OCEAN
GEOLOGY

| Quaternary | Pliocene | Miocene | Oligocene | Eocene | Paleocene | Late Cretaceous | Middle Cretaceous | Early Cretaceous | Late Jurassic |

EARTH'S DYNAMIC SYSTEMS

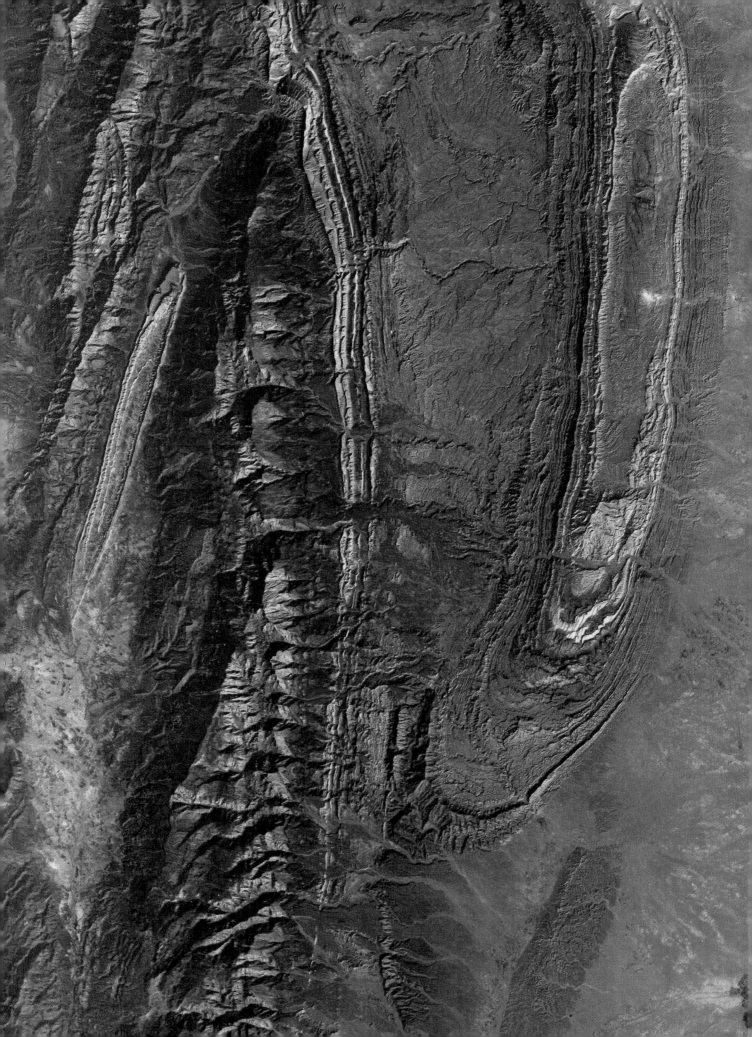

EARTH'S DYNAMIC SYSTEMS

Sixth Edition

W. KENNETH HAMBLIN

Brigham Young University, Provo, Utah

Macmillan Publishing Company
New York

Maxwell Macmillan Canada
Toronto

Maxwell Macmillan International
New York   Oxford   Singapore   Sydney

Cover: Earth's dynamic systems are vividly displayed in the Sulaiman Range of Pakistan, "the foothills of the Himalaya Mountains." The mountain belt was formed when the Indian subcontinent drifted northward and collided with Asia. This image of the Zinda Pir area shows the type of crustal deformation resulting from that collision. Folds form elliptical patterns of rock layers, as expressed by the different colors and textures in the photograph. Faults are identified by displacement of these patterns. Stream erosion dissects the folded mountain belt into a series of ridges and valleys, illustrating the effects of geologic processes operating on Earth's surface.

Cover photo: Copyright by Earth Satellite Corp., Chevy Chase, Maryland, under GEOPIC trademark.

Senior Editor: Robert A. McConnin
Developmental Editor: Madalyn Stone
Production Editor: Mary Harlan
Art Coordinator: Raydelle M. Clement
Text Designer: Cynthia M. Brunk
Cover Designer: Thomas Mack
Production Buyer: Pamela D. Bennett

This book was set in Garamond by Carlisle Communications, Ltd., with color separations and film work by H & S Graphics, Inc., and was printed and bound by R. R. Donnelley & Sons Company. The cover was printed by Lehigh Press, Inc.

Copyright © 1992 by Macmillan Publishing Company, a division of Macmillan, Inc.

Printed in the United States of America.

Earlier editions, entitled *The Earth's Dynamic Systems: A Textbook in Physical Geology*, copyright © 1989, 1985 by Macmillan Publishing Company, a division of Macmillan, Inc. Copyright © 1982, 1978, 1975 by Burgess Publishing Company.

Macmillan Publishing Company
866 Third Avenue
New York, New York 10022

Macmillan Publishing Company is part of the
Maxwell Communication Group of Companies.

Maxwell Macmillan Canada, Inc.
1200 Eglinton Avenue East, Suite 200
Don Mills, Ontario M3C 3NI

Library of Congress Cataloging-in-Publication Data

Hamblin, W. Kenneth (William Kenneth), (date)
    Earth's dynamic systems: a textbook in physical geology / by W.
Kenneth Hamblin. — 6th ed.
        p.   cm.
    Includes bibliographical references and index.
    ISBN 0-02-349490-5
    1. Physical geology.   I. Title.
QE28.2.H35 1991
550 — dc20

                                                                91 – 23832
                                                                CIP

Printing: 1 2 3 4 5 6 7 8 9   Year: 2 3 4 5

Credits for illustrations and photos appear on pages 634–36.

Preface

We are privileged to live in an unprecedented period of scientific research. Scientists around the world are unraveling the genetic code of life, probing the fundamental particles of matter, and extending the frontiers of astronomy to the beginning of the universe. Geology is also in a "golden age," a time of new enlightenment, with major discoveries and revolutionary theories giving us profound insights into the origin of our planet and how it works. New technological abilities allow us to see Earth from space and to compare our planet with other bodies in the solar system. We can now measure directly the motion of Earth's crustal plates and their effect upon rock structures and the surface topography. We can even "X-ray" the internal structure of our planet and see the hot material in the interior moving in convection cells.

With this era of enlightenment comes an increased awareness of how our planet is continually changing, and a fresh awareness of how fragile it is. Earth is a finite sphere with limited resources, so it is impossible for the population to grow indefinitely. We may find more oil, gas, and coal by improved detection methods, yet nature requires more than a million years to concentrate the oil now consumed in only 12 months. We have created nuclear waste, yet are unsure of the effects of its disposal in the geologic system. What can we do about the fact that rivers today transport more agricultural and industrial waste than natural sediment? To live efficiently on Earth, we must understand its dynamic systems and how they operate.

The sixth edition of *Earth's Dynamic Systems* introduces students to how our planet works and will help them understand and participate in the solution of these problems. It is written for freshmen-level students at both two- and four-year schools who are taking their first course in physical geology for one semester or less.

THE THEME OF THIS BOOK

Beginning geology students commonly suffocate in an avalanche of information without a unifying theme to help them focus on important concepts and grasp fundamental ideas and relationships. The core theme in this book is Earth's dynamic systems, just as the title states. There are two major energy systems that make Earth a dynamic planet: (1) the hydrologic system, the circulation of water over Earth's surface, which involves the movement of water in oceans, rivers, underground, and in glaciers; (2) the tectonic system, the movement of material powered by heat from Earth's interior. This system involves the evolution of continents, ocean basins, and mountain-building. These two systems are the focus of this book and everything discussed in it can be related to these unifying themes.

NEW TO THIS EDITION

This book has been significantly revised in an effort to make it a more effective learning tool.

1. All chapters were updated and checked for accuracy. Previous users will note we have written a new beginning chapter discussing the planet Earth in its cosmic setting. We have also combined the chapters on river systems and have added a new chapter on mass movement. However, we have retained two chapters covering the most rapidly changing field in geology—plate tectonics. This subject is introduced near the beginning of the book in Chapter 3 in order to serve as a background for all that follows. It is then treated more fully in Chapter 18 with a discussion on the history of its development and refinements in the concepts that have occurred since the last edition.

2. Special attention was given to the art program. We have expended great effort to develop line illustrations and to take photographs that visually teach the basic concepts of geology. The publisher and I have been uncompromising in our effort to make each figure an efficient learning tool. I have traveled over much of North America and many other parts of the world to take photographs specifically for this edition. Our objective was to achieve a new degree of integration of text and illustration that would help the student understand modern geology and catch a glimpse of some of its splendor. We hope the student will find that reexamining the figures and captions is one useful method of reviewing the content of the chapter.

3. We have continued to use space photography to introduce our chapters. With the high resolution of Landsat images we can view Earth as never before. By beginning each chapter with a space photograph and briefly discussing what we see, we present the student with a new and up-to-date perspective of the Earth. The twenty-four space images in the book could stand alone as a visual summary of physical geology.

4. We have rewritten every major section of each chapter in an effort to briefly state the main ideas and to enhance them with supportive discussion. The thesis statement of each topic is printed in italics and separated from the body of the text by horizontal rules.

5. We have included guiding questions for each major topic in the margin of the page. They are intended to stimulate thought and provide direction and focus for the students' reading.

IN-TEXT LEARNING AIDS

In an effort to make this book more effective and useful to the student, we have designed a number of learning aids and have incorporated them into each chapter.

Outline of Major Concepts. Students taking an introductory course in geology will probably not know enough about geology to identify all key concepts in page after page of text. To help the student focus on the key points of each chapter, we have identified them at the beginning of each chapter under the title of *Major Concepts*.

Guiding Questions. Experience has shown that the most successful students are those who read with a specific purpose—those who read to answer a question. A major pedagogical tool in this edition is the development of guiding questions which are presented in the margins adjacent to the appropriate text material. The questions are intended to guide students in their study, stimulate their curiosity, and help focus their attention.

Summary. The major concepts are further reinforced by end-of-chapter summaries that are presented in outline form. Students can also use the summary to easily locate pertinent text discussion for more information.

Key Terms. Important terms are printed in bold type when first introduced in the text. In the Key Terms section that follows the Summary at the end of each chapter, the terms are listed alphabetically, with the number of the page on which each appears. These terms are also defined in the glossary at the end of the book.

Review Questions. These discussion questions are intended to reinforce the main concepts and stimulate further investigation by pointing out some of the intriguing questions on which scientists are working.

Additional Readings. A reading list at the end of each chapter includes both periodicals and more general book references to direct the student who would like to learn more about the topic.

Glossary. At the end of the book, the student will find an illustrated glossary including definitions of approximately 800 key terms introduced in the text that are illustrated where appropriate by more than 100 full-color illustrations. The definitions of the terms are in conformance with the latest edition of the Glossary published by the American Geologic Institution.

SUPPLEMENTARY MATERIALS

1. Student Study Guide. In preparing the sixth edition, we have been acutely aware that effective learning involves more than attending lectures and reading textbooks. It involves the student's active participation, with as much one-to-one interaction with the instructor as possible. This is difficult in large introductory courses, but we have solved part of the problem by further developing the new technique of *latent image printing*, which we introduced in the fourth edition. This companion workbook–study guide utilizes an instant feedback system that is a form of personal tutor. The unique aspect of this workbook is that answers are printed next to the questions in invisible ink that can be activated by a special accompanying chemical felt-tip pen. The latent image gives students instant feedback on their work and, depending on how a student responds, provides further guidance for study. The system is similar to that used in many computer-assisted learning programs.

2. Laboratory Manual. The eighth edition of *Exercises in Physical Geology* written by Hamblin and Howard is available for laboratory work associated with a typical course in physical geology. It includes exercises on rocks, minerals, topographic maps, stereo aerial photographs, Landsat images and geologic maps.

3. Slide Set. A set of 200 slides to complement both lecture and laboratory presentations has been carefully selected as an aid to adopting instructors.

4. Overhead Transparencies. A set of 100 overhead transparencies includes four-color illustrations and photographs from the text for use in the classroom or the laboratory.

5. Instructor's Guide. This guide was prepared to help the instructor utilize the text and related supplemental material more efficiently. It contains suggestions for lecture preparation, discussion material, and a test bank of more than 1000 questions keyed to the text. This thoroughly class-tested test bank is also available on diskette for IBM or Apple personal computers.

The real test of any textbook is how well it helps the student learn. I welcome feedback from students and instructors who have used this book. Please address your comments, criticisms, and suggestions to:

W. K. Hamblin
Department of Geology
Brigham Young University

Acknowledgments

Special thanks are expressed to the following colleagues for their many helpful comments and suggestions, offered primarily while reviewing various editions of this text.

M. James Aldrich, North Carolina State University at
   Raleigh—Los Alamos National Laboratory
Burton A. Amundson, Sacramento City College
James L. Baer, Brigham Young University
David M. Best, Northern Arizona University
Myron G. Best, Brigham Young University
Willis H. Brimhall, Brigham Young University
Eric H. Christiansen, Brigham Young University
H. C. Clark, Rice University
Peter S. Dahl, Kent State University
E. Julius Dasch, Oregon State University
George H. Davis, University of Arizona
C. Patrick Ervin, Northern Illinois University
James R. Firby, University of Nevada at Reno
R. H. Grant, University of New Brunswick
Charles W. Hickcox, Emory University
Thomas M. C. Hobbs, Northern Harris County College
Roger D. Hoggen, Ricks College
Roger Hooke, University of Minnesota
Warren D. Huff, University of Cincinnati
W. Calvin James, Ohio State University
Chester O. Johnson, Gustavus Adolphus College
Cornelis Klein, University of New Mexico
Karl J. Koenig, Texas A&M University
H. Wayne Leimer, Tennessee Technical University
Paul Lowman, Goddard Space Flight Center
Erwin J. Mantei, Southwest Missouri State University
George W. Moore, Oregon State University
David A. Mustart, San Francisco State University
David Nash, University of Cincinnati
Richard S. Naylor, Northeastern University
Hallan C. Noltimier, Ohio State University

Morris S. Petersen, Brigham Young University
Arthur L. Reesman, Vanderbilt University
John R. Reid, University of North Dakota
James A. Rhodes, Stoffer Chemical Company
Steven M. Richardson, Iowa State University
W. Carl Shellenberger, Northern Montana College
Sam B. Upchurch, University of South Florida
Kenneth Van Dellen, Macomb Community College
John R. Wagner, Clemson University

Two artists who contributed to previous editions of *Earth's Dynamic Systems* deserve special recognition. William L. Chesser executed the drawings for the first edition. Many of his fine illustrations reappear here. Robert Pack prepared new figures for the third edition, many of which reappear here as well. In addition, Dale Claflin and Kim Baker have revised some of the artwork and prepared new illustrations for this book. The combined talents and imaginative work of these artists form a major contribution to this book.

My secretary, Sherrie Heywood, has my sincere thanks and appreciation for her typing and all her help on this project.

The editorial and production staff at Macmillan played a critical role in the development of this edition. I am especially indebted to Robert McConnin for his advice and supervision throughout the entire project. Madalyn Stone, the developmental editor, contributed greatly to this edition by her constructive criticism and keen insight. The production staff at Macmillan has provided much appreciated professional support. I especially appreciate the hard work, advice, cooperation, and patience of Raydelle Clement and Mary Harlan. They turned a difficult task into a pleasant experience.

W. K. H.

To the Student

OUR APPROACH TO WRITING

One of the most difficult problems you face as a student in beginning a course in a new subject is to identify fundamental facts and concepts and separate them from supportive material. This problem is often expressed by the question, *What do I need to learn?* We have attempted to overcome this problem by presenting the material in each chapter in a manner that will help you recognize immediately the essential concepts.

A brief thesis statement identifying the main ideas is presented at the beginning of each topic. This short statement expresses the facts and concepts of the subject in one all-embracing view. These may be difficult for you to comprehend fully the first time you read the "*Statement,*" but you will gain further insight from the subsequent text material in which terms are defined, illustrations are presented, and evidence supporting the statement of principles is given. If it is pertinent to an understanding of a concept, a brief history of how it developed is included. This material is designed to help you grasp the ideas presented in the statement.

With this organization you can easily recognize the major facts and concepts that are separated from the supportive discussion and examples. The great value of this system is that you can focus on the main concepts in the statement and clearly understand it by the elaboration presented in the subsequent text.

Brief Contents

Contents

1

Planet Earth

2
The Hydrologic System 27

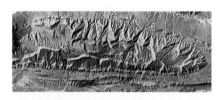

3
The Tectonic System 43

4
Minerals 57

5
Igneous Rocks 81

6
Sedimentary Rocks 109

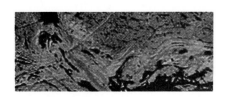

7
Metamorphic Rocks 133

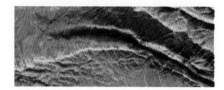

8
Structure of Rock Bodies   147

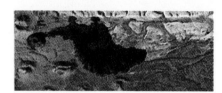

9
Geologic Time   167

10
Weathering 189

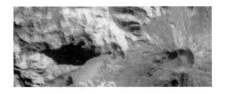

11
Mass Movement 215

12
River Systems 233

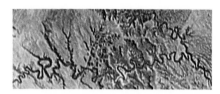

13
Evolution of Landforms   271

14
Groundwater   295

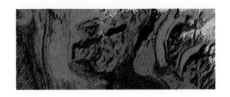

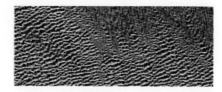

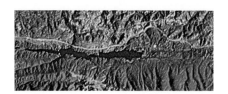

19
Earth's Seismicity 457

20
Volcanism 479

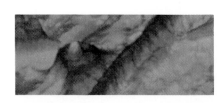

21
Evolution of Oceans 501

Summary • Key Terms • Review Questions • Additional Readings

Until recently, we obtained most of our knowledge of Earth by looking horizontally from viewpoints at or near its surface. This perspective was nearsighted and limited, and in an attempt to overcome it, local observations on its surface were plotted on maps and charts, which served as models of the real world. Rivers, mountains, shorelines, and weather patterns were surveyed and studied from hundreds of observation points, but observers were never able to see Earth in a regional, panoramic view or to see how it functions as a planet in space. Now, for the first time, we can see Earth as it really is, a tiny blue sphere suspended in the black of empty space. But that's not all. We have landed on the Moon and sampled its rocks. We have mapped the surface of Mars, tested its soil for evidence of life, and explored its huge canyons and grand volcanoes. We have surveyed the diverse landscapes on the moons of Jupiter, Saturn, Uranus, and Neptune. Every object in the solar system contains part of a record of planetary origin and evolution.

We have not only seen Earth from space, we have also seen the ocean floor, mapped its topography and structure, and gained insight into its origin and history. Until recently, this part of Earth was as inaccessible as the planets in the outer solar system. We now know that the oceanic crust is completely different from the continental crust. It is much younger, composed of different material, and has had a separate history of its own.

We have also "X-rayed" the interior of Earth and recognized that it is a highly differentiated planet whose materials are segregated into layers according to density. From this new ability to see the internal structure of Earth, we have discovered how it functions as a dynamic system.

We now know that volcanoes, earthquakes, mountain building, and continental drift are manifestations of Earth's internal heat. In our time the right tools, sophisticated technology, and decisive evidence have come together to allow us to understand how Earth works. In this chapter we will briefly describe the major geologic features of the planet Earth—features that make it unique in the solar system.

1
Planet Earth

MAJOR CONCEPTS

1. A comparison of Earth with other terrestrial planets provides an insight into the distinguishing characteristics of our planet and what makes it unique.
2. Earth's atmosphere is a thin shell surrounding the planet and is in constant motion. It is unique in that it is composed of 78% nitrogen and 21% oxygen.
3. The hydrosphere is another feature that makes Earth unique. It moves from the ocean to the atmosphere and over the surface in a great cycle.
4. The biosphere exists because of water. Although it is relatively small compared to other major layers of Earth, it has been a major geologic force operating at the surface.
5. Continents and ocean basins are the principal surface features of Earth.
6. The continents consist of three major components: (a) shields, (b) stable platforms, and (c) belts of folded mountains. All of these components show the mobility of the crust.
7. The ocean floor contains several major structural and topographic divisions: (a) the oceanic ridge, (b) the abyssal floor, (c) seamounts, (d) trenches, and (e) continental margins.
8. Earth is a differentiated planet, with its materials segregated and concentrated into layers according to density. The internal layers based on composition are (a) crust, (b) mantle, and (c) core. The major internal layers based on physical properties are (a) lithosphere, (b) asthenosphere, (c) mesosphere, and (d) core. Material within each of these units is in motion, making Earth a changing, dynamic planet.

INTRODUCTION

Geology, the science of Earth, is an incredibly fascinating subject. It concerns Earth and all that is in it: its origin, its history, and the dynamics of how it changes. Geologists study such diverse phenomena as volcanoes and glaciers, rivers and beaches, earthquakes, and even the history of life. It is a study about what happened in the past and what is happening at present—a study that increases our understanding of nature and our place in it.

But geology does much more than satisfy intellectual curiosity. As cities expand and suburbs spread across the countryside, geology is being called upon to help guide engineers in planning buildings, highways, dams, harbors, and canals, and to recognize how natural hazards such as landslides, earthquakes, floods, and beach erosion can be avoided or dealt with effectively. The discovery of natural resources is perhaps the strongest driving force in our attempt to understand Earth. All Earth materials, including water, soils, minerals, fossil fuels, and building materials, are "geologic" and are discovered, developed, and managed with the aid of geologic science. But perhaps, in the long run, comprehending nature more fully is as important as the discovery of oil fields and finding mineral deposits.

Let us begin by exploring why Earth is unique among the other planetary bodies of the solar system and how some of its important characteristics—its size, composition, atmosphere, hydrosphere, and the structure of its interior—contribute to that uniqueness.

FIGURE 1.2
Viewed from space, Earth appears as a delicate blue ball wrapped in a swirl of filmy white clouds. This is a visual embrace of our physical world and the life upon it. In a photograph we see Earth motionless, frozen in a moment of time, but in this new view of our planet, there is much more action than one might imagine. The swirling patterns of clouds that dominate the scene underline the importance of water in Earth's system. Huge quantities of water are in constant motion, in the sea, the air, and on land. Several complete cyclonic storms, spiraling over hundreds of square kilometers, can be seen pumping vast amounts of water, which erodes the land as it flows back to the sea.

Large parts of Africa and Antarctica are visible in this view, and the major climatic zones of our planet are clearly delineated. Much of the vast tropical forest of central Africa is seen beneath the discontinuous cloud cover. Also, large portions of the South Polar ice cap, a glacier more than 3000 m thick that covers the continent of Antarctica, are clearly visible. Of particular interest in this view is the rift system of East Africa, which extends in a north-south direction across most of the continent. It is mostly obscured by clouds in the equatorial region but is well expressed in the sea. The Red Sea is a large fracture in the African continent, separating the Arabian Peninsula from the rest of Africa. One of the animals that evolved from the East African rift valleys was a creature that learned to live in all of the varied landscapes of the planet. His first home was here, but he recently walked on the Moon.

5

The presence of water as a liquid on Earth's surface throughout its history has enabled life to evolve. Strange as it may seem, life profoundly changed the composition of Earth's atmosphere. Organic activity removes large quantities of carbon dioxide from the atmosphere as plants exchange oxygen for carbon dioxide. In addition, many forms of marine life also remove carbon dioxide from seawater to make their shells, which later fall to the sea floor and solidify into limestone rock. Thus, while the atmosphere of Venus became locked in a runaway greenhouse effect, it was life itself that created a new atmosphere on Earth—one with abundant oxygen and minimal carbon dioxide. Equilibrium became established by the recycling of carbon compounds through the atmosphere, oceans, and rocks in the crust. This resulted in a moderate climate throughout most of Earth's history. Perhaps, if nothing more, our studies of the diversity of compositions and conditions of the bodies in the solar system should remind us of the delicate balance that allows us to exist at all.

In contrast to Mercury, Mars, and the Moon, as you will see in the following sections, Earth continued to change as a result of its internal heat. Large planets tend to generate more internal heat and retain it longer than smaller planets. As a result of its internal heat, Earth's rigid outer layer (the lithosphere) is broken into huge fragments, or plates, like a cracked eggshell, and there is enough internal heat to cause the plates to move. The moving plates created ocean basins and continents. Earth's crust was deformed into mountain belts, and volcanic activity modified its surface. Earth thus remained a dynamic planet, continually changing as a result of its internal heat and the circulation of its surface water.

The Moon and Mercury

The Moon and Mercury are strikingly similar (Figures 1.3 and 1.4). They are relatively small planetary bodies and neither has an atmosphere, hydrosphere, or biosphere. Each is pockmarked with craters, which record the last stages of the birth of the solar system more than 4.5 billion years ago. This was a period when all of the early-formed planetary bodies were sweeping up what was left of the cosmic debris that formed the sun and its planets. The vast lava plains that form the smooth dark areas of the Moon and Mercury record a major thermal event in their histories—a time when internal heat was ignited and then died. Since then, the Moon and Mercury have not experienced significant geologic activity. We find no evidence of intense deformation of their crusts by folding or thrust faulting and no indication of major rifts. Nothing moves on the Moon except on rare occasions when a meteorite strikes the surface. No motion was observed by the *Apollo* astronauts except for what they themselves produced. The footprints they left behind will remain fresh and unaltered for literally millions of years. Both Mercury and the Moon thus remain as "fossils" of the early stages in planetary development, and both are important in that they provide a fascinating and valuable record of the first chapters in the history of our little corner of space.

Look at the surface of the Moon (Figure 1.3) and marvel at the features formed immediately after the creation of the solar system 4 billion years ago. Look again and wonder how it would be different if the Moon had water.

Mars

Beyond Earth and the Moon we come to the fascinating red planet, Mars. For years it was considered to be a planet of mystery and intrigue—a planet possibly populated with life. Telescopic observations revealed polar ice caps and shifting markings that often darkened during the Martian spring. Before the space program, some thought that life on Mars had evolved to a civilized state. Streaks were believed to be canals or vegetated land alongside canals. As it turns out, this fanciful theory was all wrong. But the Mars we have just explored

What can the surface features tell us about planetary dynamics?

FIGURE 1.3
The surface of the Moon shows two contrasting types of landforms: densely cratered highlands, called terrae, and dark, smooth areas of lava plain, called maria. We know from rock samples brought back by the *Apollo* missions that the maria resulted from great floods of lava, which filled many large craters and spread out over the surrounding area. The volcanic activity thus occurred after the formation of the densely cratered terrain. These relationships between surface features imply that the Moon's history involves three major events: (1) a period of intense bombardment by meteorites, (2) a period of volcanic activity, and (3) a subsequent period of relatively light meteorite bombardment (resulting in young, bright-rayed craters). The lunar surface has a very low level of erosion and has not been modified by wind, water, or glaciers.

FIGURE 1.4
A photomosaic of Mercury, made from photographs taken at a distance of 234,000 km, shows that Mercury and the Moon are strikingly similar. Each has a densely cratered terrain, multiringed basins, a younger area of dark plains (maria), and young rayed craters.

is even more exciting than anyone could imagine. Many of its Earth-like surface features are not only large, but gigantic. There are huge canyons, giant volcanoes, global dust storms, polar ice caps, and dry river beds (Figure 1.5). Mars has water, but at the present moment in time temperatures and atmospheric pressures are such that water can exist only as vapor or as ice. Thus, virtually all of the water on Mars is frozen as ice caps or is locked up as ground ice, like the permafrost regions on Earth. Now wind alone is the major process altering the landscape of Mars. Some dust storms grow to such proportions that at times they blanket the entire planet. The huge river channels on Mars also create an intriguing mystery. Under what ancient conditions did liquid water once flow in great floods across the surface of Mars? Why did things change on such a global scale? Do dust storms play a role in atmospheric conditions and surface temperature? Did life evolve when liquid water was abundant? Some of the answers to these questions may be suggested when we consider the size and thermal history of Mars.

Mars has about half the diameter of Earth or Venus but twice that of Mercury or the Moon. Thus, Mars has generated more internal heat, leading to more geologic activity than on Mercury and the Moon. Like all other planetary bodies in the solar system, Mars experienced an early period of intense bombardment followed by volcanic activity in which floods of lava were extruded into the lowlands of its northern hemisphere. But Mars' thermal energy continued for a time, producing subsequent volcanism that formed the giant volcanoes and causing crustal deformation that created the huge canyons.

Mars likely started out much as Venus and Earth did, developing a dense atmosphere early in its history. At one time, Mars probably had a moderate climate and abundant liquid surface water. Rainfall and flooding produced significant erosion during that period, creating stream channels and other features similar in some respects to those on Earth. But Mars is much smaller than Earth, has less internal heat, and was less geologically active. Great dust storms now rage on Mars as the major process altering its surface and atmospheric temperatures. Earth's atmosphere acts not only as a window for sunlight, but as a blanket for heat. Unlike carbon dioxide, which thickens the blanket and creates a warming greenhouse, dust thrown into the Martian atmosphere tends to close the window and block solar energy from reaching the surface below. Mars, therefore, became cold and dry. All of its water was soon locked up in ice caps or frozen in the pore spaces of rocks and soil, and stream erosion ceased.

Venus

Of all the planets, Venus is most like Earth in size, density, and distance from the sun. It probably formed at the same time as Earth and out of the same mix of iron and rocky materials that were concentrated in the inner solar system. The dense atmosphere that surrounds Venus has kept the planet's surface hidden from view. Recently, however, radar imagery from the *Magellan* spacecraft has given us a first clear glimpse of that surface. On a regional scale, Venus looks somewhat like Earth drained of its seas (Figure 1.6). Two large highlands, resembling Earth's continents, are surrounded by low smooth plains like the sea floor on Earth. High conical volcanic mountains occur on both "continents," and there are fascinating belts of folded mountains similar to those on Earth. The great difference between Venus and Earth is in the amount of liquid water on the surface and the amount of carbon dioxide in the atmosphere. But what a difference that fluid and gas make!

It is likely that Venus started out much like Earth, with massive amounts of water. But because Venus is closer to the sun it receives more solar heat, so its water soon began to evaporate and rise as vapor to the upper atmosphere. There the sun's ultraviolet rays broke the water molecules apart, freeing the hydrogen to escape into space. In this way much of the water on Venus was lost.

FIGURE 1.5
The planet Mars, as photographed from the *Viking* spacecraft, shows a surface somewhat similar to that of the Moon, yet significant differences exist. The planet has two major provinces: densely cratered highlands in the southern hemisphere and smooth plains (presumably vast floods of lava) covering the lowlands in the northern hemisphere and the floors of large basins. Unlike the Moon, however, Mars has many features indicating that its surface has been modified by atmospheric processes, recent volcanic activity, and crustal deformation. Mars, then, has experienced early events similar to those on the Moon (a period of intense meteorite bombardment and extrusion of lava floods), but it has been modified by subsequent surface processes, crustal deformation, and volcanism. Note the large volcanoes and great canyon system.

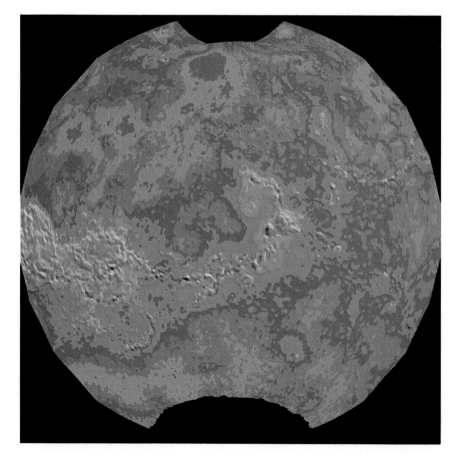

FIGURE 1.6
A shaded relief map of Venus was generated by computer from radar altimeter data collected by the *Pioneer* Venus orbiter. Two large "continents" named Ishtar and Aphrodite rise above the general surface. Circular features are probably impact structures or volcanoes.

Without water, Venus became less and less like Earth in profound ways you might at first not expect. Venus, like Earth, had a significant amount of internal heat, much more than the Moon, Mercury, and Mars; consequently, Venus has remained geologically active, with volcanic eruptions continually spewing carbon dioxide into the atmosphere. On Earth, water directly or indirectly was a means by which carbon was cycled from the atmosphere back to Earth, where it is trapped and stored in limestone rock. On Earth, therefore, much of the carbon from the early atmosphere was trapped in the crust. Because Venus lost its water early in its history it could not return carbon to the crust. As a result, the level of carbon dioxide in the atmosphere of Venus rose unchecked. This created an intense greenhouse effect: like a blanket, carbon dioxide holds the solar energy that falls on Venus, causing its surface temperature to rise to a level high enough to melt lead.

Conclusion

This then is the planet Earth in its cosmic setting—a pale blue dot in space, part of a family of planetary bodies that revolve about the sun. It is a minor planet bound to an ordinary star in the outskirts of one galaxy among billions. Yet, from a human perspective, it is a vast and complex system that has evolved over billions of years, a home we are just beginning to understand. Human beings are intelligent. But are we intelligent enough to understand how our world functions as a planetary body, to live wisely within its limits? Learning about Earth and the forces that change it—the intellectual journey upon which you are about to embark—is a journey we hope you will never forget.

EARTH'S OUTER LAYERS

The outermost layers of Earth are the atmosphere, hydrosphere, and biosphere. Their dynamics are especially spectacular when seen from space. The continents and ocean basins are Earth's major surface features.

Views of Earth from space, like that shown in Figure 1.2, reveal many features that make Earth unique and provide an insight into our planet's history and how it changes. The atmosphere is the thin, gaseous envelope that surrounds Earth, and its bright, swirling clouds are the most conspicuous features seen from space. The hydrosphere, the planet's discontinuous water layer, is seen in the vast surface of the oceans. Even parts of the biosphere—the organic realm, which includes all of Earth's living things— can be seen from space, in the dark green tropical forest of equatorial Africa. The lithosphere, the outer, solid part of Earth, is visible in continents and islands.

One of the unique features of Earth is that each of these major realms of the planet's surface is in constant motion and changes continually. The atmosphere and the hydrosphere move in ways that are dramatic and obvious. Movement, growth, and change in the biosphere are readily appreciated and easily understood, but Earth's crust is also in motion and has been throughout most of the planet's history.

The Atmosphere

The *atmosphere,* the envelope of gas surrounding Earth, constitutes an insignificantly small fraction of the planet (less than 0.01% of the mass), but it is particularly significant because it moves easily and is constantly interacting with the ocean and land. It plays a part in the evolution of most features of the landscape and is essential for life. On the scale of the illustration in Figure 1.2, most of the atmosphere would be concentrated in a layer as thin as the ink with which the photo is printed.

The atmosphere is in constant motion; its circulation patterns are clearly seen in Figure 1.2 by the shape and orientation of the clouds. At first glance, the patterns may appear confusing, but upon close examination we find that they are well organized. If we smooth out the details of local weather systems, the global atmospheric circulation becomes apparent. Solar heat, the driving force of atmospheric circulation, is greatest in the equatorial regions; it causes water in the oceans to evaporate and the moist air to rise. The warm, humid air forms an equatorial cloud belt above this low-pressure system, bordered on the north and south by relatively high-pressure zones that are cloud-free in the middle latitudes, where air descends. At higher latitudes, low-pressure systems develop where the warm air from the low latitudes meets the polar fronts. The pattern of circulation is around the resulting low pressure and produces counterclockwise winds in the Northern Hemisphere and clockwise winds in the Southern Hemisphere.

Our atmosphere is unique in the solar system. It is composed of 78% nitrogen, 21% oxygen, and minor amounts of other gases. The earliest atmosphere was much different, consisting largely of carbon dioxide and water vapor. The present atmosphere began to form as soon as organisms evolved and through photosynthesis developed the ability to extract carbon dioxide from the air and expel oxygen. Thus, the oxygen in today's atmosphere is and was produced by life.

The atmosphere is divided into several layers. The part closest to Earth is called the troposphere (Greek *tropos,* "turn" or "change") and extends from the surface to a height of about 13 km. This thin layer close to the surface contains almost all the water vapor, and therefore nearly all the clouds, storms, and precipitation. The overlying layer is the stratosphere, which extends to about 55 km above the surface. Today, jet air flight takes place mainly in the stratosphere in order to avoid the many weather hazards in the troposphere. The stratosphere is also important because it contains the ozone layer. This layer absorbs much of the sun's stronger ultraviolet rays, which would otherwise destroy exposed bacteria and severely burn animal tissue. No sharp boundary can be placed on the outer limit of the atmosphere. The density of gas molecules decreases almost imperceptibly into interplanetary space. Most scientists consider the outer boundary of the atmosphere to be about 9500 km above Earth, a distance nearly as great as the diameter of Earth itself. The atmosphere, together with the hydrosphere, is a vital agent in maintaining a suitable temperature for the majority of life in the biosphere. Movement of water in the atmosphere and its precipitation on land are responsible for sculpturing many of Earth's landforms.

Why are the atmosphere and oceans considered as much a part of Earth as solid rock?

The Hydrosphere

The *hydrosphere* is the total mass of water on the surface of our planet. About 98% of the water is in the oceans; 2% is in streams, lakes, groundwater, and glaciers. Approximately 71% of Earth is covered with water. Thus, it is for good reason that Earth has been called the water planet. It has been estimated that if all the irregularities of Earth's surface were smoothed out to form a perfect sphere, a global ocean would cover Earth to a depth of 2.25 km.

It is this great mass of water that makes Earth unique. Water permitted life to evolve and flourish. Every inhabitant on Earth is directly or indirectly controlled by it. All of Earth's weather patterns, climate, rainfall, and the extremely important carbon dioxide content of the atmosphere are influenced by the seas and oceans. The hydrosphere is in constant motion—evaporating from the oceans and moving through the atmosphere, precipitating as rain and snow, and returning to the sea in rivers, glaciers, and groundwater. As water moves over Earth's surface, it erodes, transports, and deposits weathered rock material, constantly modifying Earth's landscape. Many of the distinctive surface features of Earth are due to the hydrosphere.

The Biosphere

What is the biosphere? How does it affect Earth dynamics?

The *biosphere* may be defined as the part of the Earth where life exists. It includes the forests, the grasslands, and the familiar animals of the land, together with the numerous creatures that inhabit the sea and atmosphere. As a terrestrial covering, the biosphere is discontinuous and has an irregular shape; it is a single, interwoven web of life existing within and reacting with the atmosphere, hydrosphere, and lithosphere. It consists of more than 1.5 million described species and perhaps as many as 3 million not yet described. Each species lives within its own limited environmental setting. By far, most of the biosphere exists in a narrow zone extending from the depth to which sunlight penetrates the oceans (about 200 m) to the snowline in the tropical and subtropical mountain ranges (about 6000 m above sea level). At the scale of the photograph in Figure 1.2, the biosphere—all of the known life in the solar system—would be located in a thin layer no thicker than the paper on which the image is printed.

The main factors controlling the distribution of life on our planet are temperature, pressure, and chemistry. However, the range of environmental conditions in which life is possible is truly amazing, especially the range of environments in which microorganisms can exist. Also remarkable is the range of species that can exist in these varied environments (Figure 1.7).

Although the biosphere is relatively small compared with the other major layers of Earth (atmosphere, hydrosphere, and lithosphere), it has been a major geologic force operating at the surface. Essentially all of the present atmosphere has been produced by the chemical activity of the biosphere. The composition of the oceans is similarly affected by the activity of organisms; most marine organisms extract calcium carbonate from seawater to make their shells and hard parts. When the organisms die, their shells settle to the sea floor and accumulate as beds of limestone. The continued extraction of calcium carbonate from the ocean has a major effect upon the composition of the atmosphere as well. In addition, all of the coal, oil, and gas of Earth were

FIGURE 1.7
A global view of Earth's biosphere was produced from data derived from NASA's Nimbus-7 Coastal Zone Color Scanner and the NOAA-7 Advanced Very High Resolution Radiometer. Ocean chlorophyll concentration is shown by color hues increasing from purple to red. Land vegetation index is shown increasing from tan to green.

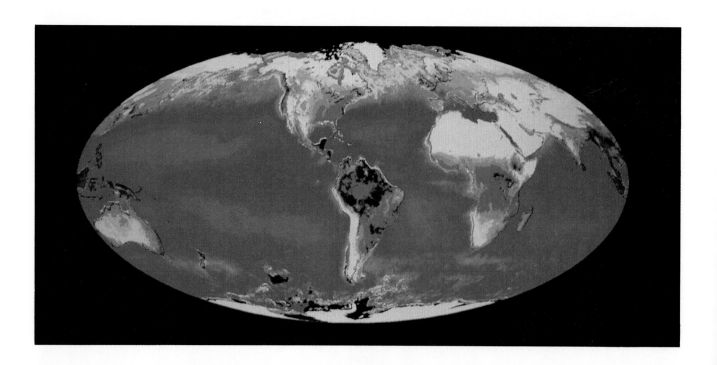

formed by the biosphere, so large parts of the rocks near the surface of Earth's crust originated in some way from organic activity.

The record of the biosphere (life on Earth) is preserved, sometimes in remarkable detail, by fossils that occur in most sedimentary rock and even in some igneous and metamorphic rocks. Indeed, the number of living species today represents only about 10% of the number of species that have existed since life first developed on Earth.

The Continents and Ocean Basins

If Earth had neither an atmosphere nor a hydrosphere, two principal regions— *continents* and *ocean basins*—would stand out as its dominant features. These major structural and topographic divisions of Earth differ markedly, not only in elevation but also in rock types, density, chemical composition, age, and history. The ocean basins, which occupy about two-thirds of Earth's surface, are characterized by a spectacular topography, most of which originated from extensive volcanic activity and Earth movements that continue today. The continents rise above the ocean basins as large platforms, but the waters of the oceans more than fill their basins and flood a large part of the continental surface. The present shoreline, so important geographically and so carefully mapped, has no special structural significance with respect to the boundary between continents and ocean basins. Indeed, shorelines have fluctuated greatly throughout Earth's entire history.

From a geological viewpoint, the elevation of the continents, which rise high above the ocean basins, is much more significant than the position of the shore. The difference in elevation of continents and ocean basins represents a fundamental difference in rock *density*. Continental rocks are less dense than the rocks of the ocean basins. That is, a given volume of continental rock weighs less than the same volume of oceanic rock. This difference causes the *continental crust* to float above the denser *oceanic crust*.

The elevation and area of the continents and ocean basins have been mapped with precision, and the data can be summarized in various forms. The data presented graphically in Figure 1.8 show that the average elevation of the continents is 0.84 km above sea level, and the average depth of the ocean is about 3.7 km below sea level. Only a relatively small percentage of Earth's surface rises significantly above the average elevation of the continents or drops below the average depth of the ocean. If the continents did not rise quite so high above the ocean floor, the entire surface of Earth would be covered with water.

Why does Earth have ocean basins and continents instead of a cratered surface like those of the Moon and Mars? By their very existence, the continents pose one of the most fundamental questions about Earth. Theoretically, a large, rotating planet like Earth with a strong gravitational field would mold itself into a smooth spheroid covered with a layer of water approximately 2.25 km deep. Continents, as they are known today, have not been found on other planets in the solar system.

If the oceans were dry, would there be any fundamental difference between the continents and ocean basins?

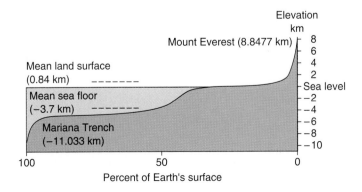

Elevation
km

Mount Everest (8.8477 km)
8
6
4
2

Mean land surface
(0.84 km)

Sea level
−2
−4

Mean sea floor
(−3.7 km)

−6
−8
−10

Mariana Trench
(−11.033 km)

100   50   0

Percent of Earth's surface

FIGURE 1.8
A graph of the elevation of the continents and ocean basins shows that the average height of the continents is only 0.84 km above sea level; the average depth of the ocean is 3.7 km below sea level. Only a small percentage of Earth's surface rises above the average elevation of the continents or drops below the average elevation of the ocean floor.

MAJOR FEATURES OF THE CONTINENTS

Continents consist of three major structural components: (1) a basement complex (shield), which is composed of completely deformed rocks eroded down to near sea level, (2) a stable platform, which is the area where the basement complex is covered with a veneer of horizontal sedimentary rocks, and (3) folded mountain belts.

The broad, flat continental masses, which rise above the ocean basins, present a great diversity of surface features, with an almost endless variety of hills and valleys, plains and plateaus, and mountains. From a regional perspective, however, the continents are remarkably flat. Most of their surface lies within a few hundred meters of sea level. Extensive geologic studies during the past 100 years have revealed several striking facts about the continents (Figure 1.9).

1. Most continents are roughly triangular in shape.
2. They are concentrated in the Northern Hemisphere.
3. Although each may seem unique, all continents have three basic components: (a) a shield, (b) a stable platform, and (c) folded mountain belts. Geologic differences between continents are mostly in size, shape, and proportions of these components.
4. Continents consist of rock that is less dense than the rock in the ocean basins.
5. The continental rocks are old, some as old as 3.8 billion years.
6. The climatic zone occupied by a continent usually determines the style and variety of landforms developed on it.

What are the fundamental structural features of continents?

The extensive flat, stable regions of the continents in which complex crystalline rocks are exposed or buried beneath a relatively thin sedimentary cover are called *cratons.* These regions have been relatively undisturbed for over a half billion years except for broad, gentle warping. The cratons include the shields, where large areas of highly deformed igneous and metamorphic rock, the *basement complex,* are exposed. The stable platforms are those regions of the craton where the basement rocks are covered with a relatively thin veneer of sedimentary rocks.

Shields

Without some firsthand knowledge of a shield, visualizing the nature and significance of this important part of the continental crust is difficult. Figure 1.10, showing part of the Canadian Shield of the North American continent, will help you to comprehend the extent, the complexity, and some of the typical features of shields. You should also study Figures 7.1 and 22.1.

What are the most significant features of shields?

Characteristically, a *shield* is a regional surface of low *relief* that generally has an elevation within a few hundred meters of sea level. The only features of relief are the resistant rocks that rise 50 to 100 m above their surroundings.

A second characteristic of shields is their complex structure and rock types. Many rock bodies were once liquid, and others have been compressed and extensively deformed. All of the rocks in the shields are crystalline and were formed several kilometers below the surface. They are now exposed only because the shields have been subjected to extensive erosion.

Stable Platforms

Large parts of the cratons are covered with a veneer of sedimentary rocks. These areas have been relatively stable throughout the last 600 million or 700

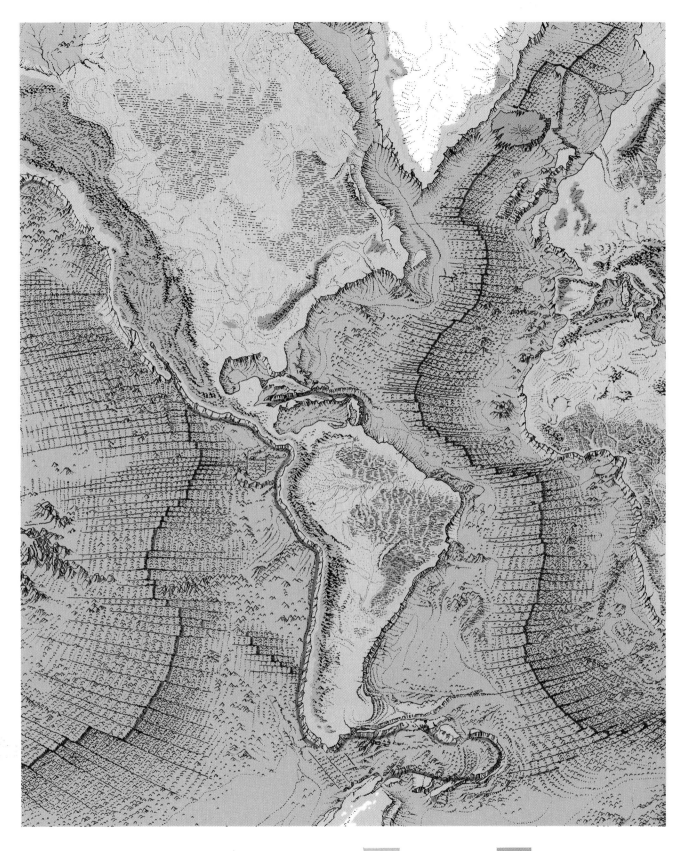

FIGURE 1.9
The major surface features of Earth reflect the structure of the crust. The continental crust rises above the ocean basins and forms continents that have as their major structural features the shields, stable platforms, and folded mountain belts. The oceanic crust, composed primarily of basalt, forms the ocean floor, the major features of which include the oceanic ridge, the abyssal floor, seamounts, and trenches.

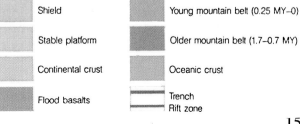

Shield

Stable platform

Continental crust

Flood basalts

Young mountain belt (0.25 MY–0)

Older mountain belt (1.7–0.7 MY)

Oceanic crust

Trench
Rift zone

15

G r a n i t i c   r o c k

L .   A l b a n e l

M e t a m o r p h i c   r o c k

L. Chibougamau

FIGURE 1.10
The Canadian Shield is a fundamental structural component of North America. It is composed of complexly deformed crystalline rock bodies, eroded down to an almost flat surface near sea level. Throughout much of the Canadian Shield, the topsoil has been removed by glaciers, and different rock bodies are etched out in relief by erosion. The resulting depressions commonly are filled with water, forming lakes and bogs, which emphasize the structure of the rock bodies. Dark tones indicate areas of metamorphic rock. Light pink tones indicate areas of granitic rock.

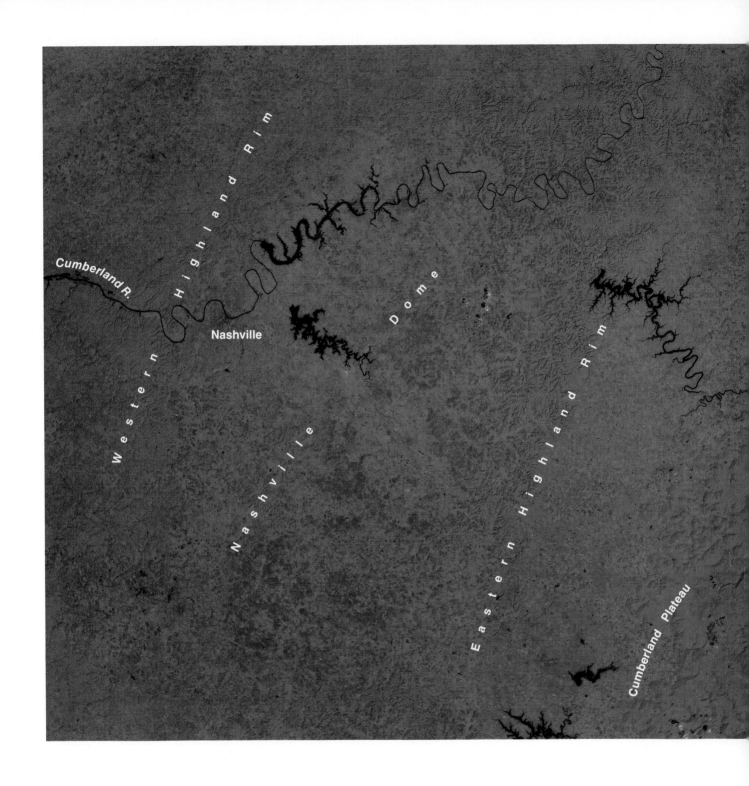

FIGURE 1.11
The stable platform in the vicinity of Nashville, Tennessee, is dominated by a broad domal upwarp, although vegetation covers many details. Erosion has removed the younger rocks from the central part of the Nashville dome, so that the eroded edges of major rock formations are exposed in large, elliptical outcrop patterns, which can be identified in this photograph by dark tones and a delicate erosional texture. These outcrops of resistant rock form a ridge, known as the Highland Rim, rising from 30 to 60 m above the adjacent area. The older rocks exposed in the center of the dome are nonresistant limestone, and the surface there has been lowered by erosion so that the region, although structurally a dome, is expressed topographically as a basin, or lowland. To the southeast, younger sedimentary rocks mark the western margins of the Cumberland Plateau, a broad flatland dissected by stream erosion.

17

What are the most significant features of stable platforms?

million years; that is, they have not been uplifted a great distance above sea level or submerged far below it—hence the term *stable platform*. In North America, the stable platform lies between the Appalachian Mountains and the Rocky Mountains and extends northward to the Lake Superior region and into western Canada. Throughout most of this area, the sedimentary rocks that cover the craton are nearly horizontal, but locally they have been warped into broad domes and basins (Figure 1.11; also see Figure 22.4).

Folded Mountains

One of the most significant features of the continents is the young, *folded mountain belts* that typically occur along their margins. Most people think of a *mountain* as simply a high, more or less rugged, landform, in contrast to flat plains and lowlands. Mountains, however, are much more than high country. To a geologist, the term *mountain belt* means a long, linear zone in Earth's crust where the rocks have been intensely deformed by horizontal stresses and generally intruded by molten rock material. The topography can be high and rugged, or it can be worn down to a surface of low relief. To a geologist, the topography of mountain belts is not as important as the extent and style of deformation. The great folds and fractures in mountain belts provide evidence that Earth's crust is, and has been, in motion.

What are the most significant characteristics of folded mountain belts?

Figure 1.12 illustrates some of the characteristics of folded mountains and the extent to which the margins of continents have been deformed. The layers of rock shown in this photograph have been deformed by compression and are folded like wrinkles in a rug. Erosion has removed the upper part of the fold, so the resistant layers form zigzag patterns similar to those that would be produced if the crests of wrinkles in a rug were cut off.

We now know that the crusts of the Moon, Mars, and Mercury lack this type of deformation. All of their impact craters, regardless of age, are circular—proof that these planets have not been strongly deformed by compressive forces. Their crusts, unlike that of Earth, appear to have been fixed and immovable throughout their histories.

MAJOR FEATURES OF THE OCEAN FLOOR

The oceanic crust is strikingly different from the continental crust with respect to rock types, structure, landforms, age, and origin. The major features of the ocean floor are (1) the oceanic ridge, (2) the abyssal floor, (3) seamounts, (4) trenches, and (5) continental margins.

What landforms and structural features characterize the ocean floor?

The ocean floor, not the continents, is the typical surface of the solid Earth. It is the ocean floor that holds the key to the evolution of Earth's crust, yet not until the 1960s did we obtain enough data about the ocean floor to get a clear picture of its regional characteristics. This new knowledge caused a revolution in geologists' ideas about the nature and evolution of the crust.

Before 1947, most geologists believed that the ocean floor was simply a submerged version of the continents, with huge areas of flat abyssal plains covered with sediment derived from the landmass. Although echo sounding was used to determine depths as early as 1922, the devices were primitive and limited to shallow water. An observer with headphones timed the interval between the transmission of a sound pulse and the return of its echo. A major breakthrough occurred in 1953 with the development of a precision depth recorder that could automatically plot a continuous profile of the ocean floor in any depth of water. Since then, millions of kilometers of profiles have been

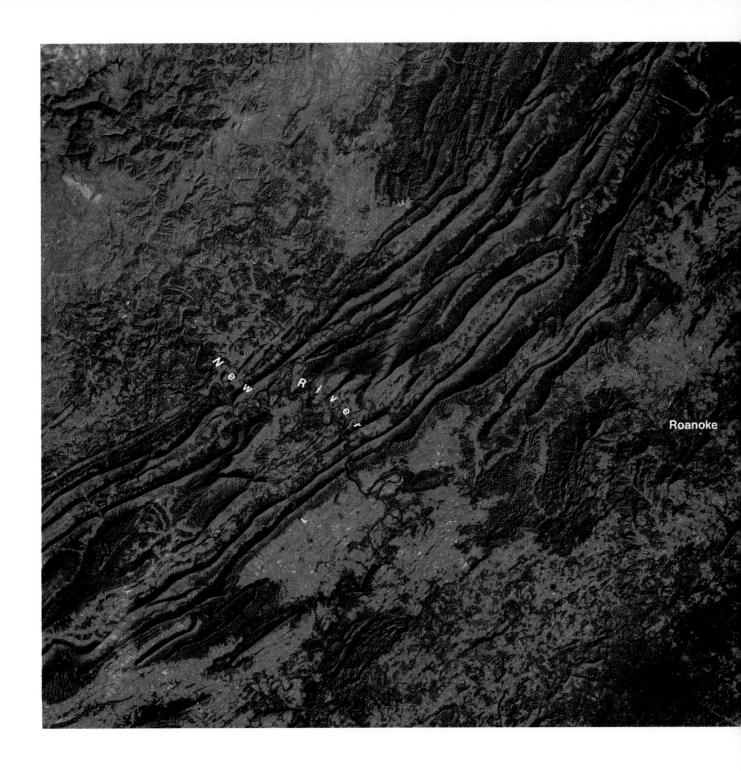

FIGURE 1.12
The Appalachian Mountains in the eastern United States show
the typical style of deformation in a folded mountain belt. The folded
strata are expressed by long, narrow ridges of resistant sandstone,
which rise about 300 m above the surrounding area. Erosion has removed the upper parts of many of the folds, so their resistant limbs
(which form the ridges) are exposed in elliptical or zigzag outcrop
patterns. This area is part of the Ridge and Valley province, one of the
physiographic provinces of North America, which extends from central
Alabama to New England. Similar styles of deformation are found in
most other mountain belts of the world. Such deformation is an important structural feature of Earth's crust and clearly shows that the
crust is mobile and has been in constant motion throughout most of
geologic time.

19

made, and the information has been used to map the surface features of the ocean basins in remarkable detail (Figure 1.9).

Profiles of the ocean floor show that submarine topography is as varied as that on the continents and, in some respects, is more spectacular. Among the most significant facts that we have learned about the oceanic crust are the following:

1. The oceanic crust is mostly basalt, a dense volcanic rock, and its major topographic features are somehow related to volcanic activity. The oceanic crust, therefore, is entirely different from the continental crust.
2. The rocks of the ocean floor are young in a geologic time frame. Most are less than 150 million years old, whereas the ancient rocks of the shields are more than 700 million years old.
3. The rocks of the ocean floor have not been deformed by compression. Their undeformed structure is in marked contrast to the complex deformation of rocks in the folded mountains and basement complex of the continents.
4. The major provinces of the ocean floor are
   a. The oceanic ridge
   b. The abyssal floor
   c. Seamounts
   d. Trenches
   e. Continental margins

Although most of the topography of the ocean floor can be seen only indirectly through profile records, some features are visible in satellite photographs, others have been photographed from deep submersible vessels, and some areas have been "seen" through radar imagery. However, *physiographic maps,* such as the one shown in Figure 1.9, provide the best visual reference for regional features of the ocean basins. You should refer to this map frequently as you study the following material.

The Oceanic Ridge

The *oceanic ridge* is perhaps the most striking and important feature on the ocean floor. It extends continuously from the Arctic basin down the center of the Atlantic Ocean, into the Indian Ocean, and across the South Pacific. The oceanic ridge is essentially a broad, fractured swell generally more than 1400 km wide. Its higher peaks rise as much as 3000 m above the ocean floor. A huge, cracklike valley, called the *rift valley,* runs along the axis of the ridge throughout most of its length. In addition, great fracture systems, some as long as 4000 km, trend across the ridge.

The Abyssal Floor

The oceanic ridge divides the Atlantic and Indian oceans roughly in half and traverses the southern and eastern parts of the Pacific. On both sides of the ridge are vast areas of broad, relatively smooth, deep-ocean basins known as the *abyssal floor.* This surface extends from the flanks of the oceanic ridge to the continental margins and generally lies at depths of about 4000 m.

The abyssal floor can be subdivided into two sections, the abyssal hills and the abyssal plains. The *abyssal hills* are relatively small hills, rising as much as 900 m above the surrounding ocean floor. They cover from 80 to 85% of the Pacific sea floor, and thus they are the most widespread landforms on Earth. Near the continental margins, land-derived sediment completely covers the abyssal hills, forming flat, smooth *abyssal plains.*

Seamounts

Isolated peaks of submarine volcanoes are called *seamounts*. Some seamounts rise above sea level and form islands, but most are completely submerged and are known only from oceanographic soundings. Although many seem to occur at random, others, such as the Hawaiian Islands, form chains along well-defined lines. Islands and seamounts testify to the extensive and continuous volcanic activity that has occurred throughout the ocean basins. They also provide an important insight into the dynamics of the inner Earth.

Trenches

The deep-sea *trenches* are the lowest areas on Earth's surface. The Mariana Trench, in the Pacific Ocean, is the deepest part of the world's oceans—11,000 m below sea level—and many other trenches are more than 8000 m deep. Trenches have attracted the attention of geologists for years, not only because of their depth but also because they represent fundamental structural features of Earth's crust. As is illustrated in Figure 1.13, the trenches are invariably adjacent to island arcs or coastal mountain ranges of the continents. We shall see in subsequent chapters how the trenches are involved in the most intense volcanic and seismic (earthquake) activity on the planet.

Continental Margins

The zone of transition between a continental mass and an ocean basin is called a *continental margin*. The submerged part of a continent is referred to as a *continental shelf*. Geologically, it is part of the continent, not part of the ocean basin. The continental shelves at present constitute 11% of the continental surface, but at times in the geologic past, shallow seas were much more extensive.

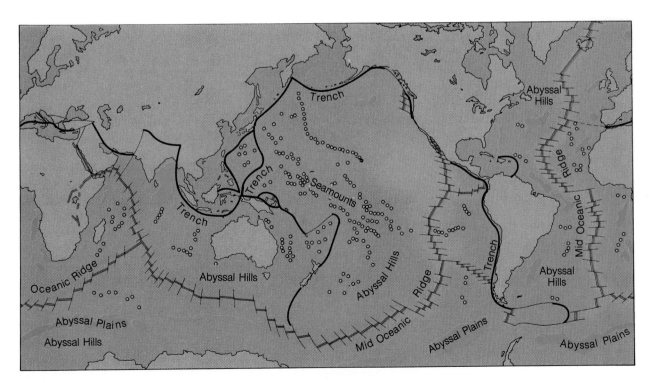

FIGURE 1.13
The major features of the ocean floor are related to plate boundaries. The oceanic ridge coincides with divergent plate margins. Trenches form where plates converge. The abyssal floor is the deep part of the ocean, off the flanks of the oceanic ridge.

The sea floor descends in a long, continuous slope from the outer edge of the continental shelf to the deep-ocean basin. This **continental slope** marks the edge of the continental rock mass. Continental slopes are found around the margins of every continent and also around smaller fragments of continental crust, such as Madagascar and New Zealand. Look at Figure 1.9 and study the continental slopes—especially those surrounding North America, South America, and Africa—and you can see that they form one of Earth's major topographic features. On a regional scale, they are by far the longest and highest slopes on Earth. Within this zone, from 20 to 40 km wide, the average relief above the sea floor is 4000 m. Along the marginal trenches, relief is as great as 10,000 m. In contrast to the shorelines of the continents, the continental slopes are remarkably straight over distances of thousands of kilometers.

In many areas, the continental slopes are cut by deep **submarine canyons,** which are remarkably similar to canyons cut by rivers into continental mountains and plateaus. As is shown in Figure 1.9, submarine canyons cut across the edge of the continental shelf and terminate on the deep abyssal floor, some 5000 or 6000 m below sea level.

THE MAJOR STRUCTURAL UNITS OF EARTH

Earth is a differentiated planet—that is, its constituent materials are separated and segregated into layers according to density. The denser materials are concentrated near the center, the less dense near the surface (Figure 1.14). The internal layers are recognized on the basis of (a) composition and (b) physical properties. Compositional layers are (1) crust, (2) mantle, and (3) core. Layers based on physical properties are (1) lithosphere, (2) asthenosphere, (3) mesosphere, and (4) core.

Layers Based on Composition

What layers of Earth are most significant to the planet's dynamics?

The Crust.   Geologists use the term "crust" in reference to the outermost layer of the *differentiated planet* that is Earth. Once Earth was thought to be completely molten in its early stages; as it cooled, a hard crust formed, enveloping the still liquid interior. Though this concept has been obsolete for nearly a century, the term "crust" is still popular. However, it has acquired another generally accepted meaning. Today the term designates the outer layer of Earth, extending from the solid surface down to the first major discontinuity in seismic wave velocity in the lithosphere. It heralds a compositional, *but not a structural,* change. Moreover, the crust of the continents is distinctly different from the crust beneath the ocean basins (Figure 1.15). The continental crust is much thicker (as much as 50 km thick) and is composed of relatively light "granitic" rock that includes the oldest rock of the crust. By contrast, the oceanic crust is only about 8 km thick and is composed of dark, dense volcanic rocks (basalt) with densities much greater than that of granite. The oceanic crust is young and relatively undeformed by folding. The differences between the continental and oceanic crust, as we shall see, are of fundamental importance in understanding Earth.

The Mantle.   The next major compositional layer of Earth, the **mantle,** surrounds or covers the core. This zone constitutes the great bulk of Earth (82% of its volume and 68% of its mass). The mantle is composed of iron and magnesium silicate rock, fragments of which have been brought to the surface by volcanic eruptions.

The upper boundary of the mantle is considered by tradition to be the first worldwide discontinuity in seismic wave velocity (the first zone below Earth's

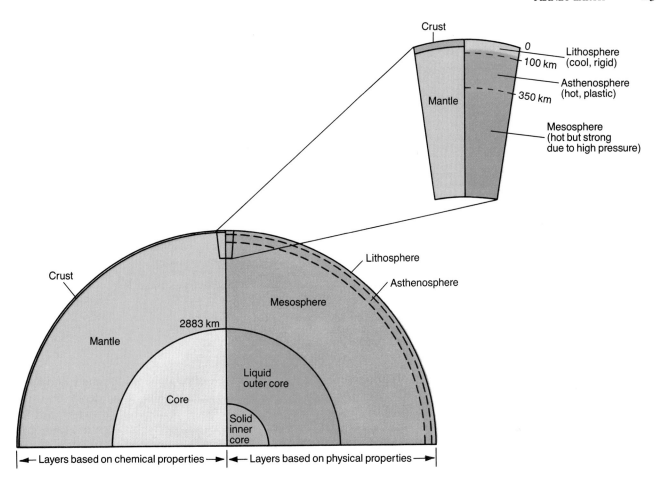

FIGURE 1.14
The internal structure of Earth is known from studies of its density, its magnetic field, and the way in which it transmits seismic waves. The left side of the diagram shows the layering based on composition: crust, mantle, and core. The right side shows the layering based upon physical properties. Note that the two do not coincide. The expanded view of the upper part of the diagram shows the nature of the outermost layers. The lithosphere is a hard, rigid shell surrounding the entire planet. The asthenosphere is a zone where increased temperature makes the rock soft, weak, and capable of plastic flow. The mesosphere below is stronger and more rigid because increase in pressure offsets the high temperature that occurs with depth.

surface where the velocity at which earthquake waves travel decreases significantly). This was discovered years ago by a Yugoslavian seismologist named Mohorovičić. The name has been shortened to Moho or M-discontinuity. The Moho is believed to be the result of a change in composition but is not a structural boundary.

The Core. The *core* of Earth is a central mass about 7000 km in diameter. Its density increases with depth but averages about 10.78 g/cm³. It constitutes only 16% of Earth's volume but accounts for 32% of Earth's mass. Indirect evidence indicates that the core is mostly iron.

Internal Layers Based on Physical Properties

Asthenosphere. During the last few decades, it has been recognized that there is a major zone within the upper mantle where temperature and pressure are at just the right balance so that part of the material melts. The rocks lose much of their strength, becoming soft plastic and flowing like warm tar. This zone of easily deformed mantle is called the ***asthenosphere*** (meaning "weak sphere"). The asthenosphere is a distinctive zone in the upper mantle and is as much as 200 km thick (Figure 1.15).

Lithosphere. The top of the asthenosphere is about 100 km below the surface. Above the asthenosphere the material is solid, strong, and rigid. This layer is called the ***lithosphere*** ("rock sphere"). The boundary between the lithosphere and the asthenosphere is distinct but does not correspond to a compositional change. The boundary is simply due to a major change in the physical properties of the rock. The lithosphere thus contains the crust and the uppermost part of the mantle.

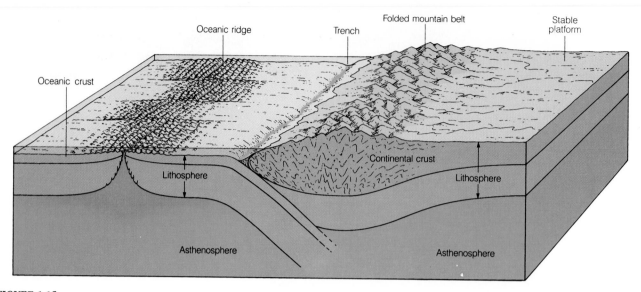

FIGURE 1.15
The outermost solid layers of Earth are the asthenosphere and the lithosphere. The asthenosphere is close to the melting point and is capable of flow. The lithosphere above it is rigid. It includes two types of crust: a thin, dense oceanic crust and a thick, lighter continental crust.

Mesosphere. The rock below the asthenosphere is stronger and more rigid than the asthenosphere because the high pressure at this depth offsets the effect of high temperature. The region between the asthenosphere and the core–mantle boundary is called the *mesosphere*. The diagram in Figure 1.15 shows how the major internal layers of Earth are related.

The Core. The core of Earth marks a change in both physical properties and composition. It is composed mostly of iron and is therefore distinctly different from the silicate (rocky) material above. On the basis of physical properties, the core has two distinct parts—a solid inner core and a liquid outer core. Heat loss from the core and the rotation of Earth probably cause the liquid outer core to circulate, and its circulation generates Earth's magnetic field.

SUMMARY

1. Perhaps the most overwhelming impression gained from seeing Earth from space and observing the major features of its surface is that Earth is a dynamic planet and has been changing throughout all geologic time. Not only are the surface fluids extremely mobile, but the crust is in motion, having been deformed by mountain-building forces, volcanic activity, and earthquakes. No such activity has occurred on the Moon or Mercury, and Mars has been much less dynamic than Earth.

2. Earth's crust and surface features are constantly changing because of its internal energy and the circulation of its surface fluids.

3. Other terrestrial planetary bodies are smaller, have less internal energy, and lack a hydrosphere. They are less dynamic and their surfaces preserve many craters that resulted from bombardment by meteorites early in the history of the solar system.

4. The atmosphere and the hydrosphere constitute a system of moving surface fluids in which tremendous volumes of water are in constant motion.

5. The two major structural and topographic features of Earth are (a) the continents and (b) the ocean basins.

6. The continents consist of three major components: (a) the shields, which are composed of complexly deformed rock eroded down to near sea level, (b) stable platforms, areas where the basement complex is covered with a veneer of horizontal sedimentary rocks, and (c) mountain belts, in which sedimentary rocks have been compressed into folds.

7. The major features on the ocean floor are (a) the oceanic ridge, (b) the abyssal floor, (c) seamounts, (d) trenches, and (e) continental margins.

8. The major structural units of Earth, based on composition, are (a) crust, (b) mantle, and (c) core.

9. The internal layers of Earth, based on physical properties, are (a) lithosphere, (b) asthenosphere, (c) mesosphere, and (d) core.

KEY TERMS

abyssal floor (p. 20)

abyssal hill (p. 20)

abyssal plain (p. 20)

asthenosphere (p. 23)

atmosphere (p. 10)

basement complex (p. 14)

biosphere (p. 12)

continent (p. 13)

continental crust (p. 13)

continental margin (p. 21)

continental shelf (p. 21)

continental slope (p. 22)

core (p. 23)

craton (p. 14)

density (p. 13)

differentiated planet (p. 22)

folded mountain belt (p. 18)

hydrosphere (p. 11)

lithosphere (p. 23)

mantle (p. 22)

mesosphere (p. 24)

mountain (p. 18)

mountain belt (p. 18)

ocean basin (p. 13)

oceanic crust (p. 13)

oceanic ridge (p. 20)

physiographic map (p. 20)

relief (p. 14)

rift valley (p. 20)

seamount (p. 21)

shield (p. 14)

stable platform (p. 18)

submarine canyon (p. 22)

trench (p. 21)

REVIEW QUESTIONS

1. Draw a diagram of the internal structure of Earth and briefly describe the core, mantle, asthenosphere, and lithosphere.

2. Why are the atmosphere and the oceans considered as much a part of Earth as solid rock?

3. What are the major differences between the continents and the ocean basins?

4. Briefly describe the distinguishing features of continental shields, stable platforms, and folded mountain belts.

5. Briefly describe the distinguishing features of the oceanic ridge, the abyssal floor, seamounts, trenches, and continental margins.

6. Study the physiographic map of Earth in Figure 1.9. Make a profile of the major structural features of Earth by tracing or sketching the continental crust, shields, stable platforms, and folded mountain belts, together with the oceanic ridge, the abyssal floor, and deep-sea trenches.

7. Study the view of Earth in Figure 1.2 and make a sketch map of the following features: (a) major patterns of atmospheric circulation, (b) low-latitude deserts, (c) the tropical belt, (d) the Red Sea rift, and (e) the Antarctic ice cap.

ADDITIONAL READINGS

Bates, D. R., ed. 1964. *The Planet Earth*. Elmsford, NY: Pergamon Press.

Bolt, B. A. 1973. The fine structure of the earth's interior. *Scientific American* 228(3):24–33.

Cloud, P. 1978. *Cosmos, Earth, and Man*. New Haven, Conn.: Yale University Press.

Motz, L., ed. 1979. *The Rediscovery of the Earth*. New York: Van Nostrand Reinhold.

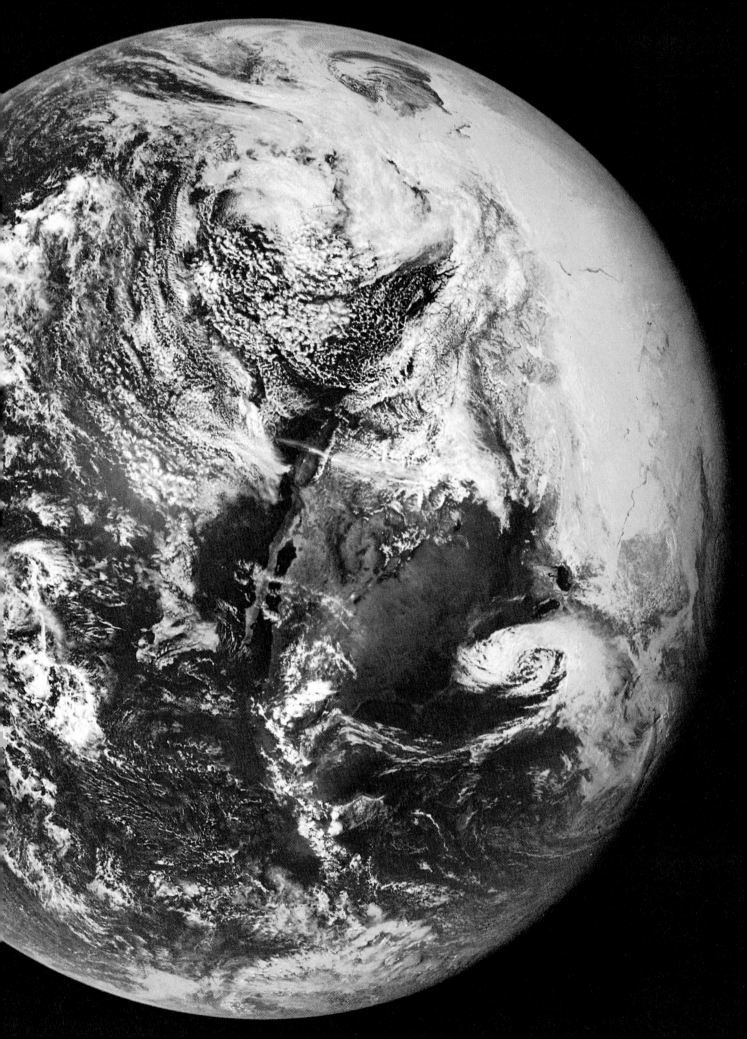

We speak of Earth as a dynamic planet because the materials of the various layers are in motion. The most obvious motion is that of the surface fluids, air and water.

Leaving Earth for the Moon, the crew of *Apollo 16* looked back and photographed North America. The picture they took is shown on the facing page. This view of Earth emphasizes the circulating atmosphere and the fact that Earth's surface is predominantly water. Indeed Earth could be called the Blue Planet. The most impressive features of Earth as seen from space are the brilliant white swirling clouds, an expression of the motion of surface fluids. Two major storm systems are seen in this view, each containing huge amounts of water pumped up from the ocean. One is in the Gulf of Alaska ready to descend on western North America. The other is over the eastern United States.

The complex cycle by which water moves from the oceans to the atmosphere, to the land, and back to the oceans again is called the hydrologic system. The energy source that drives the system is heat from the sun, which evaporates water from the oceans and causes the atmosphere to circulate. Water vapor moves with the circulating atmosphere and eventually condenses to fall as rain or snow. Acted on by the force of gravity, it then flows back to the oceans in several subsystems (rivers, groundwater, and glaciers), all of which involve gravity flow from higher to lower levels.

Water precipitated on Earth furnishes a fluid medium for many processes that shape Earth's surface features. Rivers erode the land as they flow to the sea. Groundwater percolating through the pore spaces in rocks dissolves and carries away soluble minerals. Glaciers form and spread over large parts of a continent, modifying its surface by erosion and deposition.

In this chapter we will consider the hydrologic system: the paths the water takes, the energy involved, and how Earth's surface is modified. The concepts are simple. The major difficulty you will find is becoming aware of the magnitude of the system, the volume of water involved, and the changes that result. We live in the hydrologic system. We depend on it for sustenance. Although we can alter it, most of us do not understand it. The goal of this chapter is to aid that understanding.

2
The
Hydrologic
System

1. Earth's system of moving water—the hydrologic system—involves movement of water in rivers, as groundwater, in glaciers, in oceans, and in the atmosphere.
2. Heat from the sun is the source of energy for Earth's hydrologic system.
3. The volume of water in motion is almost incomprehensibly large.
4. As water moves, it erodes, transports, and deposits its sediment, creating distinctive landforms.

DYNAMICS OF THE HYDROLOGIC SYSTEM

> *The hydrologic system is the complex cycle through which water moves from the oceans to the atmosphere, to the land, and back to the oceans again. It involves water in rivers, lakes, glaciers, and oceans, in the atmosphere, and in the pore spaces in the rocks beneath the surface.*

The Hydrologic Cycle

The complex cycle through which water moves from the oceans to the atmosphere, to the land, and back to the oceans again is called the *hydrologic system.* It operates on a global scale and unites into a single whole all possible paths of motion of Earth's surface water. The basic elements of the system can be seen from space (page 26 and Figure 2.1) and are diagrammed in Figure 2.2. The system operates as heat from the sun evaporates water from the oceans, the principal reservoir for Earth's water. Most of the water returns directly to the oceans as rain. Atmospheric circulation carries the rest over the continents, where it is precipitated as rain or snow. Water that falls on the land can take a variety of paths back to the oceans. The greatest quantity returns to the atmosphere by evaporation, but the most obvious return is by surface *runoff* in river systems, which funnel water back to the oceans. Some water also seeps into the ground and moves slowly through the pore spaces of the soil and rocks, where it is available for plants. Part of the water is used by the plants and then expelled into the atmosphere, but much of it slowly seeps into streams and lakes or migrates through the subsurface back to the oceans. In polar regions, or in high mountains, water can be temporarily trapped on a continent as glacial ice, but the glacial ice gradually moves from cold centers of accumulation into warmer areas, where melting occurs and the water returns to the oceans as surface runoff.

Water in the hydrologic system—moving as surface runoff, groundwater, glaciers, and waves and currents—erodes and transports surface rock material and deposits it as deltas, beaches, and other types of sedimentary formations. In this way, the surface material is in motion—motion that results in a continually changing landscape.

Early Concepts of the Hydrologic Cycle

An understanding of the hydrologic system did not arise as a revolutionary scientific theory through the work of a single scientist. It gradually evolved over many years, beginning in biblical time, and was based on the collective observations of many brilliant people.

Throughout history humans have wondered about the origin of rain, the source of rivers and springs, and the interactions of these waters. At first it was believed that rainfall could not account for the amount of water carried by

FIGURE 2.1
Earth's hydrologic system is especially vivid when seen from space. Indeed, until we could look at Earth from viewpoints in space, we did not fully realize how much of Earth is usually covered with clouds. Now via satellite complete storm systems can be seen in the various spiraling patterns of clouds hundreds of kilometers in diameter. The volume of water continually moving over the surface of Earth in the hydrologic system staggers the imagination. Calculations show that if the water evaporated from the oceans did not return by precipitation and by surface runoff, the ocean basins would be dry in 4000 years.

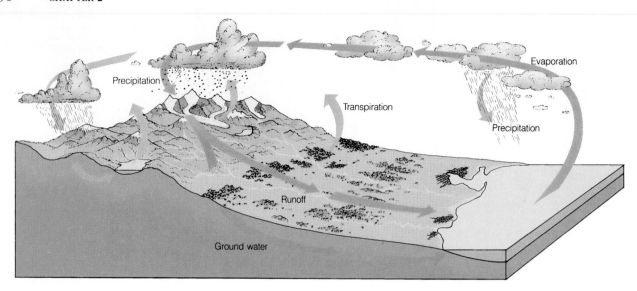

FIGURE 2.2
The circulation of water in the hydrologic system operates by solar energy. Water evaporates from the oceans, circulates with the atmosphere, and is eventually precipitated as rain or snow. Most of the water that falls on the land returns to the oceans by surface runoff and groundwater seepage. Variations in the major flow patterns of the system include the temporary storage of water in lakes and glaciers. Within this major system are many smaller cycles, or shortcuts, such as evaporation from lakes and transpiration from plants.

rivers. Part of the reason for this was that early philosophers lived in the Mediterranean area, where the headwaters of the major rivers, such as the Nile, were far away. In addition, people have always seen the quantity of water flowing in rivers they could not cross as more impressive than that in the raindrops from a passing storm. Major rivers flow continuously; storms are intermittent. Despite these disparities, however, the idea of a completed cycle in the movement of water—from the oceans to the rivers and back to the sea again—was recorded in biblical times.

> All of the rivers run to the sea, yet the sea is not full; unto the place from whence the rivers came thither they return again. (Eccles. 1:7)

Regardless of this ancient observation, prior to the sixteenth century it was still generally believed that water in rivers and springs could not be derived from rain because it was thought that rainfall occurred in insufficient amounts. Leonardo da Vinci (1452–1519), an exceptional genius in many fields, is generally credited with one of the earliest accurate descriptions of the hydrologic cycle.

> Whence we may conclude that the water goes from the rivers to the sea and from the sea to the rivers, thus constantly circulating and returning, and that all the sea and rivers have passed through the mouth of the Nile an infinite number of times. ... The conclusion is that the saltiness of the sea must proceed from the many springs of water which, as they penetrate the earth, find mines of salt, and these they dissolve in part and carry with them to the ocean and other seas, whence the clouds, the begetters of rivers, never carry it up.

But such a cycle was not demonstrated as fact until the mid-seventeenth century. The discovery was made by two French scientists, Pierre Perrault (1608–1680) and Edmé Mariotte (1620?–1684), who independently measured precipitation in the drainage basin of the Seine River and then measured the discharge into the ocean during a given interval of time. Their measurements

FIGURE 2.3
Early concepts of the hydrologic system were presented by natural theologian John Ray in the manner shown in this diagram. This model is an oversimplification, but differs little from that presented in most textbooks today.

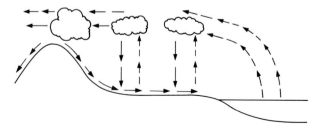

proved that precipitation alone produces not only enough water for the river to flow but also enough for springs. Precipitation thus was recognized as the basic source of all surface water. In addition, Edmund Halley (1656–1742), the English scientist after whom Halley's comet was named, estimated the evaporation from the Mediterranean Sea. He concluded it was as great as the volume of water in all the rivers flowing into the Mediterranean. These observations provided the initial basis for our understanding of the hydrologic system.

During the sixteenth and seventeenth centuries there were many heated debates between natural scientists and Christian theologians, and new ideas about nature were carefully scrutinized by the church. Although initially rejected by theologians, the hydrologic cycle gained favor when its usefulness in support of the doctrine of the "Divine Plan of Nature" became clear. This doctrine held that God created Earth expressly for man and that everything within it was part of a Great Divine Order. Early versions of the hydrologic cycle were seen as scientific verification of the Divine Plan by science. An idealized model of the hydrologic cycle was presented by John Ray, a seventeenth century theologian, and soon became generally accepted (Figure 2.3). Over the succeeding centuries a more or less standard model has evolved. This model appears in academic textbooks today in a form very similar to that presented in the seventeenth century (compare Figure 2.2 with Figure 2.3).

MOVEMENT OF WATER IN THE HYDROLOGIC SYSTEM

The movement of water in the hydrologic system begins with evaporation of seawater from the oceans. Water vapor then moves in the atmosphere and is precipitated as rain or snow. The cycle is completed when water returns to the sea by river systems. There are, however, many complex subcycles and irregularities within the system over time.

The idealized model of the hydrologic system shown in Figure 2.2 is a model of the land–ocean–atmosphere water exchange. It includes evaporation from the sea, movement of water vapor over the land, precipitation, surface runoff, infiltration into the subsurface, etc. Although the diagram is simple, in reality the flow of water through the hydrologic system is extremely complex and irregular over time and space. Today, many sophisticated measurements of almost every part of the hydrologic system have been made over most areas of the globe. These include measurements of evaporation, rainfall, stream discharge, and flow of groundwater. We now have a good idea of the quantitative aspects of the system. Most significant are (1) the volume of water within a subsystem, (2) *residence time* (average time during which a unit of water remains within the subsystem), and (3) paths of motion from one subsystem to another. Although these measurements may at first appear to be unimportant or mere statistics, we will explore their significance and what they imply about Earth's dynamics. Briefly, this is what has been found.

Components of the System

The total water of the planet is distributed in various subsystems, i.e., water in the air, on the land, and in the oceans—and in all three states: gas, liquid, and solid. Compared to the total mass of Earth, the total mass of water in the hydrosphere is extremely small, only 1 part in 4500. Yet water covers 71% of the surface of Earth and is extremely mobile, constantly moving from one place to another, often at remarkable speed. The distribution of water in the various parts of the hydrologic system is shown in Figure 2.4. The major reservoirs are the oceans and glaciers, which contain 99.5% of the world's water.

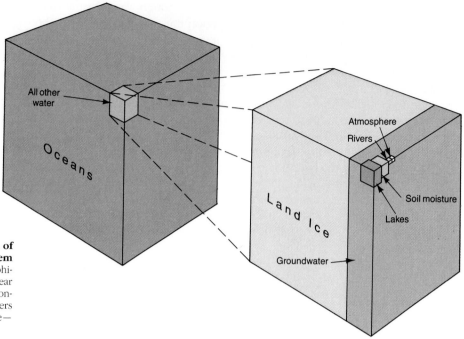

FIGURE 2.4
The relative amounts of water in each of the major parts of the hydrologic system can be best appreciated when shown graphically. More than 97% of the water on or near Earth's surface is in the oceans. Glaciers contain about 1.9%, groundwater—0.5%, rivers and lakes—0.02%, and the atmosphere—0.0001%.

What are the major components of the hydrologic system and how do they operate?

The Oceans. About 97.5% of the combined liquid water, ice, and water vapor is in the vast reservoir of the oceans. There is only one output of any importance from the oceans, that of evaporation; of the two inputs, precipitation accounts for about 90% and runoff from the land for about 10%. The residence time of water in the oceans is about 3000 years. This, of course, is only an average time because only the water at the surface of the ocean is involved in evaporation. Thus, water near the surface may have a much shorter residence time than deep water, which may be stored for very long periods, possibly hundreds of thousands of years.

Ice. Water in the form of ice constitutes about 80% of the water that is not in the oceans, or about 2% of Earth's total. Most of this form of water is located in the great glaciers that cover Greenland and Antarctica. Ice within the glaciers moves slowly from the areas where snow accumulates to the margins of the glaciers, where it melts. Water may reside in the glaciers for thousands or even millions of years. Present estimates, based on rates of loss by melting, suggest that water resides in glaciers, on the average, for 10,000 years.

The amounts of water stored in ice and in the oceans are intimately related. When glacial ice increases significantly, the sea level drops. When glacial ice decreases, the sea level rises. If the existing glaciers melted completely, the volume of water in the oceans would increase by about 2%, and the sea level would rise about 100 meters.

Groundwater. The next largest reservoir is groundwater, the water contained in the pore spaces of rock and soil. Surprisingly, about 20% of the water not in the oceans occurs as groundwater (see Figure 2.4).

Soil moisture accounts for about 0.005% of Earth's water, or about 0.5% of the water not contained in the oceans. Even at this low percentage, there is more water in the soil than is contained in river channels. There are two components of soil moisture: (1) water that is migrating down to the water table and (2) water that is retained in the soil. Water in the soil is withdrawn by evaporation from the soil surface and by plants. Residence time for water in the soil is about 1 month. Soil moisture is important because it migrates into rivers and groundwater systems, but it is also of supreme biologic importance.

Lakes. Lakes contain approximately 0.017% of Earth's water, or about 0.7% of that not contained in the ocean basins. Slightly more than half is held in freshwater lakes (lakes with outlets); the rest is in saline lakes (lakes without an outlet). About 75% of the total volume of freshwater lakes occurs in the large lakes of North America, Lake Baikal of the USSR, and the large lakes of East Africa. The Caspian Sea contains 75% of the water in closed basins with no outlet to the oceans. Water resides in the Caspian Sea for about 200 years. This relatively short residence time is due to the high rate of evaporation, which is equivalent to the discharge of a large river. Residence time for Lake Superior is almost 200 years, for Lake Erie 90 years. These figures are alarming in that serious pollution of lakes can be difficult to correct.

Atmosphere. The amount of water in the atmosphere is surprisingly small (0.0001%) when compared to that in other parts of the hydrologic system. Measurements show that if all the water held in the atmosphere condensed at a given moment and rained evenly over all Earth, it would form a layer only 2 mm deep. The volume would be roughly equivalent to the volume of water carried by the world's rivers. The daily rate of exchange between the water in the atmosphere and that on Earth includes approximately 2.5 mm that falls as rain and 2.5 mm that evaporates from the surface. Calculations show that the average residence time of water in the atmosphere is about 10 days, and there is a complete exchange of atmospheric moisture 40 times a year.

The atmosphere is a very dynamic part of the hydrologic system and a most vital link to other parts of the hydrologic system because it is the primary source of water. The dynamic aspect of atmospheric water can be appreciated by watching time-lapse movies of the regional movement of clouds on weather programs. Over a relatively short time (a few hours) the clouds may appear to explode as they move over the surface.

Rivers. At first glance, a river may seem to be a primary reservoir of water in the hydrologic system, but actually the volume of water in Earth's river systems is only about 0.0001% of the total water on Earth, or 0.005% of the water not in the oceans. The volume of water in river systems at any given time is not as much a measure of its importance as it is a measure of its residence time. Water flows through rivers at an average rate of 3 m^3/s. At this rate, it could traverse the entire length of the longest river in only 20 days. This means that although the volume of water in rivers at a given time is small, the total volume going through them over a period of time can be enormous. No other system is so universally important in shaping Earth's surface as running water.

The input of water into river systems is largely from precipitation. A considerable amount of water, however, is constantly seeping into most rivers from the ground at a slow rate. Groundwater seepage is important in most rivers to sustain perennial flow.

Living Organisms. The total amount of water stored in living organisms is extremely small when compared with amounts stored in other parts of the hydrologic system, but the volume of water circulating through living organisms and back into other parts of the hydrologic system is a different matter. Calculations indicate that in any given period of time, plants may well release as much water into the atmosphere as is discharged by all of the world's rivers combined. This is not an insignificant amount. Its importance is further emphasized when we consider that the residence time of water in organisms is very short, ranging from a few hours in warm-blooded animals to a season for most plants. Thus, organisms are involved with the flow of a significant volume of water over a period of time, and plants and animals constitute an important segment in the hydrologic system. If we continue to destroy the forests, the effect on the hydrologic system may be devastating.

Magnitude of the System

The hydrologic system is perhaps the most fundamental and significant geologic system operating on Earth's surface. Its influence on the development of surface features is sometimes subtle, sometimes dramatic. Your problem as a student new to geology will not be in understanding the process but in conceiving the worldwide scope of the system and its influence on the surface features of our planet.

One of the best ways to gain an accurate conception of the magnitude of the hydrologic system is to study the space photographs in Figure 1.2 and on page 26. These photographs provide views of the system in operation on a global scale. A traveler arriving from space would observe that Earth's surface is predominantly water (an obvious conclusion from Figure 1.2). The movement of water from the oceans to the atmosphere is expressed in the flow patterns of the clouds. This motion is one of the most distinctive features of Earth viewed from space. It stands out in marked contrast to the surfaces of Mars, and especially the Moon and Mercury, where impact craters dominate the stark landscape. Without hydrologic systems the surfaces of these planets have remained largely unmodified for 4 billion years (see Figure 1.3).

In addition, the swirling, moisture-laden clouds carry an enormous amount of energy. For instance, the kinetic energy produced by a hurricane amounts to roughly 100 billion kilowatt-hours per day. That is considerably more than the energy used by all of the people of the world in one day.

Another way to grasp the magnitude of the hydrologic system is to consider the volume of water involved. From measurements of rainfall and stream discharge, together with measurements of heat and energy transfer into bodies of water, scientists have calculated that $400,000 \text{ km}^3$ of water evaporates each year (or 1100 km^3 per day). Of this, about $336,000 \text{ km}^3$ is taken from the oceans, and $63,000 \text{ km}^3$ evaporates from bodies of water on the land. Of the total $400,000 \text{ km}^3$ of water evaporated annually, about $101,000 \text{ km}^3$ falls as rain or snow on the continents. Most of the precipitation (from 60 to 80%) returns directly to the atmosphere by evaporation. About $38,000 \text{ km}^3$ of water flows back to the oceans over the surface as rivers or beneath as groundwater.

At these rates, if the hydrologic system were interrupted and water did not return to the oceans by precipitation and by surface runoff from the continents, sea level would drop 1 m per year. All of the ocean basins would be completely dry within 4000 years. The recent glacial epoch demonstrates this point clearly: the hydrologic system was partly interrupted because much of the water that fell on the Northern Hemisphere froze and accumulated to form huge continental glaciers. This prevented the water from flowing immediately back to the oceans as surface runoff. Consequently, sea level dropped more than 100 m during the ice age.

As a final observation, consider the volume of water running off Earth's surface. The circulation of water in the hydrologic system provides a continuous supply of surface runoff over the land. At any given instant, about 1260 km^3 of water is flowing in the world's rivers. Gauging stations on the major rivers of the world indicate that, if the ocean basins were empty, runoff from the continents would fill them completely in 40,000 years. (This estimate assumes no precipitation falling directly into the oceans.) Moreover, if all of the evaporated water returned to the oceans by runoff, with no return to the atmosphere by immediate evaporation and *transpiration* (return of water to the atmosphere by plants), then the ocean basins would be filled in slightly more than 3000 years.

Gravity plays an important role in the hydrologic system. It is the force causing rain to fall on the land and move downslope. It causes rivers to flow and to transport sediment from higher to lower elevations. Indeed, the movement of water in each part of the hydrologic cycle is a type of gravity flow system. The hydrologic system eroded the Grand Canyon, carved the fantastic Himalaya Mountains, and deposited the Mississippi delta. It formed the glaciers of the last ice age, developed the Sahara Desert, and created the streams and valleys of your hometown area. Little if any surface of Earth is not affected in some way by the work of the hydrologic system. When you go for a walk in the country, it is very likely you will be walking over a surface that was formed by running water, a young surface that is still in the process of developing.

EFFECTS OF THE HYDROLOGIC SYSTEM

As water moves in each of the hydrologic subsystems (rivers, glaciers, groundwater, oceans, and winds), it erodes, transports, and deposits material and continually modifies the surface of Earth.

The hydrologic system has operated on Earth's surface throughout its entire history. Whenever we see water, we see parts of this system. The energy involved in this system—raising 400,000 km^3 of water into the air and dumping it on the surface of the land each year—is almost beyond comprehension. The surfaces of the continents have an average elevation of approximately one-half mile (2600 feet) above sea level. At this height, the energy released by the surface runoff of streams alone is equivalent to that of a waterfall one-half mile high over which a volume of water equal to 90 Mississippi Rivers pours continuously.

The enormous energy of the hydrologic system is involved with each of the subsystems—rivers, glaciers, groundwater, oceans, and wind—all of which erode, transport, and deposit material and create new landforms in the process. We will explore the details of each major geologic process in subsequent chapters, but let us now examine some of the results of water moving over Earth.

River Systems

As would be expected, the vast volumes of water that move over Earth's surface produce a landscape dominated by features formed by running water. From viewpoints on the ground, we cannot appreciate the prevalence of stream channels on Earth's surface. From space, however, we readily see that stream valleys are the most abundant landforms on the continents. In arid regions, where vegetation and soil cover do not obscure details, the intricate network of stream valleys is most impressive (Figure 2.5). Most of the surface of every continent is somehow related to the slope of a stream valley, which collects and funnels surface runoff toward the ocean.

The important point is that each continent has one or more major *river systems,* and the valley slopes formed by the network of tributary streams are the dominant landforms. All parts of the land surface are linked to the flow of water, solutes, and sediment through a drainage system. We can see from space photographs that the surface drainage of a continent is an important segment of the hydrologic system and is a vivid record of how moving water has continually modified and shaped the surface of the land.

Another important aspect of a river system is that it provides the fluid medium that transports huge amounts of sand, silt, and mud to the oceans. This material forms the great deltas of the world, which are records of the amount of material washed off the continents by rivers. The Mississippi Delta (Figure 2.6) is a classic example. The river is confined to a single channel far downstream from New Orleans. It then breaks into a series of distributaries, which

What distinctive landforms are produced by each of the five major subsystems of the hydrologic system?

FIGURE 2.5
Drainage systems are a clear record of how surface runoff has sculptured the land. They testify to the magnitude of Earth's hydrologic system, for few areas of the land are untouched by stream erosion. In this photograph of a desert region, details of the delicate network of tributaries are clearly shown. On the Moon, Mercury, and Mars, craters dominate the landscape, but on the continents of Earth, stream valleys are the most abundant landform.

FIGURE 2.6
The Mississippi Delta, like deltas of other major rivers, is a record of erosion due to the hydrologic system. Sediment eroded from the land is transported by a river system and deposited in the sea. The dynamics of delta building are displayed vividly in this photograph. The cloud of mud and silt delivered to the ocean colors the water a lighter tone around the mouth of the river. This material is deposited as banks of mud, sand, and clay over the continental shelf as the delta grows seaward at a rate of nearly 20 km per 100 years. Measurements indicate that the Mississippi River pours more than a million metric tons of sediment into the Gulf of Mexico each day. In the process of deltaic growth, the river builds up a projection of new land into the ocean. Eventually, the river finds a shorter route to the ocean and abandons its active distributary channel for the shorter course. The abandoned distributary ceases to grow and is eroded back by wave action. Abandoned river channels and inactive subdeltas can be seen clearly on each side of the present river.

FIGURE 2.7
Valley glaciers, such as these in Alaska, occur where more snow accumulates each year than is melted during the summer months. Valley glaciers originate in the snowfields of high mountain ranges and flow as large tongues of ice down preexisting stream valleys. The moving ice is an effective agent of erosion and modifies the valleys in which it flows; thus glaciers cause local modifications of the normal hydrologic system. The dark lines on the glaciers are rock debris derived from the valley walls.

FIGURE 2.8
Groundwater is a largely invisible part of the hydrologic system—it occupies pore spaces in the soil and rocks beneath the surface. It can, however, dissolve soluble rocks, such as limestone, to form complex networks of caves and subterranean passageways. As the caverns enlarge, their roofs may collapse, so circular depressions called sinkholes are formed. Sinkholes create a pockmarked surface called karst topography. The hundreds of lakes shown in this photograph of the area west of Cape Canaveral, Florida, occupy sinkholes and testify to the effectiveness of groundwater as a geologic agent.

build extensions of new land into the ocean. A large "cloud" of suspended sediment forms in the ocean where the river empties into the Gulf of Mexico and is actively building new land seaward. Ultimately, the main channel shifts its course to seek a more direct route to the ocean, and the extension of land (subdeltas) is eroded back by waves and currents. Previous courses of the Mississippi can be seen on both sides of the present river.

Glacial Systems

In cold climates, precipitation falls in the form of snow, which remains frozen and does not return immediately to the ocean as surface runoff. If more snow falls each year than melts during the summer months, huge bodies of ice build up to form *glaciers*. Figure 2.7 is an example of existing glaciers. Large valley glaciers originate from snowfall in the high country and slowly flow down valleys as rivers of ice. They melt at their lower ends and return their water to the hydrologic system as surface runoff.

At the present time, the continent of Antarctica is covered with a continental glacier, a sheet of ice from 2.0 to 2.5 km thick. It covers an area of 13,000,000 km^2—an area larger than the United States and Mexico. A part of the Antarctic glacier can be seen in Figure 15.10. The formation of a continental glacier completely modifies the normal hydrologic system because the water does not return immediately to the ocean as surface runoff but moves slowly as flowing ice. An ice sheet similar to that now on Antarctica covered a large part of North America and Europe during the last ice age and retreated only within the last 15,000 years. As the ice moved, it modified the landscape by creating numerous lakes and other landforms.

Groundwater Systems

Another segment of the hydrologic system is the water that seeps into the ground and moves slowly through the *pore spaces* in the soil and rocks. As it moves, *groundwater* dissolves soluble rocks and creates caverns and caves, which can enlarge and collapse to form surface depressions called sinkholes. This type of solution-generated landform is common in Kentucky and Florida and west Texas and is easily recognized from the air (Figure 2.8). The sinkholes are filled with water and create a pockmarked surface somewhat resembling the cratered surface of the Moon.

Shoreline Systems

The hydrologic system also operates along the shores of all continents, islands, and inland lakes by the unceasing work of waves. The oceans and lakes are bodies of mobile water subject to a variety of movements, waves, tides, and currents, all capable of eroding the coast and transporting vast quantities of sediment (Figure 2.9). The effects of shoreline processes are seen in wave-cut cliffs, shoreline terraces, deltas, beaches, bars, and lagoons. The satellite photo in Figure 2.9 shows a line of barrier islands formed by current and wave action. Suspended mud is carried beyond the shore by currents and is eventually deposited on the sea floor. The patterns of suspended sediment are most striking when seen from space through satellite photography, and sediment movements are often monitored in this fashion.

The hydrologic system expends a considerable amount of energy upon the shore. Over a period of only a few tens of years significant modification in the shape and configuration of many coasts can be seen. Hurricanes can greatly alter the sandy coastline in a single day.

Great Dismal
Swamp

Pamlico Sound

Cape
Hatteras

Cape Lookout

FIGURE 2.9
The Atlantic coastline of the southern states is dominated by offshore bars and barrier islands. A considerable amount of sediment eroded from the continent is brought seaward by rivers and reworked by wave action to form long, narrow, sandy barrier beaches, which fringe the mainland. This photograph shows part of the shoreline of North Carolina. The lagoon of Pamlico Sound and the Great Dismal Swamp, protected by the barrier island from the vigorous wave action of the Atlantic, will eventually become filled with sediment, extending the mainland seaward to the present position of the barrier beach. The edge of the continental shelf is delineated by puffy clouds, which form where the cold surface water to the east meets the warmer surface water of the shallow nearshore zone. The edge of the shelf is only about 120 m deep.

Eolian Systems

The hydrologic system also operates in the arid regions of the world. In many deserts, river valleys are still the dominant landform. There is no completely dry place on Earth. Even in the most arid regions, some rain falls, and climatic patterns change over the years. River valleys can be obliterated, however, by the great seas of windblown sand that cover the desert landscape (Figure 2.10). Air circulating in the atmosphere constitutes *eolian* (wind) *systems,* which can transport enormous quantities of loose sand and dust, leaving a distinctive record of the operation of the wind. In the broadest sense, the wind itself is part of the hydrologic system, a moving fluid on the planet's surface.

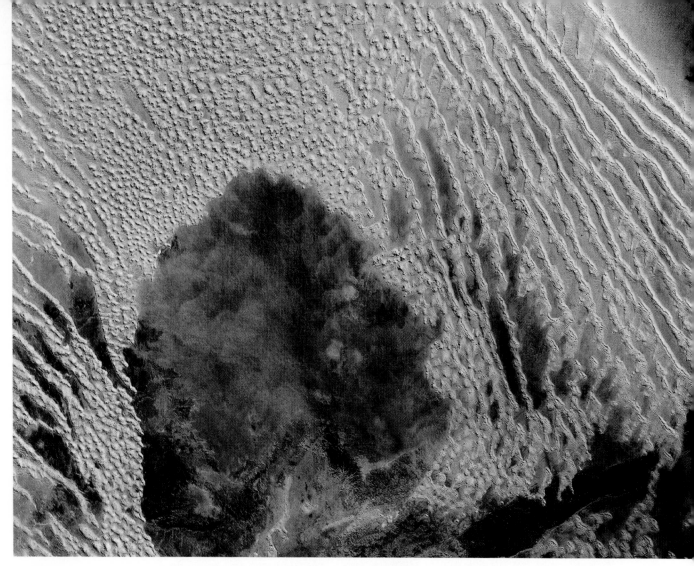

FIGURE 2.10
"Sand seas" of the great deserts of Earth form in arid low-latitude regions where there is not enough precipitation for the hydrologic system to operate in its normal manner. The vast areas of migrating sand dunes in the world's deserts illustrate the effectiveness of the circulating atmosphere as a geologic agent, continually transporting enormous quantities of sediment over the surface of Earth.

THE HYDROLOGIC SYSTEM AND THE ENVIRONMENT

The hydrologic cycle operates as a unified system, with each part or segment intimately related to all the others. Every aspect of the biosphere is interwoven with the hydrologic system and depends on it for sustenance.

Because the hydrologic cycle operates as a unified system, we cannot consider any part to be isolated and independent. The atmosphere is related to rivers and lakes, which are interrelated to the groundwater reservoir, and all are ultimately related to the ocean. Moreover, every aspect of the biosphere is interwoven with the hydrologic system and depends upon it for sustenance.

The hydrologic system is highly susceptible to human influence. We can build dams for water storage, flood control, and irrigation, and as a result we alter the system of groundwater, the balance of sediment input in a delta, and the concentration of salt in farmland.

Pollutants introduced to the hydrologic system will eventually spread and will soon enter stream networks. Increasingly, and ominously, pollutants are entering the groundwater system from a variety of sources. Because subsurface water has a long residence time, such pollutants may affect a system for thousands of years. Plants and animals are parts of the hydrologic system, and in some instances pollutants entering surface water eventually spread into plants and animals and so infect the food chain. We live within the hydrologic system. What substances we introduce into it will remain and move within the system. It is impossible to simply throw our waste away.

SUMMARY

1. The hydrologic system includes all possible paths by which water moves from the oceans, through the atmosphere, over the land, and below the ground.
2. The concept of the hydrologic system was developed in the middle of the seventeenth century from measurements of precipitation and discharge in the Seine River basin.
3. More than 99% of Earth's water is concentrated in the oceans and in the glaciers of Antarctica and Greenland.
4. Water in the soil, atmosphere, and rivers constitutes only a small part of Earth's water. It moves through these systems very rapidly and initiates significant change in Earth's surface features.
5. Rivers, glaciers, groundwater, shorelines, and wind are the major subsystems of the hydrologic system and each produces distinctive landscapes.

KEY TERMS

eolian system (p. 39)

glacier (p. 38)

groundwater (p. 38)

hydrologic system (p. 28)

pore spaces (p. 38)

residence time (p. 31)

river system (p. 35)

runoff (p. 28)

shoreline system (p. 38)

transpiration (p. 34)

REVIEW QUESTIONS

1. Diagram the paths by which water circulates in the hydrologic system.
2. What measurements prove that the source of rivers is precipitation and not springs?
3. What is the source of energy for the hydrologic system?
4. Approximately how much water evaporates from the ocean each year?
5. Describe the major landforms resulting from (a) rivers, (b) groundwater, (c) glaciers, and (d) wind.

ADDITIONAL READINGS

Baumgartner, A., and E. Reichel. 1975. *The World Water Balance*. Amsterdam: Elsevier. 179 pp.

Leopold, L. B. 1974. *Water: A Primer*. New York: Freeman.

Nace, R. L. 1967. U.S. Geological Survey Circular 536.

Peixit, J. P., and M. A. Kettani. 1973. The control of the water cycle. *Scientific American*, April.

Revelle, R. 1963. Water. *Scientific American*.

U.S. Geological Survey. 1984. *1983 National Water Summary—Hydrologic Events and Issues*. U.S. Geological Survey Water Supply Paper 2250.

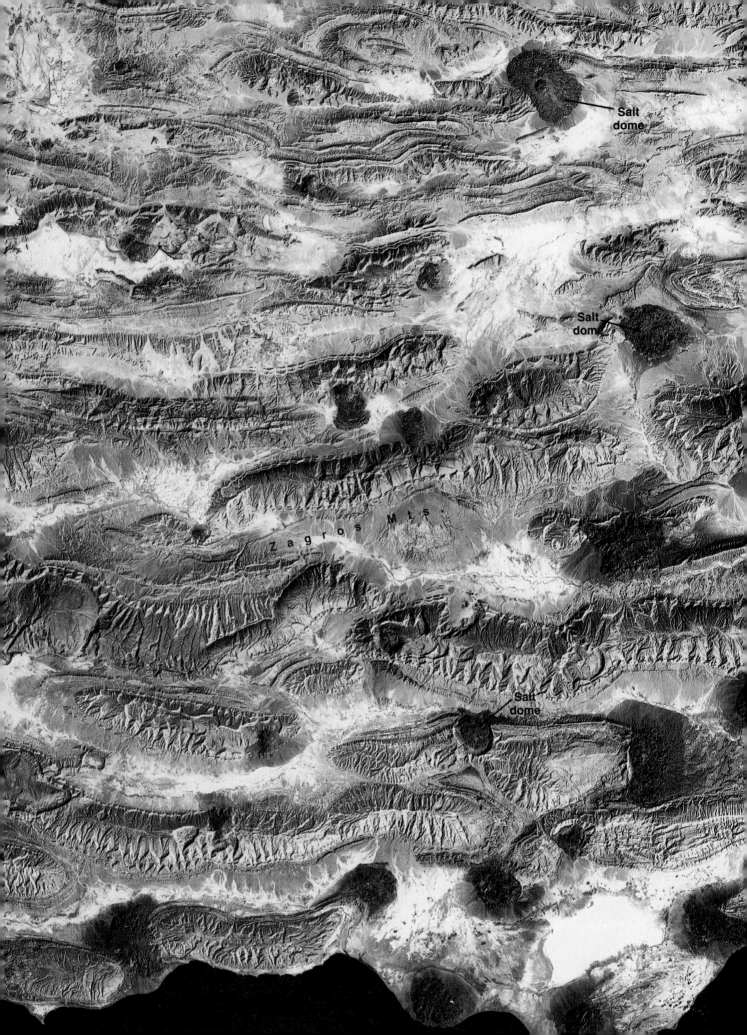

Salt
dome

Salt
dome

Z a g r o s   M t s .

Salt
dome

Even though Earth's crust appears to be fixed and stable, there is convincing evidence that it is in constant motion, and as you continue your study of geology, the profound significance of Earth's internal energy will become increasingly clear. There is now overwhelming evidence that the entire lithosphere moves, and as it moves, continents split and drift thousands of kilometers across Earth's surface.

One of the most dramatic expressions of this motion is the deformation that occurs where moving segments of the lithosphere collide and deform the crust into a series of folds. The Zagros Mountains of southern Iran, shown in the Landsat image on the facing page, are an excellent example. They were produced as a result of the collision of Arabia with Asia. The ridges that extend across the scene are folds that represent a crustal shortening of approximately 50 km. The upfolded strata are in various stages of dissection, which imparts a rough texture to the flanks of the folds.

Other expressions of mobility in the rocks in this image are the salt domes, shown as circular, dark, rugged hills. Thick layers of salt formations are less dense than the surrounding rocks and tend to flow. Many rise as teardrop-shaped domes and eventually reach the surface. Here in the extreme desert of southern Iran is one of the few places in the world where soluble salt domes form large hills.

In this chapter we will introduce you to the revolutionary theory of plate tectonics, which explains for the first time how Earth's energy systems work and how the major features of our planet can be explained by a simple system of moving plates. Later, in Chapter 18, we will consider details of the theory, how it developed, and some of the evidence upon which it is based. But for now we will be concerned with the general ideas of the theory as a necessary background for the succeeding chapters.

3
The
Tectonic
System

MAJOR CONCEPTS

1. The theory of plate tectonics explains the major structural features of Earth as a result of movements of a series of lithospheric plates.
2. Where plates move apart, hot material from the mantle wells up to fill the void and creates new lithosphere. The major features formed where plates spread apart are continental rifts, oceanic ridges, basaltic volcanism, and new ocean basins.
3. Where plates converge, one slides beneath the other and plunges down into the mantle. The major features formed at converging plate margins are (a) folded mountain belts, (b) volcanic arcs, and (c) deep-sea trenches.
4. Earth's lithosphere floats on the denser, plastic asthenosphere beneath, and it rises and sinks in attempts to maintain isostatic equilibrium.
5. The crust and upper part of the mantle form the lithosphere, about 100 km thick, that overlies the hotter and more plastic mantle of the asthenosphere, above which horizontal lithospheric displacement can take place.

THE THEORY OF PLATE TECTONICS

Plate tectonics is the theory of global dynamics in which the lithosphere is believed to be broken into a series of separate plates that move in response to convection in the upper mantle. The margins of the plates are sites of considerable geologic activity, such as sea-floor spreading, volcanic eruptions, crustal deformation, mountain building, and continental drift.

Geologists have long recognized that Earth has its own source of internal energy that is repeatedly manifested by earthquakes, volcanic activity, and folded mountain belts, but not until the middle 1960s was a unifying theory of Earth's dynamics developed. This theory, known as plate tectonics, provides, for the first time, a master plan of Earth's dynamics. The term *tectonics,* like architecture, comes from the Greek *tektonikos* and refers to building or construction. In geology, **tectonics** is the study of the formation and deformation of Earth's crust that result in large-scale structural features. The importance of the plate tectonics theory is that it provides a unifying framework in which numerous aspects of geology assume a comprehensible whole. The great power of this theory is that it can explain, with equal ease, such diverse geologic phenomena as the origin of mountains, the history of ocean basins, the chemistry of lavas, rates of sedimentation, and migration and extinction of fauna. Plate tectonics explains the dynamics of the planet Earth—the way it works—and is probably the most significant scientific breakthrough in the history of geology.

The basic elements of the plate tectonics theory are simple and can be easily understood by studying the diagram in Figure 3.1. The lithosphere, which includes Earth's crust and part of the upper mantle, is rigid, but the underlying asthenosphere yields to plastic flow. The fundamental idea of *plate tectonics* is that the segments or plates of the rigid lithosphere are in constant motion relative to one another and carry the lighter continents with them. These plates form as hot mantle material rises along midoceanic ridges, but they are destroyed in subduction zones, where one of the converging plates plunges into the hotter mantle below. Their descent is marked by deep-sea trenches that border island arcs and some continents. Where plates slide by one another, large transform faults occur. The movement and collision of

What exactly is the theory of plate tectonics?

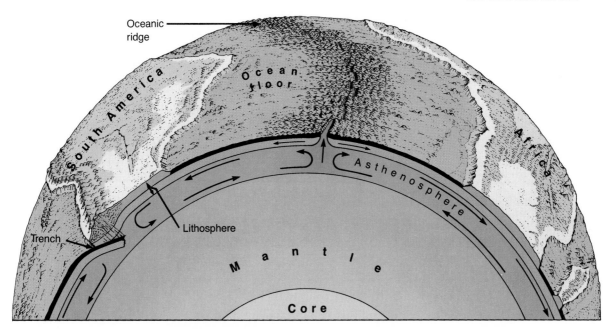

FIGURE 3.1
The tectonic system operates from Earth's internal heat. The asthenosphere is more plastic than both the overlying lithosphere and the underlying lower mantle because of an optimum balance of temperature and pressure. Above the plastic asthenosphere, relatively cool and rigid lithospheric plates split and move apart as single mechanical units. Molten rock from the asthenosphere wells up to fill the void and creates new lithosphere. Convection circulation occurs in the asthenosphere where material fills the gap between the spreading plates and moves away to areas where plates descend into the lower mantle at oceanic trenches. Some plates contain blocks of thick, low-density continental crust, which cannot sink into the mantle. As a result, where a plate carrying continental crust collides with another plate, the continental margins are deformed into mountain ranges. The plate margins are the most active areas on Earth—the sites of the most intense volcanism, seismic activity, and crustal deformation.

plates account for most of Earth's earthquakes, volcanoes, and folded mountain belts, as well as for the drift of its continents.

The theory of plate tectonics was developed during the 1960s. It resulted largely from newly acquired abilities to study the characteristics of the ocean floor, map its surface features, and measure its magnetic and seismic properties. Evidence for the revolutionary theory of crustal movement comes from many sources and includes data on the structure, topography, and magnetic patterns of the ocean floor, the locations of earthquakes, the patterns of heat flow in the crust, the locations of volcanic activity, the structure and geographic fit of the continents, and the nature and history of mountain belts. (See Chapter 18 for details.)

From the standpoint of Earth's dynamics, the boundaries of the **tectonic plates** are where the action is. As seen in Figure 3.2, plate boundaries do not necessarily coincide with continental boundaries, although some do. There are seven very large plates and a dozen or more smaller plates (not all of which are shown in Figure 3.2). Each plate may be up to several hundred kilometers thick. Plates slide over the more mobile asthenosphere below, generally at rates between 1 and 10 cm per year, although rates of interactions between adjacent plates approach 20 cm per year. Because the plates are internally rigid, they interact mostly along their edges.

The plates diverge and move apart along midoceanic ridges. Hot material from the deeper mantle wells up to fill the void. Some of this material is erupted at the surface as basaltic lava. Thus, new lithosphere is formed at the plate's trailing edge. The oceanic ridges stand high because their material is hot and, therefore, low in density. Heat flow at the ridge crest is six times greater than that of old oceanic crust beyond the flanks of the oceanic ridge.

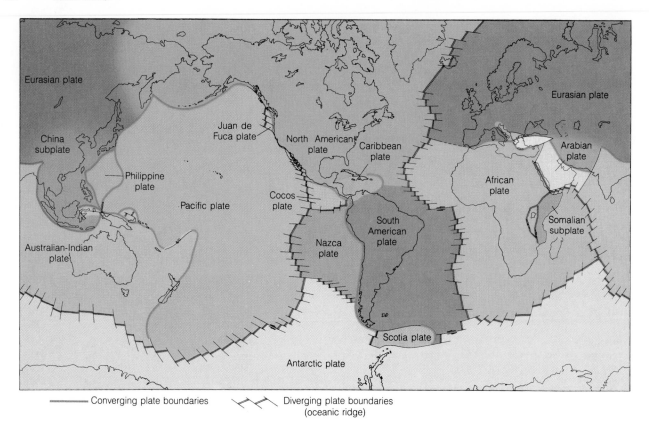

Converging plate boundaries — Diverging plate boundaries (oceanic ridge)

FIGURE 3.2
A mosaic of plates forms Earth's lithosphere, or outer shell. The plates are rigid, and each moves as a single unit. There are three types of plate boundaries: (1) the axis of the oceanic ridge, where the plates are diverging and new oceanic crust is generated, (2) transform faults, where the plates slide past each other, and (3) subduction zones, where the plates are converging and one descends into the asthenosphere.

New plates form as old ones break up and drift apart. The rift valleys of East Africa are believed to have formed along a zone of incipient *divergent plate boundaries* and are characterized by updoming, rifting, and volcanism. A more advanced stage of rifting is exemplified by the Red Sea rift, which almost completely separates the Arabian Peninsula from Africa.

The ridges are commonly offset by large faults that are also the sites of shallow earthquakes. It is here that plates slide past one another. The best known example of this type of fault is the San Andreas Fault of California (see Figure 8–15).

Within the moving plate, the continents and oceanic crust experience little tectonic or volcanic activity as they passively move away from the midoceanic ridge. Hotspots in the deep mantle, however, may create upwarps and volcanic activity in the plates' interiors.

At *convergent plate boundaries,* the geologic activity is far more varied and complicated. We now know much about the mechanics of these junctions from geophysical studies made with instruments developed for oil exploration—in particular, seismic-reflection profiles—which provide an image of plate boundaries. It is the processes at these junctions that ultimately result in folded *mountain belts,* alteration of the rock by heat and pressure, and the growth of continents. At the point where the plates converge, one tips down and slides beneath the other, a process known as *subduction.* The simplest form of the convergent junction is one where two plates with oceanic crust meet. Such plate junctions in the eastern Pacific region lie along the volcanic islands of Tonga, the Marianas, and the Aleutians. A trench, normally 5–8 km deep, is formed where the undersliding plate plunges down into the mantle. Sediment on the ocean floor and slivers of oceanic crust are scraped off against the front of the overriding plate and stacked in a telescopic sequence against it as a chaotic mass of mud and basalt, commonly referred to as a mélange. Some large slices of oceanic crust and mantle may be incorporated. The Franciscan Formation in the area of San Francisco, California, is believed to have been plastered against the continental margin as sediment was stripped from a subducting plate. Complications arise when continents or island arcs are transported to a *subduction zone.* The subducting plate is heated as it tries to descend into the mantle, and some of the material melts, rises, and may be extruded to form a string of volcanic islands (island arc) or a chain of volcanoes on a continent. It is a series of these volcanic arcs that make up the "ring of fire" around the Pacific Ocean.

The continental crust and material in the island arc are relatively buoyant and so resist subduction. The horizontal motion of the plates is transformed to intense compression. Deformation becomes extreme where two continental blocks collide, as in the case of India and Asia.

The processes occurring at island arcs and active continental margins can lead to the formation of mountain chains. Sediment along a continental margin may be deformed into a belt of thrust-faulted and folded rocks associated with volcanic rocks and igneous intrusions developed in the subduction zone. Continents may be welded together along suture zones (like the Ural Mountains of Russia), and island arcs and small continental fragments may be accreted onto a major continental mass.

The theory of plate tectonics explains much of the dynamics of Earth's crust. The concepts were proven primarily by geophysical studies and deep-sea drilling of the ocean basins, but the theory has also revolutionized our understanding of continental geology. With a better understanding of how geologic features are forming today, we can deduce how similar features in the geologic record are related to ancient plate boundaries. Geologists are now searching and recognizing ancient plate boundaries—zones where old continents have been welded together—and how small continental fragments, ancient seamounts, and island arcs have been accreted to the continents. Plate motion has dominated tectonic processes on our planet for at least the past 2500 million years, providing a fascinating record, waiting to be interpreted.

Although there are many unanswered questions, the basic source of energy for tectonic movement is believed to be Earth's internal heat, which is transferred by convection. Hot material in the mantle rises to the base of the lithosphere, where it then moves laterally, cools, and descends to become reheated, so the cycle begins again. A familiar example of convection can be seen in the heating of a pot of soup (Figure 3.3). Heat applied to the base of the pot warms the soup at the bottom, causing it to expand and become less dense. The warm fluid rises to the top and is forced to move laterally. It then cools, becomes more dense, and sinks. The regular flow circuit of rising warm fluid and sinking cold fluid is called a convection current.

In order to appreciate why the plate tectonics theory has had such an impact on the science of geology, let us consider some of the major geologic features of our planet and see how the theory explains them. Take a moment and study Figure 3.2 again. You will want to become very familiar with this map because it shows a new geography: the geography of plate tectonics. As can be seen on Figure 3.2, seven major lithospheric plates are recognized—the North American, South American, Pacific, Australian, African, Eurasian, and Antarctic plates—together with several smaller ones. The spreading axis, where the lithosphere is spreading apart, is marked by the oceanic ridge, which extends from the Arctic down through the central Atlantic and into the Indian and the

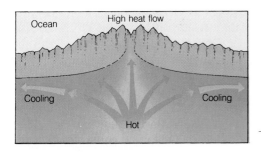

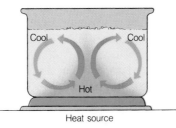

Heat source

FIGURE 3.3
Convection in the mantle can be compared to convection in a pot of soup. Heat from below causes the material to expand and become less dense. The warm material rises by convection and spreads laterally. It then cools, becomes more dense, and sinks. It is reheated as it descends, and the cycle is repeated.

Pacific oceans. Movement of the plates is away from the crest of the oceanic ridge.

The North American and South American plates are moving westward and interacting with the Pacific, Juan de Fuca, Cocos, and Nazca plates along the western coast of the Americas. The Pacific plate is moving northwestward from the oceanic ridge toward the system of deep trenches in the western Pacific basin. The Australian plate includes Australia, India, and the northeastern Indian Ocean. It is moving northward, causing India to collide with the rest of Asia and producing the Himalaya Mountains. The African plate includes the continent of Africa, plus the southeastern Atlantic and western Indian oceans. It is moving eastward and northward. The Eurasian plate, which consists of Europe and most of Asia, moves eastward. The Antarctic plate includes the continent of Antarctica plus the Antarctic Ocean. It is somewhat unique in that it is nearly surrounded by oceanic ridges.

There are many other features of both the ocean basins and continents that can be explained nicely by the plate tectonics theory. We will consider many of them in some detail in subsequent chapters, but let us first look briefly now at the major geologic features of our planet and how they fit into the tectonic system. First look at the topography of the ocean floor (see the maps inside book covers). One of the remarkable features of the sea floor is the great system of deep-sea trenches that almost surrounds the Pacific Ocean. They have been known for almost a century. They are long narrow troughs and are the lowest features on Earth. Invariably they are associated with volcanic activity and a zone of intense earthquakes. It was not known why. It is now quite obvious that the deep-sea trenches mark the zones where plates converge and one descends into the mantle. Their low topography, association with earthquakes, and lines of volcanic activity are clearly explained in light of plate tectonics. Likewise, the midoceanic ridge, the world's longest structural and topographic feature, has been known to exist for decades, but why is there a "mountain range" extending through the center of the world's ocean basins? Plate tectonics provides an answer. The midoceanic ridge marks the zone where the plate margins are uplifted and stretched apart.

Many geologic features on the continents that have been interpreted as independent phenomena are now explained in light of a single unifying theory of Earth's dynamics. Our newly acquired ability to view Earth from satellites in space permits us to see many of Earth's tectonic features in one synoptic view and to observe many phenomena that occur as a result of moving plates. For example, the great rift of the Red Sea, shown in Figure 3.4, is an extension of the spreading axis of the Indian Ocean, which splits the Sinai and Arabian peninsulas from Africa. Take the time to locate the area shown in this remarkable photograph on the tectonic map and physiographic map on the inside covers of the book. The structure of the area shown in Figure 3.4 is dominated by two long, linear fault valleys. One is occupied by the Gulf of Aqaba, the Dead Sea, and the Jordan River and the other by the Gulf of Suez. These rifts express dramatically the tensional or shear stresses in the crust, and we now see the way these stresses affect Earth's surface. The long, straight lines extending through the Jordan River valley and the Gulf of Aqaba are major fractures produced by the rift. Volcanic rocks extruded along the fracture zone are visible as a dark tone, especially evident near the Sea of Galilee, in the upper part of the photograph. Oceanographic studies confirm that the Red Sea is an extension of the oceanic-ridge system, which has split the northern part of Africa and caused the Arabian Peninsula to drift to the north. It represents the initial stages of sea-floor spreading and development of a new ocean basin. As the Arabian Peninsula moves northward, a large fracture is developing along its left flank to form the long, narrow trough of the Jordan valley and the Dead Sea.

Another type of fracture in the crust results if plates slide past each other horizontally. The great San Andreas Fault in California, parts of which are

What geologic features does the theory of plate tectonics explain?

FIGURE 3.4

The Red Sea rift, which separates Africa from Asia, is a major fracture system in Earth's crust. This feature, an important part of Earth's rift system, represents the incipient stages of crustal fragmentation and the movement of continental plates. The rift extends up the Red Sea and splits at its northern end, with one branch forming the Gulf of Suez and the other extending up the Gulf of Aqaba, into the Dead Sea, and up the Jordan valley. Movement of the Arabian plate in a northeasterly direction, away from Africa, has created a new ocean basin, which is in the initial stage of its formation. As the rift widens, the edges of the continental block break off, forming a series of steps leading down toward the depression. These steplike blocks can be seen along the Gulf of Suez as distinct parallel lines in the bedrock and as linear trends in the offshore islands.

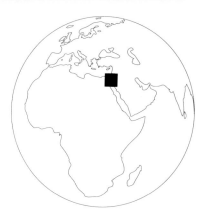

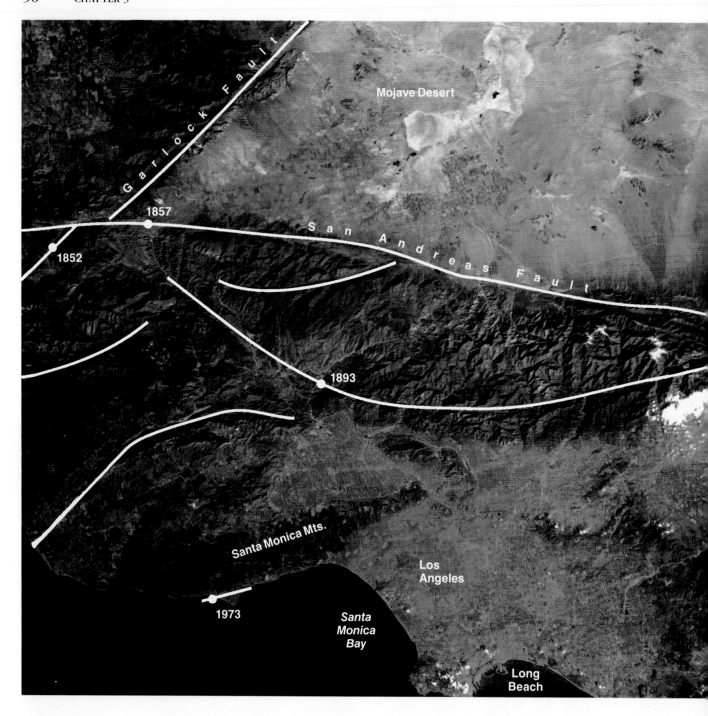

Mojave Desert

Garlock Fault

1857

1852

San Andreas Fault

1893

Santa Monica Mts.

Los Angeles

1973

Santa Monica Bay

Long Beach

FIGURE 3.5
The San Andreas Fault is one of the most significant geologic structures in California. This view shows the intersection of the San Andreas Fault and the Garlock Fault, northwest of Los Angeles. These two major structural features intersect to form the boundary of the Mojave Desert, the light-colored triangular area in the upper right-hand quarter of the photograph. At least a dozen major fault systems can be seen as linear trends in the landscape in this view. Most have been active during historical times. Movement along the San Andreas Fault is horizontal, that is, one block of Earth's crust slides laterally past the other.

clearly shown in Figure 3.5, results from this type of movement. The fault zone is marked by sharp linear landforms such as straight narrow valleys, linear ridges, and offset drainage patterns (see Figure 8.15). The San Andreas Fault is over 1000 km long and probably cuts the entire thickness of the lithosphere; its earthquakes extend vertically into Earth at least 20 km. Many smaller faults form branches off the main zone. The San Andreas fault system is an active boundary between the Pacific plate to the west and the North American plate to the east. The Pacific plate is moving at about 6 cm per year relative to the North American plate. As stress builds up between the plates, a sudden release causes the earthquakes for which California is noted. The famous San Francisco earthquake of 1906 resulted from movement that produced an offset as much as 6.4 m.

Now, let us consider the global distribution of earthquakes and volcanoes, which are dramatic expressions of Earth's internal dynamics. Earthquakes and volcanic activity are closely related in both time and space and do not occur randomly. They are concentrated along specific zones, extending around the Pacific Ocean and through the Mediterranean Sea. More recently it has been discovered that a zone of shallow earthquakes extends along the crest of the midocean ridge. When these zones of earthquakes and volcanic activity are plotted on a map, they outline with dramatic clarity the plate boundaries (Figure 3.6). Plate tectonics can thus readily explain why Japan is tormented by repeated volcanic eruptions and earthquakes (Figure 3.7)—it is created at the convergence of two tectonic plates. The same is true for Central and South America. It is equally clear that the earthquakes and volcanic eruptions in the Mediterranean area occur at plate margins. Most of the major earthquakes and volcanic eruptions you have read about in the past and will likely hear of in the future occur along the margins of tectonic plates.

These associations are enough in themselves to make the theory attractive, but they are only the beginning. The great mountain belts of the world can be related to the moving plates. Where the moving plates converge, the crust

FIGURE 3.6
Plate margins are outlined with remarkable fidelity by zones of earthquake activity and recent volcanism. Shallow earthquakes, submarine volcanic eruptions, and tensional fractures occur along the oceanic ridges, where plates are moving apart. Deep earthquakes, volcanic eruptions, and folded mountain belts occur along margins where plates converge.

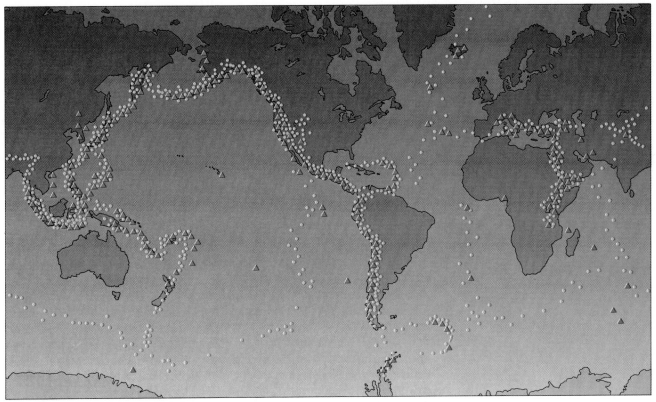

▲ Volcanoes   • Earthquakes

is compressed and folded. The space photograph in Figure 3.8 shows the type of deformation that most vividly expresses motion in the crust. The complex system of ridges and valleys is produced by folded sedimentary rocks that were deformed by collision between plates. The folded rocks now appear like wrinkles in a rug. A younger mountain belt extends from Alaska through the Rockies and the Andes of South America and is produced from the encounter of the American plates with the Pacific, Cocos, and Nazca plates. This is a young mountain system, with many parts being deformed continually as the plates continue to move. It is also apparent that the mountain systems of the Alps and Himalayas are nicely explained by plate tectonics as the result of the collisions of the African and Australian-Indian plates with Eurasia. These are young, active mountain belts, still growing, still being deformed by the present motion of plates. Older mountain belts such as the Appalachian Mountains in the eastern United States and the Ural Mountains of Russia mark zones where older plates collided long ago and were "welded" together.

GRAVITY AND ISOSTASY

Gravity plays a fundamental role in Earth's dynamics. It is intimately involved with differentiation of the planet's interior, isostatic adjustments of the crust, plate tectonics, and gravity flow of water in the hydrologic system.

Gravity is one of the great fundamental forces in the universe. It played a dominant role in the formation of the solar system, the origin of the planets, and the impact of meteorites that dominated their early history. Since then gravity has been a constant force in every phase of planetary dynamics and it is a dominant factor in all geologic processes operating on and within Earth. The differentiation and separation of Earth's materials into core, mantle, and crust is the result of gravitational force. The downward movement of rain and snow is due to gravity. Water flowing in river systems to the ocean owes its movement to gravity, and the slower movement of groundwater is due to the same force. Gravity also causes the settling of windblown dust and sand and the deposition of sediment on the floors of lakes, seas, and oceans. Likewise, glacial ice moves because of the force of gravity, as does the flow of lava erupted from a volcano.

How is gravity involved in Earth's processes?

These examples are obvious expressions of gravity familiar to everyone, but gravity also operates on a much grander scale within Earth's crust, causing lighter segments such as continents to stand higher than the denser ocean basins. Similarly, the loading of Earth's crust at one place with sediment from a river delta, glacial ice, or water into a lake will cause that region to subside, and the removal of rock from a mountain range by erosion will cause the region to be uplifted. This gravitational adjustment of Earth's crust is called *isostasy* (Greek *isos*, "equal," *stasis*, "standing"). Earth's lithosphere therefore continually responds to the force of gravity as it tries to maintain a gravitational balance.

Isostasy occurs because the crust is buoyed up by the more dense mantle beneath it, and each portion of the crust displaces the mantle according to its thickness and density (Figure 3.9). Denser crustal material sinks deeper into the mantle than less dense crustal material. Alternatively, thicker crustal material will sink deeper than thin crust of the same density.

Isostatic adjustment in Earth's crust can be compared to adjustments in a sheet of ice floating on a lake as you skate on it. The layer of ice bends down beneath you, displacing a volume of water with a weight equal to your weight. As you move ahead, the ice rebounds behind you, and the displaced water flows back.

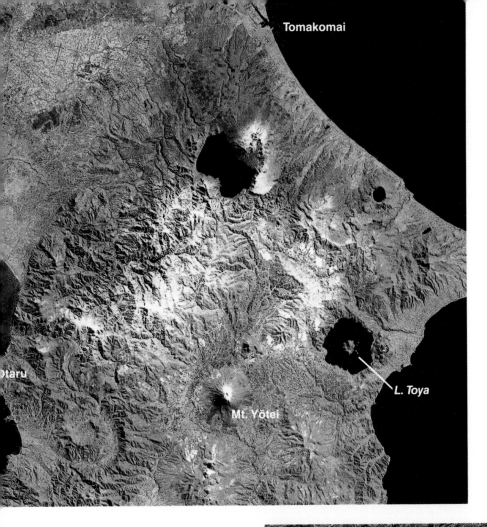

Tomakomai

Otaru

Mt. Yōtei

L. Toya

FIGURE 3.7
Volcanoes of the Japanese Island arc are formed by volcanic activity produced where two oceanic plates converge, and one is thrust under the other and is assimilated into the mantle. As the plate descends, molten rock is produced and it rises to form a chain of volcanoes. If the upper plate contains oceanic crust, the volcanic activity produces an arc of volcanic islands. This area in the vicinity of Sapporo, Japan, is dotted with both active and extinct volcanic cones built up from the sea floor. If the upper plate contains continental crust, volcanic material is extruded on a folded mountain belt. An excellent example of this type of volcanic activity occurs in the Andes Mountains of South America.

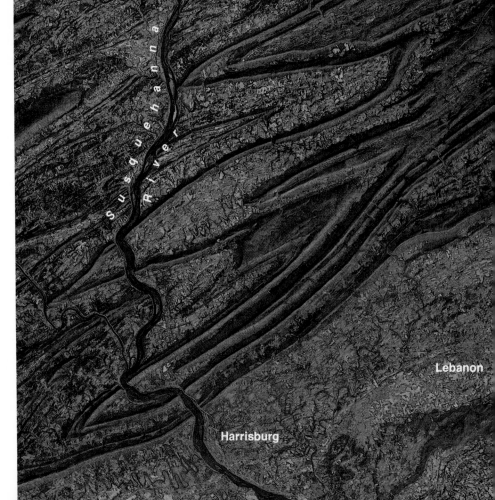

Susquehanna River

Lebanon

Harrisburg

FIGURE 3.8
The folded strata of the Appalachian Mountains were formed approximately 250 million years ago at converging plate margins. Layers of sedimentary rock, which were originally horizontal, have been compressed into tight folds that have been subsequently eroded. The resistant layers of sandstone appear as ridges, forming a zigzag pattern. Folded mountain belts like the Appalachians are one of the most significant results of converging plates.

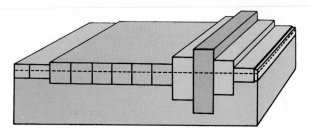

(A) Floating blocks of equal density. Thicker blocks rise higher and sink deeper.

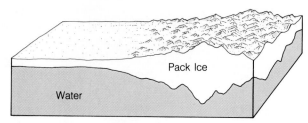

(B) Thick pack ice in the Arctic Ocean rises higher and sinks deeper than thin ice.

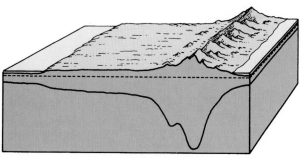

(C) High mountains are balanced by a deep root.

FIGURE 3.9
Isostasy is the universal tendency of segments of Earth's crust to establish a condition of gravitational balance. Differences in both density and thickness can cause isostatic adjustments in Earth's crust.

How do we know that isostatic adjustments occur?

As a result of isostatic adjustments, high mountains and plateaus having a great vertical thickness sink deeper into the mantle than areas of low elevation do. Any thickness change in an area of the crust—such as removal of material by erosion or addition of material by sedimentation, volcanic extrusion, or accumulations of large continental glaciers—causes an isostatic adjustment.

The concept of isostasy, therefore, is fundamental to studies of the major features of the crust, such as continents, ocean basins, and mountain ranges, and also to understanding the response of the crust to erosion, sedimentation, glaciation, and the tectonic system.

The construction of the Hoover Dam, on the Colorado River, provides an excellent, well-documented illustration of isostatic adjustment because the added weight of water and sediment in the reservoir was sufficient to cause measurable subsidence. From the time of the dam's construction in 1935, 24 billion metric tons of water, plus an unknown amount of sediment, accumulated in Lake Mead, behind the dam. In a matter of years, this added weight caused the crust to subside in a roughly circular area around the lake. Maximum subsidence was 1.7 m.

Continental glaciers are another clear example of isostatic adjustment of the crust. The weight of an ice sheet several thousand meters thick disrupts the crustal balance and causes the crust beneath to be depressed. In both Antarctica and Greenland, the weight of the ice has depressed the central part of the landmasses below sea level. A similar isostatic adjustment occurred in Europe and North America during the last ice age, when continental glaciers existed there. Parts of both continents, such as Hudson Bay and the Baltic Sea, are still below sea level. Now that the ice is gone, however, the crust is rebounding at a rate of 5 to 10 m per 1000 years.

Tilted shorelines of ancient lakes provide another means of documenting isostatic rebound. Lake Bonneville, for example, was a large lake in Utah and Nevada during the ice age but has since dried up to such small remnants as Utah Lake and Great Salt Lake. Shorelines of Lake Bonneville were level when they were formed but have been tilted in response to unloading as the water was removed. The lake was only 305 m deep and covered a much smaller area than a continental glacier, but still this relatively small weight was sufficient to depress the crust. Since it has been removed, the shorelines near the deepest part of the lake have rebounded nearly 60 m.

These and other examples illustrate several important facts about gravity and isostatic adjustments.

1. Gravity is the driving force for all isostatic adjustments. All types of loading and unloading therefore cause vertical movements. Isostasy is involved in all of the processes that shift material on Earth's surface. Some of the more obvious isostatic adjustments to be expected are as follows:
   a. In mountains and highlands, as erosion removes material, the crust should rebound.
   b. In deltaic areas, where sediment is deposited, the added weight should cause the crust to subside.
   c. In areas of volcanic activity, the added weight of extrusions should cause the crust to subside.
   d. In regions of continental glaciation, the thick ice sheets should cause the crust to subside. As the ice is removed, the crust should rebound.
2. Very small loads, such as water a few hundred meters deep, are also sufficient to cause isostatic adjustments.
3. Isostatic adjustments can occur very rapidly in a geologic time frame (60 m in less than 20,000 years).

SUMMARY

1. The theory of plate tectonics explains Earth's crustal dynamics by the movement of rigid lithospheric plates. New crust is created and heat is brought to the surface, where mantle material from below moves into the gap between the spreading lithospheric plates. At a subduction zone, cool lithosphere descends and is ultimately consumed in the hotter mantle below.
2. The major structural features of Earth are formed along plate boundaries.
3. At divergent plate boundaries the lithosphere is upwarped and spreads apart. The major features produced in this area are (a) oceanic ridges, (b) rifting of continents, (c) basaltic volcanism, and (d) new ocean basins.
4. At converging plate margins, one plate slides under the other and descends into the mantle. The major features produced in this area are (a) folded mountain belts, (b) deep-sea trenches, and (c) volcanic arcs.
5. Gravity plays a fundamental role in Earth's dynamics.

KEY TERMS

convergent plate boundary (p. 46)

divergent plate boundary (p. 46)

isostasy (p. 52)

mountain belt (p. 46)

plate tectonics (p. 44)

subduction (p. 46)

subduction zone (p. 47)

tectonic plate (p. 45)

tectonics (p. 44)

REVIEW QUESTIONS

1. Draw a diagram (cross section) showing (a) converging plates, (b) diverging plates.
2. Show on a map the plate boundaries.
3. What surface features mark plate boundaries?
4. Explain how the Andes Mountains, the midoceanic ridge, deep-sea trenches, island arcs, and volcanoes are related to plate tectonics.
5. Explain convection in Earth's interior.
6. Explain the concept of isostasy in Earth's system and give two examples of isostatic adjustment of Earth's crust in recent geologic time.

ADDITIONAL READINGS

Bird, J. M., ed. 1980. *Plate Tectonics.* Washington, DC: American Geophysical Union.

Condie, K. C. 1989. *Plate Tectonics and Crustal Evolution,* 3rd ed. New York: Pergamon Press.

Courtillot, V., and G. E. Vink. 1983. How continents break up. *Scientific American.*

Dewey, J. F. 1972. Plate tectonics. *Scientific American* 226(5): 56–68.

Hallam, A. 1973. *A Revolution in the Sciences: From Continental Drift to Plate Tectonics.* New York: Oxford University Press (Clarendon Press).

Van Andel, T. H. 1985. *New Views on an Old Planet: Continental Drift and the History of the Earth.* Cambridge: Cambridge University Press.

Igneous
intrusion

Rustenburg

Verdeford
dome

S outh Africa contains some of the richest mineral deposits in the world. Gold, diamonds, platinum, uranium, and copper have shaped its history and will undoubtedly influence its future.

The Landsat image on the facing page is an image of minerals. It shows part of the interior plateau that extends over much of the Cape Province. The prominent circular feature at the top center of the image is a mass of granite that was injected into the surrounding rock from below while it was in a liquid state, after which it slowly cooled. The Bushveld Complex is a sequence of iron- and magnesium-rich rocks, capped by granite that is nearly 2 billion years old. It appears in the upper right corner of the scene. A variety of important mineral deposits, such as chromium, vanadium, nickel, copper, and tin, occur in and around the complex. The most productive region for platinum is the famous Merensky Reef, a zone along the margins of the granite. Near the bottom of the image in the east-west bands of green-gray is the northern edge of the famous Witwatersrand Basin, South Africa's most productive area for gold and uranium.

But the importance of minerals extends far beyond their value as economic deposits. All of Earth's dynamic processes involve the growth and destruction of minerals, matter changing from one state to another. Minerals grow, melt, dissolve, or are broken and modified by physical forces. As Earth's surface weathers and erodes, some minerals are destroyed and others grow in their place. As sediments accumulate in the oceans, minerals grow from solution. Other minerals grow as lava extruded from volcanoes cools. Deep below Earth's surface, high pressure and temperature remove atoms from the crystal structures of some minerals and recombine them into new structures to form other minerals that are more stable. As tectonic plates move and continents drift, minerals are created and destroyed by a variety of processes. Some knowledge of Earth's major minerals, therefore, is essential to an understanding of Earth's dynamics.

In this chapter we survey the general characteristics of minerals and the physical properties that identify them. We then explore the silicate mineral group (the major rock-forming minerals) in preparation for a study of the major rock types in Chapters 5, 6, and 7.

4
Minerals

1. An atom is the smallest unit of an element that possesses the properties of the element. It consists of a nucleus of protons and neutrons and a surrounding cloud of electrons.
2. An atom of a given element is distinguished by the number of protons in its nucleus. Isotopes are varieties of an element, distinguished by the different numbers of neutrons in their nuclei.
3. Ions are electrically charged atoms, produced by a gain or loss of electrons.
4. Matter exists in three states: (a) solid, (b) liquid, and (c) gas. The differences among the three states are related to the degree of ordering of the atoms.
5. A mineral is a natural, inorganic solid possessing a specific internal atomic structure and a chemical composition that varies only within certain limits.
6. Minerals grow when atoms are added to the crystal structure as matter changes from the gaseous or the liquid state to the solid state. Minerals dissolve or melt when atoms are removed from the crystal structure.
7. All specimens of a mineral have well defined physical and chemical properties (such as crystal structure, cleavage or fracture, hardness, and specific gravity).
8. Silicate minerals form more than 95% of Earth's crust.
9. The most important rock-forming minerals are feldspars, micas, olivines, pyroxenes, amphiboles, quartz, clay minerals, and calcite.

MATTER

An atom is the smallest unit of an element that possesses the properties of the element. It consists of a nucleus of protons and neutrons and a surrounding cloud of electrons. There are three states of matter: gas, liquid, and solid. Each state is distinguished by completely different physical properties. Processes in Earth's dynamics mostly involve matter changing from one state to another.

To understand the dynamics of Earth and how rocks and minerals are formed and changed through time, you must have some knowledge of the fundamental structure of matter and how it behaves under various conditions. The solid materials that make up Earth's outer layers are called rocks. Most rock bodies are mixtures, or aggregates, of minerals. A mineral is a naturally occurring inorganic compound with a definite chemical formula and a specific internal structure. Because minerals, in turn, are composed of atoms, to study minerals we must understand something about atoms and the ways in which they combine.

A simplified description of modern atomic theory includes the following fundamental concepts:

1. An atom is the smallest fraction of an element that can exist and still show the characteristics of that element.
2. The main building blocks of an atom are protons, neutrons, and electrons, although many other subatomic particles have been identified in recent years.
3. A typical atom consists of a nucleus of protons and neutrons and a cloud of electrons surrounding the nucleus.

4. The distinguishing feature of an atom of a given element is the number of protons in the nucleus. The number of electrons and neutrons in an atom of a given element can vary, but the number of protons is constant.

5. Normally atoms are electrically neutral because they have one negatively charged electron for each positively charged proton.

6. Electrically charged atoms, called ions, are produced by the gain or loss of electrons.

7. Isotopes, which are varieties of a given atom (element), are produced by variations in the number of neutrons in the nucleus.

8. Atoms combine, mostly through ionic or covalent bonding, to form minerals.

Atoms

Atoms are best described by abstract models constructed from mathematical formulas involving probabilities. Recently, however, images of atoms have been made. An example is shown in Figure 4.1. In its simplest form, an *atom* is characterized by a relatively small nucleus of tightly packed protons and neutrons, with a surrounding cloud of electrons. Each proton carries a positive electrical charge, and the mass of a proton is taken as the unit of atomic mass. The neutron, as its name indicates, is electrically neutral and has approximately the same mass as the proton. The electron is a much smaller particle, with a mass approximately 1/1850 of the proton. It carries a negative electrical charge equal in intensity to the positive charge of the proton. Since the electron is so small, for practical purposes the entire mass of the atom is considered to be concentrated in the protons and neutrons of the nucleus.

Hydrogen is the simplest of all elements. It consists of one proton in the nucleus and one orbiting electron (Figure 4.2). The next heaviest atom is helium, with two protons, two neutrons, and two electrons. Each subsequently heavier element contains more protons, neutrons, and electrons. The distinguishing feature of an element is the number of protons in the nucleus of each of its atoms. The number of electrons and neutrons in an atom of a given element can vary, but the number of protons is constant.

What is the structure of an atom?

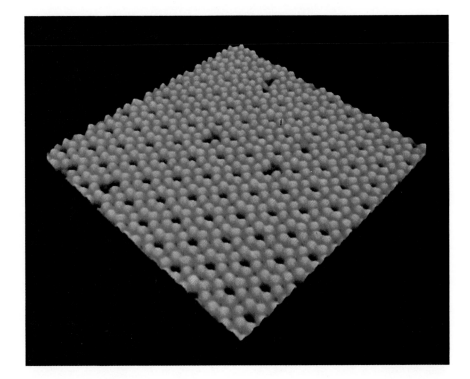

FIGURE 4.1
Image of atoms of silicon produced by a scanning tunneling microscope at the IBM Research Center, Yorkton Heights, N.Y. The blue spots are individual silicon atoms, which are arranged in a regular pattern that repeats itself across the surface. Images such as this are helpful in understanding the structure of many different minerals.

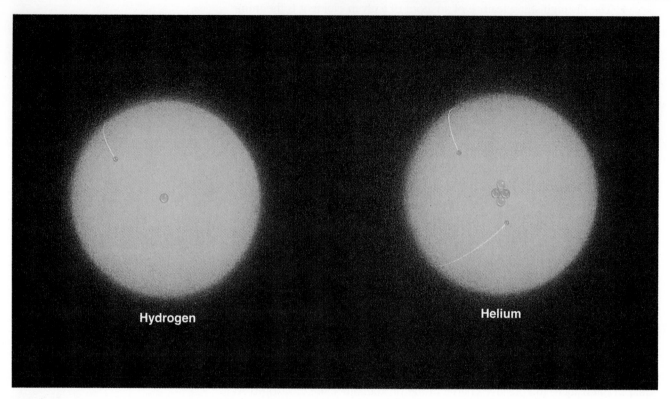

FIGURE 4.2
The atomic structures of hydrogen and helium illustrate the major particles of an atom. Hydrogen has one proton and one orbiting electron. Helium has two protons, two neutrons, and two orbiting electrons.

Atoms normally have the same number of electrons as protons and thus do not carry an electrical charge. As the number of protons increases in the succeedingly heavier atoms, the number of electrons also increases. The electrons fill a series of energy-level shells around the nucleus, each shell having a maximum capacity. The electrons in the outer shells control the chemical behavior of the element.

Isotopes

Although the number of protons in each atom of a given element is constant, the number of neutrons can vary. This means that atoms of a given element are not necessarily all alike. Iron atoms, for example, have 26 protons but can have 28, 30, 31, or 32 neutrons. These varieties of iron are examples of *isotopes*. They all have the properties of iron and differ from one another only in mass number. Most common elements exist in nature as mixtures of isotopes. Some isotopes are unstable, emitting particles and energy as they experience radioactive decay.

Ions

What are the distinguishing characteristics of an isotope? Of an ion?

Atoms that have as many electrons as protons are electrically neutral, but atoms of all elements can gain or lose electrons in their outermost shells. If this occurs, an atom loses its electrical neutrality and becomes charged. These electrically charged atoms are called *ions*. The loss of an electron results in a positively charged ion because the number of protons then exceeds the number of negatively charged electrons. If an electron is gained, the ion has a negative charge. The electrical charges of ions are important because the attraction between positive ions and negative ions is the bonding force that sometimes holds matter together.

Bonding

Atoms are most stable if their outermost shells are filled to capacity with electrons. The inner shell can hold no more than two electrons. The next shell can hold 8 electrons and is full in neon (atomic number 10). In heavier elements, the next shell can have 18 electrons, and another 32. Neon, for example, has 10 protons in the nucleus and 10 electrons, of which 2 are in the first shell and 8 are in the second shell. A neon atom does not have an electrical charge. Its two electron shells are complete, since the second shell has a limit of 8 electrons. As a result, neon does not interact chemically with other atoms. Argon and the other noble gases also have 8 electrons in their outermost shells, and they normally do not combine with other elements. Most elements, however, have incomplete outermost shells. Their atoms readily lose or gain electrons to achieve a structure like that of argon, neon, and the other inert gases, with 8 electrons in the outermost shell.

For example, an atom of sodium has only 1 electron in its outermost shell but 8 in the shell beneath (Figure 4.3). If it could lose the lone outer electron, the sodium atom would have a stable configuration like that of the inert gas neon. The chlorine atom, in contrast, has 7 electrons in its outermost shell, and if it could gain an electron, it too would attain a stable configuration. Whenever possible, therefore, sodium gives up an electron and chlorine gains one. The sodium atom thus becomes a positively charged sodium ion, and the chlorine atom becomes a negatively charged chloride ion. With opposite electrical charges, the sodium ions and chloride ions attract each other and bond together to form the compound sodium chloride (common salt, also known as the mineral halite). This type of bond, between ions of opposite electrical charge, is called an *ionic bond.*

Atoms can also attain the electronic arrangement of a noble gas, and thus attain stability, by sharing electrons. No electrons are lost or gained, and no ions are formed. Instead, an electron cloud surrounds both nuclei. This type of bond is called a *covalent bond.*

A third type of bond is called a *metallic bond.* In a metal, each atom contributes one or more outer electrons that move relatively freely throughout the entire aggregate of ions. A given electron is not attached to a specific ion pair, but moves about. This sea of negatively charged electrons holds the positive metallic ions together in a crystalline structure and is responsible for the special characteristics of metals.

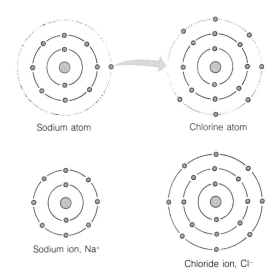

Sodium atom

Chlorine atom

Sodium ion, Na+

Chloride ion, Cl-

FIGURE 4.3
The formation of sodium and chloride ions by transfer of an electron from the outermost shell of a sodium atom to the outermost shell of a chlorine atom results in a stable outer shell for each ion.

States of Matter

The principal differences between solids, liquids, and gases involve the degree of ordering of the constituent atoms. In the typical *solid,* atoms are arranged in a rigid framework. The arrangement in crystalline solids is a regular, repeating, three-dimensional pattern called a *crystal structure,* but in *amorphous solids* the atomic arrangement is random. Each atom occupies a more or less fixed position, but has a vibrating motion. As the temperature rises, the vibration increases, and atoms move farther apart. Eventually they become free and are able to glide past one another. Melting ensues, and the solid matter passes into the liquid state.

In a *liquid,* the basic particles are in random motion, but they are packed closely together. They slip and glide past one another or collide and rebound, but they are held together by forces of attraction greater than those in gases. This explains why density generally increases and compressibility decreases as matter changes from gas to liquid to solid. If a liquid is heated, the motion of the particles increases, and individual atoms or molecules become separated as they move about at high speeds.

In a *gas,* the particles are in rapid motion and travel in straight lines until their direction is changed by collision. The individual atoms or molecules are separated by empty spaces and are comparatively far apart. This explains why gases can be markedly compressed and can exert pressure. Gases have the ability to expand indefinitely, and the continuous rapid motion of the particles results in rapid diffusion.

Water is undoubtedly the most familiar example of matter changing through the three basic states. At pressures prevailing on Earth's surface, water changes from a solid to a liquid to a gas in a temperature range of only 110°C. Other forms of matter in the solid Earth are capable of similar changes, but usually their transitions from solid to liquid to gas occur at comparatively high temperatures. At normal room temperature and pressure, 93 of the 106 elements are solids, 2 are liquids, and 11 are gases.

Most people are familiar with the effects of temperature changes on the state of matter because of their experience with water as it freezes, melts, and boils. Fewer people are familiar with the effects of pressure. Under great pressure, water will remain liquid at temperatures as high as 371°C. The combined effects of temperature and pressure on water are shown in the phase diagram in Figure 4.4. Similar diagrams, constructed from laboratory work on other minerals, provide important insight into the processes operating at the high temperatures and pressures below Earth's surface.

Why can gaseous, liquid, and solid forms of a substance have such different physical properties and still have the same composition?

FIGURE 4.4
Temperature and pressure are the major factors that determine the state in which matter exists. In this diagram, the ranges of temperature and pressure for the various phases of water are shown. The triple point is the point at which all three phases are in equilibrium. The critical temperature (T_c) and the critical pressure (P_c) cross at the critical point (the end of the gas–liquid phase line) beyond which the liquid and gas phases cannot be distinguished. Similar diagrams can be constructed for other minerals.

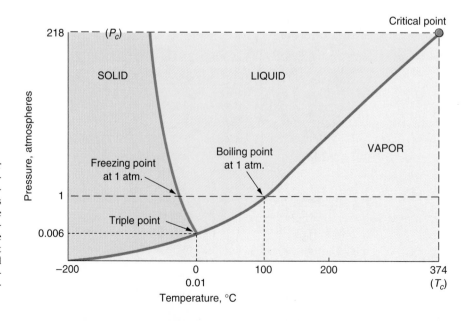

THE NATURE OF MINERALS

A mineral is a natural inorganic solid with a specific internal structure and a chemical composition that varies only within specific limits. All specimens of a given mineral, regardless of where, when, or how they were formed, share certain physical properties. Some of the more readily observable physical properties of minerals are cleavage, crystal form, hardness, specific gravity, color, and streak.

Many people commonly think of minerals only as exotic crystals in museums or as valuable gems and metals, but grains of sand, snowflakes, and salt particles are also minerals, and they have much in common with gold and diamonds.

Minerals are the major solid constituents of Earth. A precise definition is difficult to formulate, but for a substance to be considered a *mineral*, the following conditions must be met:

What is a mineral?

1. It occurs naturally as an inorganic solid.
2. It has a specific internal structure; that is, its constituent atoms are precisely arranged into a crystalline solid.
3. It has a chemical composition that varies within definite limits and can be expressed by a chemical formula.
4. It has definite physical properties (hardness, cleavage, crystal form, etc.) that result from its crystalline structure and composition.

The differences among minerals arise from the kinds of atoms they contain and the ways the atoms are arranged in a crystalline structure.

Inorganic Solids

By definition, only naturally occurring inorganic solids are minerals, that is, natural elements, or inorganic compounds in a solid state. Synthetic products, such as artificial diamonds, are therefore not minerals in the strict sense, nor are organic compounds, such as coal and petroleum, which are organic materials and not crystalline solids.

Minerals can consist of a single element, such as gold, silver, copper, diamond, and sulfur. However, because the abundant elements in Earth have a strong tendency to combine, most are compounds of two or more elements.

The Structure of Minerals

The key words in the definition of *mineral* are *internal structure*. The component atoms of a mineral have a specific arrangement in a definite geometric pattern. Every specimen of a given mineral has the same internal structure, regardless of when, where, and how it was formed. This property of minerals was suspected long ago by mineralogists who observed the many expressions of order in *crystals*. Nicolaus Steno (1638–1687), a Danish monk, was among the first to note this property. He found from numerous measurements that each mineral has a characteristic crystal form. Although the size and shape of a mineral crystal form may vary, similar pairs of crystal faces always meet at the same angle. This is known as the *law of constancy of interfacial angles*.

Later, René Haüy (1743–1822), a French mineralogist, accidently dropped a large crystal of calcite and observed that it broke along three sets of planes only, so all the fragments had a similar shape (see Figure 4.5). He then proceeded to break other calcite crystals in his own collection, plus many in the collections of his friends, and found that all of the specimens broke in exactly the same manner. All of the fragments, however small, had the shape of

FIGURE 4.5
Cleavage of calcite occurs in three planes that do not intersect at right angles. This results in a form called a rhombohedron. Each cleavage fragment is bounded by similar sets of cleavage planes.

a rhombohedron. To explain his observations, he assumed that calcite is built of innumerable infinitely small rhombohedra packed together in an orderly manner, and he concluded that the cleavage of calcite relates to the ease of parting of such units from adjacent layers. His discovery was a remarkable advance in the understanding of crystals.

Today, we know that cleavage planes are planes of weakness in the crystal structure and that they are not necessarily parallel to the crystal faces. Cleavage planes do, however, constitute a striking expression of the orderly internal structure of crystals.

With modern methods of **X-ray diffraction** (the apparent bending of X-rays as they pass through a crystalline substance), we can determine precisely a mineral's internal structure and learn much about its atomic arrangement. The technique is illustrated in Figure 4.6. If a thin beam of X-rays is passed through a mineral, it is diffracted (or dispersed) by the framework of atoms. The dispersed rays produce an array of dots on photographic film placed behind the crystal. From measurements of the relationships among the dots, the systematic orientation of planes of atoms within the crystal can be deduced and expressed in mathematical formulas. Detailed models of crystal structures can thus be constructed and analyzed. The X-ray diffraction instrument is now the most basic device for determining the internal structure of minerals, and geologists use it extensively for precise mineral identification and analysis.

To understand the importance of structure in a mineral, consider the characteristics of diamond and graphite. These two minerals are identical in

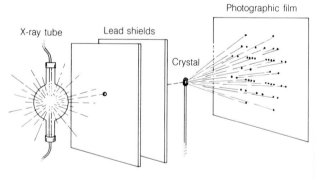

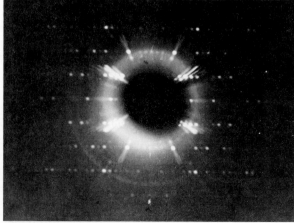

FIGURE 4.6
X-ray diffraction patterns are made by passing a thin beam of X-rays through a mineral. The X-rays are diffracted (dispersed) by the framework of atoms in the crystal, forming an orderly arrangement of dots on a photographic film placed behind the specimen.

chemical composition. Both consist of a single element, carbon (C). Their crystal structures and physical properties, however, are very different. In diamond, which forms only under high pressure, the carbon atoms are packed closely together, and the bonds between the atoms are very strong. This explains why diamonds are extremely hard—the hardest substance known. In graphite, the carbon atoms are loosely bound in a layered structure. The layers separate easily, which accounts for graphite's slippery, flaky property. Because of its softness and slipperiness, graphite is used as a lubricant and is also the main constituent of common "lead" pencils. The important point to note is that different structural arrangements of the same element produce different minerals with different properties. This ability of a specific chemical substance to crystallize with more than one type of structure is known as *polymorphism.*

The Composition of Minerals

A mineral has a definite chemical composition, in which specific elements occur in definite proportions. Thus, a precise chemical formula can be written to express the chemical composition, i.e., SiO_2, $CaCO_3$, etc. The chemical composition of some minerals can vary, but only within specific limits. In these minerals, two or more kinds of ions can substitute for each other in the mineral structure, a process called *ionic substitution.* Ionic substitution results in a chemical change in the mineral without a change in the crystal structure, so substitution can occur only within definite limits. The composition of such a mineral can be expressed by a chemical formula that specifies ionic substitution and how the composition can change.

The suitability of one ion to substitute for another is determined by several factors, the most important being the size and the electrical charge of the ions in question (Figure 4.7). Ions can readily substitute for one another if their ionic radii differ by less than 15%. If a substituting ion differs in charge from the ion for which it is substituted, the charge difference must be compensated for by other substitutions in the same structure.

Ionic substitution is somewhat analogous to substituting different types of equal-sized bricks in a wall. The substitute brick may be composed of glass, plastic, or whatever, but because it is the same size as the original brick, the structure of the wall is not affected. An important change in composition has, however, occurred, and as a result there are changes in physical properties. In minerals ionic substitution causes changes in hardness, color, etc., without changing the internal structure.

Ionic substitution is common in rock-forming minerals and is responsible for mineral groups, the members of which have the same structure but varying composition. For example, in the olivine group, with the formula $(Mg,Fe)_2SiO_4$, ions of iron (Fe) and magnesium (Mg) can substitute freely for

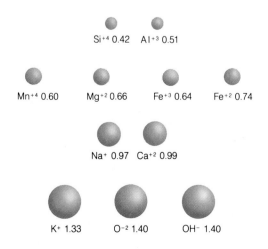

Si^{+4} 0.42   Al^{+3} 0.51

Mn^{+4} 0.60   Mg^{+2} 0.66   Fe^{+3} 0.64   Fe^{+2} 0.74

Na$^+$ 0.97   Ca^{+2} 0.99

K$^+$ 1.33   O^{-2} 1.40   OH$^-$ 1.40

FIGURE 4.7
The relative size and electrical charge of ions are important factors governing the suitability of one ion to substitute for another in a crystal structure. Iron can be replaced by magnesium, sodium by calcium, and silicon by aluminum. If the substituting ions do not have exactly the same electrical charge, the difference in charge must be balanced or compensated for by other substitutions in the structure.

one another. The total number of Fe and Mg ions is constant in relation to the number of silicon (Si) and oxygen (O) atoms in the olivine, but the ratio of Fe to Mg may vary in different samples. The common minerals feldspar, pyroxene, amphibole, and mica each constitute a group of related minerals in which atomic substitution produces a range of chemical composition and a range of physical properties within certain limits.

Physical Properties of Minerals

What are the major physical properties of minerals? What determines these properties?

Because a mineral has a definite chemical composition and internal crystalline structure, all specimens of a given mineral, regardless of when or where they were formed, have the same physical and chemical properties. If ionic substitution occurs, variation in physical properties also occurs, but since ionic substitution can occur only within specific limits, the range in physical properties also can occur only within specific limits. This means that one piece of quartz, for example, is as hard as any other piece, that it has the same specific gravity (ratio of the density of a substance to the density of water), and that it breaks in the same manner, regardless of when, where, or how it was formed.

The more significant and readily observable physical properties of minerals are crystal form, cleavage, hardness, specific gravity, color, and streak.

Crystal Form. If a crystal is allowed to grow in an unrestricted environment, it develops natural *crystal faces* and assumes a specific geometric *form.* The shape of a crystal is a reflection of the internal structure and is an identifying characteristic for many mineral specimens. If the atoms are arranged in a long chain, the crystal may be like a needle. If the atoms are arranged in a boxlike network, the crystal will likely be in the form of a cube (Figure 4.8A). Quartz, for example, typically forms elongate, hexagonal crystals (Figure 4.8B). If the space for growth is restricted, however, smooth crystal faces cannot develop.

Cleavage. *Cleavage* is the tendency of a crystalline substance to split or break along smooth planes parallel to zones of weak bonding in the crystal structure (Figure 4.8C). If the bonds are especially weak in a given direction, as in mica or halite, perfect cleavage occurs with ease, and breaking the mineral in any direction other than along a cleavage plane is difficult (Figure 4.8D). In other minerals the differences in bond strength are not great, so cleavage is poor or imperfect. Cleavage can occur in more than one direction, but the number and direction of cleavage planes in a given mineral species are always the same.

Some minerals have no weak planes in their crystalline structure, so they do not have cleavage and break along various types of fracture surfaces. Quartz, for example, characteristically breaks by *conchoidal fracture,* that is, along curved surfaces, like the curved surfaces of chipped glass.

Hardness. *Hardness* is a measure of a mineral's resistance to abrasion. This property is easily determined and is used widely for field identification of minerals. More than a century ago, Friedrich Mohs (1773–1839), a German mineralogist, assigned arbitrary relative numbers to ten common minerals in order of their hardness. He assigned the number 10 to diamond, the hardest mineral known. Softer minerals were ranked in descending order, with talc, the softest mineral, assigned the number 1. The Mohs hardness scale (Table 4.1) provides a standard for testing minerals for preliminary identification. Gypsum, for example, has a hardness of 2 and can be scratched by a fingernail (Figure 4.8E).

Specific Gravity. *Specific gravity* is the ratio of the weight of a given volume of a substance to the weight of an equal volume of water. For example,

(A) Crystals of pyrite have a cubic form because the constituent atoms are arranged in a cubic pattern.

(C) Cleavage in the mineral halite occurs in three directions at right angles. The resulting fragments have a characteristically cubic form.

(D) Perfect cleavage in one direction is illustrated by the mineral mica.

(B) Quartz crystals form prisms in which some of the crystal faces meet at 120°, regardless of the gross shape and size of the crystals or when and where they were formed. The constancy of interfacial angles is an expression of the atomic structure of the mineral. Every mineral has a characteristic crystal form as a result of its crystalline structure.

(E) Gypsum has a hardness of 2 on the Mohs hardness scale. It is a very soft mineral and can easily be scratched with a fingernail.

FIGURE 4.8
Physical properties of minerals.

TABLE 4.1
The Mohs Hardness Scale

Hardness	Mineral	Test
1	Talc	
2	Gypsum	
		Fingernail
3	Calcite	
		Copper coin
4	Fluorite	
5	Apatite	
		Knife blade or glass plate
6	K-feldspar	
7	Quartz	
8	Topaz	Steel file
9	Corundum	
10	Diamond	

a liter of solid lead weighs a little over 11 times more than a liter of water, and thus the specific gravity of lead is 11.

Specific gravity is one of the more precisely defined properties of a mineral. It depends on the kinds of atoms making up the mineral and how closely they are packed in the crystal structure. Clearly, the more numerous and compact the atoms, the higher the specific gravity. Most common rock-forming minerals have specific gravities from 2.65 (for quartz) to about 3.37 (for olivine).

With a little experience, you will be able to estimate the relative specific gravity of a mineral merely by lifting a specimen. Most metallic minerals feel heavy, whereas most common nonmetallic minerals seem relatively light.

Color. Color is one of the more obvious properties of a mineral. Unfortunately, it is *not* diagnostic. Most minerals are found in various hues, depending on such factors as subtle variations in composition and the presence of inclusions and impurities. Quartz, for example, ranges through the spec-

FIGURE 4.9
Submicroscopic crystals growing in the pore spaces between sand grains can be seen through an electron microscope. Each crystal contains all of the physical and chemical properties of the mineral.

trum from clear colorless crystals to purple, red, white, yellow, gray, and jet black.

Streak. The color of a mineral in powder form, referred to as *streak,* is usually more diagnostic than the color of a large specimen. For example, the mineral pyrite (fool's gold) has a gold color but a black streak, whereas real gold has a gold streak—the same color as that of larger grains. To test for streak, a mineral is rubbed vigorously against the surface of an unglazed piece of white porcelain. Minerals softer than the porcelain leave a streak, or line, of fine powder. For minerals harder than porcelain, a fine powder can be made by crushing a mineral fragment. The powder is then examined against a white background.

GROWTH AND DESTRUCTION OF MINERALS

Minerals are susceptible to chemical change, and they grow as matter changes from a gaseous or liquid state to a solid state. They break down as the solid changes back to a liquid or gas. All minerals came into being because of specific physical and chemical conditions, and all are subject to change as these conditions change. Minerals, therefore, provide an important means of interpreting the changes that have occurred in Earth throughout its history.

Crystal Growth

A single crystal, containing all the physical and chemical properties of the mineral, may be so small that it cannot be identified even with a high-powered electron microscope. *Crystallization* occurs by the addition of atoms to a crystal face. This is possible because the outer layers of atoms on a crystal are never completed and can be extended indefinitely. An example of exceedingly small crystals is shown in Figure 4.9.

An environment suitable for crystal growth includes (1) proper concentration of the kinds of atoms or ions required for a particular mineral and (2) proper temperature and pressure.

The time-lapse photographs in Figure 4.10 show how crystals grow in an unrestricted environment. Although the size of the crystal increases, its form

How can a mineral, which is inorganic, grow?

Time

| 0 sec. | 10 sec. | 30 sec. | 1 min. |

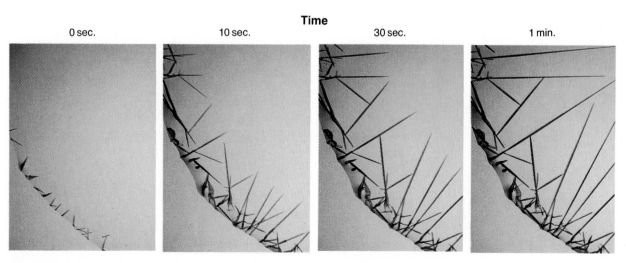

FIGURE 4.10
Crystal growth can be recorded by time-lapse photography. Each crystal grows as atoms lock onto the outer faces of the crystal structure.

and internal structure remain the same. New atoms are added to the faces of the crystal, parallel to the plane of atoms in the basic structure.

Some crystal faces, however, grow faster than others, so that in an environment where space is restricted, a crystal may not grow evenly. Where a crystal face encounters a barrier, it ceases to grow. A crystal growing in a restricted space assumes the shape of the confining area, and well-developed crystal faces do not form. This process is illustrated in Figure 4.11. The external form of the crystal can thus take on practically any shape, but the internal structure of the crystal is in no way modified. The mineral's internal structure remains the same, its composition is unaffected, and no changes in its physical and chemical properties occur. The only modification is a change in the relative sizes of the crystal faces.

The process of crystal growth in restricted space is especially important in rock-forming minerals. In a *melt* or a solution, many crystals grow at the same time and must compete for space. As a result, in the later stages of growth, crystals in rocks commonly lack well-defined crystal faces and typically interlock with adjacent crystals to form a strong, coherent mass (Figure 4.12).

Most crystals are rather small, measuring from a few tenths of a millimeter to several centimeters in diameter. Some are so small that they can be seen only when enlarged thousands of times with an electron microscope (Figure 4.9). Minerals composed of such minute crystals are said to have a ***cryptocrystalline texture***. In an unrestricted environment, however, crystals can grow to enormous sizes (Figure 4.13).

FIGURE 4.11
Crystals growing in a restricted environment do not develop perfect crystal faces.

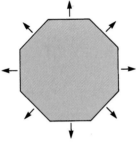

(A) Where growth is unrestricted, all crystal faces grow with equal facility, and the perfect crystal is enlarged by growth on each crystal face.

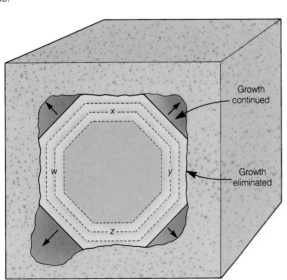

(B) In a restricted environment, growth on certain crystal faces, such as w, x, and y, is terminated at about the same time that available space is filled. Growth on face z continues but is soon inhibited by the walls of the cavity.

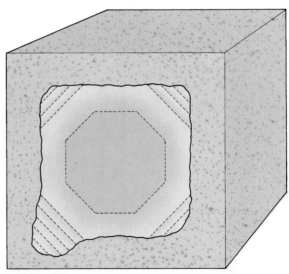

(C) The final shape of the crystal is determined by the geometry of the available space in which it grows. In the example, a normally octagonal crystal develops an irregular form. The internal (atomic) structure of the crystal, however, remains the same, regardless of the space in which the crystal grows.

Destruction of Crystals

Minerals dissolve, or melt, by removal of outer atoms from the crystal structure, so the matter returns to a fluid state (liquid or gas). The heat that causes a crystal to melt increases atomic vibrations enough to break the bonds holding an atom to the crystal structure. Similarly, atoms can be "pried" loose and carried away by a solvent, usually (in geologic processes) water. The breakdown, or *dissolution,* of a crystal begins at the surface and moves inward.

In summary, the growth and breakdown of crystals are of paramount importance in geologic processes because rocks originate and change as minerals are formed and destroyed. Crystals grow when lava cools and solidifies. They grow from solution in the sea. They grow at Earth's surface, where minerals in common rocks react chemically with elements in the atmosphere. They also form and are broken down deep within Earth's crust, where heat and pressure cause some crystal structures to collapse and new minerals with a more dense, compact atomic structure (Figure 4.14) to form in their place.

SILICATE MINERALS

More than 95% of Earth's crust is composed of the silicate minerals, a group of minerals containing silicon and oxygen linked together in tetrahedral units, with four oxygen atoms to one silicon atom. Several fundamental configurations of tetrahedral groupings are single chains, double chains, two-dimensional sheets, and three-dimensional frameworks.

Although more than 2000 minerals have been identified, 95% of the volume of Earth's crust is composed of a group of minerals called the silicates. This should not be surprising, because silicon and oxygen constitute nearly three-fourths of the mass of Earth's crust (Table 4.2) and therefore must predominate in most rock-forming minerals. Silicate minerals are complex in both chemistry and crystal structure, but all contain a basic building block called the

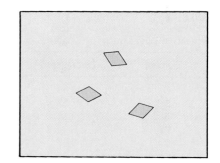
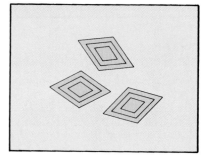
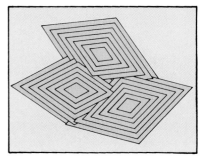

FIGURE 4.12
Interlocking texture develops if crystals grow in a restricted environment and are forced to compete for space.

FIGURE 4.13
Large crystals can form where there is ample space for growth, such as in caves. These crystals of gypsum are more than 1 m long.

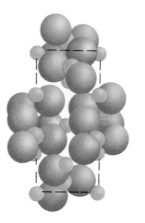
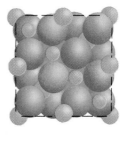

FIGURE 4.14
Under high pressure, the atomic structure of a mineral can collapse into a more dense form, in which the atoms are more closely packed.

TABLE 4.2
Concentrations of the Most Abundant
Elements in Earth's Crust (by Weight)

Element	Percentage
O	46.60
Si	27.72
Al	8.13
Fe	5.00
Ca	3.63
Na	2.83
K	2.59
Mg	2.09
Ti	0.44
H	0.14
P	0.12
Mn	0.10
S	0.05
C	0.03

After Mason, B., and C. B. Moore. 1982. *Principles of Geochemistry,* 4th ed. New York: Wiley.

silicon-oxygen tetrahedron. This is a complex ion $[(SiO_4)^{4-}]$ in which four large oxide ions (O^{2-}) are arranged to form a four-sided pyramid with a smaller silicon ion (Si^{4+}) fitted into the cavity between them (Figure 4.15). This geometric shape is known as a tetrahedron. The major groups of silicate minerals differ mainly in the arrangement of such silica tetrahedra in their crystal structures.

Perhaps the best way to understand the unifying characteristics of the *silicates,* as well as the reasons for the differences, is to study the models shown in Figure 4.16. These were constructed on the basis of X-ray studies of silicate crystals. *Silicon-oxygen tetrahedra* combine to form minerals in two ways. In the simplest combination, the oxygen ions of the tetrahedra form bonds with other elements, such as iron or magnesium. Olivine is an example. Most silicate minerals, however, are formed by the sharing of an oxygen ion between two adjacent tetrahedra. In this way, the tetrahedra form a larger ionic unit, just as beads are joined to form a necklace. The sharing of oxygen ions by the silicon ions results in several fundamental configurations of tetrahedral groups: single chains, double chains, two-dimensional sheets, and three-dimensional frameworks. These structures define the major silicate mineral groups.

1. Single chains—pyroxenes
2. Double chains—amphiboles
3. Two-dimensional sheets—micas, chlorites, and clay minerals
4. Three-dimensional frameworks—feldspars and quartz

The unmatched electrons of the silica tetrahedron are balanced by various metal ions, such as ions of calcium, sodium, potassium, magnesium, and iron. The silicate minerals thus contain silica tetrahedra linked in various patterns by metal ions.

Considerable ionic substitution can occur in the basic crystal structure. For example, sodium can substitute for calcium, or iron can substitute for magnesium. Minerals of a major silicate group can thus differ chemically among themselves but have a common silicate structure.

ROCK-FORMING MINERALS

Fewer than twenty kinds of minerals account for the great bulk of Earth's crust and most of the upper mantle. The most important rock-forming minerals are feldspars, micas, amphiboles, pyroxenes, olivines, quartz, calcite, dolomite, clay, halite, and gypsum.

Most of Earth's crust and upper mantle are composed of silicate minerals in which elements such as iron, magnesium, sodium, calcium, potassium, and aluminum combine with silicon and oxygen. But the identification of these minerals presents some special problems. Rock-forming minerals rarely are

FIGURE 4.15
The silicon-oxygen tetrahedron is the basic building block of the silicate minerals. In this figure the diagram to the right is expanded to show the position of the small silicon atom. Four large oxygen atoms are arranged in the form of a pyramid (tetrahedron) with a small silicon atom fitted into the central space between them. This is the most important building block in geology because it is the basic unit for 95% of the minerals in Earth's crust.

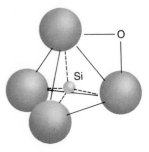

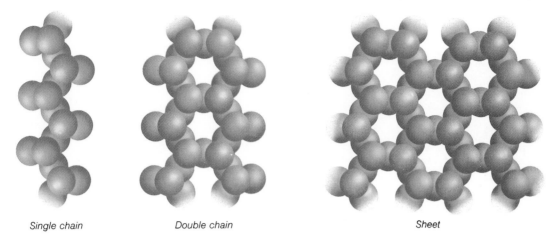

Single chain   Double chain   Sheet

FIGURE 4.16
Silicon-oxygen tetrahedral groups can form single chains, double chains, and sheets by the sharing of oxide ions among silicon ions.

well-developed because they grow from crystallization from a melt (magma) or from solution (seawater) and vigorously compete for space, or they are abraded as they are transported as sediment, or they are deformed under high temperature and pressure. In addition, most rock-forming mineral grains are small, generally less than the size of your little fingernail, so their physical properties may be difficult to see without a hand lens or microscope. Further complications arise because most rock-forming mineral groups have variable compositions as a result of ionic substitution in the crystal structure. As a result, the color, hardness, and other physical properties may be variable.

It is important for you to become familiar with the general characteristics of each of the major rock-forming mineral groups (feldspars, micas, quartz, olivine, pyroxene, amphibole, clays, calcite, dolomite, halite, and gypsum) and to know something about their physical properties, their mode of origin, the environment in which they form, and their genetic significance. You will find the following summary of each mineral group to be much more meaningful if you will examine a specimen of a rock containing the mineral while you study the written description.

A careful examination of the minerals that make up granite is a good beginning. The polished surface of a granite in Figure 4.17 shows that the rock is composed of myriads of mineral *grains* of different sizes, shapes, and colors. Although the minerals interlock to form a tight, coherent mass, each one has distinguishing properties.

Feldspar

Granite consists largely of a pink porcelainlike mineral that has a rectangular shape and a milky-white mineral that is somewhat smaller but similarly shaped. These are *feldspars* (German, "field crystals"), the most abundant mineral group, composing about 50% of Earth's crust. The feldspars have good cleavage in two directions, a porcelain luster, and a hardness of about 6 on the Mohs hardness scale.

The crystal structure permits considerable ionic substitution, giving rise to two major types of feldspars—potassium feldspar (K-feldspar) and plagioclase feldspar.

Potassium feldspar ($KAlSi_3O_8$) is commonly pink in granitic rocks. *Plagioclase* feldspar (shown in gray in the sketch) permits complete substitution of sodium (Na) for calcium (Ca) in the crystal structure, giving rise to a compositional range from $NaAlSi_3O_8$ to $CaAl_2Si_2O_8$. White plagioclase in granite is rich in sodium.

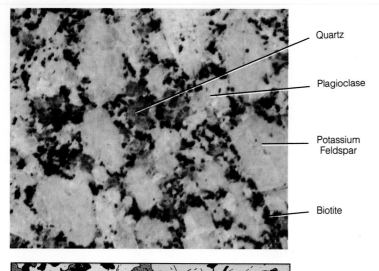

Quartz

Plagioclase

Potassium
Feldspar

Biotite

(A) A polished surface of a granite, shown at actual size, displays mineral grains of different sizes, shapes, and colors.

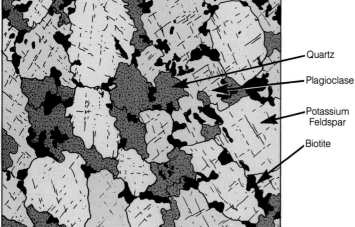

Quartz

Plagioclase

Potassium
Feldspar

Biotite

(B) A sketch emphasizing grains of individual minerals shows K-feldspar (pink), quartz (stippled), and plagioclase (gray).

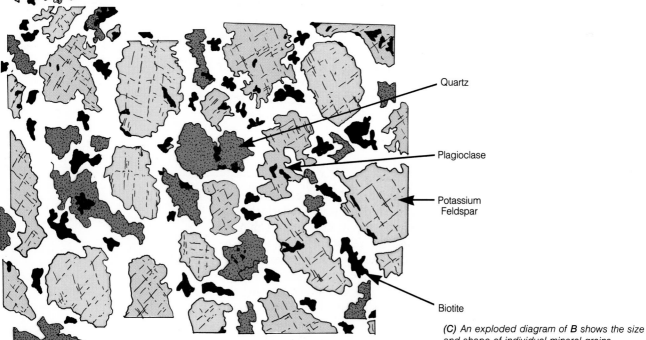

Quartz

Plagioclase

Potassium
Feldspar

Biotite

(C) An exploded diagram of B shows the size and shape of individual mineral grains.

FIGURE 4.17
Mineral grains in a granite form a tight, interlocking texture because each mineral is forced to compete for space as it grows.

Feldspars are common in most igneous rocks, in many metamorphic rocks, and in some sandstones.

What are silicate minerals? Why are they important in geology?

Micas

The tiny, black, shiny grains in Figure 4.17 are *mica*. This group of minerals is readily recognized by its perfect one-directional cleavage, which permits breakage into thin, elastic flakes. Mica is a complex silicate with a sheet structure, which is responsible for its perfect cleavage. Two common varieties occur in rocks: *biotite*, which is black mica, and *muscovite*, which is white or colorless. Mica is abundant in granites and in many metamorphic rocks and is also a significant constituent of many sandstones.

Quartz

The glassy, irregularly shaped grains in Figure 4.17 are *quartz*. It is usually the last mineral to form in a granite and is thus forced to grow in the spaces between the earlier-formed feldspars and micas. As a result, quartz in granite typically lacks well-developed crystal faces.

Quartz is abundant in all three major rock types. It has the simple composition SiO_2 and is distinguished by its hardness (7), its conchoidal fracture, and its glassy luster. Pure quartz crystals are colorless, but slight impurities produce a variety of colors. When quartz crystals are able to grow freely, their form is elongated, has six sides, and terminates in a point (Figure 4.8B), but well-formed crystals are rarely found in rocks. In sandstone, quartz is abraded into rounded sand grains. Quartz is stable both mechanically (it is very hard and lacks cleavage) and chemically (it does not react with elements at or near Earth's surface). It is therefore a difficult mineral to alter or break down once it has formed.

Ferromagnesian Minerals

Another category of silicate minerals is the *ferromagnesian minerals,* so named because they contain appreciable amounts of iron and magnesium. These minerals generally range from dark green to black in color and have a high specific gravity. Biotite is classified in this general group, together with the olivines, pyroxenes, and amphiboles. Biotite is common in granite, but the other ferromagnesian minerals are rare or absent. The ferromagnesian minerals are common, however, in basalt (volcanic rock) (Figure 4.18).

What are the distinguishing characteristics of the major minerals in igneous rocks? Sedimentary rocks?

Olivines. The only mineral clearly visible in the hand specimen in Figure 4.18 is the green, glassy mineral called *olivine*. The olivine family is a group of silicates in which iron and magnesium substitute freely in the crystal structure. The composition is expressed as $(Mg,Fe)_2SiO_4$. This hard mineral is characterized by an olive-green color (if magnesium is abundant) and a glassy luster. In rocks, it rarely forms crystals larger than a millimeter in diameter. Olivine is a high-temperature mineral and is common in basalt. It is probably a major constituent of the mantle, the material beneath Earth's crust.

Pyroxenes. *Pyroxenes* are high-temperature minerals found in many igneous and metamorphic rocks. In Figure 4.18, pyroxene occurs as microscopic crystals, but some basalt samples contain larger grains of this mineral, which typically ranges from dark green to black in color. Their internal structure shows single chains of linked tetrahedra (Figure 4.16). Pyroxene crystals commonly have two directions of cleavage that intersect at right angles.

(A) In a hand specimen, only a few large grains can be seen.

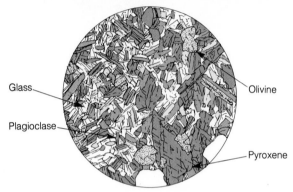

(B) Viewed under a microscope, the grains can be seen forming an interlocking texture. Plagioclase crystals typically form lathlike grains.

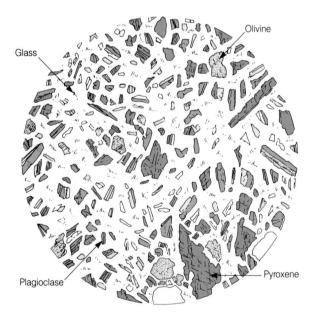

(C) An exploded diagram of B shows the size and shape of individual mineral grains.

FIGURE 4.18
Mineral grains in basalt are microscopic in size.

FIGURE 4.19
Amphibole crystals are among the first to crystallize in "granitic" rocks and therefore have well-developed crystal faces.

Amphiboles. Figure 4.19 shows a rock that is similar to granite but has less quartz and contains appreciable amounts of the mineral *amphibole*. Amphiboles have much in common with the pyroxenes. Their chemical compositions are similar, except that amphibole contains hydroxyl ions (OH^-) and pyroxene does not. The minerals also differ in structure, however. The internal structure consists of double chains of tetrahedra (Figure 4.16). The amphiboles produce elongate crystals that cleave perfectly in two directions, which are not at right angles. The color of amphibole ranges from green to black. This mineral is common in rocks closely related to granite and may be more abundant in that group than biotite. It is especially common in the metamorphic rock known as amphibolite. *Hornblende* is the most common variety of amphibole.

Clay Minerals

The *clay minerals* constitute a major part of the soil and are thus encountered more frequently than other minerals in everyday experience. Clay minerals form at Earth's surface, where air and water interact with the various silicate minerals, breaking them down to form clay and other products. Like the micas, the clay minerals are sheet silicates (Figure 4.16), but their crystals are microscopic and usually can be detected only with an electron microscope (Figure 4.9). More than a dozen clay minerals can be distinguished on the basis of their crystal structures and variations in composition.

Calcite

Calcite is composed of calcium carbonate ($CaCO_3$), the principal mineral in limestone. It can be precipitated directly from seawater and is removed from seawater by organisms as they use it to make their shells. Calcite is dissolved by groundwater and reprecipitated as new crystals in caves and fractures in rock. It is usually transparent or white in color, but the aggregates of calcite crystals that form limestone contain various impurities, which give them gray or brown hues. Calcite is common at Earth's surface and is easy to identify. It is soft enough to scratch with a knife (hardness of 3), and it effervesces in dilute hydrochloric acid. It has perfect cleavage in three directions, which are not at right angles, so that cleaved fragments form rhombohedra (Figure 4.5). Besides being the major constituent of limestone, calcite is the major mineral in the metamorphic rock marble.

Dolomite

Dolomite is a carbonate of calcium and magnesium. Large crystals form rhombohedra, but most dolomite occurs as granular masses of small crystals. Dolomite is widespread in sedimentary rocks, forming when calcite reacts with solutions of magnesium carbonate in seawater or groundwater. Dolomite can be distinguished from calcite because it effervesces in dilute hydrochloric acid only if it is in powdered form.

Halite and Gypsum

Halite and gypsum are the two most common minerals formed by evaporation of seawater or saline lake water. *Halite,* common salt (NaCl), is easily identified by its taste. It also has one of the simplest of all crystal structures—the sodium and chloride ions are united in a cubical array. Most physical properties of halite are related to this structure. Halite crystals cleave in three directions, at right angles, to form cubic or rectangular fragments (Figure 4.8C). Salt, of course, is very soluble and readily dissolves in water.

Gypsum is composed of calcium sulfate and water ($CaSO_4 \cdot 2H_2O$). It forms crystals that are generally colorless, with a glassy or silky luster. It is a very soft mineral and can be scratched easily with a fingernail. It cleaves perfectly in one direction to form thin, nonelastic plates (Figure 4.8E). Gypsum occurs as single crystals, as aggregates of crystals in compact masses (alabaster), and as a fibrous form (satin spar).

SUMMARY

1. The atomic theory is the basis for our present-day understanding of matter.
2. There are three states of matter: gas, liquid, and solid. The three states are distinguished by completely different physical properties, but matter retains the same chemical composition in all states. Processes in Earth's dynamics mostly involve matter changing from one state to another.
3. A mineral is a natural inorganic solid with a specific internal structure and a chemical composition that varies only within specific limits.
4. All specimens of a given mineral, regardless of where, when, or how they were formed, share certain physical properties. Some of the more readily observable physical properties are cleavage, crystal form, hardness, specific gravity, color, and streak, all of which aid in mineral identification.
5. Minerals are the building blocks of rocks. They grow or are broken down by chemical reaction as matter changes to and from the solid state. Minerals, therefore, provide an important means of interpreting the changes that have occurred in Earth throughout its history.
6. More than 95% of Earth's crust is composed of the silicate minerals, a group of minerals containing silicon and oxygen linked together in tetrahedral units, with four oxygen atoms to one silicon atom. Several fundamental configurations of tetrahedral groupings—single chains, double chains, two-dimensional sheets, and three-dimensional frameworks—result from the sharing of oxygen ions among silicon ions.
7. Fewer than twenty minerals form the great bulk of Earth's crust. The most important rock-forming minerals are feldspars, micas, amphiboles, pyroxenes, olivines, quartz, calcite, dolomite, clay minerals, halite, and gypsum.

muscovite (p. 75)

olivine (p. 75)

plagioclase (p. 73)

polymorphism (p. 65)

pyroxene (p. 75)

quartz (p. 75)

silicate (p. 72)

silicon-oxygen tetrahedron (p. 72)

solid (p. 62)

specific gravity (p. 66)

streak (p. 69)

X-ray diffraction (p. 64)

REVIEW QUESTIONS

1. Give a brief but adequate definition of a mineral.
2. Explain the meaning of "the internal structure of a mineral."
3. Why does a mineral have a definite chemical composition?
4. How do geologists identify minerals too small to be seen in a hand specimen?
5. Briefly explain how minerals grow and are destroyed.
6. Explain the origin of cleavage in minerals.
7. Describe the silicon-oxygen tetrahedron. Why is it important in the study of minerals?
8. Why does mica have excellent cleavage, whereas quartz and olivine lack cleavage?
9. What are the silicate minerals? List the silicate minerals that are important in the study of rocks.
10. Give three examples of how silicon-oxygen tetrahedra are arranged in silicate minerals.
11. Study Figure 4.17 and explain why most of the mineral grains in a granite have an irregular shape.

KEY TERMS

amorphous solid (p. 62)

amphibole (p. 77)

atom (p. 59)

biotite (p. 75)

calcite (p. 77)

clay mineral (p. 77)

cleavage (p. 66)

conchoidal fracture (p. 66)

covalent bond (p. 61)

cryptocrystalline texture (p. 70)

crystal (p. 63)

crystal face (p. 66)

crystal form (p. 66)

crystal structure (p. 62)

crystallization (p. 69)

dissolution (p. 71)

dolomite (p. 78)

feldspar (p. 73)

ferromagnesian mineral (p. 75)

gas (p. 62)

grain (p. 73)

gypsum (p. 78)

halite (p. 78)

hardness (p. 66)

hornblende (p. 77)

ion (p. 60)

ionic bond (p. 61)

ionic substitution (p. 65)

isotope, (p. 60)

liquid (p. 62)

melt (p. 70)

metallic bond (p. 61)

mica (p. 75)

mineral (p. 63)

ADDITIONAL READINGS

Ahrens, L. H. 1965. *Distribution of the Elements in Our Planet*. New York: McGraw-Hill.

Berry, L. G., B. Mason, and R. V. Dietrich. 1983. *Mineralogy*, 2nd ed. San Francisco: Freeman.

Deer, W. A., R. A. Howie, and J. Zussman. 1966. *An Introduction to the Rock-Forming Minerals*. New York: Wiley.

Dietrich, R. V., and B. J. Skinner. 1979. *Rocks and Rock Minerals*. New York: Wiley.

Ernst, W. G. 1969. *Earth Materials*. Englewood Cliffs, NJ: Prentice-Hall.

Klein, C., and C. S. Hurlbut, Jr. 1985. *Manual of Mineralogy (after J. D. Dana), 20th ed. New York: Wiley.*

Philips, W. J., and N. Philips. 1980. *An Introduction to Mineralogy for Geologists*. New York: Wiley.

Simpson, B. 1966. *Rocks and Minerals*. Elmsford, NY: Pergamon Press.

U.S. Geological Survey, 1971. *Atlas of Volcanic Phenomena*. Reston, VA: U.S. Geological Survey.

Walton, A. J. 1983. *Three Phases of Matter*, 2nd ed. Oxford: Oxford University Press.

Watson, J. 1979. *Rocks and Minerals*, 2nd ed. Boston, MA:

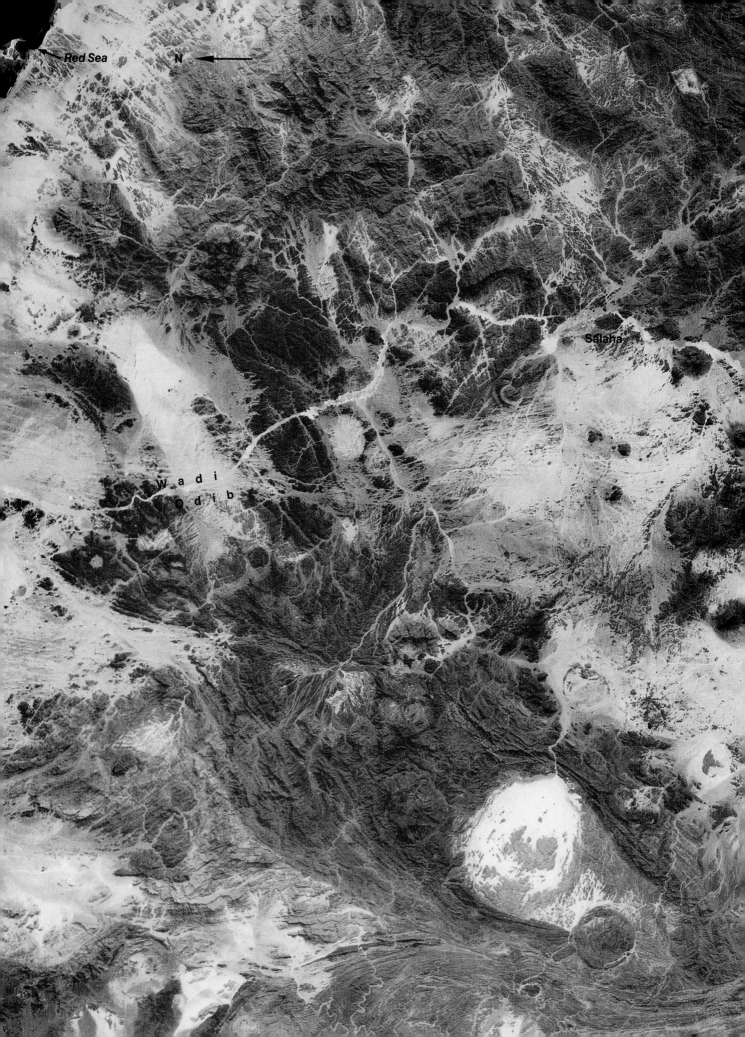

I gneous rocks are records of the thermal history of Earth. Their origin is closely associated with the movement of tectonic plates, and they play an important role in sea-floor spreading, the origin of mountains, and the evolution of continents. The best known examples of igneous activity are volcanic eruptions, where liquid rock material works its way to the surface and erupts from volcanic vents and fissures. Less spectacular, though just as important, are the tremendous volumes of liquid rock that never reach the surface, but remain trapped in the crust, where they cool and solidify. Granite is the most common variety of this type of igneous rock and is commonly exposed in eroded mountain belts and in the roots of ancient mountain systems now preserved in the shields.

The image (opposite) shows part of the shield in the northeast corner of Sudan, close to the border of Egypt. The region is part of the Sahara Desert, which has less than 4 inches of rainfall per year. There is thus little soil and vegetation cover, so the rock bodies are exceptionally well exposed. Igneous rock bodies form circular patterns ranging in size from almost 8 km across the body that is exposed in the large, white area to small structures that appear as dark circles or dots. The liquid rock that formed these bodies migrated upward through the crust like bubbles of air in water. The liquid masses were probably teardrop shaped, somewhat like a hot-air balloon, spherical at the top and tapering downward to a point at its base. The present circular exposures are horizontal cross sections through these structures. Tonal variations in the circular patterns reflect both composition and sand cover. Dark areas are probably rocks containing enough iron and magnesium to color the rock. Gold is commonly associated with such igneous rocks and has been mined in the area since the time of the ancient Egyptians.

In this chapter, we study the major types of igneous rocks and what they reveal about the thermal activity of Earth. We pay particular attention to the distinctive texture of igneous rocks and how we can read from the interlocking network of grains a fascinating history of how the hot liquid became part of the solid crust.

5
Igneous Rocks

MAJOR CONCEPTS

1. Magma is molten rock that is capable of penetrating into, or through, Earth's crust.

2. Magma originates from the partial melting of the lower crust and the upper mantle, usually at depths between 50 and 200 km below the surface.

3. Two major types of magma are recognized: (a) basaltic magma and (b) silicic magma. Intermediate types also occur.

4. The texture of a rock provides important insight into the cooling history of the magma.

5. The major textures of igneous rocks are (a) glassy, (b) aphanitic, (c) phaneritic, (d) porphyritic, and (e) pyroclastic.

6. High-silica magmas produce rocks of the granite-rhyolite family, which are composed of quartz, K-feldspar, Na-plagioclase, and minor amounts of biotite or amphibole.

7. Low-silica magmas produce rocks of the gabbro-basalt family, which are composed of Ca-plagioclase and pyroxene with lesser amounts of olivine and little or no quartz.

8. Magmas with composition intermediate between high and low silica produce rocks of the diorite-andesite family.

9. Basaltic magma is generated by partial melting of the mantle along rift zones. Silicic magma is produced at subduction zones by partial melting of oceanic and lower continental crust.

NATURE OF MAGMA

Magma is molten rock material consisting of liquid, gas, and early-formed crystals. It is mobile and is capable of penetrating into or through rocks of the crust. Two important types of magma are: (1) basaltic magma, which is typically very hot (from 900 to 1200°C) and highly fluid, and (2) silicic magma, which is cooler (less than 800°C) and highly viscous.

The term *magma* comes from the Greek word that means "kneaded mixture," like a dough or paste. In its geologic application, it refers to hot, partially molten, mobile material within Earth that is capable of penetrating into or through the rocks of the crust. This definition may be different from the concept most people have, which is derived largely from illustrations of spectacular volcanic eruptions. Popular articles about *volcanism* usually emphasize the sensational, and readers are impressed with the idea that all magma is a very fluid, red-hot liquid closely resembling molten steel (Figure 5.1).

Most *magmas,* in reality, are not entirely liquid but are a combination of liquid, solid, and gas. Early-formed crystals may make up a large portion of the mass, so a magma could be thought of more accurately as a slush, a liquid melt mixed with a mass of mineral crystals. Such a mixture has a consistency similar to that of freshly mixed concrete, slushy snow, or thick oatmeal. Movement of these magmas is slow and sluggish.

Water vapor and carbon dioxide are the principal gases dissolved in a magma, and they can constitute as much as 15% by weight. These *volatiles* (materials that are readily vaporized) are important because they strongly influence the mobility and melting point of a magma and the types of volcanic activity that can be produced. Dissolved volatiles tend to increase the fluidity of

What are the chemical and physical characteristics of molten rock?

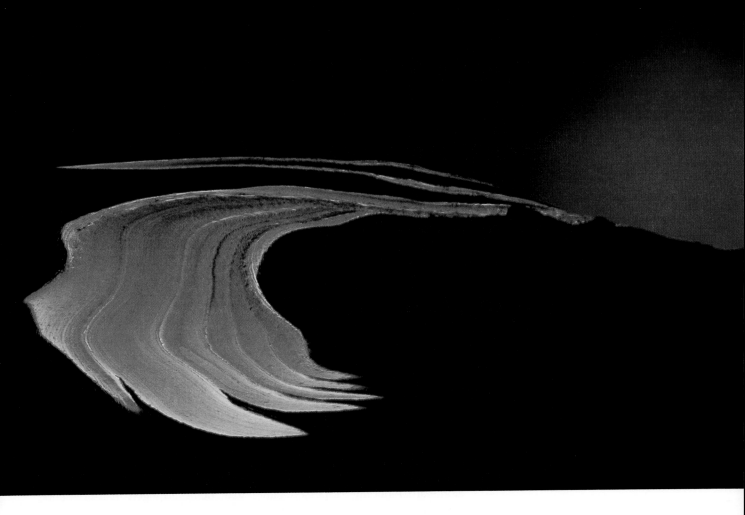

a magma. Magmas rich in volatiles also tend to erupt more violently. More than 90% of the gas emitted from volcanoes is water (H_2O) and carbon dioxide (CO_2).

The principal elements of a magma are oxygen (O), silicon (Si), aluminum, (Al), calcium (Ca), sodium (Na), potassium (K), iron (Fe), and magnesium (Mg). Two constituents—silica (SiO_2) and water (H_2O)—largely control the properties of a magma, such as its *viscosity* (the tendency for a body to resist flow) and the manner in which it is extruded.

Chemical analyses of many *igneous rocks* distinguish two principal kinds of magma: (1) *basaltic magmas,* which contain about 50% SiO_2 and have temperatures ranging from 900 to 1200°C, and (2) *silicic magmas,* which contain between 65 and 77% SiO_2 and generally have temperatures lower than 800°C. Basaltic magmas are characteristically fluid, whereas silicic magmas are thick and viscous. This is because silicic magmas have lower temperatures and greater amounts of SiO_2. The viscosity in a magma is influenced by SiO_2 content because silica tetrahedra link together even before crystallization occurs and the linkages offer resistance to flow. The higher the silica content, therefore, the greater the magma's viscosity.

Like most fluids, magma is less dense than the solid material it forms, and it tends to migrate upward through the mantle and crust. Magma can intrude into the overlying rock by forceful injection into fractures, or it can melt and assimilate the rock it invades. As it rises, it cools and eventually solidifies to form an igneous rock. When it reaches the surface and flows out over the landscape as lava, it forms *extrusive rock* bodies. Magma that solidifies below the surface forms *intrusive rock* bodies.

FIGURE 5.1
Magma is molten rock material. Much of it cools deep in Earth's crust, but some of it works its way to the surface and is extruded as lava. Lava eruptions such as this permit us to study the nature of magma, its composition, and its physical characteristics and give us an insight into the origin of igneous rocks.

IMPORTANCE OF ROCK TEXTURES

The texture of a rock refers to the size, shape, and arrangement of the constituent mineral grains. The major textures in igneous rocks are (a) glassy, (b) aphanitic, (c) phaneritic, (d) porphyritic, and (e) pyroclastic.

The texture of a rock is similar to the architecture of a building—a design of the building blocks, sized, shaped, and arranged in specific ways. The building blocks of rocks, of course, are minerals, so the **texture** of a rock is the size, shape, and arrangement of the constituent minerals. It is a characteristic separate and distinct from composition. Texture is important because the mineral grains bear a record of energy changes involved in the rock-forming process and the conditions existing at the time the rock originated.

The genetic imprint left on the texture of a rock is commonly clear and easy to read. For example, the texture of lunar rocks that were formed by the impact of meteorites is characterized by various sizes of angular rock fragments interspersed with bits and pieces of glass fused by the impact. A sandstone formed on a beach consists of well-rounded, smooth grains, all approximately the same size, which result from the abrasion and washing action of waves. In contrast, rocks formed from a cooling liquid have a texture characterized by interlocking grains.

To illustrate the importance of texture, we will consider six examples of rocks that have essentially the same chemical and mineralogical composition but different textures (Figure 5.2). In each rock, a chemical analysis would disclose about 48% O, 30% Si, 7% Al, and between 3 and 4% each of Na, K, Ca, and Fe. In glassy rocks, crystallization has not taken place, so the constituent atoms are randomly dispersed rather than being arranged in a crystalline structure. In the other rock examples, the elements occur in grains of feldspar, quartz, mica, and amphibole. On the basis of chemical composition alone, these rocks would be considered the same—the difference is in texture *only*. It is the texture that provides the most information about how the specimen was formed.

What does the texture of a rock indicate about the cooling history of a magma?

Glassy Texture

The nature of volcanic *glass* is illustrated in Figure 5.2A. The hand specimen displays a conchoidal fracture with the sharp edges typical of broken glass. No distinct grains are visible, but, viewed under a microscope, distinct flow layers are apparent. These result from the uneven concentration of innumerable, minute, "embryonic" crystals.

In the laboratory, melted rock or synthetic lava hardens to glass if it is quenched (or quickly cooled) from a temperature above that at which crystals would normally form. We can conclude that a **glassy texture** is produced by very rapid cooling. The randomness of the ions in a high-temperature melt is "frozen in" because the ions do not have time to migrate and organize themselves in an orderly, crystalline structure. Field observations of glassy rocks in volcanic regions support the hypothesis that rapid cooling produces glass. Small pieces of lava blown from a volcanic vent into the much cooler atmosphere harden to form glassy ash. A glassy crust forms on the surface of many lava flows, and glassy fragments form if a flow enters a body of water.

Aphanitic Texture

If crystal growth from a melt requires time for the ions to collect and organize themselves, then a crystalline rock indicates a slower rate of cooling than that of a glassy rock. The texture illustrated in Figure 5.2B is crystalline but ex-

(A) A glassy texture develops when molten rock material cools so rapidly that the migration of ions to form crystal grains is inhibited. Glassy texture typically forms on the crust of lava flows and in viscous magma. The sample shown here is obsidian.

(B) An aphanitic texture consists of mineral grains too small to be seen without a microscope. The sample shown here is rhyolite. Only a few grains are large enough to be seen. Most are microscopic. Aphanitic texture results from rapid cooling.

(C) A phaneritic texture consists of grains large enough to be seen with the unaided eye. All grains are roughly the same size and interlock to form a tight mass. The large crystals suggest a relatively slow rate of cooling.

(D) A porphyritic-aphanitic texture results from two separate rates of cooling. In this specimen, the larger phenocrysts formed first when the magma cooled relatively slowly. The microscopic crystals (aphanitic matrix) formed later, during a period of more rapid cooling.

(E) A porphyritic phaneritic texture results from two stages of cooling, both of which are relatively slow so all of the crystals are relatively large and can be seen without a microscope. In this specimen, the large pink crystals of K-feldspar developed during the first stage of cooling. The second stage of cooling was more rapid, so the remaining crystals are smaller.

(F) A pyroclastic texture forms when crystals, fragments of rock, and glass are blown out of a volcano as ash. Generally the particles fall to Earth and accumulate like sediment. Ash flows are very hot and move as a body close to the ground, and the individual fragments become fused together in a dense mass.

FIGURE 5.2
Textures of igneous rocks provide important information concerning rock genesis. In the examples presented here, all of the rocks have roughly the same composition, but different textures. The photographs show the actual size of the specimens.

tremely fine grained—a texture referred to as *aphanitic* (Greek *a,* "not," *phaneros,* "visible"). In hand specimens, few, if any, crystals can be detected in aphanitic textures. Viewed under a microscope, however, many crystals of feldspar and quartz can be recognized.

An aphanitic texture indicates relatively rapid cooling, but not nearly as rapid as the quenching that produces glass. Aphanitic textures are typical of the interiors of lava flows, in contrast to the glassy texture that forms on the surface, or crust.

A feature in many aphanitic and glassy rocks is the presence of numerous small spherical or ellipsoidal cavities called *vesicles.* These are produced by gas bubbles trapped in the solidifying rock. As hot magma rises toward Earth's surface, the confining pressure diminishes, and dissolved gas (mainly steam) separates and collects in bubbles. The process is similar to the effervescence of champagne and soda pop when the bottles are opened. Vesicular textures typically develop in the upper part of a lava flow, just below the solid crust, where the upward-migrating gas bubbles are trapped. Though vesicles change the outward appearance of the rock and indicate the presence of gas in a rapidly cooling lava, they do not change the basic aphanitic texture.

Phaneritic Texture

The specimen shown in Figure 5.2C is composed of grains large enough to be recognized without a microscope, a texture known as *phaneritic* (Greek *phaneros,* "visible"). The grains are approximately equal in size and form an interlocking mosaic. The equigranular texture suggests a uniform rate of cooling, and the large size of the crystals shows that the rate of cooling was very slow.

In order for cooling to take place at such a slow rate, the magma must have cooled far below the surface. Field evidence supports this conclusion, since volcanic eruptions produce only aphanitic and glassy textures. Rocks with phaneritic textures are exposed only after erosion has removed thousands of meters of covering rock.

Porphyritic Texture

Some igneous rocks have grains of two distinct sizes. The larger, well-formed crystals are referred to as *phenocrysts,* and the smaller crystals constitute the *matrix,* or the *groundmass.* This texture is called *porphyritic.* It occurs in either aphanitic or phaneritic rocks.

A porphyritic texture usually indicates two stages of cooling. An initial stage of slow cooling, during which the large grains developed, is followed by a period of more rapid cooling, during which the smaller grains formed. The aphanitic matrix shown in Figure 5.2D indicates that the cooling melt had sufficient time for all of its material to crystallize. The initial stage of relatively slow cooling produced the larger grains, and the later stage of rapid cooling, when the magma was extruded, produced the smaller grains. Similarly, a phaneritic matrix with phenocrysts indicates two stages of cooling (Figure 5.2E). An initial stage of very slow cooling was followed by a second stage, when cooling was more rapid but not rapid enough to form an aphanitic matrix.

Pyroclastic Texture

The texture shown in Figure 5.2F may appear at first to be that of a porphyritic rock with phenocrysts of quartz. Under a microscope, however, the grains are seen to be broken fragments rather than interlocking crystals. Some fragments are bent and squeezed, and shards of glass are twisted and deformed. This is a *pyroclastic texture* (Greek *pyro*, "fire," *klastos*, "broken"), produced when explosive eruptions blow ash, early-formed minerals, and glass into the air as a mixture of hot fragments. If the fragments are still hot when they are deposited, they will be welded (fused) together, or they may be cemented together later in the cooling process.

In summary, the texture of an igneous rock records considerable information about the rock's cooling history and the manner in which it was formed. All of the rocks just described have the same chemical composition. All could have been derived from the same magma, yet each texture type was formed in a different way. Geologists therefore pay particular attention to a rock's texture and use it as one basis for classification of all major rock types.

KINDS OF IGNEOUS ROCKS

Igneous rocks are classified on the basis of texture and composition. The major kinds of igneous rocks are basalt, gabbro, andesite, diorite, rhyolite, and granite.

A simple chart of the major types of igneous rocks is shown in Figure 5.3. The basis for this scheme of classification is texture and composition. Variations in composition are arranged horizontally, and variations in texture are arranged vertically. Rocks that cool below the surface are called intrusive, and those that cool at the surface are called extrusive. The rock names are printed in bold type, the size of which is roughly proportionate to the relative abundance of the rock. Rocks in the same column have the same composition but different textures. Rocks in the same horizontal row have the same texture but different compositions. The chart shows that granite, for example, has a phaneritic texture and is composed predominantly of quartz and K-feldspar. The type size indicates that it is the most abundant intrusive igneous rock. Rhyolite has the same composition as granite but is aphanitic. Basalt has an aphanitic texture and is composed predominantly of Ca-plagioclase and pyroxene. It has the same composition as gabbro but is much more abundant.

This classification attempts to show the natural, or genetic, relationships between the various rock types. As we have seen in the previous section, texture provides important information on the cooling history of the magma. Rocks that crystallize slowly are able to grow large crystals; those that cool

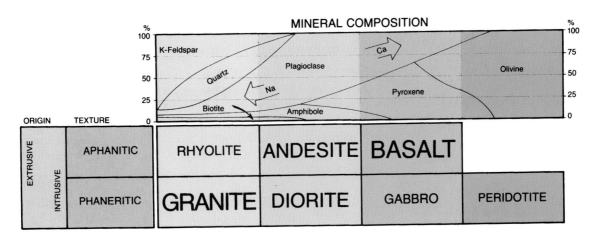

FIGURE 5.3
The classification of igneous rocks is based on texture (shown vertically on the chart) and composition (shown horizontally). The size of type in which the names of the rocks are printed is roughly proportional to their abundance.

rapidly have a fine-grained or glassy texture. The composition of a rock provides information about the nature and origin of the magma. Magmas high in iron and magnesium and Ca-plagioclase generally originate from partial melting of the mantle; they are formed in rift systems and associated with mid-oceanic ridges. Rocks high in silica, Na-plagioclase, and K-feldspar form at converging plate margins.

Rocks with Aphanitic Textures

Basalt. *Basalt* is the most common aphanitic rock. It is a very fine-grained, usually dark-colored rock that originates from the cooling lava flows. The mineral grains are so small that they can rarely be seen without a microscope. If a *thin section* (a thin, transparent slice of rock) is viewed through a microscope, the individual minerals can then be seen and studied (Figure 4.18).

Basalt is composed predominantly of calcium-rich plagioclase and pyroxene, with smaller amounts of olivine or amphibole. The plagioclase occurs as a mesh of elongate, lathlike crystals surrounding the more equidimensional pyroxene and olivine grains. In some cases, large crystals of olivine or pyroxene form phenocrysts, resulting in a porphyritic texture. Many basalts have some glass, especially near the tops of flows.

Andesite. *Andesite* is an aphanitic rock composed of plagioclase, pyroxene, and amphibole. It usually contains little or no quartz and has the same composition as diorite. The texture of andesite is generally porphyritic, with phenocrysts of feldspar and ferromagnesian minerals. The rock is named after the Andes Mountains, where volcanic eruptions have produced it in great abundance. Andesite is the next most abundant lava type after basalt and occurs most frequently along the continental margins. It is not found in the ocean basins, nor is it abundant in the continental interiors. The origin of andesite along the continental margins is probably related to partial melting of the oceanic crust and lower continental crust at convergent plate boundaries.

Rhyolite. *Rhyolite* is an aphanitic rock with the same composition as that of granite. It commonly contains a few phenocrysts of feldspar, quartz, or mica, but usually not enough to be considered porphyritic. Rhyolite lava flows are viscous. Instead of spreading in a linear flow, rhyolite typically piles up in large, bulbous domes (see Figure 5.10). Rhyolite and andesite may be difficult to distinguish without the aid of a microscope and are commonly grouped together and referred to as *felsite* (any light-colored aphanitic rock).

Rocks with Pyroclastic Textures

Tuff. Volcanic eruptions of rhyolitic and andesitic lavas commonly produce large volumes of fragmental material. The fragments range from dust-size pieces to large blocks more than a meter in diameter. The rock resulting from the accumulation of ash falls is referred to as *tuff*. Although of volcanic origin, tuff has many of the characteristics of sedimentary rock because the fragments composing tuff settle out from suspension in the air and are stratified like sedimentary rocks.

Ash-Flow Tuff. *Ash-flow tuff* is rock composed of fragments of volcanic glass, broken fragments of crystal, rock fragments, and pieces of solidified magma—all fused together in a tight, coherent mass. Typically, many of the fragments are flattened or bent out of shape. This unique texture indicates that at the time of extrusion the ash fragments were hot enough to fuse together in a rock mass.

Rocks with Phaneritic Textures

Granite. *Granite* is a coarse-grained igneous rock composed predominantly of feldspar and quartz. (Figure 5.3). K-feldspar is the most abundant mineral, and usually it is easily recognized by its pink color. Na-Ca-plagioclase is present in moderate amounts, usually distinguished by its white color and its porcelainlike appearance. Mica is conspicuous as black or bronze-colored flakes, usually distributed evenly throughout the rock. The texture of granite, together with laboratory experiments, indicates that plagioclase, biotite, and amphibole are the first minerals to crystallize from a granitic magma. A seemingly insignificant but very important property of granite is its relatively low specific gravity, about 2.7. In contrast to basalt and related rocks, which have a specific gravity of 3.2, granite is light. This fact is important in considering the nature of continents and the contrast between continental crust and oceanic crust. Granite and related rocks make up the great bulk of continental crust, whereas most of the oceanic crust is composed of basalt.

Diorite. *Diorite* is similar to granite in texture, but it differs in composition. Plagioclase feldspar is the dominant mineral, and quartz and K-feldspar are minor constituents. Amphibole is an important constituent, and some pyroxene may be present. In composition, diorite is intermediate between granite and basalt.

Gabbro. *Gabbro* is not commonly exposed at Earth's surface, but it is a major constituent of the lower part of the oceanic crust and is present in some ancient parts of the continents. It has a coarse-grained texture similar to that of granite, but it is composed almost entirely of pyroxene and calcium-rich plagioclase, with minor amounts of olivine. Gabbro is dark green, dark gray, or almost black because of the predominance of dark-colored minerals.

Peridotite. *Peridotite* is composed almost entirely of two minerals, olivine and pyroxene (Figure 5.4). It is not common at Earth's surface or within the continental crust, but it is a major constituent of the subcrustal part of Earth (the mantle). Its high specific gravity, together with other physical properties, suggest that the great bulk of Earth—the mantle—is composed of peridotite and closely related rock types. The Alps and Saint Paul's Rocks (islands in the Atlantic Ocean) are two areas where peridotite from the mantle appears to have been pushed through the crust to Earth's surface.

FIGURE 5.4
Peridotite is a phaneritic rock composed mostly of olivine, but minor amounts of pyroxene and Ca-plagioclase may be present. This specimen is composed almost entirely of olivine. It is believed that the mantle is composed of peridotite and closely related rocks.

EXTRUSIVE ROCK BODIES

Extrusive igneous rocks are those that form from magma being extruded onto Earth's surface by volcanic eruptions. The rocks include lava flows and volcanic ash. Basaltic magmas are low in silica and are relatively fluid. The lava is typically extruded quietly from fissures and fractures. Silicic magmas are thick and viscous and their eruptions are typically explosive. The lava is extruded as thick flows, bulbous domes, or ash flows.

One of the most spectacular of all geologic processes is the extrusion of lava onto Earth's surface by volcanic eruptions. Throughout recorded history more than 700 volcanoes have been active, but this is only an instant in geologic time and ignores the region of the most intense volcanic activity on Earth—the region hidden beneath the oceans, where most eruptions go unnoticed. The importance of volcanic activity is that it testifies to the continuing dynamics of Earth, provides an important window on the planet's interior, and sheds light on the processes operating below the surface in the lower crust and upper mantle.

There are two major types of magma:

1. Basaltic magmas are low in silica and are relatively fluid. Dissolved gases escape readily, so the lava is typically extruded quietly from fissures and fractures. Fissure eruptions produce a succession of thin lava flows that cover large areas.
2. Silicic magmas are thick and viscous. The escape of gas is thus retarded, and pressure builds up within the magma. These eruptions are typically violent, and the lava is extruded as thick flows, bulbous domes, or ash flows.

Basaltic Eruptions

Basaltic eruptions are probably the most common type of volcanic activity on Earth. The lava is generally extruded from fractures or fissures in the crust. Upon extrusion, it tends to flow freely downslope and spreads out to fill valleys and topographic depressions (Figure 5.5). This type of eruption occurs along fractures in the oceanic ridge, forming new oceanic crust where the tectonic plates move apart, and is the major type of eruption in the volcanic plains and plateaus of the continents. Basaltic *lavas* have been observed with eruptive temperatures ranging between 900°C and 1200°C. Before it cools and congeals, the lava can flow at speeds as high as 30 to 40 km per hour down steep slopes, but rates of 20 km per hour are considered unusually rapid. As a flow moves downslope, it loses gas, cools, and becomes more viscous. Movement then becomes sluggish, and the flow soon comes to rest.

What are the major types of volcanic eruptions? How does each originate?

There are two common types of basaltic flows, referred to by the Hawaiian terms *aa* (pronounced ah'ah') and *pahoehoe* (pronounced pa ho'e ho'e) (Figures 5.6A and 5.6B). An *aa flow* contains relatively little gas and is a slow-moving flow 3 to 10 m thick. The surface of the flow cools and forms a crust while the interior remains molten. The flow may move only a few meters per hour. As it continues to move, the hardened crust is broken into a jumbled mass of angular blocks and clinkers (Figure 5.6A). Gas within the fluid interior of the flow migrates toward the top, but it may remain trapped beneath the crust. These "fossil gas bubbles," called vesicles, make the rock light and porous.

Compared with aa flows, *pahoehoe flows* are more fluid. They are not as thick (usually less than 1 m thick) and move much faster because of their lower viscosity. As a pahoehoe flow moves, it develops a thin, glassy crust, which is

molded into billowy forms or surfaces that can resemble coils of rope. A variety of flow features (such as those shown in Figure 5.6B) can develop on the surface of the flow.

The interior of a flow may be massive and nonvesicular. As a flow cools, it contracts and may develop a system of polygonal cracks called *columnar joints,* which are similar in many ways to mud cracks (Figure 5.6C). In some flows, the sides and top freeze solid while the interior remains fluid. When pressure is great enough, the fluid interior can break through the crust and flow out, leaving a long lava tube (Figure 5.6D). More commonly, pressure from the fluid interior causes the crust of the flow to arch up in a *pressure ridge,* a blister characterized by a central crack through which gas and lava escape (Figure 5.6E).

Instead of issuing from a central vent, basaltic lava is commonly extruded from a series of fractures in the crust called *fissures.* The fluid lava usually spreads out over a large area, rather than building an isolated cone. In some places along the fissure, the rising lava may be concentrated and erupt like a lava fountain. The splashing of lava around the fountain can build up small conical-shaped mounds called *spatter cones* (Figure 5.6F).

Droplets and globs of lava blown out from a volcanic vent usually cool by the time they fall back to the ground. This material forms *volcanic ash* and dust, collectively known as *tephra* (Figure 5.6H), and larger fragments called *volcanic bombs* (Figure 5.6I). As the tephra travels through the air, it is sorted according to size. The larger particles accumulate close to the vent, or *volcanic neck,* to form a *cinder cone* (Figure 5.7), and the finer, dust-size particles are transported afar by the wind. Cinder cones, which are generally less than 200 m high and 2 km in diameter, are relatively small features compared with the large shield volcanoes and stratovolcanoes.

FIGURE 5.5
Fissure eruptions are the most common type of volcanic eruption on Earth. Lava is simply extruded through cracks or fissures in the crust. This type of eruption is typical of fluid basaltic magma and is the dominant eruptive style along the oceanic ridge. This photograph shows a recent fissure eruption on the island of Hawaii.

(A) *The surface of an aa flow* consists of a jumbled mass of angular blocks, which form when the congealed crust is broken as the flow slowly moves. Aa flows are viscous and much thicker than pahoehoe flows. This photograph shows a recent aa flow in Craters of the Moon National Monument, Idaho.

(B) *The surfaces of pahoehoe flows* are commonly twisted, ropy structures. Pahoehoe flows form on fluid lava and typically are very thin.

(C) *Hexagonal columnar joints* commonly form by contraction when a lava cools. The long axis of the column is approximately perpendicular to the cooling surface.

(D) *Lava tubes* develop where the margin of the flow cools and solidifies and the interior, molten material is drained away.

(E) *Pressure ridges* develop in lava flows when the outer crust is arched up by the pressure of gas trapped beneath. They commonly crack and release the trapped gas and lava from the interior of the flow.

(F) *Spatter cones* form in local areas along fissures where globs of lava accumulate near a major vent. This small spatter cone is along the major fissure in Craters of the Moon National Monument, Idaho.

FIGURE 5.6
A variety of surfaces develop on basaltic lava flows and reflect the manner of flow, rates of cooling, amounts of dissolved gases, and viscosity.

(G) Pahoehoe lava flows from recent eruptions on the island of Hawaii. The flow is moving away from the observer. The main flow forms a solid crust along its margins and upper surface. Hot liquid lava breaks through the crust, gradually cools, and forms a new crust. The process is then repeated downslope.

(H) Tephra is a general term referring to all pyroclastic material ejected from a volcano. It includes ash, dust, bombs, and rock fragments. It is commonly stratified.

(I) Volcanic bombs are fragments of lava ejected in a liquid or plastic state. As they move through the air they twist and turn and form spindle-shaped masses.

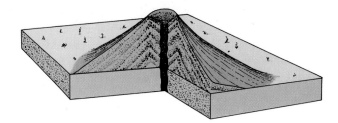

FIGURE 5.7
A cinder cone is a small volcano composed almost exclusively of ash and dust blown out from a central vent. The internal structure consists of layers of ash inclined away from the summit crater. The vent, or volcanic neck, is commonly filled with solidified lava and fragmental debris.

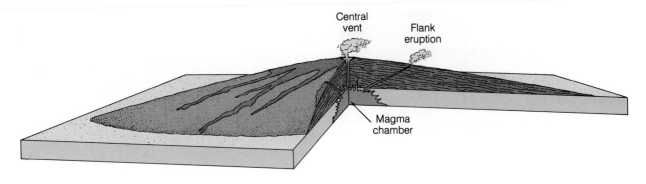

FIGURE 5.8
Shield volcanoes are the largest volcanoes on Earth. They are composed of innumerable thin basaltic flows with relatively little ash. Some of the best examples are the islands and seamounts of the Pacific. The island of Hawaii, rising approximately 10,000 m above the sea floor, is the largest volcano on Earth.

FIGURE 5.9
Pillow basalt is formed when lava extruded under water cools quickly, forming a series of ellipsoidal masses. This photograph shows pillow lava that was originally formed on the ocean floor and is now exposed in New Zealand.

If the extrusion of large quantities of fluid basaltic lava continues, a broad cone called a *shield volcano* may form around the vent (Figure 5.8). With each eruption, the fluid basaltic lava flows freely for some distance, spreading into a thin sheet, or tongue, before congealing. Shield volcanoes, therefore, have wide bases (commonly over 100 km in diameter) and gentle slopes (generally less than 10 degrees). Their internal structure consists of innumerable thin basalt flows with comparatively little ash. The Hawaiian Islands are excellent examples of shield volcanoes. They are enormous mounds of basaltic lava, rising as high as 10,000 m above the sea floor. The younger volcanoes typically have summit craters, as much as 3 km wide and several hundred meters deep, which resulted from subsidence following the eruption of magma from below.

The extrusion of basaltic lava into water produces an ellipsoidal mass referred to as *pillow lava* (Figure 5.9). The formation of pillow basalt has been observed off the coast of Hawaii, and recent undersea photographs show that it is widespread over the sea floor, where volcanic activity is common.

Silicic Eruptions

Silica-rich magmas produce granitic and associated igneous rocks. They are relatively cool, and the mechanics of eruption and flow of their lavas are quite different from that of basaltic lavas.

Some silicic magmas are so thick and viscous that small volumes hardly flow at all, but instead form massive plugs or bulbous domes over the volcanic vent (Figure 5.10). The high viscosity of silicic magmas inhibits the escape of dissolved gas, so tremendous pressure builds up. Consequently, when eruptions occur, they are highly explosive and violent and commonly produce large quantities of tephra. The alternating layers of tephra and thick, viscous lava typically produce a *composite volcano,* or *stratovolcano,* characterized by a high, steep-sided cone around the vent (Figure 5.11). This is probably the most familiar form of continental volcano, with such famous examples as Mounts Shasta, Fuji, Vesuvius, Etna, and Stromboli. A depression at the summit, the *crater,* usually marks the position of the vent.

The explosive eruptions of silicic volcanoes can blow out large volumes of ash and magma. As a result, the summit area sometimes collapses, forming a large basin-shaped depression known as a *caldera.* Crater Lake, Oregon (Figure 5.12), for example, formed when a volcano's summit collapsed after its last major eruption and the resulting caldera filled with water. Wizard Island, a small cinder cone in the caldera, resulted from subsequent minor eruptions. Krakatoa is another well-known example. The 1883 eruption was one of the largest explosions in recent history. The cone was demolished, and great quantities of volcanic ash were blown high into the atmosphere. The explosion and subsequent subsidence produced a caldera 6 km in diameter, which completely altered the configuration of the island.

A spectacular type of eruption associated with silicic magmas is the lateral flowing movement of large masses of ash and lava particles. This phenomenon

(A)

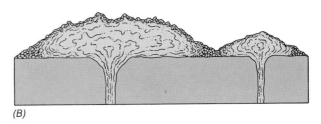

(B)

FIGURE 5.10
Domes of silicic lava form because silica-rich lava is viscous and resists flow. It therefore tends to pile up over the vent to form large, bulbous domes.

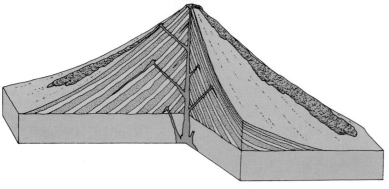

FIGURE 5.11
Composite volcanoes are built up of alternating layers of ash and lava flows, which characteristically form high, steep-sided cones.

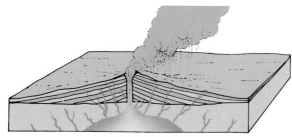

(A) Early eruptions formed the prehistoric volcano Mount Mazama.

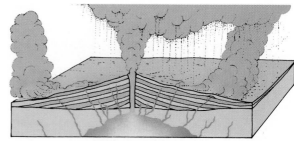

(B) Great eruptions of ash flows emptied the magma chamber, causing the top of the volcano to collapse.

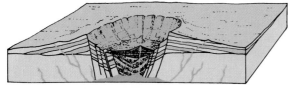

(C) The collapse of the summit into the magma chamber formed the caldera.

(D) A lake formed in the caldera, and subsequent minor eruptions produced small volcanic islands in the lake.

FIGURE 5.12
The evolution of the caldera at Crater Lake, Oregon, involved a series of great eruptions followed by the collapse of the summit into the magma chamber.

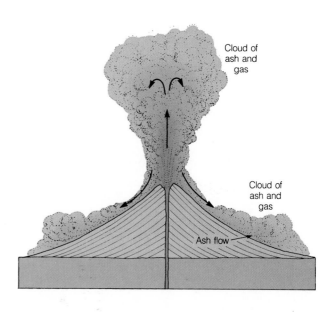

FIGURE 5.13
An ash flow is a hot mixture of highly mobile gas and ash that moves rapidly over the surface of the ground away from the vent. The ash rises into the air from the explosive force of the eruption but, being much denser than air, it moves en masse back to the surface and rushes down the slopes of the volcano as an ash flow. Less dense gas and ash continue to move upward as a cloud into the atmosphere. This ash ultimately falls back to the surface as an ash fall.

FIGURE 5.14
Devastation caused by the 1951 ash flow of Mount Lamington, New Guinea, illustrates the force of hot flowing ash.

is not a liquid lava flow or an ash fall (in which particles settle independently), but a flow consisting of gas and suspended fragments of hot mineral grains, droplets of lava, and pieces of rock suspended like a dense dust cloud, which moves rapidly close to the surface. This type of eruption is therefore called an *ash flow* (Figure 5.13). As a magma works its way to the surface, confining pressure is released and trapped gas bubbles rapidly expand. Near the surface, the magma can literally explode, ejecting pieces of lava, bits of solid rock, ash, early-formed crystals, and gas. This material is very hot, sometimes incandescent. Being denser than the air, it flows across the ground surface as a thick, dense cloud of hot ash. Ash flows can reach velocities greater than 250 km per hour because the expanding gas continually forces the cooling lava particles apart. Some have moved as much as 400 km from the vent. When an ash flow comes to rest, the particles of hot crystal fragments, glass, and ash fuse together to form *welded tuff*. Ash flows can be very large. Some form flow units more than 100 m thick and cover thousands of square kilometers. As it cools, the contracting mass can develop columnar jointing.

Ash-flow eruptions are rare and catastrophic events. A few fortunate geologists have had the opportunity to witness them from afar and to make direct observations of this type of extrusion. For example, Mount Lamington, in New Guinea, was considered extinct until it erupted in 1951. It had never been examined by geologists and was not even considered to be a volcano by the local inhabitants. When it did erupt, volcanic activity began with preliminary emissions of gas and ash, accompanied by earthquakes and landslides near the crater. Then on Sunday, January 20, 1951, a catastrophic explosion burst from the crater and produced an ash flow that completely devastated an area of about 200 km². The main eruption was observed and photographed at close quarters from passing aircraft, and a qualified volcanologist was on the spot within 24 hours. Sensitive seismographs were soon installed near the crater to monitor Earth movements, and aerial photographic records were made daily. The ash flow descended radially from the summit crater, its direction of movement controlled to some degree by the topography. As the ash flow moved down the slopes, it scoured and eroded the surface. Estimated velocities of 470 km per hour were calculated from the force required to overturn certain objects. Entire buildings were ripped from their foundations, and automobiles were picked up and deposited in the tops of trees (Figure 5.14).

INTRUSIVE ROCK BODIES

Igneous intrusions are masses of rock formed when magma cools beneath the surface. They are classified according to their sizes, shapes, and relationships to the older rocks that surround them. Important intrusive rock bodies are batholiths, stocks, dikes, sills, and laccoliths.

Magma is mobile, at times amazingly so. It rises because it is less dense than the surrounding rock. It can push aside surrounding rocks, force its way into cracks, and flow on the surface over distances of more than 100 kilometers. It can move upward in the crust by melting away surrounding rocks or wedging and prying loose large blocks of rock, which it then replaces. When a magma within the crust loses its mobility it slowly cools and solidifies, forming a mass of igneous rock called an intrusion. Igneous intrusions occur in a variety of sizes and shapes and are exposed at the surface only after the previously overlying rock has been removed by erosion.

Intrusions usually are classified according to their sizes, shapes, and relationships to the older rocks that surround them (Figure 5.15). They include batholiths, stocks, dikes, sills, and laccoliths.

Batholiths

Batholiths, which are masses of coarsely crystalline rock, generally of granitic composition, are the largest rock bodies in Earth's crust. They form almost exclusively in continents and large island arcs and do not occur in oceanic

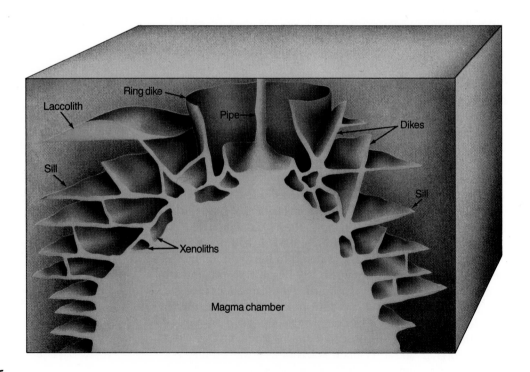

FIGURE 5.15
Magmatic intrusions may assume a variety of forms. Batholiths are large masses of coarsely crystalline rock that cools in the major magma chamber. Stocks are smaller masses and may be protrusions from a batholith. Dikes are narrow, tabular bodies formed as magma is squeezed into fractures and cools. Many dikes are related to conduits leading to volcanoes. Some radiate out from the volcanic neck; others form a circular pattern above a stock and are called ring dikes. Sills are layers of igneous rock squeezed in between sedimentary strata. Laccoliths, dome-shaped bodies with a flat floor, are formed where magma is able to arch up the overlying strata. Inclusions of the surrounding rock in the magma are called xenoliths. A pipe is a cylindrical conduit through which magma migrates upward.

islands. Many cover several thousand square kilometers. The Idaho batholith, for example, is a huge body of granite, exposed over an area of nearly 41,000 km². The British Columbia batholith, to the north, is over 2000 km long and 300 km wide and at least 30 km thick. The true three-dimensional form of batholiths is difficult to determine because of uncertainty about their shape and extension deep below the surface. Evidence from gravity measurements and from seismic studies showing the layered nature of the crust and the mantle suggests that batholiths are limited in depth to the thickness of the crust and do not extend down into the mantle. They must therefore be less than 60 km thick. The nature of the base of a batholith remains a matter of conjecture. Indirect evidence suggests that the size of batholiths may increase with depth for some distance and then taper off at still greater depth, somewhat like the root of a tooth. From these data, batholiths appear to be huge, slablike bodies, with a horizontal extent much greater than their thickness. Studies of the thickness of strata upwarped along the flanks of batholiths suggest that they were emplaced at depths of more than 7 km below the surface. The diagrams in Figure 5.16 give a rough idea of the geometric form of some of the better-known batholiths in North America.

Figure 5.17 shows a diagram of a batholith and its relationship to adjacent rock bodies. The mass of granitic rock can be elongate and elliptical or circular. It generally cuts across the surrounding rock. In some areas, however, the wall of the batholith can be parallel to the layers of the surrounding rocks. In the simplest cases, a batholith results from a single *intrusion* in which the magma

How do intrusions originate?

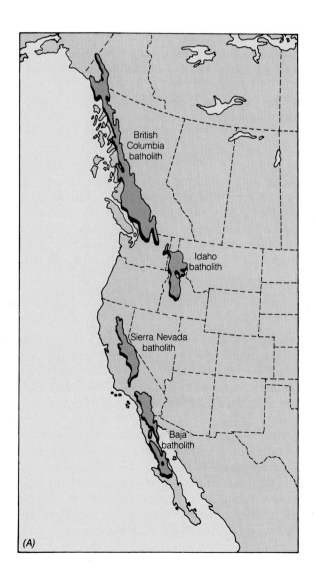

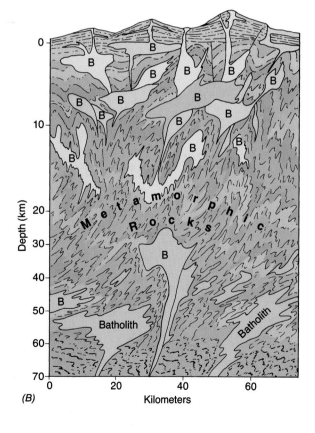

FIGURE 5.16
The shapes of batholiths can be determined by mapping the surface exposure and by studies of seismic wave velocities. Some well-known batholiths in North America, shown on the map (A), are huge tabular bodies with an irregular elliptical shape. An idealized cross section through the continental crust (B) illustrates the relative size and shape of batholiths and their relationship to the surrounding rock.

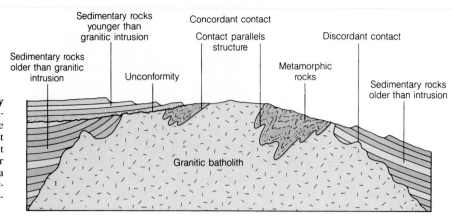

Sedimentary rocks
older than granitic
intrusion

Sedimentary rocks
younger than
granitic intrusion

Unconformity

Concordant contact

Contact parallels
structure

Metamorphic
rocks

Discordant contact

Sedimentary rocks
older than intrusion

Granitic batholith

FIGURE 5.17
The relationship between sedimentary rocks and igneous intrusions provides important information about relative ages of the rock bodies and their geologic histories. Most intrusions are discordant; that is, they cut across the surrounding older rock. If younger sedimentary rock rests upon the intrusion, a period of uplift and erosion probably occurred before the sedimentary rocks were deposited.

is injected into older rock and cools, but many batholiths are composed of several intrusions of similar composition.

Batholiths typically form in the deeper zones of mountain belts and are exposed only after considerable uplift and erosion. Some of the highest peaks of mountain ranges, such as the Sierra Nevada and the Coast Ranges of western Canada, are carved in the granite of batholiths. These rocks originally cooled thousands of meters below the surface. The trend of a batholith usually parallels the axis of the mountain range, although the intrusion can cut locally across folds within the range.

Extensive *exposures* of batholiths also are found in the shield areas of the continents (Figures 5.18 and 22.1). These exposures are considered to be the roots of ancient mountain ranges, which have long since been eroded to lowlands. The young mountain ranges and the ancient shields, therefore, are believed to expose parts of the batholith formed at different depths. In mountains, we see remnants of the roof of the magma chamber. In the shield, only the lower and deeper parts of the batholiths remain.

Stocks

What are the characteristics of the major types of intrusions?

A *stock* is an intrusive body with an outcrop area (the area exposed at the surface) of less than 10 km². Some stocks are known to be small protrusions rising from the main batholith, but the downward extent of others is unknown. Stocks are generally composed of granitic rock, and many are porphyritic with a fine-grained groundmass. Many deposits of silver, gold, lead, zinc, and copper are deposited in fractures and form veins extending from a stock into the surrounding rock.

Dikes

One of the most familiar signs of ancient igneous activity is a narrow, tabular body of igneous rock called a *dike* (Figure 5.19). A dike forms where magma squeezes into fractures of the surrounding rock and cools. The width of a dike can range from a fraction of a centimeter to hundreds of meters. The largest known example is the Great Dike of Zimbabwe, which is 600 km long and has an average width of 10 km.

The emplacement of dikes is controlled by fracture systems within the surrounding rock. Dikes, therefore, cut across sedimentary and metamorphic layers and are said to be *discordant*. They commonly radiate from ancient volcanic necks and thus reflect the stresses associated with volcanic activity. Sometimes upward pressure from a magma chamber produces circular or elliptical fracture systems, in which injected magma forms ring dikes (Figure 5.15). Larger ring dikes can be as much as 25 km in diameter and thousands of meters deep.

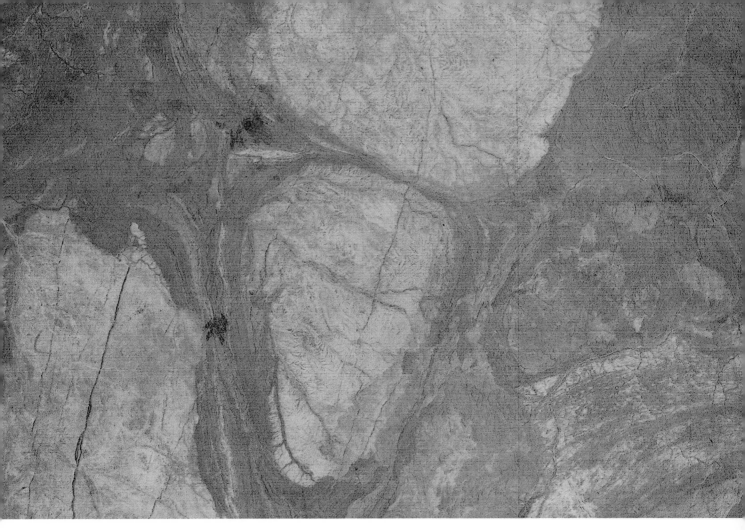

FIGURE 5.18
Batholiths in the shield of western Australia are shown on Landsat images as large, elliptical, yellowish white masses. These granitic rocks have intruded and arched up the older, greenish metamorphic rock, so the older layered structures are nearly vertical.

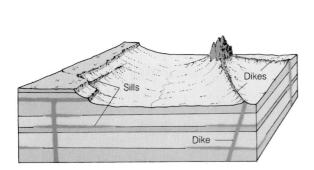

FIGURE 5.19
Dikes and sills are tabular intrusive bodies. Dikes cut across layers of the surrounding rock. Sills are injected between layers of strata.

101

After erosion, the surface expression of a dike is usually a long, narrow ridge. Dikes can also erode as fast as the surrounding rock, or even faster, and they can erode down to form long, narrow trenches.

Sills

Rising magma follows the path of least resistance. If this path includes a bedding plane, which separates layers of sedimentary rock, magma may be injected between those layers to form a *sill*—a tabular intrusive body parallel to, or **concordant** with, the layering (Figures 5.15 and 5.19). Sills range from a few centimeters to hundreds of meters thick and can extend laterally for several kilometers. A sill can resemble a buried lava flow lying within the sequence of sedimentary rock. It is an intrusion, however, squeezed between layers of older rock. Many features evident at the contact with adjacent strata can be used to distinguish between a sill and a buried lava flow. For instance, rocks above and below a sill are commonly altered and recrystallized, and a sill shows no signs of weathering on its upper surface. Sills also commonly contain **inclusions,** blocks and pieces of the surrounding rocks. A buried lava flow, in contrast, has an eroded upper surface marked by vesicles, and the younger, overlying rock commonly contains fragments of the eroded flow.

Sills can form as local offshoots from dikes, or they can be connected directly to a stock or a batholith. Characteristically, they are formed from highly fluid basaltic magma, which can be injected between the older rocks without deforming them. The more viscous silicic magma rarely forms sills.

Laccoliths

When viscous magma is injected between layers of sedimentary rock, it tends to arch up the overlying strata. The resulting intrusive body, a *laccolith,* is lens-shaped, with a flat floor and an arched roof (Figure 5.20). Laccoliths usually occur in blisterlike groups in areas of flat-lying strata. They can be several kilometers in diameter and thousands of meters thick. Typically, they are porphyritic.

Laccoliths were first discovered in the Colorado Plateau, where they occur along the margins of a central stock and appear as inflated sills. Laccoliths can also be fed through a conduit from below.

THE ORIGIN OF MAGMA

Magma is a product of the dynamics of plate margins. It is not generated in Earth's core. Basaltic magma is formed by partial melting of the mantle along spreading centers as tectonic plates move apart. Silicic magma is generated along subduction zones by partial melting of the descending oceanic crust and by partial melting associated with metamorphism in the continental crust.

The origin of magma and the mechanics by which it rises through the crust remain fundamental problems that are not completely understood. We do know that magma is not generated in Earth's lower mantle or core. Seismic evidence shows that the crust and the mantle are solid and that no permanent, worldwide reservoirs of magma exist from the surface to a depth of nearly 2900 km. Earth's core responds to seismic waves as if the core were a liquid, but gravity measurements indicate that the core is made of material far denser than any mantle rock, so the core material could not rise to the

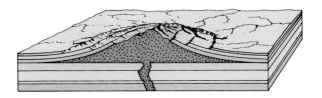

FIGURE 5.20
Laccoliths are masses of igneous rock injected between layers of the surrounding rock. They tend to deform the overlying strata in a dome-like structure.

surface. If it did reach the surface, core material would produce rocks several times denser than any other igneous rock. Contrary to popular belief, therefore, it is very unlikely that magma comes from the liquid core or even from the middle and lower parts of the mantle. Magma must originate between 50 and 200 km below the surface from the partial melting of the upper mantle and lower crust.

From the global distribution of volcanoes, batholiths, and mountain belts, we know that igneous activity is related to processes operating at active plate margins and that two fundamental types of magma apparently form in separate tectonic settings (Figure 5.21).

1. Basaltic magma is generated by the partial melting of upwelling mantle along spreading centers, where plates move apart. Basaltic volcanism dominates the igneous activity of the ocean basins.
2. Silicic magma is generated in the subduction zone by the partial melting of the oceanic crust and by the partial melting of the lower continental crust in the deeper roots of an active mountain belt (in the zone where the plates collide and temperature and pressure are high).

Critical to the understanding of the origin of magma is the concept of partial melting of rocks in the upper mantle and lower crust. You must bear in mind that most rocks are composed of more than one mineral, and thus magma does not have a specific melting temperature. Each component mineral in the rock melts at a different temperature. The major factors that influence melting are the following:

1. Temperature
2. Pressure
3. Amount of water and other volatiles present
4. Composition of the rock

A rock will begin to melt by either an increase in temperature or a decrease in pressure. If the temperature or the pressure (or both) changes in a direction to cause melting, the component minerals melt in a definite sequence such as is illustrated in Figure 5.22. At a given temperature and pressure, a rock body may thus be only partly melted, and the liquid (or melt) may have a composition quite different from that of the original rock. This process is known as *partial melting*. Conversely, as a magma cools, the various minerals crystallize in a definite sequence. When partial crystallization occurs, the remaining liquid can be separated from the crystals, so that it forms a magma quite different from the parent material. This process is called *magmatic differentiation*.

How is magma produced in Earth's tectonic system?

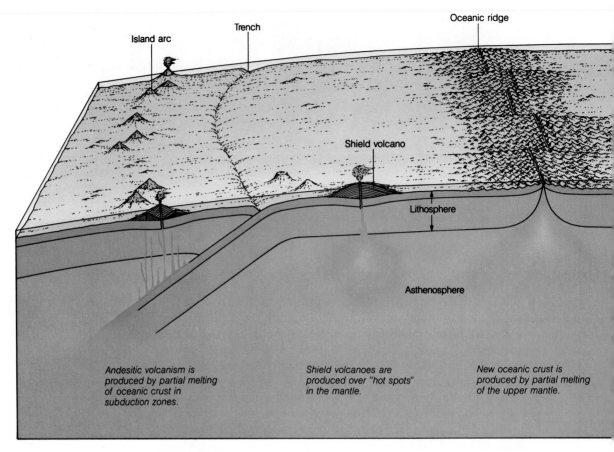

FIGURE 5.21
The origin of magma is involved with the dynamics at plate margins. Basaltic magma originates by partial melting of the upper mantle at diverging plate margins. As mantle material moves upward in a convection cell, the peridotite (major rock in the mantle) begins to melt because of a decrease in pressure. The material that melts first produces a magma of basaltic composition. Granitic magma is generated at subduction zones by partial melting of an oceanic plate and of lower continental crust. As the oceanic crust (containing basalt, oceanic sediments, and water) descends into the mantle, it is heated and begins to melt. The resulting liquid has a granitic composition and, being lighter than the surrounding rock, rises to form granitic intrusions or andesitic volcanoes.

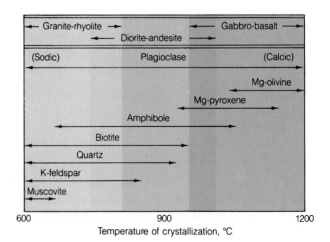

FIGURE 5.22
The order of crystallization of the common rock-forming minerals provides a key to the cooling history of a magma. Olivine, pyroxene, and Ca-plagioclase crystallize at high temperatures. Amphibole, Na-Ca-plagioclase, and biotite crystallize at intermediate temperatures. Quartz, K-feldspar, Na-plagioclase, and muscovite crystallize at low temperatures.

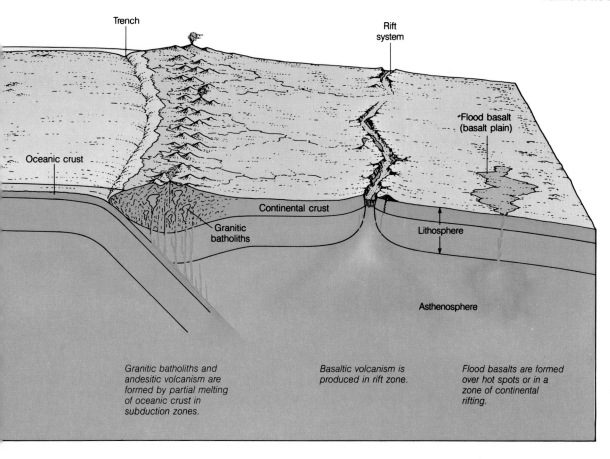

Trench

Rift system

Oceanic crust

Flood basalt (basalt plain)

Continental crust

Granitic batholiths

Lithosphere

Asthenosphere

Granitic batholiths and andesitic volcanism are formed by partial melting of oceanic crust in subduction zones.

Basaltic volcanism is produced in rift zone.

Flood basalts are formed over hot spots or in a zone of continental rifting.

Laboratory studies show that the crystallization of a silicate melt is extremely complex. Not only are temperature and pressure important, but the composition of the rock and the amount of water present also influence the process. Early-formed crystals of one mineral can react with the remaining liquid to form new minerals of different compositions. The general order of crystallization is summarized in Figure 5.22. The minerals olivine and Ca-plagioclase crystallize at the highest temperatures (between 1050 and 1200°C). They are followed by pyroxene, amphibole, and Na-plagioclase. Quartz and K-feldspar are lower-temperature minerals, crystallizing at temperatures below 900°C.

Generation of Basaltic Magma

The asthenosphere is believed to be composed of peridotite, a rock consisting of the minerals olivine and pyroxene. The balance between temperature and pressure in the asthenosphere is just about right for a slight degree of melting to occur. At greater depths in the mantle, however, pressure is too great for partial melting. The balance of temperature and pressure explains why the asthenosphere is soft and weak. As the soft asthenosphere slowly moves upward along a spreading zone between plates, the decrease in pressure causes additional partial melting. The partial melting of peridotite produces a basaltic magma, because the first minerals to melt yield basaltic constituents. Laboratory experiments on melting peridotite at high pressures indicate that basaltic magma is produced if a portion of the peridotite (ranging from less than 10 up to 30%) melts. The basaltic magma, being less dense than peridotite, rises along the oceanic ridge and is extruded as new crust in the spreading ocean basins (Figure 5.21).

Generation of Granitic Magma

At a *subduction zone,* the basaltic oceanic crust and its veneer of marine sediment descend into the mantle, where they are heated (Figure 5.21). There, melting is much more complex. In addition to basalt, the oceanic crust contains oceanic sediments with water-rich clay and silica-rich material derived from erosion of the continents. Partial melting of this material produces a magma with a high silica content, which forms andesitic or granitic rock. Where subduction zones involve a continental collision and mountain building, a greater variety of silica-rich rocks develop from the partial melting of the metamorphic rocks in the roots of the mountain systems. The lighter magma rises and collects in large bodies, producing the great granitic batholiths that characteristically form in mountain belts.

The simple model of plate tectonics discussed in Chapter 3 thus explains the major facts about most igneous rocks on Earth—facts derived from field observations, studies of composition and texture of rock types, and laboratory experiments on synthetic magma.

SUMMARY

1. Magma is molten rock material that can penetrate into or through rocks of the crust. It can include crystals that formed early in the cooling history (solids), silicate melt (liquid), and water and other volatiles (gas).
2. Two types of magma are most important: (a) basaltic magma is typically very hot (from 900 to 1200°C) and highly fluid and contains about 50% SiO_2, (b) silicic magma is cooler (less than 800°C) and highly viscous and contains between 60 and 77% SiO_2.
3. The texture of a rock refers to the size, shape, and arrangement of the constituent mineral grains.
4. The major textures in igneous rocks are (a) glassy, (b) aphanitic, (c) phaneritic, (d) porphyritic, and (e) pyroclastic.
5. The classification of igneous rocks is based on texture and composition. High-silica magmas produce rocks of the granite-rhyolite family, characterized by quartz, K-feldspar, and Na-plagioclase. Low-silica magmas produce rocks of the gabbro-basalt family, characterized by Ca-plagioclase, olivine, and pyroxene with little or no quartz. Magmas with an intermediate composition produce rocks of the diorite-andesite family, with a composition intermediate between that of granite and that of basalt.
6. The two major types of magma—basaltic and silicic—produce contrasting types of eruptions.
7. Basaltic magmas are more fluid. They are extruded as quiet fissure eruptions, producing a succession of thin flows that cover broad areas on the ocean floor, and as flood basalts on continents. Basaltic flows commonly develop columnar jointing, cinder cones, pillow lava, and shield volcanoes.
8. Silicic magmas are more viscous and contain large amounts of trapped gas. They therefore explode violently and extrude either lava flows in a thick, pasty mass or ash flows. Extrusion of silicic magma commonly produces composite volcanoes and calderas.
9. Masses of igneous rock formed by the cooling of magma beneath the surface are called intrusions. They are classified according to their size and shape and their relationship with the older rocks that surround them. The most important types of intrusions are batholiths, stocks, dikes, sills, and laccoliths.
10. Magma originates by the partial melting of the lower crust and the upper mantle, usually at depths from 50 to 200 km below the surface. It is not generated in Earth's core.
11. Basaltic magmas are formed by the partial melting of the mantle along spreading centers as tectonic plates move apart.
12. Silicic magmas are generated along subduction zones by the partial melting of oceanic crust descending into the hotter mantle and by the partial melting associated with metamorphism in the continental crust.

KEY TERMS

aa flow (p. 90)

andesite (p. 88)

aphanitic texture (p. 86)

ash flow (p. 97)

ash-flow tuff (p. 88)

basalt (p. 88)

batholith (p. 98)

caldera (p. 94)

cinder cone (p. 91)

columnar joint (p. 91)

composite volcano (p. 94)

concordant (p. 102)

crater (p. 94)

dike (p. 100)

diorite (p. 89)

discordant (p. 100)

exposure (p. 100)

extrusive rock (p. 83)

felsite (p. 88)

fissure (p. 91)

gabbro (p. 89)

glass (p. 84)

glassy texture (p. 84)

granite (p. 89)

groundmass (p. 86)

igneous rock (p. 83)

inclusion (p. 102)

intrusion (p. 99)

intrusive rock (p. 83)

laccolith (p. 102)

lava (p. 90)

magma (p. 82)

magmatic differentiation (p. 103)

matrix (p. 86)

pahoehoe flow (p. 90)

partial melting (p. 103)

peridotite (p. 89)

phaneritic texture (p. 86)

phenocryst (p. 86)

pillow lava (p. 94)

porphyritic texture (p. 86)

pressure ridge (p. 91)

pyroclastic texture (p. 87)

rhyolite (p. 88)

shield volcano (p. 94)

sill (p. 102)

spatter cone (p. 91)

stock (p. 100)

stratovolcano (p. 94)

subduction zone (p. 106)

tephra (p. 91)

texture (p. 84)

thin section (p. 88)

tuff (p. 88)

vesicle (p. 86)

viscosity (p. 83)

volatiles (p. 82)

volcanic ash (p. 91)

volcanic bomb (p. 91)

volcanic neck (p. 91)

volcanism (p. 82)

welded tuff (p. 97)

REVIEW QUESTIONS

1. Define the term *magma*.
2. What are the principal gases (volatiles) in magma?
3. What are two principal types of magma?
4. List the major types of igneous rock textures. Why is texture important in the study of rocks?
5. List the major types of igneous rocks and briefly describe their texture and composition.
6. Describe some of the common surface features of basaltic flows.
7. Why does magma tend to rise upward toward Earth's surface?
8. Draw a series of diagrams showing the form and internal structure of (a) a cinder cone, (b) a composite volcano, and (c) a shield volcano.
9. Describe the events that are typically involved in the formation of a caldera.
10. Describe the extrusion of an ash flow.
11. Describe and illustrate the major types of igneous intrusions. What is the textural difference between intrusive rocks and extrusive rocks?
12. Draw a diagram showing the following types of contacts between igneous intrusions and the surrounding rock: (a) discordant, (b) concordant, and (c) sedimentary.
13. Explain how an igneous rock can be produced from magma that does not have the same composition as the rock.
14. What is magmatic differentiation?
15. Draw a simple diagram and explain how basaltic magma originates from the partial melting of the asthenosphere.
16. Draw a simple diagram and explain how granitic and andesitic magma originate from the partial melting of the lithosphere as it descends into a subduction zone.
17. Why can't a basalt be produced by the partial melting of a granite?

ADDITIONAL READINGS

Best, M. G. 1982. *Igneous and Metamorphic Petrology.* San Francisco: Freeman.

Cox, K. G., J. D. Bell, and R. J. Pankhurst. 1979. *The Interpretation of Igneous Rocks.* London: George Allen and Unwin.

Dietrich, R. V., and B. J. Skinner. 1980. *Rocks and Rock Minerals.* New York: John Wiley.

Ehlers, E. G., and H. Blatt. 1982. *Petrology: Igneous, Sedimentary, and Metamorphic.* San Francisco: Freeman.

Ernst, W. G. 1969. *Earth Materials.* Englewood Cliffs, NJ: Prentice-Hall.

Hyndman, D. W. 1972. *Petrology of Igneous and Metamorphic Rocks.* New York: McGraw-Hill.

Maaloe, S. 1985. *Principles of Igneous Petrology.* New York: Springer-Verlag.

MacKenzie, W. S., C. H. Donaldson, and C. Guildord. 1982. *Atlas of Igneous Rocks and Their Textures.* New York: Halstead Press.

Schumann, W. 1974. *Stones and Minerals.* Elmsford, NY: Pergamon Press.

Simpson, B. 1966. *Rocks and Minerals.* Elmsford, NY: Pergamon Press.

Tindall, J. R., and R. Thornhill. 1975. *Rock and Mineral Guide.* London: Blanford Press.

The geologic processes operating on Earth's surface produce only subtle changes in the landscape during a human lifetime, but over a period of tens of thousands, or millions, of years, the effect of these processes is considerable. Given enough time, the erosive power of the hydrologic system can reduce an entire mountain range to a featureless lowland. The eroded debris is transported by rivers and deposited as new sedimentary formations.

A sequence of sedimentary rock may be thousands of meters thick. When exposed at the surface, each rock layer provides information about past events in Earth's history. Such is the case in the Grand Canyon of Arizona, where the Colorado River has cut a gorge over a mile deep clearly exposing the sequence of rock layers resting upon the older basement complex. The satellite image of the western Grand Canyon on the facing page shows that the rock layers are essentially horizontal and that the uppermost resistant limestone forms a broad plateau into which the canyon has been cut. The dark areas are young lava flows, some of which cascade over the rim of the canyon as frozen lava falls 1000 m high.

Sedimentary rock sequences preserve the record of erosion through time. Each bedding plane is a remnant of what was once the surface of Earth. Each rock layer is the product of a previous period of erosion.

To interpret the sedimentary record correctly, you must first understand something about modern sedimentary environments, such as deltas, beaches, and rivers. The study of how sediment is deposited in these areas provides insight into how ancient sedimentary rocks formed. From the sedimentary record, geologists can find the trends of ancient shorelines, map the positions of former mountain ranges, and determine the drainage patterns of ancient river systems. With this information, they are able to make paleogeographic maps of ancient landscapes, allowing us to look back through time and see the planet as it was millions of years ago.

6
Sedimentary Rocks

1. Sedimentary rocks are formed at Earth's surface by the hydrologic system. Their origin involves weathering of preexisting rock, transportation of the material away from the original site, and deposition of the eroded material in the sea or in some other sedimentary environment.
2. Two main types of sedimentary rocks are recognized: (a) clastic rocks, consisting of rock and mineral fragments, and (b) chemical rocks and organic rocks, consisting of chemical precipitates or organic material.
3. Stratification is the most significant sedimentary structure. It results from changes in erosion, transportation, and deposition of sediment during the time the rock is being formed. Other important structures include cross-bedding, graded bedding, ripple marks, and mud cracks.
4. Sedimentary processes result in sedimentary differentiation, in which material is sorted, segregated, and concentrated according to grain size and composition.
5. The major sedimentary environments are (a) fluvial, (b) alluvial-fan, (c) eolian, (d) glacial, (e) delta, (f) shoreline, (g) organic-reef, (h) shallow-marine, and (i) deep-marine systems.

NATURE OF SEDIMENTARY ROCKS

Sedimentary rocks are formed at Earth's surface by the hydrologic system. Their formation involves weathering of preexisting rock, transportation of the material away from the original site, and deposition of the eroded material in the sea or in some other sedimentary environment. Sedimentary rocks typically occur in layers, or strata, that cover large parts of the continents.

Sedimentary rocks are probably more familiar to most people than the other major rock types because they cover approximately 75% of the surface of the continents and therefore form most of the landscape. Few people, however, are aware of the true nature and extent of sedimentary rock bodies.

Sedimentary rocks are sediments that have been compacted and cemented to form solid rock bodies. The original sediment can be various substances:

1. Fragments of other rocks and minerals, such as gravel in a river channel, sand on a beach, and mud in the ocean
2. Chemical precipitates, such as salt in a saline lake and calcium carbonate in a shallow sea
3. Organic materials, such as coral reefs and vegetation in a swamp

What are the distinguishing features of sedimentary rocks?

All are sediment, and all can become sedimentary rocks.

Sedimentary rocks are important because they preserve a record of ancient landscapes, climates, and mountain ranges, as well as the history of the erosion of Earth. In addition, fossils are found in abundance in sedimentary rocks younger than 600 million years and provide evidence of changing plant and animal communities. Earth's geologic time scale was worked out using this record of sedimentary rocks and fossils.

An excellent place to study the nature of sedimentary rocks is the Grand Canyon area, where many features of sedimentary rocks are well exposed (Figure 6.1). Their most obvious characteristic is that they occur in distinct layers, or **strata** (singular **stratum**), many of which are more than 100 m thick. Rock types that are resistant to weathering and erosion form cliffs, and non-

FIGURE 6.1
The sequence of sedimentary rocks exposed in the Grand Canyon, Arizona, is almost 2000 m thick and was deposited over a period of 300 million years. Each major rock unit erodes into a distinctive landform. Formations that are resistant to weathering and erosion (such as sandstone and limestone) erode into vertical cliffs. Rocks that weather easily (such as shale) form slopes or terraces.

FIGURE 6.2
A cross section of the Grand Canyon graphically illustrates the major sedimentary formations. The sedimentary strata are essentially horizontal and were deposited on older igneous and metamorphic rocks. Graphic sections such as this are used by geologists to analyze and interpret sequences of sedimentary rock.

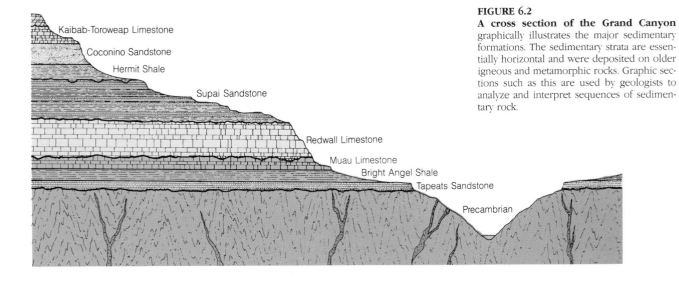

Kaibab-Toroweap Limestone
Coconino Sandstone
Hermit Shale
Supai Sandstone
Redwall Limestone
Muau Limestone
Bright Angel Shale
Tapeats Sandstone
Precambrian

resistant rocks erode into gentle slopes. From Figure 6.1, you should be able to recognize the major **formations** in the *geologic cross section* (a diagram showing a view of a vertical slice through a section of Earth's crust) shown in the diagram in Figure 6.2. The major formations exposed in the Grand Canyon can be traced across much of northern Arizona. In fact, they cover an area of more than 250,000 km². A close view of sedimentary rocks in the canyon (Figure 6.1) reveals that each formation has a distinctive texture, composition, and internal structure. The major layers of the sandstone in Figure 6.1 actually consist of smaller units separated by *bedding planes* (planes separating successive layers of rock). These bedding planes are marked by some change in composition, grain size, or color or by other physical features (Figure 6.3A). Animal and plant *fossils* are common in most of the rock units and can be preserved in great detail (Figure 6.3B). The term *fossil* is generally used to refer to any evidence of former life (plant or animal). It may be direct evidence, such as shells, bones, or teeth; or it may be indirect, such as tracks and burrows produced by organic activity. The texture of most sedimentary rocks consists of mineral grains or rock fragments that show evidence of abrasion (Figure 6.3C) or consists of interlocking grains of the mineral calcite. In addition, many layers show ripple marks (Figure 6.3D), mud cracks (Figure 6.3E), and other evidence (Figure 6.3F) of water deposition preserved in the bedding planes. All of these features indicate that sedimentary rocks form at Earth's surface in environments similar to those of present-day deltas, streams, beaches, tidal flats, lagoons, and shallow seas.

TYPES OF SEDIMENTARY ROCKS

Sedimentary rocks are classified on the basis of the texture and composition of their constituent particles. Two main groups are recognized: (1) clastic rocks, formed from fragments of other rocks, and (2) chemical rocks and organic rocks, formed by chemical or biological processes.

How are the different types of sedimentary rock distinguished and classified?

Sedimentary rocks are derived from the mechanical breakdown and chemical decay of preexisting rocks. One important category consists of particles of gravel, sand, or mud. Rocks made up of such fragmental material are called clastic rocks. The term clastic comes from the Greek word *klastos,* "broken," and describes the broken and worn particles of rock and minerals that were carried to the sites of deposition by streams, wind, glaciers, and marine currents. The other major category of sedimentary rocks consists of those precipitated chemically from water bodies such as lakes or shallow seas or derived from the growth of organisms such as coral reefs or swamp vegetation.

Clastic Rocks

Generally, *clastic* rocks are subdivided according to the grain size of the component materials. From the largest grain size to the smallest, the types of clastic rocks are conglomerate, sandstone, siltstone, and shale.

Conglomerates. A *conglomerate* consists of consolidated deposits of gravel, with various amounts of sand and mud in the spaces between the larger grains (Figure 6.4). The cobbles and pebbles usually are well-rounded fragments over 2 mm in diameter. Most conglomerates show a crude stratification and include beds and lenses of sandstone. Conglomerates are accumulating today at the bases of many mountain ranges, in stream channels, and on beaches.

(A) Stratification in sedimentary rocks consists of numerous layers from 0.5 to 1.0 m thick. Each major layer contains cross-bedding or other types of layering.

(B) Typical fossils found in sedimentary rocks include representatives of most types of marine animals.

(C) A microscopic view of sand grains in sedimentary rocks shows the effects of transportation by running water. The grains are rounded and sorted to approximately the same size.

(D) Ripple marks preserved in a sandstone indicate that the sediment was deposited by the current action of wind or water.

(E) Mud cracks form where sediment dries while it is temporarily exposed to the air. This structure is common on tidal flats, in shallow lakebeds, and on stream banks. Mud cracks in a rock layer provide important clues to the environment in which the sediment was deposited.

(F) Cross-bedding formed by currents.

FIGURE 6.3
A variety of features in sedimentary rocks indicate their origin at Earth's surface as a result of the hydrologic system.

Sandstones. *Sandstone* is probably the most familiar sedimentary rock because it is well exposed, easily recognized, and generally resistant to weathering (Figure 6.5). The sand grains range from 1/16 mm to 2 mm in diameter and can be composed of almost any material, so sandstones can be almost any color. Quartz grains, however, are usually most abundant because quartz is a common constituent in many other rock types and because it is not easily broken down by abrasion or chemical action. The particles of sand in most sandstones are cemented by calcite, silica, or iron oxide.

The composition of a sandstone provides an important clue to its history. During prolonged transportation, small rock fragments and minerals that readily decompose, such as olivine, feldspar, and mica, break down into finer particles and are winnowed out, leaving only the ultrastable quartz. Clean, well-sorted sandstone composed of well-rounded quartz grains indicates prolonged transportation or even several cycles of erosion and deposition.

Siltstones. *Siltstone* is a fine-grained clastic rock in which most of the material is 1/16 to 1/256 mm in diameter (finer than sand but coarser than mud). Siltstones commonly contain very thin layers (*laminae*) and may show evidence of burrowing by organisms. Silt is a material frequently carried in suspension by rivers and deposited in floodplains and deltas.

Shales (Mudstone). Solidified deposits of mud and clay are known as *shale*. The particles that make up the rock are less than 1/16 mm in diameter and in many cases are too small to be clearly seen and identified even under a microscope (Figure 6.6). Shale is the most abundant sedimentary rock. It usually is soft and weathers rapidly into a slope, so relatively few fresh, unweathered exposures are found. Most shale is well stratified, with thin laminae. Black shales are rich in organic material and accumulate in a variety of quiet-water environments, such as lagoons, restricted shallow seas, and tidal flats. Red shales are colored with iron oxide and indicate oxidizing conditions in the environments in which they accumulate, such as stream channels, floodplains, and tidal flats.

Chemical and Organic Rocks

Nonclastic rocks are formed from chemical precipitation and from biological activity.

Limestone. By far the most abundant nonclastic rock is *limestone*. It is composed principally of calcium carbonate ($CaCO_3$) and originates by both chemical and organic processes. Limestones have a great variety of rock textures, and many different types have been classified. Some of the major groups are skeletal limestone, oolitic limestone, and microcrystalline limestone.

Many plants and invertebrate animals extract calcium carbonate from water in their life processes and use it to construct their shells and hard parts. When these organisms die, their shells accumulate on the sea floor. Over a long period of time, the shells build up a deposit of limestone with a texture consisting of shells and shell fragments (Figure 6.7). This type of limestone, composed mostly of skeletal debris, can be several hundred meters thick and can extend over thousands of square kilometers. *Chalk,* for example, is a skeletal limestone in which the skeletal fragments are remains of microscopic plants and animals.

Some limestones are composed of small, spherical concretions of calcium carbonate called *oolites.* The individual grains, about the size of a grain of sand, can at present be observed forming in the shallow waters off the Bahamas where currents and waves are active. Evaporation and increased temperatures in the seawater raise the concentration of $CaCO_3$ until it is pre-

FIGURE 6.4
Conglomerate is a coarse-grained, clastic rock in which most of the particles are larger than 2 mm in diameter. It is, in essence, gravel that has been cemented and consolidated into a solid rock body. The fragments in a conglomerate are mostly rounded pebbles and cobbles but there is generally considerable sand and mud filling the space between pebbles. Most conglomerates are deposited by streams.

FIGURE 6.5
Sandstone is a fine-grained, clastic rock composed of fragments that may range from 1/16 to 2 mm in diameter. Most sand is composed of rounded quartz grains, but significant amounts of small rock fragments and feldspar may be present. Sandstone commonly grades into either siltstone and shale or into conglomerate. Three general types of sand are recognized: (a) **Quartz sandstone** is composed mainly of the mineral quartz. (b) **Arkose** is a sandstone rich in feldspar. Most arkoses are formed from rapid erosion of granite and gneiss. (c) **Lithic sandstone** is a sandstone in which significant amounts of the grains are small rock fragments.

FIGURE 6.6
Shale is consolidated mud. The fragments that make up a shale are less than 1/256 mm in diameter and are composed mostly of clay minerals. Coarser grains, called silt (1/256 to 1/16 mm in diameter), may make up a significant part of the rock. Shales accumulate in many different environments. The main sediment carried by the great rivers to the sea is mud and fine sand so it is not surprising that shale is the most abundant sedimentary rock. Black shales are rich in organic matter and accumulate in a variety of quiet-water environments such as lagoons and tidal flats. Red shales are colored with iron oxide and indicate an oxidizing environment.

FIGURE 6.7
Limestone is composed principally of calcium carbonate ($CaCO_3$) and is by far the most abundant nonclastic rock. It originates by both chemical and organic processes, which produce a variety of rock textures. Some of the major groups are (1) skeletal limestone (composed of fragments of fossil shells), (2) oolitic limestone, and (3) microcrystalline limestone.

cipitated. Small fragments of shells and tiny grains of $CaCO_3$ become coated with successive layers of $CaCO_3$ as they are rolled along the sea floor by waves and currents. The sorting action of the waves and the currents produces a uniform grain size. Naturally, fragments of shells and oolites accumulate together in many places, forming an oolitic-skeletal texture. Oolitic limestone is a special type of nonclastic sedimentary rock. As with clastic sedimentary rocks, the particles are carried by current action to the places where they are deposited.

In quiet water, calcium carbonate is precipitated as tiny, needlelike crystals, which accumulate on the bottom as limy mud. Soon after deposition, the grains commonly are modified by compaction and recrystallization. This modification produces microcrystalline limestone, a rock with a dense, very fine-grained texture. Its individual crystals can be seen only under high magnification. Microcrystalline limestone also is precipitated from springs and from the dripping water in caves, but the total amount of *dripstone,* compared to the amount of marine limestone, is negligible.

Dolostone.   *Dolostone* is a rock composed of the mineral dolomite, a calcium magnesium carbonate $CaMg(CO_3)_2$. It is similar to limestone ($CaCO_3$) in most textural and structural features and in general appearance. It can develop by direct precipitation from seawater, but most dolostones are apparently formed by the later substitution of magnesium in pore water for some of the calcium in limestones.

Rock Salt and Gypsum.   *Rock salt* is composed of the mineral halite (NaCl). It forms by evaporation in saline lakes (for example, the Great Salt Lake and the Dead Sea) or in restricted bays along the shore of the ocean. *Gypsum,* composed of $CaSO_4 \cdot 2H_2O$, also originates from evaporation. It collects in layers as calcium sulfate is precipitated. Since *evaporites* (rocks formed by evaporation) accumulate only in restricted basins subjected to prolonged evaporation, they are important indicators of ancient climatic and geographic conditions.

SEDIMENTARY STRUCTURES

Sedimentary rocks commonly show layering and other structures that form as sediment is moved, sorted, and deposited by currents. These features, called primary sedimentary structures, provide key information about the conditions under which the sediment accumulated. The most important sedimentary structures are stratification, cross-bedding, graded bedding, ripple marks, and mud cracks.

What are sedimentary structures? Why are they important in the study of sedimentary rocks?

Primary *sedimentary structures* are large-scale fractures of sedimentary rock such as stratification, ripple marks, and mud cracks. They are formed by a variety of processes such as currents, gravity flow, and organic activity and are important because they form as the sediment is being deposited. They provide important insight for interpreting the ancient environment in which the sediment accumulated. Some structures show the direction of the depositing currents. Others show the trends of ancient shorelines and the transgression and regression of the sea over the continents. The most important sedimentary structures are

1. Stratification
2. Cross-bedding
3. Graded bedding
4. Ripple marks and mud cracks

Stratification

One of the most obvious characteristics of sedimentary rocks is that they occur in distinct layers, which are expressed by color, texture, and the way the different rock units weather and erode. These layers are termed strata, or simply *beds.* The planes separating the layers are planes of stratification, or bedding planes. *Stratification* occurs on many scales and reflects the changes that occur during the formation of a sedimentary rock. Large-scale stratification is expressed by major changes in rock types, which can be seen in large exposures, such as the Grand Canyon (Figure 6.1), where cliffs of limestone or sandstone alternate with slopes of weaker shale. These major units are called *formations.* Within each formation, stratification, or bedding, occurs on several smaller scales, expressed by differences in the texture, color, and composition of the rock.

An important aspect of stratification is that the rock layers do not occur in a random fashion but overlie one another in definite sequences and patterns. One of the more simple and common patterns in a vertical sequence is the cycle of sandstone, shale, and limestone. This pattern is produced by the advance and retreat of shallow seas over the continental platform (Figure 6.8). In diagram A of the figure, the sea begins to expand over a lowland drained by a river system. Sand accumulates along the shore, mud is transported in suspension offshore, and lime is precipitated from solution farther offshore, beyond the mud zone. All three types of *sediment* are deposited simultaneously, each in a different environment. Stream deposits (not shown in the figure) accumulate on the floodplain of the river system.

As the sea expands over the lowland, each environment shifts landward, following the shoreline (Figure 6.8A, B, and C). Beach sands are subsequently deposited over the stream sediments, offshore mud is deposited over the previous beaches, and lime is deposited over the mud. As the sea continues to expand, the layers of sand, mud, and lime are deposited farther and farther inland.

As the sea withdraws (Figure 6.8D), the mud is deposited over the lime and the nearshore sand over the mud. The net result is a long wedge, or layer, of limestone encased in a wedge of shale, which in turn is encased in a wedge of sandstone. Below and above the marine deposits are *fluvial* (river) *sediments* deposited by the river system (not shown in the diagram). Subsequent uplift and erosion of the area reveal a definite sequence of rock (Figure 6.8E). Beginning at the base, sandstone is overlain by shale and limestone, which are in turn overlain by shale and sandstone. (See the lower rock layers in Figure 6.2.)

Cross-Bedding

Cross-bedding is a type of stratification in which the layers within a bed are inclined at an angle to the upper and lower surfaces of the bed. The formation of cross-bedding is shown in Figure 6.9. As sand grains are moved by currents (either wind or water), they form ripples or dunes called *sand waves.* These range in scale from small ripples less than a centimeter high to giant sand dunes several hundred meters high. Typically, they are asymmetrical, with the gentle slope facing in the direction of the moving current. As the particles migrate up and over the sand wave, they accumulate on the steep downcurrent face and form inclined layers. The direction of flow of the ancient currents that formed a given set of cross-strata can be determined by measuring the direction in which the strata are inclined. We can determine the patterns of ancient current systems by mapping the direction of cross-bedding in sedimentary rocks.

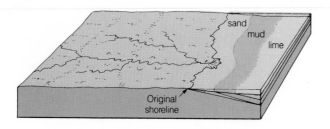

(A) The sea begins to expand over the land. The original shoreline is marked by sand deposits that grade seaward into mud and lime (CaCO₃).

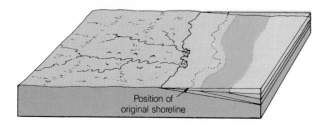

(B) The sea expands farther inland depositing a sheet of sand, overlain by mud (gray) and lime (tan).

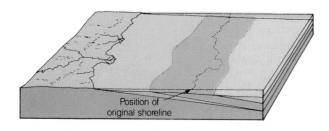

(C) With continued expansion of the sea, mud is deposited on top of the sand at the position of the original shoreline.

(D) A regression of the sea deposits shoreline sand over the offshore mud. Thus the vertical succession of sediment at the position of the original shoreline is sand, mud, lime, mud, and sand.

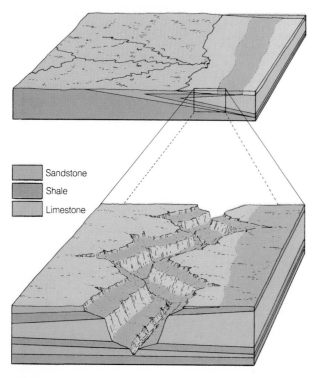

(E) The exposure of a transgressive-regressive sequence shows the cycle from bottom to top: sandstone, shale, limestone, shale, and sandstone.

FIGURE 6.8
A sequence of sediments deposited by the expansion and contraction of a shallow sea
is represented in these schematic diagrams. Sand accumulates along the beach, mud is deposited offshore, and lime is precipitated farther offshore, beyond the mud. As the sea expands, these environments move inland, producing a vertical sequence of sand-mud-lime. As the sea recedes, mud is deposited over the lime and sand is deposited over the mud. The net result is a vertical sequence of sedimentary layers: sandstone, shale, limestone, shale, and sandstone.

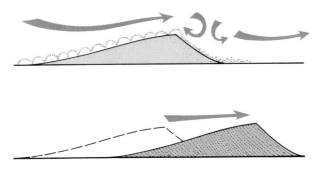

FIGURE 6.9
Cross-bedding is formed by the migration of sand waves (ripple marks or dunes). Particles of sediment, carried by currents, travel up and over the sand wave and are deposited on the steep downcurrent face to form inclined layers.

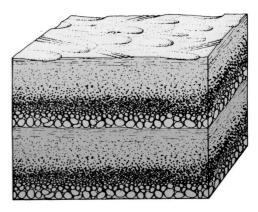

FIGURE 6.10
Graded bedding is produced by turbidity currents. It occurs in widespread layers, each layer generally less than a meter thick. Deep-marine environments commonly produce a great thickness of graded layers, which can easily be distinguished from sediment deposited in most other environments.

Graded Bedding

A distinctive type of stratification, called *graded bedding,* is characterized by a progressive decrease in grain size upward through the bed (Figure 6.10). This type of stratification commonly is produced on the deep-ocean floor by *turbidity currents,* which transport sediment from the continental slope to adjacent deep ocean. A turbidity current is generated by turbid (muddy) water, which, being denser than the surrounding clear water, sinks beneath it and moves rapidly down the continental slope (Figure 6.11).

Turbidity current flow can be easily demonstrated in the laboratory by pouring muddy water down the side of a tank filled with clear water. The mass of muddy water moves down the slope of the tank and across the bottom at a relatively high speed, without mixing with the clear water. Turbidity currents can also be observed where streams discharge muddy water into a clear lake or reservoir. The denser, muddy water moves out along the bottom of the basin and can flow for a considerable distance, even along the flat surface of the abyssal floor.

Turbidity currents can also be generated by an earthquake or a submarine landslide, during which mud, sand, and even gravel can be thrown into suspension. In 1929, one of the best documented large-scale turbidity currents was triggered by an earthquake near the Grand Banks, off Newfoundland (Figure 6.12). Slumping of a large mass of soft sediment (estimated to be 100 km^3) moved as a turbidity flow down the continental slope and onto the abyssal plain, eventually covering an area of 100,000 km^2. As the turbidity current moved downslope, it broke a series of transatlantic cables at different times. The speed of the current, determined from the intervals between the times when the cables broke, was from 80 to 95 km per hour. This mass of muddy water formed a graded layer of sediment over a large area of the Atlantic floor.

As a turbidity current moves across the flat floor of a basin, its velocity at any given point gradually decreases. The coarser sediment is deposited first, followed by successively smaller particles. After the turbid water ceases to move, the sediment remaining suspended in the water gradually settles out. One pulse of sedimentation, therefore, deposits a single layer of sediment, which exhibits continuous gradation from coarse material at the base to fine material at the top. Subsequent turbidity flows can deposit more layers of graded sediment, with sharp contacts between layers. The result is a succession

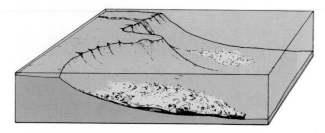

FIGURE 6.11
The movement of turbidity currents down the slope of the continental shelf can be initiated by a landslide or an earthquake. Sediment
is moved largely in suspension. As the current slows down, the coarse
grains are deposited first, followed by the deposition of successively
finer-grained sediment. A layer of graded bedding is thus produced
from a single turbidity current.

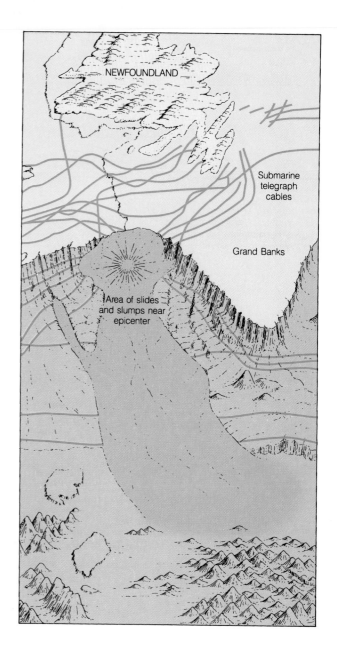

FIGURE 6.12
The Grand Banks turbidity current in 1929 was exceptionally well
documented by the rupture of submarine cables. An earthquake that
occurred south of Newfoundland caused the loose sediment on the
continental slope to move in a submarine landslide. All of the cables in
the area of the epicenter were broken instantaneously as a result of the
shock and slump. The sediment thrown into suspension by the shock
moved downslope as a turbidity current, breaking the deeper cables in
its path. The times at which the breaks occurred were determined by
the times at which transmission ceased. Cables were broken 600 km
downslope away from the shock, and the maximum speed of the turbidity current was estimated at about 95 km per hour. Approximately
100 km³ of sediment was moved by the slump and turbidity current and
was eventually spread over an area of 100,000 km² as a single layer of
graded sediment.

of widespread horizontal layers, each being a graded unit deposited by a single
pulse of sedimentation from a turbidity flow.

Ripple Marks, Mud Cracks, and Other Surface Impressions

Ripple marks are commonly seen in modern stream beds, in tidal flats, and
along the shores of lakes and the sea. Preserved in rocks, they provide information concerning the environment of deposition, such as depth of water,
ancient current directions, and trends of ancient shorelines. *Mud cracks* are
also commonly preserved in sedimentary rock and show that the sedimentary
environment occasionally was exposed to the air during the process of deposition. Mud cracks in rocks suggest that the original sediment was deposited
in shallow lakes, on tidal flats, or on exposed stream banks.

Tracks, trails, and borings of animals are typically associated with ripple
marks and mud cracks and can provide additional important clues about the
environment in which the sediment accumulated.

ORIGIN OF SEDIMENTARY ROCKS

A sedimentary environment is the place where sediment is deposited and the physical, chemical, and biological conditions that exist there. The major sedimentary environments are (a) fluvial, (b) alluvial-fan, (c) eolian, (d) glacial, (e) delta, (f) shoreline, (g) organic-reef, (h) shallow-marine, and (i) deep-marine environments. Each of these environments is characterized by certain physical, chemical, and biological conditions and therefore develops distinctive rock types and fossil assemblages.

Sedimentary rocks form at Earth's surface through interactions of the hydrologic system and the crust. Fortunately, many of these processes are in operation today, and geologists actively study rivers, deltas, and oceans of all parts of Earth. This research indicates that the genesis of sedimentary rocks involves four major processes.

1. Weathering (erosion)
2. Transportation
3. Deposition
4. Compaction and cementation

During these processes, the original rock material that is eroded from the land is sorted according to size and composition, a process called sedimentary differentiation.

Weathering

Weathering is the interaction between the elements in the atmosphere and the rocks exposed at Earth's surface. The atmosphere can mechanically break down the rock through processes such as frost action, and it can chemically decompose the rock by a variety of reactions. We will study the details of weathering in Chapter 10. For now, note that weathering is the first step in the genesis of sedimentary rock. The atmosphere breaks down and decomposes preexisting solid rock and forms a layer of loose, decayed rock debris, or soil. This unconsolidated material can then be transported easily by various agents, such as streams, wind, and glaciers.

Transportation

Running water is the most effective form of sediment transport. All rivers carry large quantities of sediment toward the sea. This fact is readily appreciated if you consider the great deltas of the world, each formed from sediment transported by rivers (see Figure 12.37). Indeed, sediment is so abundant in most rivers that a river might best be thought of as a system of water and sediment rather than simply a channel of flowing water.

As sediment is transported by a river, it is sorted and separated according to grain size and composition. Large particles accumulate as gravel, medium-size grains are concentrated as sand, and finer material settles out as mud. Dissolved minerals are carried in solution and are ultimately precipitated as limestone, salt, or other chemical deposits.

Wind and glaciers also transport sediment, although their activity is somewhat restricted to special climatic zones.

The **sorting** that occurs during transportation is an important factor in the genesis of sedimentary rock. Indeed, most **sedimentary differentiation** occurs during transportation, so when the sediment is delivered to the site where it is deposited, it is already commonly sorted and differentiated to some degree according to particle size and composition.

What are the major environments of sedimentation?

Deposition

Probably the most significant factor in the origin of sedimentary rocks is the *sedimentary environment,* that is, the physical, chemical, and biological conditions that exist at the place where the sediment is deposited. The idealized diagram in Figure 6.13 shows in a general way the regional setting of the major types of sedimentary environments. Continental environments include areas of sedimentation that occur exclusively on the land surface. Most important are major river systems, alluvial fans, desert dunes, lakes, and margins of glaciers. Marine environments include the shallow seas, which cover parts of the continental platform, *reefs,* and the floors of the deep-ocean basins. Between continental areas and marine areas are the transitional or mixed environments, which occur along the coasts and are influenced by both marine and nonmarine processes. These include *deltas, beaches, tidal flats,* and *lagoons* (Figure 6.13).

How can the sedimentary environment be interpreted from an outcrop of rock?

Each of these environments is characterized by certain physical, chemical, and biological conditions. Distinctive types of texture, composition, internal structure, and fossil assemblages are thus developed in each environment. Illustrations of modern sedimentary environments, together with examples of the rocks they produce, are shown in Figures 6.14 through 6.24.

Compaction and Cementation

The final stage in the formation of sedimentary rocks is the transformation of loose, unconsolidated sediment into solid rock. This is accomplished by compaction and cementation. The weight of overlying material, which continually accumulates in a sedimentary environment, compresses and compacts buried sediment into a tight, coherent mass. Cementation occurs as mineral matter, carried by water seeping through the pore spaces of the grains, is precipitated. This postdepositional crystallization of *cement* (the common minerals are calcite, quartz, and limonite), which holds the grains of sediment together, is a fundamental process in transforming sediment into solid rock.

Significance of Sedimentary Rocks

Sedimentary rocks are of great economic importance; so great, in fact, that they have been a controlling factor in the development of our industry, society, and culture. Without material from sedimentary rocks, humans probably could not have advanced to the neolithic age because flint and chert played such an important role in development of tools, arrowheads, and axes. Consider for a moment the obvious natural resources that are provided by sedimentary rocks. Coal, petroleum, and natural gas, upon which our industry and economy depend, originate and are contained in sedimentary rocks. Gravel, sand, and limestone are the major building materials of our culture: from them we construct buildings, freeways, dams, and so on. Clay is the basis of ceramics. Evaporites provide most of our salt and chemicals. Sandstones are exploited for rare minerals such as gold, diamonds, platinum, and uranium. Glass is made from quartz sand. (It would be difficult to develop a glass industry from materials derived only from Japan.) Most of our iron deposits are found in sedimentary rocks, and sedimentary sequences are the hosts for many metallic mineral deposits such as lead and zinc. Bricks are made from clay; the great cathedrals of Europe are made from limestone; and the statues made by the artists of ancient Greece and Rome and during the Renaissance would have been impossible without limestone.

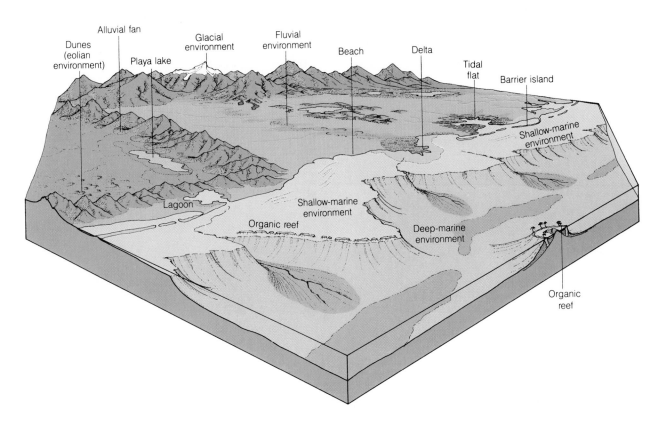

CONTINENTAL ENVIRONMENTS

Alluvial fans are fan-shaped deposits of gravel, sand, and mud that accumulate in dry basins at the bases of mountain ranges.

Eolian (wind) environments include sand seas of deserts, where sand dunes are built and transported by wind, and areas where windblown dust accumulates.

Fluvial (river) environments are the river channels, river bars, and the adjacent floodplain.

Lakes are bodies of nonmarine water, including freshwater lakes on continental lowlands and saline lakes in isolated basins.

Glacial environments are the areas where sediment is deposited by glaciers. Most obvious are the margins of the ice, where sediment carried by the glacier is dropped as the ice melts. Other subenvironments of a glacier are the lakes and meltwater streams.

SHORELINE ENVIRONMENTS (TRANSITIONAL, OR MIXED)

Deltas are deposits of mud, silt, and sand that form at the mouths of rivers, where they empty into the sea or a lake.

Beaches are shoreline accumulations of sand.

Barrier islands are linear bodies of sand built offshore by the action of ocean waves.

Lagoons are elongate bodies of seawater located between the mainland and barrier islands or reefs. Low wave energy permits the deposition of mud.

Tidal flats are shoreline areas that are covered with water at high tide and uncovered at low tide. Mud is the major type of sediment deposited.

MARINE ENVIRONMENTS

Shallow-marine environments extend from the shore to the edges of the continental shelves. Lime and mud are the principal types of sediment deposited.

Organic reefs are solid structures built from corals, algae, and shells of other marine organisms. Reefs grow in warm, shallow water near islands and continents.

Deep-marine environments characterize the deep oceans beyond the continental slopes and include deep-sea fans and abyssal plains. Turbidites are the major types of sediment deposited.

FIGURE 6.13
The major sedimentary environments are represented in this idealized diagram. Most sediment moves downslope from continental highlands toward the oceans, so the most important environments of sedimentation are found along the shores and in the shallow seas beyond. Sedimentary environments can be categorized in three groups: continental, shoreline (transitional), and marine. Their important characteristics and the types of sediment that accumulate in each are outlined below the diagram.

(A)

(B)

FIGURE 6.14
The fluvial environment. (A) Point-bar deposits in a modern river. (B) An ancient stream channel in Tertiary sediments in central Utah. The great rivers of the world are the major channels by which erosional debris is transported from the continents to the oceans. Before reaching the ocean, most rivers meander across flat alluvial plains and deposit a

considerable amount of sediment. Within this environment, sedimentation occurs in stream channels, on bars, and on floodplains. Perhaps the most significant type of sedimentation occurs on bars on the insides of meander bends (see Figure 12.25). Stream deposits are characterized by channels of sand or gravel cut into horizontal layers of silt and mud.

(A)

(B)

FIGURE 6.15
The alluvial-fan environment. (A) Modern alluvial fans in Death Valley, California. (B) Ancient alluvial-fan deposits in central Utah. In many arid regions of the world, thick deposits of sedimentary rock accumulate in a series of alluvial fans at the bases of mountain ranges. Deposition occurs because streams in arid regions do not have enough water to transport their sediment load over the flat surface of the basin.

Flash floods are an important factor in this environment. Torrents from cloudbursts pick up the loose debris on the slopes of the mountain ranges and deposit it on the basin floor. The sediment in an alluvial fan characteristically is coarse grained, and conglomerate is the most abundant rock type. In the central part of the basin, fine silt and mud can accumulate in temporary lakes and commonly are associated with the coarser fan deposits.

(A)

(B)

FIGURE 6.16
The eolian (wind) environment. (A) Modern sand dunes in the Sahara Desert. (B) Ancient dune deposits in Zion National Park, Utah. Wind is a very effective sorting agent. Silt and dust are lifted high in the air and may be transported thousands of kilometers before being deposited. Sand is winnowed out and transported close to the surface and eventually accumulates in dunes. Gravel cannot be moved effectively by wind. A major process in an eolian sedimentary environment is the migration of sand dunes. Sand is blown up and over the dunes and accumulates on the steep dune faces. Large-scale cross-strata that dip in a downwind direction are thus formed (see Figure 17.11). Ancient dune deposits are characterized by large-scale cross-strata consisting of well-sorted, well-rounded sand grains. The most significant ancient wind deposits are sandstones that accumulated in large dune fields comparable to the present Sahara and Arabian deserts and the great deserts of Australia. These sandstones are vast deposits of clean sand that preserve to an unusual degree the large-scale cross-bedding developed by migrating dunes.

(A)

(B)

FIGURE 6.17
The glacial environment. (A) The margins of the Barnes Ice Cap of Baffin Island, Canada. (B) Glacial sediments deposited during the last ice age. A glacier transports large boulders, gravel, sand, and silt, suspended together in the ice. This material is eventually deposited near the margins of the glacier as the ice melts. The resulting sediment is unsorted and unstratified, with angular individual particles. Typically, the deposits of continental glaciers are widespread sheets of unsorted debris, which rest on the polished and striated floor of the underlying rock. Fine-grained particles dominate in many glacial deposits, but angular boulders and pebbles are invariably present. Streams from the meltwaters of glaciers rework the unsorted glacial debris and redeposit it beyond the glaciers as stratified, sorted stream deposits. The unsorted glacial deposits are thus directly associated with well-sorted stream deposits from the meltwaters.

(A)

(B)

FIGURE 6.18
The delta environment. (A) A small delta formed in a lake. (B) Ancient deltaic deposits in Tertiary rocks of the Colorado Plateau. One of the most significant environments of sedimentation occurs where the major rivers of the world enter the oceans and deposit most of their sediment in marine deltas. A delta environment can be very large, covering areas of more than 36,000 km². Commonly, deltas are very complex and involve various distinct subenvironments, such as beaches, bars, lagoons, swamps, stream channels, and lakes. Because deltas are large features and include a number of both marine and nonmarine subenvironments, a great variety of sediment types accumulate in them. Sand, silt, and mud dominate. A deltaic deposit can be recognized only after considerable study of the sizes and shapes of the various rock bodies and their relationships to each other. Both marine and nonmarine fossils can be preserved in a delta.

(A)

(B)

FIGURE 6.19
The beach environment. (A) A modern gravel beach. (B) Ancient beach deposits. Much sediment accumulates in the zone where the land meets the ocean. Within this zone, a variety of subenvironments occur, including beaches, bars, spits, lagoons, and tidal flats. Each has its own characteristic sediment. Where wave action is strong, mud is winnowed out, and only sand or gravel accumulates as beaches or bars. Beach gravels accumulate along shorelines, where high wave energy is expended. The gravels are well sorted and well rounded, and commonly are stratified in low, dipping cross-strata. Ancient gravel beaches are relatively thin. They are widespread and commonly are associated with clean, well-sorted sand deposited offshore.

(A)

(B)

FIGURE 6.20
The lagoon environment. (A) A lagoon along the central Atlantic coast of the United States. (B) Lagoonal deposits of Cretaceous age in central Utah. Offshore bars and reefs commonly seal off part of the coast, forming lagoons. A lagoon is protected from the high energy of waves, so the water is relatively calm and quiet. Fine-grained sediment,

rich in organic matter, accumulates as black mud. Eventually, the lagoon may fill with sediment and evolve into a swamp. Where the bottom vegetation provides enough organic matter, a coal deposit may form. The rise and fall of sea level shift the position of the barrier bar, and thus the organic-rich mud or coal formed in the lagoon or swamp is interbedded with sand deposited on the barrier island.

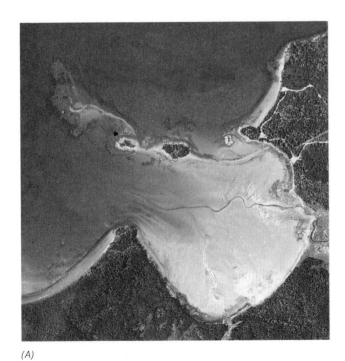

(A)

(B)

FIGURE 6.21
The tidal-flat environment. (A) A modern tidal flat in the Gulf of California. (B) Tidal-flat deposits of Triassic age in southern Utah. The tidal-flat environment is unique in being alternately covered with a sheet of shallow water and exposed to the air. Tidal currents are not strong. They generally transport only fine silt and sand and typically

develop ripple marks over a broad area of the tidal flat. Mud cracks commonly form during low tide and are subsequently covered and preserved. Ancient tidal-flat deposits are thus characterized by accumulations of silt and mud in horizontal layers with an abundance of ripple marks and mud cracks.

(A)

(B)

FIGURE 6.22
The organic-reef environment. (A) A modern organic reef. (B) Ancient reefs of Devonian age in Australia. An organic reef is a solid structure of calcium carbonate constructed of shells and secretions of marine organisms. The framework of most reefs, consisting of a mass of colonial corals, forms a wall that slopes steeply seaward. Wave action continually breaks up part of the seaward face, and blocks and fragments of the reef accumulate as debris on the seaward slope. A lagoon

forms behind the reef, toward the shore or toward the interior of the atoll (organic reef), and lime, mud, and evaporite salts may be deposited there. Gradual subsidence of the sea floor permits continuous upward growth of reef material to a thickness of as much as 1000 m. Because of their limited ecological tolerance (corals require warm, shallow water), fossil reefs are excellent indicators of ancient environments. They are commonly found in shallow-marine limestones.

(A)

(B)

FIGURE 6.23
The shallow-marine environment. (A) A modern shallow-marine environment in the Bahamas. (B) Ancient shallow-marine sediments of Pennsylvanian age in eastern Kansas. Shallow seas border most of the land area of the world and can extend to the interior of a continent, as do Hudson Bay, the Baltic Sea, and the Gulf of Carpentaria (in northern Australia). The characteristics of the sediment deposited in the shallow-sea environment depend on the supply of sediment from the land and

the local conditions of climate, wave energy, circulation of water, and temperature. If there is a large supply of land-derived sediment, sand and mud accumulate and ultimately form sandstone and shale, respectively. If sediment from the land is not abundant, limestone generally is precipitated or deposited by biological means. Ancient shallow-marine deposits are characterized by thin, widespread, interbedded layers of sandstone, shale, and limestone.

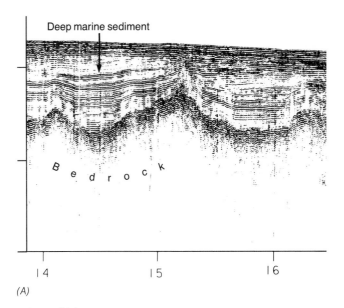

Deep marine sediment

Bedrock

| 4 | 5 | 6

(A)

(B)

FIGURE 6.24
The deep-marine environment. (A) A sketch of modern deep-sea fans off the coast of India. (B) Ancient deep-marine sediment originally deposited in a horizontal position near the southern coast of France. The deep-ocean floor lying near the continent is, for the most part, extremely smooth because of the deposition of sediment brought down the continental slope by turbidity currents. These deposits are characterized by a sequence of graded beds, with each layer extending over a large area. Such beds are thus easily distinguishable from sediment deposited in most other environments. Sediment also accumulates on the floor of the open ocean, far from the continents. This material consists of shells of microscopic organisms and fine particles of mud, which are carried in suspension and gradually sink. The most abundant sediment is a fine-grained brown or red clay. Some sediment in the open ocean apparently crystallized directly from sea water. In large parts of the South Pacific, these deposits form the bulk of the sediment.

PALEOCURRENTS AND PALEOGEOGRAPHY

Paleocurrent structures such as cross-bedding show patterns of sediment dispersal from which it is possible to deduce ancient geographic conditions at the time the sediment was deposited— locations of highlands, trends of ancient shorelines, and the direction of the regional slope down which the sediment was transported.

Paleogeography is the study of ancient landscapes and the characteristics of past conditions on Earth's surface. In this study, geologists attempt to determine the locations of ancient features such as mountain belts, drainage systems, and areas once covered by inland seas. They try to delineate the trends of ancient shorelines, to determine ancient climatic conditions, and to establish the direction of paleowinds (ancient winds).

One approach found to be successful is the study of paleocurrents (see Figure 6.9). Cross-bedding is inclined in a downcurrent direction because the sediment is deposited on the lee slope of a sand wave or sand dune. When cross-bedding directions are mapped over a broad area and are analyzed statistically, these structures may show ancient drainage directions, trends of ancient shorelines, or patterns of paleowinds.

The paleogeographic interpretation of current structure depends on the environment of sedimentation. In river deposits, the flow is in one direction only, from the highlands to the sea (or lake basin). Although local variations result as the river meanders, the average direction of current structures formed in a river is down the regional slope. This is illustrated clearly by major geographic elements. The flow of the river may be in any direction locally, but the average is downslope. A line drawn perpendicular to the average paleocurrent direction is a general contour line and is parallel to the ancient shoreline.

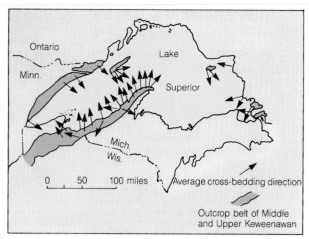

(A) The average cross-bedding direction in the Keweenawan sediments in the Lake Superior region shows a paleoslope into the area of the present lake.

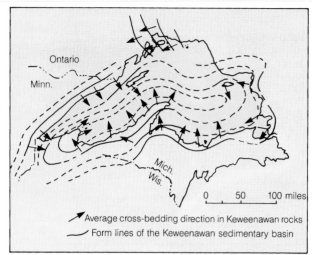

(B) The arrows indicate the average cross-bedding direction, and the lines represent generalized ancient topographic contour lines, showing the interpreted form of the ancient sedimentary basin.

FIGURE 6.25
On the basis of cross-bedding directions, geologists can reconstruct the size and shape of the ancient basin in which the Keweenawan sediments were deposited.

It is thus possible to map cross-bedding directions in a formation of river deposits and to deduce from them the location of old highlands, the direction of the regional slope down which the sediments flowed, and the general trend of the ancient shoreline at the time the sediment was deposited.

Figure 6.25 shows the results of a paleogeographic study of a series of Precambrian rocks exposed along the shores of Lake Superior. These rocks are interbedded with basaltic lava flows and consist of conglomerate and sandstone, which typically occur in channels. The composition and channeled structures in the sandstones and conglomerates and the association with ancient lava flows leave little doubt that these rocks were deposited by streams in an area subjected to basaltic volcanism. The average direction of cross-bedding in the conglomerates and sandstones is shown in Figure 6.25A. From this map, we see that the sediment was derived from both the southeast and the northwest and was deposited in an elongate basin somewhat larger than the present Lake Superior, but in approximately the same position. An outline of the basin to which the sediment was transported (Figure 6.25B) was made by drawing lines perpendicular to the average cross-bedding direction, considered to be roughly parallel to the ancient basin contour lines. Highlands must have existed in Michigan and in parts of Wisconsin during this time, as well as in parts of Minnesota and Ontario. The association of the sandstones and conglomerates with thick accumulations of basalt suggests that the elongate basin was a rift valley.

SUMMARY

1. Sedimentary rocks originate from fragments of other rocks, chemically precipitated minerals, and organic matter.
2. Sedimentary rocks typically occur in layers, or strata, which cover large parts of the continents.
3. Sedimentary rocks preserve a record of the erosional history of Earth and also a history of life on the planet over a period of at least 3.5 billion years.
4. Two main types of sedimentary rocks are recognized: (a) clastic rocks, consisting of rock and mineral fragments (clay, silt, sand, and gravel), and (b) chemical and organic rocks.
5. Four major processes are involved in the genesis of sedimentary rocks: (a) weathering, (b) transportation, (c) deposition, and (d) compaction and cementation. During weathering, transportation, and deposition, the sedimentary material is differentiated (sorted and concentrated according to grain size and composition).
6. A sedimentary environment is a place where sediment is deposited and encompasses the physical, chemical, and biological conditions that exist there. The major sedimentary environments are (a) fluvial, (b) alluvial-fan, (c) eolian, (d) glacial, (e) delta, (f) shoreline, (g) organic-reef, (h) shallow-marine, and (i) deep-marine environments. Each of these is characterized by certain physical, chemical, and biological conditions and therefore develops distinctive rock types and fossil assemblages.

KEY TERMS

alluvial-fan environment (p. 124)

beach (p. 126)

bed (p. 117)

bedding plane (p. 112)

cement (p. 122)

chalk (p. 114)

clastic (p. 112)

conglomerate (p. 112)

cross-bedding (p. 117)

deep-marine environment (p. 129)

delta (p. 126)

dolostone (p. 116)

dripstone (p. 116)

eolian environment (p. 125)

evaporite (p. 116)

fluvial environment (p. 124)

fluvial sediment (p. 117)

formation (p. 117)

fossil (p. 112)

geologic cross section (p. 112)

glacial environment (p. 125)

graded bedding (p. 119)

gypsum (p. 116)

lagoon (p. 127)

lamina (p. 114)

limestone (p. 114)

mud cracks (p. 120)

oolite (p. 114)

reef (p. 122)

ripple marks (p. 120)

rock salt (p. 116)

sandstone (p. 114)

sand waves (p. 117)

sediment (p. 117)

sedimentary differentiation (p. 121)

sedimentary environment (p. 122)

sedimentary rock (p. 110)

sedimentary structure (p. 116)

shale (p. 114)

shallow-marine environment (p. 128)

siltstone (p. 114)

sorting (p. 121)

stratification (p. 117)

stratum (p. 110)

tidal flat (p. 127)

turbidity current (p. 119)

weathering (p. 121)

REVIEW QUESTIONS

1. List the characteristics that distinguish sedimentary rocks from igneous and metamorphic rocks.
2. What is the principal mineral in sandstone? Why does this mineral dominate?
3. How does limestone differ from clastic rocks?
4. What is the mineral composition of limestone?
5. Show by a series of sketches the characteristics of stratification, cross-bedding, and graded bedding.
6. What rock types form in the following sedimentary environments: (a) delta, (b) lagoon, (c) alluvial fan, (d) eolian environment, (e) organic reef, (f) deep-marine environment?
7. How could you recognize an ancient turbidity current deposit now exposed in a mountain?
8. Abraham Werner, a leading geologist in the eighteenth century, believed that all rocks were deposited in the sea. What evidence can you cite from the sequence of rocks in the Grand Canyon that this concept is wrong?

ADDITIONAL READINGS

Blatt, H., G. V. Middleton, and R. C. Murry. 1980. *Origin of Sedimentary Rocks,* 2nd ed. Englewood Cliffs, NJ: Prentice-Hall.

Boggs, S., Jr. 1987. *Principles of Sedimentology and Stratigraphy.* Columbus, OH: Merrill Publishing Company.

Davis, R. A. 1983. *Depositional Systems.* Englewood Cliffs, NJ: Prentice-Hall.

Dunbar, C. O., and J. Rogers. 1957. *Principles of Stratigraphy.* New York: Wiley.

Laporte, L. F. 1968. *Ancient Environments.* Englewood Cliffs, NJ: Prentice-Hall.

Pettijohn. F. J., and P. E. Potter. 1964. *Atlas and Glossary of Primary Sedimentary Structures.* New York: Springer-Verlag.

Reading, H. G. 1985. *Sedimentary Environments and Facies,* 2nd ed. New York: Elsevier; Oxford, England: Blackwell.

Selley, R. C. 1978. *Ancient Sedimentary Environments,* 2nd ed. New York: Cornell University Press.

Selley, R. C. 1982. *An Introduction to Sedimentology.* New York: Academic Press.

Scholle, P. A., and D. Spearing. 1982. *Sandstone Depositional Environment.* Tulsa: Am. Assoc. Petroleum Geologists Memoir 31.

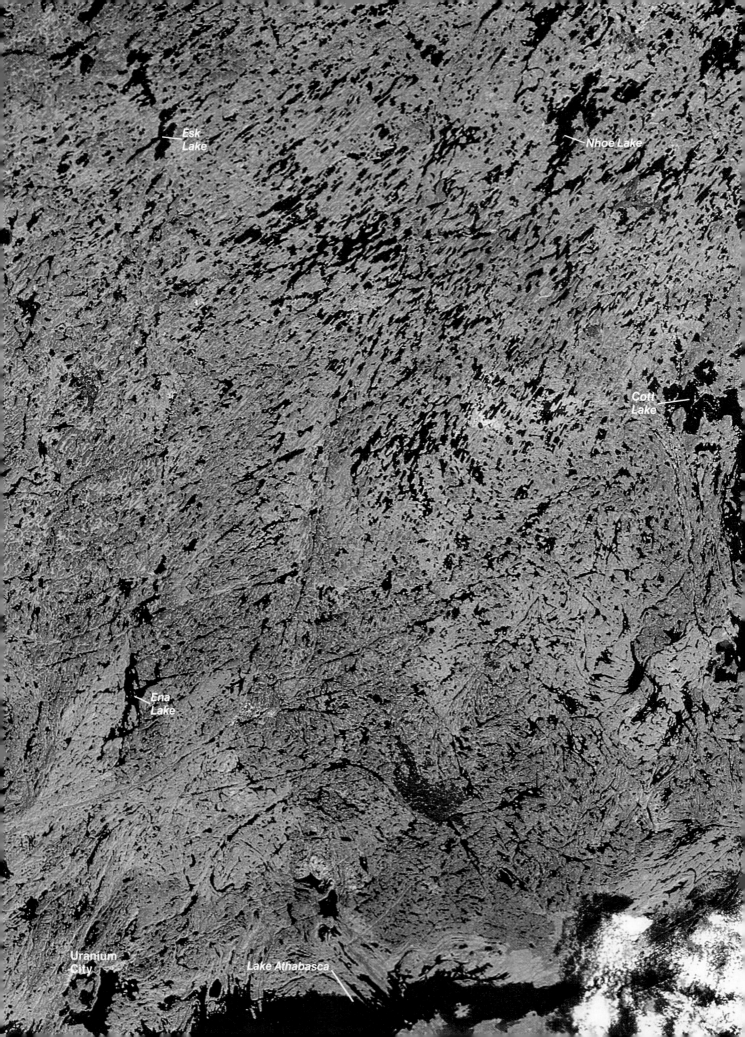

Esk
Lake

Nhoe Lake

Cott
Lake

Ena
Lake

Uranium
City

Lake Athabasca

M ost of the rocks exposed in the continental shields and in the cores of mountain belts show evidence that their original texture and composition have been changed. Many have been plastically deformed, as is indicated by contorted parallel bands of minerals, resembling the swirling layers in marble cake. Others have recrystallized and developed larger mineral grains, and the constituent minerals of many rocks have a strong fabric with a specific orientation. The mineral assemblages in these rocks are also distinctive. They are characterized by mineral types that form only under high temperature and pressure. Geologists interpret these and other lines of evidence to mean that the original rock has been recrystallized in a solid state. The result is a new rock type with a distinctive texture and fabric and, in some cases, a new mineral composition. These rocks are called metamorphic rocks.

Some of the best exposures of metamorphic rock are on the Canadian shield, where glacial erosion has removed the topsoil and thousands of square kilometers of metamorphic terrain are exposed. This Landsat image shows the northern part of Saskatchewan and the adjacent Northwest Territories. Lake Athabaska lies near the bottom of the image. Note the fabric of the terrain. The complex folds and contortions in the rock units show the degree to which metamorphic rocks have been deformed. The long linear lakes, ridges, and depressions are major fracture systems. With this image we are looking down on the roots of mountains built long before Pangaea split to form the Atlantic Ocean. Most of the rocks were formed 2 billion years ago. The glacier disrupted the preexisting drainage to produce the maze of water bodies. Small lakes, shown on the image as black patches, occur in innumerable depressions. Despite the age of the rocks on which they lie, the lakes came into existence only a few thousand years ago.

Metamorphic rocks are hosts of many important mineral deposits. They also play an important role in the evolution of continental crust. Consequently, every aspect of metamorphic rock, from the small grain to the regional fabric of a shield, points toward the same theme: metamorphic rocks dramatically show mobility of a dynamic crust.

7
Metamorphic Rocks

MAJOR CONCEPTS

1. Metamorphic rock can be formed from igneous, sedimentary, or previously metamorphosed rocks.
2. Metamorphic rocks result from changes in temperature and pressure and from changes in the chemistry of their pore fluids. These changes produce new minerals, new textures, and new structures within the rock body.
3. During metamorphism, new minerals grow in the direction of least stress, producing a planar rock structure called foliation. The three main types of foliation are (a) slaty cleavage, (b) schistosity, and (c) gneissic banding.
4. Rocks with only one mineral (such as limestone) do not develop a strong foliation but instead develop a granular texture with large crystals.
5. The major types of metamorphic rocks are slate, schist, gneiss, quartzite, marble, amphibolite, metaconglomerate, and hornfels.
6. Regional metamorphism develops in the roots of mountain belts along convergent plate boundaries. Contact metamorphism is a local phenomenon associated with thermal and chemical changes near the contacts of igneous intrusions.

NATURE AND DISTRIBUTION OF METAMORPHIC ROCKS

Metamorphic rocks are those rocks that have been altered by heat, pressure, and the chemical action of pore fluids to such an extent that the diagnostic features of the original rocks are modified or obliterated. New minerals are formed that are stable in a higher-temperature and higher-pressure environment, and a new rock texture (or fabric) develops in response to the new growth of minerals.

Many igneous and sedimentary rocks have been altered by heat, pressure, and the chemical action of pore fluids to such an extent that the diagnostic features of the original rock have been greatly modified or obliterated. Most structural and textural features in the original rock—e.g., stratification, vesicles, porphyritic textures—cannot be recognized. The rock generally becomes harder, denser, and darker. In some cases, new minerals have formed and create a new rock texture, or fabric. In others, the composition remains the same but the rock has developed much larger mineral grains. Other rocks may be highly contorted. These are metamorphic (changed-form) rocks, a major group of rocks that result from the moving tectonic plates.

Metamorphic rocks can be formed from igneous, sedimentary, or even previously metamorphosed rocks. A common factor is their crystallinity, for, like igneous rocks, they consist of a fabric of interlocking crystals, usually with a preferred grain orientation.

Many people know something about various igneous and sedimentary rocks but only vaguely understand the nature of *metamorphic rocks.* Perhaps the best way to become acquainted with this group of rocks and to appreciate their significance is to study carefully Figure 7.1. The satellite image of a portion of the Canadian shield (Figure 7.1A) shows that the rocks have been twisted and compressed. Originally, these were sedimentary and volcanic layers deposited in a horizontal position, but they have been deformed so intensely that it is difficult to determine the original base or top of the rock sequence. Light-colored granite batholiths and dikes have been injected into the metamorphic series, and the entire mass has been fractured and displaced by numerous faults.

(A) Satellite image of metamorphic rocks in the Canadian shield. Note the complex folds and fractures resulting from extensive crustal deformation.

(B) Metamorphic rocks exposed in the Black Canyon of the Gunnison River, Colorado. The dark complex metamorphic rocks have been intruded by numerous igneous dikes.

(C) Handspecimen (actual size) of a highly metamorphosed rock. Note that crystallization has concentrated light and dark minerals into alternating layers.

FIGURE 7.1
The characteristics of metamorphic rocks are shown on three different scales. Each shows features resulting from heat, pressure, and the chemical activity of pore fluids.

Figure 7.1B shows a more detailed view of metamorphic rocks. The alteration and deformation of the rock are evident in the contorted layers of dark minerals. The pattern of the distortions shows that the rock was compressed while it was in a plastic or semiplastic state.

The degree of *plastic deformation* that is possible during *metamorphism* is best seen by comparing the shape of pebbles in a conglomerate with the shape of pebbles in a rock that has been metamorphosed. In a metamorphosed rock, the original spherical pebbles in the conglomerate have been stretched into long, ellipsoidal blades (the long axis is as much as thirty times the original diameter), all oriented in the same direction. Even on a microscopic scale, distortion and deformation of the individual grains can be seen. A definite preferred orientation of the grains (Figure 7.1C) shows that they either recrystallized under stress or responded to pressure and flowed as a plastic.

It is important to note that the typical texture of metamorphic rocks does not show a sequence of formation of the individual minerals like that evident in igneous rocks. All grains in metamorphic rocks apparently recrystallize at roughly the same time, and they have to compete for space in an already solid rock body. As a result, the new minerals grow in the direction of least stress. Most metamorphic rocks thus have a layered or planar structure, resulting from recrystallization.

What are the distinguishing features of metamorphic rocks?

Metamorphic rocks constitute a large part of the continental crust. Extensive exposures (such as those shown in Figure 7.1) are found in the vast shield areas of the continents, where the cover of sedimentary rocks has been removed. Deep drilling in the stable platform indicates that the bulk of the continental crust immediately beneath the sedimentary cover is also made up of metamorphic rock. These facts permit estimates that metamorphic rocks, together with associated igneous intrusions, make up approximately 85% of the continental crust, at least to a depth of 20 km. In addition to those beneath the stable platforms of the continents and exposed in the shields, metamorphic rocks are also found in the cores of eroded mountain ranges. The nature of Earth's mantle is not known from direct observation, but it is difficult to imagine how the mantle, or at least parts of it, could be other than a type of metamorphic rock.

The widespread distribution of metamorphic rocks in the continental crust, especially among the older rocks, is highly significant evidence that Earth's crust has been subjected to repeated deformation throughout geologic time. The rocks in the continental crust have been compressed almost continually since the crust was formed. This suggests that the tectonic system has operated during most of Earth's history.

METAMORPHIC PROCESSES

The principal agents of metamorphism are changes in temperature, pressure, and chemistry. These changes occur in the solid rock.

Metamorphism is a series of changes in the texture and composition of a rock. The changes occur to restore equilibrium to rocks subjected to an environment different from the one in which they originally formed. The principal agents of metamorphism are changes in (1) temperature, (2) pressure, and (3) chemistry. These agents act in combination, so distinguishing their individual effects in a given rock body is often difficult.

The changes resulting from metamorphism occur in the solid rock—if melting occurred, a magma would be produced and the rock would be igneous.

Temperature Changes

Heat is the most important factor in metamorphism. As the rock temperature increases, minerals begin to change from the solid state to the liquid state, and the reactivity of *pore fluid* in the rocks increases. Below 200°C, only a small amount of fluid is present, and most minerals will remain unchanged for millions of years at that temperature. As the temperature rises, however, the ability of the pore fluid to react with the rocks increases considerably. Chemical reactions become more vigorous. Crystal lattices are broken down and recreated using different combinations of the same ions and different atomic structures. As a result, new mineral assemblages begin to appear. At temperatures higher than 700°C, additional components of the rock become fluid and fusible. Considerable evidence suggests that a high-temperature fluid phase, approaching the magma stage, exists in such a rock body. Extreme cases of increasing temperature probably involve layers of solid material mixed with layers of magma, which gives rise to a rock that is transitional between igneous and metamorphic rocks.

Different minerals are in equilibrium at different temperatures. The mineral composition of a rock therefore provides a key to the temperatures at which it formed. In the field, zones of differing mineral assemblages show how temperature once varied on a regional scale. These zones are particularly obvious around igneous intrusions.

Pressure Changes

High pressure within Earth's crust causes significant changes in the physical properties of many rocks. Under sufficient pressure, a rock becomes plastic and may be deformed as the constituent grains move and rotate, or the grains may be fractured and sheared. This produces a reorientation of the mineral grains and a new rock texture. Perhaps the most obvious sign of directed pressure is the distinct orientation of the grains of minerals such as mica and chlorite. High pressure tends to reduce the space occupied by the mineral components and thus can produce new minerals with closer atomic packing. An increase in pressure can result because the rocks are buried deep beneath Earth's surface, but usually the pressure involved in metamorphism results from *stress,* or directed pressure, at convergent plate boundaries.

How can the texture and composition of a rock be changed?

Chemically Active Fluids

Metamorphic changes may take place without the addition or removal of chemical constituents from the bulk rock, but recrystallization is generally accompanied by some change in chemical composition, that is, by a loss and gain of ions and atoms. Especially important is the loss of volatiles, particularly water and carbon dioxide. In the metamorphic process, low-temperature minerals (minerals that crystallize at relatively low temperatures) break down, providing a fluid medium for the migration of material. If an atom breaks loose from the crystal structure of a mineral, it may move to some other place in this fluid. Although the bulk of the rock remains solid during metamorphism, the rising temperature and pressure free many atoms from their crystal structures. These atoms migrate through fluids in the pore spaces and along the margins of grains. A constant interchange of atoms thus occurs. Original crystals break down, and new crystal structures, which are stable under the new conditions of temperature and pressure, develop. The small amount of pore fluid provides an important medium of transport, which carries material through the rock and arranges it in new mineral structures.

Other chemical reactions can occur by the introduction of ions from an external source and thus produce minerals with a different overall composition. This process is called *metasomatism.* It generally is connected with magmatic intrusions, in which new material from the magma passes through the pore fluid of the rock to create new minerals stable in the new chemical environment.

KINDS OF METAMORPHIC ROCKS

A classification of metamorphic rocks is based on texture and composition. Two major groups are distinguished: (1) rocks that possess a definite planar texture, called foliation, and (2) rocks that lack foliation and have a granular texture. The major types of foliated rocks are slate, schist, and gneiss. The major nonfoliated rocks are quartzite, marble, and amphibolite.

How are the different types of metamorphic rocks distinguished and classified?

Because of the great variety of original rock types and the variation in the kinds and degrees of metamorphism, many types of metamorphic rocks have been recognized. A simple classification of metamorphic rocks, based on texture and composition, is usually sufficient for beginning students. This classification distinguishes two major groups.

1. Rocks that possess a definite planar texture, called foliation
2. Rocks that lack foliation and have a granular texture

The foliated rocks are further subdivided on the basis of the type of foliation. The major rock names can then be qualified by adjectives describing their chemical and mineralogical compositions.

Foliated Rocks

How do rocks react to changes in pressure and temperature and to chemically active fluids? Their individual mineral grains adjust and recrystallize. An important result of these processes is the growth of new minerals in the direction of least stress, so that many metamorphic rocks tend to develop a texture in which the mineral grains have a strong preferred orientation. This orientation may impart a distinctly planar element to the rock, called *foliation* (Latin *folium,* "leaf," hence "splitting into leaflike layers"). The planar structure can result from the alignment of platy minerals, such as mica and chlorite, or from alternating layers having different mineral compositions.

   Slate. *Slate* is a very fine-grained metamorphic rock generally produced by the *low-grade metamorphism* of shale (metamorphism under conditions of relatively low temperature and low pressure). It is characterized by excellent foliation, called *slaty cleavage,* in which the planar element of the rock is a series of surfaces along which the rock can easily be split (Figure 7.2). Slaty cleavage is produced by the parallel alignment of minute flakes of platy minerals, such as mica, chlorite, and talc. The mineral grains are too small to be obvious without a microscope, but the parallel arrangement of small grains develops innumerable parallel planes of weakness, so the rock can be split into smooth slabs.

   Slaty cleavage should not be confused with bedding planes of the parent rock. It is completely independent of the original (relict) bedding and commonly cuts across the original planes of sedimentary stratification. Relict bedding can be rather obscure in slates, but it is often expressed by textural changes resulting from interbedded, thin layers of sand or silt (Figure 7.2). Excellent foliation can develop in the shale part of the sedimentary sequence, in which clay minerals are abundant and are easily altered to mica. In thick

FIGURE 7.2
Slate

layers of quartz sandstone, however, the slaty type of cleavage plane is generally poorly developed.

Phyllite. A *phyllite* is a metamorphic rock with essentially the same composition as a slate, but the micaceous minerals are larger and impart a definite luster to the plane of foliation of the rock. The large mineral grains result from higher temperatures and pressure than those that form slate.

FIGURE 7.3
Schist

Schist. *Schist* is a foliated rock ranging in texture from medium-grained to coarse-grained. Foliation results from the parallel arrangement of relatively large grains of platy minerals, such as mica, chlorite, talc, and hematite, and is referred to as *schistosity.* The mineral grains are large enough to be identified with the unaided eye and produce an obvious planar structure because of their overlapping subparallel arrangement (Figure 7.3). The foliation of schists differs from that of slate mainly in the size of the crystals. The term *schistosity* comes from the Greek *schistos,* meaning "divided" or "divisible." As the name implies, rocks with this type of foliation break readily along the cleavage planes of the parallel platy minerals.

In addition to the clay minerals, significant quantities of quartz, feldspar, garnet, amphibole, and other minerals can occur in schists. The mineral composition thus provides a basis for subdividing schists into many varieties, such as chlorite schist, mica schist, and amphibole schist.

Schists result from a higher intensity of metamorphism than the type that produces slates. They have a variety of parent rock types, including basalt, granite, shale, and tuff, and are one of the most abundant metamorphic rock types.

Gneiss. *Gneiss* is a coarse-grained, granular metamorphic rock in which foliation results from alternating layers of light and dark minerals, or *gneissic layering* (Figure 7.4). The composition of most gneisses is similar to that of granite. The major minerals are quartz, feldspar, and ferromagnesian minerals. Feldspar commonly is abundant and, together with quartz, constitutes a light-colored (white or pink) layer of interlocking grains. Mica, amphibole, and other iron-rich minerals form dark layers. Gneissic layering is usually highly contorted, and gneiss can fracture across the layers, or planes of foliation, as easily as it can do so along them (see Figure 7.4). Gneiss forms during regional *high-grade metamorphism* (metamorphism under conditions of relatively high temperature and high pressure) and in some areas appears to grade into granite.

FIGURE 7.4
Gneiss

Nonfoliated Rocks

Rocks such as sandstone and limestone are composed predominantly of one mineral that crystallizes in an equidimensional form. Metamorphism of these rocks does not result in strong foliation, although mica grains scattered through the rock can assume a parallel orientation. The minerals in nonfoliated rock can be flattened, stretched, and elongated and can show a preferred orientation, but the mass of rock does not develop strong foliation. The resulting texture is best described as granular or, simply, nonfoliated.

Quartzite. *Quartzite* is a metamorphosed, quartz-rich sandstone (Figure 7.5). It is nonfoliated because quartz grains, the principal constituents, do not form platy crystals. The individual grains commonly are deformed and fused into a tight mass, and thus the rock breaks across the grains as easily as it breaks around them. Pure quartzite is white or light colored, but iron oxide and other minerals often impart various tones of red, brown, green, and other colors.

FIGURE 7.5
Quartzite

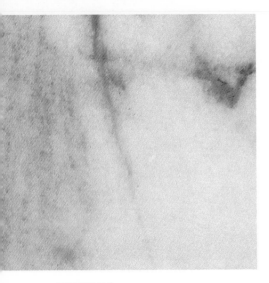

FIGURE 7.6
Marble

FIGURE 7.7
Metaconglomerate

Marble. *Marble* is metamorphosed limestone or dolomite. Calcite, the major constituent of the parent rocks, is equidimensional, so the rock is non-foliated (Figure 7.6). The grains are commonly large and compactly interlocked, forming a dense rock.

The purest marbles are snow white but many marbles contain a small percentage of minerals other than calcite that were present in the original sedimentary rock. These impurities result in streaks or bands and when abundant may impart a variety of colors to the marble. Thus marbles may exhibit a range of colors including white, green, red, brown, and black.

Most marbles occur in areas of regional metamorphism in zones between schists and phyllites.

Amphibolite. *Amphibolite* is a coarse-grained metamorphic rock composed chiefly of amphibole and plagioclase. Mica, quartz, garnet, and epidote also can be present. Amphibolites result from the metamorphism of basalt, gabbro, and other rocks that are rich in iron and magnesium. Some amphibolites develop foliation if mica or other platy minerals are abundant.

Metaconglomerate. *Metaconglomerate* is not an abundant metamorphic rock. It is important in some areas, however, and illustrates the degree to which a rock can be deformed in the solid state. Under directional stress, individual pebbles are stretched into a mass that shows distinctive linear fabric (Figure 7.7).

Hornfels. A *hornfels* is a fine-grained, nonfoliated metamorphic rock that is very hard and dense. The grains usually are microscopic and are welded into a regular mosaic. Platy minerals, such as mica, have a random orientation, and grains of high-temperature minerals are present. Hornfelses usually are dark colored, and they may resemble basalt, dark chert (flint), or dark, fine-grained limestone. They result from metamorphism around igneous intrusions, causing partial or complete recrystallization of the surrounding rock. The parent rock usually is shale, although lava, schists, and other rocks can be baked into hornfels.

The Relationship between Foliation and Larger Structures

Foliation is a response to regional stress that compresses the rocks and causes recrystallization to occur. The orientation of foliation, therefore, is closely related to the large folds and structural patterns of rocks in the field. This relationship commonly extends from the largest folds down to microscopic structures. For instance, the cleavage in slate is generally oriented parallel to

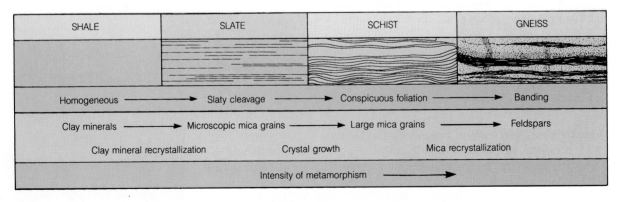

FIGURE 7.8
The metamorphism of shale can involve a series of steps, depending on the intensity of temperature and pressure. Shale can change to slate, schist, or even gneiss.

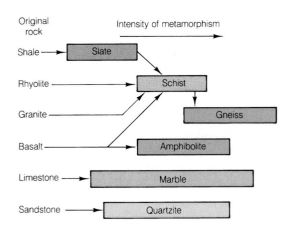

FIGURE 7.9
The origin of common metamorphic rocks is complex. In some cases, such as quartzite, marble, and metaconglomerate (not shown in diagram), the nature of the original rocks is easily determined. In other cases, such as schist and gneiss, it is difficult and sometimes impossible to determine the type of source rock. This diagram is a simplified flow chart showing the origin of some of the common metamorphic rocks.

the axial planes of the folds, which can be many kilometers apart. A slice of the rock viewed under a microscope shows small wrinkles and folds having the same orientation as the larger structures mapped in the field.

Source Material for Metamorphic Rocks

The origin of metamorphic rocks is a complicated process and presents some challenging problems of interpretation. A single source rock can be changed into a variety of metamorphic rocks, depending on the intensity or the degree of metamorphism. For example, shale can be changed to slate, schist, or gneiss (Figure 7.8). Gneiss can also form from many rocks, such as granite or rhyolite. The chart in Figure 7.9, which relates source rocks and metamorphic conditions to metamorphic rock types, gives a generalized picture of the origin of common metamorphic rocks.

METAMORPHIC ZONES

> *Regional metamorphism involves major changes in large masses of rock. Large-scale recrystallization and structural adjustments occur, and the rock commonly shows zones that reflect the difference and degree of changes in temperature and pressure. One type of zone can be defined on the basis of the occurrence of certain metamorphic minerals. Another type of metamorphic zone is defined on the basis of distinctive metamorphic rocks.*

Regional metamorphism involves large-scale changes in thick masses of rock in which major recrystallization and structural adjustments occur. These changes are not random but are systematic and generally occur in a well-defined, predictable sequence, controlled by increasing heat and pressure. Regional metamorphic rocks therefore commonly show metamorphic zones that reflect the differences in temperature and pressure. One type of zonation can be defined on the basis of the occurrence of certain index minerals. In the metamorphism of a thick sequence of shale, a typical sequence of index minerals indicating a transition from low-grade to high-grade metamorphism would be chlorite, biotite, garnet, staurolite, kyanite, and sillimanite.

Another type of metamorphic zonation is defined on the basis of a group of associated rocks, each sequence characterized by a definite set of minerals formed under specific metamorphic conditions. The distinctive group of rocks is called a *metamorphic facies* and is named after the characteristic rock type or mineral.

What features in a rock indicate zones of different degrees of metamorphism?

By mapping zones of index minerals (minerals that form over a narrow range of temperature, thus characterizing a particular degree of metamorphism) or the extent of metamorphic facies, geologists can locate the central and marginal parts of ancient mountain belts and interpret something about ancient interactions between tectonic plates.

Figure 7.10 is a diagram of the progressive change in associations of metamorphic minerals. This schematic representation shows a series of zones of increasing metamorphic grades produced in the metamorphism of shale. A metamorphic grade reflects the extent or degree of metamorphism. For example, the conversion of shale to slate would be low-grade metamorphism, whereas continued and more intense metamorphism would form a schist, or a higher grade of metamorphism. The higher the degree of metamorphism, the greater the amount of change in the rock.

Figure 7.11 is a graph showing the major types of metamorphic rocks in relation to variations in pressure (depth) and temperature. The zeolite facies represents the beginning of metamorphism at low temperature and pressure and is transitional from the changes in sediment resulting from compaction and cementation. The low temperature and pressure produce a new group of minerals called the zeolites, which are hydrous aluminosilicates of calcium and sodium with a rather large water content. With an increase in temperature and pressure, these minerals are soon altered as water is driven out. The greenschist facies then forms under moderate pressure and still fairly low temperature, commonly referred to as low-grade metamorphism. This facies is characterized by chlorite, muscovite, sodic-plagioclase, and quartz. If the pressure remains constant but the temperature increases, the amphibolite facies forms,

FIGURE 7.10
The grades of regional metamorphism are related to temperature and pressure. The arrow in diagram A shows a typical change from lower to higher grades at a given depth. The sequence of index minerals will commonly be chlorite, biotite, almandite (garnet), staurolite, kyanite, and sillimanite. Diagram B is a cross section showing the zones of different grades of metamorphism surrounding a granitic intrusion.

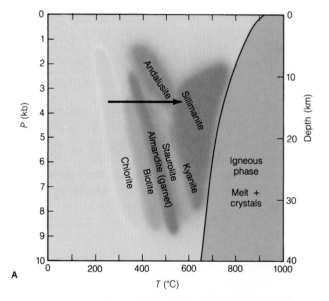

A

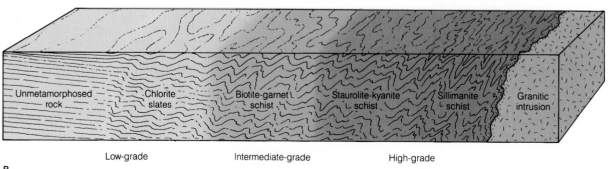

B

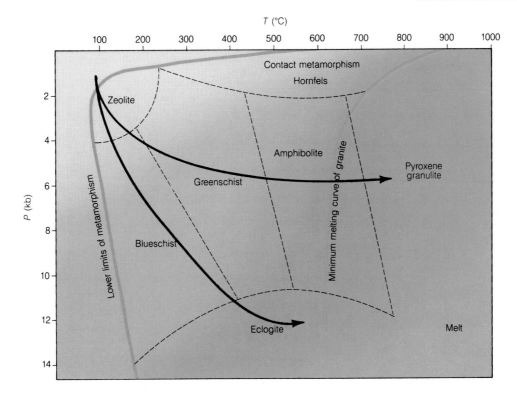

FIGURE 7.11
Metamorphic facies are related to temperature and pressure (depth). The arrows show two possible paths of evolution. At low pressure (relatively shallow depth) the sequence of changes resulting from an increase in temperature would be zeolite, greenschist, amphibolite, pyroxene granulite. If the increase in pressure is greater, changes in metamorphic facies would follow the path indicated by the lower arrow, with the formation of blueschist and then eclogite. Contact metamorphism is limited to a shallow zone of low pressure.

characterized by the presence of hornblende (a type of amphibole). Amphibolite may also contain quartz, plagioclase feldspar, garnet, and biotite, depending on the original rock composition. With a further increase in temperature, the granulite facies forms (above 650°C). The term *granulite* is used to describe the rock texture, which consists of small grains roughly the same size. Pyroxene is an important mineral in this facies, and sillimanite appears as an index mineral. Quartz and plagioclase are also common. The granulite facies represents the highest grade of metamorphism. With a still further increase in temperature, melting occurs and a new magma may be produced.

As shown in Figure 7.11, another path in the sequence of metamorphism can occur. Pressure may increase while the temperature remains relatively low (400 to 600°C) With high pressure but low temperature, the blueschist facies forms. If temperature is increased under these extremely high pressures, the blueschist facies may grade into the eclogite facies, consisting of pyroxene and garnet. Eclogites are also believed to occur in parts of Earth's mantle. In some cases eclogites form from magma and are igneous rocks.

METAMORPHIC ROCKS AND PLATE TECTONICS

Most metamorphic rocks develop deep in the roots of folded mountain belts as a result of plate collision.

We can never observe metamorphic processes in action because they occur deep within the crust. We can, however, study in the laboratory how minerals react to elevated temperatures and pressures that simulate, to some extent, the conditions under which metamorphism occurs. These laboratory studies, together with field observations and studies of texture and composition, provide the rationale for interpreting metamorphic rocks in the framework of the plate tectonics theory.

Figure 7.12 is a graphic model summarizing some of the major ideas concerning the origin of regional metamorphic rocks. According to the theory

What is the tectonic setting where metamorphic rocks are formed?

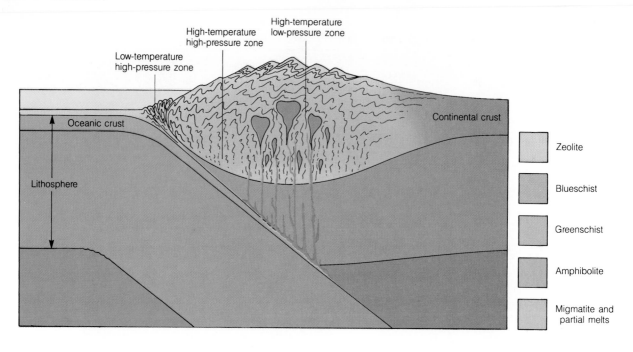

FIGURE 7.12
The origin of metamorphic rocks at convergent plate margins involves changes in pressure and temperature as a result of the collision of plates plus the chemical action of fluids generated in the subduction zone. Relatively low temperatures and high pressures act on marine sediments and oceanic crust near the continental margin. These conditions produce blueschists. In the deep mountain roots, high temperatures and high pressures occur and develop schists and gneisses, which are typical of the continental shields. Contact metamorphism develops around the margins of igneous intrusions.

What does metamorphism tell us about mobility of Earth's crust?

of plate tectonics, pressure results from the converging plates, heat is generated in the subduction zone, and shearing occurs where the plates slide past each other along fracture systems and in the subduction zone. Metamorphism is thus best developed in the deep roots of folded mountain belts, which form at convergent plate boundaries. The original material, prior to metamorphism, may be sediment derived from erosion of a continent, sediment and volcanic material derived from a volcanic arc, or deep-marine sediments and basalt from the oceanic crust. This material is squeezed at convergent plate boundaries as though it were in a vise. Recrystallization tends to produce high-angle or vertical foliation in a linear belt parallel to the margins of the converging plates. Metamorphism theoretically is most intense in the deep mountain roots, where partial melting contributes material to the rising magma generated in the subduction zone. Batholiths and dikes are thus intimately associated with zones of intense metamorphism. Different groups of metamorphic rocks are generated from different source materials—sand, shale, and limestones along continental margins, volcanic sediments and flows along island arcs, and a mixture of deep-marine sediments and oceanic basalt from the oceanic crust.

Close to the subduction zone, high pressure from the converging plates dominates the metamorphic processes. Sediments that have accumulated on the sea floor, together with fragments of oceanic crust, may be scraped off the descending plate and crushed in a chaotic mass of deep-sea sediment, oceanic basalt, and other rock types; and some material may be derived from the upper plate. This jumbled association of rock is called ***mélange*** (heterogeneous mixture). Farther away from the subduction zone, in the mountain root, high-temperature and high-pressure metamorphism occur (Figure 7.12). Metamorphism can also be produced by shearing along fracture zones.

After the stresses from the converging plates are spent, erosion of the mountain belt occurs, and the mountain roots rise because of isostasy. Ultimately, the deep roots and their complex of metamorphic rocks are exposed at the surface, forming a new segment of continental crust. The entire process takes several hundred million years. Repetition of this process causes the continents to grow larger with each mountain-building event. The belts of metamorphic rocks in the shields are thus considered to be the record of ancient continental collisions (see Figure 7.1). This topic is discussed more fully in Chapter 22.

SUMMARY

1. Metamorphic rocks result mainly from changes in temperature and pressure and changes in the chemistry of pore fluids within a rock body. These changes develop new minerals, new textures, and new structures. The diagnostic features of the original sedimentary and igneous rocks are greatly modified or obliterated.

2. Metamorphic rocks are especially significant because they constitute a large part of the continental crust and indicate that the continents have been mobile and dynamic throughout most of geologic time. They constitute an important record of ancient plate movements.

3. During metamorphism, new minerals grow in the direction of least stress, so rocks are produced with a planar element called foliation.

4. Three major types of foliation are recognized: (a) slaty cleavage, (b) schistosity, and (c) gneissic layering. Rocks originally composed of one mineral (such as limestone and sandstone) do not develop strong foliation. Instead, they develop a granular texture with large mineral grains.

5. Metamorphic rocks are classified on the basis of texture and composition. Two major groups are recognized: (a) foliated and (b) nonfoliated.

6. The major types of foliated rocks are slate, schist, and gneiss. The major types of nonfoliated rocks are quartzite, marble, amphibolite, metaconglomerate, and hornfels.

7. Heat is one of the most important factors in metamorphism. As temperatures increase, the hydrous minerals break down to form a dry solid and an aqueous fluid. The pore fluid causes new minerals to form.

8. Directed pressures reduce pore spaces and can produce new minerals with closer atomic packing.

9. Changes in chemistry can result from the decomposition of low-temperature minerals and the migration of the fluid, which subsequently recrystallizes.

10. An increase in heat and pressure can result from deep burial or from igneous intrusions.

11. Most metamorphic rocks of regional extent, however, develop deep in folded mountain belts as a result of plate collision. The metamorphic mountain roots are welded to the continent and become part of the continental crust.

KEY TERMS

amphibolite (p. 140)

foliation (p. 138)

gneiss (p. 139)

gneissic layering (p. 139)

high-grade metamorphism (p. 139)

hornfels (p. 140)

low-grade metamorphism (p. 138)

marble (p. 140)

mélange (p. 144)

metaconglomerate (p. 140)

metamorphic rock (p. 134)

metamorphism (p. 136)

metasomatism (p. 137)

phyllite (p. 139)

plastic deformation (p. 136)

pore fluid (p. 137)

quartzite (p. 139)

schist (p. 139)

schistosity (p. 139)

slate (p. 138)

slaty cleavage (p. 138)

stress (p. 137)

REVIEW QUESTIONS

1. Compare and contrast the characteristics of metamorphic rocks with those of igneous and sedimentary rocks.

2. Make a series of sketches showing the changes in texture that occur with metamorphism of (a) slate, (b) sandstone, and (c) conglomerate.

3. Define *foliation* and explain the characteristics of (a) slaty cleavage, (b) schistosity, and (c) gneissic layering.

4. Describe the major types of metamorphic rocks.

5. Make a generalized flowchart showing the origin of the common metamorphic rocks.

6. What are the agents of metamorphism?

7. Draw an idealized diagram of converging plates to illustrate the origin of regional metamorphic rocks.

8. What type of metamorphic rock would result if zeolite rocks were subjected to temperatures of about 800°C at a depth of 15 km?

ADDITIONAL READINGS

Best, M. G. 1982. *Igneous and Metamorphic Petrology.* San Francisco: Freeman.

Ehlers, E. G., and H. Blatt. 1982. *Petrology, Igneous, Sedimentary, and Metamorphic.* San Francisco: Freeman.

Hyndman, D. W. 1985. *Petrology of Igneous and Metamorphic Rocks,* 2nd ed. New York: McGraw-Hill.

Turner, F. J. 1981. *Metamorphic Petrology: Mineralogical, Field and Tectonics Aspects,* 2nd ed. New York: McGraw-Hill.

Winkler, H. G. F. 1979. *Petrogenesis of Metamorphic Rocks,* 5th ed. New York: Springer-Verlag.

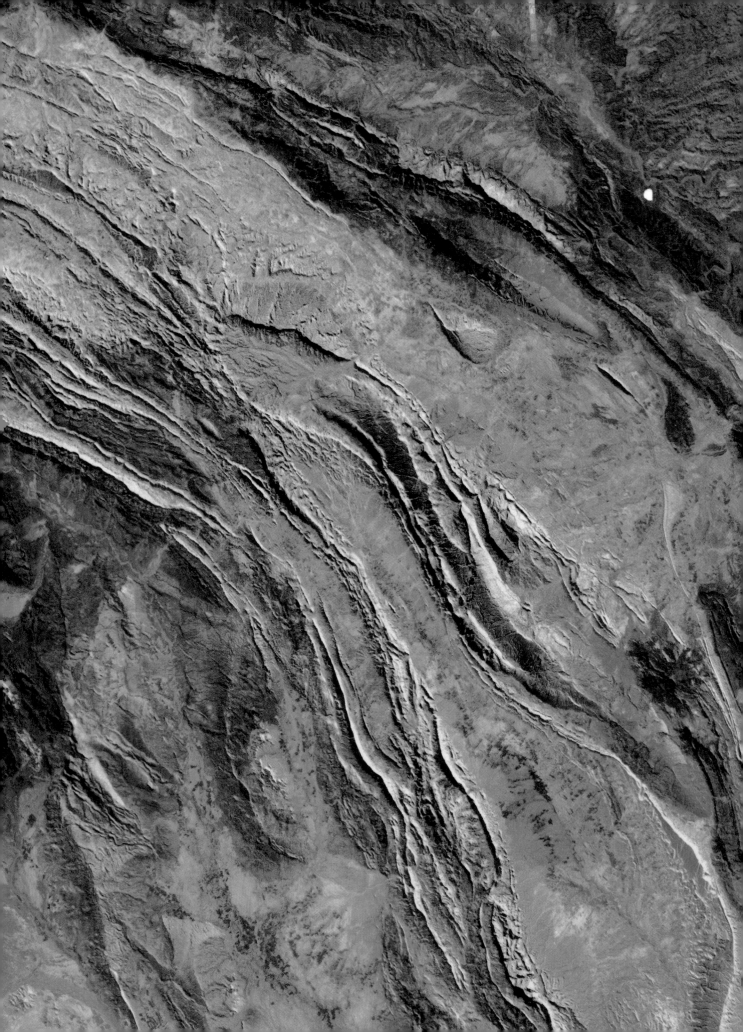

Since the beginning of geologic studies more than 150 years ago, geologists have known that rocks in certain parts of the continents have been folded and fractured on a gigantic scale. Deformation of the crust is most intense in the great mountain belts of the world, where sedimentary rocks, which were originally horizontal, are now folded, contorted, fractured, and, in some places, completely overturned. In some mountains, large bodies of rock have been thrust several tens of kilometers over younger strata. In the deeper parts of mountain belts, deformation is great enough to cause metamorphism.

One of the most dramatic examples of crustal deformation is in the mountain belts formed as India collided with Asia. The area shown in this Landsat scene is in the Sulaiman Range of Pakistan. The Sulaiman Range is oriented north–south in the western arc of the deformation belt, a large bend in the trend of the mountain range formed as the collision had to push aside the continental blocks to the east and west.

The linear ridges that dominate the scene are formed on eroded folds. The multicolored folds in this area form a picturesque and detailed pattern of sedimentary strata folded into an anticline and subsequently eroded into a series of elliptical ridges and intervening valleys.

The extensive deformation of rock bodies in the shields and mountain belts shows that Earth's tectonic system has operated throughout geologic time, with shifting plates constantly deforming the crust at their convergent margins. In this chapter, we will consider how rocks respond to tectonic forces by folding and fracturing and how these structures are expressed at Earth's surface.

8
Structure of Rock Bodies

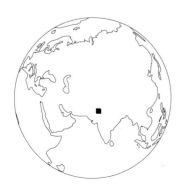

1. Deformation of Earth's crust is well documented by historical movement along faults, by raised beach terraces, and by deformed rock bodies.
2. Folds in rock strata range in size from microscopic wrinkles to large structures hundreds of kilometers long. The major types are (a) domes and basins, (b) plunging anticlines and synclines, and (c) complex flexures.
3. Faults are fractures in the crust along which slippage or displacement has occurred. The three basic types are (a) normal faults, (b) thrust faults, and (c) strike-slip faults.
4. Joints are fractures in rocks along which there is no horizontal or vertical displacement.

EVIDENCE OF CRUSTAL DEFORMATION

A great deal of evidence, both direct and indirect, indicates that the crust is in continuous motion and that it has moved on a vast scale throughout all of geologic time.

Although a casual observer might think that the crust of Earth is permanent and fixed, a great deal of evidence, both direct and indirect, indicates that the crust is in continuous motion and that it has moved on a vast scale throughout all of geologic time.

Evidence of crustal movement comes in many forms and is there for all to see. In the Mediterranean area, some ancient harbors, such as Ephesus in Asia Minor, are now high and dry several kilometers from the sea. Others have been submerged well below low tide. Evidence of more-intense crustal movement is found in sedimentary rocks that were originally deposited beneath the sea but are now found on the highest mountains on all continents.

Earthquakes are perhaps the most convincing evidence that the crust is moving. Those who experience an earthquake are certainly convinced. During earthquakes, the crust not only vibrates, but segments of it are fractured and displaced. One impressive example is the movement along the San Andreas Fault during the 1906 San Francisco earthquake, which offset fences and roads by as much as 7 m (Figure 8.1A). Another is the 1899 earthquake at Yakutat Bay, Alaska, during which a beach was uplifted 15 m above sea level. Similar displacements, well documented with photographs, occurred during the Good Friday earthquake in Alaska in 1964, the Hebgen Lake earthquake in Montana in 1959, and the Dixie Valley earthquake in Nevada in 1954.

Various topographic features also testify to Earth movements in prehistoric times. A striking example is the raised beach terrace along the coast of southern California. There, ancient wave-cut cliffs and terraces, containing remnants of beaches with barnacles, shells, and sand, rise in a series of steplike forms above the present shore (see Figure 16.23).

Many rock formations contain obvious evidence of deformation on a much larger scale. Practically every mountain range exposes deformed sedimentary strata that were originally deposited horizontally below sea level. These twisted and contorted rock layers testify to the continuing motion of the lithosphere and the deformation it produces (Figure 8.1C, D).

The folded rocks in the world's major mountain belts (the Appalachians, the Rockies, the Himalayas, the Urals, and the Alps) all exemplify this type of deformation.

How do we know that Earth's crust has been and continues to be deformed?

(A) Disruption of human structures resulting from earthquakes

(B) Uplifted shoreline in Alaska

(C) The internal structure of a mountain commonly consists of highly deformed layers of sedimentary rock, such as those shown in this photograph of the Canadian Rockies. The degree of compression and deformation may be better appreciated by studying diagram D, which traces the major beds. Note that much of the rock in the upper parts of the original folds has been removed by erosion.

FIGURE 8.1
Evidence of crustal deformation

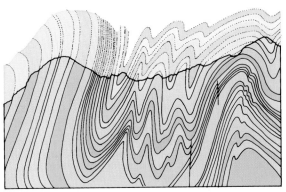

(D) Diagram of the major beds shown in photo C.

DIP AND STRIKE

The orientation in rocks of planar features such as bedding planes, faults, and joints can be defined by measurements of dip (the downward inclination of the plane) and strike (the direction or trend of the plane).

How do we measure the orientation of rocks, bodies, faults, and joints?

Many structural features of the crust are too large to be seen from any point on the ground. They are recognized only after the geometry of the rock bodies is determined from geologic mapping. At an outcrop two fundamental observations—dip and strike—describe the orientations of bedding planes, fault planes, joints, and other planar features in the rock. The ***dip*** of a plane is the angle and direction of its inclination from the horizontal. The ***strike*** is the direction, or trend, of a planar feature, such as a bedding plane or a fault plane. More precisely, it is the compass bearing of a horizontal line on the planar feature. These two measurements together define the orientation of the planar surface in space.

The concept of dip and strike can be easily understood by referring to Figure 8.2, which shows an outcrop of tilted beds along the coast. The water provides a necessary reference to a horizontal plane. The trend of the waterline along the bedding plane of the rocks is the direction of strike. The angle between the water surface and the bedding plane is the angle of dip. The photograph in Figure 8.3 shows a sequence of beds striking south (to the top of the picture) and dipping 40 degrees to the east (to the left of the picture). Another way to visualize dip and strike is to think of the roof of a building. The dip is the direction and amount of inclination of the roof, and the strike is the trend of the ridge.

Dip and strike are measured in the field with a geologic compass, which is designed to measure both direction and angle of inclination. An example of the symbols used for recording the dip and strike of bedding planes on a map is ⊥30. The long crossbar shows the strike, the short line perpendicular to it shows the direction of dip, and the number represents the angle of dip. The symbol shown here represents beds striking N 45 degrees and dipping 30 degrees to the southeast.

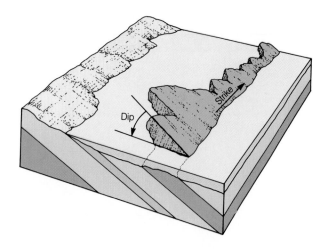

FIGURE 8.2
The concept of dip and strike can be understood by studying rock layers such as the ones shown in this photograph. The strike of a bed is the compass bearing of a horizontal line drawn on the bedding plane. It can readily be established by reference to the horizontal waterline in this example. The dip is the angle and direction of inclination of the bed, measured at right angles to the strike.

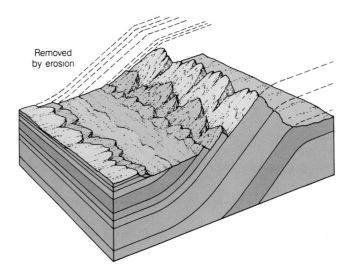

Removed
by erosion

FIGURE 8.3
A sequence of inclined beds striking south (toward the background) and dipping 40 degrees to the east is shown in this photograph of the San Rafael Swell, in Utah. The diagram shows the configuration of the flexure and the upper beds, which have been removed by erosion.

FOLDS

Folds are warps (wavelike contortions) in rock strata. They are three-dimensional structures ranging in size from microscopic crinkles to large domes and basins hundreds of kilometers wide. Broad, open folds form in the stable interiors of continents, where the rocks are only mildly warped. Complex folds develop in mountain belts, where deformation is more intense.

Three-dimensional structures that range in size from microscopic crinkles to large domes and basins hundreds of kilometers wide are known as *folds.* They are warps, or wavelike contortions, in rock strata. Small flexures are abundant in sedimentary rocks and can be seen in mountainsides and road cuts and even in hand specimens. Large folds cover thousands of square kilometers, and they can best be recognized from aerial or space photographs or from geologic mapping.

Fold Nomenclature

Three general types of folds are illustrated in Figure 8.4. *Monoclines* are folds in which horizontal or gently dipping beds are modified by simple steplike bends. *Anticlines,* in their simplest form, are uparched strata, with the *limbs* (sides) of the fold dipping away from the crest. Rocks in an eroded anticline are progressively *older* toward the interior of the fold. *Synclines,* in their simplest form, are downfolds, or troughs, with the limbs dipping toward the center. Rocks in an eroded syncline are progressively *younger* toward the center of the fold.

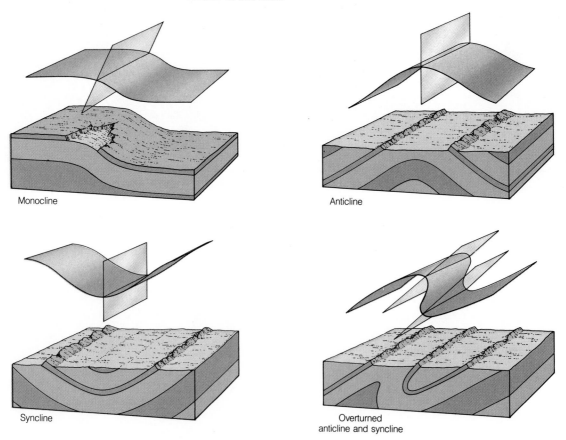

Monocline

Anticline

Syncline

Overturned
anticline and syncline

FIGURE 8.4
The nomenclature of folds is based on the three-dimensional geometry of the structure, although most exposures show only a cross section or map view.

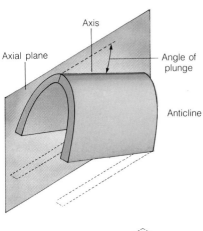

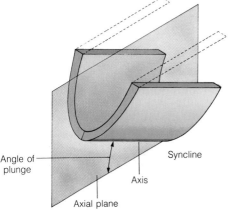

FIGURE 8.5
The axial plane of a fold is an imaginary plane that divides the fold in equal parts. The line formed by the intersection of the axial plane and a bedding plane is called the axis. The downward inclination of the axis is called the plunge.

For purposes of description and analysis, it is useful to divide a fold into two equal parts by an imaginary plane called the ***axial plane***. The line formed by the intersection of the axial plane and a bedding plane is called the ***axis***, and the downward inclination of the axis is referred to as the ***plunge*** (Figure 8.5). A plunging fold, therefore, is a fold in which the axis is inclined.

Domes and Basins

The sedimentary rocks covering much of the continental interiors have been mildly warped into broad ***domes*** and ***basins*** many kilometers in diameter. One large basin covers practically all of the state of Michigan. Another underlies the state of Illinois. An elongate dome underlies central Tennessee, central Kentucky, and southwestern Ohio. Although these flexures in the sedimentary strata are extremely large, the configuration of the folds is known from geologic mapping and from information gained through drilling. The nature of these flexures and their topographic expression are diagrammed in Figure 8.6.

The configuration of a single bed warped into broad domes and basins is shown in perspective in Figure 8.6A. If erosion cuts off the tops of the domes, the layer looks like the one shown in Figure 8.6B. The exposed rocks of both domes and basins typically have a circular or elliptical outcrop pattern. The rocks exposed in the central parts of eroded domes are the oldest rocks, whereas the rocks exposed in the centers of basins are the youngest.

A classic example of a broad fold in the continental interior is the large dome that forms the Black Hills of South Dakota (see Figure 13.10). Resistant rock units form ridges that can be traced completely around the core of the dome, and nonresistant formations make up the intervening valleys. A small dome in western Wyoming, similar to that of the Black Hills, is shown in the photograph in Figure 8.6C. It clearly illustrates the typical elliptical outcrop pattern of domes, expressed by alternating ridges and valleys that encircle the structure.

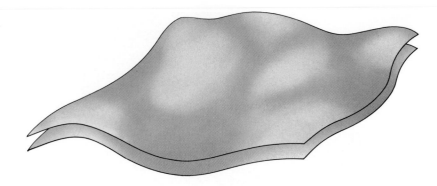

(A) A single gently folded bed is warped in a configuration of broad domes and basins.

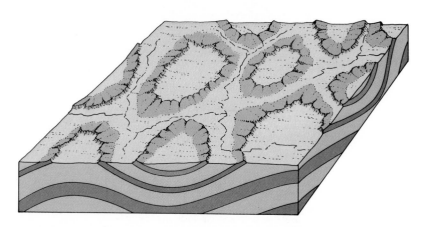

(B) As erosion proceeds, the tops of the domes are eroded first. The outcrop pattern of eroded domes and basins typically is circular or elliptical.

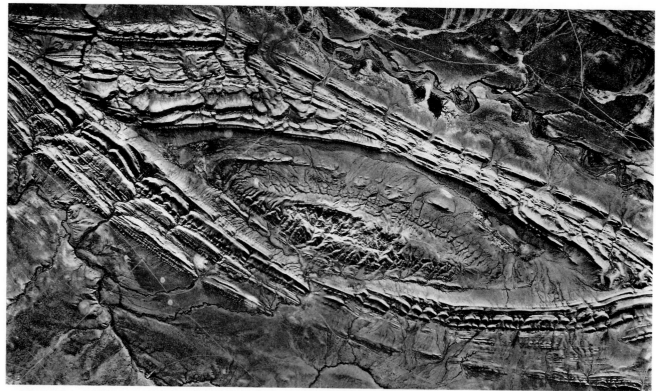

(C) The surface expression of a structural dome in western Wyoming was photographed from an altitude of approximately 6 km. The resistant layers of the structure form ridges, and the nonresistant layers are eroded into elongate valleys. Note that the oldest rocks are at the center of the structure.

FIGURE 8.6
The geometry and topographic expression of domes and basins involve broad upwarps and downwarps of layered rocks. When eroded, the exposed rock forms circular or elliptical outcrop patterns.

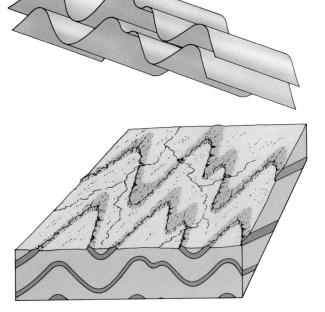

(A) *The basic form of folded strata is similar to that of a wrinkled rug. In this diagram, the strata are compressed and plunge toward the background.*

(B) *If the tops of the folded strata are eroded away, a map of the individual layers shows a zigzag pattern at the surface. Rock units that are resistant to erosion form ridges, and nonresistant layers are eroded into linear valleys. In a plunging anticline, like the one shown here, the surface map pattern of the beds forms a V pointing in the direction of plunge.*

FIGURE 8.7
A series of plunging folds forms a zigzag outcrop pattern.

Plunging Folds

Where deformation is more intense (such as, typically, in mountain belts), the rock layers are deformed into a series of tight folds. Fold geometry is not exceedingly complex. In many ways, folds resemble the wrinkles in a rug. But complexity in the exposed outcrop patterns of folds results from erosion, so folds may be difficult to recognize on aerial photographs without some previous experience in geologic observation and interpretation. The diagrams in Figure 8.7 illustrate the general configuration of *plunging folds* and their surface expressions after they have been eroded. The basic form is shown in diagram A, and diagram B shows a fold system after the upper part has been removed by erosion. The outcrop of the eroded plunging anticlines and synclines forms a characteristic zigzag pattern. The nose of an anticline forms a V-shaped pattern that points in the direction of plunge, and the oldest rocks are in the center of the fold. The nose of a syncline forms a V-shaped pattern that opens in the direction of plunge, and the youngest rocks are in the center of the fold. Together, the outcrop pattern and the relative ages of the rocks in the center of the fold make it possible to determine the structure's subsurface configuration.

What is the geometry of folded rocks?

Complex Folds

Intense deformation in some mountain ranges produces complex folding, as illustrated in Figure 8.8. Such structures commonly exceed 100 km in width, so complex folds can extend through a large part of a mountain range. Details of such intensely deformed structures are extremely difficult to work out because of the complexity of the outcrop patterns.

Figure 8.8 illustrates the geometry and surface expression of complex folds. Diagram A is a perspective drawing of a single bed in a typical complex fold. This *overturned fold* is a huge anticlinal structure with numerous minor anticlines and synclines forming digits on the larger fold. Diagram B shows the fold after it has been subjected to considerable erosion, which has removed

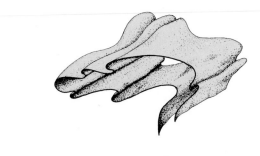

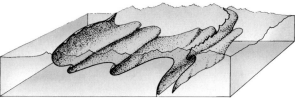

(A) Rocks that have been intensely deformed commonly consist of large overturned folds, with minor folds on the limbs.

(B) The surface outcrop of the same fold, after erosion has removed the upper surface, shows great complexity, so that its details are difficult to recognize.

(C) The topographic expression of complex folds can be a series of linear mountain ridges.

FIGURE 8.8
Complex folds produce complex outcrop patterns.

most of the upper limb. Note the cross section of the structure on the mountain front and the outcrop pattern compared with that in diagram A. The topographic expression of complex folds is variable. They usually are expressed in a series of mountains (Figure 8.8C). Complex folds are common in the Swiss Alps, but they were recognized only after more than half a century of detailed geologic studies. They are also common in the roots of ancient mountain systems and thus are exposed in many areas of the shields.

FAULTS

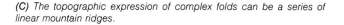

Faults are fractures in Earth's crust along which slippage or displacement has occurred. Three basic types of faults are recognized: (1) normal faults, (2) thrust faults, and (3) strike-slip faults.

Slippage or displacement along fractures in Earth's crust creates **faults**. In a road cut or in the walls of a canyon, a fault plane may be obvious, and the displaced or offset beds can easily be seen (Figure 8.9). Elsewhere, the surface expression of a fault can be very subtle, and detailed geologic mapping may be necessary before the precise location of such a fault can be established. Displacement along faults ranges from a few centimeters to hundreds of kilometers.

Faults grow by a series of small displacements, which occur as built-up stress is suddenly released. Displacement can also occur by imperceptibly slow movement called tectonic creep. Three basic types of faults are recognized:

1. Normal faults
2. Thrust faults
3. Strike-slip faults

These are illustrated in Figure 8.10.

FIGURE 8.9
The displacement of beds in a fault is often well-expressed on the side of a valley. Here the displacement of the beds can be seen along the well-defined fault plane.

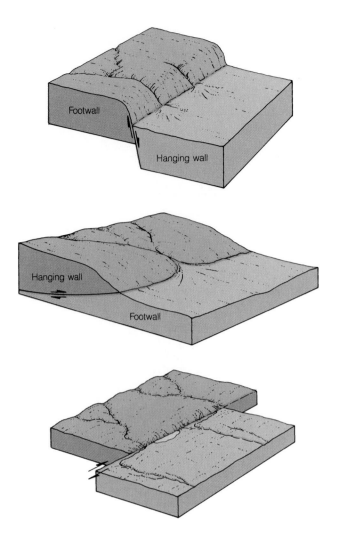

(A) In normal faults, the hanging wall moves downward in relation to the footwall.

(B) In thrust faults, the hanging wall moves upward in relation to the footwall.

(C) In strike-slip faults, the displacement is horizontal.

FIGURE 8.10
The three major types of faults are distinguished by the direction of their relative displacement.

Normal Faults

In *normal faults,* movement is mainly vertical, and the rocks above the fault plane (the *hanging wall*) move downward in relation to those beneath the fault plane (the *footwall*) (Figure 8.10). Most normal faults are steeply inclined, usually between 65 and 90 degrees. Their predominantly vertical movement commonly produces a cliff, or *scarp.*

Normal faults rarely are isolated fractures. Typically, a group of parallel normal faults develops a steplike arrangement, or a series of *fault blocks.* A narrow block dropped down between two normal faults is called a *graben* ("grave" in German), and an upraised block is a *horst* (Figure 8.11). A graben typically forms a conspicuous fault valley or basin marked by relatively straight, parallel walls. Horsts form blocklike plateaus bounded by faults.

Large-scale normal faulting is the result of major tensional stresses, which stretch and pull apart the crust. This type of stress occurs on a global basis along divergent plate margins, so that normal faults are the dominant structure along the oceanic ridge and in the several continental rift systems.

In the Basin and Range province of western North America, normal faulting produced a series of fault blocks trending north and south and extending from central Mexico to Oregon and Idaho. The horsts commonly form mountain ranges from 2000 to 4000 m high, which are considerably dissected by erosion. Grabens form topographic basins, which are partly filled with

What is the nature and origin of faults?

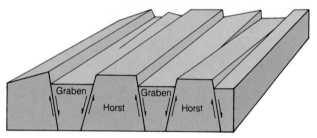

FIGURE 8.11
Horsts and grabens in Canyonlands National Park, Utah, are clearly expressed at the surface. Grabens (downdropped blocks) form elongate valleys, which are partly covered with a smooth flat veneer of sediment. Horsts (upraised blocks) form elongate ridges. Relative movement along the major faults is shown in the idealized diagram.

erosional debris from the adjacent ranges. Between the Wasatch Range, in central Utah, and the Sierra Nevada, on the Nevada–California border, Earth's crust has probably been extended some 60 km during the last 15 million years. Displacement of alluvial fans and surface soils indicates that many of the faults are still active.

The great rift valleys of Africa are another example of large-scale normal faulting produced by a zone of tension in Earth's crust. There, large grabens have been formed by a system of normal faults extending from the Zambesi River in southern Africa to northern Ethiopia, a distance of 2900 km. That distance is almost doubled if the rift system of the Red Sea and the Jordan valley (which continues northward into Syria) is added. The rift valleys of Africa are remarkably uniform in width, ranging from 30 to 45 km. They are about the same size as the grabens of the Basin and Range province and the rift valleys of the other continents.

Thrust Faults

Thrust faults are low-angle faults in which the hanging wall has moved up and over the footwall. Some geologists apply the term *thrust* to these faults only if they dip at angles less than 45 degrees, and they refer to high-angle thrusts as

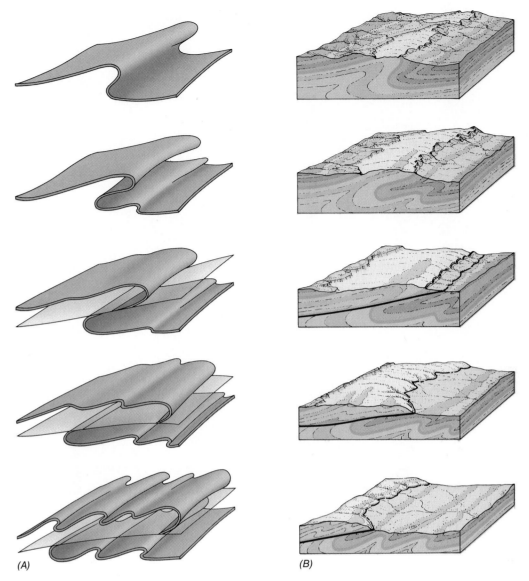

(A)

(B)

FIGURE 8.12
The evolution of thrust faults from folds is depicted in this sequence of diagrams. Diagram A shows the fault plane and the progressive development of folds into a thrust fault. Diagram B shows how the structure might be expressed at the surface as a result of contemporaneous erosion.

reverse faults. Movement on a thrust is predominantly horizontal, and displacement can be more than 50 km.

Thrust faults result from crustal shortening. They generally are associated with intense folding, caused by powerful horizontal compression in Earth's crust. They are prominent in all of the world's major folded mountain regions, commonly evolving from folds in the manner diagrammed in Figure 8.12. Where resistant rocks are thrust over nonresistant strata, a scarp is eroded on the upper plate. The scarp is not straight or smooth, as cliffs produced by normal faulting are. Rather, the outcrop of the fault surface typically is irregular in map view (Figure 8.13).

Strike-Slip Faults

Strike-slip faults are high-angle fractures in which displacement is horizontal, parallel to the strike of the fault plane. There is little or no vertical movement, so that high cliffs do not usually form along strike-slip faults. Instead, these

FIGURE 8.13
A thrust fault in the Spring Mountains in southern Nevada was formed by strong compressive stress in Earth's crust during late Cretaceous time, approximately 70 million years ago. The sharp boundary between the red and gray rocks along the top of the 700 meter cliff marks the trace of the thrust fault. The gray rocks are Cambrian limestones and were thrust 40 km eastward and rest on top of the red Jurassic sandstone. Compare this photograph with the bottom diagram in Figure 8.12.

faults are expressed topographically by a straight, low ridge extending across the surface, a feature that commonly marks a discontinuity in types of landscape.

Some of the topographic features produced by strike-slip faulting and subsequent erosion are shown in Figure 8.14. One of the more obvious is the offset of the drainage pattern. The relative movement is often shown by abrupt right-angle bends in streams at the fault line: A stream follows the fault for a short distance and then turns abruptly and continues down the regional slope. As the blocks move, some parts may be depressed to form *sag ponds*. Others buckle into low, linear ridges.

Strike-slip movement results in the juxtaposition of blocks with contrasting structures, rock types, and topographic forms (Figure 8.15). The fault line therefore often marks the boundary between distinctly different surface features and rock types. This distinction is expressed in Figure 8.14 by the contrast in the degree of dissection on the fault blocks. Faults also disrupt patterns of groundwater movement, as is reflected by contrasts in vegetation and soils and by the occurrence of springs along the fault trace.

Strike-slip faults result from shear stresses in the crust. They commonly are produced where one tectonic plate slides past another at a transform fault boundary.

160

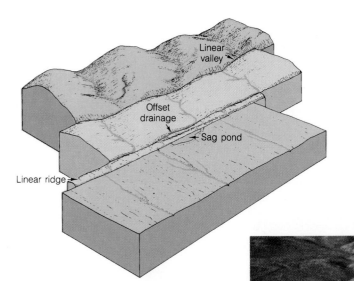

FIGURE 8.14
Strike-slip faults produce distinctive landforms. Streams are offset by recurrent movement, linear ridges and valleys form, and local sag ponds develop along the fault line.

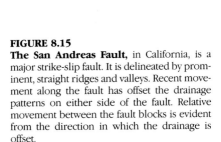

FIGURE 8.15
The San Andreas Fault, in California, is a major strike-slip fault. It is delineated by prominent, straight ridges and valleys. Recent movement along the fault has offset the drainage patterns on either side of the fault. Relative movement between the fault blocks is evident from the direction in which the drainage is offset.

Observed Movement on Faults

The movement along faults during earthquakes rarely exceeds a few meters. In the great San Francisco earthquake of 1906, the crust slipped horizontally as much as 7 m along the San Andreas Fault, so roads, fence lines, and orchards were offset. Recent vertical faults in Nevada and Idaho have produced fresh scarps from 3 to 6 m high (Figure 8.16). The Good Friday earthquake in Alaska in 1964 was accompanied by a 13 m uplift near Montague Island. The largest well-authenticated displacement during an earthquake appears to have occurred in 1899 near Yakutat Bay in Alaska, where beaches were raised as much as 15 m above sea level.

Movement along faults is not restricted to uplift during earthquakes, however. Precise surveys along the San Andreas Fault show slow shifting along the fault plane at an average rate of 4 cm per year. Such movements, called **tectonic creep,** break buildings constructed across the fault line and eventually result in considerable displacement.

The important point is that displacement on a fault does not occur in a single violent event. Rather, it is the result of numerous periods of displacement and slow tectonic creep, commonly separated by periods of tectonic stability.

FIGURE 8.16
Recent displacement along a fault, at the base of the Lost River Range in southern Idaho, has produced the fresh cliff at the base of the mountain front. Cumulative movement on the fault produced the mountain range.

JOINTS

Joints are fractures in rocks along which no appreciable displacement has occurred. They are very common in all rock bodies and are found in almost every exposure.

The most common structural features of rocks exposed at Earth's surface are simple cracks, or fractures, called *joints*. Joints do not occur at random but usually as two sets of fractures, which intersect at angles ranging from 45 to 90 degrees. They thus divide the rock body into large, roughly rectangular blocks. Joints can be related to major faults or to broad upwarps of the crust. They can form remarkably persistent patterns extending over hundreds of square kilometers.

The best areas in which to study joints are those where brittle rocks, such as thick sandstone, have been fractured and their joint planes accentuated by erosion. The massive sandstones of the Colorado Plateau are excellent examples. Joints are expressed by deep, parallel cracks, which are most impressive when seen from the air (Figure 8.17, see also Figure 8.11). In places, joints control the development of stream courses, especially secondary tributaries and areas of solution activity.

Most joints result from broad regional upwarps or from compression or tension associated with faults and folds. Some tensional joints result from erosional unloading and expansion of the rock. Columnar joints in volcanic rocks are produced by stresses that are set up as the lava cools and contracts.

Joints can have economic importance. They can provide the necessary permeability for groundwater migration and for the migration and accumulation of petroleum. Analyses of jointing have been important in both the exploration and the development of these resources. Joints also control the deposition of copper, lead, zinc, mercury, silver, gold, and tungsten ores. Hydrothermal solutions associated with igneous intrusions migrate along joint systems and precipitate minerals along the joint walls, thus forming mineral veins. Modern prospecting techniques therefore include detailed fracture analyses.

Major construction projects (such as dams) are especially affected by jointing systems within rocks, and allowances must be made for them in the project planning. For example, the Flaming Gorge Dam, on the Green River in northeastern Utah, was constructed in a small bend of the river nearly parallel to the major vertical joint system. Stresses caused by water storage therefore tended to close the fractures. Had the dam site been selected in the part of the river that trends from east to west, the joint planes would have presented major structural weaknesses in the bedrock foundation.

Joint systems can be either an asset or an obstacle to quarrying operations. Closely spaced joints severely limit the sizes of blocks that can be removed. If a quarry follows the orientation of intersecting joints, however, the expense of removing building blocks is greatly reduced, and waste is held to a minimum.

FIGURE 8.17
Joint systems in resistant sandstones in Arches National Park, Utah, have been enlarged by weathering, forming deep, narrow crevasses. The set of intersecting joints reflects the orientation of stress that deformed the rock body.

SUMMARY

1. Deformation of Earth's crust is well documented by historical movements along faults, by raised beach terraces, and by contorted rock layers.
2. The major types of rock structures are folds, faults, and joints.
3. Folds in stratified rocks range in size from microscopic wrinkles to large flexures hundreds of kilometers long.
4. In the stable platform, sedimentary rocks are mildly warped into broad domes and basins. More intense deformation occurs in mountain belts, where rocks are folded into tight, plunging anticlines and synclines. Intense deformation in parts of some mountain ranges produces complex folding.
5. Faults are fractures in Earth's crust along which slippage or displacement has occurred. Three basic types of faults are recognized: (1) normal faults, resulting from tension; (2) thrust faults, resulting from compression; and (3) strike-slip faults, resulting from shear stresses. Normal faults develop on a large scale in rift systems. Thrust faults generally are commonly formed in folded mountain belts resulting from plate collisions. Strike-slip faults result from shearing along passive plate margins.
6. Joints are fractures with no noticeable displacement. They are associated with all types of deformation.
7. The various kinds of crustal deformation record plate movement and a history of Earth's dynamics.

KEY TERMS

anticline (p. 152)	limb (p. 152)
axial plane (p. 153)	monocline (p. 152)
axis (p. 153)	normal fault (p. 157)
basin (p. 153)	overturned fold (p. 155)
dip (p. 150)	plunge (p. 153)
dome (p. 153)	plunging fold (p. 155)
fault (p. 156)	reverse fault (p. 159)
fault block (p. 157)	sag pond (p. 160)
fold (p. 152)	scarp (p. 157)
footwall (p. 157)	strike (p. 150)
graben (p. 157)	strike-slip fault (p. 159)
hanging wall (p. 157)	syncline (p. 152)
horst (p. 157)	tectonic creep (p. 161)
joint (p. 162)	thrust fault (p. 158)

REVIEW QUESTIONS

1. List evidence that Earth's crust is in motion and has moved throughout geologic time.
2. What is a mountain belt?
3. Explain the terms *dip* and *strike*.
4. Sketch a cross section of the structure of the rocks shown in Figure 8.6C.
5. Make a perspective sketch of an anticline and the adjacent syncline, and label the following features: (a) axial plane, (b) axis, (c) angle of plunge, and (d) limbs.
6. Sketch the outcrop pattern of a plunging fold.
7. Describe the development of complex alpine-type folds.
8. Draw a simple block diagram of a normal fault, a thrust fault, and a strike-slip fault, and show the relative movement of the rock bodies along each. List the defining characteristics of each type of fault.
9. What are horsts and grabens? What global tectonic features are found where horsts and grabens most commonly are formed?
10. What global tectonic features are found where thrust faults commonly are formed?
11. List some of the surface features that commonly are produced by strike-slip faults.

ADDITIONAL READINGS

Davis, G. H. 1984. *Structural Geology of Rocks and Regions.* New York: Wiley.

Dennis, J. D. 1987. *Structural Geology, an Introduction.* Dubuque, IA: Brown.

Suppe, J. 1985. *Principles of Structural Geology.* Englewood Cliffs, NJ: Prentice-Hall.

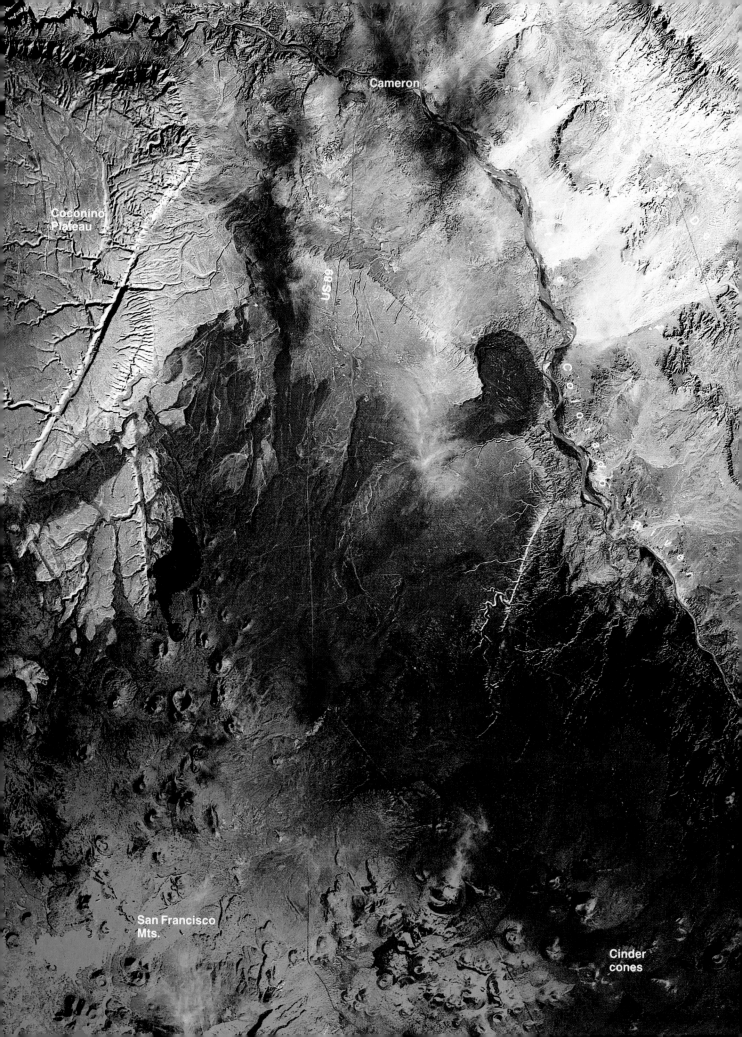

S ome sciences deal with incredibly large numbers, others with great distances, still others with infinitesimally small particles. In every field of science, students must expand their conceptions of reality, a sometimes difficult, but very rewarding, adjustment to make.

Geology students must expand their conceptions of the duration of time. Because life is short, we tend to think twenty or fifty years is a long time. A hundred years in most frames of reference is a very long time, yet in studying Earth and the processes that operate on it, you must attempt to comprehend time spans of 1 million years, 100 million years, and even several billion years.

How do scientists measure such long periods of time? Nature contains many types of time-measuring devices. Earth itself acts like a clock, rotating on its axis once every 24 hours.

Rocks are records of time, and from their interrelationships, the events of Earth's history can be arranged in proper chronological order. The area shown on this Landsat mosaic is an excellent example. The region is about 50 km south of the Grand Canyon. Flagstaff, Arizona, is just south of this image. Several major geologic events are clearly expressed in this area. The first event was the formation of an erosional surface on top of the Kaibab Limestone, the relatively smooth surface shown in tones of light green and into which stream erosion has cut deep canyons. The next event was displacement of the rocks by a series of fractures, the most prominent of which trends northeast. Volcanic activity is obviously a younger event, in that all of the lava flows rest upon, or cut across, the erosional surface and fractures. The older flows and associated volcanic cones are more weathered and eroded than the younger. The youngest event was the eruption that produced the dark black flow in the central part of the photo and the fresh cinder cone.

Fossils within a rock constitute a separate organic clock by which geologists can "tell time" and identify synchronous events in Earth's history. Rocks also contain radioactive clocks, which permit us to measure with remarkable accuracy the number of years that have passed since the minerals forming the rocks crystallized.

9
Geologic
Time

MAJOR CONCEPTS

1. The interpretation of past events in Earth's history is based on the principle that the laws of nature do not change with time.
2. Relative dating (determining the chronologic order of a sequence of events) is achieved by applying the principles of (a) superposition, (b) faunal succession, (c) crosscutting relations, and (d) inclusion.
3. The standard geologic column was established from studies of the rock sequence in Europe. It is now used worldwide. Rocks are correlated from different parts of the world on the basis of the fossils they contain.
4. Absolute time designates a specific duration of time in units of hours, days, or years. In geology, long periods of absolute time can be measured by radiometric dating.

THE DISCOVERY OF TIME

The interpretation of rocks as products and records of events in Earth's history is based on the principle of uniformitarianism, which states that the laws of nature do not change with time. Rocks are records of geologic time.

Geologic time was discovered in Edinburgh in the 1770s by a small group of scholars led by James Hutton. These men challenged the conventional thinking of their day, in which the largest unit of time was the human life span (the lives of the patriarchs), and the age of Earth was accepted to be 6000 years, as established by Bishop Ussher's summation of biblical chronology. Hutton and his friends studied the rocks along the Scottish coast and observed that every formation, no matter how old, was the product of erosion from other rocks, older still. Their discovery showed that the roots of time were far deeper than anyone had supposed. Hutton's discovery of time was based on the interpretation of rocks as products of events in Earth's history. It was perhaps the most significant discovery of the eighteenth century because it changed forever the way we look at Earth, the planets, and the stars and, as a consequence, the way we look at ourselves.

Uniformitarianism

Why are rocks considered to be records of time?

The interpretation of rocks as products and records of events in Earth's history is based on one of the fundamental assumptions of scientific inquiry: the principle of *uniformitarianism*, which states that *the laws of nature do not change with time*. We assume that the chemical and physical laws operating today have operated throughout all time. The physical attraction between two bodies (gravity) acted in the past as it does today. Oxygen and hydrogen, which today combine (under certain conditions) to form water, did so in the past under those same conditions. Although scientific explanations have improved and changed over the centuries, the natural laws and processes are constant and do not change. All chemical and physical actions and reactions occurring at present are produced by the same causes that produced similar events 100 years or 5 million years ago.

Hutton's principle of uniformitarianism was radical for the time and slow to be accepted. In the late eighteenth century, before modern geology had developed, the Western world's prevailing view of Earth's origin and history was derived from the biblical account of creation. Earth was believed to have been created in 6 days and to be approximately 6000 years old. Creation in so short a time was thought to have involved forces of tremendous violence, surpassing anything experienced in nature. This type of creation theory was

168

called *catastrophism.* Foremost among its proponents in the eighteenth century was Baron Georges Cuvier (1769–1832), a noted French naturalist. Cuvier, an able student of fossils, concluded that each fossil species was unique to a given sequence of rocks. He cited this discovery in support of the theory that each fossil species resulted from a special creation and was subsequently destroyed by a catastrophic event.

This theory was generally supported by scholars until 1785, when it was challenged by James Hutton (1762–1797). He saw evidence that Earth had evolved by uniform, gradual processes over an immense span of time, and he developed a concept that became known as the principle of uniformitarianism. According to this principle, past geologic events can be explained by natural processes we observe operating today, such as erosion by running water, volcanism, and gradual uplift of Earth's crust. Hutton assumed that these processes occurred in the distant past just as they occur now. He saw that in the vast abyss of time enormous work could be achieved by what appeared to be small and insignificant processes. Rivers could completely erase a mountain range. Volcanism and Earth movements could form new ones. On the basis of his observations of the rocks of Great Britain, he visualized "no vestige of a beginning—no prospect of an end." In a way, what Copernicus did for space, Hutton did for time. The universe does not revolve around Earth, and time is not measured by the life span of man. Before Hutton, human history was all of history. Since Hutton, we know that we are but a tiny pinpoint on an extraordinarily long time line.

Sir Charles Lyell (1797–1875) based his *Principles of Geology* (1830–1833) on Hutton's uniformitarianism. Lyell's book established uniformitarianism as the accepted method for interpreting the geologic and natural history of Earth. Charles Darwin (1809–1882) accepted Lyell's principles in formulating his theory of the origin of species and the descent of man. Lyell, however, also believed that the *rates* at which processes operate do not change with time. More recent studies indicate that changes in the rates of various processes may have occurred within certain limits.

Modern Views of Uniformitarianism (Actualism)

With the help of modern scientific instruments, geologists have studied much more of the geologic record than did Hutton and Lyell, and they have observed and measured many subtle details that earlier scientists could not measure. Modern science is making significant advances in understanding Earth, its long history, and how it was formed. By applying principles of thermodynamics, electromagnetism, chemistry, and related scientific disciplines, geologists are discovering more clues about Earth's genesis and evolution.

The assumptions of constancy in natural law are not unique to the interpretations of geologic history; they constitute the logical essentials in deciphering recorded history as well. We observe only the present and interpret past events on inferences based on present observations. We thus conclude that books or other records of history such as fragments of pottery, cuneiform tablets, flint tools, temples, and pyramids, which were in existence prior to our arrival, have all been the works of human beings, despite the fact that postulated past activities have been outside the domain of any possible present-day observations. Having excluded supernaturalism, we draw these conclusions because humankind is the only known agent capable of producing the effects observed. Similarly, in geology we conclude that ripple marks in a sandstone formation in the folded Appalachian Mountains were in fact formed by currents or wave action, or that coral shells found in limestones exposed in the high Rocky Mountains are indeed the skeletons of corals that lived in a now nonexistent sea.

Many features of rocks serve as records or documents of past events in Earth's history, and for those who listen, the rocks still echo the past. Igneous rocks are records of thermal events; the texture and composition of an igneous rock indicate if volcanic eruptions occurred or if the magma cooled beneath the surface (see Figure 5.2). Sedimentary rocks record changing environments on Earth's surface—the rise and fall of sea level, changes in climate, and changes in life forms. A layer of coal is a record of lush growth of vegetation, commonly in a swamp. Limestone composed of fossil shell debris indicates deposition in a shallow sea. Salt is precipitated from seawater or from saline lakes only in an arid climate, so a layer of salt carries specific climatic connotations. The list of examples could go on and on. For more than two centuries geologists have extracted from the rocks a remarkably consistent record of events in Earth's history: a record of time.

UNCONFORMITIES

Geologic time is continuous; it has no gaps. In any sequence of rocks, however, there are many major discontinuities (unconformities) that indicate significant interruptions in the rock-forming processes.

James Hutton was a very perceptive observer who clearly recognized the historical implications of the relationships between rock bodies. He not only recognized the vastness of time recorded in the rocks of Earth's crust, but he also recognized breaks or gaps in the record. In 1788, Hutton, together with Sir James Hall and John Playfair, visited Siccar Point in Berwickeshire, Scotland, and saw for the first time the Old Red Sandstone resting upon the upturned edges of the older strata (Figure 9.1). This exposure proved that the older rocks (primary strata) had been uplifted, deformed, and partly eroded away before the deposition of the "Secondary Strata." They soon discovered

FIGURE 9.1
Angular unconformity at Siccar Point, southeastern Scotland. It was here that the historical significance of an unconformity was first realized by James Hutton in 1788.

comparable relationships in other parts of Great Britain. This relationship between rock bodies became known as an *angular unconformity*.

To appreciate the significance of an angular unconformity, consider what the angular discordance implies by studying the sequence of diagrams in Figure 9.2. At least four major events are involved in the development of an angular unconformity: (1) an initial period of sedimentation during which the older strata are deposited in a near-horizontal position, (2) a subsequent period of deformation during which the first sedimentary sequence is folded, (3) development of an erosional surface on the folded sequence of rock, and (4) a period of renewed sedimentation and the development of a younger sequence of sedimentary rocks on the old erosional surface.

At first, only the most obvious stratigraphic breaks were recognized, but with more field observations other, more subtle, discontinuities were recognized. In Figure 9.3, for example, igneous and metamorphic rocks are overlain by flat-lying sedimentary strata. Clearly, the light-colored granite dike in the center of the photograph did not invade the overlying horizontal sandstone. Moreover, the basal layers of the sandstone contain pebbles of granite and coarse sand consisting of quartz and feldspar. These were derived from the weathering and disintegration of the granite and schist below, indicating a significant period of erosion. This relationship, in which plutonic igneous or metamorphic rocks are overlain by sedimentary shale, is called a *nonconformity*.

What are unconformities? Why are they significant in the study of geologic history?

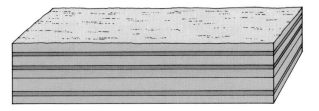

(A) Sedimentation: a sequence of rocks is deposited over time.

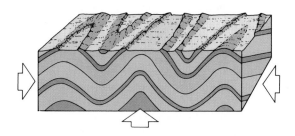

(B) Deformation: the sequence of rocks is deformed by mountain-building processes or by broad upwarps in Earth's crust, followed by erosion.

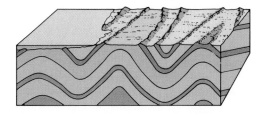

(C) Renewed sedimentation.

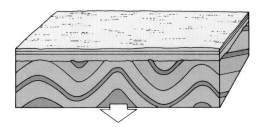

(D) Subsidence permits a new sequence of rocks to be deposited upon the eroded surface of the older deformed rocks.

FIGURE 9.2
The geologic events implied from an angular unconformity represent a sequence of major events in the geologic processes operating within the area.

FIGURE 9.3
A nonconformity is an unconformity in which sedimentary rocks were deposited on the eroded surfaces of metamorphic or intrusive igneous rocks. The metamorphic rocks and the igneous dikes shown in this photograph were formed at great depths in the crust. Subsequent uplift and erosion were necessary for them to be exposed at the surface. Younger sedimentary layers were then deposited on the eroded surface of the igneous and metamorphic terrain. The ancient erosional surface separating the horizontal sedimentary rocks from the underlying igneous and metamorphic rocks marks the nonconformity.

Why does sandstone deposited on a granite or a metamorphic rock indicate crustal mobility and a major discontinuity in the rock-forming processes? The answer becomes apparent if we consider the environment in which intrusive igneous and metamorphic rocks form. Both originate deep within Earth's crust: the granite cools slowly at great depth, and metamorphic rock recrystallizes at high temperatures and pressures far below the surface. Before these rocks can be weathered and eroded, the overlying cover of rock must be removed. Uplift must occur in order for these rocks to be uncovered and exposed; thus the deposition of sedimentary rocks on granitic or metamorphic rocks implies four major events: (1) the formation of an ancient sequence of rocks, (2) intrusion of granite and/or metamorphism, (3) uplift and erosion to remove the cover and expose the granites or metamorphic rocks at the surface, and (4) subsidence and deposition of younger sedimentary rocks on the eroded surface.

An example of another type of stratigraphic break is shown in Figure 9.4. Here the rock strata above and below the erosion surface are parallel. Erosion may strip off the top of the older sequence and may cut channels into the older beds, but there is no structural discordance between the older, eroded rock body and the younger, overlying rock. This type of discordance is referred to as a *disconformity.*

An unconformity is best seen in a vertical section exposed in a canyon wall, road cut, or quarry, where it appears as an irregular line. An unconformity is not a line, however, but a buried erosional surface. The present surface of Earth, especially the coastal plains, is an example of what an unconformable surface is like. Channels cut by streams are responsible for many irregularities, and resistant rocks protruding above the surrounding surface can cause local relief of several hundred meters.

It can be assumed that a soil profile develops over most ancient erosional surfaces, but much of the soil is eroded away as the area slowly subsides beneath the sea. Locally, however, soil profiles and minor irregularities in former landscapes are preserved in their entirety beneath the younger sediments. Erosional surfaces beneath volcanic ash or lava flows are commonly preserved in minute detail. A rare example of a buried surface preserving remarkable detail is the ancient town of Pompeii, Italy, which was covered by ash from Mt. Vesuvius in A.D. 79.

FIGURE 9.4
Disconformities do not show angular discordance, but an erosion surface separates the two rock bodies. The channel in the central part of this exposure reveals that the lower shale units were deposited and then eroded before the upper units were deposited.

CONCEPTS OF TIME

Time is measured by change. The most effective clocks for measuring long periods of time are the radioactive clocks.

We are all aware of growth and change in the physical and biological worlds. Were things unchanging and motionless, we would not be aware of time. Time is measured by change, and changes occur on many scales of time and space. Weather changes by the hour and day. Climate changes by decades and centuries. Mountains change in millions of years. There are many clocks or mechanisms for measuring time, the most fundamental of which use some periodic physical phenomenon—the swing of a pendulum, the flow of sand through an hourglass, the revolution of the Moon around Earth and Earth around the sun. For most practical and scientific purposes, Earth is our ultimate timepiece, for, as it revolves, distinct changes in the day, the season, and the year are observed and experienced. It is by these changes that we are aware of time. *Efforts to determine the age of Earth are basically efforts to determine how long Earth has been revolving around the sun.*

A clock is any mechanism that can be used for telling time. We are generally familiar only with mechanical clocks constructed to measure hours,

What is time? How is it measured?

minutes, and seconds, but a great variety of natural clocks provide as good or better measures of time than the standard "clock on the wall." Natural clocks, however, measure other intervals of time. Vibrations of atoms, for example, provide a means of measuring extremely small intervals of time with great precision. Electric waves provide another "clock" to measure intervals of time useful to us and our activities. Biological clocks measure intervals of time between various biological activities (breathing, heartbeat, hunger, menstruation, and life span or generation). Other natural clocks include layers of sediment deposited during specific seasons (a layer of sand in the spring and summer, a layer of mud in the winter when water freezes), tree rings, and growth marks on corals. For spans of time of a much greater interval, the thickness of sediment in the rock record is a measure of time much like the sand in an hourglass. The most effective clocks for measuring long periods of time, however, are the radioactive clocks, which we will discuss at the end of this chapter.

EARLY ESTIMATES OF THE AGE OF EARTH

Early attempts to estimate the age of Earth were based on (1) the salinity of the oceans, (2) the thickness of the total sequence of sedimentary rocks, and (3) heat loss from Earth. Each attempt showed evidence of a considerable period of time, but we now know that these early estimates were all too low.

How was geologic time first measured?

The first serious efforts to estimate the magnitude of geologic time were made in the late nineteenth century. Before Hutton and Lyell, few people even recognized geologic processes or thought about the age of Earth. After Hutton presented his arguments for uniformitarianism and Lyell further developed the concept, much interest was generated in the magnitude of geologic time, and scientists explored a number of ways to estimate the age of Earth.

Estimates Based on Salinity

In 1899, John Joly concluded from calculations based on the salinity of the oceans that Earth was 90 to 100 million years old. He assumed that the original ocean water was fresh and that the present salinity was a result of rivers bringing salt to the sea, where it was concentrated by evaporation. From a chemical analysis of river water, Joly simply estimated the amount of sodium delivered to the sea annually. This estimate, divided into the total amount of salt in the ocean, gave, he thought, the age of the ocean.

Joly's estimates were far too low because he failed to take into account the fact that salt is removed from the sea and deposited as part of marine sediment. Later, the salt may be exposed by uplift, eroded, and recycled back to the sea.

Estimates Based on Thickness of Sediment

Soon after sedimentary rocks were recognized as a record of erosion, transportation, and sedimentation, geologists reasoned that the time represented by the sequence of sedimentary rocks in Earth's crust could be estimated if the average rate of sedimentation in modern seas could be determined and the total thickness of ancient sedimentary rocks measured. Early measurements of the maximum thickness of the sediment preserved in the geologic record range from 25,000 to 112,000 m. With later mapping, new rock units have been discovered, and the thickness of fossiliferous rocks is now considered to be at least 150,000 m. Rates of sedimentation differ from area to area, but most estimates place the average rate of sedimentation at about 0.3 m per 1000 years. At this rate, the age of the first abundantly fossiliferous rocks is approximately 500 million years.

Two problems make this method of determining geologic time inaccurate but not completely useless. First, accurate estimates of the average rate of sedimentation are difficult to obtain because different kinds of sediments accumulate at vastly different rates. Second, many interruptions occur in the sequence of sedimentary rocks. In a given area, sedimentation will be followed by uplift and erosion, subsidence, and then renewed sedimentation. An unknown amount of time is not recorded by uplift and erosion, and during those processes, part of a previously formed record is removed.

Estimates Based on Heat Loss

Lord William Kelvin reasoned that Earth cooled from a molten state and that the entire rate of cooling could be determined by measuring the present rate of heat flow. Estimates of Earth's age based on this method ranged from 20 to 40 million years.

Kelvin's estimates were low because Earth generates heat by radioactive decay. It thus does not necessarily lose heat at a constant rate.

In summary, these and other early attempts to measure geologic time were remarkable first efforts. Each method showed that Earth was far older than anyone had supposed, but the true dimensions of time remained elusive.

> It has not been easy for man to face time. Some, in recoiling from the fearsome prospects of time's abyss, have toppled backwards into the abyss of ignorance. (Claude C. Albritton, *The Abyss of Time.*)

RELATIVE DATING

Relative dating determines the chronologic order of a sequence of events. The most important methods of relative dating are (a) superposition, (b) faunal succession, (c) crosscutting relations, and (d) inclusions.

An expanded conception of time is perhaps the main contribution of geology to the history of thought. Geologists, however, are concerned not so much with the philosophical problem of defining time, or analyzing the concept of time, as with measuring time in relation to events in the history of Earth. Two different concepts of time, and hence two different but complementary methods of dating, are used in geology: relative time and absolute time.

Relative dating is simply determining the chronologic order of a sequence of events. Geologic and recorded history are both organized in units of relative time. We speak of human events as occurring in eras B.C. or A.D., which constitute a broad division of historic time. Historians commonly use relative time when they place events in relation to ancient dynasties, reigns of kings, or major events such as wars. Geologists do the same with geologic time. Earth's history is divided into eras, periods, and ages by markers such as the existence of certain forms of life (for example, "age of dinosaurs") or major physical events (for example, Appalachian mountain building), even though the actual dates of the events may be unknown. *Relative dating* implies that no quantitative or absolute length of time in days or years is deduced. An event can only be inferred to have occurred earlier or later than another.

In studying Earth, relative dating is important because many physical events such as volcanism, canyon cutting, deposition of sediment, or upwarping of the crust can be identified and their relative ages determined. To estab-

How is geologic time currently measured?

lish the relative ages of these events is to determine their proper chronologic order. This can be done by applying several principles of remarkable simplicity and universality. The most significant of these are

1. The principle of superposition
2. The principle of faunal succession
3. The principle of crosscutting relations
4. The principle of inclusion

The Principle of Superposition

The *principle of superposition* is the most basic guide in the relative dating of rock bodies. It states that in a sequence of undeformed sedimentary rock, the oldest beds are on the bottom and the higher layers are successively younger. The *relative ages* of rocks in a sequence of sedimentary beds can thus be determined from the order in which they were deposited.

In applying the principle of superposition, we make two assumptions: (1) layers were essentially horizontal when they were deposited, and (2) the rocks have not been so severely deformed that the beds are overturned. (Rock sequences that have been overturned are generally easy to recognize by their sedimentary structures, such as cross-bedding, ripple marks, and mud cracks.)

The Principle of Faunal Succession

The *principle of faunal succession* states that groups of fossil animals and plants occur in the geologic record in a definite and determinable order and that a period of geologic time can be recognized by its characteristic fossils. Thus, in addition to superposition, the sequence of sedimentary rocks in Earth's crust is characterized by another independent element that can be used to establish the chronologic order of events.

Fossils are the actual remains of ancient organisms, such as bones and shells, or the evidence of their presence, such as trails and tracks. Their abundance and diversity are truly amazing. Some rocks (such as coal, chalk, and certain limestones) are composed almost entirely of fossils, and others contain literally millions of specimens. Invertebrate marine forms are most common, but even large vertebrate fossils of mammals and reptiles are plentiful in many formations. For example, it is estimated that over 50,000 fossil mammoths have been discovered in Siberia, and many more remain buried.

Even before Darwin developed the theory of natural selection, the principle of faunal succession was recognized by William Smith (1769–1839), a British surveyor. Smith worked throughout much of southern England and carefully studied the fresh exposures of rocks in quarries, road cuts, and excavations. In a succession of interbedded sandstone and shale formations, he noted that the several shales were very much alike, but the fossils they contained were not. Each shale had its own particular groups of fossils. By correlating types of fossils with rock sequences, Smith developed a practical tool that enabled him to predict the location and properties of rocks beneath the surface.

Soon after Smith announced that the fossil assemblages of England change systematically from the older beds to the younger, other investigators discovered the same to be true throughout the world.

How are fossils used as geologic clocks?

Fossils provide geologists with a means of establishing relative dates, in much the same way that archeologists use artifacts. Both show evolution and change with time. For example, in a city dump where refuse is buried in succession, we could recognize a period of time prior to the automobile by the remains of wagon wheels, saddles, and similar equipment. A layer containing abundant scraps of Model T Fords would be recognized as being older than a layer containing remains of the Model A, and layers containing new models such as the Porsche would be recognized as being younger, even though they might not rest on layers containing any of the older materials.

Today, the principle of faunal succession has been confirmed beyond doubt. It has been used extensively to locate valuable natural resources, such as petroleum and mineral deposits. It is also the foundation for the standard geologic column, which divides geologic time into eras, periods, epochs, and ages (see Figure 9.8).

The Principle of Crosscutting Relations

The relative age of certain events is also shown by the ***principle of crosscutting relations,*** which states that igneous intrusions and faults are younger than the rocks they cut (Figure 9.5). Crosscutting relations can be complex, however, and careful observation may be required to establish the correct sequence of events. The scale of crosscutting features is highly variable, ranging from large faults with displacements of hundreds of kilometers to small fractures less than a millimeter long.

The Principle of Inclusion

The ***principle of inclusion*** states that a fragment of a rock incorporated or included in another is older than the host rock. The relative age of intrusive igneous rocks (with respect to the surrounding rock) therefore is commonly apparent if inclusions, or fragments, of surrounding rocks are included in the intrusion (Figure 9.6). As a magma moves upward through the crust, it dislodges and engulfs large fragments of the surrounding material, which remain as unmelted foreign inclusions.

The principle of inclusion can also be applied to conglomerates in which relatively large pebbles and boulders eroded from preexisting rocks have been transported and deposited in a new formation. The conglomerate is obviously younger than the formations from which the pebbles and cobbles were derived. In areas where superposition or other methods do not indicate relative ages, a limit to the age of a conglomerate can be determined from the rock formation represented in its pebbles and cobbles.

FIGURE 9.5
Crosscutting relationships clearly indicate the relative ages of rock bodies and geologic structure. In this photograph several generations of dikes cut across the green metamorphic rock. The thick dike is the youngest because it cuts across all other rock bodies.

FIGURE 9.6
Inclusions of one rock in another provide a means of determining relative age. In this example fragments of granite are included in the basalt, clearly indicating that the granite is the older.

Succession in Landscape Development

Surface features of Earth's crust are continually being modified by erosion and commonly show the effects of successive events through time. Many landforms evolve through a definite series of stages, so the relative age of a feature can be determined from the degree of erosion. This is especially obvious in volcanic features such as cinder cones and lava flows. These features are created during a period of volcanic activity and then subjected to the forces of erosion until they are completely destroyed or buried by erosional debris.

The composite diagram in Figure 9.7 shows several kinds of crosscutting relationships as well as unconformities and superposition of major rock bodies. Although this diagram covers a large area, the relationships between rock bodies are in canyons, valley walls, and on the plateau surface. The major rock bodies, faults, and unconformities are labeled by letters arranged in alphabetical order from oldest (A) to youngest (N). In Table 9.1 the same events are listed in sequence from the youngest (N, top) to the oldest (A, bottom).

The oldest rocks in the diagram are the metamorphic rocks, A; the granite, B, intrudes these rocks and is younger; but the granite is not in contact with the tilted strata, D, so their age relationship is not certain. An erosional surface, C, developed on the metamorphic terrain, and then a sequence of sedimentary rocks, D, were deposited. These rocks were then intruded by dikes and sills, E. In order to establish the position of the granite, B, more accurately in the sequence, we would need some absolute dates based on radiometric age determination for units B and E. Faults, F, displaced the sequence D. Widespread erosion then occurred, developing the unconformity, G, which cuts

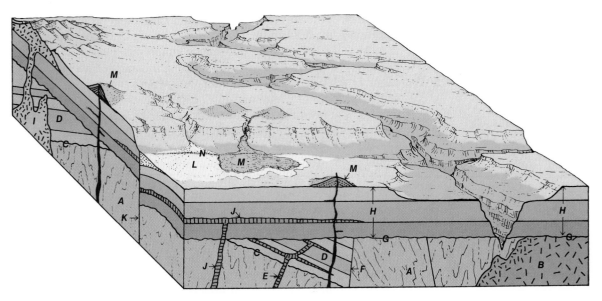

FIGURE 9.7
The sequence of geologic events can be determined by using the principles of superposition and crosscutting relations.

across all of the units A–F. The sequence of horizontal rocks, H, was then deposited. Two igneous intrusions, I and J, occurred. Intrusion I formed a laccolith, whereas J formed a dike and sill. Since we cannot tell the relative age relationship of these intrusions, we would have to obtain radiometric dates to place these events more accurately in the sequence. We do know, from crosscutting relationships, that intrusion J is older than the fault, K, and the volcanic rocks, M.

The next group of events—erosion, volcanic eruptions, and sedimentation—have surface expressions. Judging from the deep erosion of the canyon, it appears to have been initiated relatively early. The surface upon which alluvial fans, volcanic cones, and lava flows were formed is related to the erosion by the major river. Lava flow, M, is younger than the alluvial fan, L. Both are cut by recurrent movement on fault K. Note the amount of displacement along the fault of the sedimentary rocks, H, and the small amount of displacement of the fan, L, and lava flow, M. Judging from the lack of erosion on the volcanoes, it would appear that the cones are very young features.

This example is not purely hypothetical. It is a composite, idealized diagram of the Grand Canyon area, Arizona. Most of these relationships shown on the left face of the block diagram are exposed in the eastern Grand Canyon (see Figure 6.1), whereas the major fault and volcanic features are exposed in the western Grand Canyon.

TABLE 9.1
Events shown in Figure 9.7

Youngest (top) to Oldest (bottom)	
N	Recurrent movement on faults
M	Volcanism
L	Fans
K	Major faulting
J	Basaltic dikes and sills
I	Laccolithic intrusion
H	Horizontal strata
G	Unconformity
F	Fault displacing tilted strata
E	Injection into tilted strata
D	Tilted shale
C	Unconformity
B	Granite
A	Metamorphic rocks

THE STANDARD GEOLOGIC COLUMN

Using the principles of superposition and faunal succession, geologists have determined the chronological sequence of rocks throughout broad regions of every continent and have constructed a standard geologic time scale that serves as a calendar for the history of Earth.

Using the principles of superposition and faunal succession, earth scientists began to accumulate a considerable amount of knowledge concerning the strata of rock in various parts of the world. Correlation by means of fossils

How was the geologic time scale developed?

made it possible to construct a diagram called the geologic column, which showed in simple form the major units of strata in Earth's crust.

Most of the original geologic column was pieced together from sequences of strata studied in Europe during the mid-nineteenth century. Major units of rock (such as the Cambrian, Ordovician, and Silurian) generally were named after geographic areas in Europe where they are well exposed. The rock units are distinguished from each other by major changes in rock type, unconformities, or abrupt vertical changes in the fossil groups they contain. The original subdivision of the geologic column was based simply on the sequence of rock formations in their superposed order as they are found in Europe. Rocks in other areas of the world that contain the same fossil assemblages as a given part of the European succession are considered to be of the same age and commonly are referred to by the same names.

The standard *geologic column* is shown in Figure 9.8. The nomenclature of the column provides names for eras, periods, and ages in Earth's history. These names mark the highlights in the historical development of geological science over the past 200 years. Nearly every name is associated with an important scientific discovery. An understanding of the origin and meaning of these names is therefore helpful.

The Precambrian

Precambrian time is represented by a group of highly complex metamorphic and igneous rocks, which form a large volume of the continental crust. To produce these rocks, great thicknesses of sedimentary and volcanic rocks were intensely folded and faulted and were intruded with granitic rock. Precambrian rocks contain only a very few fossils of the more primitive forms of life. Arrangement of individual rock layers in their proper detailed stratigraphic sequence is therefore difficult if not impossible in this group of rocks. The structure is too complex.

The Paleozoic Era

Rocks younger than the Precambrian are much less complex and contain great numbers of fossils, permitting geologists to identify them worldwide. The term *Paleozoic* means "ancient life." *Paleozoic* rocks contain numerous fossils of marine organisms, primitive fish, and amphibians. The era is subdivided into periods distinguished largely according to the rock formations of Great Britain.

Cambrian comes from *Cambria,* the Latin name for Wales, where these rocks were first studied. In most areas of the world, Cambrian rocks rest on the highly deformed Precambrian metamorphic complex.

Ordovician is derived from the name of an ancient Welsh tribe, the Ordovices. Ordovician strata overlie the Cambrian but differ in the types of fossils they contain.

Silurian designates rocks above the Ordovician, which are exposed on the border of Wales, a territory originally inhabited by a British tribe, the Silures.

Devonian was first used to refer to rocks exposed in Devonshire, England.

Carboniferous is the name of a sequence of coal-bearing formations that lie above the Devonian rocks. These formations were first studied in England. In the United States, Carboniferous rocks are subdivided into two major units: the *Pennsylvanian* (named after the state of Pennsylvania) and the *Mississippian* (named after the upper Mississippi valley).

Permian was introduced to refer to rocks exposed over much of the province of Perm, Russia, just west of the Ural Mountains. Corresponding rocks in England lie above the Carboniferous.

EON	ERA	Duration in millions of years	Millions of years ago
PHANEROZOIC	CENOZOIC	66	66
	MESOZOIC	179	
	PALEOZOIC	325	245
			570
PRECAMBRIAN — PROTEROZOIC	LATE	330	
			900
	MIDDLE	700	
			1600
	EARLY	900	
			2500
ARCHEAN	LATE	500	
			3000
	MIDDLE	400	
			3400
	EARLY		
			4600

Era	Period	Epoch	Duration in millions of years	Millions of years ago
CENOZOIC	Quaternary	Pleistocene	1.6	1.6
	Neogene	Pliocene	3.7	5.3
		Miocene	18.4	23.7
	Tertiary — Paleogene	Oligocene	12.9	36.6
		Eocene	21.2	57.8
		Paleocene	8.6	66.4
MESOZOIC	Cretaceous		78	
				144
	Jurassic		64	
				208
	Triassic		37	
				245
PALEOZOIC	Permian		41	
				286
	Carboniferous	Pennsylvanian	34	
				320
		Mississippian	40	
				360
	Devonian		48	
				408
	Silurian		30	
				438
	Ordovician		67	
				505
	Cambrian		65	
				570
PRECAMBRIAN				

FIGURE 9.8

The standard geologic column was developed in Europe during the mid-nineteenth century on the basis of the principles of superposition and faunal succession. Later, radiometric dates provided a scale of absolute time for the standard geologic periods.

The Mesozoic Era

Mesozoic means "middle life." The term is used for this period of geologic time because the presence of fossil reptiles and a significant number of more modern fossil invertebrates dominates these rocks. The **Mesozoic** Era includes three periods: Triassic, Jurassic, and Cretaceous.

Triassic refers not to a geographic location but to the striking threefold division of the rocks overlying the Paleozoic in Germany.

Jurassic was first introduced for strata outcropping in the Jura Mountains.

Cretaceous refers to the chalk formations in France and England. The name is derived from the Latin *creta,* "chalk."

The Cenozoic Era

Cenozoic means "recent life." Fossils in these rocks include many types closely related to modern forms, including mammals, modern plants, and invertebrates. The **Cenozoic** Era has two periods: the Tertiary and the Quaternary.

Tertiary is a term held over from the first attempts to subdivide the geologic record into three divisions—Primary, Secondary, and Tertiary. The companion divisions, Primary and Secondary, have been replaced by Precambrian, Paleozoic, and Mesozoic.

Quaternary is the name proposed for very recent deposits, which contain fossils of species with living representatives.

The geologic column by itself indicates only the relative ages of the major periods in Earth's history. It tells us nothing about the specific duration of time represented by a period. With the discovery of radioactive decay of uranium and other elements, a new tool for measuring geologic time became available. This greatly enhanced our understanding of time and the history of Earth and provided benchmarks of absolute time for the standard geologic column.

RADIOMETRIC MEASUREMENTS OF ABSOLUTE TIME

> *Radiometric dating provides a method for measuring geologic time directly in terms of a specific number of years (absolute dating). It has been used extensively during the last 50 years to provide an absolute time scale for the events in Earth's history.*

How do we measure the magnitude of geologic time?

Unlike relative time, which specifies only the chronologic relationships among events, absolute time, or finite time, designates specific durations measured in units of hours, days, or years. Time can be measured by any regularly recurring event, such as the swing of a pendulum or the rotation of Earth, but throughout most of the nineteenth century there was no method of measuring long periods of time and there seemed to be little hope of finding the secret of Earth's age. Then a major breakthrough occurred when Henri Becquerel (1852–1908), a French physicist, discovered natural **radioactivity** in 1896 and opened new vistas in every field of science. Among the first to experiment with radioactive substances was the distinguished British physicist Lord Rutherford (1871–1937). After defining the structure of the atom, Rutherford made the first clear suggestion that radioactive decay could be used to date geologic events in **absolute time.**

Radioactive isotopes are unstable: their nuclei spontaneously disintegrate, transforming them into completely different atoms. In the process, radiation is given off and heat is liberated. Initially, scientists assumed that each radioactive substance disintegrates at its own rate and that for many substances the rate is extremely slow. This assumption has been proved by experiment.

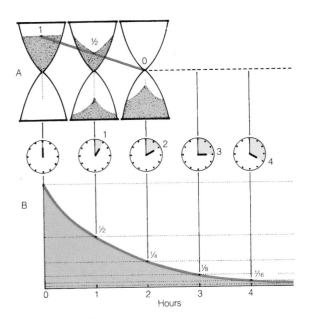

FIGURE 9.9
Rates of depletion can be linear or exponential. (A) Most processes are characterized by uniform, straight-line depletion, like sand moving through an hourglass. If half of the sand is gone in 1 hour, all of it will be gone in 2 hours. (B) Radioactive decay, in contrast, is exponential. If half is depleted in 1 hour, half of the remainder, or one-fourth, will be depleted in 2 hours, leaving one-fourth. Rates of radioactive decay are expressed in half-lives, the time required for half of the remaining amount to be depleted. In this case, the half-life is 1 hour.

The rate of radioactive decay is defined in terms of *half-life,* the time it takes for half of the nuclei in the sample to decay. In one half-life, half of the original atoms decay. In a second half-life, half of the remainder (or a quarter of the original atoms) decay. In a third half-life, half of the remaining quarter decay, and so on (see Figure 9.9). The time elapsed since the formation of a crystal containing a radioactive element can be calculated from the rate at which that particular element decays. The amount of the radioactive element remaining in the crystal (*parent isotope*) is simply compared with the amount of the disintegration product (*daughter isotope*).

There are numerous radioactive isotopes. Most decay rapidly; that is, they have short half-lives and lose their radioactivity within a few days or years. Some decay very slowly, however, with half-lives of hundreds of millions of years. These can be used as atomic clocks for measuring long periods of time. The parent isotopes that are most useful for geologic dating and their daughter products are listed in Table 9.2.

The theory of *radiometric dating* is simple enough, but the laboratory procedures are complex. The principal difficulty lies in the precise measurement of minute amounts of isotopes. The accuracy of the method also depends on the accuracy with which the half-life is known. (Measurements of the decay of uranium-235 to lead-207 are considered accurate within 2%.)

At present, the potassium–argon method of radiometric dating is of great importance. It can be used on the micas and amphiboles, which are widely distributed in igneous rocks, and can be used on rocks as young as a few thousand years or on older rocks.

Another important radioactive clock uses the decay of *carbon-14* (^{14}C), or *radiocarbon,* which has a half-life of 5730 years. Carbon-14 is produced continually in Earth's atmosphere as a result of the bombardment of nitrogen-14 (^{14}N) by cosmic rays. The newly formed radioactive carbon becomes mixed with ordinary carbon atoms in carbon dioxide gas (CO_2). The formation of radiocarbon is in balance with its decay, so the proportion of ^{14}C in atmospheric CO_2 is essentially constant.

Plants use carbon dioxide in photosynthesis, and animals eat plants. Both thus maintain a fixed proportion of ^{14}C while they are alive. After death, however, no additional ^{14}C can replenish what is lost by radioactive decay. The ^{14}C steadily reverts to ^{14}N. The time elapsed since an organism died can therefore be determined by measuring the remaining proportion of ^{14}C. The longer the time elapsed since death, the less ^{14}C remains. Because the isotope's half-life is 5730 years, the amount of ^{14}C remaining in organic matter older than 50,000 years is too small to be measured accurately. The radiocarbon method

TABLE 9.2
Radioactive Isotopes Used in Determining Geologic Time

Parent Isotope	Daughter Isotope	Half-life (billion years)
Uranium-238	Lead-206	4.5
Uranium-235	Lead-207	0.713
Thorium-232	Lead-208	13.9
Rubidium-87	Strontium-87	50.0
Potassium-40	Argon-40	1.5

is therefore useful for dating very young geologic events involving organic matter and for dating archeologic material.

To safeguard against errors, radiometric dating is subjected to constant checks. A number of radioactive isotopes are suitable for absolute dating, so one obvious test is to date a mineral or rock by more than one method. If the results agree, the probability is high that the date is reliable. If the results differ significantly, additional methods must be used to determine which date, if either, is correct. Another independent check can be made by comparing the absolute date, determined radiometrically, with the relative age of the rocks, determined from such evidence as superposition and fossils. Through this system of tests and cross-checks, many very reliable radiometric dates have been determined.

THE RADIOMETRIC TIME SCALE

Absolute dates of numerous geologic events have been determined from thousands of specimens throughout the world. These, in combination with the standard geologic column, provide a radiometric time scale from which the absolute age of other rock units and geologic events can be estimated.

Unfortunately, radiometric dates cannot be used to determine the age of every rock. Many sedimentary rocks do not contain minerals suitable for accurate radiometric dating. In others, a date indicates when a mineral within the rock formed—not the time when the sediment was deposited. Radiometric dating can be used effectively to determine when igneous rocks crystallized or when heat and pressure developed new minerals in metamorphic rocks because, in these cases, the mineral and the rock formed together. The problem in developing a reliable radiometric time scale is to accurately place the radiometric dates of igneous and metamorphic rocks in their proper positions in the relative time scale established for the relative dating of sedimentary rocks. Layers of volcanic rocks and intrusions are the most suitable time markers in the standard geologic column. These benchmarks, accurately placed in the geologic column, constitute the basis for the radiometric time scale.

Layered Volcanics

Why are radiometric dates of volcanic rocks more useful in establishing an absolute geologic time scale than radiometric dates of granite?

The best reference points for the radiometric time scale are probably volcanic ash falls and lava flows. Both are deposited instantaneously, as far as geologic time is concerned. Because they commonly are interbedded with fossiliferous sediments, their exact positions in the geologic column can be determined. They thus provide the main basis for establishing the absolute time scale within the geologic column.

Bracketed Intrusions

As is shown in Figure 9.10, molten rock can cool within Earth's crust without ever breaking out to the surface. Subsequent erosion may expose this rock at the surface. Later, younger sediments may be deposited on top. In some cases, the entire sequence of events takes only a few million years. In others, a much longer time may pass. The relative age of the igneous rock is *bracketed* between the age of the older sediments (1) and the younger sediments (4). Such a rock body is therefore known as a **bracketed intrusion**. Unfortunately, the span of time between 1 and 4 is commonly too long to permit the relative age of the intrusion to be useful in detailed geochronology. Radiometric dating of such rocks does establish the time of major igneous events, however.

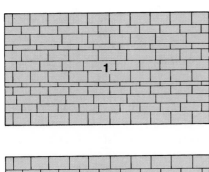

(A) A sequence of sedimentary rocks (1) is deposited.

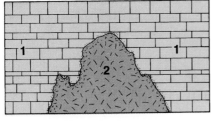

(B) The sedimentary rocks are subsequently intruded by an igneous body (2).

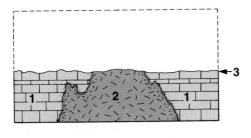

(C) Erosion (3) removes part of the sequence (1 and 2).

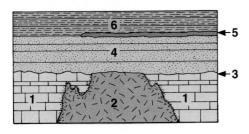

(D) Subsequent deposition of sediment (4) occurs. An extrusion of lava (5) occurs during the period of sedimentation. The lava flow is covered by deposition of younger sedimentary rocks (6).

FIGURE 9.10
Radiometric dating of igneous rocks can be used in developing an absolute time scale, but the relationship between rock bodies must be considered. In these diagrams, major geologic events are numbered.

Radiometric dating of the lava flows (5) would provide an absolute age for rocks in that position in the geologic sequence because the flows occurred as part of the normal sequence of rocks. The intrusive body (2) is more difficult to place in the column. We know only that it formed sometime after the deposition of the oldest sedimentary rocks (1), but before the episode of erosion (3) and the deposition of the next younger sediments (4).

The currently accepted *geologic time scale* is based on the standard geologic column, established by faunal succession and superposition, plus the finite radiometric dates of rocks that can be placed precisely in the column. Each dating system provides a cross-check on the other because one is based on relative time and the other on absolute time. Agreement between the two systems is remarkable, and discrepancies are few. In a sense, the radiometric dates act as the scale on a ruler, providing reference markers between which interpolation can be made. Enough dates have been established so that the time span of each geologic period can be estimated with considerable confidence. The age of a rock can be determined by finding its location in the geologic column and interpolating between the nearest radiometric time marks.

How do we know Earth is 4.5 billion years old?

From this radiometric time scale, we can make several general conclusions about the history of Earth and geologic time.

1. Present evidence indicates that the age of Earth is about 4.5 to 4.6 billion years.
2. The Precambrian constitutes more than 80% of geologic time.
3. Phanerozoic time (the Paleozoic and later) began about 570 million years ago. Rocks deposited since Precambrian time can be correlated worldwide by means of fossils, and the dates of many important events during their formation can be determined from radiometric dating.
4. Some major events in Earth's history are difficult to place in their relative positions on the geologic column but can be dated by radiometric methods.

MAGNITUDE OF GEOLOGIC TIME

The magnitude of geologic time is easier to comprehend when compared to some tangible linear time line.

Great time spans are difficult for most people to comprehend. The norms established through sensory experience are short intervals, such as the day, the week, and the changing seasons. Students of geology must continually attempt to enlarge their temporal norms to encompass the magnitude of geologic time. Without an expanded conception of time, extremely slow geologic processes, considered only in terms of human experience, have little meaning.

To appreciate the magnitude of geologic time, we will abandon large numbers for a moment and refer instead to something tangible and familiar. In Figure 9.11, the length of a football field represents the lapse of time from the beginning of Earth's history to the present. An absolute time scale and the standard geologic periods are shown on the left. Precambrian time constitutes the greatest portion of Earth's history (87 yards). The Paleozoic and later periods are equivalent to only the last 13 yards. To show events with which most people are familiar, the upper end of the scale must be enlarged. The first abundant fossils occur at the 13-yard line. The great coal swamps are at about the 6-yard line. The dinosaurs became extinct about a yard from the goal line, and the last ice age occurred an inch from the goal line. Recorded history corresponds to less than the width of a blade of grass.

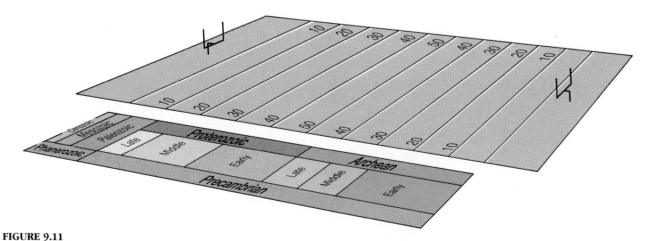

FIGURE 9.11
If the length of geologic time is compared to a football field, Precambrian time represents the first 87 yards, and all events since the beginning of the Paleozoic are compressed into the last 13 yards. Dinosaurs first appeared 5 yards from the goal line. The glacial epoch occurred in the last inch, and historic time is so short that it cannot be represented, even on the enlarged part of the figure.

SUMMARY

1. The basic assumption endorsed by essentially all geologists today in studying and interpreting the history of Earth is the principle of uniformitarianism, which states that natural laws do not change with time. This principle is not unique to geology. It is a fundamental law in all fields of science.

2. Relative dating determines the chronologic order of a sequence of events. The most important methods of relative dating are (a) superposition, (b) faunal succession, (c) crosscutting relations, and (d) inclusions.

3. The standard geologic column was established during the early and middle 1800s by means of the principles of relative dating.

4. Radiometric dating provides a method for measuring geologic time directly in terms of a specific number of years (absolute dating). It has been used extensively during the last 50 years to provide an absolute time scale for the events in Earth history.

KEY TERMS

absolute time (p. 182)

angular unconformity (p. 171)

bracketed intrusion (p. 184)

carbon-14 (p. 183)

catastrophism (p. 169)

Cenozoic (p. 182)

crosscutting relations, principle of (p. 177)

daughter isotope (p. 183)

disconformity (p. 172)

faunal succession, principle of (p. 176)

geologic column (p. 180)

geologic time scale (p. 185)

half-life (p. 183)

inclusion, principle of (p. 177)

Mesozoic (p. 182)

nonconformity (p. 171)

Paleozoic (p. 180)

parent isotope (p. 183)

Precambrian (p. 180)

radioactivity (p. 182)

radiocarbon (p. 183)

radiometric dating (p. 183)

relative age (p. 176)

relative dating (p. 175)

superposition, principle of (p. 176)

uniformitarianism (p. 168)

REVIEW QUESTIONS

1. Explain the modern concept of uniformitarianism.
2. Explain the concept of relative dating.
3. Explain how the following principles are used in determining the relative age of rock bodies: (a) superposition, (b) faunal succession, (c) crosscutting relations, (d) inclusions.
4. Discuss the sequence of events illustrated in Figure 9.2.
5. What is the standard geologic column? How did it originate?
6. Explain the meaning of half-life in radioactive decay.
7. How is the absolute age of a rock determined?
8. How is the half-life of a radioactive isotope used to determine the radiometric age of a rock?
9. Why are most rocks dated with respect to their position in the standard geologic column rather than assigned a definite numerical age, even though accurate methods of radiometric dating are well established?

ADDITIONAL READINGS

Albritton, C. C. 1980. *The Abyss of Time*. San Francisco: Freeman.

Berry, W. B. N. 1968. *Growth of Prehistoric Time Scale*. San Francisco: Freeman.

Block, J. 1976. The Bible and science on creation. *Journal of Geological Education* 24(2):58–60.

Eicher, D. L. 1968. *Geologic Time*. Englewood Cliffs, NJ: Prentice-Hall.

Faul, H. 1966. *Ages of Rocks, Planets, and Stars*. New York: McGraw-Hill.

Hallam, A. 1982. *Great Geologic Controversies*. New York: Oxford University Press.

Harbough, J. W. 1968. *Stratigraphy and Geologic Time*. Dubuque, IA: Brown.

Hume, J. D. 1978. An understanding of geologic time. *Journal of Geological Education* 26(4):141–143.

Newman, W. L. 1978. *Geologic Time*. Geological Survey Information Booklet No. 0-261-226 (9).

Toulmin, S., and J. Goodfield. 1965. *The Discovery of Time*. New York: Harper & Row.

10
Weathering

A new building gradually deteriorates. The paint chips and peels, wood dries and splits, and even bricks, building stone, and cement eventually decay and crumble. Left alone, most buildings decompose into a pile of rubble within a few hundred years. This process of natural decay is called weathering. Weathering is a general term describing all of the changes that result from the exposure of rock materials to the atmosphere.

The effects of weathering can be seen everywhere. Soil is the product of weathering, and for life on Earth it is certainly the most important. This image of part of the Snake River Plain in southern Idaho vividly shows how weathering modifies a rock body. The smooth flat surface that extends diagonally across the area was formed by floods of basalt that were extruded intermittently during the last several million years. The youngest extrusions are fresh and black and retain the original surface features of the flows. The large area of youngest flows near the left-central part of the image is Craters of the Moon National Monument. Older flows have been subjected to longer periods of weathering and have developed a thin soil that supports a sparse vegetation. These flows are bluish gray, not black, and form several irregular patches. The oldest flows from the broad, smooth plain appear as light bluish gray. Weathering has completely decomposed the surfaces of these flows and has formed a thicker soil cover. All original flow features are obliterated. Near the Snake River, along the southern margin of the plain, irrigation is possible and farmlands (here shown in rectangular patterns of red) cover most of the surface.

From a geological point of view, weathering is important because it transforms the solid bedrock into small, decomposed fragments and prepares those fragments for removal by the agents of erosion. In addition, the products of weathering form a blanket of soil over the solid bedrock, and soil is the basis for most terrestrial life. Weathering should therefore be considered a part of the geologic system and one that has tremendous ecological significance.

1. The major types of weathering are mechanical disintegration and chemical decomposition.
2. Ice wedging is the most important form of mechanical weathering.
3. The major types of chemical weathering are oxidation, dissolution, and hydrolysis.
4. Joints facilitate weathering because they permit water and gases in the atmosphere to attack a rock body at considerable depth. They also greatly increase the surface area on which chemical reactions can occur.
5. The major products of weathering are a blanket of soil (regolith) and spheroidal rock forms.
6. Climate greatly influences the type and rate of weathering. The major controlling climatic factors are precipitation and temperature.

To appreciate how geologic processes erode the surface of Earth and how the landscape evolves, it is first necessary to understand the nature of weathering—the disintegration and decomposition of rocks. Weathering involves a multitude of physical, chemical, and biological processes. By definition, weathering is different from erosion. Weathering involves only the breakdown of rock, whereas erosion involves the removal of debris produced by the breakdown. But in reality weathering and erosion are intimately involved with one another. Weathering disintegrates solid rock and produces loose debris, and the results of weathering are seen everywhere (Figure 10.1). Erosion removes the debris and exposes fresh rock, which is then weathered, and the cycle continues.

Two main types of weathering are recognized: (1) mechanical weathering and (2) chemical weathering. Mechanical weathering breaks the rock mass into small particles. It is strictly a physical process involving no change in chemical composition. Chemical weathering alters the rock by chemical reactions between elements in the atmosphere and those in the rocks. Most geologists believe that chemical weathering is most important in terms of total amount of rock breakdown, but in most places the two processes work together, each facilitating the other, so the final product results from a combination of the two processes.

It would be difficult to overemphasize the importance of weathering to humans. It is a critical base to our ecology, and our very existence depends on it. Without weathering, Earth would be forbidding indeed. The continents would be bare hard rock, for no soil cover could develop; consequently Earth would be devoid of plant and animal life. In addition to producing soil, upon which agriculture depends, weathering produces some other very practical products. Sand, gravel, and clay deposits, which we use so much in our modern culture, are the indirect results of weathering. Practically all aluminum ore, most iron ore, and some copper ore are formed and concentrated by weathering.

MECHANICAL WEATHERING

Mechanical weathering is the breakdown of rock by physical processes and involves no change in chemical composition. The most important types of mechanical weathering are ice wedging and sheeting, or unloading.

Mechanical weathering is strictly a physical process, involving no change in chemical composition of the rock. No chemical elements are added to, or

(A) Weathering is especially obvious on the old monuments in Europe. Here most of the details on the gargoyles of Notre Dame in Paris have been erased by weathering.

(B) Decomposition of solid rock is clearly seen in many old headstones in graveyards.

(C) Weathering is apparent from the fallen debris on many slopes. Here the resistant sandstone butte is shrinking as weathering separates fragments that fall and accumulate at the base of the cliff.

(D) Shattering of the rock in mountain regions results from water seeping into cracks and freezing. The fragments commonly occur in a cone at the base of the cliff.

FIGURE 10.1
The effects of weathering are seen whenever rocks are exposed. These photographs show typical examples.

subtracted from, the rock. The rock is simply broken down into small fragments by various physical stresses. The most important types of mechanical weathering are

1. Ice wedging, in which freezing water expands in cracks or bedding planes and wedges the rock apart
2. Sheeting, or unloading, in which a series of fractures is produced by expansion of the rock body itself as a result of the removal of overlying material by erosion

Ice Wedging

Figure 10.2A is a simple diagram showing how *ice wedging* breaks a rock mass into small fragments. Water from rain or melting snow easily penetrates cracks, bedding planes, and other openings in the rock. As it freezes, it expands about 9%, exerting great pressure on the rock walls, similar to the pressure produced by driving a wedge into a crack. Eventually, the fractured blocks and bedding

FIGURE 10.2
Ice wedging

(A) Ice wedging occurs when water seeps into fractures and expands as it freezes. The expanding wedge forces the rock apart and produces loose, angular fragments that move downslope by gravity and accumulate at the base of the cliff as talus cones.

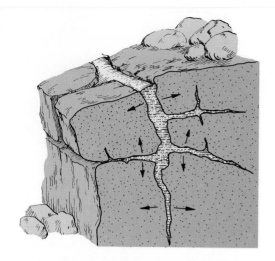

(B) The effects of ice wedging in the Teton Range of Wyoming are seen in both the rugged surface of the mountain peaks and the accumulation of fragmented debris at the base of the cliff. The rock that forms the mountain range is a massive granite cut by numerous fractures. Ice wedging, controlled in part by the fractures, produces the sharp, angular texture of the mountain peaks. The debris derived from ice wedging has accumulated in conical-shaped slopes near the base of the cliff.

How does mechanical weathering break down a solid mass of rock into small fragments?

planes are pried free from the parent material. The stress generated each time the water freezes is approximately 110 kg/cm², roughly equivalent to that produced by dropping an 8-kg ball of iron about the size of a large sledgehammer from a height of 3 m. Stress is exerted with each freeze, so that over a period of time, the rock is literally hammered apart.

Ice wedging occurs under the following conditions: (1) when there is an adequate supply of moisture, (2) where preexisting fractures, cracks, or other voids occur within the rock, into which water can enter, and (3) where temperatures frequently rise and fall across the freezing point. Temperature fluctuation above and below the freezing point is especially important because pressure is applied with each freeze. In areas where freezing and thawing occur many times a year, the ice wedging is far more effective than in excep-

FIGURE 10.3
Sheeting in granite of the Sierra Nevada occurs as erosion removes the overlying rock cover and reduces the confining pressure. The bedrock expands and large fractures develop parallel to the surface. The fractures may subsequently be enlarged by ice wedging.

tionally cold areas, where water is permanently frozen. Ice wedging thus occurs most frequently above the timberline. It is especially active on the steep slopes above valley glaciers, where meltwater produced during the warm summer days seeps into cracks and joints and freezes during the night (Figure 10.2B).

Sheeting

Rocks formed deep within Earth's crust are under great confining pressure from the weight of thousands of meters of overlying rocks. As this overlying cover is removed by erosion, the confining pressure is released, and the buried rock body tends to expand. The internal stresses set up by expansion can cause large fractures, or expansion joints, parallel to Earth's surface (Figure 10.3). This result is called *sheeting*. It can be observed directly in quarries, where the removal of large blocks is sometimes followed by the rapid, almost explosive expansion of the quarry floor. A sheet of rock several centimeters thick may burst up, and at the same time, numerous new parallel fractures will appear deeper in the rock body. The same process occasionally causes rock bursts in mines and tunnels when the confining pressure is released during the tunneling operation. It can also be seen in many valley walls and in excavations for roads, where rock slumping due to sheeting can cause serious highway problems.

Other Types of Mechanical Weathering

Animals and plants play a variety of relatively minor roles in mechanical weathering. Burrowing animals, such as rodents, mechanically mix the soil and loose rock particles, a process that facilitates further breakdown by chemical means. Pressure from growing roots widens cracks and contributes to the rock breakdown. Lichens can live on the surface of bare rock and extract nutrients from its minerals by ion exchange. This results in both mechanical and chemical alteration of the minerals. These processes may seem trivial, but the work of

FIGURE 10.4
Talus cones are piles of rock debris that accumulate at the base of a cliff as the result of
rockfalls. Most rock fragments in talus cones
are produced by ice wedging.

innumerable plants and animals over a long period of time adds significantly
to the **_disintegration_** of the rock.

Thermal expansion and contraction of the rock caused by daily or seasonal temperature changes were once thought to be an effective process of
mechanical weathering. The idea is plausible, but experiments show that
stresses developed by alternate heating and cooling over long periods of time
are insignificant in comparison with the elastic strength of rock. Even on the
Moon, where daily temperature changes are much greater than those on Earth,
the effects of thermal expansion on rock fragmentation are uncertain.

The products of mechanical weathering are best seen in the high mountain country, where ice wedging dominates and produces a large volume of
angular rock fragments. This material commonly accumulates in a pile at the
base of the cliffs from which it was derived. Because most cliffs are notched by
steep valleys and narrow ravines, the fragments dislodged from the high valley
walls are funneled through the ravines to the base of the cliff, where they
accumulate in cone-shaped deposits called **_talus cones_** (Figure 10.4).

The talus cones are built up by isolated blocks loosened by ice wedging.
The blocks commonly fall separately, as almost any mountain climber can
testify, but large masses of the material on steep slopes may be moved by an
avalanche. Earthquakes may also suddenly activate large numbers of blocks
loosened by many seasons of ice wedging.

In the example shown in Figure 10.4, all of the talus has accumulated
since the last ice age, which terminated 10–15 thousand years ago. This is a
considerable amount of material produced by ice wedging alone.

194

CHEMICAL WEATHERING

> *Chemical weathering (chemical decomposition) is the break-down of rocks by chemical alteration of the constituent miner-als. It involves several important chemical reactions between the elements in the atmosphere and those in the minerals of Earth's crust. The three main groups of chemical reactions are (1) hy-drolysis, (2) dissolution, and (3) oxidation.*

Chemical weathering involves several important chemical reactions between elements in the atmosphere and those in the rocks and minerals of Earth's crust. During chemical weathering, rocks are decomposed, the internal struc-ture of the minerals is destroyed, and new minerals are created. Thus there is a significant change in the chemical composition and physical appearance of the rock.

Water is of prime importance in chemical weathering. It takes part di-rectly in the chemical reactions. It acts as a medium to transport elements of the atmosphere to the minerals of the rocks, where reactions can occur, and it removes the products of weathering to expose fresh rock. The rate and degree of chemical weathering, therefore, are greatly influenced by the amount of precipitation.

No area of Earth's surface is continually dry. Even in the most arid deserts, some rain falls. Chemical weathering is therefore essentially a global process, but it is least effective in deserts and in climates where water is frozen the entire year.

The chemical reactions involved in the decomposition of rock are com-plex, but three main groups are recognized.

1. Hydrolysis
2. Dissolution
3. Oxidation

What are the major chemical reactions in weathering? What do they produce?

Hydrolysis

The chemical union of water and a mineral is known as *hydrolysis.* The process involves not merely absorption of water, as in a sponge, but a specific chemical change in which a new mineral is produced. In hydrolysis, ions derived from one mineral react with the H^+ or OH^- ions of the water to produce a different mineral.

A good example of hydrolysis is the chemical weathering of feldspar. As you recall from previous chapters, feldspar is an abundant mineral in a great many igneous, metamorphic, and sedimentary rocks, so it is important to understand how feldspars weather and decompose into clay minerals, which form the most abundant sedimentary rock, shale. Two substances are essential in the weathering of feldspars: (1) carbon dioxide and (2) water. The atmo-sphere and the soil contain carbon dioxide, which unites with rainwater to form carbonic acid. If K-feldspar comes in contact with carbonic acid, the following chemical reaction occurs:

What processes are involved in the weathering of feldspar?

$$2\,KAlSi_3O_8 + H_2CO_3 + H_2O \longrightarrow K_2CO_3 + Al_2Si_2O_5(OH)_4 + 4\,SiO_2$$

(K-feldspar)   (carbonic acid)   (water)   (potassium carbonate–readily soluble)   (a clay mineral)   (soluble hydrated silica or finely divided quartz)

The hydrogen ion of the H_2CO_3 displaces the potassium ion of the feldspar and thus disrupts the crystal structure. It then combines with the aluminum silicate of the feldspar to form a clay mineral. The potassium associates with the carbonate ion to form potassium carbonate, a soluble salt. Silica is also re-leased but may remain in solution. The new clay mineral does not contain

What is the origin of clay?

potassium, which was present in the original feldspar. The new mineral also has a new crystal structure, consisting of sheets of silica tetrahedra that form submicroscopic crystals.

Dissolution

Dissolution is a process whereby rock material passes directly into solution, like salt in water. Quantitatively, the most important minerals involved in dissolution are the carbonate minerals calcite and dolomite. These minerals make up the limestones of the world. Dissolution occurs because water is one of the most effective and universal solvents known. The structure of the water molecule requires the two hydrogen atoms to be positioned on the same side of the larger oxygen atom. The molecule thus has a concentration of positive charge on the side with the two hydrogen atoms, balanced by a negative charge on the opposite side. As a result, the water molecule is polar and behaves like a tiny magnet. It acts to loosen the bonds of the ions at the surface of minerals with which it comes in contact. Because of the polarity of the water molecule, practically all minerals are soluble in water to some extent.

Some rock types can be completely dissolved and *leached,* or dissolved and flushed, away by water. Rock salt is perhaps the best-known example. It is extremely soluble, surviving at Earth's surface only in the most arid regions. Gypsum is less soluble than rock salt but is also easily dissolved by surface water. Few, if any, large outcrops of these rocks occur in humid regions. Limestone is also soluble in water, especially if the water contains carbon dioxide. Where water is abundant, limestone commonly weathers into valleys, but in arid regions it forms cliffs.

The chemical analysis of river water illustrates the effectiveness of dissolution in the weathering of rocks. Fresh rainwater contains relatively little dissolved mineral matter, but running water soon dissolves the more soluble minerals in the rock and transports them in solution. Each year the rivers of the world carry about 3.9 million metric tons of dissolved minerals to the oceans. It is not surprising, then, that seawater contains 3.5% (by weight) dissolved salts, all of which were dissolved from the continents by pure rainwater.

How is limestone weathered?

Oxidation

Oxidation is the combination of atmospheric oxygen with a mineral to produce an oxide. The process is especially important in the weathering of minerals that have a high iron content, such as olivine, pyroxene, and amphibole. The iron in silicate minerals unites with oxygen to form hematite (Fe_2O_3) or limonite [$FeO(OH)$]. Hematite is deep red, and if it is dispersed in sandstone or shale, it imparts a red color to the entire rock.

Like many chemical reactions, the rate of chemical weathering increases as temperature rises. Chemical decomposition, therefore, is most intense in warm, wet, tropical climates.

Plants and bacteria are also important agents in chemical weathering because some acids and organic compounds in soils are produced by bacterial decay of plant and animal remains. Water seeping through organic remains in soils commonly becomes more acidic, increasing its effectiveness as a weathering agent.

Inasmuch as feldspars and other silicate minerals that weather into clay compose a large percentage of igneous and metamorphic rocks, an enormous amount of clay has been produced by weathering of these minerals throughout geologic time. It has been calculated that sediment and sedimentary rocks have average thicknesses of 3 km throughout the ocean basins, 5 km on the continental shelves, and 1.5 km on the continents. Because clay makes up about one-third of all sedimentary rocks, the total amount of clay would form a layer almost 2 km thick if spread uniformly over the entire surface of Earth.

A Concluding Note

We have considered mechanical and chemical weathering as separate, individual processes, but in nature these processes cannot be separated because many types of weathering processes are usually involved in the weathering of any outcrop. Mechanical fracturing of a rock increases the surface area, where chemical actions take place, and permits deeper penetration for chemical decomposition. Chemical decay in turn facilitates mechanical disintegration. One process may dominate in a given area, depending on the climate and rock composition, but mechanical and chemical weathering processes generally attack the rock at the same time.

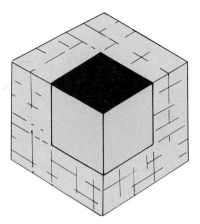

(A) A surface of bedrock, 10 m long and 10 m wide, with no joints, exposes a total area of 100 m² to weathering processes. A set of joints 10 m apart would expose a surface area of 600 m² to weathering.

GEOMETRY OF ROCK DISINTEGRATION

The size and shape of most rock fragments are inherited from patterns of joints, bedding, cleavage, and other planes of structural weakness in the parent rock material.

The breakup of a solid mass of rock into smaller particles may at first seem to be a random process in which an infinite variety of shapes may be reproduced. Careful study, however, shows that there is system and order in the process. Mechanical breakdown of rocks and the shapes of most rock fragments are inherited from patterns of joints, bedding, cleavage, and other planes of structural weakness in the parent rock material.

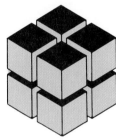

(B) Three additional joints, dividing the block into eight cubes, would increase the surface area to 1200 m².

The Importance of Joints in Weathering

Almost all rocks are broken into a system of fractures called *joints*. Joints result from strain that occurs when the rocks are uplifted, tilted, folded, or fractured by tectonic forces; from the release of confining pressure when material is removed by erosion; and from contraction produced by the cooling of lava. Joints greatly influence the weathering of rock bodies in two ways.

1. They effectively cut large blocks of rock into smaller ones and thereby greatly increase the surface area where chemical reactions take place.
2. They act as channelways through which water can penetrate to break down the rock by ice wedging.

The importance of joints in weathering processes can be appreciated by considering the amount of new surface area produced by jointing. Consider, for example, a cube of rock that measures 10 m on each side (as shown in Figure 10.5). If only the upper surface of the cube were exposed and the rock were not jointed, weathering could attack only the exposed top surface of 100 m². If the block were bounded by intersecting joints 10 m apart, however, the surface area exposed to weathering processes would be 600 m². If three additional joints cut the cube into eight smaller cubes, the surface exposed to weathering would be 1200 m². If joints 1 m apart cut the rock, 6000 m² of rock surface would be exposed. Obviously, a highly jointed rock body weathers much more rapidly than a solid one. The breakdown of a rock along a system of jointing planes is called *joint-block separation* (Figure 10.6).

Besides providing a larger surface area for chemical decomposition, joints act as a system of channels through which water can more readily penetrate the rock body. Joints thus permit mechanical and chemical weathering processes to attack the rock from several sides, even hundreds of meters below the surface.

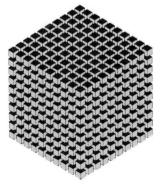

(C) If joints 1 m apart cut the rock, the surface area exposed to weathering would be increased to 6000 m².

FIGURE 10.5
A system of joints cutting a rock body greatly increases the surface area exposed to weathering.

How do joints affect weathering?

Other Planes of Weakness

Bedding planes in many sedimentary rocks form planes of weakness, which cause the rock to break into slabs or plates. Foliation in metamorphic rock is similar. Schists tend to break into small, splintery pieces with flat sides parallel to planes of foliation. Slate is an even better example of how foliation influences the way in which a rock breaks into smaller pieces.

Disintegration

Further modification of these fragments is commonly influenced by the size and shape of the mineral grains and the manner in which the grains are held together. Rocks that are very dense, such as quartzites, may rupture into irregular angular fragments. This form of disintegration is called *shattering,* as the fracture cuts across, not around, the mineral grains. A good example of shattering is seen in the rock fragments produced by blasting bedrock with explosives.

In many sandstones the quartz grains are weakly cemented together with calcite, which is readily dissolved and removed by water. Once the cementing material is removed or weakened, there is nothing to hold the grains together, and they fall apart, or disaggregate. This process of grain-by-grain breakdown is known as *granular disintegration* (Figure 10.6). Individual mineral grains simply separate from one another along their natural contacts and produce sand or gravel, in which each particle has the same shape and size as it did in the original rocks. Similarly, the quartz grains in granite are disaggregated as the feldspars are decomposed into clay. The soft clay loosens the interlocking grains and the once-solid piece of granite starts to crumble into individual grains or groups of grains. The result is that the small, irregular quartz grains that once filled the spaces between feldspar are now liberated and become the source of most sand.

The best way to appreciate how joints, bedding planes, and other planes of weakness influence the geometry of rock fragmentation is to compare and contrast outcrops of several rock types and consider the shape of the fragmented material that weathering has produced (Figure 10.6). Figure 10.6A shows a basalt flow that has been broken into hexagonal columns as it cooled. When stresses are applied, such as ice wedging, the rock comes apart along these established joints to produce a pile of jumbled columns. The bedding planes in sedimentary rocks (Figure 10.6B) commonly determine the tabular form of the fragmented material. Figure 10.6C shows how jointing in massive granites has controlled the shape of the fragments that have been produced by ice wedging. Granular disintegration is obviously the major type of rock breakdown in thick sandstone formations that are weakly cemented with calcite (Figure 10.6D). It is likewise the manner of disaggregation in many conglomerates.

In every case, regardless of the original shapes of the rock fragments, subsequent modification tends to round off the corners to produce spherical forms.

WEATHERING CHARACTERISTICS OF MAJOR ROCK TYPES

The weathering of rocks is influenced by a number of variables, such as mineral composition, texture of the rock, and the climate in which weathering occurs.

Weathering is influenced by so many factors that it is difficult to make a meaningful generalization concerning the weathering of specific rock types. Limestone, for example, may weather and erode into a soil-covered valley in a

*(A) **Joint-block separation** results when prominent fractures divide the rock into small blocks. The Devil's Post Pile in California is an excellent example where columnar joints control the geometric patterns of rock breakup.*

*(B) **Bedding planes** frequently form a prominent zone of weakness in sedimentary rocks and cause the rock to break up into slabs.*

*(C) **Jointing** is commonly the major type of structural weakness in granites and related rocks and causes the rock to break up into large blocks.*

*(D) **Granular disintegration** in sandstone commonly occurs and produces spheroidal boulders. Each grain in the disintegrated material has the same size and shape as it did in the original rock.*

*(E) **Exfoliation** occurs when the solid rock mass comes apart in a series of shells or plates that conform roughly to the configuration of the outer surface of the rock mass. Exfoliation can occur on a very large scale, such as Half Dome in Yosemite National Park, or on a very small scale, with the individual plates being only a millimeter or less thick. Exfoliation affects many rock types and is caused by both physical and chemical processes.*

*(F) **Shattering** occurs when a rock is subjected to severe stress that causes the rock to rupture into sharp, irregular, angular blocks. Ice wedging produces shattering in nature of rock outcrops; blasting bedrock with explosives produces shattering artificially.*

FIGURE 10.6
The geometric patterns of rock disintegration depend upon the composition, texture, and structure of the rock body.

humid climate, whereas the same formation would form a cliff in an arid climate. Similarly, a well-cemented quartz sandstone may be extremely resistant to weathering, whereas a sandstone with a high clay content would likely be soft and weak and would weather rapidly.

Mineral composition is of prime importance; some minerals, such as quartz, are very stable and remain essentially unaltered for long periods of time; others, such as olivine and the feldspars, are very unstable and begin to decompose almost immediately. The texture of the rock is also very significant because of its influence on *porosity* and *permeability,* which govern the ease with which water can enter the rock and attack the mineral grains. Precipitation and temperature are the chief climatic controls, but weathering will be influenced not only by the total rainfall but also by the distribution of precipitation through time, percent of runoff, rate of evaporation, and so on. Therefore, a given rock will respond to weathering in a variety of ways, but in general the major rock groups weather in characteristic fashion.

Granite

Granite is a massive homogeneous rock composed of feldspar, quartz, and mica, with minor amounts of other minerals. It forms at considerable depth and under great pressure, so it is out of equilibrium when exposed at Earth's surface. The release of pressure resulting from the erosion of the overlying rocks produces expansion joints, which aid in the development of exfoliation. Although granite is known as a rock of great strength, chemical weathering is readily apparent on most outcrops.

Feldspars weather rapidly by chemical reaction with water and are altered to various clay minerals. Calcium plagioclase is least resistant, followed by sodium plagioclase. Although potassium feldspars are most resistant, all feldspars readily break down into clay. Mica weathers somewhat more slowly than most feldspars but is easily attacked by water along its cleavage planes, and ion exchange is common. Micas alter with little change in structure to chlorite and clay minerals. Quartz, in contrast, is very resistant to both chemical and mechanical weathering and remains essentially unaltered as the other minerals are decomposed. Therefore, it constitutes the most significant particle or fragment produced by the weathering of a granite.

Basalt

Basalt is a fine-grained rock composed mostly of feldspar, olivine, and pyroxene. The surface of a basalt flow is generally vesicular and very porous, and the interior of the rock body is commonly broken by a system of columnar joints. Therefore, flows are highly permeable and susceptible to decomposition. Quartz is not present in basalt, so most minerals in this rock are eventually converted to clay and iron oxides. The ultimate weathering product is a red or brown soil.

Sandstones

Sandstones are composed mostly of quartz grains, with varying amounts of small rock fragments, feldspar, and clay minerals. The quartz in sandstone is highly resistant to chemical weathering, so chemical decomposition of the rock consists largely of an attack on the cement. The major cementing materials in sandstones are calcite, iron oxide, and quartz.

Limestone

Limestone is composed mostly of the mineral calcite, although it generally contains some clay and other impurities. It is the most soluble of the common

rock types and, except in extremely dry climates, solution is the dominant weathering process. In pure water, calcite is not very soluble; if carbon dioxide is present, it combines with water to form carbonic acid, which is capable of dissolving much more calcite than pure water. The formation of carbonic acid in water is expressed by the equation

$$H_2O \quad + \quad CO_2 \quad \longrightarrow \quad H_2CO_3$$
(water)   (carbon dioxide)   (carbonic acid)

In the solution of limestone, the acid reacts with calcite to form calcium bicarbonate, which remains in solution and is removed by groundwater. This reaction may be expressed as follows.

$$H_2CO_3 \quad + \quad CaCO_3 \longrightarrow \quad Ca(HCO_3)_2$$
(carbonic acid)   (calcite)   (calcium bicarbonate)

In most limestone regions in humid climates, solution activity enlarges joints and bedding planes and forms a network of caverns and caves; the limestone formations in such regions typically form valleys. In arid regions, where solution activity is at a minimum, limestones form cliffs.

Shales

Shales commonly weather faster than most other rocks because they are fine-grained, soft, and have the ability to absorb and expel large amounts of water.

Differential Weathering

As can be seen from the preceding brief descriptions, different rock masses or different sections of the same rock weather at different rates. This is known as *differential weathering.* It occurs on a broad scale, from the great cliffs and slopes of the Grand Canyon to thin layers of sedimentary rock. The more resistant zones stand out as ridges, and the weaker zones form depressions. Differential weathering can lead to the formation of unusual shapes and forms, such as the spindles and pinnacles in Bryce Canyon (Figure 10.7) to pits and

FIGURE 10.7
Differential weathering has produced the spectacular landforms in Bryce Canyon, Utah. Two zones of weakness occur within the rock body: (1) horizontal layers of different material and (2) a system of intersecting joints that divide the rocks into a series of rectangular columns. Rapid weathering along the joints produces a series of columns, and differential weathering of the sedimentary layers imparts an irregular form to the column.

Why do rocks weather at different rates?

201

caverns on a rock face. Differential erosion on dikes of igneous rocks can form trenches or walls, depending on whether the dike is harder or softer than the surrounding rock.

Differential weathering can be seen everywhere a rock is exposed. Study the photo in Figure 10.7, and you will notice that each layer has its own weathering characteristics. The white layers erode most rapidly and tend to form slender columns. The thicker beds of sand are more resistant, whereas the interbeds of siltstone and shale weather rapidly. Thus, the horizontal layers are etched into ridges and furrows, which are responsible for much of the beauty in this scene.

PRODUCTS OF WEATHERING

The results of weathering can be seen over the entire surface of Earth. The major products of weathering are (1) a blanket of loose, decayed rock debris, known as regolith, and (2) rock bodies modified into spherical shapes.

The results of weathering can be seen from the driest deserts and the frozen wastelands to the warm, humid tropics. The most obvious product of weathering is a blanket of loose, decayed rock debris known as regolith, which forms a discontinuous cover over the solid unaltered bedrock below. In addition, there is a universal tendency for weathering processes to form rounded or spherical surfaces on a decaying rock body.

Regolith and Soil

The term *regolith* comes from the Greek word *rego,* meaning "blanket" (blanket rock). It is a layer of soft, disaggregated rock material formed in place by the decomposition and disintegration of the bedrock that lies beneath it. Within the regolith the individual grains or small groups of mineral particles are easily separated from one another. The thickness of the regolith ranges from a few centimeters to hundreds of meters, depending on the climate, type of rock, and length of time that weathering processes have been operating. The transition from bedrock to regolith can be seen in roadcuts and stream valleys.

What are the major products of weathering?

Gravel, sand, silt, and mud deposited by streams, wind, and glaciers are sometimes referred to as transported regolith in order to distinguish them from the residual regolith produced by weathering. Many types of transported regolith, or surficial deposits, have been identified, and we will learn more about them in later chapters dealing with rivers, glaciers, and wind.

The uppermost layer of the regolith is the *soil.* It is composed chiefly of small particles of rocks and minerals, plus varying amounts of decomposed organic matter. Soil is so widely distributed and so economically important that it has acquired a variety of definitions, and you should be aware that the term, as used by engineers, geologists, farmers, and soil scientists, has somewhat different definitions.

The transition from the upper surface of the soil down to fresh bedrock is called a *soil profile* and shows a rather constant sequence of layers, or *horizons*, which are distinguished by composition, color, and texture. These are shown in Figure 10.8. The *A horizon* is the topsoil layer, which often is visibly divided into three layers: A_0 is a thin surface layer of leaf mold, especially obvious on forest floors; A_1 is a humus-rich, dark layer; and A_2 is a light, bleached layer. The *B horizon* is the subsoil, containing fine clays and colloids washed down from the topsoil. It is largely a zone of accumulation and commonly is reddish in color. The **C horizon** is a zone of partly disintegrated and decomposed bedrock. The individual rock fragments are often weathered, spheroidal boulders, which may be completely decomposed. The C horizon grades downward into fresh, unaltered bedrock.

The type and thickness of soil depend on a number of factors, the most important of which are climate, parent rock material, and topography. Climate is of major importance, because rainfall, temperature, and seasonal changes all directly affect the development of soil. For example, in deserts, arctic regions, and high mountainous regions, mechanical weathering dominates as the means of soil production, and organic matter is minimal. The resulting soil is thin and consists largely of broken fragments of bedrock (Figure 10.9A). In equatorial regions, where rainfall is heavy and temperatures are high, chemical processes dominate and thick soils develop rapidly. As a consequence, soil profiles 60 m thick are common in the tropics and subtropics. In some areas (such as central Brazil), the zone of decayed rock is more than 150 m thick.

The mineral composition of the bedrock strongly influences the type of soil because the bedrock provides the chemical elements and mineral grains from which the soil develops. Pure quartzite, for example, containing 99% SiO_2, produces a thin, infertile soil (Figure 10.9B).

Topography affects soil development because it influences the amount and rate of erosion and the nature of drainage. Flat, poorly drained lowlands develop a bog-type soil, rich in decomposed vegetation and saturated with water, whereas steep slopes permit rapid removal of regolith and inhibit the accumulation of weathered materials. Well-drained uplands are conducive to thick, well-developed soils (Figure 10.9C).

Time is important in soil development in that it takes time for mechanical and chemical processes to break down the bedrock. In Figure 10.9D, the young lava flow has a very thin, patchy soil, whereas the older flow has had time for a thick soil layer to develop.

FIGURE 10.8
A soil profile shows the transition from bedrock to regolith through a sequence of layers, or horizons, consisting of successively smaller fragments of rock.

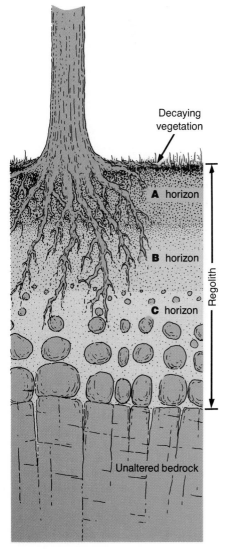

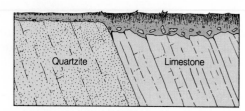

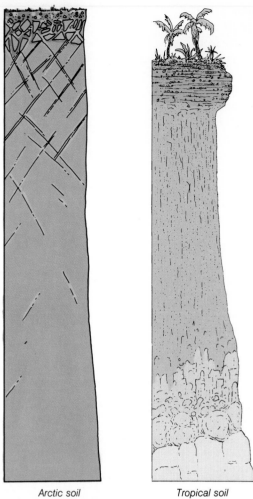

(B) *The influence of rock type is illustrated by the difference between an outcrop of quartzite and an outcrop of limestone. Quartzite resists chemical decomposition, so the soils produced from it are thin and poorly developed. Limestone is much more susceptible to chemical weathering and forms thicker soils.*

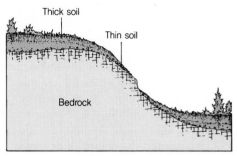

(C) *The influence of topography is apparent from the contrast between slope soils and valley soils. Thick soils can form on flat or gently sloping surfaces, but steep slopes permit only thin soils to develop.*

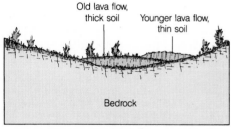

(D) *The influence of time can be seen in areas of volcanism. Thick soils have developed on old lava flows, in contrast to thin soils on younger flows.*

Arctic soil   *Tropical soil*

(A) *The influence of climate can be seen in the contrast between soils in arctic and desert regions and soils in the tropics. Thin soils, characterized by partly decomposed rock fragments, develop in arctic and desert regions. Thick soils form in tropical regions as a result of extensive chemical decomposition.*

FIGURE 10.9
The type of soil found in an area depends on a number of factors, such as climate, parent rock, topography, and time.

Spheroidal Weathering

In the weathering process, there is an almost universal tendency for rounded (or spherical) surfaces to form on a decaying rock body. A rounded shape is produced because weathering attacks an exposed rock from all sides at once, and therefore decomposition is more rapid along the corners and edges of the rock (Figure 10.10). As the decomposed material falls off, the corners become rounded, and the block eventually is reduced to an ellipsoid or a sphere. The sphere is the geometric form that has the least amount of surface area per unit of volume. Once the block attains this shape, it simply becomes smaller with further weathering. This process is known as *spheroidal weathering.*

Examples of spheroidal weathering can be seen in almost any exposure of rock (Figure 10.11). It can also be seen in the rounded blocks of ancient buildings and monuments (see also Figure 10.16). The original blocks had sharp corners and were fitted together with precision. The edges are now completely decomposed, and each block has assumed an ellipsoidal or spherical shape. In nature, spheroidal weathering is produced both at the surface and at some depth.

Why are spheroidal forms the universal result of weathering processes?

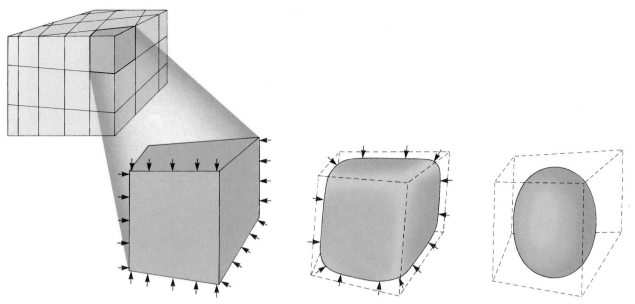

(A) Joint systems cut a rock body into angular blocks.

(B) On each block, weathering proceeds inward from the joint face.

(C) The corners of the block are soon completely decomposed, so that the weathered rock assumes a spherical or ellipsoidal shape.

FIGURE 10.10
Spheroidal weathering occurs because the edges and corners of a joint block are easily decomposed.

FIGURE 10.11
Examples of spheroidal weathering are seen in natural outcrops.

Exfoliation is a special type of spheroidal weathering in which the rock breaks apart by separation along a series of concentric shells or layers that look like cabbage leaves (Figure 10.12). The layers, essentially parallel to each other and to the surface, develop by both chemical and mechanical means. Sheeting can play an important part, because rocks like granite, if they are deeply buried, have a tendency to expand upward and outward as the overlying rock is removed. In cold climates, ice wedging along the sheeting joints helps to remove successive layers gradually. The increase in volume of mineral grains associated with the decomposition of feldspar might also promote exfoliation. Exfoliation causes massive rocks like granite to develop a spherical form characterized by a series of concentric layers ranging from boulders to a mountain.

FIGURE 10.12

Exfoliation domes are well developed in the Sierra Nevada, in California. These mountains are composed of massive granite cut by joints that separate the rock into large blocks. Exfoliation domes developed as huge slabs of rock fell off. The joints separating the rock slabs probably resulted from expansion of the rock as erosion removed the overlying rock.

A more regional view of the regolith and its relationship to bedrock is given in Figure 10.13. The photograph shows exposures of bedrock limited to certain areas of resistant limestone and sandstone strata, which form discontinuous cliffs along the upper part of the mountain front. On the steep canyon walls, little soil is retained, and bedrock is exposed from the base to the top of the canyon. The sketch in diagram B was made from the photograph and outlines the surface covered with regolith. In diagram C, the outcropping bedrock is not shown, so the regolith appears as a thin, discontinuous blanket with holes where bedrock is exposed. Sediment fills the valley in the foreground, but the regolith there is not shown in the diagram. If you carefully study the pattern of exposed bedrock areas (diagram C), you can see that the strata are warped into broad folds, shown in diagram D, which form the internal structure of the mountains.

CLIMATE AND WEATHERING

Climate is the single most important factor influencing weathering. It determines not only the type and rate of weathering but also the characteristics of regolith and weathered rock surfaces. Intense chemical weathering occurs in hot humid regions and develops thick regoliths. Chemical weathering is minimal in the deserts and polar regions.

Climate is the single most important factor influencing weathering. Evidence for this is obvious in the striking contrasts of the soil and weathered landforms in the tropics, deserts, and polar regions (Figure 10.14). As we have seen in previous sections, water is one of the most important agents in most types of weathering and influences weathering in several ways.

(A) The Wasatch Range, in central Utah, displays contrasting areas of bedrock and regolith.

(B) Outcrops of bedrock appear in cliffs and canyons. Slopes are covered with regolith.

(C) The discontinuous blanket of regolith almost completely covers some formations, while others are exposed as discontinuous cliffs. Outcrops of bedrock form "holes" in the regolith cover.

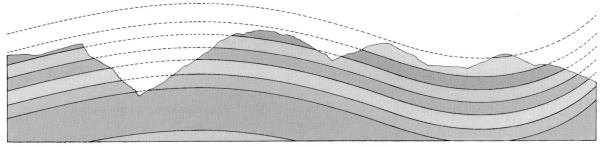

(D) The structure of the bedrock consists of rock layers warped into broad folds, some of which are cut by canyons. Compare with A.

FIGURE 10.13
The relationship between bedrock and regolith is depicted in the photograph and diagrams.

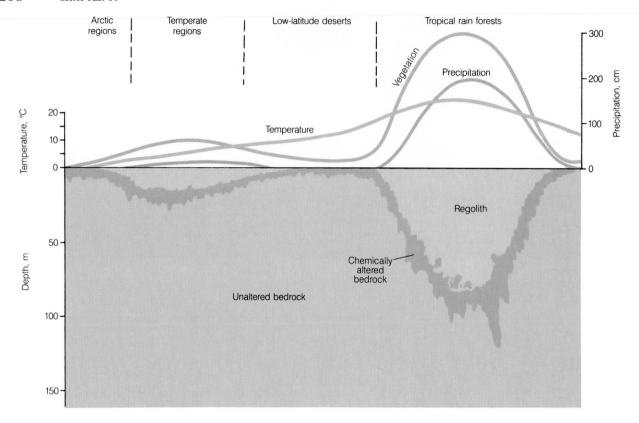

FIGURE 10.14
The type and extent of weathering vary with climate because of the combined effects of precipitation, temperature, and vegetation. (Other variables are also involved, such as those shown in Figure 10.9.) Weathering is most pronounced in the tropics, where precipitation, temperature, and vegetation reach a maximum. Conversely, a minimum of weathering is found in deserts and polar regions, where these factors are minimal.

How does climate influence weathering?

The extent and style of weathering are controlled to a considerable degree by water. The changes in volume caused by freezing and thawing of water plus the simple addition or removal of water may cause a rock to split apart or crumble. Most chemical reactions, such as hydrolysis, dissolution, and oxidation, require the presence of water, so that the total amount of precipitation in an area is clearly a major factor. The total amount of precipitation is significant, but many factors, such as the intensity of rain, seasonal variations, infiltration, runoff, and the rate of evaporation, all combine to influence weathering and weathered products in a given region. The extent and style of weathering is not controlled entirely by total water supply, and weathering may be greatly affected by other conditions. Many reactions are controlled by the hydrogen ion concentration, which is expressed as the pH value, ranging from 1 (acid) to 14 (alkaline). Iron, for example, becomes 100,000 times more soluble at pH 6 than at pH 8.5.

Temperature is also important in all aspects of weathering. In mechanical weathering perhaps the single most important temperature changes are the ones that produce a continual series of freeze–thaw changes. This results in repeated freezing and expansion of water in rock and soil and thus to mechanical fragmentation.

Temperature is important because the rate of chemical reactions (and biological activity) tends to increase as temperature increases. Commonly, a 10°C increase in temperature doubles reaction rates.

The relative importance of various types of weathering under various climatic conditions (temperature and rainfall) is shown in Figure 10.15. High temperature and high precipitation result in strong chemical weathering. Mechanical weathering dominates in regions of low temperature and low rainfall.

Perhaps the best way to appreciate the influence of climate on weathering is to consider variations in the types and thicknesses of soils from the equator to the poles, as shown in Figure 10.14. This diagram summarizes the relationships between the amount of chemical weathering and variations in precipitation and temperature.

In the humid, tropical climates, extreme chemical weathering rapidly develops thick soils to depths greater than 70 m. Under such conditions, the feldspars in granites and related rocks are completely altered to clays, and all soluble minerals are leached out. Only the most insoluble materials (such as silica, aluminum, and iron) remain in the thick, deep soil, with the result that the soil commonly is infertile. The high temperatures in tropical zones speed chemical reactions, so chemical decomposition is very rapid. Frost action, of course, is essentially nonexistent in the tropics, except on the tops of high mountains.

In the low-latitude deserts north and south of the tropical rain forests, chemical weathering is minimal because of the lack of precipitation. The soil is thin, and exposures of fresh, unaltered bedrock are abundant. Mechanical weathering is evident, however, in the fresh, angular rock debris that litters most slopes.

In the temperate regions of the higher latitudes, the climate ranges from subhumid to subarid, and temperatures range from cool to warm. Both chemical and mechanical processes operate, and the soil and regolith develop to depths of several meters. In savanna climates, where alternating wet and dry seasons occur, the type of weathering is somewhat different from weathering in regions where precipitation is more evenly distributed throughout the year.

In the polar climates, weathering is largely mechanical. Temperatures are too low for much chemical weathering, so the soil typically is thin and composed mostly of angular, unaltered rock fragments. In permafrost zones (areas where water in the pore spaces of the soil and rock is permanently frozen), the surface layer melts during the summer, but freezes again in the winter. This unique condition produces polygonal ground patterns, which result from thermal contractions and the differential thawing and freezing. We will have more to say about permafrost areas in Chapter 11.

Why are soils thick in the tropics but thin in deserts and polar regions?

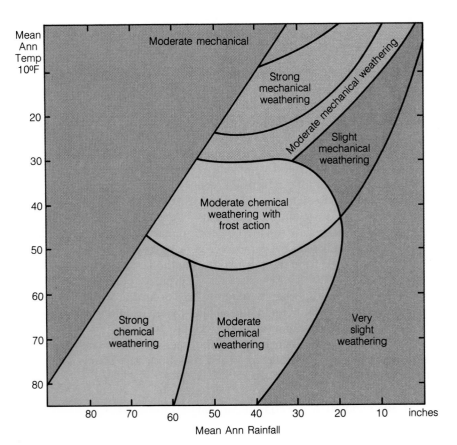

FIGURE 10.15
The relative importance of various types of weathering depends upon conditions of temperature and rainfall. This diagram shows that strong chemical weathering occurs where both temperature and precipitation are high. Mechanical weathering is strongest where the mean annual temperature is between 0 and 20° F and precipitation is between 10 and 40 inches. Weathering is at a minimum where annual precipitation is less than 10 inches.

RATE OF WEATHERING

> *The rate at which weathering processes decompose and break down a solid rock body depends on three main factors: (1) susceptibility of the constituent minerals to weathering, (2) climate, and (3) amount of surface exposed to the atmosphere.*

Why do some rocks weather more rapidly than others?

Weathering involves a multitude of physical, chemical, and biological processes, so it is difficult to make meaningful generalizations about the rate of weathering. A variety of studies, however, give us important insights into the problem, if not absolute answers. Rates of weathering can be calculated by measuring the amount of decay on rock surfaces of known age. Tombstones, ancient buildings, and monuments, for example, provide datable rock surfaces for estimating rates of weathering.

These studies show that several inches of rock can be decomposed in a few decades in some climates, whereas the same rock remains unaltered in others. A classic example of this fact is seen in the hieroglyphics cut in the monuments of ancient Egypt. Several obelisks of granite each bearing numerous deep-cut hieroglyphics stood for about 3500 years in Egypt, where the average rainfall is less than 2.5 cm per year. One was sent to New York in 1879 and now stands in Central Park. It has been so attacked by frost, water, and air rich in carbon dioxide that despite the application of shellaclike preservatives in the past 50 years the hieroglyphics are completely illegible. The obelisk that remained in Karnak, Egypt, is still unaltered by weathering.

On a longer time scale, rates of weathering have been measured from volcanic ash and basaltic lava flows that have been dated by radiometric means. On the island of St. Vincent in the West Indies, a volcanic ash deposited 4000 years ago has weathered to produce a layer of clay soil 2 m thick. Soils have also formed on the ash deposits resulting from the 1883 eruption of Krakatoa. Measurements made 45 years after the Krakatoa eruption show a new soil nearly 50 cm thick.

How do the pyramids of Egypt provide information on rates of weathering?

The Egyptian pyramids provide another interesting example of rates of weathering. The Great Pyramid of Cheops, near Cairo, was originally faced with polished, well-fitted blocks of travertine limestone. These blocks protected the rock in the pyramid core from weathering until the outer, polished layers were removed, about 1000 years ago, to build mosques in Cairo. Since then, without the rock facing, weathering has attacked all four main rock types used in the construction of the pyramid. The most durable (least-weathered) rock in the pyramid is a granite, which today remains essentially unweathered. Also resistant is a hard, gray limestone, which still retains marks of the quarry tools. The shaly limestone and fossiliferous limestone used for other blocks, however, have weathered rapidly. Many of these blocks have a zone of decayed minerals as deep as 20 cm. Most of the weathered debris remains as a talus on individual tiers and around the base of the pyramid (Figure 10.16). The volume of weathered debris produced from the pyramid during the last 1000 years has been calculated to be 50,000 m³. Near Saqqārah, Egypt, on older pyramids built nearly 4600 years ago, deeper weathering has occurred, with talus debris nearly covering the steps. Some of the smaller pyramids in the area are completely covered in their weathered debris (Figure 10.17).

In summary, we can conclude that the rates at which weathering processes decompose and break down a solid rock body vary greatly. The three factors that are most important are

1. Susceptibility of the constituent minerals to weathering
2. Climate
3. Amount of surface exposed to the atmosphere

FIGURE 10.16
The products of weathering can be observed on the Great Pyramid of Egypt. The rectangular building blocks have been modified into ellipsoidal forms by spheroidal weathering, and talus debris has accumulated on each tier.

FIGURE 10.17
Deep weathering is apparent on the ancient stepped pyramids of Egypt. Each step is completely covered with talus, and many individual blocks are weathered to spherical form.

Minerals vary widely in their ability to resist weathering. In general, minerals that crystallize at a high temperature are most susceptible to weathering. Olivine and Ca-plagioclase, therefore, weather most rapidly, followed by pyroxene and amphibole, and then by Na-plagioclase and K-feldspar. Quartz is one of the most highly resistant minerals; it resists alteration by chemical decomposition and alteration by mechanical processes.

It is clear that weathering is very rapid in warm, humid climates and slow in regions that are dry and cold. It is clear also that the weathering process slows down as the fresh rock becomes covered with the soil and regolith that forms upon it. Ultimately, a state of equilibrium may be reached.

M ass movement is the downslope transfer of material through the direct action of gravity. The processes can be rapid and devastating, as in great landslides, or imperceptibly slow, as in the creep of soil down the gentle slope of a grass-covered field. It is a major geologic process operating on a global scale on both the land and the ocean floor. It is one of the most universal of all geologic processes and operates on all planetary surfaces, modifying crater rims on the Moon and Mercury, enlarging the huge canyons on Mars, and eroding the stream valleys, sea cliffs, and mountain fronts on Earth. The net effect of mass movement is the transportation of loose rock material from the hillside into the valley floors, with the net result of modifying the landscape.

The image on the facing page shows the results of the Mount Saint Helens eruption, May 18, 1980. Destructive debris flows began within minutes of the onset of the eruption as hot pyroclastic material in the debris avalanche melted snow and glacial ice on the steep slopes of the stratovolcano, and the mixture of water and loose debris rushed down the mountain side. Such debris flows are called *lahars*, a term borrowed from Indonesia, where volcanic eruptions have produced numerous debris flows.

The largest and most destructive debris flow followed the North Fork of the Toutle River and ultimately dumped more than 65 million m^3 of sediment in the Columbia River. Other debris flows swept down the southeast flank of the volcano. On the upper steep slopes of Mount Saint Helens the debris flows traveled as fast as 200 km per hour and were more than 30 m deep.

In this chapter we will consider the various types of mass movement, some rapid and devastating such as the debris flows of Mount Saint Helens, and others imperceptibly slow. All slopes are mobile and are constantly changing under the constant pull of gravity.

11
Mass Movement

MAJOR CONCEPTS

1. Mass movement is the downslope transfer of material through the direct action of gravity. It is a major geologic process operating on all slopes.
2. The most important factors influencing mass movement are saturation of slope material with water, earthquakes, oversteepening of slopes, and freezing and thawing.
3. The major types of mass movement are creep, debris flows, and landslides.
4. Slopes are open dynamic systems in which regolith and near surface bedrock move downslope toward the main stream where they are removed through the drainage system.

FACTORS INFLUENCING MASS MOVEMENT

Gravity is the driving force for the downslope movement of material, but several factors are important in overcoming inertia, causing movement to occur. The most important are (1) saturation of material with water, (2) vibrations from earthquakes, (3) oversteepening of slopes by undercutting, and (4) alternating freezing and thawing.

Gravity pulls continually downward on all materials everywhere on Earth's surface. Bedrock is usually so strong and well supported that it remains fixed in place, but if the slope becomes too steep or if fractures or other zones of weakness occur, masses of bedrock may break free and move downslope. Soil and regolith, in contrast, are poorly held together and are much more susceptible to downslope movements. There's abundant evidence that on most slopes at least a small amount of downhill movement is occurring all the time.

Gravity is the driving force behind all slope processes. Mass movement is not limited to stream valleys but occurs on all slopes, including sea cliffs and fault-block mountain fronts, and on the slopes of craters on other planets. The force of gravity is continuous, of course, but it can move material only when it exceeds the cohesive strength of the surface material. As the products of weathering accumulate on a hillslope, the dry, loose rock fragments will tend to maintain a nearly uniform slope angle inclined at the angle of repose. The **angle of repose** is the steepest slope on which loose material such as talus will remain at rest without rolling further downslope. The angle of repose is commonly about 30 degrees but varies somewhat depending on the size, shape, and sorting of the fragments. The factors that influence mass movement can be easily understood by considering the forces acting on a rock fragment resting on a sloping surface (Figure 11.1). The weight of the object (the force caused by gravity) is directed vertically downward, in this case 1 kg. The force directed down the hillslope is only half that (1 × sine 30°). On a stable slope, the fractional cohesion of the object to the slope is greater than this downhill force and the rock fragment remains stationary. Any factor that either weakens the cohesion of the object with the surface or increases the downslope force may initiate downslope movement. Such factors include (1) saturation of the material with water, (2) vibrations from earthquakes, (3) alternating expansion and contraction of the regolith, and (4) undercutting of slopes by streams or waves. Water is an important factor in mass movement because it lubricates the unconsolidated material on slopes and adds weight to the mass, thereby promoting mobility and downslope movement. Heavy rainfalls, whether prolonged over many days or in a single storm, are particularly effective in triggering mass movement. Earthquakes, with their initial shock and aftershocks, are capable of loosening fragments of rocks on steep slopes and setting the

What factors influence mass movement?

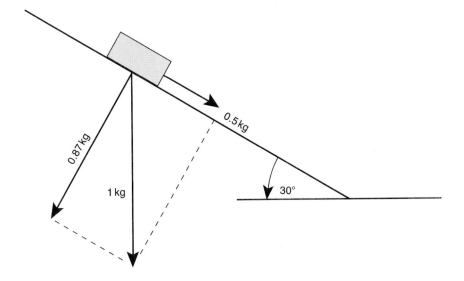

FIGURE 11.1
Diagram showing the forces acting on a rock particle on a hillslope. The force of gravity is directed vertically downward. The force directed downslope, therefore, depends on the weight of the object and angle of slope. In this example the particle will remain stable as long as the frictional cohesion of the surface material is greater than 0.5 kg. If the downslope force is increased by adding weight, increasing the slope angle, or by decreasing the cohesive strength, movement occurs.

regolith in motion. In many areas more damage is caused by mass movement than by the earthquake itself. For example, an earthquake in Guatemala in 1976 set off more than 10,000 mass movements. Most occurred on steep slopes, but some were on gentle slopes where water-saturated fine-grained regolith was mobilized and turned into mudflows. It is natural, then, for major landslides and debris flows to be especially common in active mountainous areas that are vulnerable to both earthquake shocks and major rainstorms. Earthquake shocks triggering mass movements are perhaps best known in the Andes Mountains, of South America, an active mountain range that is still forming on a convergent plate margin. For example, during the earthquake in Peru on May 31, 1970, 4000 m³ of debris roared down the slope of Mount Huascaran at a speed of more than 300 km per hour. The debris killed approximately 40,000 people as it covered the town of Yungay with as much as 14 m of mud and rock. Similar types of mass movement can be expected to occur in high, mountainous areas in the future. The effects of heavy rains are best known in the Himalaya Mountains because of their location in the Indian monsoonal system.

A significant factor in mass movement in many countries has been the modification of natural slopes to suit the needs of human communities. Farming and deforestation have brought changes in vegetation cover, soil, and drainage since prehistoric times. More recently, civil engineering works have modified coastlines, river systems, and landforms on an even larger scale. All of these changes brought about by human activity result in a new and artificial surface that is imposed on existing geological systems that had attained some degree of equilibrium. They commonly provoke unforeseen reactions that cause widespread damage. A dramatic example is the deforestation of large areas in Madagascar, resulting in tens of thousands of major slope failures and accelerated erosion (Figure 11.2).

Another example is the landslide that occurred at Vaiont Dam in northern Italy. This, the worst dam disaster in history, resulted from a huge landslide into the Vaiont Reservoir on October 9, 1963. The landslide began as a relatively slow creep over a three-year period. The rate of creep had been as much as 7 cm per week until a month before the catastrophe; at that time it increased to 25 cm per day. On October 1, animals grazing on the slopes sensed the danger and moved away. Finally, on the day before the slide, the rate of creep was about 40 cm per day. Engineers expected a small landslide and did not realize, until the day before the disaster, that a large area of the mountain slope was moving en masse at a uniform rate.

FIGURE 11.2
Accelerated mass movement resulting from deforestation in Madagascar is vividly shown in this photograph of the numerous deep gullies along the hillsides and the sediment-laden stream. Seen from space, the ocean around Madagascar is colored a thick bloody red by the silt that has been eroded from the recently deforested areas.

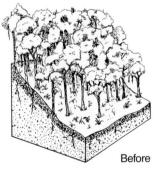

Before

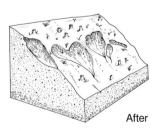

After

When the slide broke loose, more than 240,000,000 m³ of rock rushed down the hillslope and splashed into the reservoir. It produced a wave of water over 100 m high, which swept over the dam and rushed down the valley, completely destroying everything in its path for many kilometers downstream. The entire catastrophic event, including the slide and flood, lasted only 7 minutes, but it took approximately 2600 lives and caused untold property damage.

Several adverse geologic conditions contributed to the slide. The mountain rocks consisted of weak limestone interbedded with thin layers of clay. The beds were inclined steeply toward the reservoir, so there were inherent planes of weakness in the bedrock of the slope. Finally, it was the rising water level in the reservoir, saturating the adjacent soil and rock along the banks, that reduced the slope's cohesiveness and caused the slide.

The well-known and frequent landslide problems in southern California are commonly associated with the uncontrolled development of hillsides underlain by poorly consolidated rock material. Landslides would naturally be common in the area, but as slopes are artificially modified for building sites and roads, the magnitude and frequency of mass movement increase greatly. This results in millions of dollars of property loss each year. Careful geological investigations and proper land-use planning could greatly reduce these losses.

TYPES OF MASS MOVEMENT

Many types of mass movement may be recognized on the basis of behavior of the material and mechanics of movement. The most important are (1) creep, (2) debris flows, and (3) landslides.

Mass movements include all types of slope failure. Because of their potential for destruction, such movements have been studied extensively by engineers as well as geologists. As a result, they have been classified in various ways, depending on the type of motion, type of material involved, and rate of movement (Figure 11.3). In general, three main types of mass movements are recognized: (1) creep, (2) debris flows, and (3) landslides. Creep is an extremely slow, downslope movement involving minor readjustments of the individual particles of regolith. Debris flows are a flow phenomenon with no definite plane along which slippage occurs. Landslides involve movement of a mass of rock or regolith along a definite plane.

Creep

Creep is an extremely slow, almost imperceptible downslope movement of soil and rock debris that results from the constant minor rearrangements of the constituent particles. The motion is so slow that it generally is difficult to observe directly, but it is expressed in a variety of ways (Figure 11.4). On weakly consolidated, grass-covered slopes, evidence of creep can be seen as bulges, or low, wavelike swells in the soil. In road cuts and stream banks, creep can be expressed by the bending of steeply dipping strata in a downslope direction or by the movement of blocks of a distinctive rock type downslope from their outcrop. Additional signs of creep include tilted trees and posts, deformed roads and fence lines, and tilted retaining walls (Figure 11.4). The slow movement of large blocks of bedrock (blockslides) can be considered a type of creep.

Many factors combine to cause creep, but the heaving process that results from the alternating expansion and contraction of the loose rock fragments in the regolith is probably the most important. The heaving process is accom-

How is creep expressed on a hillslope?

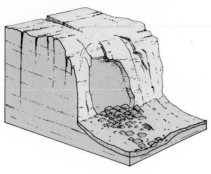

(A) A rockfall is the free fall of rock from steep cliffs.

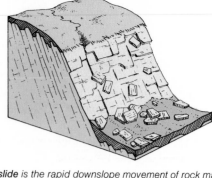

(B) A rockslide is the rapid downslope movement of rock material along a bedding plane, joint, or other plane of structural weakness.

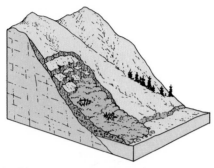

(C) A debris slide is the rapid movement of soil and loose rock fragments. The mass can be dry or moderately wet.

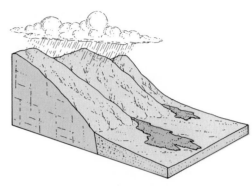

(D) A debris flow is the rapid flow of a mixture of rock fragments, soil, mud, and water. The mixture generally contains a large proportion of mud and water. Mudflows are common.

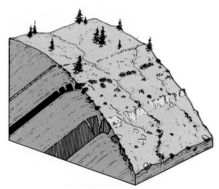

(E) Creep is the slow downslope migration of soil and loose rock fragments, resulting from a variety of processes, including frost heaving.

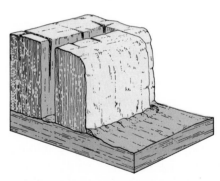

(F) A blockslide is the slow movement of large blocks of material over a layer of weak, plastic material (such as clay or shale).

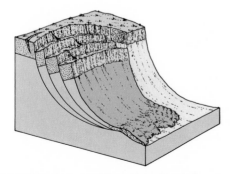

(G) Slump is the slow or moderately rapid movement of a coherent body of rock along a curved rupture surface. Debris flows commonly occur at the end of a slump block.

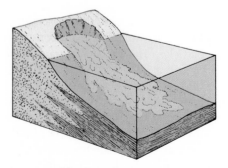

(H) A subaqueous sand flow is the flow of saturated sand or silt beneath the surface of a lake or an ocean.

FIGURE 11.3
Mass movement takes various forms, all of which produce slope retreat and enlarge valleys.
Examples of various types of mass movement are illustrated in the diagrams.

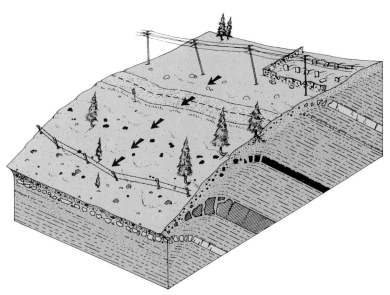

(A) Creep is expressed in various ways, such as the downslope displacement of fence lines, roads, telephone poles, headstones in cemeteries, and rock outcrop debris.

FIGURE 11.4
Creep

(B) Surface expressions of creep. One of the most vivid expressions of creep is seen in exposures along the side of a hillslope where the upper edges of stratified rock bend downslope.

plished in two principal ways: (1) by wetting and drying and (2) by freezing and thawing. In both instances the regolith expands and shifts upward perpendicular to the hillslope. When it contracts, it settles back vertically. With each cycle of expansion and contraction, each particle of rock comes to rest slightly downslope from its original position (Figure 11.5). Repeated expansion and contraction cause the particles to move downslope in a zigzag path. Freeze-thaw cycles will be most numerous in regions where the temperature regularly crosses the freezing point. Therefore, creep is facilitated by cold climates. Cycles of wetting and drying will occur in greatest number where heavy precipitation alternates with periods of desiccation. Wetting and drying will produce much greater volumetric changes where the regolith contains a high proportion of clay because clay minerals have the ability to absorb large amounts of water and, as a result, they expand or swell. Freezing and thawing is similarly affected by grain size, but in this case a relatively high proportion of silt-size particles promotes the greatest degree of expansion and contraction. Thus creep is affected by both climate and the composition of the regolith.

Many other factors also contribute to creep. Growing plants exert a wedgelike pressure between rock particles in the soil and thus cause them to be displaced downslope. Burrowing organisms also displace particles, and with each movement, however slight, the force of gravity pulls them downslope. In addition, creep can be facilitated by undercutting by rain runoff and streams, the increased load of rainwater and snow, and the disturbance of slope surfaces by earthquakes and construction by humans.

Studies in various parts of the world indicate that the rate of creep is highly variable, but some general patterns have been discovered. On moderately steep slopes (10–15 degrees), a rate of 1–2 mm per year is common in humid temperate regions. In semiarid regions with cold winters, creep reaches an average of 5–10 mm per year.

Solifluction (soil flowage) is a special type of creep. It is common in polar regions, where groundwater in the pore spaces of soil and rock is permanently frozen (Figure 11.6). The layer of permanently frozen ground is called the permafrost layer. It ranges from less than a meter to several hundred meters thick and occupies some 20% of the world's land. The presence of permafrost presents some special conditions for the downslope movement of

What produces creep?

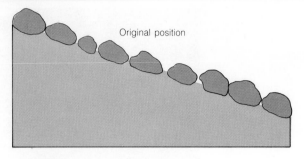

(A) Water seeps into the pore spaces between fragments of loose rock debris.

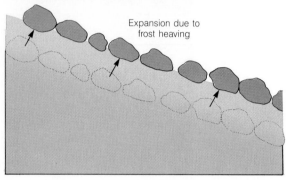

(B) As the water freezes and expands, the soil and rock fragments are lifted perpendicular to the ground surface.

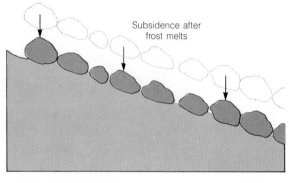

(C) As the ice melts, gravity pulls the particles down vertically, displacing them slightly downhill.

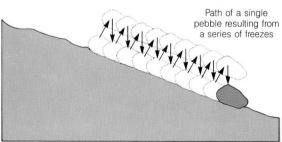

(D) Repeated freezing and thawing cause a significant net displacement downslope.

FIGURE 11.5
Creep can result from repeated expansion and contraction of the regolith. With each cycle, there is a net downslope displacement of all loose material.

regolith. During the spring and early summer, the ground begins to thaw from the surface downward. Because the meltwater cannot percolate downward into the impermeable permafrost layer, the upper zone of soil becomes completely saturated and large areas of the regolith will flow slowly down even the most gentle slopes. Solifluction can also occur in temperate regions in nonfrozen soil if a sufficient amount of water accumulates in the upper soil.

Debris Flows

Debris flows consist of mixtures of rock fragments, mud, and water that flow downslope as a viscous fluid. Movement can range from a flow that is similar

FIGURE 11.6
Solifluction is a major type of mass movement in cold polar regions and in some high mountains. In this view of a hill in Alaska, the water-saturated regolith moves slowly downslope like a viscous liquid.

to the flow of freshly mixed concrete to that of a stream of mud in which rates of flow are nearly equal to those of running water. The consequences of a debris flow can be catastrophic if human habitation lies in its path. The reason for the high velocity of flow is the presence of large amounts of water that penetrate and soak into the regolith. Water acts as a lubricant by decreasing the friction between grains and adds weight to the mass because it replaces the air in the open spaces between the fragments. Therefore, the more water present, the greater the speed of the flow (Figure 11.7).

Mudflows are a variety of debris flow consisting of a large percentage of silt and clay-size particles. They almost invariably result from an unusually heavy rain or a sudden thaw. Their water content can be as much as 30%. As a result of the predominance of fine-grained particles and the high water content, mudflows typically follow stream valleys. They are common in arid and semiarid regions and typically originate in steep-sided gullies where there is abundant loose, weathered debris (Figure 11.8). If they reach the open country at a mountain front, they spread out in the shape of a large lobe, or fan. Because of their density, mudflows can transport large boulders by "floating" them over slopes as gentle as 5 degrees and have been known to move houses and barns from their foundations. Many of the disastrous landslides in southern California are really mudflows that move rapidly down a valley for considerable distances. Mudflows vary in size and rate of flow, depending on water content, slope angle, and available debris. Many are over 100 m thick; some can be as long as 80 km.

The combination of conditions most favorable for rapid flow is the presence, on steep slopes, of unconsolidated or poorly consolidated materials that become water-soaked through abnormal precipitation. The angle of repose for

What factors influence the development of debris flows?

FIGURE 11.7

A debris flow in Spanish Fork Canyon, Utah, was mobilized in April 1983, as a result of high rainfall combined with rapid melting of a thick snowpack. The mass of debris blocked the canyon and formed a lake upstream that completely flooded the town of Thistle. Four million m^3 of debris flowed downslope and cut off railway and highway access to a large area of the western United States. Concerns that the lake would overflow, destroy the dam, and flood the cities and towns downstream prompted engineers to quickly construct a diversion tunnel to control the lake level. The "natural" dam still remains. Damages were estimated at $250 million, making the Thistle debris flow the most costly in U.S. history. More than 90 other debris flows occurred along the steep front of the Wasatch Mountains, some crashing through towns, burying homes.

FIGURE 11.8
Mudflows typically originate on steep slopes where there is an abundance of unconsolidated shale, volcanic ash, or thick regolith. They commonly mobilize during periods of high precipitation and spread out in the valley, forming a large fan, or lobe. These mudflows in the Wasatch Mountains, Utah, resulted from a period of heavy rain in 1983. More than 90 mudflows occurred along the Wasatch front during a period of 2 months.

wet material is much less than for the same material when dry; therefore, dry soil or loosely consolidated rock standing at the angle of repose for dry material is no longer in equilibrium when it becomes wet.

The great Slumgullion mud flow, in the San Juan Mountains of Colorado, moved more than 10 km and finally dammed a fork of the Gunnison River, creating Lake San Cristobal. The flow started in weathered and water-soaked volcanic rocks high in the mountains and moved down the steep valley of a tributary of the Gunnison River.

Illustrations of a special type of mudflow are found in the St. Lawrence valley of eastern Canada and in various parts of Scandinavia. In both regions, marine mud deposited near the margins of receding glaciers has a remarkable property known as "sensitive clay" or "quick clay." As in quicksand, the sediment particles are loosely packed together and consequently have a high water content. With only a slight disturbance, the material can become liquefied or "quick" (transformed from a weak solid to a viscous fluid) without a change in water content. The material can flow rapidly, even on very gentle slopes, once this change takes place. Disastrous debris flows of quick clay have affected a number of settlements along the valley of the St. Lawrence in recent years. In these areas liquefaction appears to be progressive, triggered by some localized internal shearing and spreading rapidly through the body of clay. Large masses of clay become liquefied completely and flow as fast as a river.

Another special type of debris flow is a volcanic mudflow, for which the Indonesian word *lahar* is used. The abundant loose pyroclastic material that accumulates on the flanks of a volcanic cone comes to rest at the angle of repose, so it is inherently unstable. During explosive eruptions the material commonly becomes saturated by rain, or by snow melting to water by the escaping volcanic heat, or from water expelled from a crater lake. The size and speed of such flows can be extremely great. The explosive eruption of Mount

Why are lahars common on volcanoes?

225

Saint Helens triggered several large mudflows that flowed many miles down the Toutle River. Other lahars from prehistoric eruptions of Mount Rainier traveled more than 80 km. The eruption of Vesuvius in A.D. 79 created a flow that engulfed the town of Herculaneum in a slurry up to 20 m thick. More recently, on November 13, 1985, a lahar in Armero, Colombia, raced down the slopes of the ice-capped Andean volcano called Nevado del Ruiz at speeds of more than 150 km an hour and buried more than 25,000 people alive—90% of the residents.

A new type of debris flow is created by mine dumps where unstable slopes of mine tailings accumulate at the angle of repose for dry rock fragments. Perhaps the best known, and certainly one of the most tragic, was the disastrous flow of 1966 that destroyed the village school at Aberfan (South Wales), with the loss of 140 lives. The debris flow had its origin in a large mine dump 40 m high. During a period of heavy rain, the debris became saturated with water and flowed down into the valley, covering the school and adjacent buildings. This example illustrates the need for thorough investigation of slope conditions whenever they are modified by construction or mining.

Landslides

Although the vague term *landslide* has been applied to almost any kind of slope failure, true landslides involve movement along a well-defined slippage plane. Landslides, therefore, differ from creep and debris flows in their mechanics of movement. A landslide block moves as a unit (or series of units) along a definite fracture (or system of fractures), with much of the material moving as a large *slump block* (Figure 11.9). The detached block leaves behind a distinct curved incision, or scar. The slippage plane is typically spoon-shaped. As the block moves downward and outward, it commonly rotates in such a way that bedding or other identifiable surfaces are tilted backward toward the source (Figure 11.9). In the lower part of the slump block, part of the displaced material may move as a debris flow. Several slippage planes commonly develop in the same slide, so the top of the slump block is broken into a series of steps, or small terraces. The characteristic scar, tilting of bedding or other surfaces, and jumbled, poorly drained hillocks formed by previous slides serve to identify terrains that are prone to landslides.

Landslides are fairly common phenomena and occur on a small scale nearly everywhere. Large slides are less numerous but commonly develop on steep slopes of weak shale. They can move in a matter of seconds or slip gradually over a period of weeks and months.

How does a landslide differ from a debris flow?

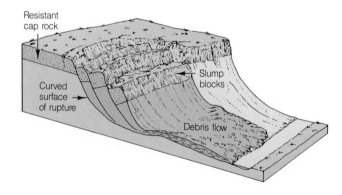

FIGURE 11.9
Landslides occur along rather well-defined slippage surfaces. Large blocks will slump and rotate downslope and many will grade into debris flows at their lower margins. The photograph shows the upper part of a landslide in California where homes were displaced along the curved rupture surface of the slump block.

Many landslides come to rest on a valley floor and in many instances dam the stream flowing through the valley, forming a lake behind them. Such lakes are temporary because the impounded water soon overflows the barrier and rapidly erodes through the unconsolidated rock debris. This may result in catastrophic flooding downstream as the lake is almost instantly drained.

Many landslides are started by earthquakes. An outstanding example is the series of slides that occurred after an earthquake in December 1920 in the loess deposits of the province of Kansu in the interior of China. About 100,000 lives were lost by engulfment as great masses of earth slid off the hills into the valleys. Many villages were completely buried.

Other Types of Mass Movement

Rockslides.   The term *rockslide* is used to denote the rapid movement of a large block of rock along a bedding plane, joint, or other plane of structural weakness (Figure 11.3B). A large block may move en masse for a short distance, but generally there is some disintegration as the body moves downslope and tends to break into smaller blocks of rubble. On one hand the rockslide grades into rockfalls; on the other it grades into landslides, with the entire mass moving as a coherent unit. Joint systems are critical in the development of rockslides because they are continuous fractures through massive rock that ultimately weaken the structure and lead to failure. Once the stress exceeds the cohesive strength along any plane in a rock, mass movement will be initiated. The failure will tend to be progressive because weakening along one joint will direct additional stress on others.

FIGURE 11.10
A recent rockfall in the Swiss Alps is a good example of mass movement and slope retreat. The process will continue until the mountain is completely consumed.

FIGURE 11.11
The Frank rockfall, Alberta, Canada, involved a huge mass of rock that broke away from the mountain face completely burying the mining town of Frank, Alberta. Among the factors that contributed to this slope failure were the steepness of the mountain front, the dip of the bedding planes parallel to the mountain face, and underlying weak shale and coal beds. Mining activity may have triggered the movement.

FIGURE 11.12
A rock glacier in the Canadian Rockies, British Columbia, illustrates many features of this type of mass movement. Note the long tongues of moving rock debris and the wrinkled surface resulting from this flow of rock fragments and ice.

What causes rock glaciers to move?

Rockslides usually occur on steep mountain fronts, but they can develop on slopes with gradients as low as 15 degrees. They are among the most catastrophic of all forms of mass movement. Sometimes millions of metric tons of rock plunge down the side of a mountain in a few seconds (Figure 11.10).

Rockfalls. *Rockfalls* include the free fall of a single fragment ranging from a small grain upward to huge blocks (Figure 11.3A). Over time, great quantities of small- to moderate-size fragments (a few centimeters to a few meters long) shower down from the face of a cliff and accumulate at the base as talus. Ice wedging is a major process in dislodging the fragments. Some rockfalls are much larger—a whole hillside may break off from the face of a mountain (Figure 11.10).

The huge rockfall that buried the town of Frank, Alberta, Canada, in 1903 was among the largest of such mass movements in the world. A gigantic wedge of limestone 400 m high, 1200 m wide, and 160 m thick crushed down from Turtle Mountain at 4:10 A.M., April 29, 1903, and destroyed much of the town of Frank. Seven million tons of rock fell off the mountain face and swept over the valley floor, burying numerous homes, mines, railways, and 3200 acres of farm land to a depth of 30 m. The fall occurred in 100 seconds, dammed the Crowsnest River, and created a small lake at the base of Turtle Mountain (Figure 11.11).

Rock Glaciers. *Rock glaciers* are long tonguelike masses of angular rock debris that resemble a glacier in general outline and form. The surface of a rock glacier is typically furrowed by a series of parallel flow ridges similar to those in an advancing lava flow (Figure 11.12). Evidence of movement includes concentric ridges within the body, the rock glacier's lobate form, and its steep front. Measurements indicate that rock glaciers move as a body downslope at rates ranging from 5 cm (2 inches) a day to 1 m (3 feet) per year. Rock glaciers commonly occur at the heads of glaciated valleys and are fed by a continuous supply of rock fragments produced by ice wedging on the cirque wall. Excavations into rock glaciers reveal ice in the pore spaces between the rock fragments. Presumably, the ice is responsible for much of the flow movement. With a continuous supply of rock fragments from above, the constantly increasing weight causes the ice in the pore spaces to flow. Favorable conditions for the development of rock glaciers thus include steep cliffs and a cold climate. The steep cliffs supply coarse rock debris with large spaces between fragments in which ice can form, and the cold climate keeps the ice frozen. Some rock glaciers may be debris-covered, formerly active glaciers.

Subaqueous Mass Movement. A variety of *subaqueous mass movements* affect large areas of the sea floor and are probably as common as those that occur on land. They are especially active near delta areas, where sediment accumulates rapidly over the continental slope. The soft, water-saturated sediment may slide or flow downslope if the slope becomes unstable. Slump blocks are common along the continental slope, and exceptionally large blocks may move en masse. Sand flows and turbidity flows commonly move farther downslope and out across the abyssal plain, where they can damage submarine cables and other installations and are potential dangers to offshore oilfields. Submarine mass movements are often triggered by earthquakes or large storms.

Subsidence. *Subsidence* is the downward movement of earth material lying at or near the surface. It differs from other types of mass movement in that movement is essentially vertical; there is little or no horizontal component. The primary force producing subsidence is gravity. Before gravity can act, however, other processes must operate in order to create space into which the earth material can sink. Both natural and human agencies can create conditions that will cause subsidence. The formation of caves by solution of minerals of soluble rock by groundwater is a major natural cause of subsidence and will be discussed in Chapter 14. As we will see, subsidence into caves has created a distinctive landscape called karst topography. The natural burning of combustible materials such as peat and coal in the subsurface will also reduce the volume of rock, as will melting of isolated blocks of glacial ice covered with glacial sediment. Subsidence into these large voids is inevitable. Lava tunnels formed by the draining off of liquid lava from the core of a lava flow after the top and sides of the flow have solidified are areas of potential subsidence. Subsidence may also result from compaction of soil and regolith that results from overloading of the surface or from removal of water by evaporation or groundwater seepage.

Examples of subsidence that result from human activity are varied and numerous. Where subsurface mining activity has removed large quantities of rock such as coal and ore, subsidence into the abandoned workings beneath may be so widespread that entire towns built above the underground workings are abandoned. Subsidence may also follow the removal of fluids such as water, oil, or gas from the subsurface. A notable example is the subsidence of buildings in Mexico City as a result of the excessive pumping of groundwater from aquifers below. A similar situation is beginning to occur in Phoenix, Arizona, because a tremendous amount of groundwater is being removed to provide for the rapidly expanding metropolitan area. At Long Beach, California, pumping from the Wilmington oil field caused the surface to subside 10 m in 30 years. Pipelines, bridges, roads, and harbor facilities had to be modified to counter the effect of subsidence. The injection of water into the petroleum reservoir rock has now minimized subsidence in the area by raising the fluid pressure in the subsurface rock.

What geologic conditions are conducive to subsidence?

SLOPE SYSTEMS

Slopes are open, dynamic systems in which the effects of weathering, mass movement, and erosion of minor gully tributaries combine to transport rock material downslope to the main stream.

Although we commonly perceive the surface in most regions to be stable and unchanging from year to year, or even throughout a lifetime, it is obvious from the previous discussion that mass movement of material downslope occurs. The movement is commonly imperceptibly slow, but at times may be rapid and devastating. Is mass movement an abnormal event that occurs only rarely and under special conditions? The answer is a definite no! A *slope* is perhaps best thought of as an open dynamic system that has an input of coarse rock material at the base of a cliff and an output of fine rock fragments into a stream (Figure 11.13). Weathering, mass movement, and the headward erosion of small gully tributaries all combine to transport regolith and loose rock material downslope to the main stream. Mass movement may be the most dramatic demonstration of the transport of rock material downslope, but remember that headward erosion plays a significant role. Numerous tributaries and small gullies operate on essentially all slopes, collecting and funneling debris to the main stream. These small tributaries work headward, undercutting the resistant cliff and thus making it more susceptible to rockfalls and slumping. During heavy storms, much of the loose rock debris on the slope is moved in the tributary system. Debris flows commonly follow the tributary stream channels. In the slope system, therefore, mass movement and headward erosion of minor tributaries operate together. Each facilitates and reinforces the other in transporting debris and causing slope retreat. As the system operates, the valley slope gradually recedes from the stream channel.

How does mass movement influence the slope of stream valleys?

To appreciate the importance of mass movement and headward erosion of minor tributaries in the process of valley development, try to visualize what a valley would look like without these processes. If downcutting of the stream channel were the only process in operation, a deep, vertical-walled canyon would result (Figure 11.14A). This does occur where rocks are strong enough to resist the force of gravity, such as in the deep canyons cut in resistant sandstones and limestones in the Colorado Plateau. More commonly, slope retreat occurs simultaneously with downcutting, to produce sloping valley walls (Figure 11.14B).

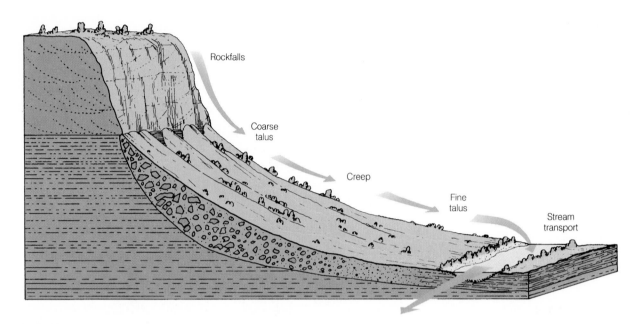

FIGURE 11.13
Slopes are dynamic systems in which material is continually moving downslope and into a drainage channel. Talus produced by frost action and rockfalls accumulates at the base of the cliff. It is then transformed into smaller and smaller particles by mechanical and chemical weathering and moves downslope by creep and other types of mass movement. Some of the debris may be collected by minor tributaries and moved to the main stream by running water. The remaining fine-grained debris continues to move downslope by gravity. It is eventually fed into a stream, which carries it out through the drainage system.

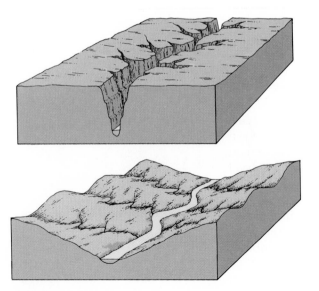

(A) Where rocks are very resistant, mass movement cannot keep pace with downcutting. As a result, a vertical-walled canyon develops.

(B) If slope processes can keep pace with downcutting, smooth rolling hills and valleys develop.

FIGURE 11.14
Slope processes affect the development of landscapes in important ways.

SUMMARY

1. Mass movement is the transfer of material downslope through the direct action of gravity.
2. Gravity is the driving force for all types of mass movement. In order for mass movement to occur, however, the force of gravity must exceed the cohesive strength of the surface material.
3. The main factors that aid or initiate mass movement are (1) the saturation of material with water, (2) vibrations from earthquakes, (3) the alternating expansion and contraction of the regolith, and (4) the undercutting of slopes by streams and waves.
4. The main types of mass movement are (1) creep, (2) debris flows, and (3) landslides. Other types of mass movement include solifluction, rockfalls, lahars, and subaqueous slides and flows.
5. Creep is an extremely slow downslope movement of soil and rock produced primarily by the expansion and contraction of the surface materials as a result of cycles of wetting and drying or of alternating freezing and thawing.
6. Debris flows consist of a mixture of rock fragments and water that flows downslope as a viscous fluid.
7. A lahar is a special type of debris flow composed of volcanic ash.
8. Landslides are a type of mass movement in which the material moves as a unit or block along definite slippage planes.
9. Subsidence differs from other types of mass movement in that it has essentially vertical motion. Voids in the subsurface that create subsidence are created by groundwater, lava tunnels, combustion of peat and coal, compaction, and removal of subsurface material by mining and by pumping of water, oil, and gas.
10. Slopes are dynamic systems of moving material that consist of an input of rock debris near the top of the slope and an output of finer rock fragments into a stream.

KEY TERMS

angle of repose (p. 216)
creep (p. 219)
debris flow (p. 222)
lahar (p. 225)
landslide (p. 226)
mass movement (p. 219)
mudflow (p. 223)
rockfall (p. 228)
rock glacier (p. 228)

rockslide (p. 227)
slope system (p. 229)
slump (p. 220)
slump block (p. 226)
solifluction (p. 221)
subaqueous mass movement (p. 228)
subsidence (p. 228)

REVIEW QUESTIONS

1. List the factors that affect mass movement.
2. Why does deforestation cause unstable slopes and accelerate mass movement?
3. Describe four types of rapid mass movement.
4. List the types of mass movement that are dominantly slow.
5. List five ways in which creep is expressed on a hillslope.
6. What factors promote debris flows?
7. What causes creep?
8. What is solifluction?
9. Explain why slopes are considered open systems.

ADDITIONAL READINGS

Bolt, B. A., W. L. Horn, G. A. MacDonald, and R. F. Scott. 1975. Hazards from Landslides, in *Geological Hazards,* chapter 4. New York: Springer-Verlag.

Fleming, R. W., and F. A. Taylor. 1980. Estimating the Cost of Landslide Damage in the United States. *U.S. Geological Survey Circular* 832.

Radbrich-Hall, D. H., R. B. Colton, W. E. Davies, B. A. Skipp, I. Lucchita, and D. J. Varnes. 1976. Preliminary Landslide Overview Map of the Conterminous United States. *U.S. Geological Survey Miscellaneous Field Studies Map* MF-771.

Schuster, R. L., and R. J. Krizek, eds. 1978. Landslides, Analysis and Control. National Research Council, Transportation Research Board Special Report 176.

Schuster, R. L., D. J. Varnes, and R. W. Fleming. 1981. Landslides, in *Facing Geologic and Hydrologic Hazards,* ed. W. W. Hays, pp. 55–65. *U.S. Geological Survey Professional Paper* 1240-B.

Utgard, R. O., G. D. McKenzie, and D. Foley. 1978. *Geology in the Urban Environment.* Minneapolis: Burgess.

Zaruba, Q., and V. Mencl. *Landslides and Their Control.* 1969. New York; Elsevier.

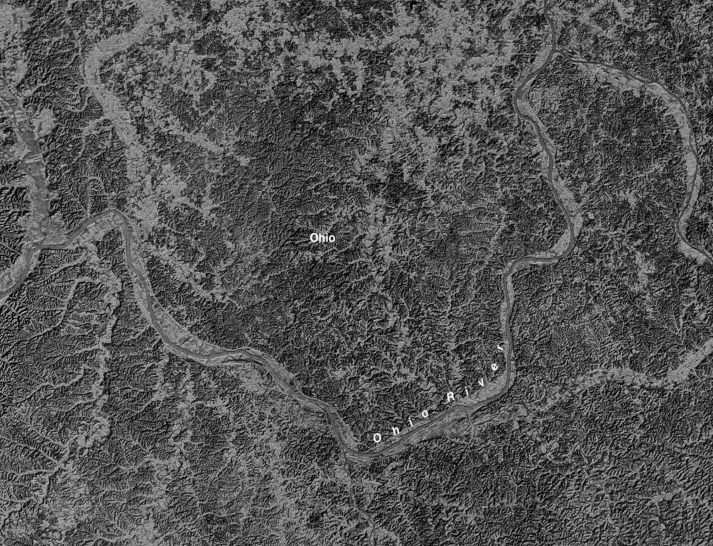

Ohio

Ohio River

West Virginia

Kentucky

12
River
Systems

A river system is a network of connecting channels through which water, precipitated on the surface, is collected and funneled back to the ocean. At any given time, about 1300 km³ of water flows in the world's rivers. As it moves, it picks up weathered rock debris and carries it to the oceans.

This Landsat mosaic shows part of the western Appalachian Plateau near the common border of West Virginia, Kentucky, and Ohio. The region is underlain by flat-lying shale, siltstone, and sandstone, which are easily eroded in the humid temperate climate to an intricate system of tributary stream valleys. The entire area is dissected. No surface remains untouched by erosion. The entire landscape is a system of valley slopes formed by stream action.

The Ohio River is the major trunk stream that collects runoff from the area. It flows through a wide, flat valley from the north, southward to the central part of the photo, and then turns abruptly and flows to the northwest. An ancient river channel, now mostly abandoned, appears as a wide valley without a river. This is part of the pre–ice age drainage system from which the present Ohio River evolved. Studies of sediment within the ancient valley indicate that it was abandoned about 700,000 years ago. Just by looking at the network of stream valleys, you can begin to appreciate the importance of running water in forming the landscape.

1. Running water is part of Earth's hydrologic system and is the most important agent of erosion. Stream valleys are the most abundant and widespread landforms on the continents.
2. A river system consists of a main channel and all of the tributaries that flow into it. It can be divided into three subsystems (1) a collecting system, (2) a transporting system, and (3) a dispersing system.
3. The most important variables in stream flow are (1) discharge, (2) velocity, (3) gradient, (4) sediment load, and (5) base level.
4. The variables in a stream constantly adjust toward a state of equilibrium.
5. Rivers erode by (1) removing regolith, (2) downcutting, and (3) headward erosion.
6. As a river develops a low gradient, it deposits part of its load on point bars, on natural levees, and across the surface of its floodplain.
7. Most of a river's sediment is deposited where the river empties into a lake or ocean. This deposition commonly builds a delta at the river's mouth. In arid regions, many streams deposit their loads as alluvial fans at the base of steep slopes.
8. The origin and evolution of the major rivers of the world are controlled by the tectonic and hydrologic systems.

GEOLOGIC IMPORTANCE OF RUNNING WATER

Running water is by far the most important agent of erosion. Other agents, such as groundwater, glaciers, and wind, are locally dominant but affect only limited parts of Earth's surface.

An attempt to appreciate the significance of streams and stream valleys in the regional landscape of Earth presents a problem of perspective, much like trying to appreciate the abundance of craters on the Moon from viewpoints on the lunar surface. To an astronaut on the Moon, the surface appears to be an irregular, broken landscape cluttered with rock debris. The crater systems and circular terrain patterns, so striking when viewed from space, are not at all apparent from vantage points on the Moon's surface. Indeed, crater rims appear only as rounded hills. Without the aid of maps or space photographs, some of the larger craters may not be recognizable as circular forms when they are seen only from *Apollo* landing sites.

Viewed from the ground, Earth's stream valleys also may appear to be only irregular depressions between rolling hills and plains. Viewed from space, however, stream valleys are seen to dominate most continental landscapes of Earth, much as craters dominate the landscape of the Moon.

What is the most common landform on Earth's surface?

The ubiquitous stream valleys on Earth's surface and the importance of running water as the major agent of erosion can best be appreciated by considering a broad, regional view of the continents and their major river systems, as seen through high-altitude photography. As the photographs in Figure 12.1 show, the surface, throughout broad regions of the continents, is little more than a complex of valleys created by stream erosion (see also the Landsat image on page 232.). Even in the desert, where it sometimes does not rain for decades, the network of stream valleys commonly is the dominant landform. No other landform on the continents is as abundant and significant.

Look again at the space photograph in Figure 2.5. Is there any part of the terrain that is not influenced by stream erosion?

THE MAJOR CHARACTERISTICS OF A RIVER SYSTEM

A river system consists of a main channel and all of the tributaries that flow into it. A typical river system can be divided into three subsystems: (1) a collecting system, (2) a transporting system, and (3) a dispersing system.

Although rivers and the valleys through which they flow are the most familiar of all nature's phenomena, it is difficult to define the word river precisely because of the great variety of physical characteristics rivers exhibit. There are big rivers, such as the Mississippi, Amazon, and Nile, and there are little rivers, sometimes referred to as streams, creeks, or brooks. Some rivers in hot and dry regions flow only after a heavy rain and then dry up, whereas rivers in the Arctic are frozen two-thirds of the year. From the viewpoint of geology, it is perhaps most useful to consider a river not as a natural channel through which water flows but as a system. A *river system*, or a *drainage basin*, consists of a main channel and all of the tributaries that flow into it. It is bounded by a *divide* (ridge), beyond which water is drained by another system. Within a river system, the surface of the ground slopes toward the network of tributaries, so the *drainage system* acts as a funneling mechanism for removing surface runoff and weathered rock debris. A typical river system can be divided into three subsystems:

1. A collecting system
2. A transporting system
3. A dispersing system

A map of a typical river system is shown in Figure 12.2. Although the boundaries between the three subsystems are somewhat gradational, the distinguishing characteristics of each subsystem on a regional scale are readily apparent.

The Collecting System

The *collecting system* of a river, consisting of a network of *tributaries* in the headwater region, collects and funnels water and sediment to the main stream. It commonly has a *dendritic* (treelike) *drainage* pattern, with numerous branches that extend upslope toward the divide. Indeed, one of the most remarkable characteristics of the collecting system is the intricate network of tributaries, shown in the enlargement in Figure 12.2, which was made by plotting all visible streams shown on an aerial photograph. That is not, however, the entire system. Each of the smallest tributaries shown in Figure 12.2 has its own system of smaller and smaller tributaries, so the total number becomes astronomical. From the details in Figure 12.2, it is apparent that most of the land's surface is part of some drainage basin.

The Transporting System

The *transporting system* is the main trunk stream, which functions as a channelway through which water and sediment move from the collecting area toward the ocean. Although the major process is transportation, this subsystem also collects additional water and sediment. Deposition occurs where the channel meanders back and forth and at times when the river overflows its banks during a flood stage. Erosion, deposition, and transportation thus occur in the transporting subsystem of a river.

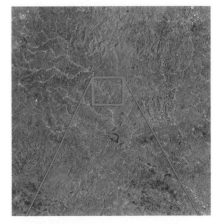

(A) A Landsat image of an area in the Ozark Plateau in Missouri shows the regional patterns of a river system and its valleys.

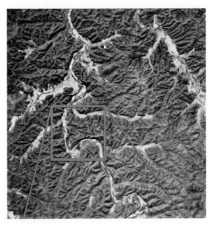

(B) A high-altitude aerial photograph of a portion of the area shown in A reveals an intricate network of streams and valleys within the tributary regions of the larger streams.

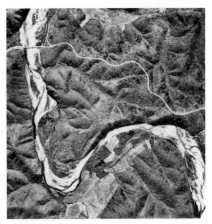

(C) A low-altitude aerial photograph of part of the area shown in B reveals many smaller streams and valleys in the drainage system.

FIGURE 12.1
Erosion by running water is the dominant process in the formation of the landscape. The Landsat image was taken from an altitude of 650 km. The aerial photographs were taken from 12 km and 3 km.

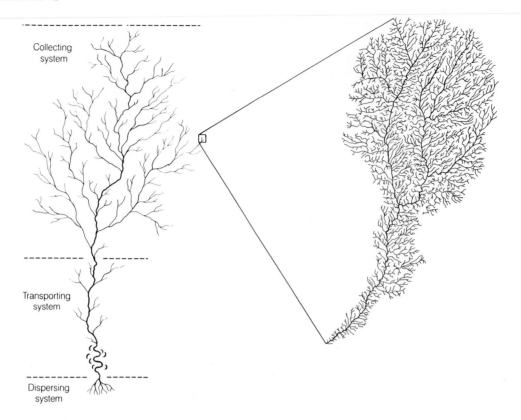

Collecting
system

Transporting
system

Dispersing
system

FIGURE 12.2
The major parts of a river system are characterized by different geologic processes. The tributaries in the headwaters constitute a subsystem that collects water and sediment and funnels them into a main trunk stream. Erosion is dominant in this headwater area. The main trunk stream is a transporting subsystem. Both erosion and deposition can occur in this area. The lower end of the river is a dispersing subsystem, where most sediment is deposited in a delta or an alluvial fan and water is dispersed into the ocean. Deposition is the dominant process in this part of the river.

How do the characteristics of a river change from the headwaters to the sea?

The Dispersing System

The ***dispersing system*** consists of a network of ***distributaries*** at the mouth of a river, where sediment and water are dispersed into an ocean, a lake, or a dry basin. The major processes are the deposition of the coarse sediment load and the dispersal of fine-grained material and river waters into the basin.

Order in Stream Systems

It is apparent from Figures 12.1 and 12.2 that a stream does not occur as a separate, independent entity. Every stream, every river, and every gully and ravine are part of a drainage system, with each tributary intimately related to the stream into which it flows and to the streams that flow into it. Every stream has tributaries, and every tributary has smaller tributaries, extending down to the smallest gully. Studies of drainage systems show that when a stream system develops freely on a homogeneous surface, definite mathematical ratios characterize the relationships between the tributaries and the size and gradient of the stream and of the stream valley. Some of the more important relationships and generalizations are the following:

1. The number of stream segments (tributaries) decreases downstream in a mathematical progression.
2. The length of tributaries becomes progressively greater downstream.
3. The gradient, or slope, of tributaries decreases exponentially downstream.
4. The stream channels become progressively deeper and wider downstream.
5. The size of the valley is proportional to the size of the stream and increases downstream.

These relationships constitute the basis for the assumption that streams erode the valleys through which they flow.

To appreciate stream order and system in a drainage basin, carefully study Figure 12.3, which is a graphic model of an idealized stream. The decrease in the gradient of the tributaries is represented by the curve on the side of the block diagram. The curve can be expressed by a mathematical formula. As is shown in the diagram, the stream channel becomes progressively deeper and wider downstream. Similarly, the sizes of the valleys increase downstream in proportion to the sizes of the streams that flow through them. The lengths of the tributaries also increase downstream. Numerous statistical studies support these observations, showing that a number of definite mathematical relationships exist between a stream and the valley through which it flows.

If valleys were ready-made by some process other than stream erosion, such as faulting or other earth movements, these relationships would be "infinitely improbable." You can easily confirm the high degree of order in streams by studying the aerial photographs in Figure 12.1 and the Landsat image on page 232. Does each tributary have a steeper gradient than the stream into which it flows? Does each tributary flow smoothly into a larger stream without an abrupt change in gradient? Are the tributary valleys smaller than the valleys into which they drain?

Stream erosion has been studied in great detail over the last 100 years, and geologists have been able to observe and measure many aspects of stream development and erosion by running water. The origin of valleys by erosion is well established, and running water is clearly the most significant agent of erosion on Earth's surface.

How do we know streams have eroded the valleys through which they flow?

THE DYNAMICS OF STREAM FLOW

Rivers are highly complex systems influenced by a number of variables and, as is the case with so many natural systems, if one variable is changed it produces a change in the others. The most important variables are (1) discharge, (2) velocity, (3) gradient, (4) sediment load, and (5) base level.

Anyone who has watched the fascinating flow of water in a river realizes that the process is complex. The water moves down the stream channel through the force of gravity, and the velocity of flow increases with the slope or gradient of the stream bed. The flow, however, is not in a straight line downslope, but is turbulent, with numerous secondary eddies and swirls in addition to the main downstream current. It is obvious to the observer that the energy of a river increases with increasing gradient and increasing volume of water. In most streams, moreover, there is a considerable load of sediment carried

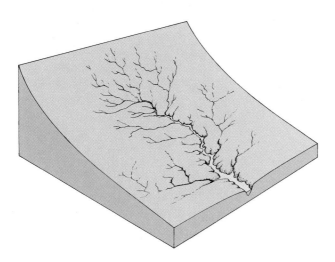

FIGURE 12.3
The characteristics of a river change systematically downstream. The gradient decreases downstream, and the channel becomes larger. Other downstream changes include an increase in the volume of water and an increase in the size of the valley through which the stream flows.

downstream by the flowing water. The flow of water in natural streams is variable, fluctuating from hour to hour and from place to place. Some of the most important variables in stream dynamics are (1) discharge, (2) velocity, (3) gradient, (4) sediment load, and (5) base level. As we will see, these variables are intimately related and a change in one will bring about a compensating change in another.

What are the major factors influencing the flow of water in river systems?

Discharge

Discharge is the amount of water passing a given point during a specific interval of time. It is usually measured in cubic meters per second. The discharges of most of the world's major drainage systems have been monitored by gauging stations for years and clearly indicate that the water for a river system comes from both surface runoff and seepage of groundwater into the stream channels. Groundwater seepage is important because it can maintain the flow of water throughout the year. Continual seepage establishes *permanent streams*. If the supply of groundwater is depleted seasonally, streams become temporarily dry. Such streams are called *intermittent streams*.

Velocity

The velocity of flowing water is not uniform throughout the stream channel. It depends on the shape and roughness of the channel and on the stream pattern. The velocity usually is greatest near the center of the channel and above the deepest part, away from the frictional drag of the channel walls and floor (Figure 12.4). As the channel curves, however, the zone of maximum velocity shifts to the outside of the bend, and a zone of minimum velocity forms on the inside of the curve (Figure 12.5). This flow pattern is an important cause of the lateral erosion of stream channels and the migration of stream patterns.

How does a channel of a meandering river change its course?

The velocity of flowing water is proportional to the gradient of the stream channel. Steep gradients produce rapid flow, which commonly occurs in high-mountain streams. Where slopes are very steep, waterfalls and rapids develop, and the velocity approaches that of free fall. Low gradients produce slow, sluggish flow, and where a stream enters a lake or an ocean, its velocity is soon reduced to zero. The velocity of flowing water in a given channel also depends on the volume. The greater the volume, the faster the flow.

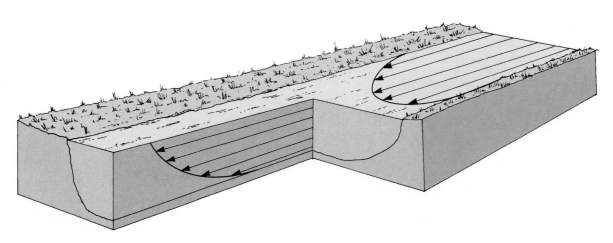

FIGURE 12.4
Variations in the velocity of flow in natural stream channels occur both horizontally and vertically. Friction reduces the velocity along the floor and sides of the channels. The maximum velocity in a straight channel is near the top and center of the channel.

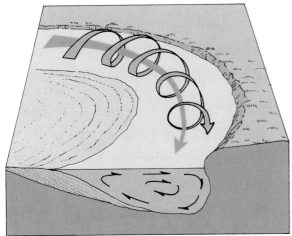

(A) Flow in a curved channel follows a corkscrew pattern. Water on the outside of the bend is forced to flow faster than that on the inside of the curve. This difference in velocity, together with normal frictional drag on the channel walls, produces a corkscrew flow pattern. As a result, erosion occurs on the outer bank, and deposition occurs on the inside of the bend. These processes produce an asymmetrical channel, which slowly migrates laterally.

FIGURE 12.5
Stream flow in a curved channel

(B) Variations in stream flow, around a meander bend, in a river in eastern Canada. Note the sharp steep bank on the outside of the meander bend where velocity is greatest and erosion occurs. On the inside of the meander velocity is at a minimum and deposition occurs to form a point bar. Note the scars on the point bar which mark the previous position of the stream channel.

Stream Gradient

Certainly one of the most obvious factors controlling stream flow is the *gradient,* or slope, of the stream channel. The gradient of a stream is steepest in the headwaters and decreases downslope. The *longitudinal profile* (a cross section of a stream from its headwaters to its mouth) is a smooth, concave-upward curve that becomes very flat at the lower end of the stream. The gradient usually is expressed in the number of meters the stream descends for each kilometer of flow. The headwater streams that drain the Rockies can have gradients of over 50 m/km; the lower reaches of the Mississippi River have a gradient of only 1 or 2 cm/km.

Sediment Load

Flowing water in natural streams provides a fluid medium by which loose, disaggregated regolith is picked up and transported to the ocean. The capacity of a stream to transport sediment increases to a third or fourth power of its velocity; that is, if the velocity is doubled, the stream can move from 8 to 16 times as much sediment. Running water is the major agent of erosion, not only because it can abrade and erode its channel, but because of its enormous power to transport loose sediment produced by weathering. Within a stream system, sediment is transported in three ways (Figure 12.6):

1. Fine particles are moved in suspension (suspended load).
2. Coarse particles are moved by traction along the stream bed (bed load).
3. Dissolved material is carried in solution (dissolved load).

Suspended Load. The *suspended load* is the most obvious, and generally the largest, fraction of material moved by a river. In most major streams, silt and clay-size particles remain in suspension most of the time and move downstream at the velocity of the flowing water, to be deposited in an ocean, in a lake, or on a floodplain.

How do rivers in the United States transport an average of 1.3 million tons of sediment per day to the oceans?

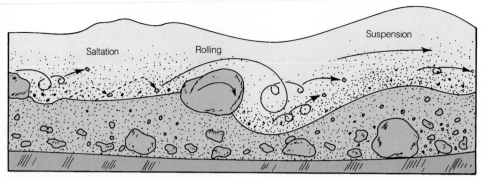

FIGURE 12.6
Movement of the sediment load in a stream is accomplished in a variety of ways. Mud and silt are carried in suspension. Particles that are too large to remain in suspension are moved by sliding, rolling, and saltation. Increases in discharge, due to heavy rainfall or spring snowmelt, can flush out all of the loose sand and gravel, so the bedrock is eroded by abrasion.

Bed Load. Particles of sediment too large to remain in suspension collect on the stream bottom and form *bed load,* or traction load. These particles move by sliding, rolling, and *saltation* (short leaps). The bed load moves only if there is sufficient velocity to move the large particles. It thus differs fundamentally from the suspended load, which moves constantly. There is not always a sharp distinction between the largest particles of the suspended load and the smallest particles of the bed load because the velocity of a stream fluctuates constantly; part of the bed load can suddenly move in suspension, or part of the suspended load can settle. The bed load can constitute 50% of the total load in some rivers, but it usually ranges from 7 to 10% of the total sediment load. The movement of the bed load is one of the major tools of stream abrasion because, as the sand and gravel move, they abrade (wear away) the sides and bottom of the stream channel.

Dissolved Load. The *dissolved load* is matter transported in the form of chemical ions and is essentially invisible. All streams carry some dissolved material, which is derived principally from the groundwater that emerges from seeps and springs along the river banks. The most abundant materials in solution are calcium and bicarbonate ions, but sodium, magnesium, chloride, ferric, and sulfate ions are also common. Various amounts of organic matter are present, and some streams are brown with organic acids derived from the decay of plant material.

Flow velocity, which is so important to the transportation of the suspended and traction loads, has little effect on a river's capacity to carry dissolved material. Once mineral matter is dissolved, it remains in solution, regardless of velocity, and is precipitated and deposited only if the chemistry of the water changes.

Chemical analysis shows that most rivers carry a dissolved load of less than a thousand parts per million. Some streams in arid regions, however, carry several thousand parts per million and are distinctly salty in taste. Although these amounts of dissolved material seem small, they are far from trivial. Sampling indicates that, in some rivers, when rainfall is abundant, more than half of all the material carried to the ocean is in solution.

Base Level

The *base level* of a stream is the lowest level to which the stream can erode its channel. It is a key feature in the study of stream activity. The base level is, in effect, the elevation of the stream's mouth where the stream enters an ocean, a lake, or another stream. A tributary cannot erode lower than the level of the stream into which it flows. Similarly, a lake controls the level of erosion for the entire course of the river that drains into it. The levels of tributary junctions and lakes are temporary base levels: lakes can be filled with sediment or drained, and streams can then be established across the former lakebed. For all practical purposes, the ultimate base level is sea level because the energy of a river is quickly reduced to zero as it enters the ocean. (Exceptions are fault valleys like Death Valley, which is faulted below sea level.) Even sea level can change, however, and as base level or sea level changes, the longitudinal profile of the river changes, and the stream adjusts to the new condition.

THRESHOLD VELOCITY FOR SEDIMENT TRANSPORT

The ability of running water to erode and transport sediment is largely dependent on stream velocity.

The results of experimental studies of water's capacity to erode, transport, and deposit sediments are summarized in Figure 12.7. The velocity at which a particle of a given size is picked up and moved is shown on the upper curve. This curve is a zone on the graph instead of a line because the erosional velocity varies according to the characteristics of the water and the shape and density of the grain to be moved. (Shape has a definite effect on the suspension of flat mica flakes, for example.) Moreover, erosional velocities vary with the depth and density of the water. The lower curve shows the *settling velocity*—the velocity at which a particle settles out of suspension. Small particles, such as fine silt and clay, require a relatively high velocity to lift them into suspension because they tend to stick together when they are deposited. Once in suspension, however, fine particles remain suspended with a minimum velocity.

The graph in Figure 12.7 shows also that sediment particles are deposited according to size if the velocity of the water decreases below the particle's critical settling velocity. On gentle slopes, where the stream's velocity is reduced, a significant part of the sediment load is thus deposited along the channel or on the floodplains. Most of the remaining sediment is deposited where the river enters a lake or the ocean because velocity is reduced there.

The time of greatest erosion and transportation of sediment is, of course, during floods. The increase in discharge results in an increase in both the maximum size of the transported material and the size of the total load. An exceptional flood can cause a large amount of erosion and transportation during a brief period. In arid regions, many streams carry little or no water during most of the year, but during a cloudburst they can move great quantities of sediment.

How is the velocity of a river related to erosion and the transport and deposition of sediment?

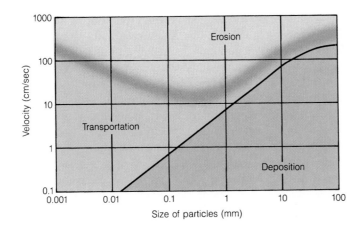

FIGURE 12.7
The threshold velocity for sediment transport provides important insight into processes of stream erosion, transportation, and deposition. The upper curve shows the range of minimum velocities at which a stream can pick up and move a particle of a given size. This threshold velocity is represented by a zone on the graph, not a line, because of variations resulting from stream depth, particle shape and density, and other factors. The lower curve indicates the velocity at which a particle of a given size settles out and is deposited. Note that fine particles stay in suspension at velocities much lower than those required to lift them from the surface of the stream bed.

EQUILIBRIUM IN RIVER SYSTEMS

A river system functions as a unified whole—any change in one part of the system affects the other parts. The major factors that determine stream flow (discharge, velocity, channel shape, gradient, and load) constantly change toward a balance, or equilibrium, so eventually the gradient of the stream is adjusted to accommodate the volume of the water available, the channel's characteristics, and the velocity necessary to transport the sediment load.

What major factors of a river system continually change in an effort to establish equilibrium? What happens when a river's equilibrium is disturbed?

We have repeatedly emphasized the fact that a stream does not occur as a separate, independent entity. One of the most important characteristics of a river system is that it functions as a unified whole—any change in one part of the system affects the other parts. The major factors that determine stream flow (discharge, velocity, channel shape, gradient, base level, and load) constantly change toward a balance, or equilibrium, so eventually the gradient of a stream is adjusted to accommodate the volume of water available, the channel's characteristics, and the velocity necessary to transport the sediment load. A change in any of these factors causes compensating adjustments to restore equilibrium within the entire drainage system. A river is in equilibrium if its channel form and gradient are balanced so that neither erosion nor deposition occurs; rivers are constantly adjusting to approach this ideal condition. This adjustment is important to an understanding of the natural evolution of the landscape. It also has practical considerations: if we continually modify rivers to suit our needs, we should know how river systems respond to artificial modifications.

The concept of equilibrium in a river system can be appreciated by considering a hypothetical stream in which equilibrium has been established. In Figure 12.8A the variables in the stream system (discharge, velocity, gradient, base level, and load) are in balance, so neither erosion nor sedimentation occurs along the stream's profile. There is just enough water to transport the available sediment down the existing slope. Such a stream is in equilibrium and is known as a **graded stream**. In Figure 12.8B the stream's profile is displaced by a fault that creates a waterfall. The increased gradient across the falls greatly increases the stream's velocity at that point, so rapid erosion occurs, and the waterfall (or the rapid) begins to migrate upstream. The eroded sediment added to the stream segment on the dropped fault block is more than the stream can transport because the system was already in equilibrium before faulting occurred. The river therefore deposits part of its load at that point, thus building up the channel gradient (the shaded areas in Figure 12.8C–E) until a new profile of equilibrium is established.

An example of the adjustments just described occurred in Cabin Creek, a small tributary of the Madison River, north of the Hebgen Dam in Montana. In 1959, a 3-m fault scarp formed across the creek during the Hebgen Lake earthquake. By June 1960, erosion by Cabin Creek had erased the waterfall at the cliff formed by the fault, and only a small rapid was left. By 1965, the rapid was completely removed, and equilibrium was reestablished.

Equilibrium in a river system is also illustrated by the results of dam construction. In the reservoir behind a dam, the gradient is reduced to zero. Hence, where the stream enters the reservoir, its sediment load is deposited as a delta and as layers of silt and mud over the reservoir floor (Figure 12.9). Because most sediment is trapped in the reservoir, the water released downstream has practically no sediment load. The clear water in the river downstream of the dam is therefore capable of much more erosion than the previous river, which carried a sediment load adjusted to its gradient. As a result, extensive scour and erosion commonly result downstream from a new dam.

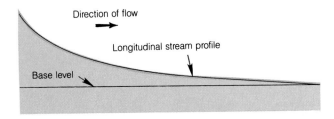

(A) *Initially, when the stream profile is at equilibrium, the velocity, load, gradient, and volume of water are in balance. Neither erosion nor deposition occurs.*

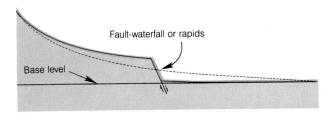

(B) *Faulting disrupts equilibrium by decreasing the gradient downstream and increasing the gradient at the fault line.*

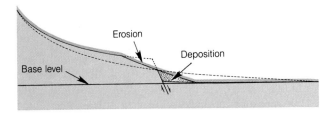

(C) *Erosion proceeds upstream from the fault, and deposition occurs downstream.*

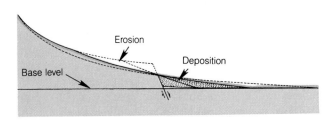

(D) *Erosion and deposition continue to develop a new stream profile at which the velocity, load, gradient, and volume of water will be in balance so that neither erosion nor deposition occurs.*

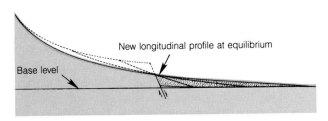

(E) *A new profile of equilibrium, in which neither erosion nor deposition occurs, is eventually reestablished.*

FIGURE 12.8
Adjustments of a stream to reestablish equilibrium are illustrated by profile changes after disruption by faulting.

How have humans modified river systems? What are the results?

The Aswan High Dam of the Nile provides a good example of the many consequences of modifying a river system that has approached equilibrium. For centuries, the Nile River has been the sole source of life in Egypt. The principal headwaters of the Nile are located in the high plateaus of Ethiopia. Once a year, for approximately a month, the Nile used to rise to flood stage and cover much of the fertile farmland in the Nile Delta area. The Aswan High Dam, completed in the summer of 1970, was intended to provide Egypt with water to irrigate 1 million acres of arid land and to generate 10 billion kilowatts of power. This, in turn, was to double the national income and permit industrialization. The dam, however, destroyed the equilibrium of the Nile, and many unforeseen adjustments in the river resulted. This is what happened.

FIGURE 12.9
The volume of sediment transported by a stream is illustrated by the Mono Reservoir in California, which has been completely filled with sand and mud.

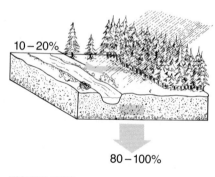

FIGURE 12.10
Surface runoff vs. infiltration under natural conditions. From 80 to 100% of the surface water filters into the subsurface, and from 0 to 20% flows through the drainage system.

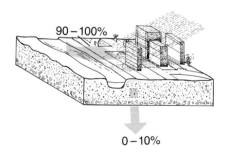

FIGURE 12.11
Surface runoff vs. infiltration in urban areas. From 0 to 10% filters into the subsurface, and 90 to 100% moves as surface runoff.

The Nile is not only the source of water for the delta; it is also the source of sediment. When the dam was finished and began to trap sediment in a reservoir (Lake Nasser), the physical and biological balance in the delta area was destroyed. Without the annual "gift of the Nile," the delta coastline is now exposed to the full force of marine currents, and wave erosion is eating away at the delta front. Some parts of the delta are receding several meters a year.

The sediment previously carried by the Nile was an important link in the aquatic food chain, nourishing marine life in front of the delta. The recent lack of Nile sediment has reduced plankton and organic carbon to a third of the former levels. This change either killed off or drove away sardines, mackerel, clams, and crustaceans. The annual harvest of 16,000 metric tons of sardines and a fifth of the fish catch have been lost.

The sediment of the Nile also naturally fertilized the floodplain. Without this annual addition of soil nutrients, Egypt's 1 million cultivated acres need artificial fertilizer.

The water discharged from the reservoir is clear, free of most of its sediment load. Without its load, the discharged water flows downstream much faster and is vigorously eroding the channel bank. This scouring process has already destroyed three old barrier dams and more than 500 bridges built since 1953. Ten new barrier dams must be built between Aswan and the ocean at a cost equal to one-fourth the cost of the Aswan Dam itself.

The annual flood of the Nile was also important to the ecology of the area because it washed away salts that formed in the arid soil. Soil salinity has already increased, not only in the delta, but throughout the middle and upper Nile areas. Unless corrective measures are taken (at a cost of over $1 billion), millions of acres will revert to desert within a decade.

The change in the river system has permitted double cropping, but there are no periods of dryness. The dry seasons previously helped limit the population of bilharzia, a blood parasite carried by snails, which infects the intestinal and urinary tracts of humans. One out of every two Egyptians now has the infection, and it causes a tenth of the deaths in the country.

Problems have also occurred in the lake behind the dam. The lake was to have reached a maximum level in 1970, but it might actually take 200 years to fill. More than 15,000,000 m³ of water annually escape underground into the porous Nubian Sandstone, which lines 480 km of the lake's western bank. The sandstone is capable of absorbing an almost unlimited quantity of water. Moreover, the lake is located in one of the hottest and driest places on Earth, and the rate of evaporation is staggering. A high rate was expected, but additional losses from transpiration by plants growing along the lakeshore and increased evaporation caused by high winds have brought the total loss of water from the lake to nearly double the expected rate. This loss equals half the total amount of water that once was "wasted," flowing unused to the ocean.

Another way that equilibrium in river systems has been disrupted is through urbanization. The construction of cities may at first seem unrelated to the modification of river systems, but a city significantly changes the surface runoff, and the resulting changes in river dynamics are becoming serious and costly. Water that falls to Earth as precipitation usually follows several paths in the hydrologic system. Generally, from 54 to 97% returns to the air directly by evaporation and transpiration, from 2 to 27% collects in stream systems as surface runoff, and from 1 to 20% infiltrates the ground and moves slowly through the subsurface toward the ocean. Urbanization disrupts each of these paths in the normal hydrologic system. It changes the nature of the terrain and consequently affects the rates and percentages of runoff and infiltration. Roads, sidewalks, and roofs of buildings render a large percentage of the surface impervious to infiltration. Not only does the volume of surface runoff increase, but runoff is much faster because water is channeled through gutters, storm drains, and sewers. As a result, flooding increases in intensity and frequency (see Figures 12.10 and 12.11).

PROCESSES OF STREAM EROSION

> *River systems erode the landscape by three main processes: (1) the removal of regolith, (2) downcutting of the stream channel by abrasion, and (3) headward erosion.*

Erosion of the land is one of the major effects of the hydrologic system. It has occurred throughout all of geologic time and will continue as long as the system operates and land is exposed above sea level. Evidence of erosion is ubiquitous and varied. We see it in the development of gullies on farmlands and in the cutting of great canyons. We see it in the thick layers of sedimentary rocks that cover large parts of the continents and bear witness to erosion during past ages. But exactly how does a river system erode the land? How can a relatively small stream like the Colorado River form the Grand Canyon, which is more than a mile deep and fifteen miles wide? What processes are involved in erosion? How do river systems evolve? Answers to these basic questions have eluded scientists until recent times and even today some details remain controversial. But we now know that erosion by running water and the evolution of a river system is accomplished by three basic processes: (1) removal of regolith, (2) downcutting of the stream channel by abrasion, and (3) headward erosion.

Removal of Regolith

One of the most important processes of erosion is the removal and transport of rock debris (regolith) produced by weathering. The process is simple but important. Loose rock debris is washed downslope into the drainage system and is transported as sediment load in streams and rivers. In addition, soluble material is carried in solution. The net result is that the blanket of regolith over Earth's surface is continually being removed and transported to the sea by stream action. As it is removed, it is also continually being regenerated by weathering of the fresh bedrock below. Measurements of the amount of sediment carried by rivers indicate that the surface of the land is being lowered on the average of 6 cm (2.4 in.) per 1000 years.

Downcutting of Stream Channels

Downcutting is a fundamental process of erosion in all stream channels, whether small hillsides, gullies, or great canyons of major rivers. The process is accomplished by the ***abrasion*** of the channel floor by sand and gravel as they are swept downstream by the flowing water. It is similar in many respects to the action of a wire saw used in quarries to cut and shape large blocks of stone (Figure 12.12). An abrasive such as garnet, corundum, or quartz, dragged across a rock by a wire, can cut through a stone block with remarkable speed.

Some dramatic examples of the power of streams to cut downward are the steep, nearly vertical gorges in many canyons in the southwestern United States (Figure 12.13). Although the bed load of sand and gravel on the channel floor is stationary much of the time, during spring runoff and periodic flash floods it moves with the flowing water. This material is an effective abrasion tool and can cut the stream channel to a profile of equilibrium in a short time.

An effective and interesting type of abrasion of the channel floor is the drilling action of pebbles and cobbles trapped in a depression and swirled around by currents. The rotational movement of the sand, gravel, and boulders acts like a drill and cuts deep holes called ***potholes***. As the pebbles and cobbles are worn away, new ones take their place and continue to drill into the bedrock of the stream channel. Some potholes are several meters in diameter and more than 5 m in depth (Figure 12.14).

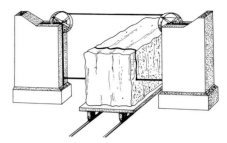

FIGURE 12.12
Sketch of a wire saw commonly used in quarry operations. The wire is pulled across the rock, dragging abrasives as it moves. When the rock is raised (or the wire lowered) the abrasives slice through the rock.

FIGURE 12.13
The tools of erosion are sand and gravel. Transported by a river, they act as powerful abrasives, cutting through the bedrock as they are moved by flowing water. The abrasive action of sand and gravel cut this vertical gorge through resistant limestone in the Grand Canyon, Arizona.

FIGURE 12.14
Potholes are eroded in a stream bed by sand, pebbles, and cobbles whirled around by eddies.

As the pebbles and cobbles are carried by flowing water, they themselves are worn down by striking one another and the channel bottom. Their corners and edges are chipped off, and the particles become smaller, smoother, and more rounded. Large boulders that have fallen into a stream and are transported only during a flood are thus slowly broken and worn down to smaller fragments. Ultimately, they are washed away as grains of sand.

Another important factor in the downcutting of a stream channel is the upstream migration of waterfalls and rapids. Here again, the process is simple but important. It can be appreciated by considering the erosion of Niagara Falls (Figure 12.15). The increased velocity of the falling water sets up strong turbulence at the base of the falls, causing rapid erosion of the underlying weak, nonresistant rock layers. The cliff is gradually undermined, and the falls retreat upstream.

Headward Erosion

In the process of stream erosion and valley evolution, streams have a universal tendency to erode headward, or upslope, and to increase the lengths of their valleys until they reach the divide. *Headward erosion* can be analyzed by reference to Figure 12.16. The reason erosion is more vigorous at the head of a valley than on its sides is apparent from the relationships between the valley and the regional slope. Above the head of a valley, water flows down the regional slope as a sheet (sheet flow), but the water starts to converge to a point where a definite stream channel begins. As the water is concentrated into a channel, its velocity and erosive power increase far beyond those of the slower-moving sheet of water on the surrounding ungullied surface. The additional volume and velocity of the channel water erode the head of the valley much faster than sheet flow erodes ungullied slopes or the valley walls. In addition, groundwater moves toward the valley, so the head of the valley is a favorable location for the development of springs and seeps. These, in turn, help to undercut overlying resistant rock and cause headward erosion to occur much faster than retreat of the valley walls. The head of the valley is thus extended upslope.

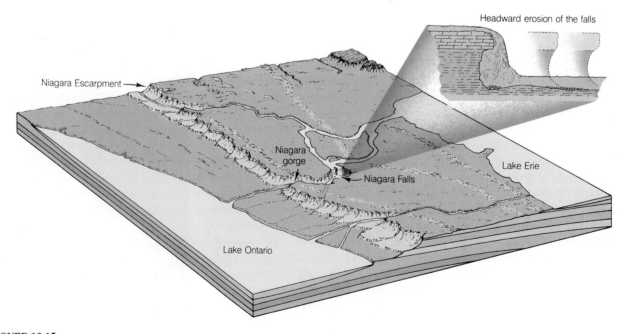

FIGURE 12.15
Retreat of Niagara Falls upstream occurs as hydraulic action undercuts the weak shale below the limestone. The Niagara River originated as the last glacier receded from the area and water flowed from Lake Erie to Lake Ontario over the Niagara cliffs. Erosion causes the waterfalls to migrate upstream at an average rate of 1.3 m per year.

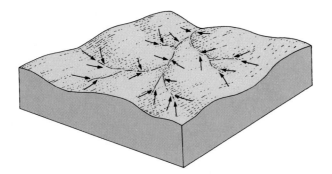

FIGURE 12.16
Headward erosion is the basic mechanism by which a drainage system is extended upslope. Water flows as a sheet down the regional slope (away from the foreground of the figure). As it converges toward the head of a tributary valley, its velocity is greatly increased, so its ability to erode also increases. The tributary valley is thus eroded headward, up the regional slope.

With the universal tendency for headward erosion, the tributaries of one stream can extend upslope and intersect the middle course of another stream, thus diverting the headwater of one stream to the other. This process, known as *stream piracy*, or *stream capture*, is illustrated in Figure 12.17. Stream piracy is most likely to occur if headward erosion of one stream is favored by a steeper gradient or by a course in more easily eroded rocks. Some of the most spectacular examples occur in the folded Appalachian Mountains, where nonresistant shale and limestone are interbedded with resistant quartzite formations. The process of stream capture and the evolution of the region's drainage system are shown in the series of diagrams in Figure 12.18. The original streams flowed in a dendritic pattern (a branching, treelike pattern) on horizontal sediments that once covered the folds. As uplift occurred, erosion removed the horizontal sediments, and the dendritic drainage pattern became superposed, or placed upon the folded rock beneath. The *superposed stream* thus cuts across weak and resistant rocks alike. As the major stream cuts a valley across the folded rocks, new tributaries rapidly extend headward along the nonresistant formations. By headward erosion, these new streams progressively capture the superposed tributaries and change the dendritic drainage pattern to a *trellis drainage pattern* (a pattern in which the tributaries join the main stream at right angles).

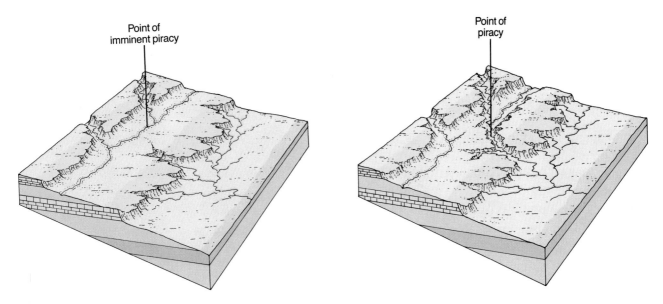

FIGURE 12.17
Stream piracy occurs where a tributary with a high gradient rapidly erodes headward and captures the tributary of another stream.

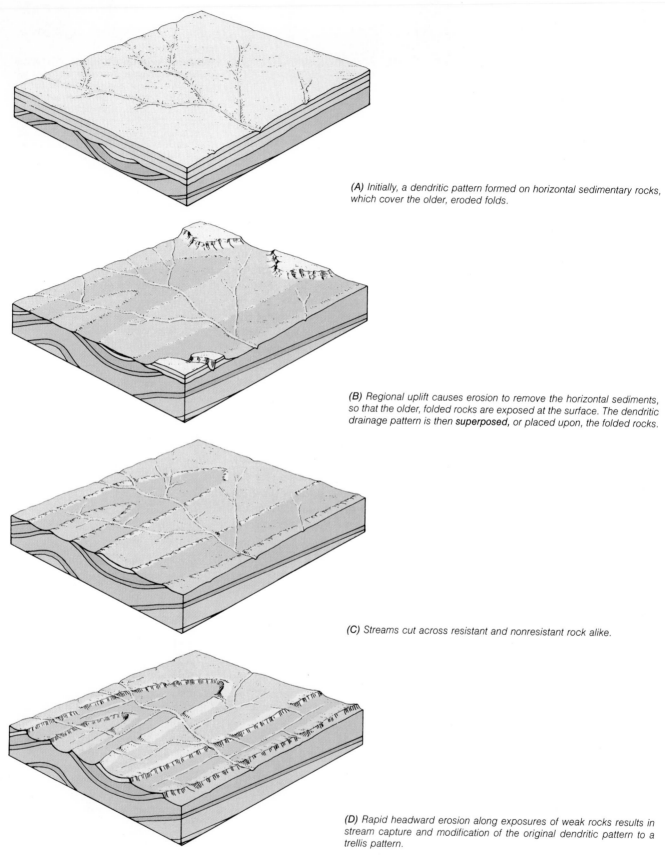

(A) Initially, a dendritic pattern formed on horizontal sedimentary rocks, which cover the older, eroded folds.

(B) Regional uplift causes erosion to remove the horizontal sediments, so that the older, folded rocks are exposed at the surface. The dendritic drainage pattern is then **superposed**, or placed upon, the folded rocks.

(C) Streams cut across resistant and nonresistant rock alike.

(D) Rapid headward erosion along exposures of weak rocks results in stream capture and modification of the original dendritic pattern to a trellis pattern.

FIGURE 12.18
A dendritic drainage pattern superposed on a series of folded rocks evolves into a trellis pattern as headward erosion proceeds along nonresistant rock formations.

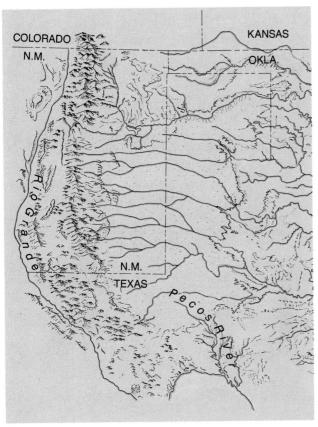

(A) Prior to the development of the Pecos Valley, drainage is believed to have been eastward from the Rocky Mountains across the Great Plains.

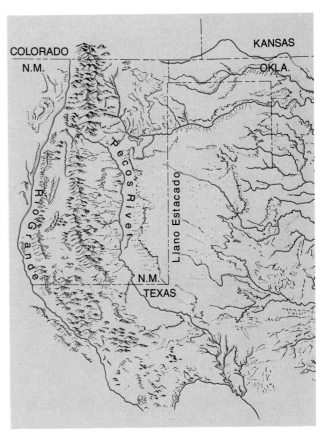

(B) Headward erosion of the Pecos River northward along the nonresistant rocks of the Pecos plains captured the headwaters of the eastward-flowing streams.

FIGURE 12.19
The Pecos River evolved as headward erosion extended the drainage network northward and captured the eastward-flowing streams.

Extensive stream piracy and development of a trellis drainage pattern can be seen almost any place where folded rocks are exposed at the surface. In the folded Appalachian Mountains, the major streams that flow to the Atlantic (such as the Susquehanna and the Potomac) are all superposed across the folded strata. Their tributaries, however, flow along the nonresistant rocks parallel to the geologic structure and have captured many superposed streams.

Another example is the Pecos River in New Mexico. By extending itself headward to the north along the weak shale and limestone, which crop out in a north–south zone parallel to the Rocky Mountain front, it has captured a series of eastward-flowing streams that once extended from the Rockies across the Great Plains. The original eastward drainage (shown in Figure 12.19A) resulted from the uplift of the Rocky Mountains. Now the headwaters of most of the original streams have been captured by the Pecos River. Water that once would have flowed across the Llano Estacado (the High Plains of Texas) now flows down the Pecos valley (Figure 12.19B).

In summary, headward erosion is one of the most important processes by which a drainage pattern evolves. It continues to extend the drainage network upslope until the tributaries reach the crest of the divide (the ridge separating two drainage systems). There it is effectively eliminated, because no undissected slope remains. Subsequent erosion is restricted to downcutting and valley widening. A drainage system thus grows in a specific way. Once sheet flow is concentrated in a channel, the future development of the drainage pattern is somewhat predictable.

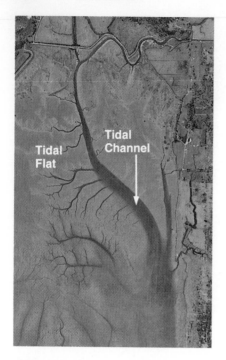

FIGURE 12.20
Tidal channels develop a characteristic dendritic drainage pattern between high tide and low tide because the sediment is homogeneous and the tidal flat slopes gently seaward.

Extension of Drainage Systems Downslope

In addition to downcutting and headward erosion, a drainage system can grow in length simply by extending its course downslope as sea level falls or as the landmass rises. This process is probably fundamental in determining the original course of many major streams, especially in the interior lowlands, where the oceans once covered much of the continents and then slowly withdrew. As the oceans retreated, drainage systems were extended down the newly exposed slopes. Later they were modified by headward erosion and stream piracy.

Tidal channels (major channels formed by tidal currents extending from offshore well into the tidal flat) along coastal plains are an example of the beginning of a new segment of a drainage system as a result of a fall in sea level. The pattern of land and tidal drainage is shown in Figure 12.20. It is characteristically dendritic because the material on which this drainage is established consists of recently deposited horizontal sediments. If the slope is pronounced, however, the tributaries, as well as the major streams, flow parallel for a long distance. If sea level were to drop, the streams would continue to flow downslope, following the courses established by tidal channels, as is shown in Figure 12.21. Major streams would extend their drainage patterns over the deltas that they deposited. Most of the streams in the Gulf and Atlantic coastal plains originated in this way. If sea level is falling, the youngest parts of a river are therefore near the shoreline and upslope, where headward erosion develops new channels.

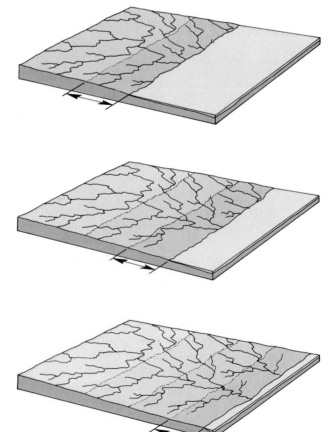

(A) In the original position of the shore, tidal channels develop between high tide and low tide.

(B) As the sea level falls and the shoreline recedes, tidal channels become part of the permanent drainage system.

(C) With each successive retreat of the shoreline, new tidal channels develop and drainage is extended farther downslope. A dendritic drainage pattern typically is produced on the homogeneous tidal-flat material.

FIGURE 12.21
Extension of a drainage system downslope occurs as a shoreline recedes. This downslope extension commonly results in a dendritic pattern.

The Grand Canyon, A Model of Stream Erosion

Because the evolution of a drainage system may require tens of millions of years, we can study the origin of stream valleys only indirectly. One approach is to study the interaction of downcutting and slope retreat by means of a computer model of the Colorado River's erosion of the rock sequence in the Grand Canyon area. Variables of a drainage system that affect various rock formations, such as rates of downcutting and slope retreat, were analyzed. This study produced a series of hundreds of computer-calculated profiles of the Grand Canyon, showing changes that have occurred between the time the Colorado River began cutting through the Colorado Plateau and the present. Four of these profiles are shown in Figure 12.22. Erosion of the model canyon was not uniformly fast or slow, but occurred in a series of pulses. Downcutting of the main stream was extremely rapid and was largely a function of the rate of uplift. In areas of little uplift, erosion was slower. In areas of rapid uplift, erosion was correspondingly faster, and the stream maintained a smooth, gently curving longitudinal profile.

On minor intermittent tributaries, downcutting was slow where resistant rocks were encountered, and rapid erosion occurred on nonresistant strata. The main river, however, cut through resistant and nonresistant rocks with almost equal ease. The rate of slope retreat was shown to be intimately related to the rate of downcutting.

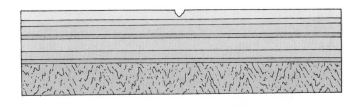

(A) The original undissected surface is a flat, stripped surface formed on horizontal sedimentary rock.

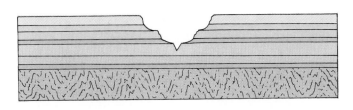

(B) Initial dissection and slope retreat occur as a result of uplift. Slope retreat causes nonresistant rocks to recede from the river, so that a terrace is left on the resistant rock layers.

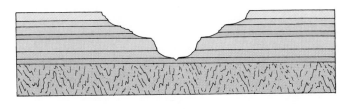

(C) Downcutting accelerates as the river breaks through a resistant formation and rapidly erodes the weaker underlying rocks.

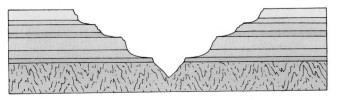

(D) Downcutting continues as differential slope retreat produces alternating cliffs and terraces.

FIGURE 12.22
A series of profiles representing the evolution of the Grand Canyon has been constructed from a computer model.

Although this model cannot be verified directly, you can get a glimpse of the stages of canyon development by studying the canyon longitudinally (Figure 12.23). Upstream, near Lees Ferry, the river has not cut through the Kaibab Limestone. Here, uplift has been minimal, and the entire sequence of strata exposed farther downstream in the Grand Canyon is below the surface. The river cuts only a shallow, narrow gorge in the Kaibab Limestone, which forms the upper rim of the Grand Canyon downstream. Farther downstream (in the foreground of the figure), uplift permitted the river to cut much deeper into the rock sequence, and the sequence of profiles across the canyon is similar to the one developed by the computer model. Evidence of the evolution of slope morphology from the canyon itself thus supports the findings of the computer model.

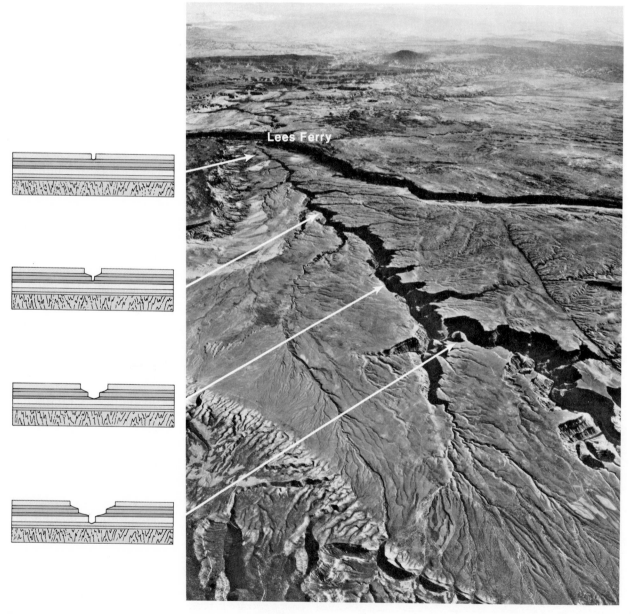

FIGURE 12.23
The effects of erosion of the eastern Grand Canyon are seen in this high-altitude photograph. At Lees Ferry, in the eastern Grand Canyon, the river is just beginning to cut through the rock sequence and has produced a profile like the one in Figure 12.22A. Downstream, uplift has permitted the river to cut deeper, and it has produced a sequence of profiles similar to the ones in Figures 12.22B, C, and D.

PROCESSES OF STREAM DEPOSITION

In the lower parts of the drainage system (transporting and dispersing systems) the gradient of the rivers is very low. As a result, the stream's velocity is reduced and deposition of the sediment load occurs, creating distinctive landforms. Foremost among these are (1) floodplains, (2) alluvial valleys, (3) deltas, and (4) alluvial fans.

Rivers transport huge volumes of sediment which is clearly indicated in practical problems of silting of reservoirs and maintenance of navigable channels and harbors. Most large rivers are always muddy; in some rivers the amount of sediment sometimes exceeds the amount of water. Sediment is deposited when the velocity of the current falls below the minimum velocity required to keep the particles of a particular size in motion (Figure 12.7). Thus, if a river carrying silt, sand, and gravel is slowed by a more gentle gradient, or by entering a lake or the sea, the coarsest particles of the load are deposited first, and progressively finer particles are deposited as the velocity of the current continues to decrease. Deposition of the sediment load in the lower transporting and dispersal segments of a river creates prominent and distinctive landforms such as (1) floodplains, (2) alluvial valleys, (3) deltas, and (4) alluvial fans.

Floodplains

On the gentle slopes of the shields and stable platform, most stream valleys are covered with large quantities of sediment that make up a flat surface over which the stream flows. This surface is called the *floodplain,* and during high floods it may be completely covered with water. Rivers that flow across floodplains are characterized by channels that either meander in sinuous loops or braid in interweaving multiple channels. These differences in channel configurations reflect variations in the type of sediment load and fluctuations in the volume of water. A schematic diagram showing the features commonly developed on a river floodplain is shown in Figure 12.24. It serves as a simple graphic model of floodplain sedimentation.

What major geologic processes operate on a floodplain?

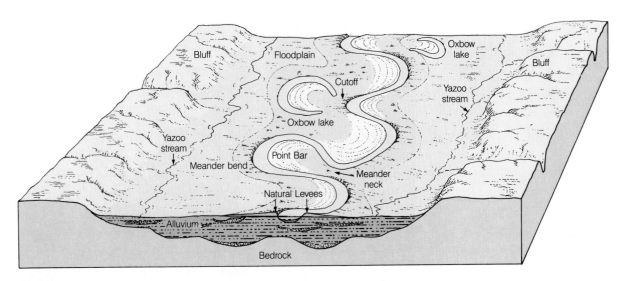

FIGURE 12.24
The major features of a floodplain include meanders, point bars, oxbow lakes, natural levees, backswamps, and streams. A stream flowing around a meander bend erodes the outside curve and deposits sediment on the inside curve to form a point bar. The meander bend migrates laterally and is ultimately cut off, to form an oxbow lake. Natural levees build up the banks of the stream, and backswamps develop on the lower surfaces of the floodplain. Yazoo streams have difficulty entering the main stream because of the high natural levees and thus flow parallel to it for considerable distances before becoming tributaries.

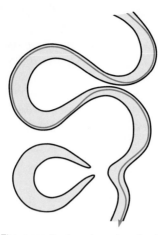

(A) Stream flow is deflected by an irregularity and moves to the opposite bank, where erosion begins.

(B) Once the bend begins to form, the flow of water continues to impinge on its outside curve, so that a meander loop develops. At the same time, deposition occurs on the inside of the bend as a result of the lower stream velocities in that area.

(C) The meander is enlarged and migrates laterally, with the contemporaneous growth of a point bar. There is a general downslope migration of meanders as they grow larger and ultimately cut themselves off to form oxbow lakes.

FIGURE 12.25
The evolution of stream meanders results because erosion occurs on the outside of a curve in the stream channel, where velocity is greatest, and deposition occurs on the inside of the curve, where velocity is least.

Meanders and Point Bars. All rivers naturally tend to flow in a sinuous pattern, even if the slope is relatively steep. This is because water flow is turbulent, and any bend or irregularity in the channel deflects the flow of water to the opposite bank. The force of the water striking the stream bank causes erosion and undercutting, which initiate a small bend in the river channel. In time, as the current continues to impinge on the outside of the channel, the bend grows larger and is accentuated, and a small curve ultimately grows into a large *meander* bend (Figure 12.25). On the inside of the meander, velocity is at a minimum, so some of the sediment load is deposited. This type of

FIGURE 12.26
Floodplain features include meander bends, point bars, natural levees, oxbow lakes, and backswamps.

deposit occurs on the point of the meander bend and is called a ***point bar***. The two major processes around a meander bend—erosion on the outside and deposition on the inside—cause meander loops to migrate laterally.

Because the valley surface slopes downstream, erosion is more effective on the downstream side of the meander bend, and thus the meander also migrates slowly down the valley (Figure 12.25). As a meander bend becomes accentuated, it develops an almost complete circle. Eventually, the river channel cuts across the meander loop and follows a more direct course downslope. The meander cutoff forms a short but sharp increase in stream gradient, causing the river to completely abandon the old meander loop, which remains as a crescent-shaped lake called an ***oxbow lake*** (Figure 12.26).

Natural Levees.   Another key process operating on a floodplain is the development of high embankments, called ***natural levees***, on both sides of the river. Natural levees form when a river overflows its banks during flood stage and the water is no longer confined to a channel but flows over the land surface in a broad sheet. This unchanneled flow significantly reduces the water's velocity, and some of the suspended sediment settles out. The coarsest material is deposited close to the channel, where it builds up a natural levee. Natural levees grow with each flood. Some grow high enough that the river channel is higher than the surrounding area (Figure 12.27).

Backswamps.   As a result of the growth and development of natural levees, much of the floodplain may be lower than the river flowing across it.

Why does a river build its own natural levees?

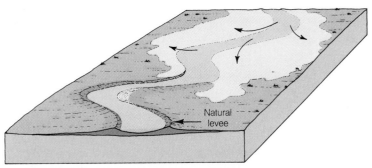

(A) Block diagram showing how natural levees are formed.

FIGURE 12.27
Natural levees are wedge-shaped deposits of fine sand, silt, and mud that taper away from the stream banks toward the backswamp. They form during flood stages because, as the stream overflows its banks, the velocity of the water is reduced and silt is deposited. As the levees grow higher, the stream channel also rises, and thus the river can be higher than the surrounding floodplain.

(B) Flood stage of the Sevier River in central Utah. Note the natural levees, which are expressed by vegetation growing on the high ground next to the river channel.

FIGURE 12.28
Radar image of the Mississippi River near Baton Rouge, Louisiana. Note the flat floodplain, the yazoo stream, and oxbow lakes. The natural levees are commonly the highest and driest areas in the region and are farmed in fields laid out perpendicular to the river. This is especially apparent in the oxbow lake near the lower right of the image.

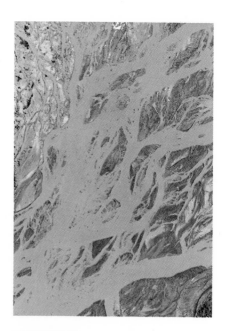

FIGURE 12.29
A braided stream pattern commonly results if a river is supplied with more sediment than it can carry. Deposition occurs, causing the river to develop new channels.

This area, known as the ***backswamp,*** is poorly drained and commonly is the site of marshes and swamps. Tributary streams in the backswamp are unable to flow up the slope of the natural levees, so they are forced either to empty into the backswamp or to flow as ***yazoo streams,*** streams that run parallel to the main stream for many kilometers. Strangely enough, then, the highest parts of the floodplain may be along the natural levees immediately adjacent to the river.

The lower Mississippi River is well known for its floodplain features (Figure 12.28). Between Cairo, Illinois, and the Gulf of Mexico, the Mississippi meanders over a broad floodplain, forming high natural levees, oxbow lakes, and backswamps. The dynamics of the river and the changes that it can bring about by deposition are illustrated by the fact that, during the period from 1765 to 1932, the river cut off 19 meanders between Cairo, Illinois, and Baton Rouge, Louisiana. Now the level of the Mississippi is controlled by dams and artificial levees, which have modified its hydrology, much as the Nile and Colorado rivers have been artificially manipulated. What do you think will happen to the river in the future?

Braided Streams. If streams are supplied with more sediment than they can carry, they deposit the excess material on the channel floor as sand and gravel bars. This deposition forces a stream to split into two or more channels, so the stream pattern forms an interlacing network of braided channels and islands (Figure 12.29). The braided pattern is best developed in rivers that carry coarse sand and gravels and fluctuate greatly in the volume of water that they discharge. These conditions commonly occur in arid or semiarid regions, where the amount of water in a stream varies greatly from season to season. Fluctuation in seasonal discharge will cause the river to deposit much of its sediment load, especially the coarse material, as channel bars, resulting in a **braided stream** pattern. Melting ice caps and glaciers also produce favorable conditions for braided streams because the streams in front of the melting ice

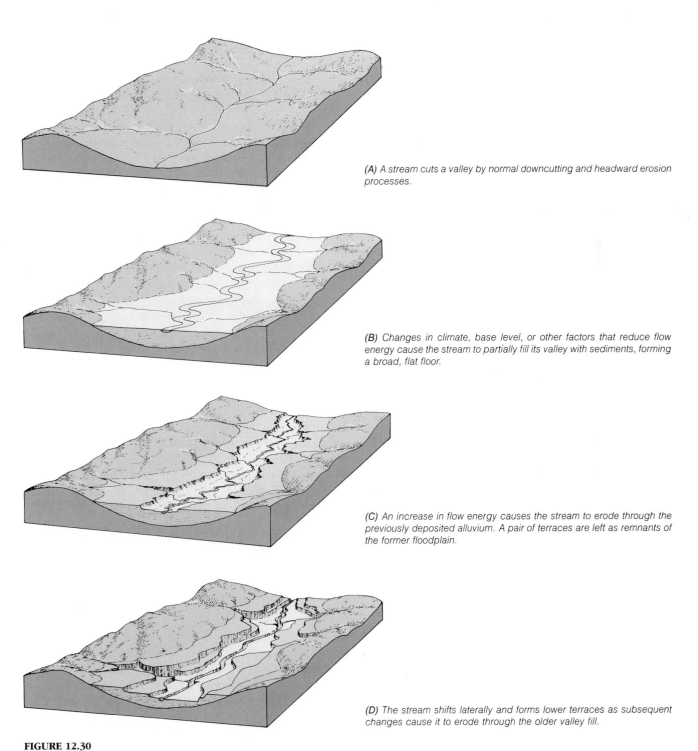

(A) *A stream cuts a valley by normal downcutting and headward erosion processes.*

(B) *Changes in climate, base level, or other factors that reduce flow energy cause the stream to partially fill its valley with sediments, forming a broad, flat floor.*

(C) *An increase in flow energy causes the stream to erode through the previously deposited alluvium. A pair of terraces are left as remnants of the former floodplain.*

(D) *The stream shifts laterally and forms lower terraces as subsequent changes cause it to erode through the older valley fill.*

FIGURE 12.30
The evolution of stream terraces involves the deposition of sediment in a stream valley, subsequent change in the stream's gradient, and renewed downcutting. These changes can be initiated by various factors that affect a stream's capacity to transport sediment, such as changes in climate, changes in base level, or regional uplift.

cannot transport the exceptionally large load of sediment deposited by the glaciers. Moreover, the cold climate near glaciers causes most rivers to freeze over during the winters, so the volume of water discharged fluctuates from almost nothing to spring floods.

Alluvial Valleys

Many streams fill part of their valleys with sediment during one part of their history and then cut through the sediment fill during a subsequent period. This fluctuation in stream processes commonly produces *stream terraces.* Deposition can be initiated by any change that reduces a stream's capacity to transport sediment. These changes include (1) a reduction in discharge (as a result of climatic change or of a loss of water volume to stream piracy), (2) a change in gradient (caused by a rise in base level or by regional tilting), and (3) an increase in sediment load.

The basic steps in the evolution of stream terraces are shown in Figure 12.30. In block A, a stream cuts a valley by downcutting and slope retreat. In block B, changes such as regional tilting of the land or rise of base level cause the stream to deposit part of its sediment load and build up a floodplain, which forms a broad, flat valley floor. In block C, subsequent changes (such as uplift or increased runoff) cause renewed downcutting into the easily eroded floodplain deposits, so a single set of terraces develops on both sides of the river. Further erosion can produce additional terraces (block D) by the lateral shifting of the meandering stream.

During the last ice age, the hydrology of most rivers changed significantly and produced numerous stream terraces. Stream runoff was increased greatly by the melting ice, and large quantities of sediment deposited by the glaciers were reworked by the streams, causing many to become overloaded. In addition, the climatic changes accompanying the ice age caused a general worldwide increase in precipitation. As a result, many streams filled part of their valleys with sediment during the ice age, and now they are cutting through that sediment fill to form stream terraces.

Deltas

As a river enters a lake or the ocean, its velocity suddenly diminishes, and most of its sediment load is deposited to form a *delta.* The growth of a delta can be complex, especially for large rivers depositing huge volumes of sediment. Two major processes, however, are fundamental to the formation and growth of a delta:

1. The splitting of a stream into a distributary channel system, which extends into the open water in a branching pattern.
2. The development of local breaks, called crevasses, in natural levees, through which sediment is diverted and deposited as splays in the area between the distributaries.

The diagrams in Figure 12.31 illustrate the development of distributaries. As a river enters the ocean (or a lake) and the flowing water is no longer confined to a channel, the currents flare out, rapidly losing velocity and flow energy. The coarse material carried by the stream is deposited in two specific areas: (1) along the margins of quiet water on either side of the main channel (these deposits build up subaqueous natural levees) and (2) in the channel at the river mouth, where there is a sudden loss of velocity (these deposits build a bar at the mouth of the channel). These two deposits effectively create two smaller channels (distributary channels), which extend seaward for some distance. The process is then repeated, and each new distributary is divided into two smaller distributaries. In this manner, a system of branching distributaries builds seaward in a fan-shaped pattern.

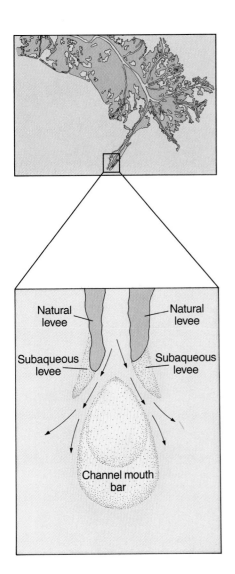

FIGURE 12.31
Distributaries develop where a stream enters a lake or the sea. A bar forms at the mouth of a river channel where water that was previously confined to the channel loses velocity, resulting in deposition as the stream enters the ocean. Subaqueous natural levees also form below water level. The bar then diverts the water coming from the main stream into two distributary channels, which grow seaward. This process is repeated, forming branching distributaries.

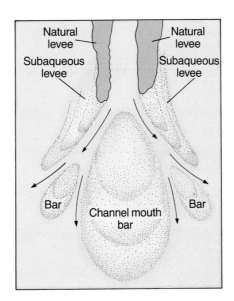

Figure 12.32 shows how the area between distributaries is filled with sediment. A local break in the levee, called a *crevasse,* forms during periods of high runoff and diverts a significant volume of water and sediment from the main stream. The escaping water spreads out and deposits its sediment to form a *splay,* which is essentially a small delta with small distributaries and systems of subsplays.

The sediment deposited by distributaries is vulnerable to erosion and transportation by marine waves and tides. The growth of a delta is therefore influenced by the balance between the rate of input of sediment by the river and the rate of erosion by marine processes. If waves or tides are strong, the development of distributaries is limited, and the sediment is reworked into bars, beaches, and tidal flats.

A major phenomenon in the construction of deltas is the shifting of the entire course of a river. Distributaries cannot extend indefinitely into the ocean because the gradient and capacity of the river to flow gradually decrease. The river, therefore, is eventually diverted to a new course, which has a higher gradient. This diversion generally happens during a flood. The river breaks through its natural levee, far inland from the active distributaries of the delta, and develops a new course to the ocean. The new course shifts the site of sedimentation to a different area, and the abandoned segment of the delta is attacked by wave and current action. The new, active, delta builds seaward, developing distributaries and splays, until it also is eventually abandoned, and another site of active sedimentation is formed. The shifting back and forth of

(A) Splays form where a break in a natural levee permits part of the stream to be diverted to the backswamp. This diversion reduces the velocity and causes deposition of sediment in a fan-shaped splay. Like the main river, a splay has distributaries and a series of smaller subsplays.

(B) A splay in the Mississippi River Delta

FIGURE 12.32
The formation of splays

What processes are involved in building major deltas? Why is an understanding of these processes important to the people who live in New Orleans?

the main river channel is thus a major way in which sediment is dispersed and a delta grows (Figure 12.33).

Several types of deltas are illustrated in Figure 12.34. Each shows a different balance between the forces of stream deposition and the forces re-working the sediment (waves and tides). In the Mississippi Delta, processes of river deposition dominate (Figure 12.34A). The delta is fed by the extensive Mississippi River system, which drains a large part of North America and discharges an annual sediment load of approximately 454 million metric tons. The river is confined to its channel throughout most of its course except during high floods. Most of the sediment reaches the ocean through two or three main distributary channels and has rapidly extended the delta far into the Gulf of Mexico. This extension is known as a ***bird-foot delta***.

If a bird-foot extension grows too far out to sea, the gradient of the river becomes too low for the river to flow. The river then shifts its entire course by breaking through its natural levee far upstream, and the process is repeated.

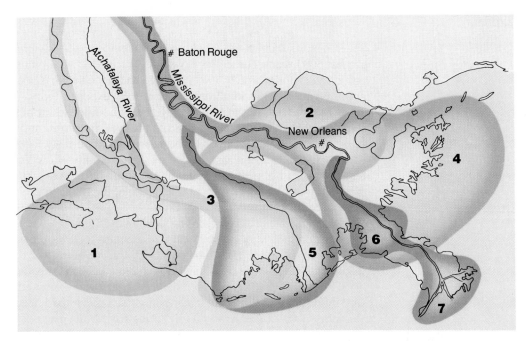

FIGURE 12.33
The history of the Mississippi Delta involves repeated shifting of the main channel, which has formed seven subdeltas. The flow of the Mississippi River is relatively constant throughout most of the year, and the water is confined to the main distributary. Most of the sediment is deposited in a small sector of the delta front. The gradient of the river where it enters the sea is very low, so the extension of the delta seaward cannot continue indefinitely. A major break in the natural levee upstream eventually diverts the entire flow to some other sector, and the process is repeated. Wave action then erodes the inactive bird-foot deltas. Previous subdeltas are indicated by numbers (1 through 6) according to age. Number 7 is the present subdelta. The active distributary system (6 and 7) has built a major bird-foot delta during the last 500 years.

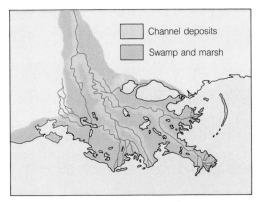

(A) The Mississippi Delta is dominated by fluvial processes, which produce a bird-foot extension.

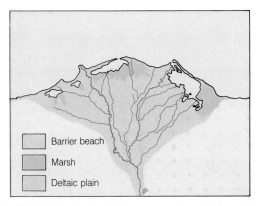

(B) The Nile Delta is dominated by wave action, which produces an arcuate delta front.

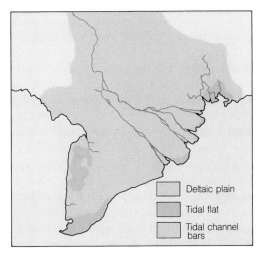

(C) The Mekong Delta is dominated by tidal forces, which produce wide distributary channels.

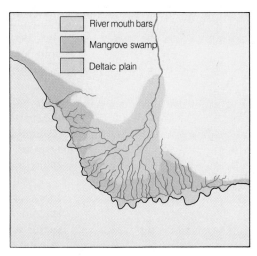

(D) The Niger Delta has formed where stream deposition, wave action, and tidal forces are about equal. An arcuate delta front and wide distributary channels are thus produced.

FIGURE 12.34
The shape of a delta depends on the balance between fluvial and marine processes.

261

Seven major subdeltas have been constructed by the Mississippi River during the last 5000 years. These are shown in Figure 12.33. The oldest lobe (1) was abandoned approximately 4000 years ago and since then has been eroded back and inundated. Only small remnants remain exposed today. The successive lobes, or subdeltas (2 through 7), have been modified to various degrees. The abandoned channels of the Mississippi are well preserved and can be recognized on satellite photographs. The currently active delta lobe (7) has been constructed during the last 500 years. River studies show that the present bird-foot delta has been extended as far as the balance of natural forces permits. Without continued human intervention, the Mississippi will shift to the present course of the Atchafalaya River.

The Nile Delta differs from the Mississippi Delta in several ways. Instead of being confined to one channel, the Nile begins to split up into distributaries at Cairo, Egypt, more than 160 km inland, and fans out over the entire delta (Figure 12.34B). Before construction of the Aswan High Dam, the Nile's annual flood briefly covered much of the delta each year and deposited a thin layer of silty mud. In the past 3000 years, 3 m of sediment have accumulated near Memphis. Two of the large distributaries have built major lobes extending beyond the general front of the delta, but strong wave action in the Mediterranean redistributes the sediment at the delta front. The reworked sediment forms a series of arcuate barrier bars, which close off segments of the ocean to form lagoons. The lagoons in turn form a subenvironment, which soon becomes filled with fine sediment. The difference between the Nile Delta and the Mississippi Delta is due largely to different balances between the influx of sediment, which builds bird-foot deltas, and the strength of wave action, which redistributes sediment to form barrier bars.

The Mekong Delta, along the southern coast of Vietnam, is dominated by tidal currents, which redistribute sediment in the river channels and along the delta fronts (Figure 12.34C). The distributaries branch into two main courses near Phnom Penh, about 500 km inland. Sediment carried by the river is reworked by tidal currents, forming broad distributary channels and a series of elongate shoals.

The Niger Delta (Figure 12.34D) is a good example of a delta in which the important energy systems are nearly in equilibrium. Stream deposition, wave action, and tidal currents are more evenly balanced here than in the other delta types, so the Niger Delta is remarkably symmetrical.

Alluvial Fans

How does an alluvial fan differ from a delta?

Alluvial fans are stream deposits that accumulate in dry basins at the base of a mountain front. Such areas usually have a large quantity of loose, weathered rock debris on the surface, so when rain falls, the streams transport huge volumes of sediment. Deposition results from the sudden decrease in velocity as a stream emerges from the steep slopes of the upland and flows across the adjacent basin with its gentle gradient. The channel soon becomes clogged with sediment and the stream is forced to seek a new course. In this manner, the stream shifts from side to side and builds up an arcuate, fan-shaped deposit (Figure 12.35). As several *fans* build basinward at the mouths of adjacent canyons, they eventually merge to form broad slopes of *alluvium* at the base of the mountain range (Figure 12.36).

Although alluvial fans and deltas are somewhat similar, they differ in mode of origin and internal structure. In deltas, stream flow is checked by standing water, and sediment is deposited in a body of water. The level of the ocean or lake effectively forms the upper limit to which the delta can be built. In contrast, a fan is deposited in a dry basin, and its upper surface is not limited by water level. The coarse, unweathered, poorly sorted sands and gravels of an alluvial fan also contrast with the fine sand, silt, and mud that predominate in a delta.

FIGURE 12.35
Alluvial fans form in arid regions where a stream enters a dry basin and deposits its load of sediment.

FIGURE 12.36
Alluvial slopes develop as fans grow and merge together. This photograph of part of the Sierra Nevadas shows large alluvial slopes, which cover much of the dry basin.

RIVER SYSTEMS AND PLATE TECTONICS

The evolution of the major rivers of the world is influenced directly and indirectly by plate tectonics.

In previous sections of this chapter we have considered river systems on a local basis: how they erode, transport, and deposit material. The major rivers of the world have other features, of a much larger order of magnitude, that relate to the global patterns of the tectonic and hydrologic systems. This subject is just beginning to attract the attention of geologists and has stimulated many provocative questions:

1. What controls the origin of the major drainage systems of the world?
2. How do river systems originate and evolve?
3. How is a river system destroyed or terminated?
4. What conditions are necessary for long-lasting rather than short-lived river systems?
5. Were the major rivers in the past similar to those of today, or were there larger rivers on Pangaea?

These questions cannot all be satisfactorily answered at the present time, but it does seem clear that rivers evolve in a systematic way and are controlled by both the hydrologic and tectonic systems.

Characteristics of Major River Systems

A glance at a drainage map of the world may give a first impression that the drainage of the continents is haphazard and unsystematic. Rivers appear to flow in any direction in an almost unlimited variety of patterns. Upon further study, however, some system becomes apparent in the locations of the major rivers, their tributaries, and the patterns they form. Figure 12.37 is a map showing the

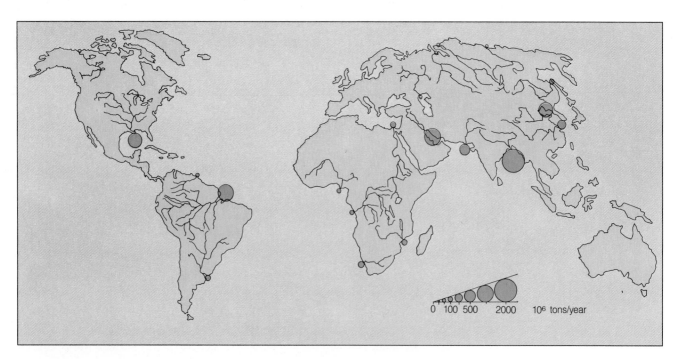

FIGURE 12.37
The world's largest drainage basins transport vast volumes of sediment, most of which is deposited as huge deltas. This map shows the annual discharge of the suspended sediment in tons for the 20 major drainage systems of the world.

locations of the major rivers of the world and the relative sizes of their deltas. From this map we see that the largest rivers in the world (those with the highest discharge) are found on sloping terrains where there is a maximum rainfall. The Amazon (discharge: 104,083 m^3 per second) and the Congo (discharge: 44,082 m^3 per second) illustrate this point very well.

It is apparent also from Figure 12.37 that there are large areas of the continents that do not have major river systems. The arid and semiarid low-latitude deserts such as the Sahara of North Africa, the Kalahari, in South Africa, and the great desert of Australia are most obvious. Sand seas cover much of these areas.

Low surface runoff is not confined to arid regions, however. Larger river systems are not found in the polar regions of North America and Europe because these areas were covered with glaciers during the ice age, which ended only a few thousand years ago. Previous drainage systems were obliterated as the glaciers expanded over the region.

Humid areas underlain by porous limestone also have poor drainage because solution activity has developed a network of subterranean caverns and enlarged fractures, which divert the drainage to the subsurface. Many of these areas have no integrated drainage system despite the humid climate. Parts of Kentucky, Florida, and the Yucatan Peninsula of Mexico are in this category.

Drainage Patterns and Plate Tectonics

In Earth's tectonic system there is a fundamental tendency for a continent to develop a high mountain range on its active margin, with a regional slope toward the passive margin. This results in an asymmetry distinctive to the watershed of a continent. The drainage pattern of an idealized major river would consist of large collecting systems in the mountain belt and a trunk stream flowing across the stable platform or shield and emptying into the passive margin of the continent. The Amazon River of South America is a classic example of this basic pattern. The preglacial drainage of North America was similar, with the major drainage system flowing across the Canadian shield into the North Atlantic. Indeed, the 28 largest rivers of the world all drain to the passive margins of the continents or to marginal seas. This, therefore, seems to be the fundamental pattern created by the tectonic system; if river systems did not continually evolve and adjust to structural and climatic conditions, most major rivers of the world would clearly have this distinctive pattern.

Modification of Basic River Pattern

Many exceptions and modifications to this basic pattern are also influenced by tectonics. Continental rifting will effectively behead or dismember a previously established river system. Also, if rifting or subsidence occurs in the shield or platform, it will tend to focus and orient the trunk stream of the drainage (examples are the lower Niger, Amazon, Paraná, and lower Mississippi).

Indirectly, the tectonics of a continent will influence drainage patterns because the folded rocks produced by crustal deformation will produce zones of alternating hard and soft rock parallel to the trend of the mountain belt. Headward erosion will follow the zones of weakness and will modify the pattern so that large segments of a river will flow parallel to the structural trends of the folded mountain belt. The Mekong and Irrawaddy rivers of Southeast Asia are excellent examples.

How does plate tectonics control the origin and evolution of the major rivers of the world?

Volcanic activity is another method by which the tectonic system modifies a drainage pattern. Extrusion of flood basalts can obliterate the preexisting drainage system. A new pattern is then established on the volcanic surface or along the margins of the flows.

The drifting of a continent into a new climatic zone is yet another way in which tectonic activity can modify a river system. As a continent drifts into the low latitudes, precipitation is greatly reduced, and windblown sand can completely cover large parts of the previously established drainage. Proof of this was recently discovered in radar images of parts of the Sahara made during the space flight of *Columbia* (Figure 12.38). These images show a large and extensive ancient drainage system now buried beneath the sand.

The drifting of continents into cold climatic zones may cause similar destruction to a substantial part of the drainage system if glaciers develop and cover large areas of the continent. Continental glaciation will obliterate the drainage system beneath it and force the major rivers to establish a new course along the margins of the ice. After the continental ice sheet retreats, a new and complex drainage pattern is reintegrated through a system of overflowing ponds and lakes.

Types of Rivers

In light of the concepts just described, it is possible to distinguish five major types of rivers in relation to their tectonic setting:

1. Rivers that have their headwaters in a mountain range at converging margins and flow across the stable platform or shield. The Amazon and preglacial drainage of North America are examples.
2. Rivers that have their headwaters in a mountain range at converging plate margins, but that have a main trunk stream that flows parallel to the range, probably as a result of associated subsidence. The Ganges of India and the Paraná of South America are examples.
3. Rivers that flow within a mountain belt parallel to the major structural trends. The Mekong, Brahmaputra, and Irrawaddy rivers are examples.
4. Rivers that flow across the stable platform or shield but do not have their headwaters in a folded mountain belt. The Nile, Niger, and Zambezi rivers are examples. The headwaters are located in highlands of volcanic rock or uplifted parts of the shield. This pattern was probably more common in periods of the geologic past when continental blocks were not bordered by subduction zones.
5. Rivers that flow in a superposed course across the structural grain of a mountain belt. The Danube, Columbia, and Snake rivers are examples.

Deltas of Major Rivers

The size of the deltas formed at the mouths of the major rivers of the world (Figure 12.37) is controlled by the size of the drainage basin, elevation of the land, and climate (amount of surface runoff). Maximum sediment load occurs in large rivers that drain mountainous topography in a humid climate. The largest deltas of the world are built by the Amazon, Tigris-Euphrates, Ganges, Mekong, and Hwang (Yellow) rivers. Twenty-five of the world's largest deltas are found on the passive margins of continents. In addition, there is a common association of many rivers with a submarine canyon and a huge submarine fan built out onto the abyssal plains in the deep-ocean basin. The submarine fan, like the delta built at the river mouth, is an indication of the vast amount of erosion accomplished by the work of river systems.

Age of Rivers

Important characteristics of a river system are its age and history, but these determinations are extremely complicated. We may consider the time of the origin of a river to be the earliest date at which a continuous system drained the region in question. We may consider a river to date from the last marine regression, the last significant tectonic uplift, the termination of lava extrusion, or the waning of an ice sheet. All produce new surfaces upon which a drainage

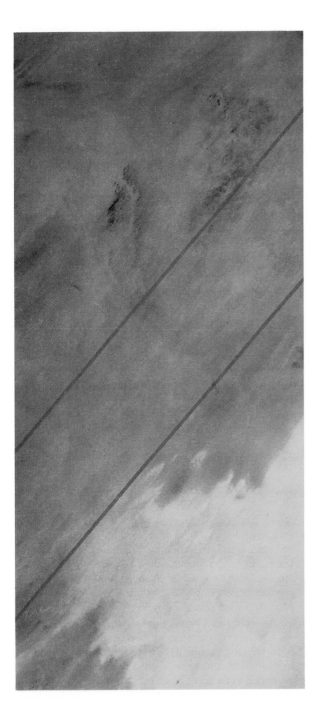

FIGURE 12.38
Ancient river systems in the Sahara Desert are now buried under sand, but are revealed by radar imagery made during the space shuttle flight of *Columbia.* On Landsat images, the present sand desert surface is yellowish orange, and the surface beneath, eroded by old river systems, is black and white. The radar image cutting diagonally across the picture on the right covers an area approximately 50 km wide and 300 km long. The largest valley on the image is as wide as the present Nile River valley and represents millions of years of erosion when the Sahara had a much wetter climate.

system may evolve. A river could be terminated by a new marine invasion, new tectonism, glaciation, volcanic extrusion, or expansion of a sand sea, but it is not that simple. Various parts of a river system originate and evolve at different times and in different ways, so we cannot establish a precise time when an entire river system originated. Very few rivers (and certainly no major ones) begin or end without some relationship to the drainage system that preceded them. Instead, a drainage system continually evolves by headward erosion and stream capture, adjustment to structure, and adjustments and modifications related to marine transgressions, continental glaciation, desert sand, and continental rifting. As the system continually evolves, each period of its history inherits something from the preceding conditions. The reason rivers continue to evolve is that the hydrologic system is continuous. Uplift of a mountain belt cannot divert or change the course of a river because a river has the capacity to downcut its channel much faster than uplift occurs.

In summary, the orientation of major drainage systems depends to a considerable extent on the tectonic history of the continent. The general tendency for continents to be asymmetrical is the direct result of plate tectonics, so there is a strong tendency for major river systems to flow from the mountain range across the stable platform and shield and to empty into the sea along the passive margin of the continent. Variations of this basic pattern commonly reflect local rifting and subsidence within the continent or disruptions of the basic pattern as a result of glaciation, development of deserts, or volcanism. Major river systems therefore depend to a considerable degree on tectonics for orientation, size, and longevity.

Some of the major river systems today date back to the early Cenozoic Era, 40 to 50 million years ago. Segments of some rivers may predate the breakup of Pangaea. The ancestral Congo, for example, could have flowed across South America. In all probability, the great Amazon is not the largest river the world has seen. Larger rivers probably drained Gondwanaland and Pangaea, the ancient land masses that existed before the present continental masses were outlined and drifted apart.

SUMMARY

1. A river system is a network of connecting channels that collects surface water and funnels it back to the ocean. It can be divided into three subsystems: (a) tributaries, which collect and transport water and sediment into the main stream, (b) a main trunk stream, which is largely a transporting system, and (c) a dispersing system at the river's mouth.

2. Studies of river systems show that there is a high degree of order among the tributaries in relation to the size and gradients of their valleys. These relationships indicate that streams erode the valleys through which they flow.

3. The major variables in a river system are (a) discharge, (b) velocity, (c) gradient, (d) base level, and (e) load.

4. A river system is best thought of as a system of moving water and sediment because an enormous volume of sediment is constantly being transported to the ocean by running water. The sediment is moved in suspension, by traction, and in solution. It is visible in most rivers, especially where it is being deposited as a delta near the river's mouth. The capacity of running water to erode and transport sediment is largely dependent on the stream's velocity.

5. A river system functions as a unified whole, adjusting its profile to establish equilibrium among the factors that influence flow. If one factor changes, the river system adjusts to reestablish equilibrium.

6. Running water is the most important eroding agent on this planet, and stream valleys are the most characteristic landforms on continental surfaces.

7. Most sediment carried by rivers is derived from the regolith. Weathering therefore plays an important role in preparing solid rock for erosion.

8. Streams and rivers erode the landscape by (a) removal of the regolith, (b) downcutting of stream channels, (c) slope retreat, and (d) headward erosion.

9. In the transporting segment of a river system, sediment is deposited to form (a) point bars, (b) natural levees, and (c) backswamps.

10. Braided stream patterns tend to develop in rivers that carry a heavy load of coarse sediment (sand and gravel) and that fluctuate greatly in the volume of water they discharge.

11. Stream terraces form by deposition of sediment in a stream valley, followed by rejuvenation of erosion. The renewed erosion may result from uplift, climatic changes, or changes in base level.

12. Most of the sediment load carried by a river is deposited as a delta where the river enters a lake or ocean. The major delta-forming processes are (a) development of distributaries and (b) development of splays.

13. Alluvial fans form in dry basins at the base of a mountain front. They form mostly in arid regions where stream flow is intermittent.

14. Plate tectonics plays a major role in the evolution of a river system.

KEY TERMS

abrasion (p. 245)

alluvial fan (p. 262)

alluvium (p. 262)

backswamp (p. 256)

base level (p. 240)

bed load (p. 240)

bird-foot delta (p. 260)

braided stream (p. 257)

collecting system (p. 235)

crevasse (p. 259)

delta (p. 258)

dendritic drainage pattern (p. 235)

discharge (p. 238)

dispersing system (p. 236)

dissolved load (p. 240)

distributary (p. 236)

divide (p. 235)

drainage basin (p. 235)

drainage system (p. 235)

fan (p. 262)

floodplain (p. 253)

graded stream (p. 242)

gradient (p. 239)

headward erosion (p. 246)

intermittent stream (p. 238)

longitudinal profile (p. 239)

meander (p. 254)

natural levee (p. 255)

oxbow lake (p. 255)

permanent stream (p. 238)

point bar (p. 255)

pothole (p. 245)

river system (p. 235)

saltation (p. 240)

settling velocity (p. 241)

splay (p. 259)

stream piracy (p. 247)

stream terrace (p. 258)

superposed stream (p. 247)

suspended load (p. 239)

transporting system (p. 235)

trellis drainage pattern (p. 247)

tributary (p. 235)

yazoo stream (p. 256)

REVIEW QUESTIONS

1. Explain the reasons for concluding that stream action (running water) is the most important process of erosion on Earth.
2. Describe and illustrate the three major subsystems of a river.
3. Define stream gradient, stream load, stream capacity, and competence.
4. Draw a diagram showing the general nature of transportation of (a) bed load, (b) suspended load, and (c) dissolved load.
5. Explain the concept of threshold velocity in the transportation and deposition of stream sediment.
6. Explain the concept of equilibrium in river systems and cite several examples of how streams adjust to attain equilibrium.
7. How does urbanization affect surface runoff?
8. Explain how stream action is able to cut a valley through solid bedrock.
9. What is headward erosion? Why does it occur?
10. Explain the process of stream piracy and cite examples of how it modifies a drainage system.
11. How does a stream system grow longer?
12. Name and describe the important landforms associated with floodplain deposits.
13. Describe the steps involved in the growth of a stream meander and the formation of an oxbow lake.
14. How does a point bar develop?
15. Explain the origin of natural levees.
16. What conditions are conducive to the development of braided streams?
17. Describe and illustrate the steps in the development of stream terraces.
18. Explain how a delta is built where a stream enters a lake or the sea.
19. Outline the history of the Mississippi Delta.
20. Make a series of sketches to show the form of a delta in which (a) fluvial processes dominate, (b) wave processes dominate, and (c) tidal processes dominate.
21. Explain how an alluvial fan is built.

ADDITIONAL READINGS

Chorley, R. J., S. A. Schumm, and D. E. Sugden. 1984. *Geomorphology*. London: Methuen.

Knighton, D. 1984. *Fluvial Forms and Processes*. London: Edward Arnold.

Leopold, L. B., and W. B. Langbein. 1966. River meanders. *Scientific American* 214(6):60–70.

Leopold, L. B., M. G. Wolman, and J. P. Miller. 1964. *Fluvial Processes in Geomorphology*. San Francisco: Freeman.

Morisawa, M. 1968. *Streams: Their Dynamics and Morphology*. New York: McGraw-Hill.

Ollier, C. D. 1981. *Tectonics and Landforms*, Chapter 12. London and New York: Longman.

Richards, K. 1982. *Rivers*. New York: Methuen.

Ritter, D. F. 1985. *Process Geomorphology*. Dubuque, IA: Brown.

Schumm, S. A. 1977. *The Fluvial System*. New York: Wiley.

In the preceding chapter, we learned how a stream valley evolves through downcutting, headward erosion, and slope retreat. These processes operate as a system, so a landscape erodes in a systematic way. The processes of landscape development, however, are extremely complex because tectonism produces a variety of structural settings upon which erosion occurs, and climate influences the rates and processes that operate within a region.

This Landsat mosaic of part of the Colorado Plateau illustrates this point. At first glance, this appears to be an image of intricate detail with many complex shapes and forms. The landscape may appear at first to be overwhelming and incomprehensible. But there are system and beauty in the evolution of the land that promise a rewarding satisfaction for those who take time to understand and appreciate.

The Colorado River and two of its major tributaries flow through the region and cut deep canyons into the flat-lying, colorful formations of sandstone, shale, and limestone. All of the terrain—every landform from the small valleys to the high mountain peaks—is related in some way to the erosion of this river system. The evolution of the landscape is thus closely related to the evolution of the river system. Follow the river pattern with your eye. Note the pronounced, entrenched meanders of the Colorado and Green rivers upstream from their confluence and the famous Goosenecks of the San Juan. Note the canyons of Canyonlands National Park, also near the confluence of the Colorado and Green rivers. Study the drainage patterns of the tributaries to the Colorado River and see how they form the plateaus, mesas, and buttes. Note the ridges of resistant white sandstone etched into relief by erosion along adjacent nonresistant beds. Note the several prominent, high mountain peaks that stand out in brilliant red and seem to be completely out of place in the plateau country. These are classic laccolithic domes covered with forests, which appear red on a Landsat image. Even these igneous structures are eroded by tributaries to the Colorado.

You can see from this image that the landscape has much to tell about its history and origin, but you must learn to read a new language, a language written in the rivers, canyons, and rocks.

13
Evolution of Landforms

1. The surfaces of the continental shields evolve through the erosion of mountain belts.
2. The process involves both erosion and isostatic adjustment; ultimately a surface of low relief near sea level is produced on the complex metamorphic and igneous rocks formed in the roots of the mountain belt.
3. The rate of erosion of a mountain belt rapidly decreases with decrease in elevation. However, erosion rates are high enough that a mountain belt can be reduced to a planar surface near sea level in 60 to 100 million years.
4. Subsequent tectonic movements may result in subsidence and expansion of the sea over the shield or in broad upwarps of the shield or the stable platform and renewed stream erosion.
5. The pattern of a river system continually adjusts to the structure of the underlying rock by differential erosion. Headward erosion of tributaries is the major way in which this adjustment is accomplished.

EROSION AND PLATE TECTONICS

A model of landscape development set in the framework of plate tectonics involves (1) the evolution of shields as a result of the erosion and subsequent isostatic adjustments of mountain belts and (2) broad upwarps of shields (and stable platforms), causing renewal of stream erosion, or subsidence, allowing expansion of shallow seas over the continent.

Compared to most other planetary bodies in the solar system, Earth is unique because of its infinite variety of constantly changing surface features. Mountain building, volcanism, erosion, and sedimentation constantly modify Earth's surface at an astonishing rate. Indeed, most of Earth's surface was formed during the last 2 million years, and little, if any, is older than Tertiary. On a local scale, many distinctive landforms are obviously the result of a specific surface process. A sand dune is formed by the wind, a delta by rivers, and a beach by wave action. But what about the regional features of Earth? Did they form in a systematic way? Is each continent unique, or do they have some features common to all?

Evolution of Shields

To answer these questions, we must consider the evolution of the major structural components of the continents—the craton with its shields and the stable platform. A general model showing how the basement rock evolves from mountain building is shown in Figure 13.1. Two major factors control this process: (1) erosion of the mountain belt by running water and (2) continual isostatic adjustment of the mountain belt as a result of the removal of material by erosion. Both erosion and isostatic adjustment continue until equilibrium is reached.

In Figure 13.1A a new mountain belt has been formed by plate collision. You will note on the diagram that there are significant changes in the dominant structural features of the mountain belt from the surface down to the deep roots. Andesitic volcanism occurs at the surface. At shallow depths, where there is little confining pressure, the rocks are relatively brittle, and the compression that formed the mountain belt has developed thrust faults. At greater depth, the

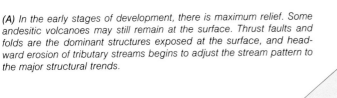

(A) In the early stages of development, there is maximum relief. Some andesitic volcanoes may still remain at the surface. Thrust faults and folds are the dominant structures exposed at the surface, and headward erosion of tributary streams begins to adjust the stream pattern to the major structural trends.

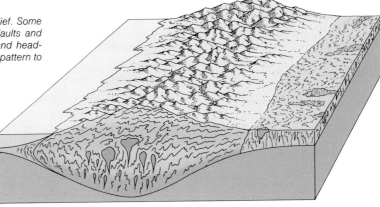

(B) As erosion removes material from the mountain belt, isostatic equilibrium is upset and the mountain root rebounds. Tight folds formed in the deeper part of the mountain system are exposed at the surface, and headward erosion adjusts the stream pattern to the folded rocks. Many tributaries flow parallel to the structural grain of the mountain belt.

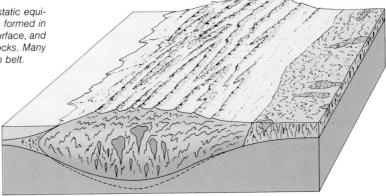

(C) Continued erosion and isostatic adjustment reduce the topographic relief and the size of the root of the mountain belt below. Complex folds and granitic igneous intrusions, originally formed in the deeper part of the mountain belt, are now exposed at the surface. The stream patterns adjust to the new structure and rock types and develop a different style of topography. Local relief and rates of erosion are greatly reduced.

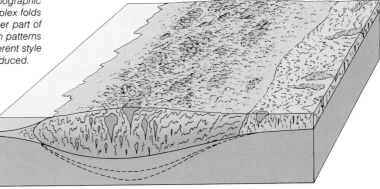

(D) Isostatic equilibrium is ultimately reestablished. The rocks formed deep in the mountain root are exposed to the surface, and local relief is only a few meters. Stream patterns adjust to the structural trends in the metamorphic terrain. At this stage, the mountain belt constitutes a new segment of the shield.

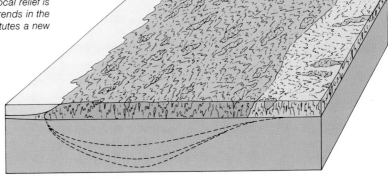

FIGURE 13.1
A model of landscape development shows how a folded mountain belt evolves into a new segment of shield. Erosion occurs during crustal deformation, so by the time mountain building terminates, the mountain range is already carved into a rugged terrain. After deformation, a mountain belt is in isostatic equilibrium; a mountain root extends down into the lithosphere to compensate for the high topography. Note that the style of structure in a mountain belt changes with depth. Andesitic volcanic features may dominate at the surface. Thrust faults and folds occur at shallow depths. Tight folds and granitic intrusions occur at intermediate depths. In the deeper roots, metamorphic rocks intruded by small granitic bodies dominate.

273

rocks are under greater confining pressure, and they yield to plastic flow, which produces tight folds. At still greater depths, complex folds are formed. Silicic magma, generated in the lower crust and by partial melting in the subduction zone, rises because it is less dense than the surrounding rock and reaches a level where it spreads out to form large batholiths. In the deeper roots of the mountain belt, metamorphic rocks dominate and are intruded by smaller bodies of granite. Each of these major zones is shown on the diagrams in Figure 13.1 by a different tone and color: faulting in green, tight folds in tan, complex folds in yellow, and metamorphic rocks in reddish-brown. The topography in Figure 13.1A is young. It is controlled by andesitic volcanism, thrust faulting, and folds (green). The topographic relief is high, and headward erosion is beginning to adjust the drainage pattern to the structural trends of the mountain belt.

In Figure 13.1B the upper segment of the mountain belt has been removed by erosion, but isostatic rebound causes the mountain belt to continually rise. The dominant structure, now exposed at the surface, is a series of tight folds (tan), which are etched into relief. Ridges formed on resistant rock zigzag across the landscape, forming a valley-and-ridge type of topography, typically developed on folded rocks.

At a later stage (Figure 13.1C), erosion has removed the zone of folded strata, but isostatic adjustment continues to elevate the mountain belt. Note the position of the root, or base, of the mountain belt in diagram C, compared to its former position in diagrams A and B, shown in dashed lines. Isostatic rebound is less because the mountain root is not so deep, and topographic relief is consequently not as great as before. Complex folds and granitic intrusions (tan), which were formed deep in the mountain belt, are now exposed at the surface, and the landforms are controlled by these features. The topographic relief and rate of erosion at this stage are much less than in stages A and B.

In Figure 13.1D erosion and isostatic adjustments have reached a state of equilibrium. There is no mountain root extending down as a bulge below the mountain topography. Metamorphic rocks and igneous intrusions (reddish brown) are now at the surface and control the types of surface features to be developed. The entire area is eroded close to sea level. The mountain root is in isostatic equilibrium. The area is stable, and a new segment of the basement complex is formed. Slight changes in sea level may cause the sea to spread across the region and deposit shallow-marine rocks over the shield to form a stable platform.

How does a mountain system evolve into a segment of a shield?

Uplift

The rate of uplift during mountain building is variable and difficult to measure. However, estimates can be made on the basis of the age of strata found high in present mountain ranges together with data obtained from precise geodetic surveys across active mountain belts. A number of such observations give a general rate of uplift of 6 m per 1000 years. If erosion did not occur contemporaneously with uplift, a mountain summit could rise 6 km in 1 million years. If these measurements are correct, a full-size mountain belt could be created in a relatively short time (5 to 10 million years).

Rates of erosion are based on measurements of the volume of sediment carried by streams. In mountainous areas, rates range from 1 to 1.5 m per 1000 years. From these data we can draw a generalized graph showing rates of uplift and the relation of rates of erosion to elevation (Figure 13.2). The rate of uplift is roughly 5 to 10 times faster than the maximum rate of erosion. The relatively rapid rate of deformation and uplift is shown by the steep line on the graph and can occur in a time span of 5 million years. Erosion would occur during uplift, so by the time deformation and uplift end, the mountain range would be carved into a rugged terrain and perhaps as much as 1 km of rock would already have been removed. The main idea that this graph emphasizes is not the absolute rates of erosion, but the rapid decrease in the rate of erosion with a decrease in elevation. From the regional viewpoint, rates of erosion are a function of the height of the landmass above sea level.

As erosion removes material from the mountain belt, the isostatic equilibrium is upset, and the mountain root rebounds in a broad upwarp in an attempt to reestablish the balance. In the early stages of erosion the removal of 500 m of rock is generally compensated for by an isostatic uplift of about 400 m, so there is a net lowering of only 100 m of the mountain surface. It we assume that isostatic adjustment occurs constantly at a ratio of 4:5, the initial rate of net lowering of the surface will be 0.2 m per 1000 years. As shown at the top of the curve in Figure 13.2, the initial net rate of erosion is 200 m per million years. In contrast, at the end of 15 million years, the net rate of lowering of the surface has been reduced to approximately 100 m per million years.

Erosion and isostatic adjustments continue to reduce the topographic relief. By the end of 30 million years the elevation and rate of erosion are again halved, to one-quarter of the initial value. Approximately three-quarters of the original landmass has now been removed, and the structures of the deep mountain roots are exposed. The regional surface is a broad, nearly flat plain. Local relief of a few tens of meters is produced by *differential erosion* of belts of different metamorphic and igneous rock types. Erosion of the surface and the associated isostatic adjustment have declined at a rapid rate, probably exponential, so that a near balance is reached. The original mountain's topography is eroded almost to sea level. At this stage, the mountain belt constitutes a new segment of the continental shield.

The subsequent evolutionary history of the landscape is intimately related to sea level. Sea level is important in our model of erosion because it is

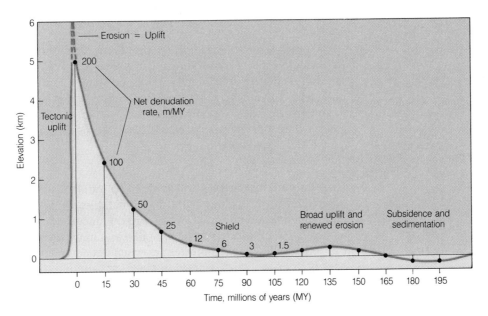

FIGURE 13.2
Rates of erosion of a mountain belt decrease exponentially with time. The period of mountain building is shown by the steep line at the beginning of the graph. The tectonic deformation is shown to last 5 million years, but most of the uplift may occur within 2 million years. Erosion proceeds contemporaneously with uplift, increasing in intensity with increase in elevation. By the time deformation ends, uplift of 6 km had occurred, but the surface is already eroded down to 5 km. The initial rate of erosion is 1 m per 1000 years, but isostatic adjustment occurs at a ratio of 4:5. The initial rate of new lowering of the surface is thus 200 m per million years. In 15 million years, one-half of the mass is removed, and the net rate at which the surface is lowered is reduced to 100 m per million years. After 30 million years, only one-quarter of the mass remains, and the average altitude is 1.25 km. In 60 million years, the mountain belt is reduced to a new segment of the shield.

the ultimate level to which stream erosion can effectively lower the continental surface. Both erosion and isostatic adjustment combine to produce a flat slab of continental crust that is eroded to near sea level. There are, however, a number of reasons why the continental lithospheric plate can be expected to move both up and down with respect to sea level. For example, the top of the asthenosphere is not a perfectly smooth surface, but undulates in swells and depressions. Bulges in the upper surface of the asthenosphere may also result from hotspots in the mantle. As the plate moves over the highs and lows in the asthenosphere, the continental crust is upwarped or depressed with respect to sea level. Changes in sea level can also result from changes in the rate of convection in the asthenosphere. If convection is rapid, the oceanic ridge swells and is arched upward, reducing the volume of the ocean basins and causing expansion of the sea over the flat continental surface. Slow convection would deflate the oceanic ridge, causing the sea to withdraw. The important point is that any change in sea level, regardless of the cause, will affect the erosional processes on the continent. If sea level is lowered (or the continent is upwarped), the processes of erosion will be renewed and the land will be rejuvenated. An increase in elevation of the land brings a rapid increase in the rate of erosion. An uplift of 1 km would be followed by accelerated erosion lasting probably 5 to 10 million years. If sea level rises (or a continent is depressed), the sea will advance over the land. Erosion terminates, and shallow marine sediments are deposited on the eroded surface of the shield.

We must add to our basic model the concept of repeated broad uplifts of the shield and stable platform: uplifts that would cause rejuvenation of erosional forces. Figure 13.3 is a series of block diagrams that show in a simplified way how a new uplifted segment of the shield or platform may evolve through a series of stages until it is reduced to a surface of low relief near sea level. We assume that uplift was rapid and that the surface was elevated about 500 m above sea level. The major processes involved in sculpturing this landscape are headward erosion, downcutting, and slope retreat.

How do moving tectonic plates cause different landscapes to develop?

Initial Stage. Figure 13.3A shows the area immediately after uplift. Downcutting and headward erosion are renewed. These processes have been in progress during uplift, but there may be some relatively large areas that are undissected. In this initial stage of downcutting, valley walls extend right down to the stream bank. Tributaries develop rapidly as the drainage system is extended up the regional slope by headward erosion, and new tributaries form on the valley walls of the major rivers.

Intermediate Stage. In Figure 13.3B tributaries have increased in number and length, so the original surface is completely dissected. Local relief (the difference between the valley floor and the divide summit) is at a maximum. Slope retreat becomes a dominant process. Streams begin to meander across the wider valley floor. Ultimately, downcutting is greatly reduced, and thereafter both erosion and deposition combine to reduce local relief.

Late Stage. The landscape in Figure 13.3C is reduced to a low, almost featureless surface near sea level. This surface is considered to be a *peneplain,* over which the stream meanders, depositing a significant amount of sediment. Subsequent modifications, if conditions remain stable, include deep weathering and fluvial deposition. Oxbow lakes, meander scars, and natural levees are common. Locally isolated erosional remnants of resistant bedrock may rise above the peneplain surface.

After a new segment of the basement complex is formed, its subsequent erosional history would involve periods of broad upwarps and renewed downcutting by streams or of subsidence and deposition.

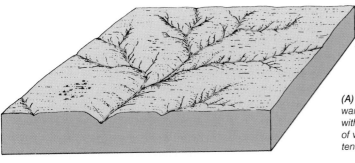

(A) Initial stage. Large areas are undissected. Downcutting and head-ward erosion are the dominant processes. Streams have steep gradients, with rapids and waterfalls. Only a small percentage of the area consists of valley slopes. The streams have not developed floodplains or an extensive meandering pattern.

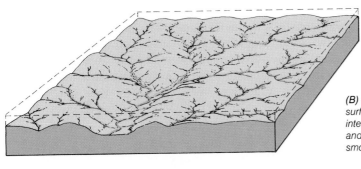

(B) Intermediate stage. The area is completely dissected, so most of the surface consists of valley slopes. Relief is at a maximum, and a well-integrated drainage system is established. The main streams meander, and a floodplain begins to develop. The topography is characterized by smooth, rolling hills.

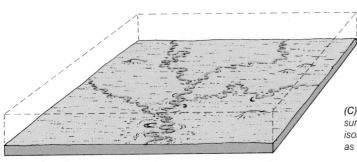

(C) Late stage. The landscape has been eroded to an almost featureless surface near sea level. Rivers meander over a broad floodplain. Some isolated erosion remnants remain, but deposition is nearly as important as erosion.

FIGURE 13.3
A model of landscape development in a humid climate shows how a newly uplifted area may evolve through a definite series of stages until it is reduced to a surface of low relief near sea level.

INTERRUPTIONS IN LANDSCAPE DEVELOPMENT

Interruptions in the general trend of landscape development can result from (1) tectonic uplift or subsidence, (2) climatic changes, and (3) volcanic activity.

The idealized model we have just described shows the basic trends by which a regional landscape would likely evolve according to the plate tectonics theory. Various forces, however, can interrupt and modify this development. Most of these forces are either directly or indirectly controlled by tectonics.

Perhaps the most significant interruptions in landscape development are due to Earth movements. These can be broad, regional uplift, subsidence of the coast, or the intense folding and faulting associated with mountain building. Each results in a change in the energy of the stream system, with a corresponding change in erosion or deposition.

Uplift, for example, increases the elevation of the land, causing an increase in stream energy. As a result, many aspects of landscape development return to the initial stage, and the cycle begins again. If uplift occurs in the late stage, the gradient of a meandering stream pattern is increased, and the stream begins to cut down its channel. As a result, the meandering stream pattern, which is characteristic of the late stage in erosion, can take the form of an ***entrenched meander,*** cutting a deep, winding canyon. Such a river, however, has many characteristics of streams in the initial stage of erosion, such as a steep gradient and rapids. Classic examples of entrenched meanders are found throughout the Colorado Plateau (Figure 13.4).

Climatic changes can also disrupt the processes of stream erosion and drastically alter the landscape. For example, as a continent migrates into the arid low-latitude regions, there is little precipitation to maintain stream erosion. Wind would then become a significant geologic agent, developing extensive deserts. Seas of sand may cover the landscape previously formed by running water. If a continental plate migrated to the polar regions, glaciers could form and cover the landscape. Previous river systems would be obliterated, and the surface features would be significantly modified by glacial erosion and deposition. Volcanic activity can also occur, especially where continents are rifted apart, and floods of lava may bury the previous topography. In addition, continental rifting can dismember a previously established drainage system and modify the way it functions.

FIGURE 13.4
An entrenched meander in the Colorado River, Utah. This area is known as the Bowknot because of the two nearly symmetrical meander bends. The entrenchment illustrates rejuvenation of stream erosion. The meandering pattern of the river was formed during an early period when the river's gradient was much lower than at present. Recent uplift of the region increased the gradient of the river and its power to erode. Downcutting of the stream channel resulted in the entrenchment of the meanders.

DIFFERENTIAL EROSION AND STRUCTURE

Factors such as climate and mineralogy cause some rocks to be hard and resistant and others to be soft and weak. This results in differential erosion on all scales, from ridges and valleys in a mountain belt to the thin laminae within a single bed. Differential erosion on horizontal, tilted, folded, and faulted rock bodies results in a variety of landforms, such as plateaus, mesas, buttes, hogbacks, and strike valleys.

The model of landscape development just described explains the evolution of the major surface features of Earth within our present understanding of Earth's dynamics. On a smaller scale, we see that different rock types erode at vastly different rates (differential erosion) and give rise to a fascinating variety of local landforms. Factors such as climate and mineralogy cause some rocks to be hard and resistant and others to be soft and weak. The structure of the rocks is critical to differential erosion and the style of landforms to be developed. The term *structure* is used here in the broadest sense and includes horizontal rocks, tilted or inclined strata, domes and basins, folds, and faults. Differential erosion occurs on all scales, from ridges and valleys in a mountain belt to thin laminae within a single bed. In general, the structural features of a rock are older than the landforms developed on them, so structure is a dominant control in the evolution of many landscapes. As a result of differential erosion, drainage systems have a universal tendency to adjust to the structure of the rocks over which they flow. Headward erosion of tributaries is the basic process by which the drainage pattern of a stream adjusts to the structure.

Let us consider briefly some of the landforms that typically develop on some of the more important structural features of the continents.

The Stable Platform — Landforms Developed by Erosion of Horizontal Strata

The stable platform is that part of the basement complex that is covered with sedimentary strata (see Chapter 1). The rocks were deposited mostly in shallow seas that spread over the flat eroded surface of the complex. The total thickness of strata rarely exceeds 2000 to 3000 m, so on a regional basis the cover of sedimentary strata is only a thin veneer overlying the igneous and metamorphic basement rock. Subsequent broad upwarps of the crust involving little or no tilting can expose these rocks to the processes of erosion.

Some of the landforms that typically develop by erosion of horizontal strata (in an arid climate) are shown in Figure 13.5. The expansion and contraction of a shallow sea will typically produce interbedded sand and shale or interbedded limestone and shale (see Figure 6.8). The shale is typically weak and nonresistant, whereas the sandstone and limestone layers are relatively resistant. As erosion removes a sequence of rocks, the resistant layers of sandstone or limestone will commonly form a protective cover, or *cap rock*, resulting in a nearly flat plateau. Erosion and slope retreat along the edge of the plateau form alternating cliffs and slopes on the alternating weak and resistant layers. This is probably the most widespread example of differential erosion. The height of the cliffs and the width of the slopes are largely functions of the thickness of the formations involved and the rates of erosion.

As erosion progresses, a large portion of a plateau can be detached to form a *mesa,* which then shrinks by slope retreat on all sides to form groups of smaller *buttes.* These continue to waste away into *pillars* and *pinnacles,* which ultimately disappear (Figure 13.5).

A given layer of sedimentary rocks is essentially homogeneous in all directions. Such formations have no appreciable structural control over the

Why are alternating cliffs and slopes so common in areas of horizontal, tilted, and folded rocks?

Why isn't there a "Grand Canyon" in Kansas?

(A) **Monument Valley, Utah,** a classic area of plateaus, mesas, buttes and pinnacles.

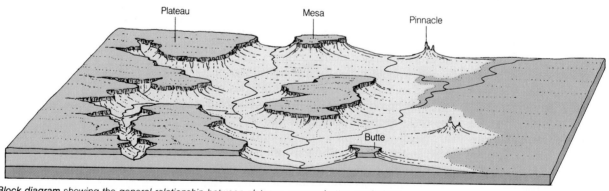

(B) **Block diagram** showing the general relationship between plateaus, mesas, buttes, and pinnacles.

FIGURE 13.5
Differential erosion of horizontal strata

drainage system. All tributaries are free to grow with equal ease, so a dendritic drainage pattern typically develops on horizontal sedimentary rocks.

Excellent examples of erosion landforms developed on horizontal strata are found throughout the stable platform, especially in Kansas, Texas, and the Colorado Plateau.

Local Features Developed by Differential Erosion

Differential erosion of the receding cliffs and slopes may develop a fascinating variety of landforms. Examples are the columns, pinnacles, and pillars developed on sandstones. If a massive cliff has well-developed joints, vertical columns may form, like those shown in Figure 13.6A. Where stratification produces alternating layers of hard and soft rock, additional detail may be developed by etching. Jointing commonly plays an important role, for it permits weathering processes to attack a rock body from many sides at once. The columns and pillars in Bryce Canyon National Park, Utah, for example, result from differential weathering and erosion along a set of intersecting joints (Figure 13.6). The joints divide the rock into columns. Nonresistant shales separate the more-resistant sandstone and limestone, and differential erosion forms recesses in the columns, producing fascinating forms.

Natural arches (Figure 13.7) are another interesting product of differential erosion on receding cliffs. The best examples are carved in the massive sandstones of the Colorado Plateau in the western United States, where more than 90% of the natural arches in the world are found. The diagrams in Figure 13.7 illustrate one way in which natural arches may be formed. The arid western parts of the United States receive little precipitation, and much of the surface water seeps into the thick, porous sandstone. Subsurface water is most abundant beneath dry stream channels, and its movement follows the general lines of surface drainage. Groundwater, emerging as a seep in a cliff beneath a dry waterfall, dissolves the cement that holds the sand grains together. Loose sand grains are washed or blown away, so an *alcove* soon develops at the base of the normally dry waterfall. If the sandstone is cut by joints, a large block can be separated from the cliff as the joints are enlarged by weathering processes. The alcove and joints continue to enlarge, and an isolated arch is eventually produced. Weathering then proceeds inward from all surfaces until the arch is destroyed, leaving only the columns standing.

Erosion of Inclined Strata

Where a sequence of alternating resistant and nonresistant strata is tilted, the nonresistant units are eroded into long valleys or lowlands trending parallel to the *strike,* or trend, of the rocks. The resistant layers are left standing as long, asymmetric ridges. Ridges formed on gently inclined strata are called *cuestas.* Sharp ridges formed on steeply inclined rocks are called *hogbacks* (Figure 13.8).

In the Coastal Plains of the Atlantic and Gulf Coast states, the alternating layers of sandstone and shale deposited during the Mesozoic and Cenozoic eras have a persistent gentle *dip* (inclination) seaward. The topography of the Coastal Plains therefore consists of a series of low cuestas (formed on the sandstone layers) and broad, low valleys (formed on the soft shales) (Figure 13.9). As the area emerged from below sea level, the streams draining the area were extended downslope.

With continued uplift, headward erosion proceeded rapidly along the nonresistant shales to form strike valleys, leaving the more resistant sandstone as cuestas. The drainage pattern developed by erosion of the Coastal Plains is a trellis pattern.

Erosion of Structural Domes and Basins

In many areas of the stable platform, the sedimentary rocks are warped into broad structural *domes* and *basins.* On the flanks of these structures the strata

(A) Eroded columns provide spectacular scenery in Bryce Canyon National Park, Utah. Rapid erosion along intersecting joint systems separates the columns from the main cliff. Differential erosion, accentuating the difference between rock layers, produces the fluted columns.

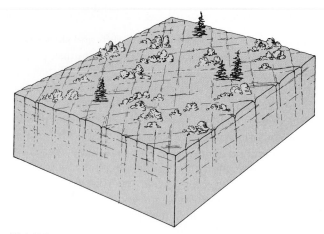

(B) **Initial stage.** Intersecting joints separate the rocks into columns.

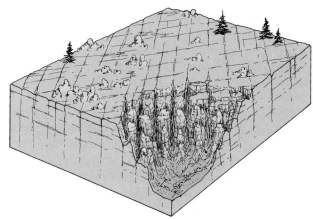

(C) **Intermediate stage.** Weathering and erosion along the joints accentuate the columns, which erode into various forms as a result of alternating hard and soft layers.

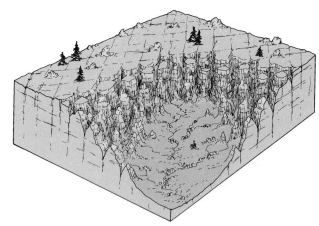

(D) **Final stage.** As weathering and erosion proceed, the cliff retreats. Old columns are completely destroyed, but new ones are continually created.

FIGURE 13.6
The evolution of columns by differential erosion along a receding cliff commonly is controlled by intersecting joint systems.

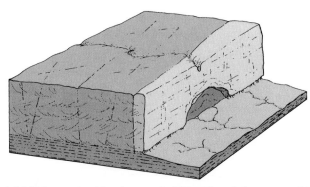

(A) *Initial stage.* In arid regions much of the surface drainage seeps into the subsurface below a stream channel. This water may then move laterally above an impermeable layer and eventually emerge as a seep near the base of a cliff. The cement that holds the sand grains together is soon dissolved in this area of greatest moisture, and the sand grains fall away, so that a recess, or alcove, forms beneath the dry falls from the intermittent stream above.

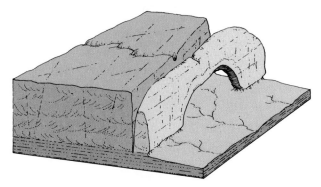

(B) *Intermediate stage.* If a joint system in the sandstone is roughly parallel to the cliff face, the joints can be enlarged by weathering, which separates a slab from the main cliff.

(D) *The initial stage in the development of a natural arch* can be observed in Zion National Park, Utah. In this photograph, a well-developed alcove has formed beneath a dry waterfall. Weathering along a large joint will soon separate the alcove from the cliff to produce a natural arch.

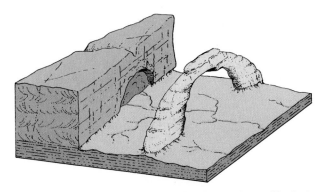

(C) *Final stage.* An arch is produced as the alcove enlarges. Weathering then proceeds inward from all surfaces until the arch collapses.

FIGURE 13.7
Natural arches develop in massive sandstone formations by selective solution activity.

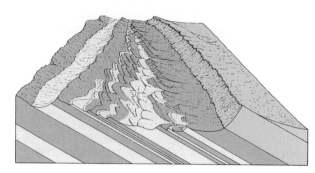

FIGURE 13.8
Differential erosion of a sequence of tilted strata produces ridges called hogbacks on resistant formations and long, narrow valleys in nonresistant rocks.

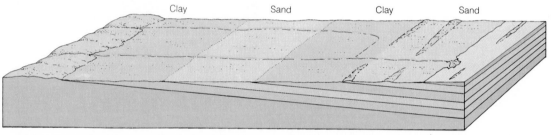

Clay   Sand   Clay   Sand

(A) As the coastal plains emerge above sea level, the drainage system is simply extended downslope directly to the new shoreline. These are called consequent streams.

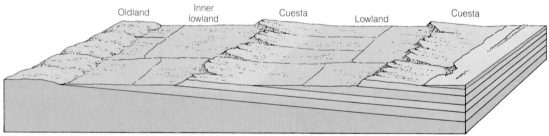

Oldland   Inner lowland   Cuesta   Lowland   Cuesta

(B) Headward erosion of tributary streams along the nonresistant shale units produces linear lowlands (strike valleys). The more-resistant sandstone units remain as linear ridges called cuestas.

FIGURE 13.9
The Atlantic and Gulf Coastal Plains consist of Tertiary and Cretaceous rocks that are inclined toward the sea. The evolution of landforms developed on the inclined strata is shown in idealized block diagrams.

may dip at angles of 20 or 30 degrees or more. As erosion progresses, strata are removed from the top of a dome, which is eroded outward to form a series of sharp-crested cuestas or hogbacks with intervening ***strike valleys***. The drainage pattern developed by erosion along strike valleys is commonly circular. If the older rocks in the center of the dome are nonresistant, the center of the uplift may be eroded into a topographic lowland bordered by inward-facing ***escarpments*** (cliffs) formed on the younger, resistant units. If the older rocks are more resistant, the center of the dome remains high, forming a dome-shaped hill or ridge. Large structural domes have inward-facing cliffs, whereas large

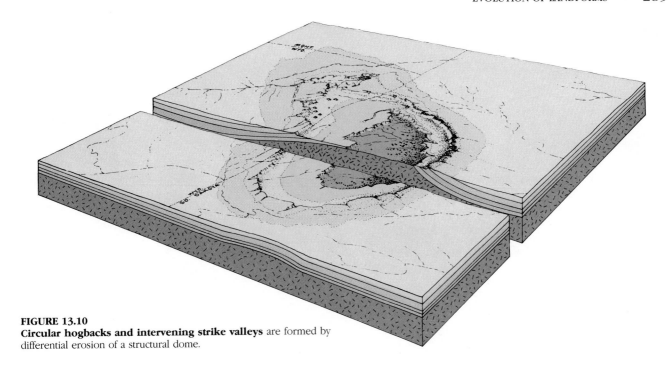

FIGURE 13.10
Circular hogbacks and intervening strike valleys are formed by differential erosion of a structural dome.

basins have outward-facing escarpments. Excellent examples of domal structures in the stable platform include the Black Hills of South Dakota (Figure 13.10), the Ozark Mountains of Missouri, and the Nashville dome of Tennessee.

Erosion on Folded Strata

Landforms that typically develop on folded strata are illustrated in Figure 13.11. In the initial stages of erosion, the uparched folds, or **anticlines,** may form ridges, and the downwarped folds, or **synclines,** may form long valleys. Some major streams from previous cycles of erosion may be superposed across the anticlinal ridges, as is shown in Figure 12.18. As erosion proceeds, the crests of the anticlines are cut by the narrow valleys that grow along the flanks of the ridge. As the crest of the anticlinal ridge is breached, the "anticlinal valleys" are enlarged and deepened so that the crest of the anticlinal ridge becomes completely opened along its length.

As erosion proceeds rapidly headward along the nonresistant formations, the crests of the anticlines are eroded away, and the surface topography bears little or no resemblance to the underlying folded structure. Differential erosion effectively removes the weak rock layers to form long strike valleys. Resistant rock bodies stand up as narrow hogback ridges. The ridges (mountains) thus mark the limbs of the folds. The pattern of topography is typically one of alternating valleys and ridges, and the drainage system generally forms a trellis pattern.

The Ridge and Valley province of the eastern United States is a classic example of a landscape formed on folded and thrust-faulted strata (Figure 13.11A). It extends from Pennsylvania to central Alabama, a distance of more than 2000 km. In Alabama, the folded structures plunge beneath the younger sediments of the Coastal Plain. Similar topography is exposed again in the Ouachita Mountains in Arkansas and Oklahoma and in the Marathon Mountains of west Texas. A large segment of the folded Appalachian mountain belt is therefore still buried beneath younger sediments. The deformation that produced the folds occurred during the late Paleozoic era (more than 200 million years ago). The mountain belt was then deeply eroded and most of the region was covered by Cretaceous and possibly Tertiary sediments. Regional uplift ensued, and renewed erosion began to remove the sedimentary cover.

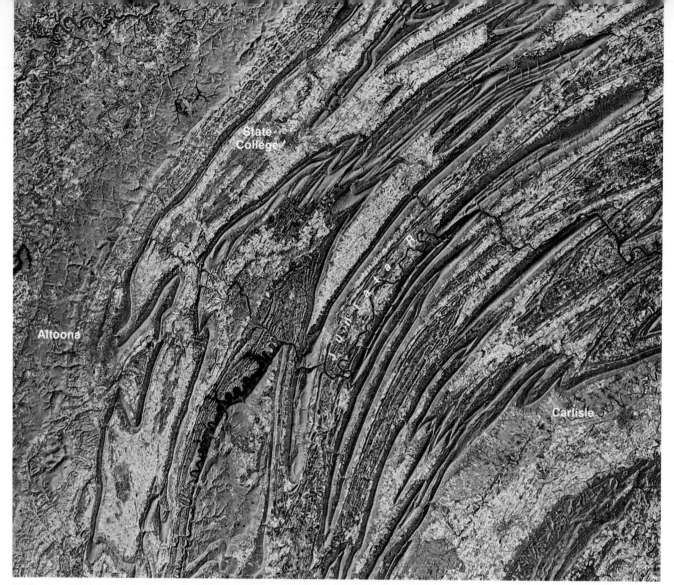

(A) *Erosion of folded strata in the Appalachian Mountains,* in the eastern United States, produces a series of zigzag ridges and valleys.

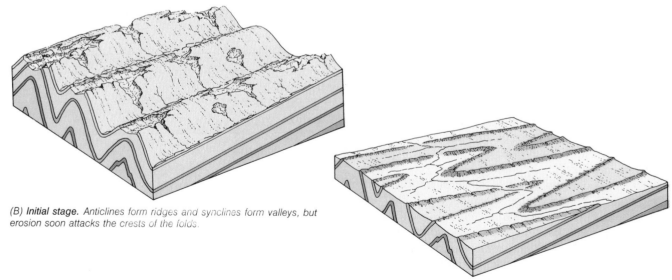

(B) *Initial stage.* Anticlines form ridges and synclines form valleys, but erosion soon attacks the crests of the folds.

(C) *Later stages.* The tops of the folds are eroded away, so that hogback ridges are left along the flanks of the folds. These ridges commonly form a zigzag pattern.

FIGURE 13.11
Differential erosion of folded rocks typically forms alternating ridges and valleys.

The major east-flowing streams were superposed on the northeastward-trending structures, and differential erosion began to cut the extensive valley system between the resistant sandstone ridges (Figure 12.18). Extension of the drainage system by more rapid erosion of the nonresistant shale and limestone formations resulted in stream capture and the development of a trellis drainage pattern. (Note that in humid regions limestones are eroded rapidly, whereas in drier regions they are resistant.)

Erosion on Fault Blocks

A *rift zone* is a region where the crust has been arched upward and pulled apart. The dominant structures in this tectonic setting are parallel systems of faults with large vertical displacements. Typically, the faults produce large, elongate, downdropped blocks called *grabens* (rift valleys) and associated uplifted blocks called *horsts*. The major landform produced by block faulting is a steep cliff, or *fault scarp,* which is soon dissected by erosion. The evolution of erosional landforms developed on a fault block is shown in Figure 13.12. As soon as uplift occurs, erosion begins to dissect the cliff and produces a series of triangular faces called *faceted spurs*. Erosion also begins to form gullies along the blunt face of the faceted spurs, so the cliff produced by faulting is considerably modified. Recurrent movement along the fault can produce a fresh scarp at the base of the older faceted spurs, but it is also rapidly modified by gullying to form a series of compound faceted spurs. When movement on the fault ceases, the cliff continues to erode down and back from the fault line.

The idealized stages in landscape development by *block faulting* in an arid region are illustrated in Figure 13.13. In the initial stage (Figure 13.13A), maximum relief is produced by uplift of *fault blocks.* Relief diminishes throughout subsequent stages, unless major uplift recurs and interrupts the evolutionary trend by producing greater relief during later stages. In arid regions, depressions between mountain ranges generally do not fill completely with water. Large lakes do not form because of the low rainfall and great evaporation. Instead, shallow, temporary lakes, called *playa lakes,* form in the central parts of basins and fluctuate considerably in size as a result of wet and dry periods. They may be completely dry for many years and then expand to cover a large part of the valley floor during years of high rainfall. When rainfall is high enough to maintain a permanent body of water, the lake is commonly saline due to the lack of an outlet. Great Salt Lake, Utah, is an example. Most basins in the Basin and Range province have dry playas during most of the year, but during the cooler, wetter climatic periods that accompanied the Pleistocene glaciers, lakes with surface areas of thousands of square kilometers developed in many of these valleys.

Weathering in the uplifted mountain mass produces more sediment than can be carried away by the intermittent streams, which may flow only during spring runoff. The overloaded streams commonly deposit much of their load where they emerge from the mountain front. This deposited sediment accumulates in a broad alluvial fan. Throughout the history of an intermontane basin, mountain ranges are eroded, and the debris is deposited in the adjacent basin.

As geologic processes continue (Figure 13.13B), the mountain mass becomes dissected into an intricate network of canyons and is worn down to a lower level. At the same time, the mountain shrinks farther as the front recedes through the process of slope retreat. The fans along the mountain front grow and merge to form a large alluvial slope called a *bajada.* As the mountain front retreats, an erosion surface called a *pediment* develops on the underlying bedrock. It expands as the mountain shrinks. Pediments are generally covered by a thin, discontinuous veneer of alluvium. As shown in Figure 13.13C, pediments may become completely buried.

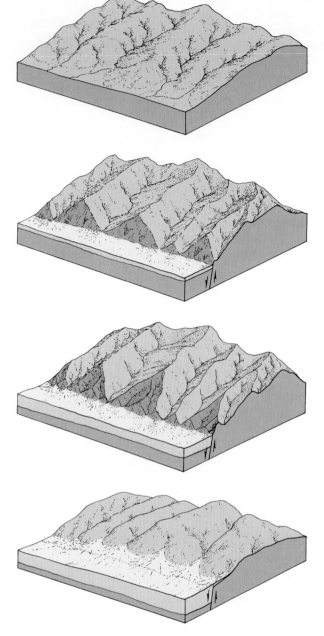

(A) The original dissected upland, before faulting, consists mostly of valley slopes.

(B) The first major period of faulting is accompanied by accelerated stream erosion. Valleys are cut through the scarp produced by faulting to form triangular faceted spurs.

(C) Recurrent movement along the fault can produce a series of fresh scarps, which are subsequently dissected by stream erosion. Older faceted spurs recede and are worn down.

(D) Once faulting ceases, erosion removes most of the fault scarp and faceted spurs, and the cliff recedes from the fault line.

FIGURE 13.12
Erosion of a fault-block mountain follows a series of stages in which faceted spurs are a prominent landform.

The final stage (Figure 13.13C), according to this model, is characterized by small, islandlike remnants of the mountains, surrounded by an extensive erosion surface, the pediment, much of which is buried by alluvial fans. The pediment shows little relief and is largely covered with erosional debris.

This erosional model, or some variation of it, effectively explains much of the landscape in the Basin and Range province of the western United States. The Basin and Range is a large area where the crust has been arched upward and pulled apart, forming a complex rift system that extends from northern Mexico to southern Idaho and Oregon. The block faulting, resulting from tensional stresses, has produced alternating mountain ranges and intervening fault-block basins. There are over 150 separate ranges in this province. Some are simple tilted fault blocks that are asymmetrical in cross section, the steeper

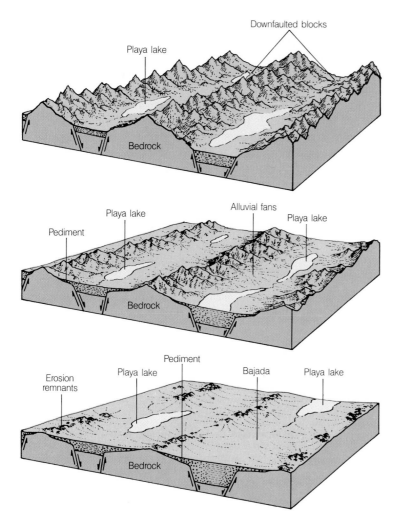

(A) Initial stage. Faulting produces maximum relief. Initially, some areas in the mountains are undissected. Playa lakes may develop in the central parts of the basins.

(B) Intermediate stage. The mountain range is completely dissected, and the mountain front retreats as a pediment develops. Alluvial fans spread out into the valley.

(C) Late stage. The basins become filled with sediment. Erosion wears down the mountain ranges to small, isolated remnants. The pediments expand and are buried by the alluvial fans, which merge to form bajadas. Most of the surface is an alluvial slope.

FIGURE 13.13
A model of landscape development in the Basin and Range province of the United States shows that fault-block mountains evolve through a series of stages until the mountain is consumed.

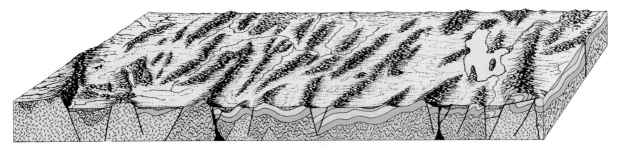

FIGURE 13.14
Erosion of fault blocks in the western United States produces alternating mountain ranges and intervening basins filled with sediment.

side marking the side along which faulting occurred. Others are faulted on both sides. The internal structure of the fault blocks is complex, with folds, faults, and igneous intrusions recording an earlier history of crustal deformation (Figure 13.14).

To the north, throughout much of western Utah and all of Nevada, the area is in the initial stage of development. The basins occupy about half of the total area, and the pediments are small. The relief of the mountain ranges is high, with most fans just beginning to coalesce into a broad alluvial slope. Recurrent movement along many of the fault systems is indicated by the complex faceted spurs, similar to those illustrated in Figure 13.12C. Continued movement in recent times is clear from faulted alluvial fans and recurring earthquakes in the area.

In Arizona and Mexico, erosion in the Basin and Range has proceeded much farther, and the area is in the late stage of development. The ranges are eroded down to small remnants of their original size. Extensive bajadas, spreading over approximately four-fifths of the area, cover wide pediments, through which isolated remnants of bedrock protrude.

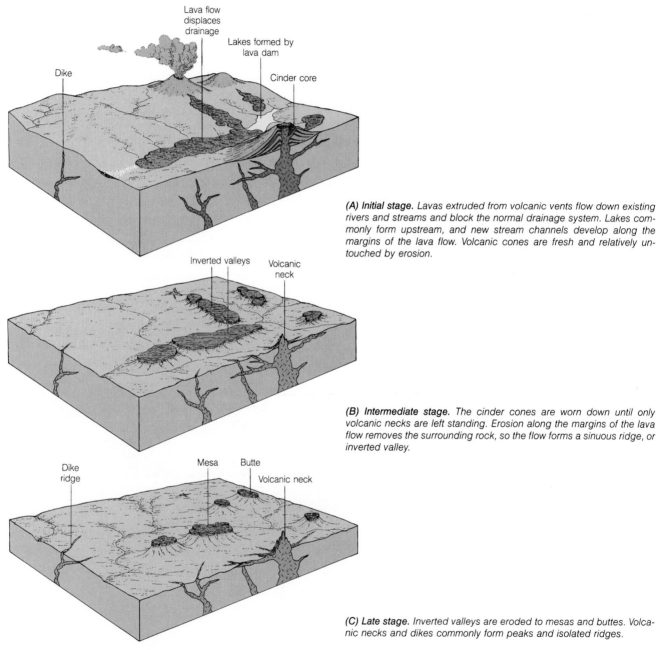

(A) Initial stage. Lavas extruded from volcanic vents flow down existing rivers and streams and block the normal drainage system. Lakes commonly form upstream, and new stream channels develop along the margins of the lava flow. Volcanic cones are fresh and relatively untouched by erosion.

(B) Intermediate stage. The cinder cones are worn down until only volcanic necks are left standing. Erosion along the margins of the lava flow removes the surrounding rock, so the flow forms a sinuous ridge, or inverted valley.

(C) Late stage. Inverted valleys are eroded to mesas and buttes. Volcanic necks and dikes commonly form peaks and isolated ridges.

FIGURE 13.15
A model of landscape development in an area of local volcanism shows how inverted topography may develop.

Erosion on Basaltic Plains

The extrusion of lava provides a new surface, which is also eroded and modified through a series of stages. Lava flows follow previously established drainage systems, displacing the river channels. In small areas of volcanic eruptions, the lava may be confined to the stream valleys. Lava flows generally are very resistant to erosion, and thus, when the stream channel is displaced, erosion proceeds along the margins of the flow. This results in a long, sinuous ridge called an *inverted valley*. As the name implies, the old stream valley beneath the lava is left standing higher than the surrounding area.

The sequence of landforms resulting from the erosion of an area of minor volcanic activity is illustrated in Figure 13.15. In the initial stages (Figure 13.15A), volcanic flows enter the drainage system, following the river channels and partly filling the valleys. They disrupt drainage in two principal ways: lakes are impounded upstream, and the river is displaced and forced to flow along the margins of the lava flow. Subsequent stream erosion is then concentrated along the lava margins.

As erosion is initiated in the displaced drainage system (Figure 13.15B), new valleys are cut along the flow margins and become gradually deeper and wider with time. Cinder cones, formed during the initial volcanic activity, are soon obliterated because the unconsolidated ash is easily eroded. Only the conduit through which the lava was extruded remains as a resistant *volcanic neck*. With time, the area along the margins of the lava flow is eroded, so an inverted valley, where the old stream valley was flooded with lava, is left standing higher than the surrounding area. In the final stage of erosion (Figure 13.15C), the inverted valley is reduced in size and ultimately becomes dissected into isolated mesas and buttes.

In regions of extensive volcanism, large areas may be completely buried by lava flows (Figure 13.16). In the initial stage (Figure 13.16A), drainage usually is displaced to the margins of the lava plain, but some rivers may migrate across the plain. When extrusion ceases (Figure 13.16B), stream erosion begins to dissect the lava plain and eventually cuts it into isolated plateaus and mesas (Figure 13.16C). These, in turn, are ultimately eroded away.

An example of landforms developed on volcanic plains is the Columbia Plateau in Washington, Oregon, and Idaho, which covers a region of over 400,000 km^2 and represents one of the world's major accumulations of lava on a continent (Figure 13.17). Approximately 100,000 km^3 of basaltic lava have been extruded in this area through fissure eruptions. The lava was very fluid and completely covered a previously mountainous topography, which had a *local relief* of over 750 m. The age of the basalts ranges from early Tertiary to Recent, with the youngest flows in Craters of the Moon National Monument being less than 1000 years old. Individual lava flows vary from 5 to 7 m in thickness, and in places a sequence over 4 km thick has built up.

The extrusion of the lavas produced a new surface, which is currently being eroded and modified by the Columbia and Snake river systems. The erosional model in Figure 13.16 shows the general way in which the landscape in this region developed. In southern Idaho the region is still in the initial stage of development, with large areas untouched by erosion. The Snake River meanders across the new basalt plain and in western Idaho has cut a deep canyon in the basalts. Hell's Canyon, along the Idaho–Oregon border, is deeper than the Grand Canyon of the Colorado River. In the youngest part of the province in eastern Idaho, the Snake River flows on the lava surface and has virtually no tributaries from the north because the surface water seeps underground in the porous layers of basalt.

The Columbia Plateau in Oregon and Washington is older and has a more complex history because of the catastrophic flooding that occurred in the area during the late stages of the ice age.

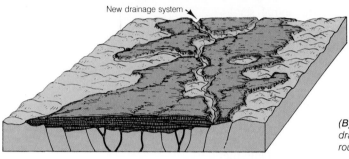

(A) Initial stage. Extensive lava flows fill the preexisting drainage system and flood the surrounding area. New drainage patterns develop along the margins of the volcanic field, as well as across the new surface of lava.

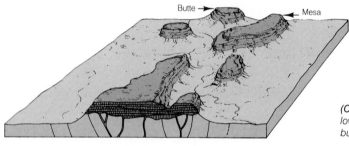

(B) Intermediate stage. The lava begins to be dissected by the new drainage system. Igneous rock usually is very resistant, so the surrounding area is eroded faster, and the lava field is left as a plateau.

(C) Late stage. The area surrounding the volcanic flow is reduced to a lowland. Remnants of lava form cap rock on isolated high mesas and buttes.

FIGURE 13.16
A model of landscape development in an area of regional volcanism shows the evolution of basaltic plateaus.

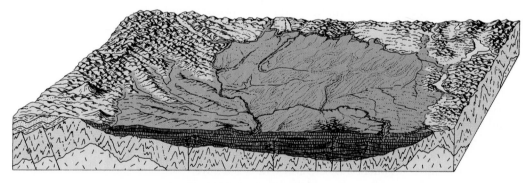

FIGURE 13.17
The Columbia Plateau, in Washington, Oregon, and Idaho, is composed of extensive floods of basalt, which are in the initial stage of dissection by the major river systems.

SUMMARY

1. A model of landscape evolution can be constructed in the framework of the plate tectonics theory. This model involves the erosion and isostatic adjustments of a mountain range with the ultimate development of a surface of low relief carved on the complex igneous and metamorphic rocks of the mountain roots. The folded mountain belt thus evolves into a new segment of the shield.

   Using the best available measurements of erosion rates, we conclude that this process can be accomplished in approximately 60 to 100 million years. Rates of erosion decrease rapidly (probably exponentially) with decrease in elevation.

2. The structure of the rocks in a region is a dominant factor in controlling the characteristics of erosional landforms. Through the processes of headward erosion and stream capture, stream patterns constantly adjust to rock structure.

3. Erosion of horizontal strata typically produces a series of alternating cliffs and slopes and the development of plateaus, mesas, and buttes. Small-scale features resulting from differential erosion on a receding cliff include rock columns or pinnacles and natural arches.

4. Erosion of inclined strata develops alternating cuestas or hogbacks and intervening strike valleys. The Coastal Plains of the Atlantic and Gulf states are examples.

5. Erosion of structural domes and basins produces a series of circular hogbacks and strike valleys. These landforms are common in several regions of the stable platform such as the Black Hills of South Dakota, the Ozark Mountains of Missouri, and the Nashville dome of Tennessee.

6. Erosion on folded strata produces ridges and valleys on the resistant and nonresistant strata, respectively. The topography typically forms a zigzag pattern, following the surface exposures of the strata in the plunging folds.

7. Erosion on fault blocks commonly produces a series of faceted spurs. In addition to the faceted spurs, major landforms include alluvial fans, bajadas, playa lakes, and pediments.

KEY TERMS

alcove (p. 281)

anticline (p. 285)

bajada (p. 287)

basin (p. 281)

block faulting (p. 287)

butte (p. 279)

cap rock (p. 279)

cuesta (p. 281)

differential erosion (p. 275)

dip (p. 281)

dome (p. 281)

entrenched meander (p. 278)

escarpment (p. 284)

faceted spur (p. 287)

fault block (p. 287)

fault scarp (p. 287)

graben (p. 287)

hogback (p. 281)

horst (p. 287)

inverted valley (p. 291)

local relief (p. 291)

mesa (p. 279)

natural arch (p. 280)

pediment (p. 287)

peneplain (p. 276)

pillar (p. 279)

pinnacle (p. 279)

playa lake (p. 287)

rift zone (p. 287)

strike (p. 281)

strike valley (p. 284)

syncline (p. 285)

volcanic neck (p. 291)

REVIEW QUESTIONS

1. Describe the model of evolution of a mountain belt into a new segment of the basement complex.
2. Describe the model of erosion of the shield, or platform, after rapid uplift of approximately 500 m.
3. List the landforms developed on horizontal rocks.
4. Explain the origin of alternating cliffs and slopes as a result of differential erosion.
5. Explain the origin of columns and pillars such as those in Bryce Canyon National Park.
6. How are natural arches formed?
7. Describe and illustrate, by means of a cross section, the landforms that typically develop on a stable platform.
8. What changes would be necessary for deep canyons to form in Kansas?
9. Describe and illustrate, by means of a cross section, the landforms that typically develop by erosion of folded sedimentary rocks.
10. Describe and illustrate by means of a cross section the origin of landforms in a rift system.

ADDITIONAL READINGS

Bloom, A. L. 1978. *Geomorphology.* Englewood Cliffs, NJ: Prentice-Hall.

Bradshaw, M. J., et al. 1978. *The Earth's Changing Surface.* New York: Wiley.

Easterbrook, D. J. 1969. *Principles of Geomorphology.* New York: McGraw-Hill.

Garner, H. F. 1974. *The Origin of Landscapes.* New York: Oxford University Press.

Hunt, C. G. 1973. *Natural Regions of the United States and Canada.* San Francisco: Freeman.

Pirkle, E. C., and W. H. Yoho. 1982. *Natural Landscapes of the United States.* Dubuque, IA: Kendall/Hunt.

Ritter, D. F. 1978. *Process Geomorphology.* Dubuque, IA: Brown.

Thornbury, W. D. 1969. *Principles of Geomorphology,* 2nd ed. New York: Wiley.

Tuttle, S. D. 1970. *Landforms and Landscapes.* Dubuque, IA: Brown.

The movement of water in the pore spaces of rocks beneath Earth's surface is a geologic process that is not easily observed and therefore not readily appreciated, yet groundwater is an integral part of the hydrologic system and a vital natural resource. Groundwater is not a rare or unusual phenomenon. It is distributed everywhere beneath the surface. It occurs not only in humid areas but also beneath desert regions, under the frozen Arctic, and in high mountain ranges. In many areas, the amount of water seeping into the ground equals or exceeds the surface runoff.

One such area is just west of Cape Canaveral, Florida, shown in this high-altitude infrared aerial photograph. Here groundwater has dissolved a network of caves and subterranean passageways. As groundwater dissolves the limestone formations, the caves enlarge and their roofs commonly collapse, so circular depressions called sinkholes are formed. In areas such as this, sinkholes dominate, and a unique landscape evolves. There is no integrated drainage in the area, and most of the water falling on the surface as rain becomes part of the groundwater system. The numerous lakes shown in this photograph occupy sinkholes and testify to the effectiveness of groundwater as a geologic agent.

In this chapter we will study the groundwater system, how water moves through various types of pore spaces in the rock and how it forms karst topography. We will then consider how we use groundwater, this most precious natural resource, and how we attempt to cope with the environmental problems that result when we modify and manipulate this part of the hydrologic system.

14
Groundwater

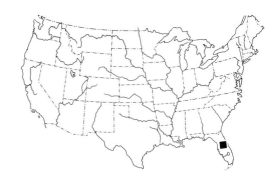

1. The movement of groundwater is controlled largely by the porosity and permeability of the rocks through which it flows.
2. The water table is the upper surface of the zone of saturation.
3. Groundwater moves slowly (percolates) through the pore spaces in the rocks by the pull of gravity. In artesian systems, it is moved by hydrostatic pressure.
4. The natural discharge of groundwater is generally into streams, marshes, and lakes.
5. Artesian water is water that is confined under pressure, like water in a pipe. It occurs in permeable beds bounded by impermeable formations.
6. Erosion by groundwater produces karst topography, which is characterized by sinkholes, solution valleys, and disappearing streams.

POROSITY AND PERMEABILITY

Two physical properties of a rock largely control the amount and movement of groundwater. One is porosity, the percentage of the total volume of the rock consisting of voids. The other is permeability, the capacity of a rock to transmit fluids.

Water can infiltrate the subsurface because solid bedrock, as well as loose soil, sand, and gravel, contains pore spaces. The pores, or voids, within a rock can be spaces between mineral grains, cracks, solution cavities, or vesicles. Two physical properties of a rock largely control the amount and movement of groundwater. One is porosity, the percentage of the total volume of the rock consisting of voids. Porosity determines how much water a rock body can hold. The second property is permeability, the capacity of a rock to transmit fluids. Permeability depends on such factors as the size of the voids and the degree to which they are interconnected. Porosity and permeability are not identical. Some very porous rocks (such as shale) have such small pore spaces that it is difficult for water to move through them. Even though they have high porosity, such rocks are impermeable.

Contrary to popular belief, underground rivers are uncommon and occur only in cavernous limestone and in some lava tunnels in volcanic terrain. Most groundwater occurs in the pore spaces between grains, in fractures, and in other small voids.

How can water flow through solid rock?

Porosity

There are four main types of *pore spaces,* or voids, in rocks (Figure 14.1): (1) spaces between mineral grains, (2) fractures, (3) solution cavities, and (4) vesicles. In sand and gravel deposits, pore space can constitute from 12 to 45% of the total volume. If several grain sizes are abundant and the smaller grains fill in the space between larger grains, or if a significant amount of cementing material fills in the spaces between grains, the *porosity* is greatly reduced. All rocks are cut by fractures, and in some dense rocks (such as granite), fractures constitute the only significant pore spaces (Figure 14.1). Solution activity, especially in limestone, commonly removes soluble material, forming pits and holes (Figure 14.1). Some limestones thus have high porosity. As water moves along joints and bedding planes in limestone, solution activity enlarges fractures in the rock and develops passageways, which may grow to become caves. In basalts and other volcanic rocks, vesicles formed by trapped

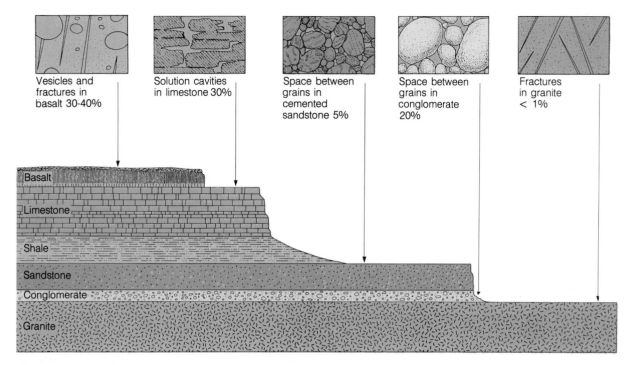

FIGURE 14.1
Various types of pore spaces in rocks permit the flow of groundwater. Porosity resulting from vesicles is common in basalt. Porosity resulting from solution activity is common in limestones. Porosity resulting from spaces between grains is exemplified in sandstone and conglomerate. Porosity resulting from fractures occurs in almost all rocks.

gas bubbles significantly affect porosity (Figure 14.1). Vesicles commonly are concentrated near the top of a lava flow and form zones of very high porosity, which can be interconnected by columnar joints or by cinders and rubble at the top and base of the flow.

Permeability

Permeability, the capacity of a rock to transmit a fluid, varies with the fluid's viscosity, **hydrostatic pressure,** the size of openings, and particularly, the degree to which the openings are interconnected. A rock can have high porosity but low permeability.

Rocks that commonly have high permeability are conglomerates, sandstones, basalt, and certain limestones. Permeability in sandstones and conglomerates is high because of the relatively large, interconnected pore spaces between the grains. Basalt is permeable because it often is extensively fractured by columnar jointing and because the tops of most flows are vesicular. Fractured limestones are also permeable, as are limestones in which solution activity has created many small solution cavities. Rocks that have low permeability are shale, unfractured granite, quartzite, and other dense, crystalline metamorphic rocks.

Water moves through the available pore spaces, twisting and turning through the tiny voids. Regardless of the degree of permeability, groundwater flows extremely slowly in comparison with the turbulent flow of rivers. Whereas the flow velocity of water in rivers is measured in kilometers per hour, the flow velocity of groundwater commonly ranges from 1 m per day to 1 m per year. The highest rate of percolation measured in the United States, in exceptionally permeable material, is only 250 m per day. Only in special cases, such as the flow of water in caves, does the movement of groundwater even approach the velocity of slow-moving surface streams.

How can a rock be highly porous and have low permeability?

THE WATER TABLE

The water table is the upper surface of the zone of saturation.

What are the major zones of subsurface water?

As water seeps into the ground, gravity pulls it downward through two zones of soil and rock. In the upper zone, the pore spaces in the rocks are only partly saturated, and the water forms a thin film, clinging to grains by surface tension. This zone, in which pore space is filled partly with air and partly with water, is called the *zone of aeration.* Below a certain level, all of the openings in the rock are completely filled with water (Figure 14.2). This area is called the *zone of saturation.* The water table, which is the upper surface of the zone of saturation, is an important element in the groundwater system. It may be only a meter or so deep in humid regions, but it might be hundreds or even thousands of meters below the surface in deserts. In swamps and lakes, the water table is essentially at the land surface.

Although the *water table* (the upper surface of zone of saturation) cannot be observed directly, it has been studied and mapped with data collected from wells, springs, and surface drainage. In addition, the movement of groundwater has been studied by means of radioactive isotopes, dyes, and other tracers, so extensive knowledge of this invisible body of water has been acquired.

What is the general configuration of the water table?

Several important generalizations can be made about the water table and its relation to surface topography and surface drainage. These are diagrammed in Figure 14.3. In general, the water table tends to mimic the surface topography. In flat country, the water table is flat. In areas of rolling hills, it rises and falls with the surface of the land. The reason for this is that groundwater moves very slowly, so the water table rises in the areas beneath the hills during periods of greater precipitation and tends to flatten out during droughts.

Where impermeable layers (such as shale) occur within the zone of aeration, the groundwater is trapped above the general water table, forming a local *perched water table.* If a perched water table extends to the side of a valley, springs and seeps occur.

The water table is at the surface in lakes, swamps, and most streams, and water moves in the subsurface toward these areas, following the general paths shown in Figure 14.3. In arid regions, however, most streams lie above the water table, so they lose much of their water through seepage into the subsurface.

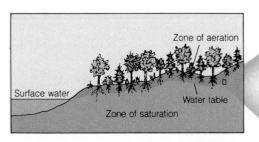

(A) Water seeps into the ground through pore spaces in the rock and soil. It passes first through the zone of aeration, in which the pore spaces are occupied by both air and water, and then into the zone of saturation, in which all of the pore spaces are filled with water. The depth of the water table varies with climate and amount of precipitation.

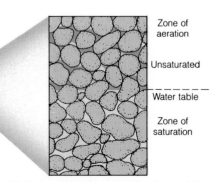

(B) Zones of aeration and saturation and the water table are shown in microscopic view.

FIGURE 14.2
The water table is the upper surface of the zone of saturation.

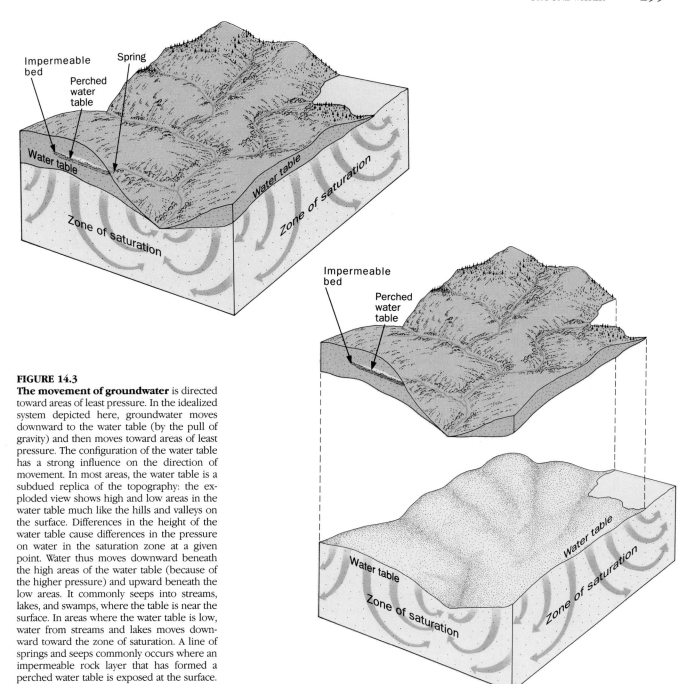

FIGURE 14.3
The movement of groundwater is directed toward areas of least pressure. In the idealized system depicted here, groundwater moves downward to the water table (by the pull of gravity) and then moves toward areas of least pressure. The configuration of the water table has a strong influence on the direction of movement. In most areas, the water table is a subdued replica of the topography: the exploded view shows high and low areas in the water table much like the hills and valleys on the surface. Differences in the height of the water table cause differences in the pressure on water in the saturation zone at a given point. Water thus moves downward beneath the high areas of the water table (because of the higher pressure) and upward beneath the low areas. It commonly seeps into streams, lakes, and swamps, where the table is near the surface. In areas where the water table is low, water from streams and lakes moves downward toward the zone of saturation. A line of springs and seeps commonly occurs where an impermeable rock layer that has formed a perched water table is exposed at the surface.

The Movement of Groundwater

The difference in elevation between parts of the water table is called the ***hydraulic head***. It causes the water to follow the paths illustrated in Figure 14.3. If we could trace the path of a particle of water, we would find that gravity slowly pulls it through the zone of aeration to the water table. When a particle encounters the water table, it continues to move downward by the pull of gravity along curved paths from areas where the water table is high toward areas where it is low (lakes, streams, and swamps). The path of groundwater movement is *not* down the slope of the water table, as one might first suspect. The explanation for this seemingly indirect flow is that the water table is not a solid surface like the ground surface. It is a surface of liquid, which in some ways resembles the surface of a wave. Water at any given point below the water table beneath a hill is under greater pressure than water at the same elevation below the lower water table in a valley. Groundwater therefore moves downward and toward points of lesser pressure.

Although these paths of groundwater movement may seem indirect, they conform to the laws of fluid physics and have been mapped in many areas by tracing the movement of dye injected into the system. The movement of the dye reveals a continual slow circulation of groundwater from infiltration at the surface to seepage into streams, rivers, and lakes.

The groundwater body is not stagnant and motionless. Rather, it is a dynamic part of the hydrologic system, in constant motion, and is intimately related to surface drainage. At considerable depths, all pore spaces in the rocks are closed by high pressure and there is no free water. This is the lower limit, or base, of the groundwater system.

Like other parts of the hydrologic system (rivers and glaciers), the groundwater system is an open system in which water enters the system by surface water infiltrating into the ground, moves through the system by percolating through the pore spaces of the rock, and ultimately leaves the system by seeping into streams or lakes.

What is the general direction of groundwater movement?

NATURAL AND ARTIFICIAL DISCHARGE

Natural discharge of the groundwater reservoir occurs wherever the water table intersects the surface of the ground. Generally such places are inconspicuous, typically occurring in the channels of streams and on the floors and banks of marshes and lakes. This discharge is the major link between groundwater reservoirs and other parts of the hydrologic system.

The natural discharge of groundwater into streams, lakes, and marshes is the major link between groundwater reservoirs and other parts of the hydrologic system. If it were not for groundwater discharge, many permanent streams would be dry during parts of the year. Most natural discharge is near or below the surfaces of streams and lakes and therefore usually goes unnoticed. It is detected and measured directly by comparing the volume of precipitation with the volume of surface runoff.

Artificial discharge results from the extraction of water from wells, which are made by simply digging or drilling holes into the zone of saturation. Many thousands of wells have been drilled, so in some areas artificial discharge has modified the groundwater system. Indeed, in some areas more water is removed from the groundwater system by artificial discharge than is added by natural recharge, and the level of the water table drops.

Natural Discharge

What geologic conditions produce natural springs?

Several geologic conditions that produce natural discharge in the form of *seeps* and *springs* are shown in Figure 14.4. If permeable beds alternate with impermeable layers (Figure 14.4A), the groundwater is forced to move laterally to the outcrop of the permeable bed. Conditions such as this usually are found in mesas and plateaus where permeable sandstones are interbedded with impermeable shales. The spring line commonly is marked by a line of vegetation. Figure 14.4B shows a limestone terrain in which springs occur where the base of the cavernous limestone outcrops. The Mammoth Cave area in Kentucky is a good example. Figure 14.4C shows lava formations that outcrop along the sides of a canyon. Springs develop because groundwater migrates readily through the vesicular and jointed basalt. Note that surface drainage disappears

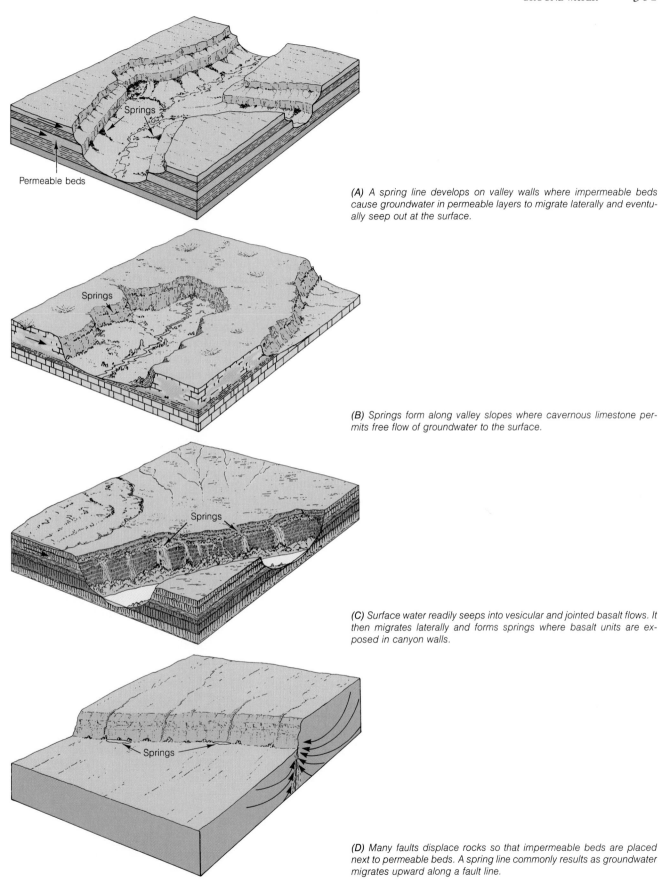

Permeable beds

(A) A spring line develops on valley walls where impermeable beds cause groundwater in permeable layers to migrate laterally and eventually seep out at the surface.

(B) Springs form along valley slopes where cavernous limestone permits free flow of groundwater to the surface.

(C) Surface water readily seeps into vesicular and jointed basalt flows. It then migrates laterally and forms springs where basalt units are exposed in canyon walls.

(D) Many faults displace rocks so that impermeable beds are placed next to permeable beds. A spring line commonly results as groundwater migrates upward along a fault line.

FIGURE 14.4
Springs can be produced under a variety of geologic conditions, some of which are illustrated in the block diagrams here. They are natural discharges of the groundwater reservoir and introduce a significant volume of water to surface runoff.

as the water flows over the lava plain. An excellent example of this type of discharge is found in the Thousand Springs area in Idaho, where numerous springs occur along the sides of the Snake River valley (see also Figure 14.7). Figure 14.4D shows springs along a fault that produces an avenue of greater permeability. Faults frequently displace strata for significant distances; thus, impermeable beds, which block the flow of groundwater, may be displaced along a fault so that they are positioned against permeable rocks. The water then moves up along the fault plane and forms springs along the fault line.

In brief, springs occur wherever the water table intersects the ground surface or where confined water seeps up to the surface along joints or faults. Generally, springs occur along valley walls, where downward-eroding streams have cut a valley below the level of the regional water table.

Wells

Ordinary wells are made simply by digging or drilling holes through the zone of aeration into the zone of saturation, as shown in Figure 14.5. Water then flows out of the pores into the well, filling it to the level of the water table. When a well is pumped, the water table is drawn down around the well in the shape of a cone, called the ***cone of depression.*** If water is withdrawn faster than it can be replenished, the cone of depression continues to grow, and the well ultimately goes dry. The cone of depression around large wells, such as those used by cities and industrial plants, can be many hundreds of meters in diameter. All wells within the cone of depression are affected (Figure 14.5). This undesirable condition has been the cause of "water wars," fought physically as well as in the courts. Because groundwater is not fixed in one place, as mineral deposits are, it is difficult to determine who owns it. Many disputes are now being arbitrated using computer models that simulate subsurface conditions such as permeability, direction of flow, and level of the water table. The models predict what changes will occur in the groundwater system if given amounts of water are drawn out of a well over specified periods of time.

Extensive pumping can lower the general surface of the water table. This effect has had serious consequences in some metropolitan areas of the southwestern United States, such as Phoenix, Arizona, where the water table has fallen hundreds of meters. The supply of groundwater is limited. Although the groundwater reservoir is being continually replenished by precipitation, the migration of groundwater is so slow that it can take hundreds of years to raise a water table to its former position of balance with the hydrologic system.

Although groundwater is largely invisible and only isolated glimpses of its presence can be seen in one place, a dramatic example of its movement can be seen in the Snake River Plain in Idaho (Figure 14.6). This region is a vast lava

What is the cone of depression? How is it produced?

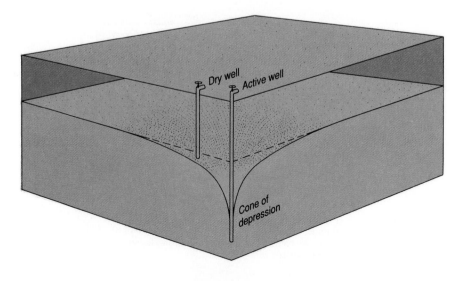

FIGURE 14.5
A cone of depression in a water table results if water is withdrawn from a well faster than it can be replenished. The cone can extend out for hundreds of meters around large deep wells and effectively lower the water table over a large area. Shallow wells nearby then run dry because they lie above the lowered water table.

FIGURE 14.6
Drainage in the Snake River Plain, Idaho, is influenced to a considerable extent by the high porosity and permeability of the basaltic bedrock. The major tributaries coming from the north lose their entire volume of water by seepage into the subsurface, and the rivers simply terminate. Much of the groundwater reappears to form the spectacular Thousand Springs approximately 400 km downstream.

plain extending across the entire southern part of the state. It was built up by innumerable floods of basaltic lava, with some interbedded coarse sand and gravel deposited in streams and lakes that occupied the region during the intervals between volcanic eruptions. The porosity and permeability of these basaltic rocks are remarkably high. Porosity is produced by columnar joints and vesicular texture in the basalt and by pore space in layers of ash and rubble between the basalt flows. In addition, porosity is naturally high in lava tubes and in layers of unconsolidated coarse sand and gravel between some of the flows. In terms of permeability, the rock sequence is almost like a sieve.

The Snake River is located near the southern margin of the plain, so tributary streams coming from the mountains to the north are forced to flow across the plain before they can join the Snake (Figure 14.6). Only one river actually completes the short journey! The rest terminate after flowing a short distance across the plain; they lose their entire volume of water by seepage into the subsurface. Two of the main would-be tributaries are known, respectively, as Big Lost and Little Lost rivers.

However, the groundwater returns to the surface in a series of spectacular springs approximately 200 km downstream. The largest and best known are the Thousand Springs just west of Twin Falls, Idaho. These springs clearly show the tremendous movement of groundwater as they discharge about 1500 cubic meters (nearly 37,000 gallons) per second. The visible springs issue from a layer of vesicular basalt 50 m above the river. However, the volume of water that seeps into the Snake River in a less spectacular fashion below the banks is no doubt many times as great. So much water comes from the Thousand Springs area that an electric power plant has been built on the site to utilize the energy (Figure 14.7).

FIGURE 14.7
The Thousand Springs, Snake River Canyon, Idaho, issue from the north wall of the canyon and are fed by water from the mountains approximately 400 km to the northeast.

ARTESIAN WATER

Artesian water is confined in an aquifer between impermeable beds. It is under pressure, like water in a pipe; where a well or fracture intersects it, the aquifer water rises in the opening, producing a flowing well or an artesian spring.

A very important type of groundwater occurs in a permeable rock confined between impermeable beds, where it is under pressure. A well drilled into such an aquifer will commonly allow water to rise and flow freely without pumping. The name "artesian" was originally applied to flowing wells for a French province along the English Channel, where artesian conditions are common.

The necessary geologic conditions for artesian water, illustrated in Figure 14.8, include the following:

1. The rock sequence must contain interbedded permeable and impermeable strata. This sequence occurs commonly in nature as interbedded sandstone and shale. Permeable beds usually are called aquifers.
2. The rocks must be tilted and exposed in an elevated area where water can infiltrate into the aquifer.
3. Sufficient precipitation and surface drainage must occur in the outcrop area to keep the aquifer filled.

The water confined in an aquifer bed behaves much like water in a pipe. Hydrostatic pressure builds up, so where a well or fracture intersects the bed, water rises in the opening, producing a flowing well or artesian spring.

What produces flowing wells?

The height to which *artesian water* rises above the *aquifer* is shown by the dashed colored line in Figure 14.8. The surface defined by this line is called the *artesian-pressure surface.* You might expect it to be a horizontal surface, but actually, an artesian-pressure surface slopes away from the *recharge* area (the area where water is absorbed into the ground and into the aquifer, usually an area where a stream or river crosses an exposed aquifer). Pores in the aquifer provide resistance to flow, and pressure is lost through fractures (leaks) in the underground plumbing system. If a well were drilled at location A or C in Figure 14.8, water would rise in the well, but it would not flow to the surface because the artesian-pressure surface is below the ground surface. Water in a well at location B or D, where the artesian-pressure surface is above the ground surface, would flow to the surface. Nonetheless, all of these wells are artesian wells; that is, the water is under artesian pressure and rises above the top of the aquifer.

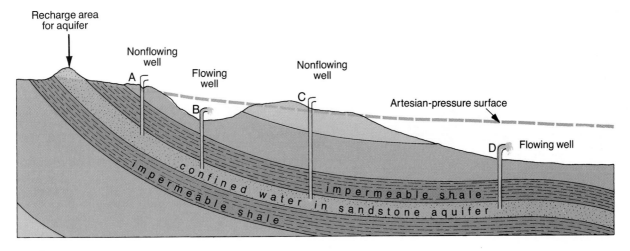

FIGURE 14.8
Necessary geologic conditions for an artesian system include the following: (1) a permeable bed (aquifer) must be confined between impermeable layers, (2) rocks must be tilted so the aquifer can receive infiltration from surface waters, and (3) adequate infiltration must exist to fill the aquifer and create hydrostatic pressure. The diagram shows an idealized artesian system. All the wells shown are artesian wells (that is, water rises in them under pressure). Flowing wells occur only when the top of the well is below the artesian-pressure surface.

Artesian water is commonplace in most areas underlain by sedimentary rocks because the necessary geologic conditions for an artesian system are present in sedimentary rocks in a variety of ways. Examples of well-known areas in the United States should help you appreciate how artesian systems result from different geologic settings.

One of the better-known artesian systems underlies the Great Plains states (Figure 14.9A). The sequence of interbedded sandstones, shales, and limestones is nearly horizontal throughout most of Kansas, Nebraska, and the Dakotas, but it is upwarped along the eastern front of the Rockies and the margins of the Black Hills. Several sandstone formations are important aquifers. Water confined in them under hydrostatic pressure gives rise to an extensive artesian system in the Great Plains states. The recharge area is along the foothills of the Rockies.

Figure 14.9B illustrates a regional artesian system in the inclined strata of the Atlantic and Gulf Coast plains. The rock sequence consists of permeable sandstone and limestone beds alternating with impermeable clay. Surface water, flowing toward the coast, seeps into the beds where they are exposed at the surface. It then moves slowly down the dip of the permeable strata.

A third example is from the western states, where the arid climate makes artesian water an important resource (Figure 14.9C). In this region, the subsurface rocks in the intermontane basin consist of sand and gravel deposited in ancient alluvial fans. Farther into the basin, in the playa lakes, these deposits become interbedded with layers of clay and silt. The playa deposits act as confining layers between the permeable sand and gravel. Water seeping into the fan deposits becomes confined as it moves away from the mountain front.

Artesian systems also underlie some of the world's great desert regions. Natural discharge from them is largely responsible for oases. Part of the Sahara system is shown in Figure 14.9D. Oases occur where artesian water is brought to the surface by fractures or folds or where the desert floor is eroded down to the top of the aquifer. You will note that each example shown in Figure 14.9 has the same basic geologic conditions necessary for artesian water: (1) there is a sequence of interbedded permeable and impermeable strata, and (2) the sequence of strata is tilted so that the strata are exposed in an elevated area, enabling surface water to infiltrate into the aquifer. The main difference in each area is in the details of the rock structure and sequence of strata.

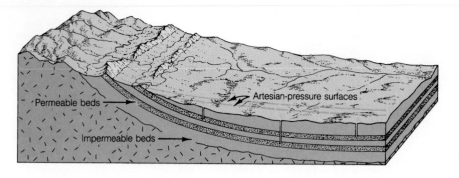

(A) The Great Plains states are underlain by permeable Cretaceous sandstones, which are warped up along the front of the Rocky Mountains, where they receive infiltration. This structure forms a widespread artesian system in Kansas, Nebraska, and the Dakotas.

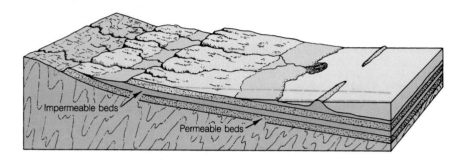

(B) The Atlantic and Gulf Coast states are characterized by Tertiary and Cretaceous rocks dipping uniformly toward the ocean. Water enters permeable beds where they are exposed and becomes confined downdip to form a large artesian system.

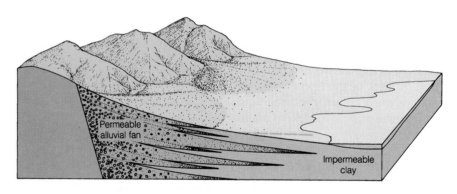

(C) The intermontane basins in the western United States contain permeable sand and gravel (originally deposited in alluvial fans) interfingered with impermeable clay deposits (originally deposited in playa lakes). Water seeps into the lenses of buried fan deposits and is confined by the clay to form an artesian system.

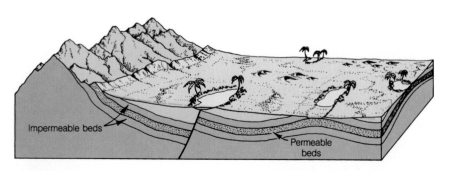

(D) The Sahara Desert is underlain mostly by gently warped permeable beds, which receive water where they are exposed at the base of the Atlas Mountains. The artesian water forms oases by natural discharge through fractures or at sites where the rock is exposed.

FIGURE 14.9
Artesian systems develop under a variety of geologic conditions, some of which are illustrated in these block diagrams. All of them satisfy the basic conditions necessary for an artesian system, as illustrated in Figure 14.8. The main difference is in the geometry of the rock structures in each area.

THERMAL SPRINGS AND GEYSERS

In areas of recent igneous activity, rocks associated with old magma chambers can remain hot for hundreds of thousands of years. Groundwater migrating through these areas of hot rocks becomes heated and, when discharged to the surface, produces thermal springs and geysers.

The most spectacular manifestation of groundwater is in the areas of thermal springs and geysers, where scalding water and steam commonly erupt high into the air. Geysers and thermal springs are usually the result of groundwater migrating through areas of hot, but not molten, igneous rock.

The three most famous regions of hot springs and *geysers* (a hot spring that intermittently erupts jets of hot water and steam) are Yellowstone National Park, Iceland, and New Zealand. All are regions of recent volcanic activity, so the rock temperatures just below the surface are quite high. Although no two geysers are alike, all require certain conditions for their development:

1. A body of hot rocks must lie relatively close to the surface.
2. A system of irregular fractures must extend downward from the surface.
3. A relatively large and constant supply of groundwater must be present.

Eruptions of Geysers

Geyser eruptions occur when groundwater pressure in fractures, caverns, or layers of porous rock builds up to a critical point where the temperature-pressure balance is such that any increase in temperature will cause the water to change instantly into steam (Figure 14.10). Because the water at the base of the fracture is under greater pressure than the water above, it must be heated to a higher temperature before it boils. Eventually, a slight increase in temperature or a decrease in pressure (resulting from the liberation of dissolved gases) causes the deeper water to boil. The expanding steam throws water from the underground chambers high into the air. After the pressure is released, the caverns refill with water and the process is repeated.

This process accounts for the periodic eruption of many geysers. The interval between eruptions is the amount of time required for water to percolate into the fracture and be heated to the critical temperature. Geysers like Old Faithful in Yellowstone National Park erupt at definite intervals because the rocks are permeable and the "plumbing system" refills rapidly. Other geysers, which require more time for water to percolate into the chambers, erupt at irregular intervals because the water supply over a longer period of time can fluctuate.

Why do geysers erupt in cycles?

Geothermal Energy

The thermal energy of groundwater, called *geothermal energy,* offers an attractive source of energy for human uses. At present, it is used in various ways in local areas of the United States, Mexico, Italy, Japan, and Iceland. Recent estimates show that from 1 to 2% of our current energy needs could be met by geothermal sources.

In Iceland geothermal energy has been used successfully since 1928. The plan is simple. Wells are drilled in geothermal areas, and the steam and hot water are piped to storage tanks and then pumped to homes and municipal buildings for heating and hot water. The cost of this direct heating is only about 60% of that of fuel-oil heating and about 75% of the cost of the cheapest

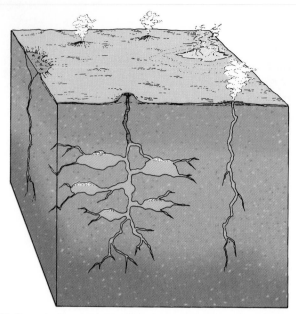

(A) Groundwater circulating through hot rocks in an area of recent volcanic activity collects in caverns and fractures. As steam bubbles rise, they grow in size and number and tend to accumulate in restricted parts of the geyser tube.

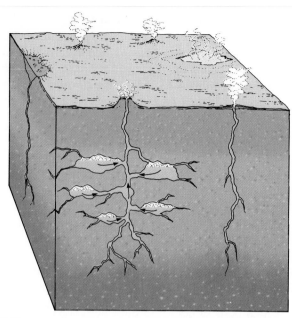

(B) The expanding steam forces water upward until it is discharged at the surface vent. The deeper part of the geyser system then becomes ready for the major eruption.

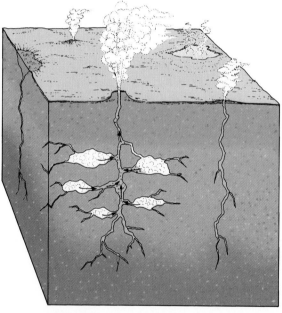

(C) The preliminary discharge of water reduces the pressure on the water lower down. Water from the side chambers and pore spaces begins to flash into steam, forcing the water in the geyser system to erupt.

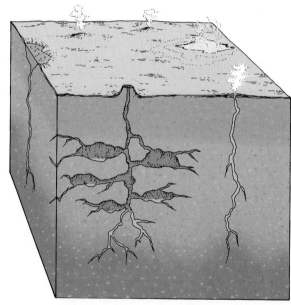

(D) Eruption ceases when the pressure from the steam is spent and the geyser tubes are empty. The system then begins to fill with water again, and the eruption cycle starts anew.

FIGURE 14.10
The origin of geysers is depicted in this series of diagrams. A geyser can develop only if (1) a body of hot rock lies relatively close to the surface, (2) a system of irregular fractures extends down from the surface, and (3) there is a constant supply of groundwater.

method of electrical heating. Steam from geothermal energy is also used to run electric generators, producing an easily transported form of energy. Corrosion is a problem, however, because most geothermal waters are acidic and contain undesirable dissolved salts. As a result, the expense of periodically replacing the plumbing system may make geothermal energy uneconomical in some areas.

EROSION BY GROUNDWATER

Slow-moving groundwater can dissolve huge quantities of soluble rock and carry it away in solution. In some areas, it is the dominant agent of erosion and produces karst topography, which is characterized by sinkholes, solution valleys, and disappearing streams.

Groundwater is capable of accomplishing erosion on an enormous scale, but unlike streams, groundwater erodes only by dissolving soluble rocks such as limestone, rock salt, and gypsum. It then transports the dissolved mineral matter and either discharges it into other parts of the hydrologic system or deposits it in the pore spaces within the rock. Groundwater erosion starts with water percolating through joints, faults, and bedding planes and dissolving the soluble rock. In time, the fractures enlarge to form a subterranean network of caves, which can extend for many kilometers. The caves grow larger until ultimately the roof collapses, and a craterlike depression called a *sinkhole* is produced. Solution activity then enlarges the sinkhole to form a solution valley, which continues to grow until the soluble rock is removed completely.

As we saw in chapter 10, the most important dissolution reagent in groundwater is carbonic acid (H_2CO_3). This acid forms readily as carbon dioxide in the atmosphere and soil dissolves in rainwater. Sulfuric acid, formed from sulfur compounds abundant in organic sediments such as coal, peat, and liquid petroleum, is also commonly present in groundwater. Together with various more complex organic acids generated from the soil, these dilute acids react with the minerals in rocks and remove them in solutions.

The rate of erosion of limestone terrains by the chemical action of groundwater has been measured in several different ways. One method is to precisely measure the weight of small limestone tablets and place them in different climatic conditions in several parts of the world. Once the weight loss of the limestone during a specific period of time is known, the average rate at which limestone terrains are being lowered by chemical processes alone can be calculated. Other methods include measuring the amount of mineral matter in water dripping through caves and the dissolved mineral matter carried by a river system in limestone regions. More recently, precise measurements of dissolution rates have been obtained by a microerosion meter, an instrument capable of measuring the erosion of a rock surface to the nearest 0.005 mm. The results of these measurements indicate that in temperate regions the landscape is being lowered at an average rate of 10 mm per 1000 years. In areas of greater rainfall, rates may be as high as 300 mm per 1000 years. These averages may seem small, but they indicate that erosion by groundwater in some areas can be greater than the average erosion of a surface by running water. It is clear from these measurements and from the characteristics of limestone terrains that groundwater accomplishes erosion on a grand scale.

Rock salt and gypsum, the most soluble rocks, are eroded rapidly by solution activity. They are relatively rare, however, and are not widely distributed on any of the continents. Limestone, which is a common rock type, is also fairly soluble, and solution activity plays an important role in eroding limestone terrains in humid regions.

Caves

Perhaps the best way to appreciate the significance of solution activity is to consider the nature of a cave system and the amount of rock removed by solution. The origin and evolution of *caves* are shown in Figure 14.11. Rainwater dissolves atmospheric carbon dioxide and forms a weak acid. It then percolates through the fractures and bedding planes, slowly dissolving the

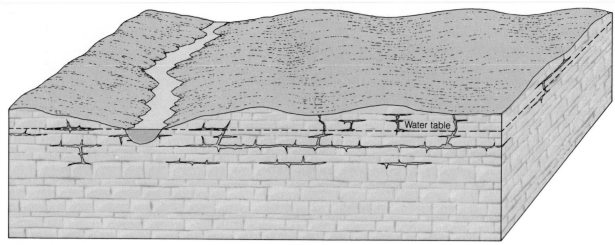

(A) In the early stages, water seeps through the fractures and bedding planes of limestone formations. It seeps downward to the water table and then moves toward the surface streams.

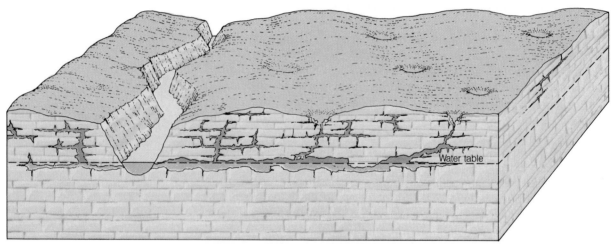

(B) As the surface streams erode the valley floor, the water table drops. The surface water seeping through the zone of aeration enlarges the existing joints and caves. Movement of water toward the surface stream develops a main system of horizontal caverns.

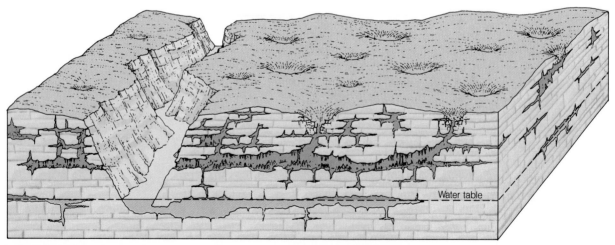

(C) As the river erodes a deeper valley, the water in the main underground channel seeks a new path to the lower river level. A new, lower, system of horizontal caverns develops. The older, higher, caverns may continue to enlarge and ultimately collapse to form sinkholes, or they may be filled with fallen rubble or cave deposits.

FIGURE 14.11
The evolution of a cave system is shown schematically in these diagrams.

limestone and enlarging the openings. The general direction of groundwater motion is downward, toward the water table. The water then moves toward a natural outlet, such as a river system, and as it moves, it dissolves the limestone. In time, a main subterranean channelway is developed, which transports the solution to the main streams. If the water table drops (usually by downcutting of the river), water in the main subterranean channel begins once again to seep downward to a new level. Eventually, the old horizontal channel drains, and water dripping from the old channel ceiling begins to deposit calcite, in time creating major dripstone deposits in the open cave.

As caves grow larger, they become unstable and tend to collapse. The fallen rubble occupies about one-third more volume than it did as intact rock on the cave ceiling. Small caves may completely backfill themselves (or fill up as a result of debris falling from the ceiling), but larger caverns tend to migrate upward as their roofs collapse and bury their floors. Larger or shallower caves ultimately break through to the surface to become sinkholes. A map of a cave system may show long, winding corridors, with branched openings that enlarge into chambers, or a maze of interlacing passageways and channels, controlled by intersecting joint systems (Figure 14.12). Where a sequence of limestone formations occurs, several levels of cave networks may exist. Mammoth Cave, for example, has over 50 km of continuous subterranean passages.

Karst Topography

Karst topography is a distinctive type of terrain resulting largely from erosion by groundwater. In contrast to a landscape formed by surface streams, which is characterized by an intricate network of stream valleys, karst topography lacks a well-integrated drainage system. Sinkholes are generally numerous and, in many karst regions, they dominate the landscape. Where sinkholes grow and enlarge, they merge and form elongate or irregular closed depressions called *solution valleys.* Small streams commonly flow on the surface for only a short distance and then disappear down a sinkhole, becoming *disappearing streams.* There the water moves slowly through a system of caverns and caves, sometimes as sluggish underground streams. Springs, which are common in karst areas, return water to the surface drainage.

In tropical areas where dissolution is at a maximum because of the abundance of water from heavy rainfall, a particular type of karst topography, called *tower karst,* develops. Tower karst is characterized by steep, cone-shaped hills rather than sinkholes and solution valleys (Figure 14.13). The towers are largely residual landforms left after most of the rock has been

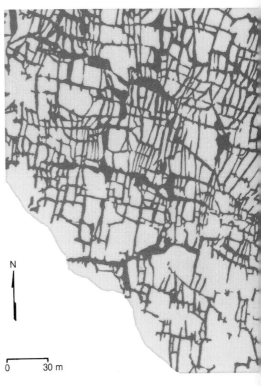

N

0   30 m

FIGURE 14.12
A map of Anvil Cave, Alabama, shows the extent to which caverns are controlled by a fracture system. The joints occur in two intersecting sets, one trending nearly north and south, and the other east and west. Solution activity along the joints has produced the network of caverns.

(A) Tower karst near Guilin, China

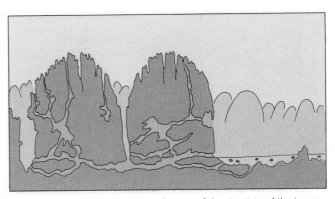

(B) Diagram showing the general nature of the structure of the towers

FIGURE 14.13
The tower karst topography of central China forms some of the most spectacular landforms in the world. The towers are largely residual landforms. Many are laced with caves and caverns.

What unique landforms are produced by groundwater activity?

removed by collapse of caverns and enlargement of solution valleys. They are the remnants of a once continuous layer of rock that covered the area.

Thus, in detail karst topography is highly diverse, ranging from fantastic landscapes such as that shown in Figure 14.13 to the low relief of a plane pitted with small depressions, as shown in Figure 14.14. What is common to all karst terrains is that their landforms are caused by the unusually great solubility of certain rock types. Humid climate is a very important factor in developing karst topography. The more water that is moving through the system, the more solution activity will occur. Karst topography is, therefore, largely restricted to humid and temperate climatic zones. In desert regions, where little rain falls, karst topography will not develop.

A model of the evolution of karst topography is shown in the block diagrams in Figure 14.15. Initially, water follows surface drainage until a large river cuts a deep valley below the limestone layers. Groundwater then moves through the joints and bedding surfaces in the limestone and emerges at the river banks. As time goes on, the passageways become larger and caverns develop. Surface waters disappear into solution depressions. The roofs of caves collapse, so numerous sinkholes are produced (Figure 14.15A). Springs commonly occur along the margins of major stream valleys. Sinkholes proliferate and grow in size as the limestone formation is dissolved away. The cavernous terrain of central Kentucky, for example, is marked by over 60,000 holes. As solution processes continue, sinkholes increase in number and size. Some merge to form larger depressions with irregular outlines. This process ultimately develops solution valleys (Figure 14.15B). Most of the original surface is finally dissolved, with only scattered hills remaining (Figure 14.15C). When the soluble bedrock has been removed by groundwater solution, normal surface drainage patterns reappear.

The best-known karst topography in the United States is the sinkhole country of Florida, Kentucky, and southern Indiana (Figure 14.16). These regions consist of plains pockmarked by innumerable small, isolated depressions

FIGURE 14.14
Karst topography in the limestone region of Kentucky is dominated by sinkholes and solution valleys. Note the absence of an apparent drainage system in this region.

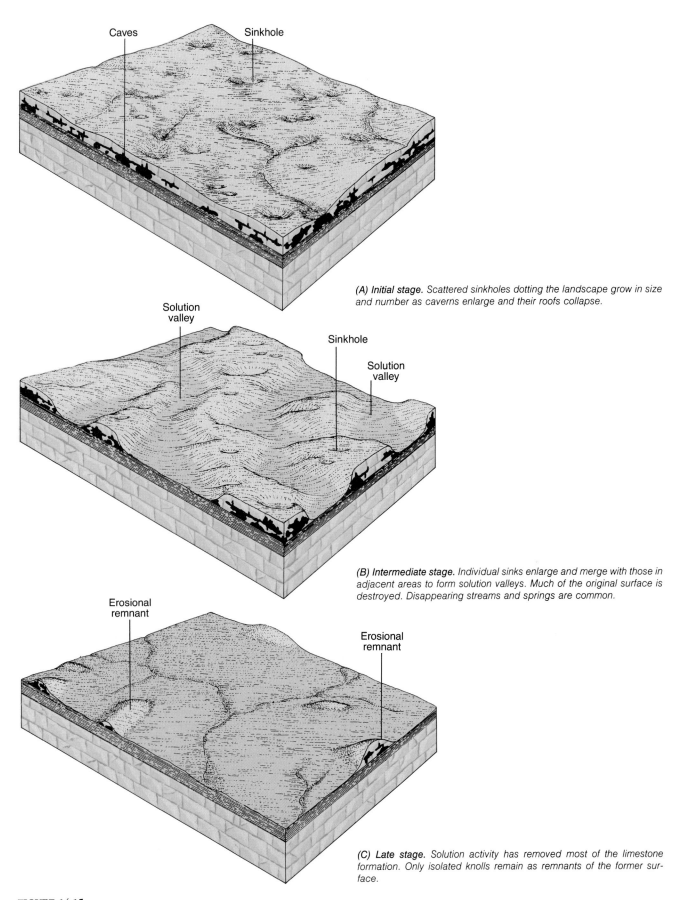

(A) Initial stage. Scattered sinkholes dotting the landscape grow in size and number as caverns enlarge and their roofs collapse.

(B) Intermediate stage. Individual sinks enlarge and merge with those in adjacent areas to form solution valleys. Much of the original surface is destroyed. Disappearing streams and springs are common.

(C) Late stage. Solution activity has removed most of the limestone formation. Only isolated knolls remain as remnants of the former surface.

FIGURE 14.15
The evolution of karst topography involves these major processes: (1) the enlargement of caves and the development of sinkholes, (2) the enlargement of sinkholes and the development of solution valleys, and (3) the enlargement of solution valleys until the original limestone terrain is completely destroyed.

313

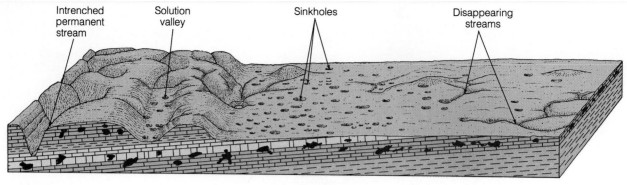

Intrenched permanent stream   Solution valley   Sinkholes   Disappearing streams

FIGURE 14.16
The karst area of southern Indiana is characterized by sinkholes and solution valleys. The style of karst landforms depends to a considerable degree on the characteristics of the limestone bedrock.

(Figures 14.14 and 14.16). Surface streams of any significant length are extremely rare. In Kentucky, the Green River is the only major stream to cross the limestone terrain. In southern Indiana, the White River is the sole river to cross the karst area. Statistical studies of "sinkhole plains" indicate that the size and frequency of the sinks are closely related to the characteristics of specific types of limestone.

In striking contrast to the sinkhole plains of Indiana and Kentucky, the tower karst topography of southern China presents some of the most spectacular limestone scenery on Earth (Figure 14.13). Here an area of thousands of square kilometers, once covered by thick limestone formations, is in an advanced stage of dissection by groundwater. Only remnants between sinkholes and solution basins remain, standing like clusters of towers above the surrounding terrain. These strange mountains, shaped like upended loaves of French bread, form an intricate system of precipitous slopes and overhanging cliffs, with caves, arches, and strange landforms made by solution activity.

The small-scale solution features in this area are almost as impressive as the larger features. Evidence of solution activity is everywhere. Most outcrops look like Swiss cheese because of the maze of interconnected cavities dissolved and enlarged by groundwater.

Classical Chinese art is noted for portraying these bizarre and exotic landforms, which appear unreal to the foreign eye. Western artists believed that the Chinese masters who painted these landforms were impressionists, but anyone fortunate enough to visit the region realizes that the artists were not visionaries; the shapes they painted were nature's own.

DEPOSITION BY GROUNDWATER

The mineral matter dissolved by groundwater can be deposited in a variety of ways. The most spectacular deposits are stalactites and stalagmites, which are found in caves. Less obvious are the deposits in permeable rocks such as sandstone and conglomerates. Here groundwater commonly deposits mineral matter as a cement between grains.

The chemical processes that cause groundwater to dissolve soluble material are easily reversed, and the minerals are precipitated in the pore spaces, voids, and caves within the rock. The change from solution to precipitation is commonly caused by lowering of the water table. The main solution processes occur in the zone of saturation; precipitation occurs in the zone of aeration after the caves and pore spaces are drained. The deposits formed in caves are some of nature's fancy work—the endless variety of cave deposits is familiar to

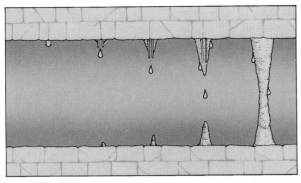

(A) *Diagram showing, left to right, the evolution of stalactites, stalagmites, and columns.*

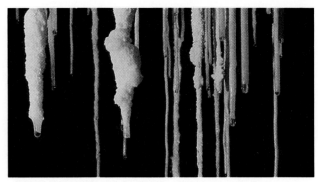

(B) *Photograph of long slender stalactites (soda straws), which grow as the drop of water suspended at the end evaporates.*

FIGURE 14.17
Dripstone originates on the ceilings of caves. Water seeps through a crack and partially evaporates. This causes a small ring of calcite to be deposited around the crack. The ring grows into a tube, which commonly acquires a tapering shape as water seeps from adjacent areas and flows down its outer surface.

almost everyone. Most originate in a similar way, however, and they are referred to collectively as *dripstone.* The process of dripstone formation is shown in Figure 14.17. As water enters the cave (usually from a fracture in the ceiling), part of it evaporates, and a small amount of calcium carbonate is left behind. Each succeeding drop adds more calcium carbonate, so that eventually a cylindrical or cone-shaped projection is built downward from the ceiling. Many beautiful and strange forms result, some of which are shown in Figures 14.18 and 14.19. Icicle-shaped forms growing down from the ceiling are called *stalactites.* These commonly are matched by deposits growing up from the floor, called *stalagmites,* because the water dripping from a stalactite precipitates additional calcium carbonate on the floor directly below. Many stalactites and stalagmites eventually unite to form columns. Water percolating from a fracture in the roof may form a thin, vertical sheet of rock called a *drip curtain.* Pools of water on the cave floor flow from one place to another, and as they evaporate, calcium carbonate is deposited on the floor, forming *travertine terraces.*

Although cave deposits are a spectacular expression of deposition by groundwater, they are trivial compared to the amount of material deposited in the pore spaces of rock. In sandstones and conglomerates, precipitation of silica and calcium carbonate cement the loose grains together into a hard, strong rock body. In some formations the cementing minerals deposited by groundwater may exceed 20% of the volume of the original rock.

Mineral precipitation by groundwater action is a slow process, in some cases involving the slow removal—one at a time—of atoms or molecules of organic matter and their simultaneous replacement by other mineral ions carried by the groundwater. One example of this process is petrified wood. Perhaps the best-known deposit of petrified wood is the Petrified Forest National Park in eastern Arizona. Here great accumulations of petrified logs buried in ancient river sediments are now being uncovered by weathering and erosion (Figure 14.20). The Petrified Forest is not really a forest at all, but rather a great collection of driftwood. This driftwood washed down from adjacent highlands about 230 million years ago and accumulated as logjams in ancient river bars and floodplains. It was subsequently covered with hundreds of meters of younger sediments. While the driftwood was covered with sediment, groundwater percolating through the strata replaced the cellular structure of the wood with silica. This transformed the wood to agate, a variety of silicon dioxide (SiO_2).

What are the major features formed by mineral matter deposited by groundwater?

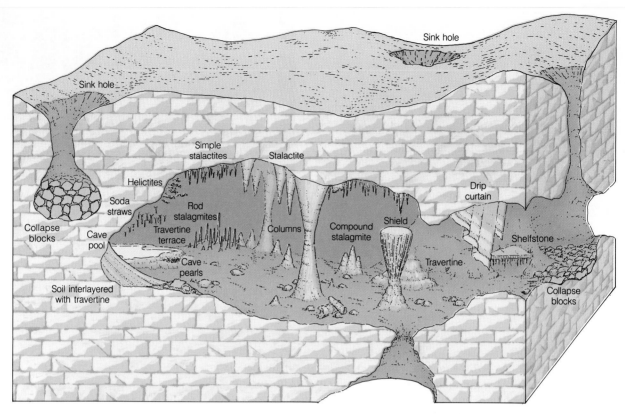

FIGURE 14.18
Many varieties of cave deposits have been recognized. All of them are composed of calcite
deposited by water that seeps into the cave and then evaporates.

FIGURE 14.19
Carlsbad Caverns, New Mexico, show many of the forms of dripstone illustrated in Figure 14.18.

FIGURE 14.20
Petrified trees litter the area, piled like giant jackstraws about a rolling landscape on the Petrified Forest Member of the Chinle Formation, Arizona.

Geodes are another common example of the result of the action of groundwater deposition. A geode is a roughly spherical, hollow rock mass with its central cavity lined with mineral crystals. They are common in limestones, but also occur in shales. The formation of geodes can be explained as a two-stage development. First, a cavity is formed in the rock by groundwater solution activity. Then, under different conditions, the mineral matter carried by groundwater is precipitated on the walls of the rock cavity. Quartz, calcite, and fluorite are the most common minerals precipitated. They accumulate very slowly and form perfect crystals, pointing toward the center of the cavity. Subsequent erosion removes the material around the geode cavity, but the mineral-lined walls of the cavity are resistant, so geodes remain; they are found as boulderlike remnants, left from the weathering of the parent rock material. In a sense, a geode is a fossil cavity.

Another expression of deposition by groundwater is the mineral deposits formed around springs. Mammoth Hot Springs in Yellowstone National Park is one of the most spectacular examples (Figure 14.21).

FIGURE 14.21
Mammoth Hot Springs, Yellowstone National Park, was formed by the deposition of travertine ($CaCO_3$) as the warm spring water cooled and evaporated.

Saltwater Encroachment

On an island or a peninsula where permeable rocks are in contact with the ocean, a lens-shaped body of fresh groundwater is buoyed up by the denser saltwater below, as is illustrated in Figure 14.23A. The fresh water literally floats on the saltwater and is in a state of balance with it. If excessive pumping develops a large cone of depression in the water table, the pressure of the fresh water on the saltwater directly below the well is decreased, and a large cone of *saltwater encroachment* develops below the well, as is shown in Figure 14.23B. Continued excessive pumping causes the cone of saltwater to extend up the well and contaminate the fresh water. It is then necessary to stop pumping for a long time to allow the water table to rise to its former position and depress the cone of saltwater. Restoration of the balance between the freshwater lens and the underlying saltwater can be hastened if fresh water is pumped down an adjacent well (Figure 14.23C).

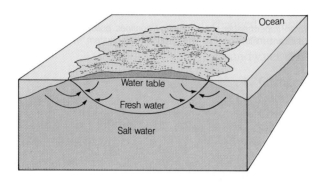

(A) A lens of fresh groundwater beneath the land is buoyed up by denser saltwater below.

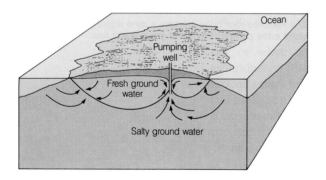

(B) Excessive pumping causes a cone of depression in the water table on top of the freshwater lens and a cone of saltwater encroachment at the base of the freshwater lens.

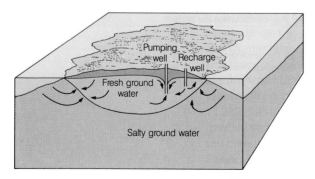

(C) Fresh water pumped down an adjacent well can raise the water table around the well and lower the interface between the fresh water and the saltwater.

FIGURE 14.23
The relationship between fresh water and saltwater on an island or a peninsula is affected by the withdrawal of water from wells. Excessive pumping causes a cone of saltwater encroachment, which limits the usefulness of the well.

Changes in the Position of the Water Table

The water table is intimately related to surface runoff, the configuration of the landscape, and the ecological conditions at the surface. The balance between the water table and surface conditions, established over thousands or millions of years, can be completely upset by changes in the position of the water table. Two examples illustrate some of the many potential ecological problems.

In southern Florida, fresh water from Lake Okeechobee has flowed for the past 5000 years as an almost imperceptible "river" only a few centimeters deep and 64 km wide. This sheet of shallow water created the swampy Everglades. The movement of the water was not confined to channels. It flowed as a sheet in a great curving swath for more than 160 km (Figure 14.24A). The surface of the Everglades slopes southward only 2 cm per km, but this gradient was enough to keep the water moving slowly to the coast and to prevent saltwater from invading the Everglades and the subsurface aquifers along the coast. In effect, the water table in the swamp was at the surface, and the ecology of the Everglades was in balance with the water table.

Today many canals have been constructed to drain swamp areas for farmland, to help control flooding, and to supply fresh water to the coastal megalopolis (Figure 14.24B). The canals diverted the natural flow of water across the swamp, in effect lowering the water table, in some places as much as 0.5 m below sea level. This change in position of the water table produced many unforeseen and often unfortunate results. As the water table was lowered, saltwater encroachment occurred in wells all along the coast. Some cities had to move their wells far inland to obtain fresh water.

How does human activity alter the groundwater systems?

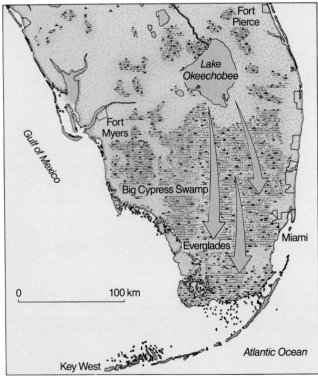

(A) Natural drainage of southern Florida in 1871 spreads southward from Lake Okeechobee in a broad sheet only a few centimeters deep. This maintained swampy conditions in the Everglades and established a water table very close to the surface.

(B) Canals diverted the natural flow of surface water across the Everglades. The water table was lowered, the swamp was destroyed in some areas, and saltwater encroached in wells along the coast.

FIGURE 14.24
Modification of the natural drainage system of southern Florida

The most visible effects, however, involve the ecology of the swamp. In the past, the high water table could maintain a marsh during periods of natural drought. Now the surface is dry during droughts. Forest fires ignite the dry organic muck, which burns like peat, smoldering long after the surface fires die out. This effectively destroys the ecology of the swamp. The lowering of the water table also caused the muck to compact, so that it subsided in places as much as 2 m. In addition, muck exposed to the air oxidizes and disappears at a rate of about 2.5 cm per year. Once the muck is gone from the swamp, it can be replaced only by nature.

Raising the water table can also modify many surface processes. An example is found in the environmental changes caused by irrigation in the Pasco basin of Washington. This area, which lies in the rain shadow of the Cascade Mountains, receives only 15 to 25 cm of precipitation a year. The basin's surface conditions (the soil cover, slope of land, drainage network, and so on) developed in response to an overall increase in aridity over the last several million years. The surface material, slope, and vegetation developed a balance with an arid climate and a low water table.

In recent years, extensive irrigation has caused the water table to rise, introducing many changes in the surface conditions. Today, from 100 to 150 cm of water are applied each year to the ground by irrigation, which simulates the effect of a climatic change of considerable magnitude. The higher water table has rapidly developed large springs along the sides of river valleys. The springs are now permanent, reflecting saturation of much of the ground. Erosion is accelerated, and many farms and roads have been damaged severely. Landslides present the most serious problems. Slopes that were stable under arid conditions are now unstable because they are partly saturated from the high water table and from the formation of perched water bodies.

In many areas, it is imperative that people modify the environment by reclaiming land or by irrigation, but unless we are careful, the detrimental effects of our modifications may outweigh the advantages. Before we seriously modify an environment, we must attempt to understand the many consequences of altering the natural systems.

Subsidence

Surface *subsidence* related to groundwater can result from natural Earth processes, such as the development of sinkholes in a karst area, or from the artificial withdrawal of fluids.

An ever-present hazard in limestone terrains is the collapse of subterranean caverns and sinkhole formation. Buildings and roads have frequently

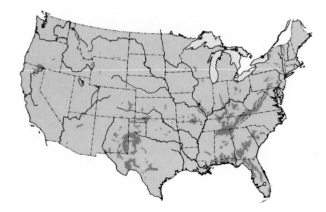

FIGURE 14.25
The major areas of karst topography in the continental United States are restricted to the central and southern states, where outcrops of limestone occur in humid climatic conditions. Limestone outcrops in many areas in the western states, but there the climate is in most cases too arid to develop a typical karst topography.

been damaged by sudden collapses into previously undiscovered caverns below. In the United States, important karst regions occur in central Tennessee, Kentucky, southern Indiana, Alabama, Florida, and Texas (Figure 14.25). The problem of potential collapse is difficult to solve. Important construction in karst regions should be preceded by test borings to determine whether subterranean cavernous zones are present. Wet concrete can be pumped down into caves and solution cavities, but such remedies can be very expensive.

Compaction and subsidence also present serious problems in areas of recently deposited sediments. In New Orleans, for example, large areas of the city are now 4 m below sea level, a drop due largely to the pumping of groundwater. As a result, the Mississippi River flows some 5 m above parts of the city, and rainwater must be pumped out of the city at considerable cost. Also, as the surface subsides, water lines and sewers are damaged.

Where groundwater, oil, or gas is withdrawn from the subsurface, significant subsidence can also occur, damaging construction, water supply lines, sewers, and roads. Long Beach, California, has subsided 9 m as the result of 40 years of oil production from the Wilmington oil field. This subsidence resulted in almost $100 million worth of damage to wells, pipelines, transportation facilities, and harbor installations. Parts of Houston, Texas, have subsided as much as 1.5 m as a result of the withdrawal of groundwater.

Probably the most spectacular example of subsidence is Mexico City, which is built on a former lakebed. The subsurface formations are water-saturated clay, sand, and volcanic ash. The sediment compacts as groundwater is pumped for domestic and industrial use, and slow subsidence is widespread. The opera house (weighing 54,000 metric tons) has settled more than 3 m, and half of the first floor is now below ground level. Other large structures are noticeably tilted, as is illustrated in Figure 14.26.

Another type of groundwater problem is shown in Figure 14.27. In western Wyoming a dam was built in the tilted strata of the Madison Limestone Formation for the purpose of storing irrigation water. The limestone, however, was so porous and permeable that all of the water that was supposed to be stored in the reservoir seeped into the subsurface and was lost. The reservoir never filled and the project was abandoned.

FIGURE 14.26
Subsidence of buildings in Mexico City resulted from compaction after groundwater was pumped from unconsolidated sediment beneath the city. Subsidence has caused this building to tilt and sink more than 2 m.

FIGURE 14.27
**A dam constructed on permeable lime-
stone** in western Wyoming never functioned
because the surface water seeped into the subsurface.

We are using and altering the groundwater system at an ever-increasing
rate. Approximately 20% of all water used in the United States is pumped from
the subsurface. This amounts to more than 83 billion gallons a day, almost
three times as much as in 1950. How much groundwater will be needed in the
future? What effect will this have on the environment?

SUMMARY

1. Groundwater is an integral part of the hydrologic system, and it is intimately related to surface water drainage.
2. The movement of groundwater is very slow, controlled largely by the porosity and permeability of the rock. At some depth below the surface, all pore spaces are filled with water.
3. The upper surface of this saturated zone is called the water table.
4. Movement of groundwater below the water table is closely associated with surface runoff, and groundwater commonly discharges into streams, lakes, and swamps.
5. Artesian water is water confined between impermeable beds and is under hydrostatic pressure.
6. Groundwater erodes by dissolving soluble rocks, such as limestone, rock salt, and gypsum. Such erosion produces karst topography, characterized by sinkholes, solution valleys, and disappearing streams.
7. Under normal conditions, groundwater is essentially in balance, or in equilibrium, with surface runoff, surface topography, and saltwater in the ocean.
8. Alteration of the groundwater system can produce many unforeseen problems, as does alteration of any segment of the hydrologic system.

KEY TERMS

artesian-pressure surface (p. 304)

artesian water (p. 304)

aquifer (p. 304)

cave (p. 309)

cone of depression (p. 302)

disappearing stream (p. 311)

drip curtain (p. 315)

dripstone (p. 315)

geothermal energy (p. 307)

geyser (p. 307)

hydraulic head (p. 299)

hydrostatic pressure (p. 297)

karst topography (p. 311)

leach (p. 318)

leachate (p. 318)

perched water table (p. 298)

permeability (p. 297)

pore space (p. 296)

porosity (p. 296)

recharge (p. 304)

saltwater encroachment (p. 320)

seep (p. 300)

sinkhole (p. 309)

solution valley (p. 311)

spring (p. 300)

stalactite (p. 315)

stalagmite (p. 315)

subsidence (p. 322)

tower karst (p. 311)

travertine terrace (p. 315)

water table (p. 298)

zone of aeration (p. 298)

zone of saturation (p. 298)

REVIEW QUESTIONS

1. Define porosity and permeability.
2. Describe and illustrate the major types of pores, or voids, in rocks.
3. What rock types are generally impermeable or nearly impermeable?
4. Describe the major zones of subsurface water and explain how water moves through each zone.
5. Explain some of the ways in which springs originate.
6. What effects are produced in the water table by excessive and rapid pumping?
7. Explain how artesian flow occurs.
8. Explain the origin of geysers.
9. What is the source of heat for hot springs and geysers?
10. Describe the evolution of a landscape in which groundwater is the dominant agent of erosion.
11. Explain how stalagmites and stalactites originate.
12. Describe the forms and processes of groundwater pollution.
13. Describe the relationship between salty groundwater and fresh groundwater beneath an island or a peninsula.
14. What undesirable effects can result from withdrawing an excessive amount of groundwater from wells close to the ocean?
15. Explain how the alteration of the natural drainage system in southern Florida has affected the Everglades.
16. How can subsidence of the land result from the withdrawal of groundwater? Give examples.
17. Describe the occurrence of groundwater in the area where you live.

ADDITIONAL READINGS

Bowen, R. 1980. *Ground Water.* London: Applied Science Publishers.

Davis, S. N., and R. J. M. DeWiest. 1966. *Hydrogeology.* New York: Wiley.

Dunne, T., and L. B. Leopold. 1978. *Water in Environmental Planning.* New York: Freeman.

Ford, R. S. 1978. Ground water—California's priceless resource. *California Geology* 31(2):27–32.

Hunt, C. A., and R. M. Garrels. 1972. *Water: The Web of Life.* New York: Norton.

Leopold, L. B. 1974. Surface water and ground water, in *Water: A Primer.* San Francisco: Freeman.

Mather, J. R. 1984. *Water Resources.* New York: Wiley.

Peixit, J. P., and M. A. Kettani. 1973. The control of the water cycle. *Scientific American.* April.

Price, M. 1985. *Introducing Groundwater.* London: Allen and Unwin.

Solley, W. B., E. B. Chase, and W. B. Man IV. 1983. Estimated use of water in the United States in 1980, in U.S. Geological Survey Circular 1001.

Trudgill, S. 1985. *Limestone Geomorphology.* White Plains, NY: Longman.

US Geological Survey. 1984. National water summary 1983—*hydrologic events and issues,* in U.S. Geological Survey Water Supply Paper 2250.

Atlantic
Ocean

No event in recent geologic history has had such a profound effect upon Earth as the last, great ice age. Its impact extended far beyond the margins of the ice itself and influenced almost every aspect of the physical and biological world. For example, the present sites of many northern cities, such as Chicago, Detroit, Montreal, and Toronto, were buried beneath thousands of feet of glacial ice as recently as 15,000 to 20,000 years ago.

The satellite image on the facing page shows the Vatnajökull glacier of Iceland. The glacier completely buries the underlying surface and spreads out over the terrain to the lower left as large lobes. In the more rugged terrain to the upper right, tongues of ice advance through the valleys toward the sea and tint the nearshore water with outwash sediment. A glacier is a system of flowing ice. Ice flows through the system and leaves by melting and evaporation at the lower margins. As it moves, it erodes and transports considerable amounts of rock debris, shown here by the dark ribbons of soil and rock in the white ice.

When glaciation occurs, many geologic processes are interrupted or modified significantly. Much precipitation becomes trapped in glaciers instead of flowing immediately back to the ocean. Consequently, sea level drops and the hydrology of streams is greatly altered. As the great ice sheets advance over the continents, they obliterate preexisting drainage networks. The moving ice scours and erodes the landscape and deposits the debris near its margins, covering the preexisting topography. The crust of Earth is pushed down by the weight of the ice, and meltwater commonly collects and forms lakes along the ice margins. As the glaciers melt, new drainage systems are established to accommodate the large volume of meltwater. Far beyond the margins of the glaciers, stream systems are modified by changing climatic patterns. Even in arid regions, the imprint of climatic changes associated with glaciation is seen in the development of large lakes in closed basins.

In this chapter we will study how glaciers operate as a system of flowing ice and how they modify the landscape. We will then consider the cause of an ice age, which remains a tantalizing question, partly unanswered.

15
Glacial Systems

1. Glaciers are systems of flowing ice that form where more snow accumulates each year than melts.
2. As ice flows, it erodes the surface by abrasion and plucking. Sediment is transported by the glacier and deposited where the ice melts.
3. The two major types of glaciers—continental and valley glaciers—produce distinctive erosional and depositional landforms.
4. The major effects of an ice age include the rise and fall of sea level, isostatic adjustments of land, modification of drainage systems, creation of numerous lakes, and migration and selective extinction of plant and animal species.
5. The cause of glaciation is not completely understood, but it may be related to several simultaneously occurring factors, such as astronomical cycles, plate tectonics, and ocean currents.

GLACIAL SYSTEMS

A glacier is an open system of flowing ice. Water enters the system as snow, which is transformed into ice. The ice then flows through the system under the pressure of its own weight and leaves the system by evaporation and melting. The balance between the rate of accumulation and the rate of melting determines the size of the glacial system.

Ice is a brittle substance and when struck sharply it will fracture and break; in large masses, however, it behaves like a viscous liquid and flows. The ice in many valley glaciers is over 300 m thick, and in continental glaciers it is commonly 3 km thick. Gravity is the fundamental force acting upon the ice, causing it to flow, and flow structures are extremely obvious in most glaciers. Thus, a glacier may be defined as a system of flowing ice that originates on land as a result of compaction and recrystallization of snow.

The necessary conditions for the development of a glacier are simple; more snow must fall each year than is lost by melting and evaporation. Under these conditions, a new layer of snow accumulates each year, and, over many years, the mass of ice eventually becomes thick enough to flow under its own weight. Perennial snowfields that do not move are not considered glaciers, nor is pack ice formed from seawater in polar latitudes.

Glaciers are open systems and have much in common with other gravity flow systems, such as rivers and groundwater. Water enters the system primarily in the upper parts of the glacier, where snow accumulates and is transformed into ice. The ice then flows out of the zone of accumulation, generally moving a few centimeters per day. At the lower end (or terminus) of the glacier, ice leaves the system by melting and evaporating. There are two principal types of glaciers:

1. Valley glaciers are streams of ice that originate in the snowfields of high mountain ranges.
2. Continental glaciers are ice sheets thousands of meters thick that spread out and cover large parts of continents.

Valley Glacial Systems

Valley glaciers, or *alpine glaciers,* occur in some of the most scenic mountain ranges of the world and are relatively accessible for direct observation. As a result, they have been studied for many years, and their general characteristics are well understood. Figure 15.1 is an idealized diagram showing how a glacial system operates. Valley glaciers are long, narrow rivers of ice that originate in the snowfields of high mountain ranges and flow down preexisting stream valleys. They range from a few hundred meters to more than a hundred kilometers in length. In many ways they resemble river systems. They receive an input of water (in the form of snow) in the higher reaches of the mountains. They have a system of tributaries leading to a main trunk system. The flow direction is controlled by the valley the glacier occupies, and as the ice moves, it erodes and modifies the landscape over which it flows.

The essential parts of the system are (1) the *zone of accumulation,* where there is a net gain of ice, and (2) the *zone of ablation,* where ice leaves the system by melting and evaporating. In the zone of accumulation, snow is transformed into glacial ice through a process identical to rock metamorphism. Freshly fallen snow consists of delicate, hexagonal ice crystals or needles, with as much as 90% of the total volume as empty space. As snow accumulates, the ice at the points of the snowflake melts from the pressure of snow buildup and migrates toward the center of the flake, eventually forming an elliptical granule of recrystallized ice approximately 1 mm thick. The accumulation of these particles packed together is called *firn,* or *névé.* With repeated annual deposits, the loosely packed névé granules are compressed by the weight of the over-lying snow. Meltwater, which results from daily temperature fluctuations and the pressure of the overlying snow, seeps through the pore spaces between the grains; when it freezes, it adds to the recrystallization process. Air in the pore spaces is driven out. When the ice reaches a thickness of approximately 30 or 40 m, it can no longer support its own weight and yields to plastic flow. The

How does solid ice flow?

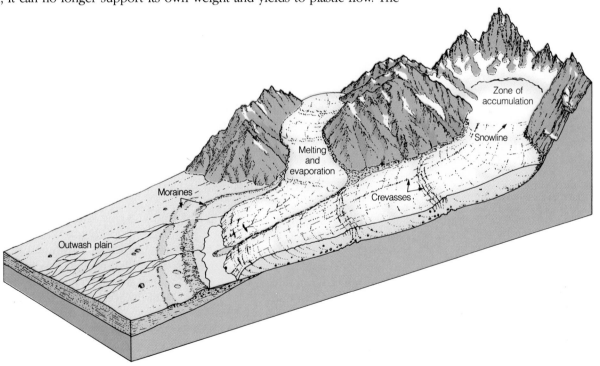

FIGURE 15.1
A glacial system is an open system of ice that flows under the pull of gravity. Snow enters the system by precipitation and is transformed into ice. The ice flows outward from the zone of accumulation under the pressure of its own weight. The ice leaves the system by evaporating and melting in the zone of ablation. The balance between the rate of accumulation and the rate of melting determines the size of the glacial system. The major parts of a valley glacial system are the zone of accumulation, where snow enters the system, and the zone of ablation, where ice leaves the system by melting and evaporation. The boundary between these zones is approximated by the snowline. As the ice moves through the system, it erodes the preexisting stream valley and deposits sediment at the end of the glacier. Meltwater can rework much of the glacial sediment and redeposit it downstream.

FIGURE 15.2
Valley glaciers in Switzerland show many of the features of the glacial system illustrated in the diagram of Figure 15.1. Note the snow-covered zone of accumulation, the rough ice in the zone of ablation, the flow structures in the glacier, and the sediment transported and deposited at the end of the glacier.

upper part of a glacier is thus rigid and tends to fracture, but the ice beneath moves by plastic flow.

As is shown in Figures 15.1, 15.2, and 15.3, the boundary between the zone of accumulation and the zone of ablation is approximated by the *snowline.* Above the snowline, the surface of the glacier is smooth and white because more snow accumulates than is lost by melting, and any irregularities are soon covered and filled with snow. Below the snowline, melting and evaporation exceed snowfall. There, the surface of the ice is rough and pitted and commonly is broken by open *crevasses.*

At the terminal margin of a glacier, the loss of ice by melting and evaporation combined exceeds the rate of accumulation. This margin is the major exit boundary of the system.

It is important for you to understand that the margins of a glacier constitute the boundaries of a system of flowing ice, much as the banks and mouth of a river constitute the boundaries of a river system. If more snow is added in the zone of accumulation than is lost by melting or evaporation at the end of the glacier, the ice mass increases, and the glacial system expands. If the accumulation of ice is less than ablation, there is a net loss of mass and the size of the glacial system is reduced. If accumulation and ablation are in balance, the mass of ice remains constant, the size of the system remains constant, and the terminus of the ice remains stationary. Nevertheless, *ice within the glacier*

FIGURE 15.3
Valley glaciers on Bylot Island just off the northern end of Canada's Baffin Island originate in the snow fields that almost completely cover the mountain peaks. Note that the snowline extends down almost to sea level. The main glaciers extend down from the highland as tongues of ice (blue). Note that glaciers, like river systems, consist of a main trunk stream and an intricate system of branching tributaries.

continually flows toward the **terminus,** *or terminal margins,* regardless of whether the terminal margins are advancing, retreating, or stationary.

The behavior of a glacial system is determined by the balance between the rate of input and the rate of output of ice. The two major variables in this balance are temperature and precipitation. A glacier can grow or shrink with an unchanging rate of precipitation if the temperature varies enough to increase or decrease the rate of melting (rate of output). The length of a glacier in no way represents the amount of ice that has moved through the system, just as the length of a river does not represent the volume of water that has flowed through it. Length simply shows the amount of ice currently in the system.

An example from the last ice age illustrates this point. A glacial valley 20 km long in the Rocky Mountains was eroded 600 m deeper than the original stream valley. This large amount of erosion was not accomplished by 20 km of ice moving down the valley. It was the result of many thousands of kilometers of ice flowing through the valley. If the ice occupied the valley during each glacial epoch and moved 0.3 m per day, a total of approximately 72,000 km of ice would have moved down the valley. The enormous abrasion caused by such a long stream of ice would be able to wear down the valley to a depth of 600 m.

The movement of the ice within a glacier is well illustrated in Figure 15.3 and also in Figure 15.2, where large tongues of ice flow from the mountain tops and spread out onto the valley floor.

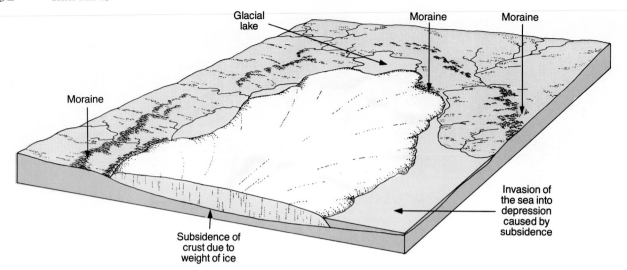

FIGURE 15.4
A continental glacial system covers a large part of a continent and causes a number of significant changes in the regional physical setting. The weight of the ice depresses the ground surface, so the land commonly slopes toward the glacier. Consequently, glacial lakes form in the depressions along ice margins, or an arm of the ocean may invade the depression. The original drainage system is greatly modified, as the streams that flow toward the ice margins are impounded to form lakes. The glacier advances more rapidly into lowlands, so the margins are not smooth but are typically irregular or lobate. As the system expands and contracts, ridges of sediment are deposited along the margins, and a variety of erosional and depositional landforms develop beneath the ice. The balance between the rate of accumulation and rate of melting determines the size of the glacial system.

Continental Glacial Systems

In terms of their effect on the landscape and on Earth's hydrologic system, *continental glaciers* are by far the most important type of glacial system. These large ice sheets form in some of the most rigorous and inhospitable climates on Earth. Nonetheless, teams of scientists from various countries have used modern technology to study existing continental glaciers in Canada, Greenland, and Antarctica. From these studies we can construct a reasonably accurate model of an idealized continental glacial system and analyze how it operates (Figure 15.4).

The basic elements of a continental glacier are much the same as those in a valley glacier. Both systems have a zone of accumulation, where there is a net gain of ice from snowfall. The ice flows out from the zone of accumulation to the zone of ablation, where it leaves the system through melting and evaporation. As shown in Figure 15.4, a continental glacier is a roughly circular or elliptical plate of ice, rarely more than 3000 m thick. Ice does not have the strength to support the weight of an appreciably thicker accumulation. If more ice is added by increased precipitation, the glacier simply flows out from the centers of accumulation more quickly.

The weight of such a huge ice mass causes Earth's crust to subside, so the surface of the land commonly slopes toward the glacier. Subsidence creates a lowland along the ice margin, which traps meltwater to form large lakes. If the margin of the glacier is near the coast, an arm of the sea may flood the depression, as is shown in Figure 15.4.

Preexisting drainage systems are modified or completely obliterated. Rivers that flow toward the ice margins are impounded, forming lakes, which may overflow and develop a river channel parallel to the ice margin. Drainage systems covered by the glacier are destroyed. Thus, when the ice melts, no established, integrated drainage system exists, so numerous lakes form in the natural depressions.

The margins of continental glaciers commonly form large lobes. These develop because, where the ice is not confined to a system of valleys, it moves more rapidly into preexisting lowlands. The sediment deposited at the ice margins is therefore arcuate or lobate in map pattern.

The Barnes Ice Cap of Baffin Island, Canada (Figure 15.5), is one of the last remnants of the glacier that covered much of Canada and parts of the northern United States only 14,000 years ago. This example illustrates the relationship between the continental glacier and the regional landforms. As shown on the map, the glacier is elliptical, with irregular or lobate margins. The ice is thickest in the central part and thins toward the edges. The presence of the glacier has caused an isostatic subsidence of the crust, so the land slopes

FIGURE 15.5
The Barnes Ice Cap, Baffin Island, Canada, is a remnant of the last continental glaciers that covered large parts of North America and shows many features typically produced by continental glaciation. Isostatic adjustment of the crust causes the surface of the land to slope toward the ice, so lakes form along the ice margins. Drainage coming from the north is blocked by the ice and also contributes to lake formation near the ice margins. Erosional debris, carried to the margins of the glacier, is deposited as moraines. Irregularities in the surface over which the ice flows cause the ice margins to be uneven or lobate.

toward the ice margins. In addition, the glacier has completely disrupted the former drainage system. Meltwater has therefore accumulated along the ice margins, forming a group of lakes. A photograph of the southern margins of the ice cap (Figure 15.6) shows the large, gently arched surface of the glacier, sediment deposited along the ice contact, and stream channels formed by meltwater on the glacier's surface.

The ice cap that covers nearly 80% of Greenland is much larger than the remnant on Baffin Island. In cross section, the glacier is shaped like a drop of water on a table (Figure 15.7). Its upper surface is a broad, almost flat-topped arch and is typically smooth and featureless. The base of the glacier is relatively flat. The Greenland glacier is over 3000 m thick in its central part, but it thins toward the margins. The zone of accumulation is in the central part of the island, where the ice sheet is nourished by snowstorms moving from west to east. The snowline lies from 50 to 250 km inland; thus, the area of ablation constitutes only a narrow belt along the glacial margins.

In rugged terrain, especially in areas close to the margins, the direction of ice movement is greatly influenced by mountain ranges, and the ice moves through mountain passes in large streams called *outlet glaciers* (Figure 15.8). These resemble valley glaciers in that they are confined by the topography. Pressure builds up in the ice behind a mountain range and forces outlet glaciers through mountain passes at relatively high speeds. Measurements in Greenland show that the main ice mass advances at approximately 10 to 30 cm per day. Outlet glaciers, however, can move as fast as 1 m per hour. In some places, you can actually see the ice move.

The glacier of Antarctica is similar to that of Greenland in that it covers essentially the entire landmass (Figures 15.9 and 15.10). Antarctica, however, is much larger than Greenland, and its glacier contains more than 90% of Earth's ice. Much of the glacier is over 3000 m thick, and its weight has depressed large parts of the continent's surface below sea level. Parts of Antarctica (mostly near the continental margins) are mountainous, with the higher peaks and ranges protruding above the ice. In the mountains, outlet glaciers funnel ice from the interior to the coast. Two large ice shelves and numerous small ones occur

FIGURE 15.6
The margins of the Barnes Ice Cap are marked by ridges of sediment deposited as the ice melts. The glacier's upper surface is gently arched, and meltwater has formed small meandering streams. The landscape of the Great Lakes region must have appeared something like this 20,000 years ago.

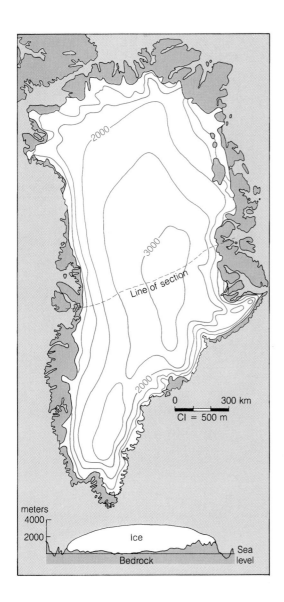

FIGURE 15.7
The Greenland Ice Sheet covers nearly 80% of the island. In this diagram, the thickness of the glacier is shown by contour lines. The upper surface of the glacier is a broad, almost flat-topped arch and is typically smooth and featureless. The base of the glacier is relatively flat. The Greenland glacier is over 3000 m thick in its central part, but it thins toward the margins. The zone of accumulation is in the central part of the island, where the ice sheet is nourished by snowstorms moving from west to east. The snowline lies from 50 to 250 km inland; thus, the zone of ablation constitutes only a narrow belt along the glacial margins. Note from the cross section that the central part of Greenland has been depressed below sea level by the weight of the ice. CI = contour interval.

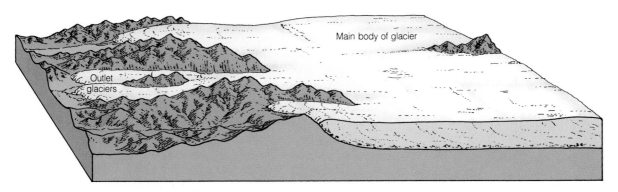

FIGURE 15.8
Outlet glaciers are segments of a continental glacier that advance rapidly through a mountain pass. A mountain range is a physical barrier to the movement of a continental glacier, and great pressure builds up in the ice behind the range. This pressure causes the ice in an outlet glacier to move very rapidly in comparison with the main body of the glacier.

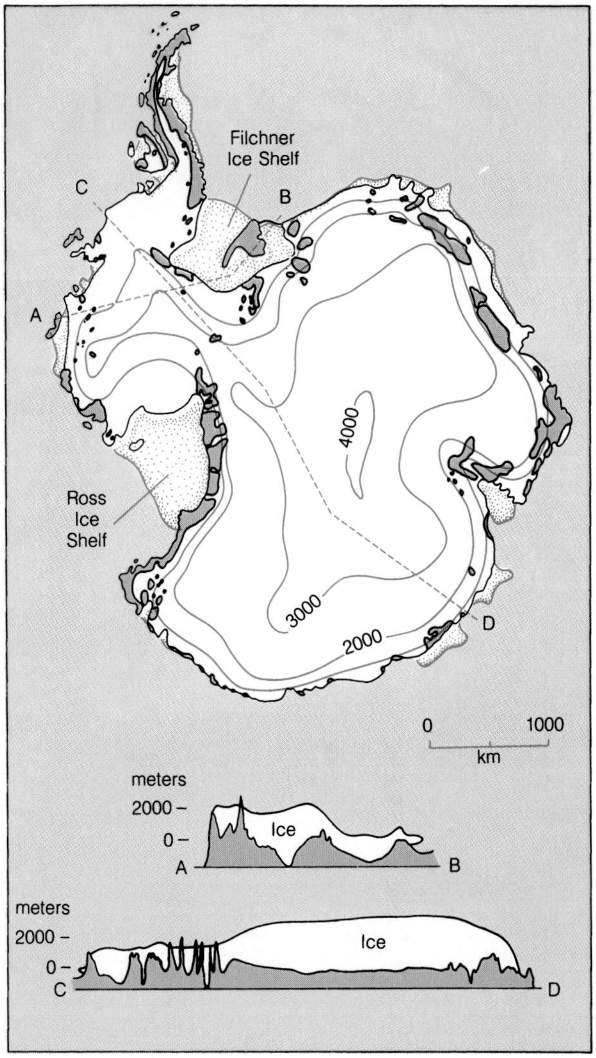

FIGURE 15.9
The Antarctic glacier is over 3000 m thick in many areas and has a general cross-sectional shape similar to that of the Greenland glacier. The tremendous weight of the ice has depressed large parts of the continent's surface below sea level. Parts of Antarctica (mostly near the continental margins) are mountainous, with the higher peaks and ranges protruding above the ice. In the mountains, outlet glaciers funnel ice from the interior to the coast. Two large ice shelves and numerous small ones occur along the coast. These break up, in a process called *calving,* to form large icebergs in the South Atlantic Ocean.

Both Greenland and Antarctica are surrounded by water, so there is an ample supply of moisture to feed their glaciers. In contrast, Siberia is cold enough for glaciers to exist, but does not have sufficient precipitation for ice to accumulate.

along the coast. These break up, in a process called *calving,* to form large icebergs in the South Atlantic Ocean.

Both Greenland and Antarctica are surrounded by water, so there is an ample supply of moisture to feed their glaciers. In contrast, Siberia is cold enough for glaciers to exist, but does not have sufficient precipitation for ice to accumulate.

EROSION, TRANSPORTATION, AND DEPOSITION BY GLACIERS

Glaciers erode bedrock in two principal ways: (1) by abrasion and (2) by glacial plucking. The eroded material is then carried suspended in the ice and is ultimately deposited near the margins of the glacier, where the ice melts.

As glacial ice flows over the land, it erodes, transports, and deposits vast amounts of rock material and greatly modifies the preexisting landscape. Glaciers erode bedrock in two ways: by glacial plucking and by abrasion (grinding). The eroded material is then carried in suspension in the ice and is deposited near the margins of the glacier, where melting dominates (Figure 15.11).

FIGURE 15.10
The Antarctic glacier was photographed by the *Galileo* spacecraft during its Earth flyby on December 8, 1990, as part of its journey to Jupiter. It was two weeks before Antarctic midsummer at the time, and nearly the entire continent was sunlit, permitting this unusual view of a continental glacier. The small dark spots are the tops of the mountain peaks of the Transantarctic Range. The ice of the Ross Ice Shelf is easily distinguished from the glacier by its smooth surface and blue tinge. The faint blue line along the curvature of Earth marks our planet's atmosphere. A scene such as this existed over much of North America only 15,000 years ago.

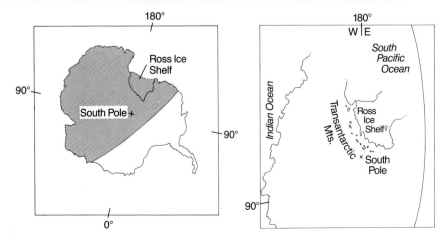

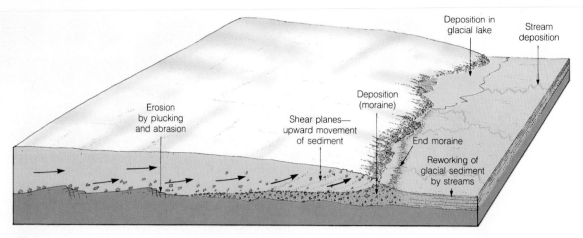

FIGURE 15.11
Glaciers erode, transport, and deposit rock material in the manner illustrated in this diagram. Erosion by glacial plucking occurs as blocks of rock, separated by joints, freeze to the base or sides of the glacier and are lifted from the outcrop by the moving ice. These fragments, frozen in the ice, then act as abrasives and wear down the surface by grinding. Near the end of the glacier, the ice is commonly stagnant and no longer moves. Much of the sediment load is thus forced upward along shear planes and is concentrated at the surface of the glacier. As the ice melts, this sediment accumulates in an end moraine.

Erosion

Glacial plucking is the lifting out and removal of fragments of bedrock by a glacier. It is one of the most effective ways in which a glacier erodes the land. The process involves ice wedging. Beneath the glacier, meltwater seeps into joints or fractures, where it freezes and expands, wedging blocks of rock loose. The loosened blocks freeze to the bottom of the glacier and are plucked, or quarried, from the bedrock, becoming incorporated in the moving ice. The process is especially effective where the bedrock is cut by numerous joints and where the surface is unsupported on the downstream side.

Abrasion is essentially a filing process. The angular blocks plucked and quarried by the moving ice freeze firmly into the glacier, where they act as tools that grind and scrape the bedrock. Aided by the pressure of the overlying ice, the angular blocks are very effective agents of erosion, capable of wearing away large quantities of bedrock. The rock fragments incorporated in the glacial ice are themselves abraded and worn down as they grind against the bedrock surface. As a result, they usually develop flat surfaces that are deeply scratched.

Evidence of the distinctive abrasive and quarrying action of glaciers can be seen on most bedrock surfaces over which glacial ice has moved. Small hills of bedrock (*roches moutonnée*) commonly are streamlined by glacial abrasion. Their upstream sides typically are rounded off, while the downstream sides are made steep and rugged by glacial plucking (Figure 15.12). Glacial *striations,* such as those illustrated in Figure 15.13, range from hairline scratches to large, deep furrows more than a kilometer long.

Glacial Transport

Why do glacial deposits differ from those formed by rivers?

As with streams, the rock fragments transported by a glacier are collectively referred to as its *load.* The manner in which a glacier transports its load, however, differs from stream transport in a very important way. The load in a glacier is carried in suspension, and large blocks are transported side by side with small grains, without the sorting or separation of the material according to size that occurs in streams. As a result, the deposits of a glacier are unsorted and unstratified, and thus they differ markedly from stream deposits. The load of a glacier is concentrated near the contact between the ice and the bedrock, from which it was derived.

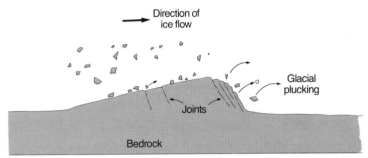

FIGURE 15.12
Roche moutonneé is an erosional feature that forms as ice moves over bedrock, eroding it into a streamlined shape. Glacial plucking commonly produces a ragged edge on the downcurrent side.

FIGURE 15.13
Glacial striations result from the abrasive action of a glacier and clearly show the direction in which the ice moved.

Most of the particles carried by a glacier are fresh and unweathered and have angular, jagged surfaces. The grinding action of the moving ice further crushes the grains to produce an abundance of fine particles known as *rock flour.*

Deposition

Most of the particles transported by a glacier are deposited near the terminus, where melting dominates. There, the ice may become stagnant. This commonly forces the active ice upstream to move up and over the "dead ice" at the terminus, so that much of the load is carried upward along shear planes and is concentrated on the surface of the glacier (Figure 15.11). When melting is completed, this material accumulates as a ridge, marking the former margins of the glacier.

Some of the sediment deposited by a glacier is picked up and reworked by meltwaters and redeposited beyond the margins of ice by streams. The sediment, deposited directly by the glacier or indirectly in glacial lakes and streams, is referred to as glacial **drift** (the name is held over from the time when such deposits were believed to have "drifted" to their present resting place during the flood of Noah). Both stratified and unstratified glacial drift occurs. The term **till** is used to designate unsorted, unstratified drift, deposited directly by the glacier.

Rates of Ice Flow

Ice flow in a glacier may seem extremely slow in comparison to the flow of water in rivers, but the movement is continuous, and over the years, vast quantities of ice can move through a glacier. Measurements show that some of Switzerland's large valley glaciers move as much as 180 m per year. Smaller glaciers move from 90 to 150 m per year. Some of the most rapid rates have been measured on the outlet glaciers of Greenland, where ice is funneled through mountain passes at a speed of 8 km per year. From these and other measurements, flow rates of a few centimeters per day appear common, and velocities of 3 m per day are exceptional. The most rapid movement occurs in outlet glaciers, where pressure built up behind a mountain range helps to force the ice through a mountain pass.

Recent close observations of the movements of valley glaciers show that occasionally a glacier can surge forward more than several hundred meters per day. Now that movements can be monitored by satellite photography, these surges are known to be fairly common. (In the past, only a few surges had been observed because the flow is so short-lived.) A glacier in the Himalayas, for instance, advanced 11 km in 3 months. In 1966, the Steele Glacier, in the Yukon Territory, advanced approximately 8 km within a matter of weeks. Glacial surges apparently result from sudden slippages along the bases of glaciers, caused by the buildup of extreme stress upstream. Stagnant or slow-moving ice near the terminus can act as a dam for the faster-moving ice upstream. If this happens, stress builds up behind the slow-moving ice, and a surge occurs when a critical point is reached. Surges can also be caused by a sudden addition of mass to the glacier, such as a large avalanche or landslide on its surface.

In considering the flow of ice in a glacier and the erosion it can cause, remember that a glacier is an open system. Material enters the system in the zone of accumulation, flows through, and then exits at the distal margins. Ice within a glacier flows continuously through the system, regardless of whether the terminal margin is advancing, retreating, or stationary. The length of a valley glacier is therefore no indication of the amount of erosion it can accomplish or has accomplished. Erosion by a glacier is a function of the duration of the process, the thickness of the ice, and the velocity at which it flows.

LANDFORMS DEVELOPED BY VALLEY GLACIERS

Erosion and deposition by valley glaciers produce many distinctive landforms, the most important of which are (1) U-shaped valleys, (2) cirques, (3) hanging valleys, (4) horns, (5) moraines, and (6) outwash plains.

Valley glaciers are responsible for some of the most rugged and scenic mountainous terrain on Earth. The Alps, the Sierra Nevadas, the Rockies, and the Himalayas were all greatly modified by glaciers during the last ice age, and the shapes of their valleys, peaks, and divides retain the unmistakable imprint of erosion by ice.

The idealized diagram in Figure 15.14 and the photograph in Figure 15.15 illustrate the major erosional landforms resulting from valley glaciation. Figure 15.14 permits a comparison and contrast of landscapes formed only by running water with those that have been modified by valley glaciers. Figure 15.14A shows the typical topography of a mountain region being eroded by streams. A relatively thick mantle of soil and weathered rock debris covers the slopes. The valleys are V-shaped in cross section and have many bends at tributary junctions, so ridges and divides between tributaries appear to overlap if you look up the valley. In Figure 15.14B, the valleys are shown occupied by glaciers. The growing glaciers expand down the tributary valleys and merge to form a major glacier. During glaciation, several thousand kilometers of ice might flow down the valley. This enormous quantity of moving ice can erode the valley as much as 600 m below its original level.

Ice wedging is a major process in the cold regions of glaciers. It effectively sharpens the mountain summits and causes glacial plucking. A valley glacier commonly fills more than half of the valley length, and as it moves, it modifies the former V-shaped stream valley into a broad U-shaped, or troughlike, form. The head of the glacier is enlarged by plucking and grows headward toward the mountain crest to form a *cirque*. Where two or more cirques approach the summit crest, they sculpt the mountain crest into a sharp pyramid-shaped peak called a *horn*. The projecting ridges and divides between glacial valleys are subjected to rigorous ice wedging, abrasion, and mass movement. In contrast to the rounded topography developed by stream erosion, these processes produce sharp, angular crests and divides (called *arêtes*).

Note that where tributaries enter the main glacier, the upper surfaces of the glaciers are at the same level. The main glacier, however, is much thicker, and it therefore erodes its valley to a greater depth than that of the tributary valleys. When the glaciers recede from the area, the floors of the tributary valleys will be higher than the floor of the main valley, and thus the tributary valleys are called *hanging valleys*.

Part of a valley glacier's load consists of rock fragments that avalanche down the steep valley sides and accumulate along the glacier margins. Frost action is especially active in the cold climate of valley glaciers and produces large quantities of angular rock fragments. This material is transported along the surface of the glacial margins, forming conspicuous dark bands called lateral moraines (Figure 15.2). Where a tributary glacier enters the main valley, the lateral moraine of the tributary glacier merges with a lateral moraine of the main glacier to form a medial moraine in the central part of the main glacier. In addition to transporting the load near its base, a valley glacier thus acts as a conveyor belt and transports a large quantity of surface sediment to the terminus. At the terminus, ice leaves the system through melting and evaporation, and the load is deposited as an end moraine. End moraines commonly block the ends of the valleys, so meltwater from the ice accumulates and forms ponds and lakes. Downstream from the glacier, meltwater reworks the glacial sediments and redeposits them to form an outwash plain.

What unique landforms result from valley glaciation?

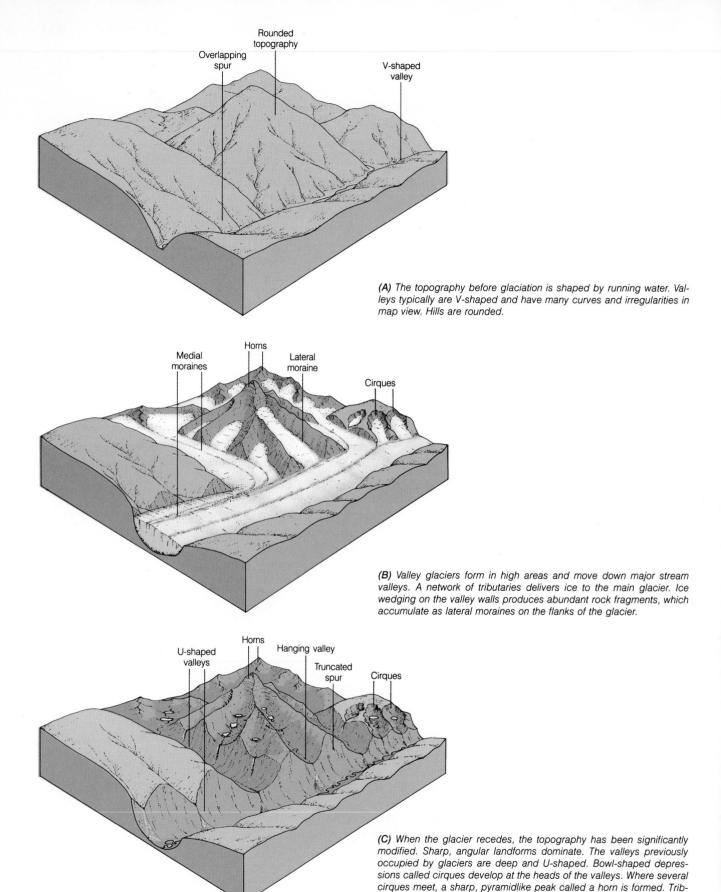

(A) The topography before glaciation is shaped by running water. Valleys typically are V-shaped and have many curves and irregularities in map view. Hills are rounded.

(B) Valley glaciers form in high areas and move down major stream valleys. A network of tributaries delivers ice to the main glacier. Ice wedging on the valley walls produces abundant rock fragments, which accumulate as lateral moraines on the flanks of the glacier.

(C) When the glacier recedes, the topography has been significantly modified. Sharp, angular landforms dominate. The valleys previously occupied by glaciers are deep and U-shaped. Bowl-shaped depressions called cirques develop at the heads of the valleys. Where several cirques meet, a sharp, pyramidlike peak called a horn is formed. Tributaries form hanging valleys, which can have spectacular waterfalls.

FIGURE 15.14
Landforms produced by valley glaciers constitute some of the most spectacular scenery in the world. In these diagrams, an idealized landscape formed by stream erosion is shown as it might appear before, during, and after glaciation.

FIGURE 15.15
Glaciated topography in the Alps of Switzerland shows most of the classic landforms produced by valley glaciers. Note the major U-shaped valley and its tributaries. Cirques and horns dominate the landscape in the background where remnants of valley glaciers still exist. Compare the landforms shown here with those in Figure 15.14C.

Figures 15.14C and 15.15 show the region after the glaciers have disappeared. The most conspicuous and magnificent landforms developed by valley glaciers are the long, straight, U-shaped valleys, or troughs. Many are several hundred meters deep and tens of kilometers long. The heads of glacial valleys terminate in large amphitheater-shaped or bowl-like cirques, which commonly contain small lakes.

The diagrams in Figure 15.14 do not show the landforms that develop at the terminus of a valley glacier. These are illustrated in Figures 15.1 and 15.2. Three types of *moraines* are immediately obvious: (1) *lateral moraines,* along the margins of the glacier, (2) *medial moraines,* formed where two lateral moraines join, and (3) a *terminal moraine,* or *end moraine.* The terminal moraine characteristically extends in a broad arc conforming to the shape of the terminus of the ice. It commonly traps meltwater and forms a temporary lake. If periods of stabilization occur during the recession of ice, *recessional moraines* may form behind the terminal moraine.

The great volume of meltwater released at the terminus of a glacier reworks much of the previously deposited moraine and redeposits the material in an *outwash plain* beyond the glacier. Outwash sediment has all of the characteristics of stream deposits, and the sediment is typically rounded, sorted, and stratified.

LANDFORMS DEVELOPED BY CONTINENTAL GLACIERS

Continental glaciers greatly modify the entire landscape they cover. The flowing ice removes the soil and commonly erodes several meters of the underlying bedrock. Material is transported long distances and deposited near the ice margins, producing depositional landforms such as moraines, drumlins, eskers, kettles, lake sediment, and outwash plains. The preexisting drainage is disrupted or obliterated, so numerous lakes form after the ice melts.

From viewpoints on the ground, the landforms developed by continental glaciers are relatively inconspicuous and not nearly as spectacular as those produced by valley glaciers. On a regional basis, however, continental glaciation modifies the entire landscape, producing many important and distinctive surface features.

Perhaps the best way to approach the study of landforms produced by continental glaciers is to study the block diagrams in Figure 15.16 and the photographs in Figure 15.17. Figure 15.16A shows the margins of an ice sheet. Debris (till) transported by the glacier accumulates at the ice margin as a terminal moraine. Beneath the ice is a variable thickness of till, transported by the glacier and deposited as a *ground moraine.* This material, together with outwash-plain sediment, can be reshaped by subsequent advances of ice to produce streamlined hills called *drumlins.*

Streams of meltwater flow in tunnels within, and beneath, the ice and carry a large bed load, which is ultimately deposited to form a long, sinuous ridge called an *esker.* Debris-laden meltwater forms braided streams that flow from the glacier over the outwash plain, where they deposit much of their load. During the retreat of the glaciers, meltwater forms subglacial channels and tunnels, which open into the outwash plain. Temporary lakes can develop where meltwater is trapped along edges of the glacier, and deltas and other shoreline features form along the lake margins. Deposits on the lake bottom typically are stratified in a series of alternating light and dark layers called *varves* (Figure 15.18). The coarse, light-colored material accumulates during spring and summer runoff. During the winter, when the lake is frozen over, no

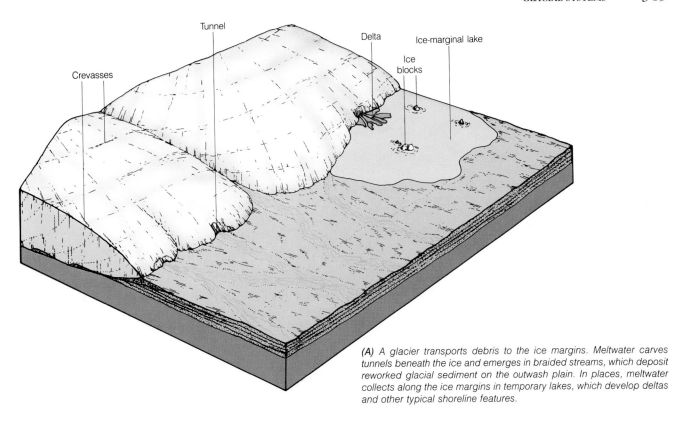

(A) A glacier transports debris to the ice margins. Meltwater carves tunnels beneath the ice and emerges in braided streams, which deposit reworked glacial sediment on the outwash plain. In places, meltwater collects along the ice margins in temporary lakes, which develop deltas and other typical shoreline features.

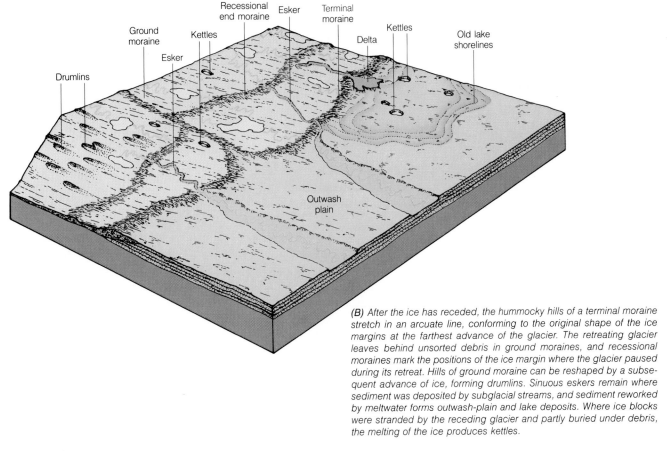

(B) After the ice has receded, the hummocky hills of a terminal moraine stretch in an arcuate line, conforming to the original shape of the ice margins at the farthest advance of the glacier. The retreating glacier leaves behind unsorted debris in ground moraines, and recessional moraines mark the positions of the ice margin where the glacier paused during its retreat. Hills of ground moraine can be reshaped by a subsequent advance of ice, forming drumlins. Sinuous eskers remain where sediment was deposited by subglacial streams, and sediment reworked by meltwater forms outwash-plain and lake deposits. Where ice blocks were stranded by the receding glacier and partly buried under debris, the melting of the ice produces kettles.

FIGURE 15.16
Landforms developed by continental glaciers commonly are related to the position of the ice margin or the direction of the flow.

(A) Eskers on the Canadian shield form long, sinuous ridges composed of sand and gravel deposited by streams that flowed beneath the glacier.

(B) Drumlin fields in the Canadian shield are streamlined hills that were shaped by the movement of the glacier and show the direction in which the ice flowed.

(C) Moraines in Alberta, Canada, form a distinctive topography of rolling hills and numerous closed depressions.

(D) Glacial till resting on horizontal limestone in Iowa is responsible for much of the rich farmland in that area.

FIGURE 15.17
Features formed by continental glaciers in various parts of Canada, photographed from an altitude of approximately 9 km, and till deposited on bedrock in Iowa show the profound imprint the ice made on the landscape during the Pleistocene ice age.

new sediment enters the lake, and the fine mud settles out of suspension to form the thin, dark layers.

Ice blocks left behind by the retreating glacier front can be partly or completely buried in the outwash plain or in moraines. Where an isolated block of debris-covered ice melts, a depression called a **kettle** is formed.

Figure 15.16B shows the area after the glacier has disappeared completely. The end moraine appears as a belt of hummocky hills, which mark the former position of the ice. An end moraine can be several kilometers wide, with local relief from 100 to 200 m. From the ground, it probably would not be recognized by an untrained observer as anything more than a series of hills. Mapped over a large area, however, it can be seen to have an arcuate pattern, conforming to the lobate margin of the glacier. Many small depressions occur throughout the moraine, some of which may be filled with water, forming small lakes and ponds.

PLEISTOCENE GLACIATION

The Pleistocene ice age was one of the most significant events in recent Earth history. The major effects of the ice age were (a) glacial erosion and deposition over large parts of the continents, (b) isostatic adjustment of the crust, (c) changes in sea level, (d) modification of the drainage systems, (e) creation of numerous lakes, (f) catastrophic flooding in Washington, and (g) stress on most life forms. The severe climatic fluctuations had a drastic impact on many life forms, and numerous species became extinct.

The glacial and interglacial periods that began between 2 and 3 million years ago constitute one of the most significant events in the recent history of Earth. During this time, the normal hydrologic system was interrupted completely throughout large areas of the world and was considerably modified in other areas. The evidence of such an event in the recent past is overwhelmingly abundant. Over the last century, field observations have provided incontestable evidence that continental glaciers covered large parts of Europe, North America, and Siberia. These ice sheets disappeared only between 15,000 and 20,000 years ago (Figure 15.19). The general extent of glaciation in the Northern Hemisphere is shown in Figure 15.20, and a more detailed map of glacial features in the eastern United States is given in Figure 15.21. These maps were compiled after many years of fieldwork by hundreds of geologists who mapped the location and orientation of drumlins, eskers, moraines, striations, and glacial stream channels.

Four major periods of Pleistocene glaciation in the United States are recorded by broad sheets of till and complex moraines separated by ancient soils and layers of windblown silt. Striations, drumlins, eskers, and other glacial features show that all of Canada, the mountain areas of Alaska, and the eastern and central United States down to the Missouri and Ohio rivers were covered with ice (Figures 15.19, 15.20, and 15.21). There were three main zones of accumulation. The largest was centered over Hudson Bay. Ice advanced radially from there, northward to the Arctic islands and southward into the Great Lakes area. A smaller center was located in the Labrador Peninsula. Ice spread southward from this center into what are now the New England states. In the Canadian Rockies to the west, valley glaciers coalesced into ice caps. These grew into a single ice sheet, which then moved westward to the Pacific shores and eastward down the Rocky Mountain foothills until it merged with the large sheet from Hudson Bay.

FIGURE 15.18
Varves are thin, alternating layers of light and dark sediment deposited in a glacial lake. A layer of relatively coarse-grained, light-colored sediment accumulates during the spring and summer runoff. During the winter, when the lake is frozen over, fine, dark mud settles to form a dark layer. Each set of light and dark layers therefore represents a year's accumulation.

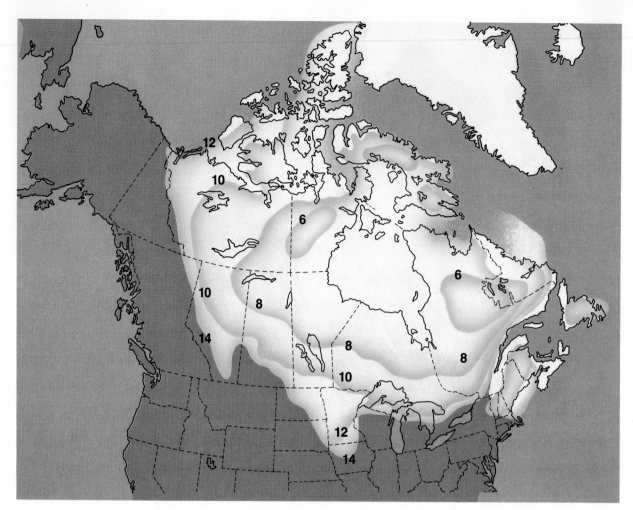

FIGURE 15.19
Successive positions of the ice front during the recession of the last ice sheet have been mapped from data collected by many geologists in Canada and the United States. Contours indicate the position and age of the ice front in thousands of years before the present.

FIGURE 15.20
Pleistocene glaciers covered large areas in North America, Europe, and Asia as well as many high mountain regions. Parts of Alaska and Siberia were not glaciated because those areas were too dry. They were cold enough, but not enough precipitation fell for glaciers to develop.

Throughout much of central Canada, the glaciers eroded from 15 to 25 m of regolith and solid bedrock. This material was transported to the glacial margins and accumulated as ground moraine, end moraines, and outwash in a broad belt from Ohio to Montana (Figure 15.21). In places, the glacial debris is over 300 m thick, but the average thickness is about 15 m. Meltwater carried sediment down the Mississippi River, and much of the fine-grained sediment was transported and redeposited by wind.

The presence of so much ice on the continents had a profound effect on Earth's entire hydrologic system. Sea level dropped more than 100 m. River patterns beneath the ice were obliterated. Elsewhere, river discharge and sediment loads were modified. In many areas, glacial erosion and deposition replaced running water as the dominant geologic process. Even areas far removed from the ice were affected by the modification of drainage systems.

Rhythms of Climatic Changes (Substages of the Ice Age)

Even before the theory of worldwide glaciation was generally accepted, many observers recognized that more than a single advance and retreat of the ice occurred during the *Pleistocene Epoch*. Extensive evidence now shows that a number of periods of growth and retreat of continental glaciers occurred during the ice age. The interglacial periods of warm climate are represented by buried soil profiles, peat beds, and lake and stream deposits separating the unsorted, unstratified deposits of glacial debris.

Radiometric dating shows that the ice first began to advance between 2 and 3 million years ago, and the last glacier began to retreat between 15,000 and 20,000 years ago. Remnants of these last glaciers, now occupying about 10% of the world's land surface, still exist in Greenland and Antarctica. Indeed, we may be living in an interglacial period of an unfinished ice age.

What evidence suggests multiple advances and retreats of the glaciers during the ice age?

Studies of the details of the advance and retreat of the ice are significant because they provide a surprising amount of climatic data. This evidence may be helpful in determining whether Earth is now in a minor interglacial substage or whether it is in the beginning of an ice-free period, such as is typical of most geologic history. Although we cannot predict in the short term whether the climate will be warmer or colder, we can, from the patterns of climatic change in the past, say that significant changes are likely to occur in the future. Earth is near the end of a glacial cycle. A return to normal (warmer) climates would lead to the melting of existing ice caps and the submergence of all major seaports and lowland cities of the world. A return to an ice age, with another major advance of continental glaciers, would also drive people from much presently habitable land. Whatever happens, it is not likely that stability will be established under the present conditions, with ice covering approximately 10% of Earth's land surface. If the past is any indication, this ice will disappear before climatic conditions can be considered normal.

Isostatic Adjustment

Major isostatic adjustment of Earth's crust resulted from the weight of the ice, which depressed the continents during Pleistocene glaciation. In Canada, a large area around Hudson Bay was depressed below sea level, as was the area around the Baltic Sea in Europe. The land has been rebounding from these depressions ever since the ice melted. The former sea floor around Hudson Bay has risen almost 300 m and is still rising at a rate of about 2 cm per year. The land must rise an additional 80 m before it regains its preglacial level and reestablishes isostatic balance.

The tilting of Earth's crust as it rebounds from the weight of the ice can be measured by mapping the elevations of shorelines of ancient lakes (Figure 15.22). The shorelines were level when they formed but were tilted as the crust rebounded from the unloading of the ice. In the Great Lakes region, old

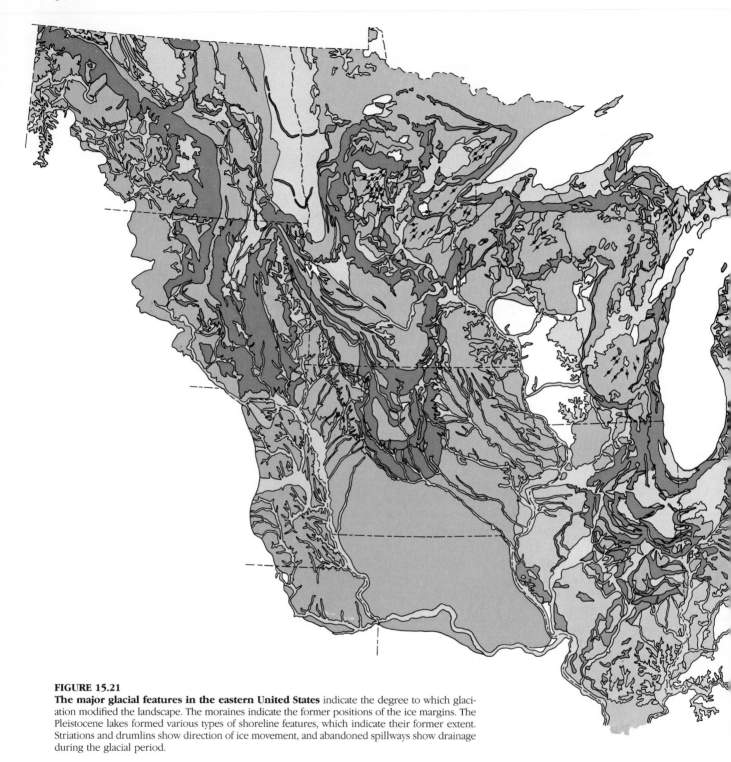

FIGURE 15.21
The major glacial features in the eastern United States indicate the degree to which glaci-
ation modified the landscape. The moraines indicate the former positions of the ice margins. The
Pleistocene lakes formed various types of shoreline features, which indicate their former extent.
Striations and drumlins show direction of ice movement, and abandoned spillways show drainage
during the glacial period.

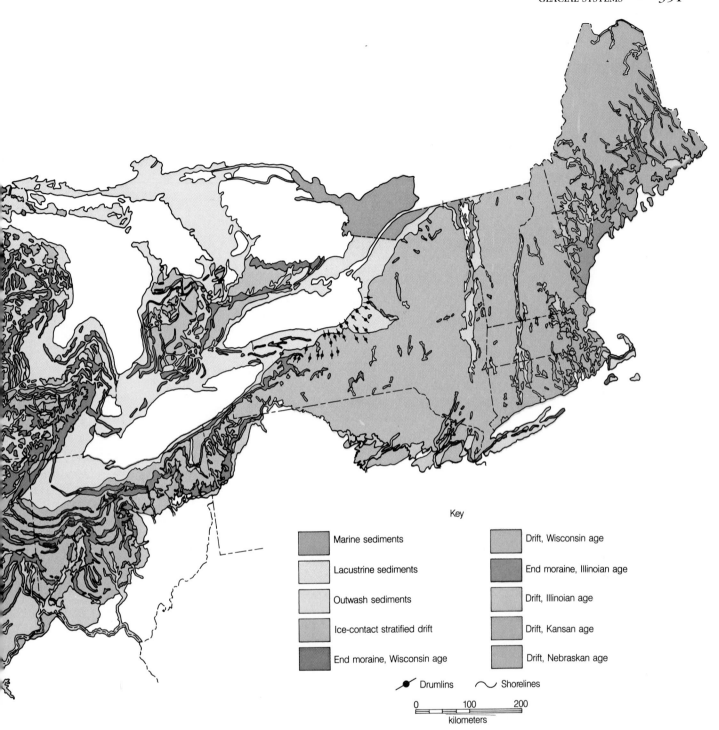

Key

Marine sediments

Lacustrine sediments

Outwash sediments

Ice-contact stratified drift

End moraine, Wisconsin age

Drift, Wisconsin age

End moraine, Illinoian age

Drift, Illinoian age

Drift, Kansan age

Drift, Nebraskan age

Drumlins   Shorelines

0   100   200
kilometers

South
Beaches
North

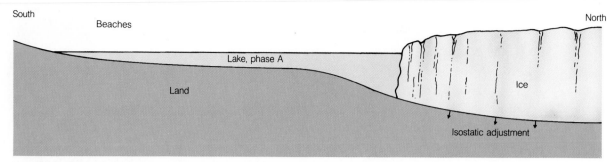

Lake, phase A

Land

Ice

Isostatic adjustment

(A) When a lake develops along a glacier's margins, the shoreline features, such as beaches and bars, are horizontal.

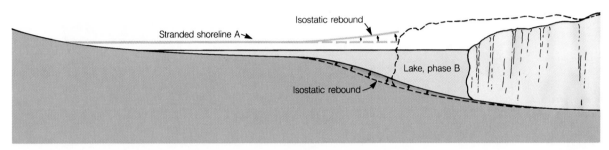

Isostatic rebound

Stranded shoreline A

Lake, phase B

Isostatic rebound

(B) As the ice recedes, isostatic rebound occurs. The shoreline features formed during phase A are tilted away from the ice. Younger horizontal shoreline features are formed by the lake during phase B.

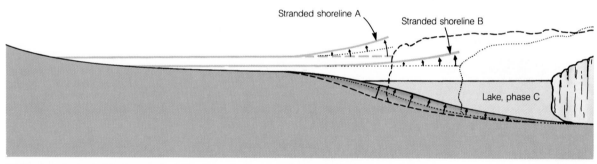

Stranded shoreline A

Stranded shoreline B

Lake, phase C

(C) Continued retreat of the ice causes further isostatic rebound and tilting of both shorelines A and B, which converge away from the glacier.

FIGURE 15.22
The tilted shorelines of glacial lakes can be used to measure the rate and extent of isostatic adjustment of the crust after the ice recedes.

shorelines slope downward to the south, away from the centers of maximum ice accumulation, indicating a rebound of 400 m or more.

Changes of Sea Level

One of the most important effects of Pleistocene glaciation was the repeated worldwide rise and fall of sea level, a phenomenon that corresponded to the retreat and advance of the glaciers. During a glacial period, water that normally returned to the ocean by runoff became locked on the land as ice, and sea level was lowered. When the glaciers melted, sea level rose again. The amount of change in sea level can be calculated because the area of maximum ice coverage is known in considerable detail, and the thickness of the ice can be estimated from the known volumes of ice in the glaciers of Antarctica and Greenland. The Antarctic ice sheet alone contains enough water to raise sea level throughout the world by about 70 m.

The dates of sea-level changes are well documented by radiocarbon dates from terrestrial organic matter and from nearshore marine organisms obtained by drilling and dredging off the continental shelf. These dates show that about 35,000 years ago, the sea was near its present position. Gradually, it receded. By 18,000 years ago, it had dropped nearly 137 m. It then rose rather rapidly to within 6 m of its present level. These fluctuations caused the Atlantic shoreline to recede between 100 and 200 km, so that vast areas of the continental shelf were exposed. Early humans probably inhabited large parts of the shelf that are now more than 100 m below sea level.

The glaciers extended far across the exposed shelf of the New England coast, as is evidenced by unsorted morainal debris and the remains of mastodons dredged from the sea floor in those areas. In the oceans off the central and southern Atlantic states, depth soundings reveal drainage systems and eroded stream valleys that extended across the shelf.

Modification of Drainage Systems

Before glaciation, the landscape of North America was eroded mainly by running water. Well-integrated drainage systems collected runoff and transported it to the ocean. Much of North America was drained by rivers flowing northeastward into Canada because the regional slope throughout the north-central part of the continent was to the northeast. The preglacial drainage patterns are not known in detail. Various features of the present system, however, together with segments of ancient stream channels now mostly buried by glacial sediments, suggest a pattern similar to that shown in Figure 15.23. Before glaciation, the major tributaries of the upper Missouri and Ohio rivers were part of a northeastward-flowing drainage system. This system also included the major rivers draining the Canadian Rockies, such as the Saskatchewan, Athabasca, Peace, and Liard rivers. It emptied into the Arctic Ocean, probably through Lancaster Sound and Baffin Bay, and an eastern drainage was out of the Saint Lawrence River.

As the glaciers spread over the northern part of the continent, they effectively buried the trunk streams of the major drainage systems, damming up the northward-flowing tributaries along the ice front. This damming created a series of lakes along the glacial margins. As the lakes overflowed, the water drained along the ice front and established the present courses of the Missouri and Ohio rivers. A similar situation created Lake Athabasca, Great Slave Lake, and Great Bear Lake and their drainage through the Mackenzie River. This established the present drainage pattern over much of North America (Figure 15.24). Compare this diagram with Figure 15.5, which shows a drainage system currently undergoing similar modifications as a result of the Barnes Ice Cap.

We can clearly see extensive and convincing evidence of these changes in South Dakota. There, the Missouri River flows in a deep, trenchlike valley, roughly parallel to the regional contours. All important tributaries enter from the west. East of the Missouri River, preglacial valleys are now filled with glacial debris, marking the remnants of preglacial drainage. The pattern of preglacial drainage is also supported by recent discoveries of huge, thick, deltaic deposits in the mouth of Lancaster Sound and in Baffin Bay. These deposits are difficult or impossible to explain as results of the present drainage pattern because no major drainage system currently empties into that area.

Beyond the margins of the ice, the hydrology of many streams and rivers was profoundly affected by the increased flow from meltwater or by the greater precipitation associated with the glacial epoch. With the appearance of the modern Ohio and Missouri rivers, water that formerly emptied into the Arctic and Atlantic oceans was diverted to the Gulf of Mexico through the Mississippi River.

How was the landscape of North America modified by the last ice age?

(A) Drainage of central North America before the ice age was northeast-ward, from the northern and central Rocky Mountains, into the St. Lawrence Bay, Hudson Bay, and the Arctic area.

FIGURE 15.23
Glacial modification of North American drainage

(B) The present drainage patterns show major modifications. Preglacial drainage was impounded against the glacial margins and developed new outlets to the ocean through the Missouri, Ohio, and Mackenzie rivers. The drainage system beneath the ice was obliterated. The present drainage in most of Canada is deranged, consisting of numerous lakes, swamps, and unintegrated meandering streams.

Other streams became overloaded and partly filled their valleys with sediment. Still others became more effective agents of downcutting as a result of glacial sediment, and they deepened their valleys. Although the history of each river is complex, the general effect of glaciation on rivers was to produce thick alluvial fill in their valleys; the fill is now being eroded to form stream terraces.

Lakes

Pleistocene glaciation created more lakes than all other geologic processes combined. The reason is obvious if we recall that a continental glacier completely disrupts the preglacial drainage system. The surface over which the glacier moved was scoured and eroded by the ice, leaving a myriad of closed, undrained depressions in the bedrock. These depressions filled with water and became lakes (Figure 15.24).

Farther south in the north-central United States, lakes formed in a different manner. There, the surface was covered by glacial deposits of ground moraine and end moraines. Throughout Michigan, Wisconsin, and Minnesota, these deposits formed closed depressions which soon filled with water to form tens of thousands of lakes. Many of these lakes still exist. Others have been drained or filled with sediment, leaving a record of their former existence in peat bogs, lake silts, and abandoned shorelines.

During the retreat of both the European and the North American ice sheets, several conditions combined to create large lakes along the glacial margins. We can envision their formation with the help of the basic model of continental glaciation shown in Figure 15.4. The ice on both continents was about 3000 m thick near the centers of maximum accumulation, but it tapered toward the glacier margins. Crustal subsidence was greatest beneath the thickest accumulation of ice. In parts of Canada and Scandinavia, the crust was depressed more than 600 m. As the ice melted, rebound of the crust lagged behind, so a regional slope toward the ice was produced. This slope formed basins that have lasted for thousands of years. These basins became lakes or were invaded by the ocean. The Great Lakes of North America and the Baltic Sea of northern Europe were formed primarily in this way.

Although the origin of the Great Lakes is extremely complex, the major elements of their history are known and are illustrated in the four diagrams of Figure 15.25. The preglacial topography of the Great Lakes region was influenced greatly by the structure and character of the rocks exposed at the surface. A geologic map of this region shows that the major structural feature is the Michigan basin, which exposed a broad, circular belt of weak Devonian shale and Silurian salt formations surrounded by the more-resistant Silurian limestone. Preglacial erosion undoubtedly formed a wide valley or lowland along the shale, and escarpments developed on the resistant limestone.

As the glaciers moved southward into this area, large lobes of ice advanced down the great valleys, eroding them into broad, deep basins. Lakes Michigan, Huron, and Erie were scoured from the belt of weak Devonian shale by these lobes of ice. Figure 15.25A shows the Great Lakes area as it probably appeared at the time the Wisconsin glaciers began to recede, about 16,000 years ago. Meltwaters flowed away from the glacier margin to the south. As the glaciers receded, lower land was uncovered, and meltwaters became impounded in front of the ice margins to form the ancestral Great Lakes (Figure 15.25B). Drainage was still to the south through various ancient channels that joined the Mississippi River. As deglaciation continued, an eastern outlet was established (Figure 15.25C) through the Mohawk–Hudson valleys. Finally, as the ice receded farther (Figure 15.25D), a new outlet was developed through the St. Lawrence estuary. Niagara Falls came into existence at this time, when water from Lake Erie flowed across the Niagara escarpment into Lake Ontario. The exposed sequence of rock consists of a resistant limestone formation underlain by a weak shale. Undercutting of the shale below the limestone causes the falls to retreat upstream about 1.3 m per year. Since their inception, the falls have retreated 11.2 km upstream (see Figure 12.15).

To the northwest, another group of lakes formed in much the same way, but they have since been reduced to small remnants of their former selves (Figure 15.23B). The largest of these marginal lakes, known as Lake Agassiz, covered the broad flat region of Manitoba, northwestern Minnesota, and the eastern part of North Dakota (Figure 15.26). It drained into the Mississippi River and then, at lower stages, developed outlets into Lake Superior. Later, when the ice dam retreated, it drained into Hudson Bay. Remnants of this vast lake include Lake Winnipeg, Lake Manitoba, and Lake of the Woods. The sediments deposited on the floor of Lake Agassiz provided much of the rich soil for the wheatlands of North Dakota, Manitoba, and the Red River valley of Minnesota. Even now, ancient shorelines of Lake Agassiz remain, marking its former margins.

Northward, along the margin of the Canadian Shield, Lake Athabasca, Great Slave Lake, and Great Bear Lake are remnants of the other great ice-marginal lakes.

In northern Europe, the recession of the Scandinavian ice sheet caused similar depressions along the ice margins, and the large lakes that were thus produced ultimately connected with the ocean to form the Baltic Sea.

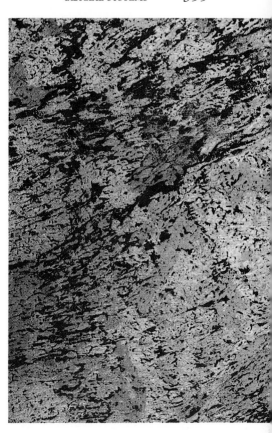

FIGURE 15.24
Lakes created by continental glaciation in the shield area of North America were photographed from a height of approximately 918 km. More lakes were created by glaciation than by all other geologic processes combined.

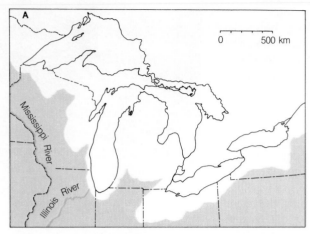

(A) Approximately 16,000 years ago, the ice front extended beyond the present Great Lakes. The ice advanced into lowlands surrounding the Michigan basin, with large lobes extending down from the present sites of Lake Erie and Lake Michigan.

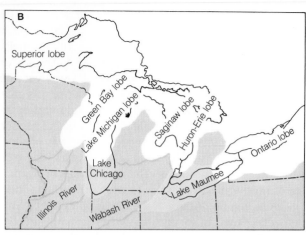

(B) The ancestral Great Lakes appeared about 14,000 years ago, as the ice receded. The northern margins of the lakes were against the retreating ice. Drainage was to the south, to the Mississippi River.

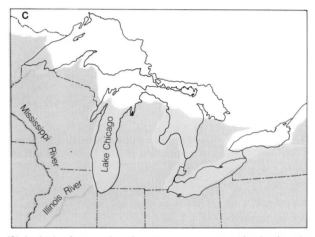

(C) As the ice front continued to retreat, an eastern outlet developed to the Hudson River, but the western lakes still drained into the Mississippi. The lakes began to assume their present outlines about 10,500 years ago.

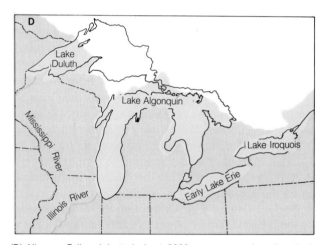

(D) Niagara Falls originated about 8000 years ago, when the glacier receded past the Lake Ontario basin, and water from Lake Erie flowed over the Niagara Escarpment into Lake Ontario.

FIGURE 15.25
The evolution of the Great Lakes can be traced from their origin along the ice margins about 16,000 years ago. The sequence of events and modifications of the landscape are inferred from numerous studies of glacial features in the Great Lakes area.

Pluvial Lakes

The climatic conditions that caused glaciation had an indirect effect on arid and semiarid regions far removed from the large ice sheets. The increased precipitation that fed the glaciers also increased the runoff of major rivers and intermittent streams, resulting in the growth and development of large *pluvial lakes* (Latin *pluvia,* "rain") in numerous isolated basins in nonglaciated areas throughout the world. Most pluvial lakes developed in relatively arid regions where, prior to the glacial epoch, there was insufficient rain to establish an integrated, through-flowing drainage system to the sea. Instead, stream runoff in those areas flowed into closed basins and formed playa lakes. With increased rainfall, the playa lakes enlarged and sometimes overflowed. They developed a variety of shoreline features—wave-built terraces, bars, spits, and deltas—now recognized as high-water marks in many desert basins. Pluvial lakes were most extensive during glacial intervals. During interglacial stages, when less

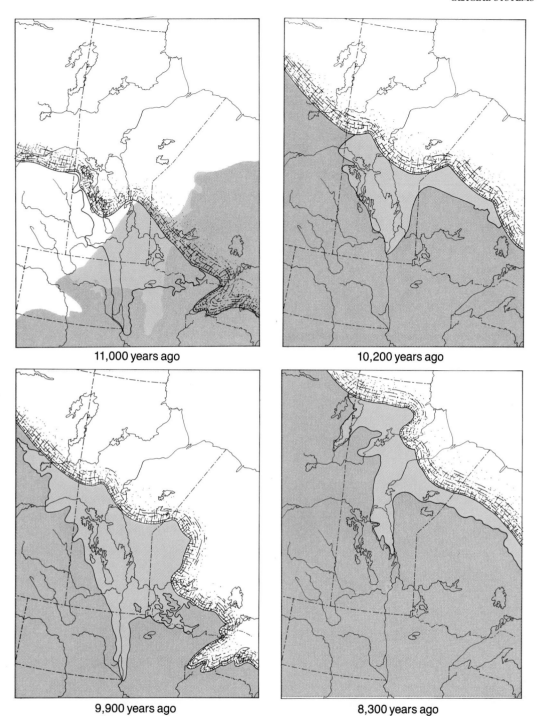

11,000 years ago

10,200 years ago

9,900 years ago

8,300 years ago

FIGURE 15.26
Lake Agassiz was the largest glacial lake in North America. Its former shorelines are now marked by beach ridges, spits, and bars. The dry lake bed now forms the fertile soils of Manitoba and North Dakota. Remnants of this former glacial lake include Lake Winnipeg, Lake Manitoba, and Lake of the Woods.

precipitation fell, the pluvial lakes shrank to form small salt flats or dry, dusty playas.

The greatest concentration of pluvial lakes in North America was in the northern part of the Basin and Range province of western Utah and Nevada. The fault-block structure there has produced more than 140 closed basins, many of which show evidence of former lakes or former high-water levels of existing lakes. The distribution of the former lakes is shown in Figure 15.27.

FIGURE 15.27
Pluvial lakes were formed in the closed basins of the western United States as a result of climatic changes associated with the glacial epoch. Most are now dry lakebeds because of the arid climate. Former shorelines of the pluvial lakes are well marked along the basin margins (see Figure 15.28). Lake Bonneville, in western Utah, was the largest. Its present remnants are Great Salt Lake, Utah Lake, and Sevier Lake.

How did glaciation affect the desert regions of the world?

Lake Bonneville was the largest by far and occupied a number of coalescent intermontane basins. Remnants of this great body of fresh water are Great Salt Lake, Utah Lake, and Sevier Lake. At its maximum extent, Lake Bonneville was approximately the size of Lake Michigan, covering an area of 50,000 km², and was 300 m deep. The principal rivers entered the lake from the high Wasatch Range, to the east. They built large deltas, shoreline terraces, and other coastal features along the mountain front that are now high above the valley floors. The most conspicuous feature is a horizontal terrace high on the Wasatch mountain front (Figure 15.28).

FIGURE 15.28
Shoreline features of Lake Bonneville include deltas, beaches, bars, spits, and wave-cut cliffs. The horizontal terrace shown in the photograph marks the high-water mark of the lake.

As the level of the lake rose to 300 m above the floor of the valley, it overflowed to the north into the Snake River and thence to the ocean. The outlet, established on unconsolidated alluvium, rapidly eroded down to bedrock, 100 m below the original pass. The level of the lake was then stabilized, fluctuating only with the pluvial epochs associated with glaciation. Some valley glaciers from the Wasatch Range extended down to the shoreline of the old lake, and some of their moraines were carved by wave action. This wave erosion shows conclusively that glaciation was contemporaneous with the high level of the lake. As the climate became drier, the lake dried up, leaving faint shorelines at lower levels.

The Channeled Scablands

The continental glacier in western North America moved southward from Canada only a short distance into Washington, but it played an important role in producing a strange complex of interlaced deep channels, a type of topography found nowhere else on Earth. This area, the *Channeled Scablands,* covers much of eastern Washington and consists of a network of braided channels from 15 to 30 m deep. The term *scabland* is appropriately descriptive because, viewed from the air, the surface of the area has the appearance of great wounds or scars (Figure 15.29). Many of the channels have steep walls and dry waterfalls or cataracts. In addition, there are sediment deposits with giant ripple marks and huge bars of sand and gravel (Figure 15.30). These features attest to extreme erosion by running water, a catastrophic flooding by normal standards, yet today the area does not have enough rainfall to maintain a single permanent stream.

How did glaciation produce catastrophic flooding in Washington?

Briefly, the scablands were eroded by the following process. A large lobe of ice advanced southward across the Columbia Plateau and temporarily blocked the Clark Fork River, one of the major northward-flowing tributaries of the Columbia River (Figure 15.31). The impounded water backed up to form glacial Lake Missoula, a long, narrow lake extending diagonally across part of western Montana. Sediments deposited in this lake now partly fill the long, narrow valley. As the glacier receded, the ice dam failed, releasing a tremendous flood over the southwestward-sloping Columbia Plateau. The enormous discharge, barely diverted by the preexisting shallow valleys, spread over the basalt surface, scouring out channels and forming giant ripple marks, bars, and other sediment deposits. Estimates suggest that, during the flood, as much as 40 km^3 of water per hour may have been discharged from Lake Missoula. Because the glaciers advanced several times into the region, such catastrophic flooding probably occurred many times. Lake Missoula formed each time the ice front advanced past the Clark Fork River and then flooded the scablands with each recession of the ice and subsequent dam failure.

Biological Effects of the Ice Age

The severe climatic changes during the ice age had a drastic impact on most life forms. With each advance of the ice, large areas of the continents (the areas beneath the ice) became totally depopulated, and plants and animals retreating southward in front of the advancing glacier were under tremendous stress. The most severe stresses resulted from drastic climatic changes, reduced living space, and a curtailed food supply. As the glaciers advanced, most species were displaced, along with their environments, across distances of approximately 3200 km. As the ice retreated, some new living space became available in deglaciated areas, but the formerly exposed continental shelves were inundated by the rising sea. During the major glacial advances, when sea level was lower, new routes of migration opened from Asia to North America, because much of Alaska and Siberia were not glaciated (see Figure 15.20), and from Southeast Asia to the islands of Indonesia. Land plants were forced to migrate

FIGURE 15.29
The Channeled Scablands of Washington consist of a complex of deep channels cut into the basalt bedrock. The scabland topography is completely different from that produced by a normal drainage system. It is believed to have been produced by "catastrophic" flooding.

FIGURE 15.30
Giant ripples were formed in the surface of a gravel bar by "catastrophic" flooding in the Channeled Scablands.

with the climatic zones in front of the glaciers. As the glaciers pushed cold-weather belts southward, displaced storm tracks and changes in precipitation affected even the tropics.

Many life forms could not cope with the repeated and overwhelming environmental changes brought about by the cycles of advancing and retreating ice. Numerous species, particularly giant mammals, became extinct. During glaciation, the now-extinct imperial mammoth, 4.2 m high at the shoulders,

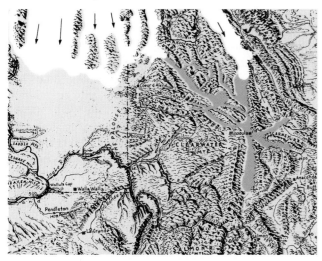

(A) The ice sheet in northern Washington blocked the drainage of the northward-flowing Clark Fork River to form Lake Missoula, a long, deep lake in northern Idaho and western Montana.

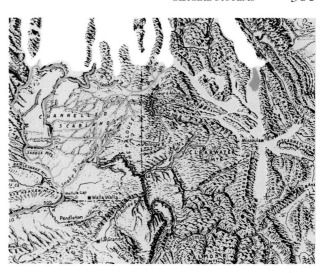

(B) As the glacier receded, the ice dam that formed Lake Missoula failed, and water from the lake quickly flowed across the scablands, eroding deep channels. Repeated advance and retreat of the glacier probably produced several ice dams that failed as the ice melted, each time causing "catastrophic" flooding.

FIGURE 15.31
The origin of the Channeled Scablands is attributed to "catastrophic" flooding, on a magnitude apparently unique in Earth's history. The flood resulted when the ice dam that formed glacial Lake Missoula failed as the glacier receded.

roamed much of North America. The saber-toothed tiger became extinct about 14,000 years ago. Fossils of the giant beaver, as large as a black bear, and the giant ground sloth, which measured 6 m tall standing on its hind legs, have been found in Pleistocene sediments. In Africa, fossil sheep 2 m tall have been found, in addition to pigs as big as a present-day rhinoceros. In Australia, giant kangaroos and other marsupials thrived during the Pleistocene.

Effects of Winds

The presence of ice over so much of the continents greatly modified patterns of atmospheric circulation. Winds near the glacial margins were strong and unusually persistent because of the abundance of dense, cold air coming off the glacier fields. These winds picked up and transported large quantities of loose, fine-grained sediment brought down by the glaciers. This dust accumulated as *loess* (windblown silt), sometimes hundreds of meters thick, forming an irregular blanket over much of the Missouri River valley, central Europe, and northern China.

Sand dunes were much more widespread and active in many areas during the Pleistocene. A good example is the Sand Hills region in western Nebraska, which covers an area of about 60,000 km². This region was a large, active dune field during the Pleistocene, but today the dunes are largely stabilized by a cover of grass.

The Oceans

Pleistocene glaciation affected the waters of all of the oceans to some extent. Besides changing the sea level so that shorelines were altered and much of the continental shelves were exposed, the glacial periods cooled the ocean waters. The lower temperatures affected the kind and distribution of marine life and also influenced seawater chemistry. Furthermore, patterns and strengths of oceanic currents were changed. Circulation was significantly restricted by glacially formed features such as the Bering Strait, extensive pack ice, and exposed shelves.

Even the deep-ocean basins did not escape the influence of glaciation. Where glaciers entered the ocean, icebergs broke off and rafted their enclosed load of sediment out into the ocean. As the ice melted, debris ranging from huge boulders to fine clay settled on the deep-ocean floor, resulting in an unusual accumulation of coarse glacial boulders in fine oceanic mud. Ice-rafted sediment is most common in the Arctic, the Antarctic, the North Atlantic, and the northeastern Pacific.

In the warmer reaches of the oceans, the glacial and interglacial periods are recorded by alternating layers of red clay and small calcareous shells of microscopic organisms. The red mud accumulated during cold periods, when fewer organisms inhabited the colder water. During the warmer interglacial periods, life flourished, and layers of shells mixed with mud were deposited.

RECORDS OF PRE-PLEISTOCENE GLACIATION

Glaciation has been a rare event in Earth's history, but there is evidence of widespread glaciation during late Paleozoic time (200 to 300 million years ago) and during late Precambrian time (600 to 800 million years ago).

Before the great ice age, which began 2 to 3 million years ago and terminated only a few thousand years ago, Earth's climate was typically mild and uniform for long periods of time. This climatic history is implied by the types of fossil plants and animals and by the characteristics of sediments preserved in the stratigraphic record. There are, however, widespread glacial deposits—unsorted, unstratified debris containing striated and faceted cobbles and boulders—recording several major periods of ancient glaciation. These glacial deposits commonly rest on striated and polished bedrock, and they are associated with varved shales and with sandstones and conglomerates that are typical of outwash deposits. Such evidence implies several periods of glaciation in the remote history of Earth.

How do we know that glaciation occurred a number of times prior to the last ice age?

The best-documented record of pre-Pleistocene glaciation is found in late Paleozoic rocks (formed 200 to 300 million years ago) of South Africa, India, South America, Antarctica, and Australia. Exposures of ancient glacial deposits are numerous in these areas, many resting on a striated surface of older rock (Figure 15.32).

Deposits of even older glacial sediment exist on every continent but South America. These indicate that two other periods of widespread glaciation occurred during late Precambrian time.

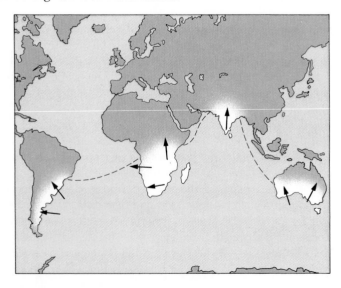

FIGURE 15.32
Late Paleozoic glaciation is well documented in southern continents by deposits of moraine, striated bedrock surfaces, and other glacial features. This map shows the areas covered by ice during the late Pennsylvanian and the Permian periods.

Small bodies of glacial sediment from other geologic periods have been found in local areas, but they are not nearly as well documented or as widespread. Glaciation, therefore, has been a relatively rare phenomenon and has not occurred in regular cycles. One feature seems to be common to all periods of glaciation: the continents appear to have been relatively high and undergoing periods of mountain building. Mountain-building episodes were numerous in the geologic past, but often without accompanying glaciation. This fact suggests that an elevated landmass is not the primary cause of glaciation, but only one of several prerequisites. Glacial epochs must require a special combination of conditions, which has occurred only a few times in the 4.5 billion years of Earth's history.

CAUSES OF GLACIATION

No completely satisfactory theory has been proposed to account for Earth's history of glaciation. The cause of glaciation may be related to several simultaneously occurring factors such as astronomical cycles, plate tectonics, and ocean currents.

Although the history of Pleistocene glaciation is well established and the many effects of glaciation are clearly recognized, we do not know with complete certainty why climates change and why glaciation takes place. For over a century, geologists and climatologists have struggled with this problem, but it remains unsolved. An adequate theory of glaciation must account for the following facts:

1. During the last ice age, repeated advances of the ice in North America and northern Europe were separated by interglacial periods of warm climate. Glaciation, therefore, is not related to a slow process involving long-term cooling.
2. Glaciation is an unusual event in Earth's history. Widespread glaciation also occurred at the end of the Paleozoic era, 200 to 300 million years ago, and during late Precambrian time, approximately 700 million years ago.
3. Throughout most of Earth's history, the climate has been milder and more uniform than it is now. A period of glaciation would require a lowering of Earth's present average surface temperature by about 5°C and, perhaps, an increase in precipitation.
4. Continental glaciers grow on elevated or polar landmasses that are situated so that storms bring moist, cold air to them. Glaciers can move into lower latitudes, but they originate in highlands or in high latitudes. Greenland and Antarctica provide favorable topographic conditions today, as do the Labrador Peninsula, the northern Rocky Mountains, Scandinavia, and the Andes Mountains.
5. Precipitation is critical to the growth of glaciers. A number of areas are cold enough at present to produce glaciers but do not have sufficient snowfall to develop glacial systems.

It has been known for some time that Earth's orbit around the sun changes periodically, forcing Earth to be slightly closer to the sun during some epochs than others. M. Milankovitch, a Yugoslavian geophysicist, convincingly calculated that these irregularities in Earth's orbit could cause climatic cycles. The main period of the cycle is about 100,000 years. In addition, the inclination of Earth's axis varies periodically between 22 and 24 degrees. The tilt of Earth's axis, of course, causes the seasons: the greater the tilt, the greater the contrasts between summer and winter temperatures. The changes in the angle of inclination occur every 41,000 years. Also, Earth wobbles on its axis like a giant top and completes one wobble every 21,000 years. According to the Milankovitch theory, these astronomical factors cause a periodic cooling of Earth, with the coldest part in the cycle occurring about every 40,000 years.

Milankovitch worked out the ideas of climatic cycles in the 1920s and 1930s, but it was not until the 1970s that a sufficiently long and detailed chronology of the Pleistocene was worked out to test the theory adequately. A correspondence between astronomical cycles and late Cenozoic climatic fluctuations now seems clear. Furthermore, studies of deep-sea cores indicate that the fluctuation of climatic cycles is remarkably close to that predicted by Milankovitch.

A problem with this theory is that the astronomical cycles have been in existence for billions of years. One might expect that glaciation would have been a cyclic event throughout geologic time instead of a rare occurrence. Other factors must also be involved. Some scholars have proposed that variations in solar energy may possibly be related to sunspots. Others have suggested variations in atmospheric carbon dioxide. Still others argue that volcanic dust injected into the atmosphere would shield Earth from the sun's rays and initiate an ice age.

A more plausible explanation for the erratic occurrence of the conditions necessary for glaciation relates to the position of the continents relative to the poles and to the circulation of the oceans and atmosphere. Here again the theory of plate tectonics helps to explain how Earth's systems operate.

Throughout most of geologic time, the polar regions appear to have been broad, open oceans that allowed major ocean currents to move unrestricted. Equatorial waters were spread over the polar regions, warming them with water from more temperate latitudes. This unrestricted circulation produced mild, uniform climates, which persisted throughout most of geologic time.

The large North American and South American continental plates moved westward from the Eurasian plate throughout Tertiary time. This drift culminated in the development of the Atlantic Ocean, trending north–south, with the North Pole in the small, nearly landlocked basin of the Arctic Ocean. Meanwhile, by late Miocene time, the Antarctic continent had drifted over the South Pole, and glaciation began on that continent. Evidence from deep-sea cores in the southern oceans strongly suggests that glaciation in the Antarctic began long before the Pleistocene and has continued ever since. By the beginning of Pleistocene time, the present location and configuration of the continents and ocean basins had been established and areas near the polar regions that had adequate precipitation were at the glacial threshold.

It appears that in order for an ice age to occur, there must be the following conditions:

1. Special geographic conditions, that is, continents must be in polar regions.
2. Adequate precipitation.
3. Temperatures low enough so that more snow falls in the winter than melts in the summer.

An important point concerning the question of the origin of the ice age is that there are many variables in the interactions of the climate and the hydrologic system, so no single causative agent can be identified. Apparently, an ice age occurs as a result of several simultaneously occurring factors.

SUMMARY

1. A glacier is a system of flowing ice that originates on land through the accumulation and recrystallization of snow.
2. There are two principal types of glaciers: (1) valley glaciers, which originate in snowfields of high mountain ranges, and (2) continental glaciers, which are sheets of ice thousands of meters thick that cover large parts of continents.
3. Glaciers erode, transport, and deposit vast amounts of rock material and greatly modify the preexisting landscape.
4. The major type of landforms resulting from valley glaciation are (a) U-shaped valleys, (b) cirques, (c) horns, (d) hanging valleys, (e) moraines, and (f) outwash plains.
5. The major landforms resulting from continental glaciation are (a) moraines, (b) drumlins, (c) eskers, (d) outwash plains, (e) kettles, and (f) lake deposits.
6. The Pleistocene ice age began 2 to 3 million years ago and terminated in most areas about 15,000 years ago. During the ice age there were a number of glacial and interglacial epochs.
7. The major effects of the ice age were (a) glacial erosion and deposition over large parts of the continents, (b) isostatic adjustment of the crust, (c) changes in sea level, (d) modification of drainage systems, (e) creation of numerous lakes, (f) catastrophic flooding in Washington, and (g) stress on most life forms.
8. Glaciation has been a relatively rare event in Earth's history, but there were glacial periods during the late Paleozoic, Ordovician, and Precambrian periods.
9. The exact causes of glaciation are not fully understood. In order for glaciation to occur, there must be adequate precipitation and cool temperatures. Many aspects of glaciation may be explained in terms of plate tectonics, oceanic circulation, and astronomical cycles.

REVIEW QUESTIONS

1. Describe the processes by which snow is transformed into glacial ice.
2. Draw a cross section of a typical valley glacier and explain how a valley glacial system operates.
3. Sketch a model of a continental glacial system and explain how it operates.
4. Explain the processes by which glaciers erode the surface over which they flow.
5. Name and describe the landforms produced by valley glaciers.
6. Make a sketch map of North America showing the extent of the ice sheet during Pleistocene time.
7. Briefly describe the major effects, both direct and indirect, of Pleistocene glaciation.
8. Explain the origin of the Channeled Scablands.
9. Compare and contrast the origins of Lake Michigan and the Great Salt Lake.
10. List several hypotheses to explain the cause of continental glaciation.
11. Explain the origin of Hudson Bay.
12. Why did sea level change during each period of advance and retreat of the ice?
13. Explain the origin of the Missouri River.
14. Study Figure 15.21 and explain why the terminal moraines occur in a series of lobate patterns rather than in a straight or gently curving line.
15. How do geologists measure isostatic adjustments of the crust that result from glaciation?
16. Why did a large number of lakes develop in the arid part of the western United States during each major advance of the ice during the Pleistocene ice age?
17. List the periods of major pre-Pleistocene glaciation that are well documented in the geologic record.

KEY TERMS

abrasion (p. 338)

alpine glacier (p. 328)

arête (p. 341)

Channeled Scablands (p. 359)

cirque (p. 341)

continental glacier (p. 332)

crevasse (p. 330)

drift (p. 340)

drumlin (p. 344)

end moraine (p. 344)

esker (p. 344)

glacial plucking (p. 338)

ground moraine (p. 344)

hanging valley (p. 341)

horn (p. 341)

kettle (p. 347)

lateral moraine (p. 344)

loess (p. 361)

medial moraine (p. 344)

moraine (p. 344)

outlet glacier (p. 334)

outwash plain (p. 344)

Pleistocene Epoch (p. 349)

pluvial lake (p. 356)

recessional moraine (p. 344)

roche moutonnée (p. 338)

snowline (p. 330)

striation (p. 338)

terminal moraine (p. 344)

till (p. 340)

valley glacier (p. 328)

varve (p. 344)

zone of ablation (p. 329)

zone of accumulation (p. 329)

ADDITIONAL READINGS

Covey, C. 1984. Earth's orbit and the ice ages. *Scientific American* 8:15.

Denton, G. H., and T. J. Hughes. 1981. *The Last Great Ice Sheets.* New York: Wiley.

Embleton, C., and C. A. M. King. 1975. *Glacial Geomorphology.* London: Edward Arnold.

Flint, R. F. 1971. *Glacial and Quaternary Geology.* New York: Wiley.

Imbrie, J., and K. Imbrie. 1979. *Ice Age: Solving the Mystery.* Short Hills, NJ: Enslow.

Matsch, C. L. 1976. *North America and the Great Ice Age.* New York: McGraw-Hill.

Paterson, W. S. B. 1981. *The Physics of Glaciers,* 2nd ed. Elmsford, NY: Pergamon Press.

Post, A., and E. R. LaChapelle. 1971. *Glacier Ice.* Seattle: University of Washington Press.

Price, R. J. 1972. *Glacial and Fluvioglacial Landforms.* New York: Hafner.

Sharp, R. P. 1960. *Glaciers.* Eugene, OR: University of Oregon Press.

Turekian, K. K., ed. 1971. *The Late Cenozoic Glacial Ages.* New Haven, CT: Yale University Press.

Atlantic Ocean

Boston

Cape Cod

Cape Cod Bay

Providence

New Bedford

Newport

Martha's Vineyard

Nantucket Island

W ater in oceans and lakes is in constant motion. It moves by wind-generated waves, tides, tsunami (seismic sea waves), and a variety of density currents. As it moves, it constantly modifies the shores of all the continents and islands of the world, reshaping coastlines with the ceaseless activity of waves and currents. Shoreline processes can change in intensity from day to day and from season to season, but they never stop.

The present shorelines of the world, however, are not the result of present-day processes alone. Nearly all coasts were profoundly affected by the rise in sea level caused by the melting of the Pleistocene glaciers between 15,000 and 20,000 years ago. The rising sea flooded large parts of the low coastal areas, and shorelines moved inland over landscapes formed by continental processes. The configuration of a given coastline may, therefore, be largely the result of processes other than marine and may originally have been shaped by stream erosion or deposition, glaciation, volcanism, Earth movements, or even the growth of organisms.

The view on the facing page is dominated by the familiar hook of Cape Cod. Boston, Massachusetts, and its suburbs cover many square kilometers along the northern coast, and Providence, Rhode Island, Newport, Rhode Island, and New Bedford, Massachusetts, can be seen along the southern coast. Why did a sandy coast develop on Cape Cod? Why is it so different from the coast around Providence? How will the coast change with time?

In this chapter we will consider these and other questions of coastal dynamics as part of the hydrologic system. Shorelines are especially important to our society because of the concentration of population on or near the coasts. To live in harmony with this rapidly changing environment, we must understand its history and dynamics.

16
Shoreline
Systems

1. Wind-generated waves provide most of the energy for shoreline processes.
2. Wave refraction concentrates energy on headlands and disperses it in bays.
3. Longshore drift, generated by waves advancing obliquely toward the shore, is one of the most important shoreline processes.
4. Erosion along a coast tends to develop sea cliffs by the undercutting action of waves and longshore currents. As a cliff recedes, a wave-cut platform develops, until equilibrium is established between wave energy and the configuration of the coast.
5. Sediment transported by waves and longshore current is deposited in areas of low energy to form beaches, spits, and barrier islands.
6. Erosion and deposition along a coast tend to develop a straight or gently curving shoreline that is in equilibrium with the energy expended upon it.
7. Reefs grow in a special environment and form coasts that can evolve into atolls.
8. The worldwide rise in sea level associated with the melting of the Pleistocene glaciers drowned many coasts between 15,000 and 20,000 years ago.
9. Tides are produced by the gravitational attraction of the Moon and exert a major local influence on shorelines.
10. Tsunami are waves generated by Earth movements that cause disturbance of the sea floor.

WAVES

Shorelines are dynamic systems involving the energy of waves and currents. Wind-generated waves provide most of the energy for erosion, transportation, and deposition of sediment, but tides and tsunami can be locally important.

Most shoreline processes are directly or indirectly the result of wave action. An understanding of wave phenomena is therefore fundamental to the study of shoreline processes. All waves are a means of moving some form of energy from one place to another. This is true of sound waves, radio waves, and water waves. All waves must be generated by some source of energy. By far the most important types of ocean waves are generated by wind.

As wind moves over the open ocean, the turbulent air distorts the surface of the water. Gusts of wind depress the surface where they move downward, and as they move upward, they cause a decrease in pressure, elevating the water surface. These changes in pressure produce an irregular, wavy surface in the ocean and transfer part of the wind's energy to the water. In a stormy area, waves tend to be choppy and irregular, and wave systems of different sizes and orientations may be superposed on each other. As the waves move out from their place of origin, however, the shorter waves move more slowly and are left behind, and the wave patterns develop some measure of order.

Wave Motion in Water

Water waves are described in the same terms as those applied to other wave phenomena. These are illustrated in Figure 16.1. The **wavelength** is the horizontal distance between adjacent **wave crests** or adjacent **wave troughs**. The

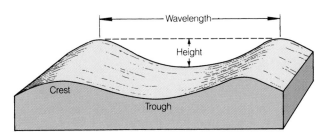

FIGURE 16.1
The morphology of a wave can be described in terms of its length (the distance from crest to crest), height (the vertical distance between crest and trough), and period (the time between the passage of two successive crests).

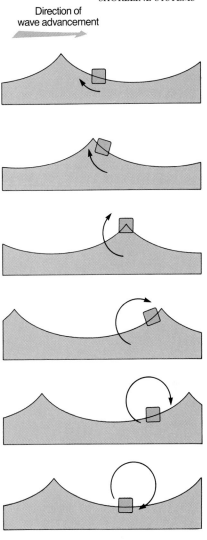

Direction of
wave advancement

FIGURE 16.2
The motion of a water particle as a wave advances is indicated by the movement of a floating object. As the wave advances (from left to right), the object is lifted up to the crest and then returns to the trough. The wave form advances, but the water particles move in an orbit, returning to their original position.

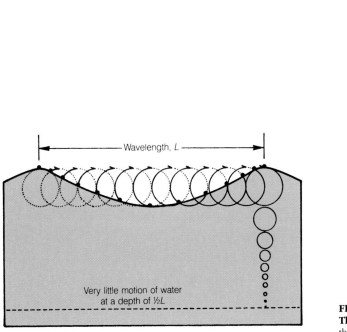

Wavelength, L

Very little motion of water
at a depth of ½L

FIGURE 16.3
The orbital motion of water in a wave decreases with depth and dies out at a depth equal to about half the wavelength.

wave height is the vertical distance between wave crest and wave trough. The time between the passage of two successive crests is called the *wave period,* or frequency.

Wave motion can easily be observed by watching a floating object move forward as the crest of a wave approaches and then sink back into the following trough. Viewed from the side, the object moves in a circular orbit with a diameter equal to the wave height (Figure 16.2). Beneath the surface this orbital motion dies out rapidly, becoming negligible at a depth equal to about one-half the wavelength. This level is known as the *wave base* (Figure 16.3). The motion of water in waves is therefore distinctly different from the motion in currents, in which water moves in a given direction and does not return to its original position.

The energy of a wave depends on its length and height. The greater the wave height, the greater the size of the orbit in which the water moves. The total energy of a wave can be represented by a column of water in orbital motion.

What is the nature of the motion of water in wind-generated waves?

Breakers

Why do breakers occur only along coasts and not in the open ocean?

Wave action produces little or no net forward motion of the water because the water moves in an orbital path as the wave form advances. As a wave approaches shallow water, however, some important changes occur (Figure 16.4). First, the wavelength decreases because the wave base encounters the ocean bottom, and the resulting friction gradually slows down the wave. Second, the wave height increases as the column of orbiting water encounters the sea floor. As the wave form becomes progressively higher and the velocity decreases, a critical point is reached at which the forward velocity of the orbit distorts the wave form. The wave crest then extends beyond the support range of the underlying column of water, and the wave collapses, or breaks. At this point, all of the water in the column moves forward, releasing its energy as a wall of moving, turbulent surf called a *breaker*.

After a breaker collapses, the *swash* (turbulent sheet of water) flows up the beach slope. The swash is a powerful surge, which causes a landward movement of sand and gravel on the beach. After the force of the swash is dissipated against the slope of the beach, the water flows down the beach slope as *backwash*, although some seeps into the permeable sand and gravel.

In summary, waves are generated by the wind on the open ocean. The wave form moves out from the storm area, but the water itself moves in a circular orbit with little or no forward motion. As a wave approaches the shore, it breaks, and the energy of the forward-moving surf is expended on the shore, causing erosion, transportation, and deposition of sediment.

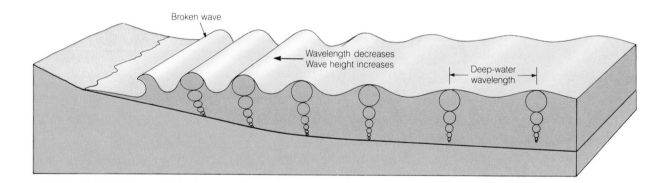

FIGURE 16.4
A wave approaching the shore undergoes several significant changes as the water in orbital motion encounters the sea floor. (1) The wavelength decreases due to frictional drag, and the waves become crowded together as they move closer to shore. (2) The wave height increases as the column of water, moving in an orbit, stacks up on the shallow sea floor. (3) The wave becomes asymmetrical, because of increasing height and frictional drag on the sea floor, and ultimately breaks. The water then ceases to move in an orbit and rushes forward to the shore. Note the change in wave morphology in the time-lapse photos (right to left).

WAVE REFRACTION

Waves approaching a shore are bent, or refracted, so that energy is concentrated on headlands and dispersed in bays.

Wave refraction is a key factor in shoreline processes because it influences the distribution of energy along the shore as well as the direction in which coastal water and sediment move. It occurs because the part of a wave in shallow water begins to drag the bottom and slows down, while the segments of the same wave in deeper water move forward at normal velocity. As a result, the wave is bent, or refracted, so the crest line tends to become parallel to the shore. Wave refraction thus concentrates energy on headlands and disperses it in bays.

To appreciate the effect of wave refraction on the concentration and dispersion of energy, consider the energy in a single wave. In Figure 16.5, the unrefracted wave is divided into three equal parts (AB, BC, and CD), each having an equal amount of energy. As the wave moves toward the shore, segment BC, in front of the *headland,* interacts with the shallow floor first and is slowed down. Meanwhile, the rest of the wave (segments AB and CD) moves forward at normal velocity. This difference in velocity causes the crest line of the wave to bend as it advances shoreward. The wave energy between points B and C is concentrated on a relatively short segment (B′C′) of the headland, whereas the equal amounts of energy between A and B and between C and D are distributed over much greater distances (A′B′ and C′D′). Breaking waves are thus powerful erosional agents on the headlands, but are relatively weak in bays, where they commonly deposit sediment to form beaches. Where major wave fronts are refracted around islands and headlands, the refraction patterns are obvious from the air (Figure 16.6).

How does wave refraction influence erosion and deposition along a shoreline?

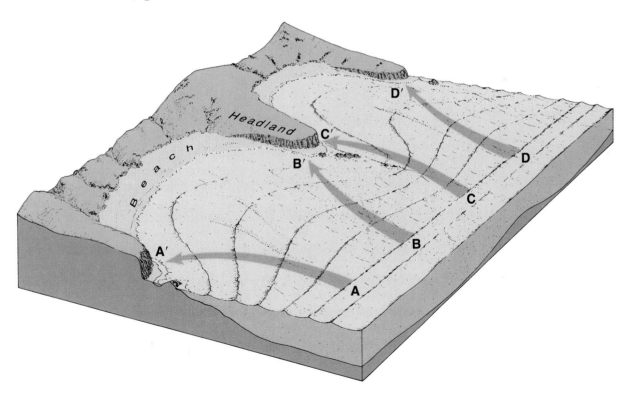

FIGURE 16.5
Wave refraction concentrates energy on headlands and disperses it across bays. Each segment of the unrefracted wave, AB, BC, and CD, has the same amount of energy. As the wave approaches shore, segment BC encounters the sea floor sooner than AB or CD and moves more slowly. This difference in the velocities of the three segments causes the wave to bend, so that the energy contained in segment BC is concentrated on the headland, while the energy contained in AB and CD is dispersed along the beach.

FIGURE 16.6
Wave refraction around headlands and islands is clearly shown in aerial photographs taken along the coast of Oregon. The energy concentrated on the headlands has reduced some of them to offshore islands.

LONGSHORE DRIFT

Longshore drift is generated as waves strike the shore at an angle. Water and sediment move obliquely up the beach face but return with the backwash directly down the beach, perpendicular to the shoreline. This results in a net transport parallel to the shore. As a result an enormous amount of sediment is constantly moving parallel to the shore.

Longshore drift is one of the most important shoreline processes. It is generated as waves advance obliquely to the shore (see Figure 16.7). As a wave strikes the shore at an angle of less than 90 degrees, water and sediment moved by the breaker are transported obliquely up the beach in the direction of the wave's advance. When the wave's energy is spent, the water and sediment return with the backwash directly down the beach, perpendicular to the shore. The next wave moves the material obliquely up the shore again, and the backwash returns it again directly down the beach slope. A single grain of sand is thus moved in an endless series of small steps, with a resulting net transport parallel to the shore. This process is known as **beach drift**. A similar process that develops in the breaker zone is called a **longshore current;** thus, longshore movement occurs in two zones. One is along the upper limits of wave action and is related to the surge and backwash of the waves. The other is in the surf and breaker zone, where material is transported in suspension and by saltation. Both processes work together, and their combined action is called **longshore drift.**

Why is a beach commonly referred to as a river of sand?

Longshore drift results in the movement of an enormous volume of sediment. A beach can be thought of as a river of sand, moving by the action of beach drift. If the wave direction is constant, longshore drift occurs in one direction only. If waves approach the shore at different angles during different seasons, longshore drift is periodically reversed. Longshore currents can pile significant volumes of water on the beach, which return seaward through the breaker zone as a narrow **rip current**. These currents can be strong enough to be dangerous to swimmers.

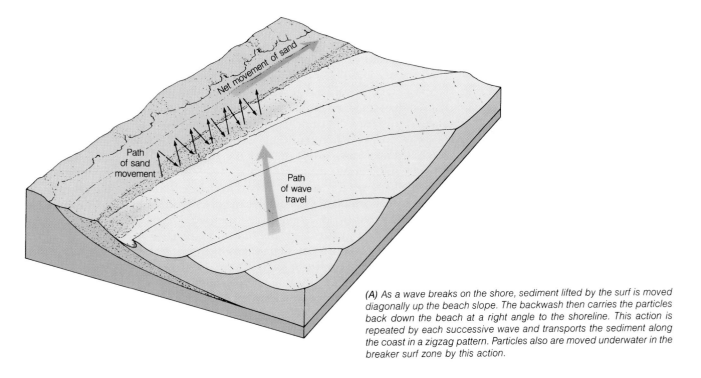

(A) As a wave breaks on the shore, sediment lifted by the surf is moved diagonally up the beach slope. The backwash then carries the particles back down the beach at a right angle to the shoreline. This action is repeated by each successive wave and transports the sediment along the coast in a zigzag pattern. Particles also are moved underwater in the breaker surf zone by this action.

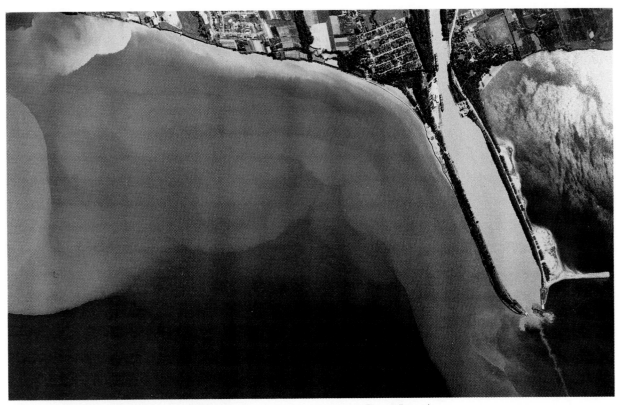

(B) Longshore drift can be seen in patterns of sediment on aerial photographs. The sediment is moved parallel to the shore in a series of waves.

FIGURE 16.7
Longshore drift occurs where waves strike the beach at an oblique angle.

One of the best ways to appreciate the process of longshore drift is to consider how it has influenced human affairs. A good example occurred in the area of Santa Barbara, along the southern coast of California, where data have been collected over a considerable period. Santa Barbara is a picturesque coastal town at the base of the Santa Ynez Mountains. It is an important educational, agricultural, and recreational area, and the people there wanted a harbor that could accommodate deep-water vessels. Studies by the United States Army Corps of Engineers indicated that the site was unfavorable because of the strong longshore currents, which carry large volumes of sand to the south (Figure 16.8A). Rivers draining the mountains of the coastal ranges supply new sediment to the coast at a rate of 592 m^3 per day. Longshore drift continually moves the sand southward from beach to beach. The currents are so strong that boulders of 0.6 m in diameter can be transported. Ultimately, the sand transported by longshore drift is delivered to the head of a submarine canyon and then moves down the canyon to the deep-sea floor.

In spite of reports advising against the project, a breakwater 460 m long was built and a deep-water harbor was constructed in 1925, at a cost of $750,000. This breakwater was not tied to the shore. Sand, moved by longshore drift, began to pour through the gap and fill the harbor, which was protected from wave refraction and longshore currents by the breakwater (Figure 16.8B). To stop the filling of the harbor, it was necessary to connect the breakwater to the shore. Sand then accumulated behind the breakwater, at its southern end.

(A) The Santa Barbara coast had significant longshore drift before the breakwater was built.

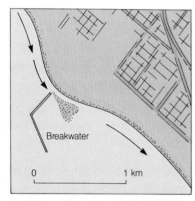

(B) The initial breakwater prevented the generation of longshore currents in the protected area behind it, and therefore the harbor filled with sand.

(C) After the breakwater was connected to the shore, longshore currents moved sand around the breakwater and filled the mouth of the harbor. Sand is now dredged from the harbor and pumped down the coast.

(D) Photo of Santa Barbara Harbor.

FIGURE 16.8
The effect of a breakwater on longshore drift in Santa Barbara, California, is documented by a series of maps of the coast from 1925 to 1938.

Soon a smooth, curving beach developed around the breakwater, and long-shore drift carried sand around the breakwater and deposited it inside the harbor (Figure 16.8C). This produced two disastrous effects: first, the harbor became so choked with sand that it could accommodate only vessels with very shallow draft, and second, the beaches downcoast were deprived of their source of sand and began to erode. Within 12 years, more than $2 million worth of damage had been done to property down the coast from Santa Barbara as the beach in some areas was cut back 75 m. To solve the problem, a dredge was installed in the Santa Barbara harbor to pump out the sand and return it to the longshore drift system on the downcurrent side of the harbor. Most of the beaches have been partly replenished, but dredging costs exceed $30,000 per year.

EROSION ALONG COASTS

Erosion along coasts results from the abrasive action of sand and gravel moved by the waves and currents and, to a lesser extent, from solution and hydraulic action. The undercutting action of waves and currents typically produces sea cliffs. As a sea cliff recedes, a wave-cut platform develops. Minor erosional forms associated with the development of sea cliffs include sea caves, sea arches, and sea stacks.

Coastal regions are sculpted in many shapes and forms, such as rocky cliffs, low beaches, quiet bays, tidal flats, and marshes. The topography of a coast results from the same basic forces that shape other land surfaces: erosion, deposition, tectonic uplift, and subsidence.

Wave action is the major agent of erosion along coasts, and its power is awesome during storms. When a wave breaks against a sea cliff, the sheer impact of the water can exert a pressure exceeding 100 kg/m^2. Water is driven into every crack and crevice of the rocks, compressing the air within. The compressed air then acts as a wedge, widening the cracks and loosening the blocks.

Solution activity also takes place along the coast and is especially effective in eroding limestone. Even noncalcareous rocks can be weathered rapidly by solution activity because the chemical action of seawater is stronger than that of fresh water.

The most effective process of erosion along coasts, however, is the abrasive action of sand and gravel moved by the waves. These tools of erosion operate like the bed load of a river. Instead of cutting a vertical channel, however, the sand and gravel moved by waves cut horizontally, forming wave-cut cliffs and wave-cut platforms.

To understand the nature of wave erosion and the principal features of its forms, first consider what happens along a profile at right angles to a shore (Figure 16.9). Where steeply sloping land descends beneath the water, waves act like a horizontal saw, cutting a notch into the bedrock at sea level (Figure 16.10). This undercutting produces an overhanging *sea cliff,* or *wave-cut cliff,* which ultimately collapses. The fallen debris is broken up and removed by wave action, and the process is repeated on the fresh surface of the new cliff face. As the sea cliff retreats, a *wave-cut platform* is produced at its base, the upper part of which commonly is visible near shore at low tide. Sediment derived from the erosion of the cliff and transported by longshore drift may be deposited in deeper water to form a *wave-built terrace.* Stream valleys that formerly reached the coast at sea level are shortened and left as *hanging valleys* when the cliff recedes.

Why are most shorelines undergoing vigorous erosion? Can coastal erosion be stopped?

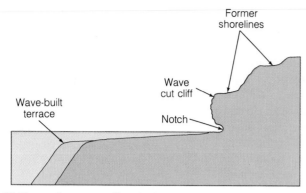

(A) *Wave action operates like a horizontal saw cutting at the base of the cliff. The cliff is undermined and collapses. The debris is soon removed by wave action, and undercutting continues. As erosion continues, the cliff recedes farther, and a gently sloping wave-cut platform is left. Some sediment eroded from the shore can be deposited in deeper water to form a complementary wave-built terrace.*

(B) *Photo of a wave-cut platform along the California coast. Note the flat surface of the uplifted wave-cut platform and the new platform being eroded at sea level.*

FIGURE 16.9
A profile of a wave-cut platform shows the features commonly produced by wave erosion.

FIGURE 16.10
Wave erosion along a coast of Mexico has produced this notch and overhanging cliff. Collapse appears imminent. The process will then be repeated, causing retreat of the sea cliff and development of a wave-cut platform.

As the platform is enlarged, the waves break progressively farther from shore, losing much of their energy by friction as they travel across the shallow platform. Wave action on the cliff is consequently greatly reduced. Beaches can then develop at the base of the cliff, and the cliff face is gradually worn down, mainly by weathering and mass movement. Wave-cut platforms effectively dissipate wave energy and thus limit the size to which they can grow. Some volcanic islands, however, appear to have been truncated completely by wave action and slope retreat, so that only a flat-topped platform is left near low tide.

Sea Caves, Sea Arches, and Sea Stacks

The rate at which a sea cliff erodes depends on the durability of the rock and the degree to which the coast is exposed to direct wave attack. Zones of weakness (such as outcrops with joint systems, fault planes, and beds of shale between harder sandstones) are loci of accelerated erosion. If a joint extends across a headland, wave action can hollow out an alcove, which may later enlarge to a **sea cave.** Because the headland commonly is subjected to erosion from two sides, caves excavated along a zone of weakness can join to form a **sea arch** (Figures 16.11 and 16.12). Eventually, the arch collapses, and an isolated pinnacle called a **sea stack** is left in front of the cliff. In the erosion of a shoreline, marine and terrestrial agents operate together to produce erosion above wave level. The seepage of groundwater, ice wedging, wind, and mass movement all combine with the undercutting action of waves to erode the coast.

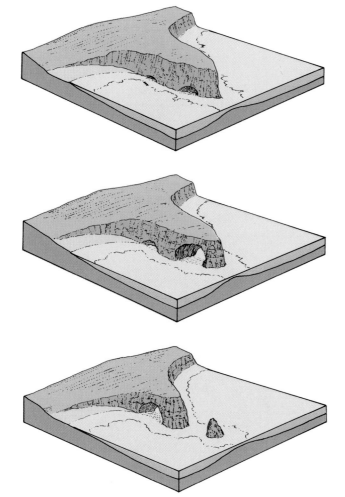

(A) Wave energy is concentrated on a headland as a result of wave refraction. Zones of weakness, such as joints, faults, and nonresistant beds, erode faster, so sea caves develop in those areas.

(B) Sea caves enlarge to form a sea arch.

(C) Eventually, the arch collapses, leaving a sea stack. A new arch can develop from the remaining headland.

FIGURE 16.11
The evolution of sea caves, sea arches, and sea stacks is associated with differential erosion of a headland.

(A) Sea arches and stacks along the coast of southern California, 1969.

(B) Same area as shown in (A), 1987.

FIGURE 16.12
Sea arches and sea stacks along the coast of southern California. The same area was photographed in 1969 (left) and 1987 (right). Erosion eliminated the small sea stack and caused the collapse of the large sea arch.

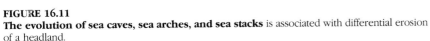

The Evolution of a Wave-Cut Platform

The evolution of erosional features along a coast is shown in Figure 16.13. In the initial stage (Figure 16.13A), sea level rises over a stream-eroded landscape, and an irregular shoreline is formed. Wave action develops a small notch, and abrasion of the platform begins. Continued wave erosion enlarges the platform and develops a high wave-cut cliff (Figure 16.13B). Minor features, such as sea caves, sea arches, and sea stacks, form by differential erosion in weak places in the bedrock. These are continually being formed and destroyed as the sea cliff recedes. In the advanced stage of development (Figure 16.13C), the platform is so enlarged that it absorbs most of the wave energy. Weathering, mass movement, and stream erosion subdue the cliff, and a beach develops as a result of the low energy level along the coast. The net result is a broad wave-cut platform.

In summary, coastal erosion is a natural process that has altered the world's shorelines ever since the oceans were first formed at least 3 to 4 billion years ago. Every day the surging action of waves, the movement of longshore currents, and the pounding of storms erode shorelines dramatically. In addition, sea level is constantly changing; with each rise or fall, a new coastline is formed and the process of reshaping the shore begins anew. Because of the recent rise in sea level due to the melting of the glaciers, many shorelines of the world are several hundred meters higher than they were 30,000 years ago. Vigorous erosion by waves and currents will continue.

The rate at which wave action cuts away at the shore is extremely variable. It depends on the configuration of the coast, the size and strength of the

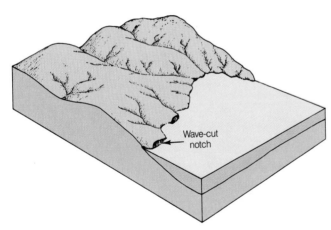

(A) Initial stage. Wave action begins to develop a notch at sea level, which evolves to form a wave-cut cliff.

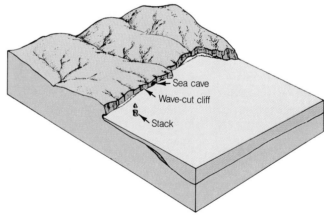

(B) Intermediate stage. Continued wave erosion causes the cliff to recede, and a wave-cut platform develops. Sea stacks, sea arches, and sea caves result from differential erosion along zones of weakness.

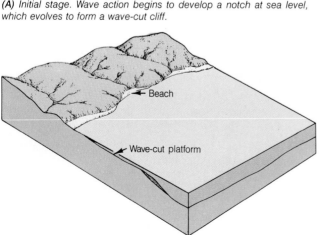

(C) Late stage. The wave-cut platform grows so large that wave energy is dissipated across it. Erosion along the shore is greatly reduced. Beaches develop, and the sea cliff retreats through mass movement.

FIGURE 16.13
The evolution of a shoreline involves a series of stages in which the configuration of the coast is modified by both erosion and deposition until, finally, only a minimum of energy is expended on it. The most stable shoreline form has a smooth, straight coast and a wide platform.

FIGURE 16.14
Erosion of the sea cliff along the coast of southern California is graphically illustrated in this photograph taken near San Diego. Since World War II dozens of homes and hundreds of attendant structures have been lost or damaged as a result of sea cliff recession.

waves, and the physical characteristics of the bedrock. In poorly consolidated material, such as glacial moraines, stream deposits, or sand dunes, the rate of cliff retreat may be as much as 30 m per year, but rates of erosion along most coasts are much slower.

An interesting example of rates of coastal erosion over a longer period of time is documented by maps made by the ancient Romans when they conquered Britain. These maps show that in approximately 2000 years, parts of the British coast have been eroded back more than 5 km, and the sites of many villages and landmarks have been swept away. Other examples of rapid wave erosion are found on new volcanic islands, such as Surtsey, near Iceland. The newly formed volcanic ash that makes up such islands can be completely planed off by wave action in a matter of only a few decades.

The reality of coastal erosion is made painfully clear by the passion of Americans to live and vacation on the seashore. Development projects unwittingly put more and more people and property on the shore, an area that by its very nature is dynamic and mobile (Figure 16.14). About 86% of California's coast is receding at an average rate of .15 to 0.75 m per year. Parts of Monterey Bay lose as much as 2 m to 3 m per year. Cape Shoalwater, Washington, about 100 km west of Olympia, has been eroding at a rate of more than 30 m per year. Parts of Chambers County, Texas, have lost 3 m of coast in nine months. In North Carolina erosion in one year has cut into beach-front property up to 25 m in places.

To combat these losses, sea walls and breakwaters have been erected, but these are local and temporary solutions at best. A sea wall or jetty (a long concrete or rock structure that juts out into water to restrain waves and currents) may protect threatened property near it, but it often hastens erosion in other areas. There appears to be no simple answer. Coasts are dynamic features, and by their very nature are mobile. In our battle with nature, retreat might be the ultimate solution.

DEPOSITION ALONG COASTS

Sediment transported along the shore is deposited in areas of low wave energy and produces a variety of landforms, including beaches, spits, tombolos, and barrier islands.

What is the source of sand on beaches?

A shoreline is a dynamic system that involves input of sediment from various sources, transportation of the sediment, and ultimate deposition. Much of the sediment is derived from the land and delivered to the sea by major rivers. The sediment is then transported by waves and longshore currents and is deposited in areas of low energy, where it builds a variety of landforms, including beaches, spits, tombolos, and barrier islands. Shoreline material is in constant motion, and the configuration of the coast is continually modified by both erosion and deposition. Changes continue until the configuration of the coast is at equilibrium with the available wave energy. The final configuration is usually a smooth and straight, or gently curving, coastline.

Figure 16.15 shows some of the important elements in a coastal system. The primary sources of sediment for beaches and associated depositional features are the rivers that drain the continents. Sediment from the rivers is transported along the shore by longshore drift and is deposited in areas of low energy. Erosion of headlands and sea cliffs is also a source of sediment. In tropical areas, the greatest source of sand commonly is shell debris derived from wave erosion of nearshore coral reefs. Sediment can leave the system by landward migration of coastal sand dunes and by transportation into deep areas of the ocean floor, where it accumulates as submarine deposits.

Beaches

A **beach** is a shore built of unconsolidated sediment. Sand is the most common material, but some beaches are composed of cobbles and boulders and others of fine silt and clay. The physical characteristics of a beach (such as slope, composition, and shape) depend largely on wave energy, but the supply and size of available sediment particles are also important. Beaches composed of fine-grained material generally are flatter than those composed of coarse sand and gravel.

Spits

In areas where a straight shoreline is indented by **bays** or **estuaries,** longshore drift can extend the beach from the mainland to form a **spit.** A spit can grow far out across the bay as material is deposited at its end (Figure 16.16). Eventually, it may extend completely across the front of the bay, forming a **bay-mouth bar.**

Tombolos

Beach deposits can also grow outward and connect the shore with an offshore island to form a **tombolo.** This feature commonly is produced by the island's effect on wave refraction and longshore drift (Figure 16.17). An island near shore can cause wave refraction to such an extent that little or no wave energy strikes the shore behind it. Longshore drift, which moves sediment along the coast, is not generated in this wave shadow zone. Sediment carried by longshore currents is therefore deposited behind the island. The sediment deposit builds up and eventually forms a tombolo, a bar or beach connecting the shore to the island. Longshore currents then move uninterrupted along the shore and around the tombolo.

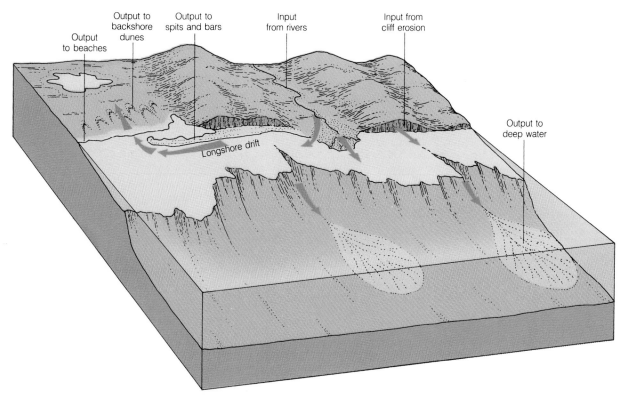

FIGURE 16.15
A shoreline is a dynamic system of moving sediment. Most of the sediment in a shoreline system is supplied by rivers bringing erosional debris from the continent and by the erosion of sea cliffs by wave action. This material is transported by longshore drift and can be deposited on growing beaches, spits, and bars. Some sediment, however, leaves the system either by transportation to deeper water or by the landward migration of coastal sand dunes.

FIGURE 16.16
Curved spits develop as longshore drift moves sediment parallel to the shoreline.

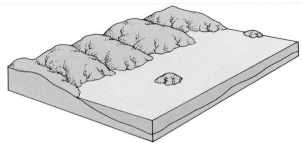

(A) An offshore island acts as a breakwater to incoming waves and creates a wave shadow along the coast behind it.

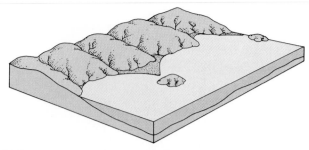

(B) Sediment moved by longshore drift is trapped in the shadow zone.

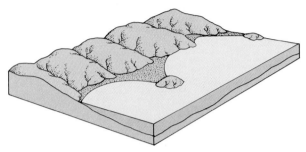

(C) The zone of sediment deposit eventually grows until it connects with the island. Longshore drift will then move sediment along the shore and around the tombolo.

FIGURE 16.17
A tombolo is a bar or beach that connects an island to the mainland. It forms because the island creates a wave shadow zone along the coast, in which longshore drift cannot occur.

(D) An aerial photograph of a tombolo.

Barrier Islands

Barrier islands are long, low offshore islands of sediment trending parallel to the shore (Figure 16.18). Almost invariably, they form long shorelines adjacent to gently sloping coastal plains, and they typically are separated from the mainland by a lagoon. Most barrier islands are cut by one or more tidal inlets.

The origin of barrier islands has been a controversial subject for many years. One theory contends that they result from the growth of spits across irregularities in the shore (Figure 16.19). Certainly, many barrier islands develop in this manner. Others may form by the shoreward migration of offshore bars. Another hypothesis contends that barrier islands formed when the worldwide rise of sea level drowned early-formed beaches and isolated them offshore. This process is illustrated in Figure 16.20. During a period when sea level was lower, such as during the ice age, a beach formed along a low coastal plain. Beach sand was piled up in a dune ridge parallel to the shore. As sea level rose with the melting of the ice, the dune ridge was drowned and became the nucleus for a barrier island. Later, wave action and longshore drift reworked, enlarged, and modified the dune ridge into a barrier island.

Transportation and deposition along many coasts can be measured using historical monuments, maps, and sequences of aerial photography. In northern France a dike built at the shoreline in 1597 is now more than 3 km inland from the present shore, indicating an average rate of spit migration of about 1 km per 100 years.

Other rates of spit migration based on dated maps include the western end of Fire Island (off the south coast of Long Island, New York), the Rockaway spit (western Long Island), and Sandy Hook (along the coast of New Jersey). The rate of lateral migration for both Long Island spits is 65 m per year, or 6.5 times the rate of migration along the French coast. Sandy Hook has migrated about 12 m per year, nearly the same rate as on the French coast.

FIGURE 16.18
A barrier island along the Atlantic coast has a smooth seaward face where wave action and longshore drift actively transport sediment. A tidal inlet forms a break in the island, and sediment transported through it is deposited as a tidal delta in the lagoon.

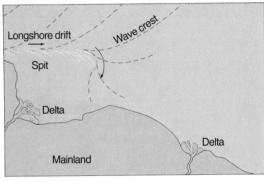

(A) Sediment moving along the shore is deposited as a spit in the deeper water near a bay.

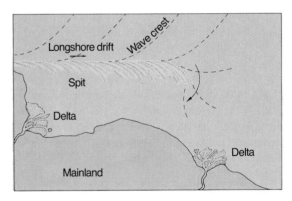

(B) The spit grows parallel to the shore by longshore drift.

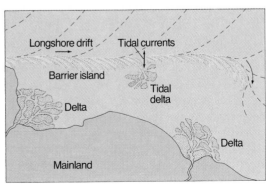

(C) Tidal inlets cut the spit, which is then long enough to be considered a barrier island.

FIGURE 16.19
A barrier island may form by migration of a spit.

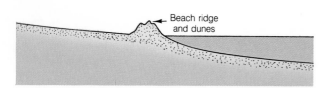

(A) On a gently inclined coast, wave action builds a beach ridge at sea level. The ridge can grow to be over 30 m high as sand is piled up by wind action.

(B) A rise in sea level due to the melting of glaciers drowns the beach ridge and produces a barrier island, which is then modified by wave action.

FIGURE 16.20
A barrier island may also develop as sea level rises over beach ridges.

EVOLUTION OF SHORELINES

> *Processes of shoreline erosion and deposition tend to develop a long straight or gently curving coastline. Headlands are eroded back, and bays and estuaries are filled with sediment. The configuration of the shoreline evolves until wave energy is distributed equally along the coast and neither large-scale erosion nor deposition occurs.*

If Earth is so old, why are shorelines still changing?

All of the coastlines throughout the world are constantly changing. In many areas changes are rapid, and within only a few decades the local configuration of a shoreline can be significantly modified. Over a longer period of time, regional variations in coastal configuration occur. This constant and rapid change in our coasts is due in part to the recent rise in sea level that accompanied the melting of glaciers. Other changes in the coast result from uplift or subsidence of the land or expansion or contraction of the sea. Thus, the configuration of most coastlines is far from being at equilibrium with the wave energy expended upon them. The general trend is for headlands to be eroded back and bays and estuaries to become filled with sediment. The change in the configuration of the shoreline is always in the direction such that energy is equally distrbuted along the shore and neither large-scale erosion nor deposition occurs.

The energy of the waves and longshore drift is just sufficient to transport the sediment that is supplied. A shoreline with such a balance of forces is called a shoreline of equilibrium. As is the case with a stream profile of equilibrium, a delicate balance is maintained between the landforms and the geologic processes operating on them.

What is the shape of a coastline that has reached equilibrium?

We can construct a simple conceptual model of a shoreline's evolution toward equilibrium and show the types of changes that would be expected to occur as the processes of erosion and deposition operate. One such model is shown in Figure 16.21. Diagram A shows an area originally shaped by stream erosion and subsequently partly drowned by rising sea level. River valleys are invaded by the sea to form irregular, branching bays, and some hilltops form peninsulas and islands. Next, as is shown in Figure 16.21B, marine erosion begins to attack the shore. The islands and headlands are eroded into sea cliffs. As erosion proceeds (Figure 16.21C), the islands and headlands are worn back, and the cliffs increase in height. A wave-cut platform enlarges, reducing wave energy, so a beach forms at the base of the cliff. In a more advanced stage of development (Figure 16.21D), the islands are eroded away and bays become sealed off, partly by the growth of spits, forming lagoons. The shoreline then becomes straight and simple. In the final stages of marine development (Figure 16.21E), the shoreline is cut back beyond the limits of the bay. Sediment moves along the coast by longshore drift, but the wave-cut platform is so wide that it

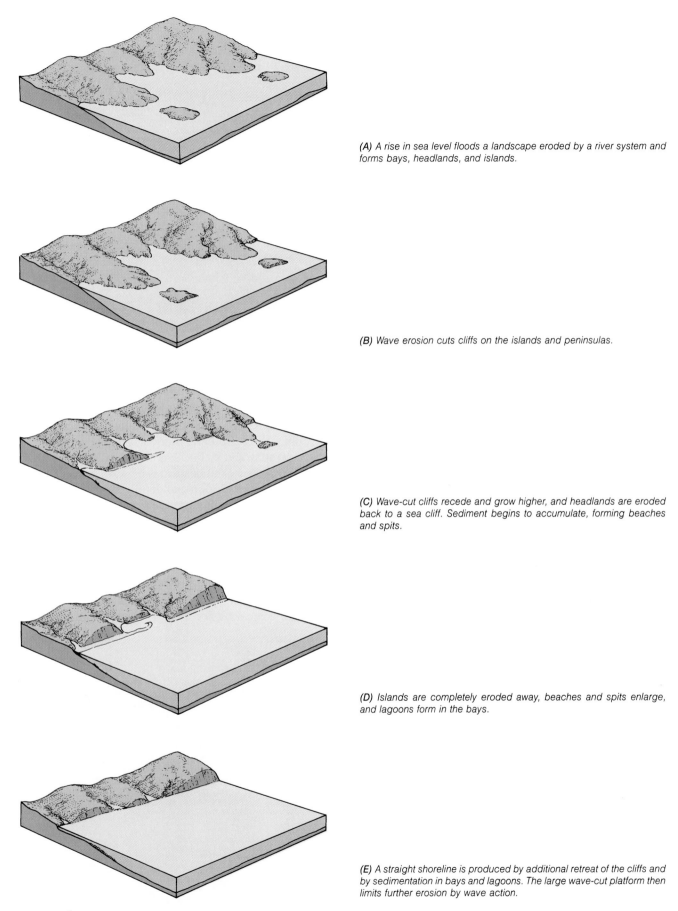

(A) A rise in sea level floods a landscape eroded by a river system and forms bays, headlands, and islands.

(B) Wave erosion cuts cliffs on the islands and peninsulas.

(C) Wave-cut cliffs recede and grow higher, and headlands are eroded back to a sea cliff. Sediment begins to accumulate, forming beaches and spits.

(D) Islands are completely eroded away, beaches and spits enlarge, and lagoons form in the bays.

(E) A straight shoreline is produced by additional retreat of the cliffs and by sedimentation in bays and lagoons. The large wave-cut platform then limits further erosion by wave action.

FIGURE 16.21
The evolution of a shoreline of equilibrium from an embayed coastline involves changes due to both erosion and deposition. Eventually, a smooth coastline is produced, and the forces acting on it are essentially at equilibrium, so neither erosion nor deposition occurs on a large scale.

FIGURE 16.22
The effect of a hurricane along a coastal region is dramatically shown in this photograph taken along the Texas Gulf Coast. Huge quantities of sand were washed over the barrier island and deposited in the lagoon.

effectively eliminates further erosion of the cliff by wave action. The *shoreline of equilibrium* is straight and essentially in equilibrium with the energy acting on it. Further modification of the cliffs results from weathering and stream erosion; also, hurricanes can cause temporary modification (Figure 16.22).

Naturally, the development of a shoreline is also affected by special conditions of structure and topography and by fluctuations of sea level or tectonics. However, the process of erosion of the headlands by wave action and the straightening of the shoreline by both erosion and deposition follow the general sequence of this idealized model, although actual shorelines rarely proceed through all of these stages because fluctuations of sea level upset the previously established balance.

The development of a shoreline is interrupted in many areas by tectonic uplift, which abruptly elevates sea cliffs and wave-cut platforms above the level of the waves. When this happens, wave erosion begins at a new, lower level, and the *elevated marine terraces,* stranded high above sea level, are attacked and eventually obliterated by weathering and stream erosion (Figure 16.23).

FIGURE 16.23
A series of elevated beach terraces resulted from tectonic uplift along the southern coast of California and the offshore islands. This photograph of San Clemente Island was taken with the sun at low angle to emphasize the sequence of terraces.

REEFS

Reefs form a unique type of coastal feature because they are of organic origin. Modern reefs are built by a complex community of corals, algae, sponges, and other marine invertebrates. Most grow and thrive only in the warm, shallow waters of semitropical and tropical regions.

In many regions of the ocean, coral reefs grow and flourish to such an extent that they significantly modify, if not control, the configuration of a coastline. This is especially true in the warm tropical waters of the South Pacific, where reefs are a major influence along the coasts of most islands (Figure 16.24). Reefs are constructed from invertebrate colonial animals, which instead of building separate isolated shells, build enormous "apartment houses" in which thousands of individuals live. When the animals die, the shell structure remains intact and subsequent generations build their apartments upon the abandoned homes of their predecessors; soon a reef develops into a "rocky" coast. The most impressive modern reef is Australia's Great Barrier Reef, which stretches a distance of 1242 miles (2000 km). Others are noted throughout the South Pacific as barrier reefs and atolls surrounding ancient volcanic islands.

Reef Ecology

The marine life that forms a reef can flourish only under strict conditions of temperature, salinity, and water depth. Most modern *coral* reefs occur in warm

FIGURE 16.24
A barrier reef in the Society Islands, French Polynesia, is typical of the intermediate stage in the evolution of an atoll. Note the outer margin of the reef, where growth of organisms is most active. The shallow lagoon inside the reef, shown in light blue, is mostly calcareous sand formed by erosion of the reef. The remnant volcano in the center is highly dissected by stream erosion, indicating a long period of time has elapsed since the volcano was active.

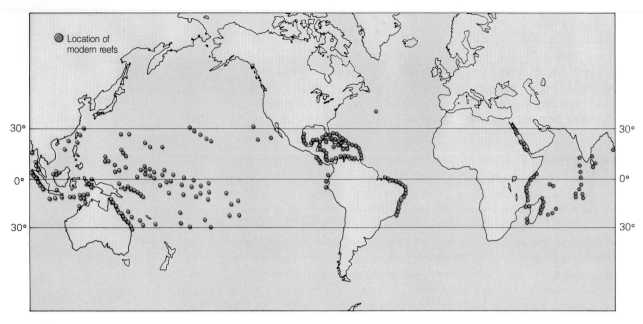

FIGURE 16.25
The distribution of coral reefs is restricted to the low-latitude areas where the average water temperature exceeds 18 °C throughout the year. Reefs are widely developed throughout the Pacific and Indian oceans.

tropical waters between the limits of 30 degrees south latitude and 30 degrees north latitude (Figure 16.25). Colonial corals need sunlight, and they cannot live in water deeper than about 76 m. They grow most luxuriantly just a few meters below sea level. Dirty water inhibits rapid, healthy growth because it cuts off sunlight and the suspended mud chokes the organisms that filter feed. Corals are therefore absent or stunted near the mouths of large, muddy rivers. They can survive only if the salinity of the water ranges from 27 to 40 parts per thousand; thus, a reef can be killed if a flood of fresh water from the land appreciably reduces the salinity. Coral reefs are remarkably flat on top, the upper surface being positioned at the level of the upper third of the tidal range. They usually are exposed at low tide but must be covered at high tide. In summary, corals thrive only in clear, warm, shallow oceans where wave action brings sufficient oxygen and food.

Reefs can grow upward with rising sea level if the rate of rise is not excessive. They can also grow seaward over the flanks of reef debris.

The fact that reefs form in such restricted environments makes them especially important as indicators of past climatic, geographic, and tectonic conditions.

Types of Reefs

The most common types of reefs in present oceans are fringing reefs, barrier reefs, and atolls (Figure 16.26).

Fringing reefs, generally ranging from 0.5 to 1 km wide, are attached to such landmasses as the shores of volcanic islands. The corals grow seaward, toward their food supply. They are usually absent near deltas and mouths of rivers, where the waters are muddy. Heavy sedimentation and high runoff also make some tropical coasts of continents unattractive to fringing reefs.

Barrier reefs are separated from the mainland by a lagoon, which can be more than 20 km wide. As seen from the air, the barrier reefs of islands in the South Pacific are marked by a zone of white breakers. At intervals, narrow gaps occur, through which excess shore and tidal water can exit. The finest example of this type is the Great Barrier Reef, which stretches for 800 km along the northern shore of Australia, from 30 to 160 km off the Queensland coast.

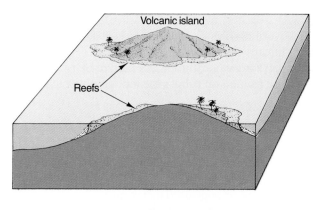

(A) *A reef begins to grow along the coast of a newly formed volcanic island.*

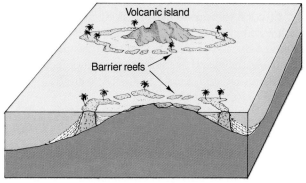

(B) *As the island subsides, the reef grows upward and develops a barrier that separates the lagoon from open water.*

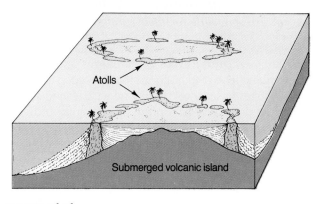

(C) *Further subsidence completely submerges the island, but if subsidence is not too rapid, the reef continues to grow upward to form an atoll.*

FIGURE 16.26
The evolution of an atoll from a fringing reef was first recognized by Charles Darwin. The theory assumes that continued slow subsidence of the ocean floor allows the reef to continue growing upward.

Atolls are roughly circular reefs that rise from deep water, enclosing a shallow lagoon in which there is no exposed central landmass. The outer margin of an atoll is naturally the site of most vigorous coral growth. It commonly forms an overhanging rim, from which pieces of coral rock break off, accumulating as submarine talus on the slopes below. A cross-sectional view of a typical atoll shows that the lagoon floor is shallow and is composed of calcareous sand and silt with rubble derived from erosion of the outer side (see the foreground of Figure 16.26).

Atolls are by far the most common type of coral reef. Over 330 are known, of which all but 10 lie within the Indo-Pacific tropical area. Drilling into the coral of atolls tends to confirm the theory that atolls form on submerged volcanic islands (see the following section, "The Origin of Atolls"). In one instance, coral extends down as much as 1400 m below sea level, where it rests on a basalt platform carved on an ancient volcanic island. Because coral cannot grow at that depth, it presumably grew upward as the volcanic island sank. A reef this thick probably accumulated over 40 or 50 million years.

What types of shorelines are built by living organisms?

Platform reefs grow in isolated oval patches in warm, shallow water on the continental shelf. They were apparently more abundant during past geologic periods of warmer climates. Most modern platform reefs seem to be randomly distributed, although some appear to be oriented in belts. The latter feature suggests that they were formed on submarine topographic highs, such as drowned shorelines.

The Origin of Atolls

In 1842, Charles Darwin first proposed a theory to explain the origin of atolls. As is indicated in Figure 16.26, the theory is based on the continued relative subsidence of a volcanic island. Darwin suggested that coral reefs are originally established as fringing reefs along the shores of new volcanic islands. As the island gradually subsides, the coral reef grows upward along its outer margins. The rate of upward growth essentially keeps pace with subsidence. With continued subsidence, the area of the island becomes smaller, and the reef becomes a barrier reef. Ultimately, the island is completely submerged, and the upward growth of the reef forms an atoll. Erosional debris from the reef fills the enclosed area of the atoll to form a shallow lagoon.

TYPES OF COASTS

On a global scale, coasts can be classified on the basis of their tectonic setting. On a local scale, coasts are classified on the basis of the process most responsible for their configuration.

Why are there so many different types of coastlines?

Nearly all coasts are complex, both in the types of landforms and in their geologic history. All are dependent upon the landforms that preceded them, all are subject to the effects of changes in sea level during the ice age, and all are influenced by the operation of present coastal processes. In spite of these complexities, insight into the understanding of coasts can be gained by considering them in light of the processes responsible for their configuration. To do this effectively, we must consider coastal features on two different scales:

1. Features on a global scale—those associated with tectonic plates
2. Features on a local scale—those that are mainly associated with processes of erosion and deposition, which modify the regional features

In tectonic terms, on a global scale, coasts can be classified as follows:

1. Convergence coasts
   a. Continental convergence
   b. Island arc convergence
2. Passive-margin coasts
3. Marginal sea coasts

On a smaller scale, coastal features are often associated with local geologic processes—erosion, deposition, volcanism, and so on. Two principal subdivisions are recognized:

1. Primary coasts shaped mainly by terrestrial processes (stream erosion, deposition, and so on)
2. Secondary coasts shaped by marine processes

Classification Based on Plate Tectonics

Convergence Coasts. The classification of the world's coastlines on the basis of tectonics is shown in Figure 16.27. The convergence coasts are all relatively straight and mountainous and are further characterized by sea cliffs,

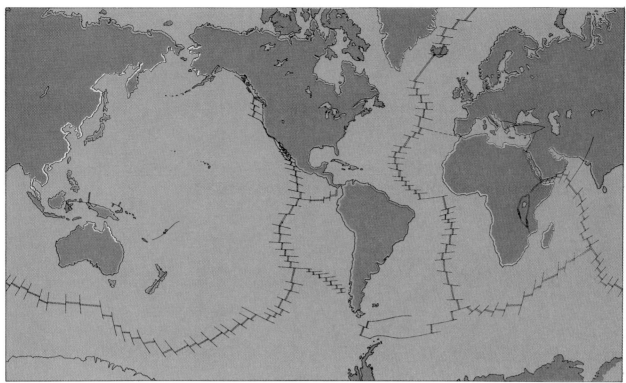

FIGURE 16.27
The tectonic classification of coasts is based upon the tectonic setting of the continental margins. Passive margins characterize the eastern coasts of North and South America, Africa, and Australia. Mountainous coasts are typical of converging plate margins.

Convergence coast
Passive-margin coast — American type
Passive-margin coast — African type
Marginal seas

raised marine terraces, and narrow continental shelves. They typically contain zones of active volcanism and seismicity. The west coasts of North and South America are excellent examples of continental convergence coasts. The Aleutian Islands and Japan are typical convergence coasts of island arcs.

Passive-Margin Coasts. Passive-margin coasts are formed by rifting. As the oceans spread and the continents move apart, the coasts are tectonically passive. In the initial stages of rifting, the coasts may be topographically high and marked by steep escarpments. With time, the coast subsides as it moves off the uparched spreading center. Sediment derived from erosion of the mountain belt on the active margin is transported to the passive margin, where it is deposited and helps to construct a broad continental shelf. Typically, passive-margin coasts are low and have been modified by marine deposits.

The east coasts of the Americas are typical passive-margin coasts. They have low relief and broad coastal plains and are bordered by a wide continental shelf. The coasts of the Red Sea and Baja California are also passive-margin coasts, but they were formed by recent uplift and rifting and, as a result, have high cliffs and narrow continental slopes. The passive-margin coasts of Africa, India, and Greenland are distinct in that both the east and west coasts face spreading centers, and the coasts have relatively high relief.

Marginal Sea Coasts. Marginal sea coasts have the greatest diversity. They are protected from the open ocean by island arcs and are frequently modified by large rivers and their deltas; Vietnam and southern China are examples.

Classification Based on Geologic Processes

The physical characteristics of second-order coastal features (lengths of 100 km) modify the characteristics of the configuration produced by tectonics.

They are highly diverse and are commonly controlled by local erosion and deposition or by the growth of reefs. These effects are superposed on the first-order tectonic features. In some cases, such as glacial erosion or the formation of large deltas, second-order features may predominate, completely masking the first-order effects. The particular climatic region of Earth is a major controlling factor influencing second-order coasts because it controls glacial systems, locations of major deltas, and the growth and destruction of reefs.

Primary Coasts

What factors determine the configuration of a shoreline?

The configuration of *primary coasts* is largely the result of subaerial geologic agents, such as streams, glaciers, volcanism, and Earth movements. These produce highly irregular coastlines characterized by bays, estuaries, fiords, headlands, peninsulas, and offshore islands. The landforms can be either erosional or depositional but are only slightly modified by marine processes. Some of the more common types are illustrated in Figure 16.28.

Stream Erosion Coasts.   If an area eroded by running water is subsequently flooded by the rise of sea level, the landscape becomes partly drowned. Stream valleys become bays or estuaries, and hills become islands. The bays extend up the tributary valley system, forming a coastline with a dendritic pattern. Chesapeake Bay is a well-known example.

Stream Deposition—Deltaic Coasts.   At the mouths of major rivers, fluvial deposition builds deltas out into the ocean. The deltas dominate the configuration of the coast. They can assume a variety of shapes and are locally modified by marine erosion and deposition.

Glacial Erosion Coasts.   Drowned glacial valleys usually are called *fiords.* They form some of the most rugged and scenic shorelines in the world. Fiords are characterized by long, troughlike bays that cut into mountainous coasts, extending inland as much as 100 km. In polar areas, glaciers still remain at the heads of many fiords. The walls of fiords are steep and straight. Hanging valleys with spectacular waterfalls are common.

Glacial Deposition Coasts.   Glacial deposition dominates some coastlines in the northern latitudes, where continental glaciers once extended beyond the present shoreline, over the continental shelf. The ice sheets left drumlins and moraines, to be drowned by the subsequent rise in sea level. Long Island, for example, is a partly submerged moraine. In Boston Harbor, partly submerged drumlins form elliptical islands.

As erosion and deposition continue, marine processes ultimately control the coastal configuration, and primary coasts become secondary coasts.

Secondary Coasts

Secondary coasts are shaped by marine erosion and deposition. They are characterized by wave-cut cliffs, beaches, barrier islands, spits, and (in some cases) sediment deposited through the action of biological agents, such as marsh grass, mangroves, and coral reefs. Marine erosion and deposition smooth out and straighten shorelines and establish a balance between the energy of the waves and the configuration of the shore. The most common types of secondary coasts are illustrated in Figure 16.29.

Wave Erosion Coasts.   Wave erosion begins to modify primary coasts as soon as the landscape produced by other agents is submerged. Wave energy is concentrated on the headlands, and a wave-cut platform develops slightly below sea level. Ultimately a straight cliff is formed, with hanging

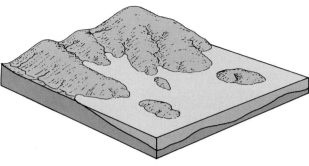

(A) Stream erosion produces an irregular, embayed coast with offshore islands.

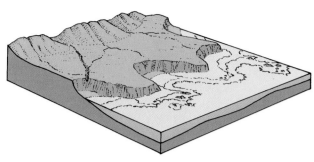

(A) Marine erosion produces wave-cut cliffs.

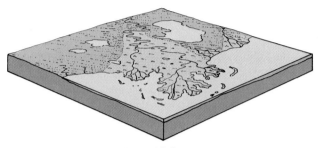

(B) Stream deposition produces deltaic coasts.

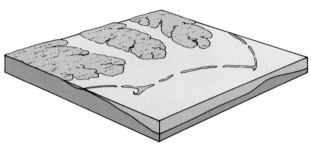

(B) Marine deposition produces barrier islands and beaches.

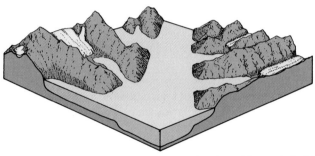

(C) Glacial erosion produces long, narrow, deep bays (drowned glacial valleys) called fiords.

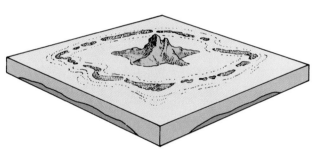

(C) The growth of coral reefs produces barrier reefs and atolls.

FIGURE 16.28
Primary coasts are those in which the configuration of the shoreline is produced by nonmarine processes. Rivers and glaciers are the most important processes forming this type of coast.

FIGURE 16.29
Secondary coasts are those in which the configuration of the shoreline is produced by marine processes, including erosion, deposition, and the growth of organisms.

stream valleys and a large wave-cut platform. The cliffs of Dover, England, are a prime example.

Marine Deposition Coasts. Where abundant sediment is supplied by streams or ocean currents, marine deposits determine the characteristics of the coast. Barrier islands and beaches are the dominant features. The shoreline is modified as waves break across the barriers and transport sand inland. The barriers also increase in length and width as sand is added. The lagoons behind the barriers receive sediment and fresh water from streams; thus, they are often capable of supporting dense marsh vegetation. Gradually, a lagoon fills with stream sediment, with sand from the barrier bar (which enters through tidal deltas), and with plant debris from swamps. The barrier coasts of the southern Atlantic and Gulf Coast states are excellent examples.

Coasts Built by Organisms. Coral reefs develop a type of coast that is prominent in the islands of the southwestern Pacific. The reefs are built up to the surface by corals and algae, and they can ultimately evolve into an atoll.

Another type of organic coast that is prevalent in the tropics is formed by intertwined root systems of mangrove trees, which grow in the water, particularly in shallow bays.

As an example of coastal classification based on tectonics and morphology, let us consider the coasts of North America and how they fit into this general scheme (Figure 16.27).

The west coast from the Aleutians to Central America is a convergent coast. The Aleutian segment forms an island arc in which volcanic islands dominate, but from Alaska to Mexico, the coast is one of continental collision. The coast is tectonically active, involving volcanic eruptions and earthquakes, and it is mountainous, so on a regional scale the coast rises rapidly out of the sea. From Alaska to northern Washington, the west coast has been significantly influenced by glaciation. Throughout Canada and southern Alaska, the coast is mountainous, and the imprint of valley glaciation dominates. Long fiords reflect a network of U-shaped glacial valleys that have not been significantly modified by wave action.

Southward, beyond the influence of glaciation, in Washington, Oregon, and California, stream and wave erosion are the dominant processes responsible for the local configurations of the shoreline. Old sea cliffs and wave-cut terraces have been elevated high above the present sea level, in places as much as 400 m.

The Gulf of California is a local segment of the west coast that is not tectonically a convergent coast. It is a passive-margin coast recently formed by the rifting of Baja California from Mexico. The region has high relief and is protected from strong waves, so small deltaic coasts develop locally.

The east coast of North America stands out in strong contrast to the west coast. Throughout its entire length it is a classic passive-margin coast. In the Gulf of Mexico the coast is restricted from the Atlantic by the Yucatan Peninsula and the island arc of the Caribbean and is therefore considered to be a marginal sea coast. The Mississippi Delta has had a profound influence on the local configuration of the coast throughout much of this region. In Louisiana it forms a deltaic coast with numerous bays, distributaries, and splays. During historic times, the Mississippi Delta shows the greatest change of any shoreline in North America, as the active distributary lobe advances seaward and the inactive lobes are reworked by wave action. In places the shoreline is changing at a rate of more than 20 m per year.

The influence of the delta extends far beyond the immediate vicinity of the distributaries as sediment delivered by the Mississippi River is transported by longshore currents west and south along the Texas coast. The sediment is deposited in a series of remarkably smooth barrier islands, which are separated from the mainland by lagoons.

Southward in Mexico, reefs are abundant and form an organic, controlled coast. A similar climatic situation exists in southern Florida, where the Florida Keys form a string of organic reefs. The western part of Florida is also influenced by organic activity because the coast consists alternately of raggedly indented swamps and long, sandy beaches. The swamps, mostly mangrove, form a distinctive type of organically controlled coast.

The Atlantic coast of North America is low, relatively flat, and is bordered by a wide continental shelf. South of New York, the coast is deeply *embayed* as major river valleys are drained by the recent rise in sea level. The drowned river valleys form Delaware Bay, Chesapeake Bay, and Pamlico Sound. These estuaries are now in the process of being filled with sediment carried by rivers that drain the land. South of Norfolk, Virginia, long, sandy barrier islands, separated from the mainland by a lagoon, form the shores. The lagoons are partly filled with marshes and swamps. Southward, some of the lagoons are completely filled, so the beach is tied to the mainland. Well-known seaside resorts such as Atlantic City, New Jersey; Ocean City, Maryland; and Virginia Beach, Virginia, are built on these barriers.

North of New York City, the coast is greatly influenced by glaciation. From New York to Cape Cod, many of the islands are built by glacial deposition. Long Island is a segment of a moraine; the islands in Boston Harbor are slightly modified drumlins. These islands are composed of unconsolidated glacial sediment and are especially vulnerable to wave erosion.

North of Cape Cod, the coast has been modified by glacial erosion, which is responsible for the irregular, rugged, low, rocky coast, characterized by numerous long bays and small, rocky islands. In the highlands of the Labrador Peninsula and on Baffin Island, valley glaciation has cut U-shaped valleys, and fiords dominate the coast.

TIDES

Tides are produced by the gravitational attraction of the Moon and the centrifugal force of the Earth-Moon system. They affect coasts in two major ways: (1) by initiating a rise and fall of the water level and (2) by generating tidal currents.

On most shorelines throughout the world, the sea advances and retreats in a regular rhythm twice in approximately 24 hours. These changes are called *tides,* and their cause has intrigued people for thousands of years. In the Mediterranean, tides are almost imperceptible; in the Bay of Fundy, they are more than 15 m high. Ignorance of tides has had an impact on history. Caesar's war galleys were devastated on the British shore because he failed to pull them high enough out of the water to avoid the returning tide. King John of England (1167–1216) was caught in a high tide, lost his treasure and part of his army, and was so enraged he died a week later. The origin of tides was not known until Newton showed how tides arise from gravitational attraction of the Moon and Earth.

The diagram in Figure 16.30 illustrates on a highly exaggerated scale the principal forces that produce tides. The gravitational force exerted by the Moon tends to pull the oceans facing the Moon into a bulge. Another tidal bulge, on the side of Earth opposite the Moon, is caused by centrifugal force. Earth and the Moon rotate around a common center of mass, which lies approximately 4500 km from the center of Earth on a line directed toward the Moon. The eccentric motion of Earth as it revolves around the center of mass of the Earth–Moon system creates a large centrifugal force, which forms the

What are the major effects of the rise and fall of tides?

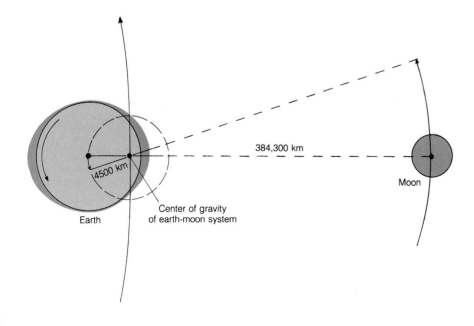

384,300 km

Moon

Center of gravity
of earth-moon system

4500 km

Earth

FIGURE 16.30
Ocean tides are caused by the gravitational attraction of the Moon and the centrifugal force of the Earth–Moon system. On the side of Earth facing the Moon, the gravitational attraction is greater, forming a tidal bulge in the ocean's water. On the other side of Earth, the centrifugal force is greater, causing another tidal bulge.

second tidal bulge. Earth rotates beneath the bulges, so the tides rise and fall twice every 24 hours.

The major effect of the rise and fall of tides is the transportation of sediment along the coast and over the adjacent shallow sea floor. Extremely high tides are produced in shallow seas where the rising water is funneled into bays and estuaries. For example, in the Bay of Fundy, between New Brunswick and Nova Scotia, the tide range (the difference in height between high tide and low tide) is as much as 21 m. Where fine-grained sediment is plentiful and the tide range is great, the configuration of the coast is greatly influenced by tides and tidal currents.

TSUNAMI

Movement of the ocean floor by earthquakes, volcanic eruptions, or submarine landslides frequently produces an unusual wave called a tsunami, which has a long wavelength and travels across the open ocean at high speeds. As tsunami approach shore, their wavelengths decrease and their wave heights increase; therefore, tsunami can be formidable agents of destruction along shorelines.

Large waves known as seismic sea waves or by the Japanese word *tsunami* originate from disturbances within Earth's crust. They are also commonly called *tidal waves,* which implies they originate from tides. They have no relationship to tides, however. Tsunami are usually caused by earthquakes or volcanic eruptions and by submarine landslides. For example, the eruption of Krakatoa in 1883 created a tsunami that killed 36,000 people in the East Indies, and Japan lost 27,000 lives from a tsunami in 1896. Tsunami are most frequent in the Pacific Ocean; to help minimize loss of life, an effective seismic sea wave warning system has been established. Tsunami have long wavelengths and travel across the open ocean at high speeds. As they approach shore, the wavelength decreases and the wave height increases, so tsunami can be formidable agents of destruction along the shoreline.

Tsunami differ from wind-produced ocean waves in that energy is transferred to the water from a sea-floor disturbance, so the entire ocean depth of water participates in the wave motion. The wave front travels from its point of origin at high speeds, ranging from 500 to 800 km per hour, and can traverse an entire ocean. The wave height is only from 30 to 60 cm, and the wavelength ranges from 55 to 200 km. In an open ocean, therefore, a tsunami can pass unnoticed. As the wave approaches the shore, however, important changes take place. The energy distributed in the deep column of water becomes concentrated in an increasingly shorter column, resulting in a rapid increase in wave height. Waves that were less than 60 cm high in the deep ocean can build rapidly to heights exceeding 15 m in many cases and well over 30 m in rare instances. They exert an enormous force against the shore and can inflict serious damage and great loss of life.

A number of tsunami have been well documented by seismic stations and coastal observers. For example, the tsunami that hit Hawaii on April 1, 1946, originated in the Aleutian Trench off the island of Unimak. The waves moving across the open ocean were imperceptible to ships in their path because the wave height was only 30 cm. Moving at an average speed of 760 km per hour, they reached the Hawaiian Islands, 3200 km away, in less than 5 hours. Because the wavelength was 150 km, the wave crests arrived about 12 minutes apart. As the waves approached the island, their height increased at least 17 m and thus produced an extremely destructive surf, which swept inland and demolished houses, trees, and almost everything else in its path.

Why is a tsunami (tidal wave) so small that it is imperceptible in the open ocean yet may be over 30 m high when it reaches a coastline?

SUMMARY

1. Shorelines are dynamic systems involving the forces of waves and currents. Wind-generated waves provide most of the energy for erosion, transportation, and deposition of sediment, but tides and tsunami can be locally important.
2. Waves approaching a shore are bent, or refracted, so that energy is concentrated on headlands and dispersed in bays.
3. Longshore drift, one of the most important shoreline processes, is generated as waves strike the shore at an angle. Water and sediment move obliquely up the beach face but return with the backwash directly down the beach, perpendicular to the shoreline. This results in a net transport parallel to the shore.
4. Erosion along coasts results from the abrasive action of sand and gravel moved by the waves and currents and, to a lesser extent, from solution and hydraulic action. The undercutting action of waves and currents typically produces sea cliffs. As a sea cliff recedes, a wave-cut platform develops. Minor erosional forms associated with the development of sea cliffs include sea caves, sea arches, and sea stacks.
5. Sediment transported by waves and longshore drift is deposited in areas of low energy. It produces a variety of landforms, including beaches, spits, tombolos, and barrier islands.
6. Erosion and deposition processes tend to develop a straight or gently curving coastline where neither large-scale erosion nor large-scale deposition occurs. A shoreline with such a balance of forces is called a shoreline of equilibrium.
7. Reefs are a special type of coastal feature because the active process is organic. They commonly evolve through a series of types, from fringing reefs to barrier reefs to atolls.
8. The worldwide rise in sea level associated with the melting of the Pleistocene glaciers drowned many coasts. Coasts are classified on the basis of the process that has been most significant in developing their configurations.
9. Tides are caused by gravitational attractions of the Moon and have a significant effect on shorelines.
10. Tsunami are waves generated by earthquakes or submarine mass movement. They grow higher as they approach the shore and may be extremely destructive.

KEY TERMS

atoll (p. 389)

backwash (p. 370)

barrier island (p. 382)

barrier reef (p. 388)

bay (p. 380)

baymouth bar (p. 380)

beach (p. 380)

beach drift (p. 372)

breaker (p. 370)

coral (p. 386)

elevated marine terrace (p. 386)

estuary (p. 380)

fiord (p. 392)

fringing reef (p. 388)

hanging valley (p. 375)

headland (p. 371)

longshore current (p. 372)

longshore drift (p. 372)

platform reef (p. 390)

primary coast (p. 392)

rip current (p. 372)

sea arch (p. 377)

sea cave (p. 377)

sea cliff (p. 375)

sea stack (p. 377)

secondary coast (p. 392)

shoreline of equilibrium (p. 386)

spit (p. 380)

swash (p. 370)

tide (p. 395)

tombolo (p. 380)

tsunami (p. 396)

wave base (p. 369)

wave-built terrace (p. 375)

wave crest (p. 368)

wave-cut cliff (p. 375)

wave-cut platform (terrace) (p. 375)

wave height (p. 369)

wavelength (p. 368)

wave period (p. 369)

wave refraction (p. 371)

wave trough (p. 368)

REVIEW QUESTIONS

1. Describe the motion of water in a wind-generated wave.
2. Explain how wave refraction alters the form of a coastline.
3. Explain the origin of longshore drift.
4. Describe the stages in the evolution of a sea cliff and wave-cut platform.
5. Name the major depositional landforms along a coast and explain the origin of each.
6. What effect would the construction of dams on major rivers have on beaches along the coast?
7. How are marine terraces formed?
8. What conditions are necessary for the formation of a coral reef?
9. Explain the origin of atolls.
10. Describe six common types of shoreline.
11. Explain how ocean tides are generated.
12. Explain the origin of tsunami.

ADDITIONAL READINGS

Bascom, W. 1964. *Waves and Beaches.* New York: Doubleday, Anchor Books.

Bird, C. F., and M. L. Schwartz, 1985. *The World's Coastlines.* New York: Van Nostrand Reinhold.

Davis, J. L. 1980. *Geographical Variations in Coastal Development,* 2nd ed. New York: Longman.

Fisher, J. S., and R. Dolan, eds. 1977. *Beach Processes and Coastal Hydrodynamics.* New York: Academic Press.

Moore, J. R., ed. 1971. *Oceanography: Readings from Scientific American.* San Francisco: Freeman.

Snead, R. E. 1982. *Coastal Landforms and Surface Features: A Photographic Atlas and Glossary.* Stroudsburg, PA: Hutchinson Ross.

Stowe, K. 1983. *Ocean Science.* New York: Wiley.

Oil Well
Flare

Longitudinal
dunes

Transverse dunes

Barchanoid
dunes

In the low-latitude deserts of the world, where
precipitation is low and evaporation is high, the
principal agent of erosion and deposition may be
the wind, and a variety of sand dunes may completely
cover the surface. One such area is the Empty Quarter
(Rub' al Khali) of Saudi Arabia, which is a sea of sand
covering an area of almost 650,000 km². This image lies
on the northern flank of the Empty Quarter and shows
part of the United Arab Emirates. The Persian Gulf lies
just 24 km north of the scene's edge. This area seems
unearthly, notably lacking any signs of life except for
the great black plume of smoke rising from an oil flare
and spreading across the upper part of the picture.
Beneath the tan sand dunes lies the Bu Hasa oil field,
one of the United Arab Emirates' largest onshore
petroleum resources. A pipeline connects this field
with the coast, and a faint, dark line of a road trends to
the northeast.

The prevailing winds that form this sea of sand are
from the northwest, and the dune patterns all exist in
relation to this direction. Dune shapes differ, however,
because of differences in wind velocity, source of sand,
and nature of the terrain. Several types of dune
patterns are distinct in this region. Crescent-shaped
dunes, called barchans, occur in the lower right.
Between many of these large dunes are dry ponds
containing salt and clay. These appear as small,
triangular, bluish patches. Wavelike dunes with ridges
perpendicular to the prevailing wind cover the central
part of the area. These are transverse dunes. To the left
of the image the sand forms long, linear ridges parallel
to the prevailing wind. These are the great seif, or
longitudinal, dunes, formed by strong, converging
winds and relatively limited sand supply.

In addition to forming dunes in the great deserts of
the world, wind activity is important in the formation
of dunes along many coastal areas and in smaller
"rain-shadow" deserts. Also, deposits of windblown
dust, called loess, blanket millions of square kilometers
of the middle-latitude continents.

Wind is probably the least effective agent of erosion,
although many erosional landforms are mistakenly
attributed to it. Even in the desert most erosional
landforms are the products of running water, and the
greatest effect of wind is forming shifting sand dunes.

17
Eolian
Systems

1. Wind is not an effective agent in eroding the landscape, but it can transport loose, unconsolidated fragments of sand and dust.
2. Wind transports sand by saltation and surface creep. Dust is transported in suspension, and it can remain high in the atmosphere for long periods.
3. Sand dunes migrate as sand grains are blown up and over the windward side of the dune and accumulate on the lee slope. The internal structure of a dune consists of strata inclined in a downwind direction.
4. Various types of dunes form, depending on wind velocity, sand supply, constancy of wind direction, and characteristics of the surface over which the sand migrates.
5. Windblown dust (loess) forms blanket deposits, which can mask the older landscape beneath them. The source of loess is desert dust or the fine rock debris deposited by glaciers.

WIND AS A GEOLOGIC AGENT

The great deserts of the world form in low-latitude regions in a zone roughly 30 degrees north and south of the equator. Wind is an effective local geologic agent, capable of lifting and transporting loose sand and dust, but its ability to erode solid rock is limited. The main effects of wind as a geologic agent are the transportation and deposition of sand and dust.

Formerly, geologists thought that wind, like running water and glaciers, had great erosional power—power to abrade and wear down Earth's surface. It is apparent, however, that few major topographical features are formed by wind erosion. Winds vary considerably in strength from one moment of time to the next and are thus able to lift and transport rock debris for only short periods. At wind speeds of 30 km per hour, sand grains can be moved, but extremely rare gusts of 120 km per hour are needed to roll pebbles along the surface. Even the lightest winds are sufficient to lift dust into the air, however, and keep it in suspension for long periods of time. Abrasion by wind-transported sand thus aids in eroding and shaping some rock surfaces, but most are small-scale features. Even in the desert, where water is not an obvious geologic agent, wind is not the major agent of erosion. Most erosional landforms in deserts were produced by weathering and by running water in times of wetter climate. They are, in a sense, "fossil" landscapes, formed by processes that are no longer active. Even minor alcoves and niches, or "wind caves," and certain topographic features called pedestal rocks, which are often thought to be caused by wind erosion, are actually produced by differential weathering.

Although it is relatively insignificant as an erosional agent, wind is effective in transporting loose, unconsolidated sand, silt, and dust. It is responsible for the formation of great "seas of sand" in the Sahara, the Arabian, and other deserts, as well as the blankets of windblown dust covering millions of square kilometers in China, the central United States, and parts of Europe. It is estimated that windblown dust covers one-tenth of the land surface. This fact is important because soils from these deposits constitute some of Earth's richest farmland.

What is the principal cause of wind?

Prevailing wind patterns are determined by variations of solar radiation with latitude, the *Coriolis effect* (deflection due to Earth's rotation) (Figure 17.1), the configuration of continents and oceans, and the location of mountain ranges. The world's great deserts, such as the Sahara and the deserts of Asia, are mostly located in low-latitude belts (Figure 17.2). The tropical regions are the

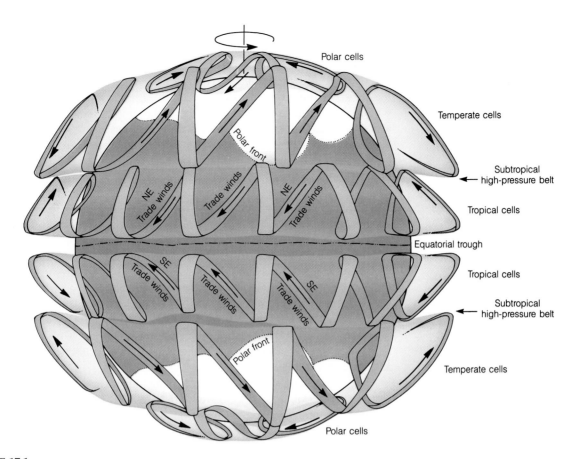

FIGURE 17.1
Atmospheric circulation and prevailing wind patterns are generated by solar radiation and by Earth's rotation. Air heated in the equatorial regions rises in convection columns and moves toward higher latitudes. There it is cooled, compressed, and forced to descend, forming subtropical high-pressure belts. The air then moves toward the equator as trade winds. In the Northern Hemisphere, this air is deflected by Earth's rotation to flow southwestward. In the Southern Hemisphere, it is deflected to flow northwestward. Cold polar air tends to wedge itself toward the lower latitudes and forms polar fronts.

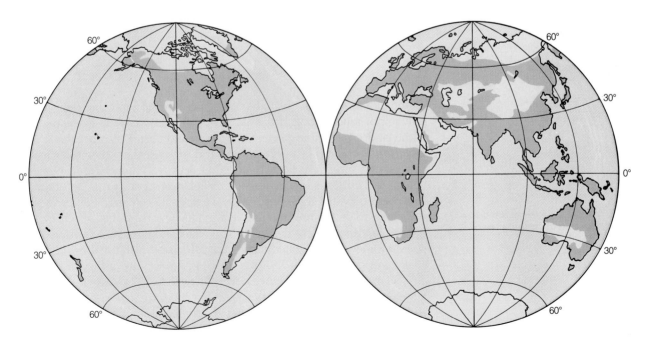

FIGURE 17.2
The major desert areas of the world, such as the Sahara, the Arabian, the Kalahari, and the deserts of Australia, all lie between 10 and 30 degrees north or south of the equator. These areas are under almost constant high atmospheric pressure and are characterized by subsiding dry air and low humidity. Desert and near-desert areas cover nearly one-third of the land surface.

birthplace of the winds. This is because equatorial air, heated by solar radiation, rises in convection cells, which descend again about 30 or 35 degrees north and south of the equator. This hot band encircles the planet and powers the wind. As the air rises to higher altitudes, it cools and releases moisture, which falls as tropical rains in the equatorial regions. The air is now much drier as it continues to move in convection cells northward and southward. The dry air descends to the surface near latitude 30 degrees north and south of the equator. It is capable of holding more moisture, so it rarely releases any as precipitation. As a result, evaporation of the surface moisture, rather than precipitation, occurs in the low latitudes where the convection cells descend. When the air reaches the equator, it again rises and cools, releasing this moisture as rain.

Other deserts lie in the *wind shadow* behind higher mountain ranges, which intercept the moisture-laden air. As the air is forced to rise over a mountain range, it cools and precipitates its moisture. Here, too, the dry descending air is heated by compression. A good example of an arid region in the rain shadow of a high mountain range is in Nevada and Utah, which lie in the rain shadow of the Sierra Nevada.

Wind action is most significant in desert areas, but it is not confined to them. Many coasts are modified by winds that pick up loose sand on the beach and transport it inland.

The impact of wind action on the planet Mars was dramatically recorded by the space probes of 1971, 1972, and 1976 (Figure 24.11). Because of the absence of running water and vegetation, wind action is probably the dominant process on the surface of Mars at the present time. In former geologic periods here on Earth, when climatic conditions were much different from what they are now, the wind may have been a more significant force in various regions.

What controls the origin and evolution of deserts?

WIND EROSION

Wind erosion acts in two ways: (1) by deflation, the lifting and removal of loose sand and dust particles from Earth's surface, and (2) by abrasion, the sandblasting action of windblown sand.

Deflation

Although wind erodes only when it is transporting sand, and then produces only minor sandblasting effects on a rock surface, wind can lower the surface of the landscape by transporting loose sand and dust. This process is known as *deflation*. It commonly occurs in semiarid regions where the protective cover of grass and shrubs has been removed by activity of humans and animals. The results are broad shallow depressions called *deflation basins*. Deflation basins also commonly develop where calcium carbonate cement in sandstone formations is dissolved by groundwater, leaving loose sand grains, which are picked up and transported by the wind. Large deflation basins, covering areas of several hundred square kilometers, are associated with the great desert areas of the world, particularly in North Africa in the vicinity of the Nile Delta.

Generally, wind can move only sand and dust-size particles, so deflation leaves concentrations of coarser material called *lag deposits,* or *desert pavements* (Figure 17.3). These striking desert features of erosion stand out in contrast to deposits in dune fields and playa lakes.

Deflation occurs only where unconsolidated material is exposed at the surface. It does not occur where there are thick covers of vegetation or layers of gravel. The process is therefore limited to areas such as deserts, beaches, and barren fields.

Why is the erosive power of wind less effective than that of running water?

FIGURE 17.3
Wind selectively transports sand and fine sediment, leaving the coarser gravels to form a lag deposit called desert pavement. The protective cover of lag gravel limits future deflation. During the 1991 Persian Gulf War, military construction, heavy vehicular traffic, and intensive bombing caused thousands of craters in the desert pavement. This widespread disruption of the protective surface was predicted to lead to increased sand dune formation and migration in areas of Kuwait, northeastern Saudi Arabia, and southern Iraq.

Abrasion

Wind abrasion is essentially the same process as the artificial sandblasting used to clean building stone. Some effects of wind abrasion can be seen on the surface of the bedrock in most desert regions (Figure 17.4A). In areas where soft, poorly consolidated rock is exposed, wind erosion can be both spectacular and distinctive. Some pebbles, called *ventifacts* (literally meaning "wind-made"), are shaped and polished by the wind (Figure 17.4B and 17.5). Such pebbles are commonly distinguished by two or more flat faces that meet at

(A) Pitted and polished surfaces

(B) Faceted pebbles and cobbles called ventifacts

(C) Erosion of rock outcrops into elongate ridges called yardangs

FIGURE 17.4
Features produced by wind abrasion are apparent on the bedrock surface in most deserts of the world.

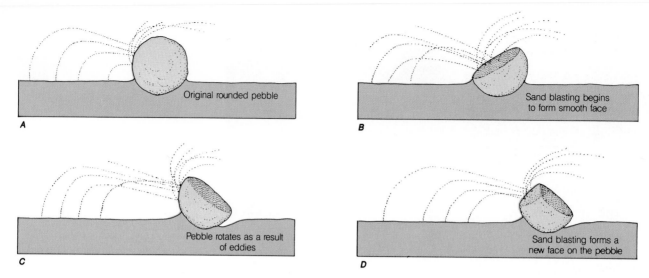

FIGURE 17.5
Ventifacts are pebbles shaped and polished by wind action. They commonly have two or more well-polished sides, which are formed as the sandblasting action of the wind reshapes one side of the pebble and then another. In diagram B a facet (smooth face) is formed by sandblasting on the side facing the prevailing wind. Removal of the sand by deflation causes the pebble to rotate so that other faceted and polished surfaces are produced on another side of the pebble (D).

sharp ridges and are generally well polished. Other ventifacts have a variety of shapes. Some have surface irregularities and grooves aligned with the wind direction (Figure 17.4A).

Perhaps the best example of wind erosion in the United States is in the Great Plains, especially the High Plains of Colorado, Kansas, and Texas. In this area, innumerable deflation basins, ranging from small dimples 30 cm deep and 3 m in diameter to larger basins 15 to 20 m deep and more than a kilometer across, are scattered across the landscape (Figure 17.6). Many have permanent or intermittent lakes in them. Although some of these depressions may be the result of collapse and local subsidence, deflation has played a major role in their development. During wet periods, water collecting in small depressions will dissolve the calcareous cement in the horizontal sandstone that covers the area. Thus, many of the individual grains in the sandstone formation are loose and free to move about. During dry periods, wind will pick up the loose grains and blow them away. This creates a larger basin, which collects more water, which in turn dissolves more cement to produce more loose grains. The process is therefore self-perpetuating. Many of these depressions are enlarged by the activity of animals. This was especially true of the time when great herds of buffalo thronged to the temporary ponds for water. After wading and wallowing, the herds would carry mud away on their bodies and destroy the surrounding vegetation, producing conditions that favored further wind erosion. The depressions have thus been referred to as "buffalo wallows."

Larger landforms produced by wind abrasion are less common, but in some desert regions distinctive, linear ridges, called **yardangs,** are produced by wind erosion. These features were first discovered in the Taklimakan Desert of China. The name is derived from the Turkistani word *yar,* meaning "ridge" or "bank." Typical yardangs have the form of an inverted boat hull (Figure 17.4C) and commonly occur in clusters, oriented parallel to the prevailing wind that formed them. Theoretically, they may be formed in any rock type, but they are best developed in soft, unconsolidated, fine-grained sediment that is easily sculpted but is cohesive enough to retain steep slopes. Yardangs evolve a streamlined shape that offers minimum resistance to the moving air, thus their resemblance to the shape of a boat hull. In a way, they are analogues to drumlins, which are shaped by moving ice. To attain this shape, erosion may combine with deposition to shape the flanks or end of the

What are the major features produced by wind erosion?

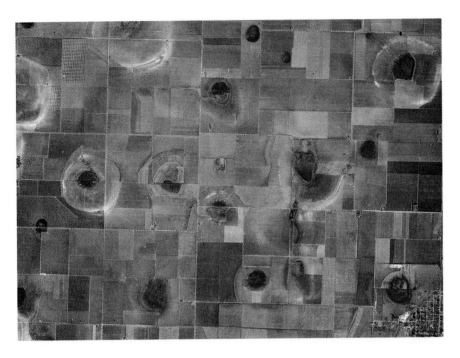

FIGURE 17.6
Deflation basins in the Great Plains are produced where solution activity in the layers of horizontal bedrock dissolves the cement that binds the sand grains together. The loose sand is removed by the wind, and a basin is formed. Water trapped in the basin dissolves more cement and the basin is enlarged. The main roads form squares 1 mile on each side.

yardang. Differences in the rock body, such as bedding, may produce differential erosion, and rock layers within the yardang may be etched out in relief. In China, resistant, horizontal layers of sand erode into flat-topped yardangs.

Yardangs are generally restricted to the most arid parts of the deserts, which are relatively sand-poor and are areas where vegetation and soil are minimal. There is some indication that the Sphinx of Egypt was constructed out of yardang.

The most spectacular wind erosional features on Earth are the great yardangs of the Lūt Desert of Iran (Figure 17.7). Here ridges almost 150 m high and over 300 m long are carved by the wind out of lake sediments consisting of sand, silt, clay, and gypsum. The ridges are separated by troughs 100 m or more wide. There are no regional erosional channels between the yardangs, and no evidence of water erosion can be seen on the floor of the yardang field, although the slopes of the yardangs have been gullied by periodic storms.

FIGURE 17.7
Yardangs in the Lūt Desert, southeastern Iran, are carved out of fine-grained, horizontally bedded, silty clay.

TRANSPORTATION OF SEDIMENT BY WIND

Wind transports sand by saltation and surface creep. Silt and dust-size particles are carried in suspension.

Field observations and wind-tunnel experiments indicate that windblown sand grains move by skipping or bouncing into the air (a process called saltation) and by rolling or sliding along the surface (called surface creep). Fine silt and dust are carried in suspension over great distances and settle back to the ground only after the turbulent wind ceases.

Movement of Sand

Although both wind and water transport sand by **saltation,** the mechanics of motion involved are different because the viscosity of water is so much greater than that of air. In water, the process involves hydraulic lift. In air, saltation results from impact and elastic bounce (Figure 17.8). The energy that initially lifts sand grains into the air comes from collision with other grains. If the wind reaches a critical velocity, the grains begin to roll or slide. As one grain moves along, it strikes others. The impact can cause one or more of the colliding grains to bounce into the air, where they are driven forward by the stronger wind above the ground surface. Gravity soon pulls them back, and the grains strike the ground at an angle generally ranging from 10 to 16 degrees. If the sand is moving over solid rock, the grains bounce back into the air. If the surface is loose sand, the impact of a falling grain can knock other grains into the air, setting up a chain reaction, which eventually sets in motion the entire sand surface.

Some sand grains, too large to be ejected into the air, move by **surface creep** (rolling and sliding). These large grains are moved by the impact of saltating grains, not directly by the wind. Approximately one-fifth to one-fourth of the sand moved by a sandstorm travels by rolling and sliding. Particles with a diameter greater than 1 cm are rarely moved by wind.

Movement of Dust

Dust storms are a major process in deserts. They are capable of transporting thousands of tons of sediment hundreds of kilometers. Although they are rarely seen, dust storms are a major dynamic feature of the planet, and they subtly, but constantly, change its surface.

What is the difference in the way wind and water transport sediment?

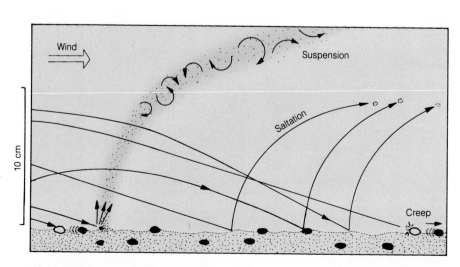

FIGURE 17.8
The transportation of sediment by wind is accomplished by surface creep, saltation, and suspension. Coarse grains move by impact from other grains and slide or roll (surface creep). Medium grains move by skipping or bouncing (saltation). Fine silt and dust move in suspension.

Throughout human history, dust storms have been a major cause of soil erosion. References to dust storms were recorded in 1150 B.C. in China and in biblical times in the Middle East. However, because of their regional extent and violent nature, it has been difficult to study them in a meaningful way. Our understanding of dust storms, however, has been greatly enhanced with the advent of satellite photography, which reveals intricate patterns of dust movement and associated dune fields and has clarified the behavior and mechanics of dust storms.

Dust storms are initiated by the downdraft of cool air from a cumulonimbus cloud. When such a cloud develops to the point that rain begins to fall from it, the rain cools the air as it falls. Because the cool air is denser than the surrounding air, it descends in a downdraft. As the heavy cooled air reaches the ground, it is deflected forward and moves in a large tongue-shaped pattern. It flows across the ground as a density current, that is, a body of moving air that is heavier than the surrounding air because it is cooler. As the dense air moves across the dry surface, it sweeps up dust and sand by the churning action of its turbulent flow. Dust storms of this type are called haboobs, from the Arabic word for violent wind (Figure 17.9).

Great dust storms sometimes reach elevations of 2500 m and advance at speeds of up to 200 m per second. It has been estimated that 500×10^6 tons of windblown dust are carried from the deserts each year. (This is only slightly less than the amount of sediment deposited each year by the Mississippi River.) Some is deposited downwind from the desert such as in China (Figure 17.2), but because of the prevailing wind pattern in the Sahara, Australia, and South America, large quantities of windblown dust are carried out to sea. Some of the larger dust storms in the Sahara have even carried dust across the Atlantic to the eastern coast of South America.

FIGURE 17.9
Dust storm in the Blue Nile area—Sudan, Africa, results when cool air descends and moves laterally over the surface as a density current. As the dense, cool air moves across the surface it sweeps up dust and sand by its turbulent flow, creating a dust storm, or "haboob."

MIGRATION OF SAND DUNES

Windblown sand commonly accumulates in dunes that migrate downwind.

Sand dunes migrate relentlessly downwind and may completely modify the landscape and damage or obliterate almost anything in their path. Forests have been entombed by advancing dunes, streams diverted, and villages completely covered. One example is found in England and France, where entire towns were overwhelmed by advancing dunes so that nothing was seen but the church spire. Then the dunes marched on, leaving behind a devastated countryside of dead trees and collapsed buildings. But why do dunes form and why do some grow so large? How do they move?

In many respects, dunes are similar to ripple marks (formed either by air or water) and to the large sand waves or sandbars found in many streams and in shallow-marine water.

Many *dunes* originate where an obstacle, such as a large rock, a clump of vegetation, or a fence post, creates a zone of quieter air behind it (Figure 17.10). As sand is blown up or around the obstruction and into the protected area (the wind shadow), its velocity is reduced and deposition occurs. Once a small dune is formed, it acts as a barrier itself, disrupting the flow of air and causing continued deposition downwind. Dunes range in size from a few meters high to as much as 200 m high and 1 km wide.

The movement of sand in a typical dune is diagrammed in Figure 17.11. Dunes are asymmetrical, with a gently inclined windward slope and a steeper downwind slope, called the *lee slope,* or *slip face.* The steep slip face of the dune indicates the direction of the prevailing wind. Dunes migrate grain by grain as wind transports sand by saltation and surface creep up the windward slope. The wind continues upward past the crest of the dune, creating divergent airflow and eddies just over the lee slope. Beyond the crest, the sand drops out of the wind stream and accumulates on the slip face. As more sand is transported from the windward slope and accumulates on the lee slope, the dune migrates downwind. The internal structure of a migrating dune consists of cross-bedding formed as the saltating grains accumulate on the inclined downwind slope of the dune (Figures 17.10 and 17.11). Strata formed on the lee slope are therefore inclined in a downwind direction. Geologists map the directions of ancient eolian systems by measuring the directions in which the cross-strata of windblown sandstone are inclined (Figure 17.12).

How can the wind move an entire sand dune?

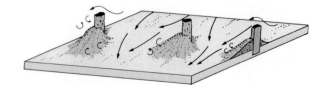

FIGURE 17.10
Sand dunes commonly originate in wind shadows. Any obstacle that diverts the wind, such as a bush or a fence post, creates eddies and reduces wind velocity. Windblown sand is deposited in protected areas, and eventually enough sand accumulates in the wind-shadow area to form a dune. The dune itself then acts as a barrier, making its own wind shadow, and thus causes additional accumulation of sand.

 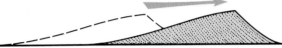

FIGURE 17.11
A sand dune migrates as sand grains move up the slope of the dune and accumulate in a protected area on the downwind face. The dune slowly moves grain by grain. As the grains accumulate on the downwind slope, they produce a series of layers (cross-beds) inclined in a downwind direction.

FIGURE 17.12
Cross-bedding in the Navajo Sandstone in Zion National Park, Utah, is excellent evidence that the rock formed in an ancient desert. The inclination of the strata shows that the wind blew from north to south (from left to right, in the photograph) during most of the time when this formation was deposited.

TYPES OF SAND DUNES

A variety of types of sand dunes result from the variations in sand supply, wind direction, and velocity. The most significant types of dunes are (a) transverse dunes, (b) barchan dunes, (c) longitudinal dunes, (d) star dunes, and (e) parabolic dunes.

What determines the shape of the various types of sand dunes?

Sand dunes vary greatly in size and shape, and although they form bleak and barren wastelands they impart a sense of beauty and mystique to the areas they cover. The largest dune fields are found in the Middle East and North Africa and are so vast that they are called "sand seas." One of the most spectacular sand seas is called the Rub' al Khali, or Empty Quarter (see photo on page 398) where dunes cover an area of about 400,000 km² of land.

The shape of a sand dune depends on such factors as the following:

1. Sand supply
2. Wind velocity
3. Variability of wind direction
4. Characteristics of the surface over which the sand moves

The most common varieties are transverse dunes, barchan dunes, longitudinal dunes, star dunes, and parabolic dunes (Figure 17.13).

Transverse Dunes

Transverse dunes typically develop where there is a large supply of sand and a constant wind direction (Figure 17.13A). These dunes cover large areas and develop a wavelike form, with sinuous ridges and troughs perpendicular to the prevailing wind. Transverse dunes commonly form in desert regions where exposed ancient sandstone formations provide an ample supply of sand. They usually cover large areas known as sand seas, so called because the wavelike dunes produce a surface resembling a stormy sea.

Barchan Dunes

Where the supply of sand is limited and winds of moderate velocity blow in a constant direction, crescent-shaped *barchan dunes* tend to develop (Figure 17.13B). Typically, they are small, isolated dunes from 1 to 50 m high. The tips (or horns) of a barchan point downwind, and sand grains are swept around them as well as up and over the crest. With a constant wind direction, beautiful symmetrical crescents form. With shifts in wind direction, however, one horn can become larger than the other. Although barchans typically are isolated dunes, they may be arranged in a chainlike fashion extending downwind from the source of sand.

Longitudinal Dunes

Longitudinal dunes, also called *seif dunes* (Arabic, "sword"), are long, parallel ridges of sand, elongate in a direction parallel to the vector resulting from two slightly different wind directions (Figure 17.13C). They develop where strong prevailing winds converge and blow in a constant direction over an area having a limited supply of sand. Many longitudinal dunes are less than 4 m high, but they can extend downwind for several kilometers. In larger desert areas, they can grow to 100 m high and 120 km long, and they are usually spaced from 0.5 to 3 km apart. Longitudinal dunes occupy a vast area of central Australia called the Sand Ridge Desert. They are especially well developed in some desert regions of North Africa and the Arabian Peninsula.

Star Dunes

A *star dune* is a mound of sand having a high central point from which three or four arms, or ridges, radiate (Figure 17.13D). This type of dune is typical of parts of North Africa and Saudi Arabia. The internal structure of these dunes suggests that they were formed by winds blowing in three or more directions.

(A) Transverse dunes develop where the wind direction is constant and the sand supply is large.

(B) Barchan dunes develop where the wind direction is constant but the sand supply is limited.

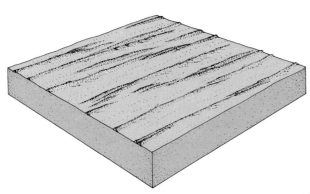

(C) Longitudinal dunes are formed by converging winds in an area with a limited sand supply.

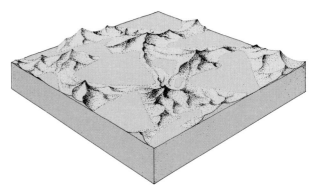

(D) Star dunes develop where the wind direction is variable.

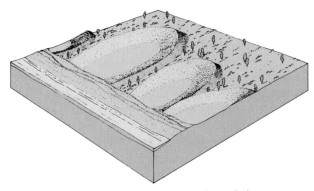

(E) Parabolic dunes are formed by strong onshore winds.

FIGURE 17.13
Each of the major types of sand dunes represents a unique balance between sand supply, wind velocity, and variability in wind direction.

Parabolic (Blowout) Dunes

Parabolic dunes (blowouts) typically develop along coastlines where vegetation partly covers a ridge of windblown sand transported landward from the beach (Figure 17.13E). Where vegetation is absent, small deflation basins are produced by strong onshore winds. These blowout depressions grow larger as more sand is exposed and removed. Usually, the sand piles up on the lee slope of the shallow deflation hollow, forming a crescent-shaped ridge. In map view, a parabolic dune is similar to a barchan, but the tips of the parabolic dune point upwind. Because of their form, parabolic dunes are also called hairpin dunes.

Sand Seas

Although Earth is commonly referred to as the water planet, several continents have vast areas where precipitation is rare and the surface is covered with windblown sand. These areas are referred to as **sand seas** or **ergs** (North Africa). It has been calculated that 99.8% of all windblown sand is in the great sand seas of the world. The largest are in Africa, Asia, and Australia (Figure 17.14). In Africa, about 800,000 km^2 (or one-ninth of the entire area of the Sahara) is covered by stable or active sand dunes. One-third of Saudi Arabia, approximately 1 million km^2, is covered by dunes, and in the vast Empty Quarter (the Rub' al Khali desert), dunes may be 213 m high. The Australian sand seas are mainly in the western and central portions of the continent.

Desertification

The great deserts of the world were formed by natural processes over long periods of time as continents migrated into dry climates, which were produced in low-latitude, high-pressure zones. During development, a desert may expand and shrink in response to cyclic climatic fluctuations. The margins of deserts, therefore, have always been transitional or gradational to the adjacent, more humid environments.

The desert fringes have very delicately balanced ecosystems. Sparse vegetation serves to inhibit wind erosion, but when it is destroyed, the desert expands. Along the desert margins human activity is commonly superimposed upon the natural processes that cause expansion and contraction of the deserts. Grazing livestock, the pounding of soil by hooves, and even collection of firewood by man can reduce the plant cover and stress the ecosystem beyond its tolerance. This results in a degradation of the land and, in many cases, expansion of the desert. This process, in which productive land is converted to desert, is referred to as *desertification*. When the general climatic trend is toward increasing aridity, desertification can occur with remarkable speed.

Desertification does not occur in a broad, even swath that can easily be mapped along the desert fringe. Deserts advance erratically, forming patches on their borders, and areas far from the desert may quickly degrade into barren rock and sand.

Desertification presents an enormous problem for human existence. About one-third of Earth's land is arid or semiarid, but only about one-half of this area is so dry that it cannot support human life. More than 600 million people live in the dry areas, and about 80 million live on land that is nearly useless because of desertification. The most severe problems are in Africa and Asia.

FIGURE 17.14 (opposite)
Sand seas in North Africa have been mapped with the use of satellite photography. The main sand seas in Algeria, Tunisia, and western Libya are located in structural and topographic basins, separated by plateaus and low mountain ranges. A variety of dune types are formed as a result of variations in wind velocity and direction, supply of sand, and the nature of the surface over which the sand moves.

LOESS

Loess is a deposit of windblown silt (dust) that accumulates slowly and ultimately blankets large areas, often masking pre-existing landforms. It covers one-tenth of the world's present land surface. The dust is derived either from nearby deserts or from rock flour near margins of recently glaciated areas.

What geologic features are produced by the windblown dust?

Windblown dust, known as *loess,* is found in thick deposits that blanket the surface in a number of regions throughout the world. It is a yellowish brown color and is composed principally of small angular grains of quartz, feldspar, and clay. Observations indicate that loess can be dispersed high into the atmosphere and carried great distances by the wind. This accounts for not only the widespread nature of loess but also the fact that, unlike dune deposits, loess deposits tend to blanket the landscape, covering hills and valleys alike. It may cover as much as one-tenth of the world's land surface and is particularly widespread in semiarid regions along the margins of great deserts (Figure 17.2). The equatorial tropics are free from loess because it is washed away by the heavy rainfall as soon as it is deposited. Areas formerly covered by continental glaciers are also loess-free because, until only a few thousand years ago, they were covered with ice.

FIGURE 17.15
A deposit of loess, composed of fine, closely packed silt particles, commonly erodes into nearly vertical cliffs. Most of the loess in the United States is found in the Great Plains and in the Mississippi valley region. This area is in the lower Mississippi valley.

Loess is a distinctive sedimentary deposit. Its particles are very similar, if not identical, to the dust in the air at the present time. Normally, loess deposits lack stratification and erode into vertical cliffs (Figure 17.15). Large loess deposits are derived from (1) desert regions and (2) glacial deposits.

Desert Loess

In northern China, extensive loess deposits consist primarily of disintegrated rock material brought by the prevailing westerly winds from the Gobi Desert, in central Mongolia. The yellow-colored loess, which reaches a thickness of more than 60 m, blankets a large area. It is easily eroded and transported in suspension by running water and is responsible for the characteristic color of the Hwang Ho (the Yellow River) and the Hwang Hai (the Yellow Sea). In the nearly vertical walls of loess deposits, a great number of Chinese have excavated caves in which they live (Figure 17.16).

Loess in the eastern Sudan, in North Africa, probably originated from a desert, most likely the Sahara to the west. Similarly, the loess deposits of Argentina are derived from arid regions to the west rather than from glacial deposits to the south.

FIGURE 17.16
Loess deposits in China are thick and widespread and have been excavated into an elaborate system of chambers for dwellings.

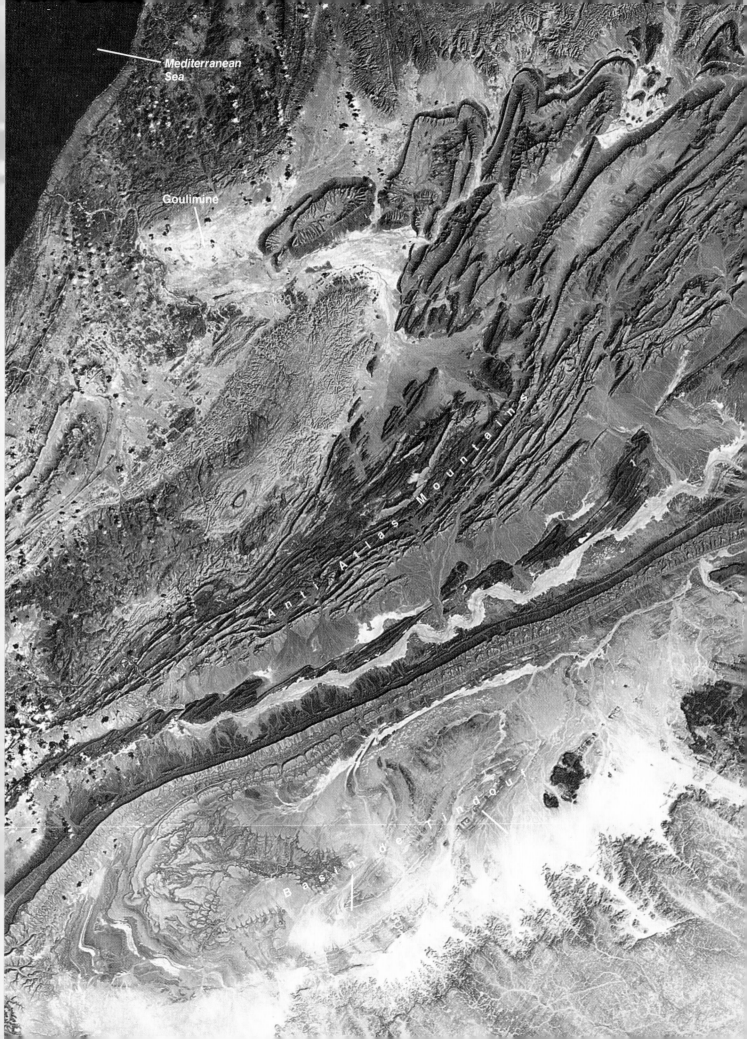

Mediterranean
Sea

Goulimine

Anti Atlas Mountains

Basin de Tindouf

18
Plate
Tectonics

The spectacular fold belt of the Anti Atlas Mountains in Morocco, North Africa, is one of the great surface expressions of plate tectonics. The deformed strata form dark ridges that zigzag across the terrain, showing the style of deformation produced as Africa drifted northward and impinged against the Eurasian plate. Careful study of this image is well worth your time because from it you can begin to develop an accurate concept of what Earth's tectonic system does.

Plate tectonics has done much more than explain the deformation of a mountain belt in this region. It explains how the Atlas Mountains are related to the Mediterranean Sea and how they are related to the volcanism in the Canary Islands several hundred kilometers west of the Moroccan coast. It explains how the Atlas Mountains are related to the Alps and the far-off mountains of the Himalayas. In brief it provides a single unifying theory of Earth's dynamics.

How did scientists develop the revolutionary theory of plate tectonics? Less than three decades ago most geologists believed that continents and ocean basins were fixed, permanent features on Earth and the theory of continental drift was considered a radical idea. What changed the minds of the entire science?

In this chapter, we will consider how the theory of plate tectonics developed and the evidence upon which it is based. We will then consider the nature of the lithosphere plates, what causes them to move, and how we measure rates and direction of plate motion. We will also look into the dynamics of plate interiors, where deformation is not closely associated with the activity of plate margins.

1. The theory of continental drift was proposed in the early 1900s and was supported by a variety of geologic evidence. Without a knowledge of the nature of the oceanic crust, however, a complete theory of Earth dynamics could not have been developed.
2. A major breakthrough in the development of the plate tectonics theory occurred in the early 1960s, when the topography of the ocean floors was mapped and magnetic and seismic characteristics were determined.
3. Most tectonic activity occurs along plate boundaries.
4. Divergent plate boundaries are zones where the plates split and spread apart. The major geologic processes are (a) rifting and (b) generation of basaltic volcanism and new oceanic lithosphere.
5. Convergent plate boundaries are zones where plates collide. Important geologic processes include (a) subduction, (b) generation of granitic magma, (c) mountain building, and (d) metamorphism.
6. Transform fault boundaries are zones where plates slide horizontally past each other.
7. Plate motion can be described in terms of a pole of rotation. The direction of relative motion of plates is indicated by (a) trend of oceanic ridge and associated fracture zones, (b) seismic data, (c) magnetic stripes on the sea floor, and (d) ages of chains of volcanic islands and seamounts.
8. The major forces acting on plates are (a) slab-pull, (b) ridge-push, (c) basal drag, and (d) friction along transform faults and in subduction zones.
9. Convection in the mantle is believed to be the fundamental process responsible for plate motion.

CONTINENTAL DRIFT

> *The theory of continental drift was proposed in the early 1900s and was supported by a variety of impressive geologic data. Without a knowledge of the nature of the oceanic crust, however, a complete theory of Earth dynamics could not have been developed.*

The theory of plate tectonics wrought a sweeping change in our understanding of Earth and the forces that shape it. Some scientists consider this conceptual change as profound as those that occurred when Darwin reorganized biology in the nineteenth century or when Copernicus, in the sixteenth century, determined that Earth is not the center of the universe, and yet the concept of continental drift is an old idea.

Soon after the first reliable world maps were made, scientists noted that the continents, particularly Africa and South America, would fit together like a jigsaw puzzle if they could be moved. Antonio Snider-Pelligrini, a Frenchman, was one of the first to study the idea in some depth. In his book *Creation and Its Mysteries Revealed* (1858), he showed how the continents looked before they separated (Figure 18.1). He cited fossil evidence in North America and Europe but based his reasoning on the catastrophe of Noah's flood. The idea seemed too farfetched for science or the general public and it was forgotten, not to be revived for 50 years. The theory was first considered seriously in 1908, when an American geologist, Frank B. Taylor, pointed out a number of geologic facts that could be explained by continental drift.

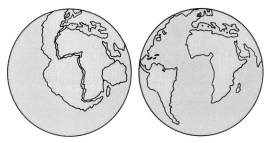

(A) Maps made by Antonio Snider-Pelligrini in 1858.

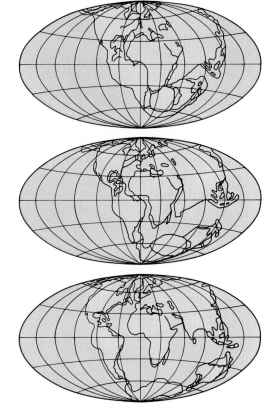

FIGURE 18.1
Continental drift was first illustrated by Antonio Snider-Pelligrini in 1858 when he published these maps in his book *La Création et ses Mystères dévoiles* (A). The idea seemed too farfetched to the public and scientific communities of the time and was forgotten, not to be revived for 50 years. Wegener later published his series of maps in 1915 (B). His evidence, most of which was quite valid, was drawn from all of the sciences. Wegener called the original landmass Pangaea ("all lands") and believed that the continents somehow plowed through the oceanic crust as they drifted.

(B) Maps made by Alfred Wegener in 1915.

But Alfred Wegener, a German meteorologist, was the first to make an exhaustive investigation of the idea of continental drift and to convince others to take it seriously. In his book *The Origin of the Continents and Oceans* (1915), Wegener based his theory not only on the shapes of the continents, but also on geologic evidence such as similarities in the fossils found in Brazil and Africa. He drew a series of maps showing three stages in the drift process, beginning with an original large landmass, which he called Pangaea (meaning "all lands"). Wegener believed that the continents, composed of light silicic rock, somehow plowed through the denser rocks of the ocean floor, driven by forces related to the rotation of Earth (Figure 18.1).

Most geologists and geophysicists rejected Wegener's theory, although many scientific observations supporting it were known at the time. A few noted scholars, however, seriously considered the theory. Alexander L. du Toit, from South Africa, compared the landforms and fossils of Africa and South America and further expounded the theory in his book *Our Wandering Continents* (1937). Arthur Holmes, of England, later developed it in his textbook *Principles of Physical Geology* (1944).

The early arguments concerning the breakup of the supercontinent *Pangaea* and the theory of **continental drift** were supported by some important and imposing evidence, most of which resulted from regional geologic studies.

Paleontological Evidence

The striking similarity of certain fossils found on the continents on both sides of the Atlantic is difficult to explain unless the continents were once connected. The fossil record indicates that a new species appears at one point and disperses outward from there. Floating and swimming organisms could migrate in the ocean from the shore of one continent to another, but the Atlantic Ocean would present an insurmountable obstacle for the migration of land-dwelling

What evidence indicates that continents split and drift apart?

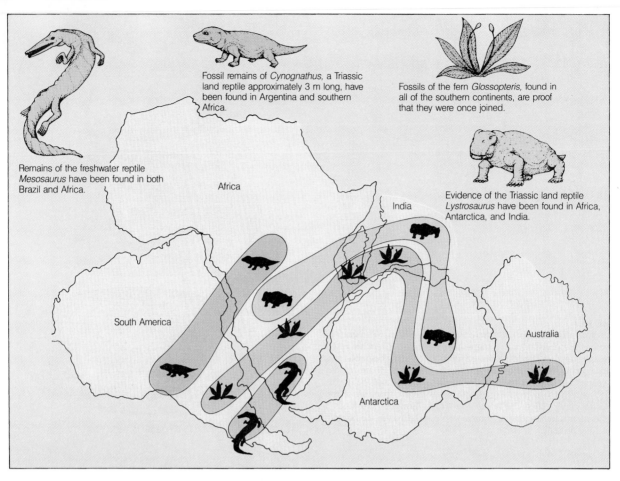

FIGURE 18.2
Paleontologic evidence of continental drift can be appreciated by considering the distribution of some fossil plants and animals found in South America, Africa, Madagascar, India, Antarctica, and Australia. *Mesosaurus,* a Permian freshwater reptile, is found in both Brazil and South Africa. *Glossopteris,* a fossil fern, is found in all of the southern continents in the zone shown on the map. *Lystrosaurus,* a Triassic land reptile, is found in South Africa, South America, India, and Antarctica. *Cynognathus,* an older Triassic reptile, is found in Argentina and South Africa.

animals, such as reptiles and insects, and certain land plants. Consider the following examples (Figure 18.2).

Fossils of *Glossopteris,* a fernlike plant, have been found in rocks of the same age from South America, South Africa, Australia, and India and within 480 km of the South Pole, in Antarctica. Mature seeds of this plant were several millimeters in diameter, too large to have been dispersed across the ocean by winds. The simultaneous presence of *Glossopteris* on all of the southern continents, therefore, is strong supporting evidence that the continents were once connected.

The distribution of Paleozoic and Mesozoic reptiles provides similar evidence, as fossils of several species have been found in the now separated southern continents. An example is a mammal-like reptile belonging to the genus *Lystrosaurus.* This creature was strictly a land dweller. Its fossils are found in abundance in South Africa, South America, and Asia, and in 1969 a United States expedition discovered them in Antarctica. This genus thus inhabited all of the southern continents. Clearly, these reptiles could not have swum thousands of kilometers across the Atlantic and Antarctic oceans, so some previous connection of the continents must be postulated. A former land bridge, similar to present-day Central America, would explain the presence of *Lystrosaurus* in distant parts of the world, but surveys of the ocean floor show no evidence of such a submerged land bridge.

Evidence from Structure and Rock Type

A number of geologic features end abruptly at the coast of one continent and reappear on the facing continent across the Atlantic (Figure 18.3). Folded mountain ranges at the Cape of Good Hope, at the southern tip of Africa, trend from east to west and terminate sharply at the coast. An equivalent structure, of the same age and style of deformation, appears near Buenos Aires, Argentina.

The folded Appalachian Mountains are another example. The deformed structures of the mountain belt extend northeastward across the eastern United States and through Newfoundland and terminate abruptly at the ocean. They reappear at the coasts of Ireland and Brittany.

Other examples could be cited, but the important point is that the continents on both sides of the Atlantic fit together, not only in outline, but in rock type and structure. They are related much like matching pieces of a torn newspaper (Figure 18.4). The jagged edges fit, and the printed lines (structure

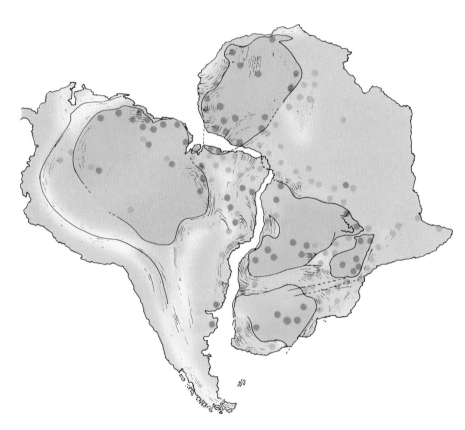

FIGURE 18.3
South America and Africa fit together, not only in outline, but in rock types and geologic structure. The green areas represent the shields of metamorphic and igneous rocks, formed at least 2 billion years ago. The gray areas represent younger rock, much of which has been deformed by mountain building. Structural trends such as fold axes are shown by dashed lines. Most of the deformation occurred from 450 million to 650 million years ago. Several fragments of the African shield are stranded along the coast of Brazil. Green dots represent rocks that are more than 2 billion years old. Orange dots represent younger Precambrian rocks.

FIGURE 18.4
The continents fit together like a jigsaw puzzle or pieces of a torn newspaper. Not only do the outlines of the torn pieces fit together, but the printing on them (analogous to the ages and structural features of the continents) also matches across the edges of the separate pieces.

and rock types) join together in a coherent unit. One important point needs emphasis. The geologic similarities on opposite sides of the Atlantic are found only in rocks older than the Cretaceous period, which began about 137 million years ago. The continents are believed to have split and begun drifting apart in Jurassic time, about 200 million years ago.

Evidence from Glaciation

During the latter part of the Paleozoic Era (about 300 million years ago), glaciers covered large portions of the continents in the Southern Hemisphere. The deposits left by these ancient glaciers can be readily recognized, and striations and grooves on the underlying rock show the direction in which the ice moved (Figure 18.5A). Except for Antarctica, all of the continents in the Southern Hemisphere now lie close to the equator. In contrast, the continents in the Northern Hemisphere show no trace of glaciation during this time. In fact, fossil plants indicate a tropical climate in that area. This evidence is difficult to explain in the context of fixed continents because the climatic belts are determined by latitude.

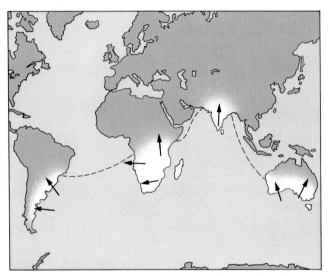

(A) The distribution of late Paleozoic glacial deposits is restricted to the Southern Hemisphere (except for India). Arrows indicate that the direction of ice movement was from the sea toward the land, which is impossible because glaciers flow from centers of accumulation on the continents outward toward the sea. These areas are now close to the tropics. The present-day cold latitudes in the Northern Hemisphere show no evidence of glaciation during the late Paleozoic.

(B) If the continents were restored to their former positions according to Wegener's theory of continental drift, and if the former South Pole were located approximately where South Africa and Antarctica meet, the location of late Paleozoic glacial deposits and the directions in which the ice flowed would be explained nicely.

FIGURE 18.5
Distribution and flow direction of late Paleozoic glaciers provide further evidence of continental drift.

Even more difficult to explain is the direction in which the glaciers moved. Regional mapping of striations and grooves indicates that, in South America, India, and Australia, the ice moved inland from the oceans. Such movement would be impossible unless there was a landmass where the oceans now exist.

If the continents were grouped together as Wegener proposed, the glaciated areas would have comprised a neat package near the South Pole (Figure 18.5B), and Paleozoic glaciation could be explained nicely. The pattern of glaciation was considered strong evidence of continental drift, and many geologists who worked in the Southern Hemisphere became ardent supporters of the theory because they could see the evidence with their own eyes.

Evidence from Other Paleoclimatic Records

Other evidence of striking climatic changes tends to support the drift theory. Great coal deposits in Antarctica show that abundant plant life once flourished on that continent, now mostly covered with ice.

Why did most geologists reject the idea of continental drift?

On the other continents, salt deposits, formations of windblown sandstone, and coral reefs provide additional clues that permit us to reconstruct the climatic zones of the past. The paleoclimatic patterns are baffling with the continents in their present positions, but if they are grouped together in their predrift positions, the patterns are easily explained (Figure 18.6).

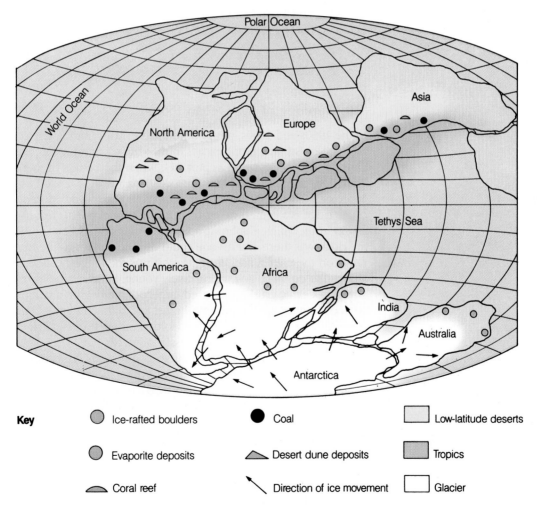

Key

- ⬤ Ice-rafted boulders
- ⬤ Evaporite deposits
- ◠ Coral reef
- ● Coal
- ◣ Desert dune deposits
- ╲ Direction of ice movement
- ▢ Low-latitude deserts
- ▢ Tropics
- ▢ Glacier

FIGURE 18.6
Paleoclimatic evidence for continental drift includes deposits of coal, desert sandstone, rock salt, windblown sand, gypsum, and glacial deposits. Each indicates a specific climatic condition at the time of its formation. The distribution of these deposits is best explained if we assume that the continents were once grouped together as shown in this diagram.

The evidence for the theory of continental drift was considered and debated for years. Wegener was criticized for failing to explain what forces would permit continents of granite to plow through oceans of rock. The idea of a moving lithosphere was yet to come. In the absence of a reasonable mechanism for drift, there was little further development of the theory until after World War II. An explosion of knowledge then provided renewed support for the drift hypothesis and also led to the discovery of a possible mechanism.

DEVELOPMENT OF THE THEORY OF PLATE TECTONICS

The plate tectonics theory was developed during the early 1960s, when new instrumentation permitted scientists to map the topography of the ocean floor and to study its magnetic and seismic characteristics.

Although the theory of continental drift was supported by some convincing evidence, the data upon which it was originally based came only from the continents because, prior to the 1950s, there was no effective means of studying the ocean floor. Before 1950, therefore, geologists faced an almost total absence of data about the geology of three-fourths of Earth's surface; then, in the 1950s and 1960s, a burst of new data and new ideas resulted from research efforts in two areas:

1. The geology of the ocean floor
2. Paleomagnetism

The Geology of the Ocean Floor

In the 1950s and 1960s, newly developed echo-sounding devices enabled marine geologists and geophysicists to map the topography of the ocean floor in considerable detail (see Chapter 21). When the results of these studies were compiled, they revealed that the ocean basins are divided by a great ridge approximately 64,000 km long and about 1500 km wide. Moreover, at the crest of the ridge is a central valley, from 1 to 3 km deep. This feature appears to be a *rift valley*, which is splitting apart under tension. No one could imagine why the ridge was there, but no one could dispute that it was the longest mountain range on the planet, and along its crest was the longest valley. Other evidence shows that ocean basins are relatively young. Seismic studies have established that the oceanic crust (composed largely of basalt) has a completely different composition from the continental crust and is much thinner. Furthermore, the oceanic crust is not deformed into folded mountain structures and apparently is not subjected to strong compressional forces.

In 1960, H. H. Hess, a noted geologist from Princeton University, proposed a theory of *sea-floor spreading,* which took into account the new data from echo soundings and also suggested a possible mechanism for continental drift. Hess postulated that the ocean floors are spreading apart, propelled by convection currents in the mantle, and are moving symmetrically away from the oceanic ridge. According to his theory, this continuous spreading produces fractures in the rift valley, and magma from the mantle is injected into these fractures to become new oceanic crust. The convection currents in the mantle carry the continents away from the oceanic ridge and toward deep-sea trenches. There, the oceanic crust descends into the mantle with the descending convection current and is reabsorbed. In this way, the entire ocean floor is completely regenerated in 200 million or 300 million years.

Hess thus elaborated on the theory of continental drift in the light of fresh knowledge and redefined it in the scheme of sea-floor spreading. A test of his ideas, using new studies in paleomagnetism, was soon to follow.

Paleomagnetism

The study of rock magnetism developed during the 1950s with the perfection of new, highly sensitive magnetometers. Certain rocks, such as basalt, are fairly rich in iron and become weakly magnetized by Earth's magnetic field as they cool. In a sense, the mineral grains in the rock become "fossil" magnets, oriented with respect to Earth's magnetic field at the time when the rock was formed, and thus preserve a record of *paleomagnetism.* Similarly, the iron in grains of red sandstone becomes oriented in Earth's magnetic field as the sediment is deposited, so red sandstone also can indicate the orientation of the paleomagnetic fields. These rocks therefore retain an imprint of Earth's magnetic field at the time of their formation.

The magnetic field of Earth resembles that of a simple bar magnet with its axis inclined 11 degrees from Earth's geographic axis (Figure 18.7A). The mantle and core of Earth are far too hot to retain a permanent magnetic field. Earth's magnetism, therefore, must be generated electromagnetically. The electromagnetic, or dynamo, theory postulates that the outer core of liquid iron slowly rotates with respect to the surrounding mantle. Such motion would generate electrical currents, which would establish a magnetic field (Figure 18.7B).

The theory of plate tectonics is a simple, straightforward explanation of how Earth works. Why wasn't it discovered earlier?

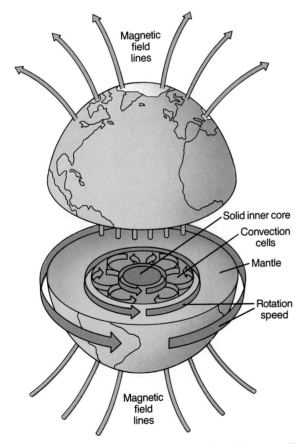

(A) Lines of force in Earth's magnetic field are shown by arrows. If a magnetic needle were free to move in space, it would be deflected by Earth's magnetic field. Close to the equator, the needle would be horizontal and would point toward the poles. At the magnetic poles, the needle would be vertical.

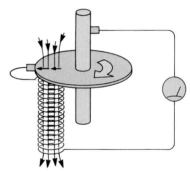

(B) Theoretically, convection in Earth's core can generate an electrical current (in a manner similar to the operation of a dynamo), which produces a magnetic field.

FIGURE 18.7
Earth's magnetic field is like that of a simple bar magnet. The temperature in the core and mantle, however, is far too high for permanent magnetism. Earth's magnetism must therefore be generated electromagnetically.

How does polar wandering support the plate tectonics theory?

Apparent Polar Wandering. Studies of paleomagnetism in European rocks of widely different ages demonstrate that Earth's north magnetic pole apparently has steadily changed its position with time. As is shown in Figure 18.8, the pole appears to have slowly migrated northward and westward to its present position. The change in position was systematic, not random. A similar migration of the magnetic pole was found from paleomagnetic work in North America, and, although the path of migration was systematically different, it paralleled that of the European shift. These observations could be explained nicely by drifting continents, so students of paleomagnetism became leading proponents of the theory of continental drift. Soon results collected from the southern continents were reported. Again, a systematic change in the position of the magnetic pole through time was documented—but with different paths for different continents.

It is impossible that there were numerous magnetic poles migrating systematically and eventually merging. The most logical explanation is that there has always been only one magnetic pole, which has remained fixed while the continents moved with respect to it. The results of paleomagnetic studies make sense if the continents were once arranged as shown in Figure 18.8B and then drifted to their present positions. This discovery brought renewed interest in the theory of continental drift and lent support to the conclusion that the Atlantic Ocean opened relatively recently.

Patterns of Magnetic Reversals on the Sea Floor. Recent studies of the magnetic properties of numerous samples of basalt from many parts of the world demonstrate that Earth's magnetic field has been reversed many times over the last 70 or 80 million years. Epochs of *normal polarity* (that is, periods when the magnetic field was oriented as it is today and the magnetic poles were thus close to their present location), lasting from 1 to 3 million years, have been followed by similar periods during which the north magnetic pole and the south magnetic pole were reversed. At least nine magnetic reversals

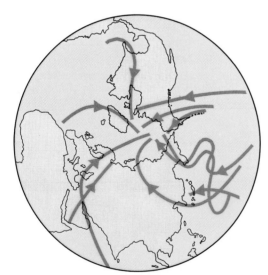

(A) The magnetic properties of rocks in North America suggest that the north magnetic pole has apparently migrated in a sinuous path over the last several hundred million years. Evidence from other continents shows similar migration, but along different paths. How could different continents show different paths of polar migration? The paleomagnetic evidence implies that, if the continents had remained fixed, different continents would have had different magnetic poles at the same time, but that would be impossible.

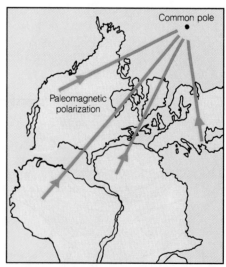

(B) The question can be answered if the pole has remained fixed while the continents have drifted. If, for example, Europe and North America were previously joined, the paleomagnetic field preserved in their rocks would indicate a single pole location until they drifted apart. The sequence of rocks on each continent would then show the pole taking a different path of migration to its present position.

FIGURE 18.8
Changes in the position of the magnetic poles during the geologic past are indicated by paleomagnetic studies of rocks.

have occurred in the last 4.5 million years. The present period of normal polarity began about 700,000 years ago. It was preceded by a period of *reversed polarity,* which began 2.5 million years ago and lasted about 2 million years. That period of generally reversed polarity contained two short periods of normal polarity.

The major intervals of alternating polarity (about 1 million years apart) are termed *polarity epochs,* and the intervals of shorter duration are termed *polarity events.* The pattern of alternating polarities has been clearly defined, and evidence of the occurrence of polarity epochs has been found in widely separated parts of Earth. From the sequence of *magnetic anomalies* and their radiometric ages, a reliable chronology of *magnetic reversals* has been established for the last 4 million years (Figure 18.9). In addition, extrapolation as far back as 76 million years reveals a sequence of at least 171 reversals.

In 1963, Fred Vine and D. H. Matthews saw a way to test the idea of sea-floor spreading put forth by Hess. If sea-floor spreading has occurred, they suggested, it should be recorded in the magnetism of the basalts in the oceanic crust. (The same idea was developed independently by L. W. Morley.) If Earth's magnetic field reversed intermittently, new basalt forming at the crest of the oceanic ridge would be magnetized according to the polarity at the time when it cooled. As the ocean floor spreads, a symmetrical series of magnetic stripes, with alternating normal and reversed polarities, would be preserved in the crust along either side of the oceanic ridge. Subsequent investigations have conclusively proved this theory proposed by Vine and Matthews and by Morley.

To better understand the origin of these magnetic patterns, consider how the sea floor could have evolved during the last few million years. Figure 18.10A shows the sea floor as it is considered to have been about 2.75 million years ago, during the Gauss normal polarity epoch (named for the German mathematician Karl Friedrich Gauss). Basalt, injected into the fractures of the oceanic ridge, formed dikes or was extruded over the sea floor as submarine flows. As it solidified, it became magnetized in the direction of the existing (normal) magnetic field, and thus basalt extruded along the oceanic ridge formed a zone of new crust with normal magnetic polarity. As the sea floor spread, this zone of crust split and migrated away from the ridge but remained parallel to it. About 2.5 million years ago, Earth's magnetic polarity was reversed. New crust generated at the oceanic ridge was magnetized in the opposite direction (Figure 18.10B), producing a zone of crust with reverse polarity. When the polarity changed to normal again, the newest crustal material was magnetized in the normal direction. In this way, the sequence of polarity reversals became imprinted as magnetic stripes on the oceanic crust.

Note that the patterns of magnetic stripes on the ocean floor on either side of the ridge match the patterns found in a sequence of recent basalts on the continents (Figure 18.10A and B; see also Figure 18.9C); that is, the crest of the ridge shows normal polarity and is flanked on either side by a broad stripe of rocks with reversed polarity (formed during a reversed epoch) and containing two narrow bands of rocks with normal polarity (formed during normal epochs). Then follows a stripe with normal polarity containing one narrow band with reversed polarity, and so on. In brief, the patterns of magnetic reversals away from the rest of the ridge are the same as those found in a vertical sequence of rocks on the continents, from youngest to oldest. These data provide compelling evidence that the sea floor is spreading and that continents drift.

An important aspect of these reversal patterns is that they enable us to determine rates of plate movement. Magnetic reversals in rock sequences on the continents have been radiometrically dated. These studies show that the present normal polarity has existed for the last 700,000 years and was preceded by the pattern shown in Figure 18.9C. Because the same pattern exists in the oceanic crust, we can assign provisional ages to the magnetic anomalies on the ocean floor.

What are magnetic reversals?

How do patterns of magnetic reversals support the plate tectonics theory?

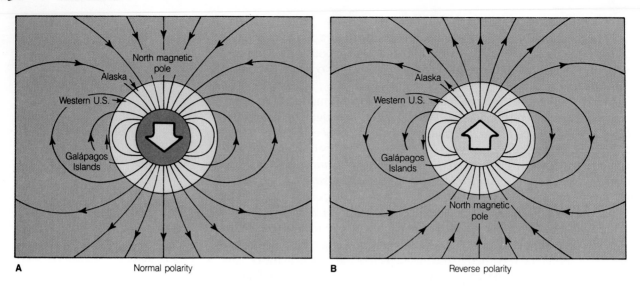

A Normal polarity   **B** Reverse polarity

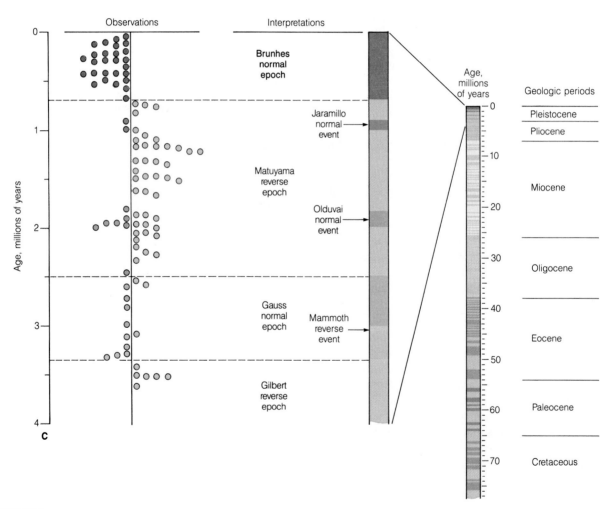

FIGURE 18.9
Reversals of lines of force in Earth's magnetic field are documented by paleomagnetic studies of numerous rock samples from throughout the world. Lines of force with normal polarity are shown in diagram **A.** With reverse polarity (diagram **B**), the lines of force are oriented in the opposite direction. Diagram **C** shows the patterns of changing polarity with time. The pattern of change during a period of 1 million or 2 million years is distinctive, and it can be used to help establish the age of a rock sequence.

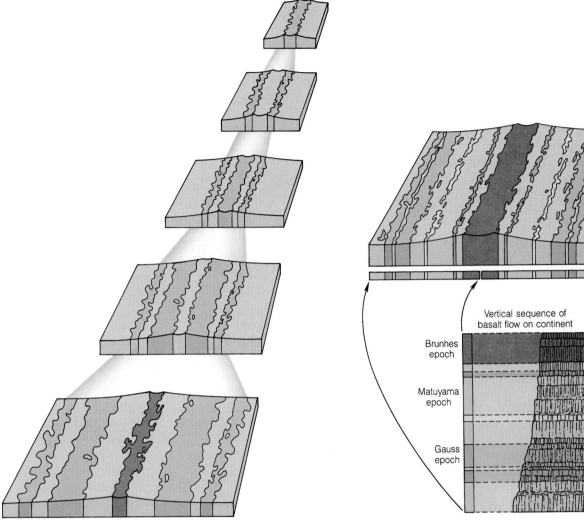

(A) As magma cools and solidifies along the ridge in dikes and flows, it becomes magnetized in the direction of the magnetic field existing at that time (normal polarity). As sea-floor spreading continues, the magnetized crust formed during earlier periods separates into two blocks. Each block is transported laterally away from the ridge, as though on a conveyor belt. New crust, formed at the ridge, becomes magnetized in the opposite direction. Note that the pattern of magnetic reversals away from the ridge is the same as the pattern found in a sequence of basalt flows on the continents.

(B) Patterns of magnetic reversals in a vertical sequence of basalts on the continents. The youngest (upper) continental rocks correlate with the youngest oceanic crust (at the center of the oceanic ridge).

FIGURE 18.10
Specific patterns of magnetism are preserved in the newly formed crust, which is generated at the oceanic ridge as the lithosphere moves laterally. The patterns of magnetic reversals away from the ridge are identical to the patterns of magnetic reversals in a vertical sequence of rocks on the continents.

Magnetic surveys have now determined patterns of magnetic reversal for much of the ocean floor, and from these patterns, the age of various segments of the sea floor has been established (Figure 18.11). These studies show that most of the deep-sea floor was formed during Cenozoic time (during the last 65 million years). It now seems probable that very little or none of the present ocean basin was formed before the Jurassic. From the pattern of magnetic reversals, the rate of sea-floor spreading appears to range from 1 to 17 cm per year.

Iceland, because it is essentially a large exposure of the *mid-Atlantic ridge,* offers a unique opportunity to study the physical mechanism of sea-floor

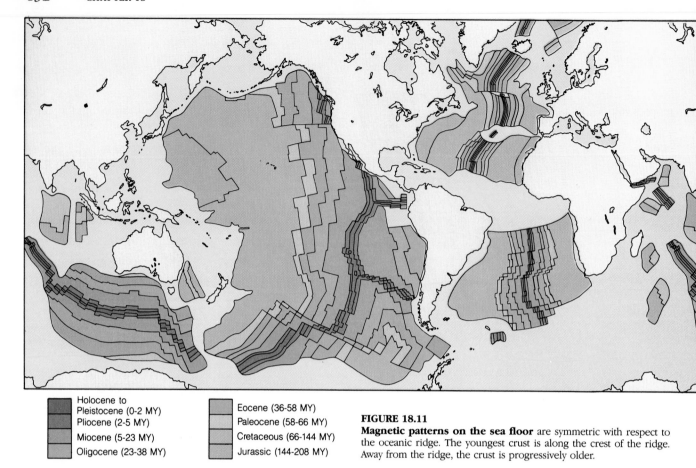

Holocene to
Pleistocene (0-2 MY)
Pliocene (2-5 MY)
Miocene (5-23 MY)
Oligocene (23-38 MY)

Eocene (36-58 MY)
Paleocene (58-66 MY)
Cretaceous (66-144 MY)
Jurassic (144-208 MY)

FIGURE 18.11
Magnetic patterns on the sea floor are symmetric with respect to
the oceanic ridge. The youngest crust is along the crest of the ridge.
Away from the ridge, the crust is progressively older.

spreading. Geologic studies there show that the island is being pulled apart by
the spreading crust beneath. The tension causes faults and fissures parallel to
the axis of the ridge. Volcanic eruptions occur through these fissures, and
swarms of parallel vertical dikes are injected into them with each increment of
crustal extension. The aggregate width of these dikes is about 400 km, which
corresponds to the total amount of crustal extension since the beginning of
Tertiary time, about 65 million years ago. A geologic map of Iceland (Figure
18.12) shows that the oldest rocks are at the extreme eastern and western ends
of the island. Rocks become progressively younger toward the center of the
island, where most of the present-day fissures and volcanism occur.

FIGURE 18.12
A geologic map of Iceland shows that the
oldest rocks are along the eastern and western
margins, and the youngest rocks are near the
center of the island. The pattern is identical to
that of rocks on either side of the oceanic
ridge.

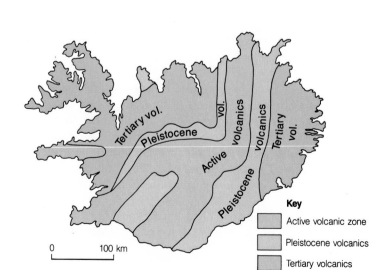

Key

Active volcanic zone

Pleistocene volcanics

Tertiary volcanics

0   100 km

Evidence from Sediment on the Ocean Floor

To many geologists, some of the most convincing evidence for the plate tectonics theory comes from recent drilling in the sediment on the ocean floor. The Deep-Sea Drilling Project is a truly remarkable expedition in scientific exploration. It began in 1968 with the *Glomar Challenger,* a special ship designed by a California offshore drilling company. The *Challenger* can lower more than 6100 m of drilling pipe into the open ocean, bore a hole in the sea floor, and bring up bottom cores and samples. The project was funded by the National Science Foundation and was planned by JOIDES—Joint Oceanographic Institutions for Deep Earth Sampling—under the direction of the Scripps Institution of Oceanography. Since 1968, the *Challenger* has drilled hundreds of holes in the sea floor and has penetrated more than a kilometer into the oceanic crust. This project has provided many data in support of the theory of plate tectonics.

Deep-sea drilling confirms the conclusions drawn from paleomagnetic studies by providing samples of the fossils that first accumulated on different portions of the ocean floor. As is predicted by the plate tectonics theory, the youngest sediment resting on the basalt of the ocean floor is found near the oceanic ridge, where new crust is being created. Away from the ridge, the sediments that lie directly above the basalt become progressively older, with the oldest sediment nearest the continental borders.

Measurements of rates of sedimentation in the open ocean show that about 3 mm of red clay accumulates every 1000 years. If the present ocean basins were old enough to have existed since Cambrian time, for example, the sediments would be 1.5 km thick (Figure 18.13A), but the average thickness of deep-ocean sediments measured to date is only 300 m, suggesting that the ocean basins are younger geologic features (Figure 18.13B). In fact, the oldest sediments yet found on any ocean floor are only about 200 million years old. In contrast, the metamorphic rocks of the continental shields are as much as 3.8 billion years old.

Not only do the thickness and age of the deepest sediments increase away from the crest of the oceanic ridge, but certain types of sediment also indicate sea-floor spreading. For example, plankton thrive in the upwelling, warm,

How does sediment on the ocean floor support the plate tectonics theory?

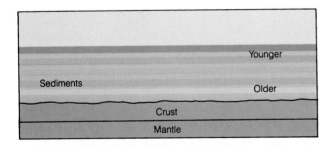

(A) With no sea-floor spreading, the entire ocean floor would be covered with a thick sequence of oceanic sediment, with alternating polarity preserving a record of Earth's magnetism since the Precambrian.

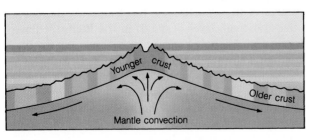

(B) With sea-floor spreading, the blanket of oceanic sediment thins progressively toward the crest of the oceanic ridge and is almost nonexistent in the rift valley. The edge of each layer of magnetized sediment lies upon basaltic crust, which was generated at the spreading center during the same time interval when the sediment was deposited.

FIGURE 18.13
The thickness of sediment and magnetic reversals on the oceanic ridge confirm the theory of sea-floor spreading.

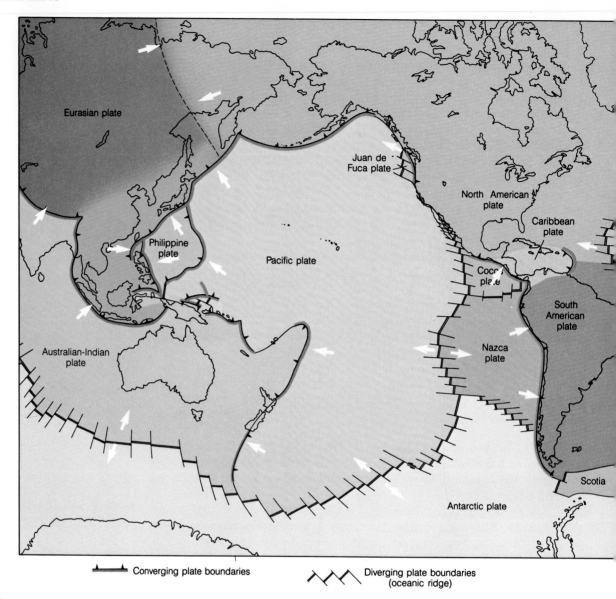

Converging plate boundaries

Diverging plate boundaries
(oceanic ridge)

FIGURE 18.14
The major tectonic plates are delineated by the major tectonic features of the globe: (1) the oceanic ridge, (2) deep-sea trenches, and (3) young mountain belts. Plate boundaries are outlined by earthquake belts and volcanic activity. Most plates (such as the North American, African, and Australian) contain both continental and oceanic crust. The Pacific, Cocos, and Nazca plates contain predominantly oceanic crust.

nutrient-rich water of the Pacific equatorial zone. As the creatures die, their tiny skeletons rain down unceasingly to build a layer of soft, white chalk on the sea floor. The chalk can form only in the equatorial belt, as plankton do not flourish in the colder waters of higher latitudes; yet drilling by the *Glomar Challenger* has shown that the chalk layer at depth in the Pacific sediments extends north of today's equator. The only logical conclusion is that the Pacific sea floor has been migrating northward for at least 100 million years.

The theory of plate tectonics is now firmly established and accepted as the fundamental theory of Earth's dynamics. It was first used to explain the meaning of features on the ocean floor. Now the emphasis has switched to the continents, and most previous geologic observations of the continents are being reexamined in light of plate tectonics theory.

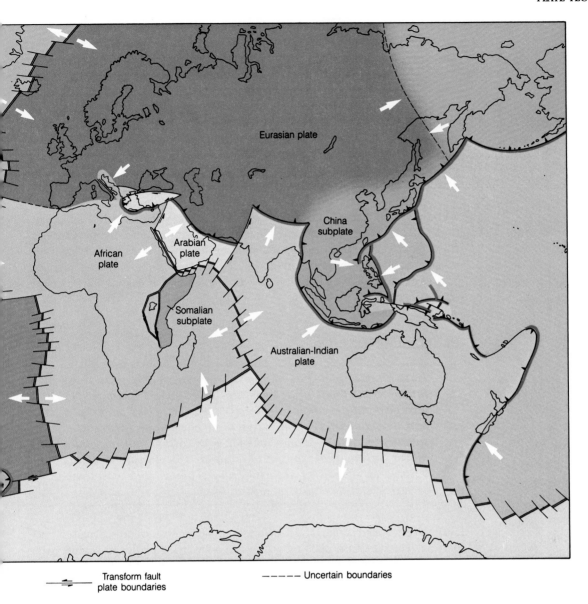

| ⇌ | Transform fault plate boundaries | – – – – – Uncertain boundaries |

PLATE GEOGRAPHY

Plate boundaries are the most significant structural elements of the planet Earth because they reflect the internal dynamics of the planet.

The shorelines of the continents are major geographic features but have little significance from the standpoint of Earth's tectonics. Plate boundaries are the most significant structural elements of the planet, and to understand plate tectonics, you must learn a new geography—the geography of plate boundaries. This should not be difficult, because plate boundaries generally are marked by major topographic features. Understanding requires only that you focus your attention on the structural features of Earth, rather than on the boundaries between land and ocean.

The new geography of *tectonic plates* is illustrated in Figure 18.14. The outer rigid layer of Earth—the lithosphere—is divided into a mosaic of seven major plates and a number of smaller subplates. The major plates are outlined

How does the geography of tectonic plates differ from classical physical geography? Why is it more important geologically?

by oceanic ridges, trenches, and young mountain systems. These include the Pacific, Eurasian, North American, South American, African, Australian, and Antarctic plates. The largest is the Pacific plate, which is composed almost entirely of oceanic crust and covers about one-fifth of Earth's surface. The other large plates contain both continental crust and oceanic crust. No major plate is composed entirely of continental crust. Smaller plates include the China, Philippine, Arabian, Juan de Fuca, Cocos, Caribbean, and Scotia plates, plus a number of others that have not yet been precisely defined. The smaller plates appear to form near convergent boundaries of major plates where collision between continents occurs or between a continent and an island arc. The smaller plates are thus characterized by rapid and complex movement.

Individual plates are not permanent features. They are in constant motion and continually change in size and shape. Plates that do not contain continental crust can be completely consumed in a subduction zone. Plate margins are not fixed. A plate can change its shape by splitting along new lines, by welding itself to another plate, or by accretion of new oceanic crust along its passive margin. The movement and modification of a plate margin can change the size and shape of the entire plate.

PLATE BOUNDARIES

Three kinds of plate boundaries are recognized and define three fundamental kinds of deformation and geologic activity: (1) divergent plate boundaries—zones of tension, where plates split and spread apart, (2) convergent plate boundaries (also called subduction zones)—zones where plates collide and one plate moves down into the mantle, and (3) transform fault boundaries—zones of shearing, where plates slide past each other without diverging or converging.

Each tectonic plate is rigid and moves as a single mechanical unit; that is, if one part moves, the entire plate moves. It can be warped or flexed slightly as it moves, but relatively little change occurs in the middle of a plate. Nearly all major tectonic activity occurs along the plate boundaries, and thus geologists and students of geology focus their attention on the plate margins, the ones that are active as well as the ancient plate boundaries preserved on the continents.

Three kinds of plate boundaries are recognized (Figure 18.15):

FIGURE 18.15
Types of plate margins are depicted in this idealized diagram. Constructive margins (divergent plate boundaries) occur along the oceanic ridge, where plates move apart. Destructive margins (convergent plate boundaries) occur along the deep trenches. Margins with no change in sea-floor area during displacement occur along transform faults.

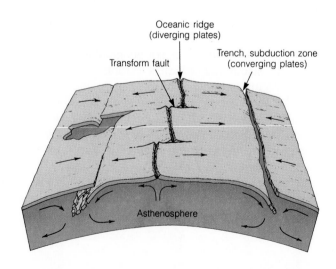

Oceanic ridge
(diverging plates)

Transform fault

Trench, subduction zone
(converging plates)

Asthenosphere

1. Divergent plate boundaries
2. Convergent plate boundaries (also called subduction zones)
3. Transform fault boundaries

Processes at Divergent Plate Boundaries

Divergent plate boundaries, or **spreading axes,** form where a plate splits and is pulled apart. Where a zone of spreading extends into a continent, rifting occurs, and the continent splits (Figure 18.16). The separate continental fragments drift apart with the diverging plates, so a new and continually enlarging ocean basin is formed at the site of the initial rift zone. Divergent plate boundaries are thus characterized by tensional stresses that produce block faulting, fractures, and open fissures along the margins of the separating plates. Basaltic magma derived from the partial melting of the mantle is injected into the fissures or extruded as fissure eruptions. The magma then cools and becomes part of the moving plates. Divergent plate boundaries are some of the most active volcanic areas on Earth, but are generally characterized by unspectacular, quiet fissure eruptions, most of which are concealed beneath the sea. The importance of volcanism along this zone is underlined by the fact that more than half of Earth's surface has been created by volcanic activity along divergent plate boundaries during the past 200 million years.

What major geologic processes occur at the three types of plate boundaries?

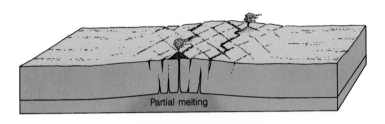

(A) Continental rifting begins when the crust is uparched and stretched, so that block faulting occurs. Continental sediment accumulates in the depressions of the downfaulted blocks, and basaltic magma is injected into the rift system. Flood basalt can be extruded over large areas of the rift zone during this phase.

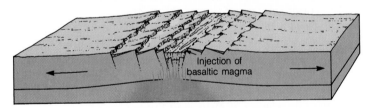

(B) Rifting continues, and the continents separate enough for a narrow arm of the ocean to invade the rift zone. The injection of basaltic magma continues and begins to develop new oceanic crust.

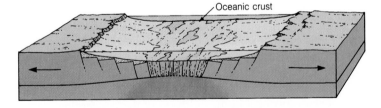

(C) As the continents separate, new oceanic crust and new lithosphere are formed in the rift zone, and the ocean basin becomes wider. Remnants of continental sediment can be preserved in the downdropped blocks of the new continental margins.

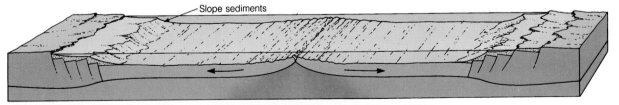

(D) As spreading continues, the ocean basin grows larger. The continents move off from the uparched spreading zone, and parts of the continental crust can be covered by the ocean.

FIGURE 18.16
Stages of continental rifting are shown in this series of diagrams. The major geologic processes at divergent plate boundaries are tensional stress, block faulting, and basaltic volcanism.

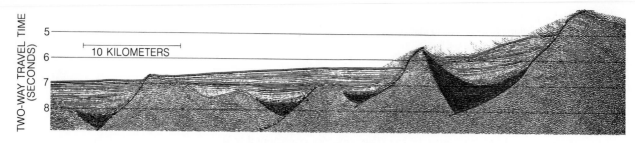

FIGURE 18.17
Seismic images of divergent plate margins show major fault blocks produced when the brittle, upper layers of the continental crust are stretched and arched upward. The fault blocks slide down curved fault planes and produce a series of ridges. The troughs are filled with sediment eroded off the fault blocks and deposited in lake basins. Younger marine sediments cover the fault blocks and older lake sediments.

Except for a few rift zones in Africa and western North America, essentially all present divergent plate margins are submerged beneath the sea, and these regional characteristics cannot be observed. Seismic-reflection techniques have recently provided images of divergent plate boundaries that were previously unavailable. One such image is shown in Figure 18.17. This image clearly shows a series of tilted fault blocks produced by stretching and thinning of the lithosphere. Curved fault planes have developed as a result of tension, and the resulting fault blocks were tilted as they rotated, producing a series of ridges and troughs. The troughs are partly filled with sediment produced by erosion and deposition occurring contemporaneously with fault displacement.

Examples of continental rifting in various stages are found in various parts of the world. The initial stage is represented by the system of great rift valleys in East Africa. The long, linear valleys, partly occupied by lakes, are huge, downdropped fault blocks, which result from the initial tensional stress. Magma rising from the mantle into the rift zone produces volcanism, exemplified by the great volcanoes of Mount Kenya and Mount Kilimanjaro. The Red Sea illustrates a more advanced stage of rifting. The Arabian Peninsula has been completely separated from Africa, and a new linear ocean basin is just beginning to develop. The Atlantic Ocean represents a still more advanced stage of continental drift and sea-floor spreading. The American continents have become separated from Africa and Europe by thousands of kilometers. The mid-Atlantic ridge is the boundary between the diverging plates, with the American plates moving westward in relation to Africa and Europe.

Processes at Convergent Plate Boundaries

Convergent plate boundaries, or *subduction zones,* where the plates collide and one moves down into the mantle, are areas of complicated geologic processes, including igneous activity, crustal deformation, and mountain building. The specific processes that are active along a convergent plate boundary depend on the types of crust involved in the collision of the converging plates.

If both plates at a convergent boundary contain oceanic crust, one is thrust under the margin of the other, in a process called *subduction.* The subducting plate descends into the asthenosphere, where it is heated and ultimately absorbed in the mantle. If one plate contains a continent, the lighter continental crust always resists subduction and overrides the oceanic plate. If both converging plates contain continental crust, neither can subside into the mantle, although one can override the other for a short distance. Both continental masses are instead compressed, and the continents are ultimately "fused" or "welded" together into a single continental block, with a mountain range marking the line of suture.

The zone of convergence between two plates is a zone of deformation, mountain building, and metamorphism. If the overriding plate contains continental crust, compression deforms the margins into a folded mountain belt, and the deep roots of the mountains are metamorphosed.

In subduction zones, the interaction between the two converging plates can easily be seen on seismic-reflection profiles. In many subduction zones, like that shown in Figure 18.18, the oceanic crust consists of three easily identifiable layers: (1) unconsolidated sediment, (2) lithified sediment, and (3) basalt. It is apparent from the seismic profile in Figure 18.18 that the unconsolidated sediment layer is being scraped off the descending plate by the overriding plate and piled up, forming what is called an accretionary prism.

The consolidated sediment and the underlying basalt continue intact for about 50 km past the base of the ridge, indicating that some sedimentary rock is subducted. In contrast, the soft sediment is being deformed into a chaotic mass, or mélange, with thrust faults that generally dip toward the subduction zone.

The major processes and geologic phenomena that characterize convergent plate margins are shown in Figure 18.19. A subduction zone (or zone of underthrusting) usually is marked by a deep-sea trench, and the movement of the descending plate generates an inclined zone of seismic activity.

As cold, wet oceanic crust is subducted into the hot asthenosphere, water is driven off by metamorphic dehydration reactions at elevated temperatures. The water was originally incorporated into the oceanic crust on its path from the ridge to the subduction zone. As the light, water-rich fluids rise into the overlying wedge of mantle, they act as fluxes and lower the melting temperature of the mantle sufficiently to produce the distinctive magmas of subduction zones. The characteristic magmas of subduction zones are andesites, but more-silicic magmas are found there as well. Some geologists think that the magmas of subduction zones are products of direct melting of the oceanic crust as it becomes hot in the subduction zone. In any case, these hot magmas rise into the crust and there digest crustal materials as they crystallize, becoming more silicic in the process.

Some of this magma is extruded at the surface as lava and forms an island arc or a chain of volcanoes in the mountain belt of the overriding plate. Usually most of the magma intrudes into the deformed mountain belt to produce batholiths. Both extrusion and intrusion add new material to the continental

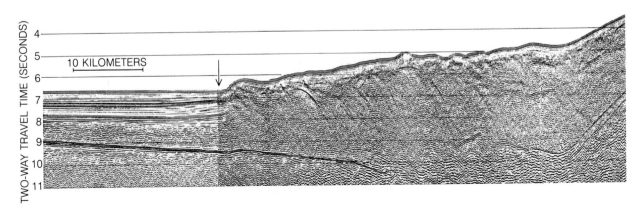

FIGURE 18.18
Seismic images of convergent plate boundaries show the zone where two lithospheric plates converge and one plunges under the other into the asthenosphere. Unconsolidated sediment on the descending plate may be scraped off and piled up, forming what is called an accretionary prism.

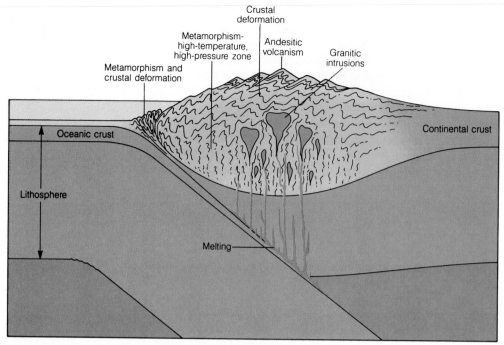

FIGURE 18.19
The major geologic processes at convergent plate boundaries include the deformation of
continental margins into folded mountain belts, metamorphism due to high temperatures and high
pressures in the mountain roots, and partial melting of the descending plate, which produces
granitic intrusions and andesitic volcanism on the overriding plate.

plate, and thus continents grow by accretion. This is an important mechanism
in the differentiation of Earth, whereby less-dense material, enriched in elements such as Si, Al, K, and Na, is concentrated in the upper layers of the planet.

Examples of the three types of convergent plate boundaries are illustrated in Figure 18.20. The convergence of the Pacific and the Eurasian and
Philippine plates has formed the island arc of Japan. Along the west coast of
South America, the convergence of the Nazca and South American plates has
produced the South American Trench and the Andes Mountains. The collision
of the Australian plate with Asia has formed the Himalaya Mountains.

Back-arc spreading (extension and spreading of the sea floor behind the
island arc) may also result at convergent plate margins, presumably as a result
of complex convective eddies in the asthenosphere above the subducting plate
or by pulling away of the adjacent plate (Figure 18.21). Either process could
cause regional tension, which would result in the formation of a back-arc basin,
characterized by crustal thinning and block faulting. The Sea of Japan and the
Lau Basin behind the Tonga arc are examples.

The back-arc region is somewhat similar to a major spreading axis, yet
there are substantial differences as well. Numerous indicators show that the
floors of the basins are young. Sediment is generally thin, and exposed rocks
include fresh basalt. Heat flow is high, but there is no well-defined ridge or rift
valley, and magnetic anomalies appear jumbled and unorganized. It appears,
therefore, that back-arc spreading is the product of a type of secondary sea-
floor spreading. Such phenomena are considered to have been common
throughout much of Earth's history.

Chapter 3 emphasized that continents are composed of lighter silicic
material that rides passively on the moving plates. Except for the small amount
of continentally derived sediment subducted back into the mantle, this material
is so buoyant that it cannot sink into the denser mantle material below. It is

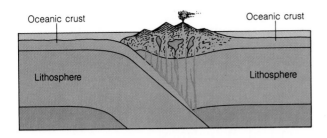

Oceanic crust   Oceanic crust

Lithosphere   Lithosphere

(A) The Japanese Islands represent the convergence of two oceanic plates.

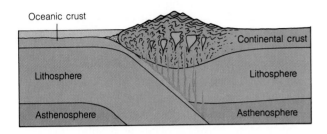

Oceanic crust

Lithosphere   Continental crust

Asthenosphere   Lithosphere

Asthenosphere

(B) South America represents the convergence of an oceanic plate and a continental plate.

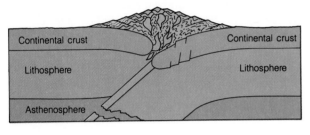

Continental crust   Continental crust

Lithosphere   Lithosphere

Asthenosphere

(C) The Himalaya Mountains represent the convergence of two continental plates.

FIGURE 18.20
Examples of the main types of convergent plate boundaries can be found today in various parts of the world.

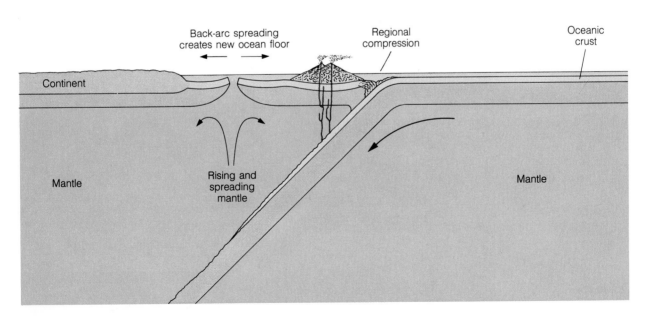

Back-arc spreading
creates new ocean floor

Regional
compression

Oceanic
crust

Continent

Rising and
spreading
mantle

Mantle

Mantle

FIGURE 18.21
Back-arc spreading may result at convergent plate margins caused by convection eddies in the asthenosphere above the subducting plate. Regional tension and normal faulting result from the rising mantle material.

simply not dense enough. The continents, therefore, move with the plates, sometimes colliding and sometimes splitting. They are much older than the ocean basins. Because continents are not consumed back into the mantle, they preserve records of plate movements in the early history of Earth—records in the form of ancient faults, old mountain belts, granitic batholiths, and sediment deposited along ancient continental margins.

Processes at Transform Fault Boundaries

Transform fault boundaries are zones of shearing where plates slide past each other without diverging or converging and without creating or destroying lithosphere. These boundaries occur along a special type of fault called a **transform fault,** which is simply a **strike-slip fault** between plates (that is, movement along it is horizontal and parallel to the fault). The term *transform* is used because the kind of motion between plates is changed—transformed—at the ends of the active part of the fault. For example, the diverging motion between plates at an oceanic ridge can be transformed along the fault to the converging motion between plates at a subduction zone (Figure 18.22).

A seismic image of a transform-fault boundary, which dissects a midocean ridge, is shown in Figure 18.23. Blocks of crust with slightly different ages occur on each side of the fracture zone. Note that near the fracture zone the crust becomes thinner.

Transform faults connect convergent and divergent plate boundaries in various combinations (Figure 18.22). Where segments of the oceanic ridge

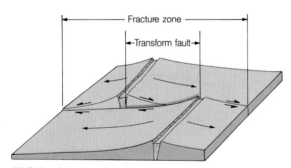

(A) Ridge-ridge transform fault.

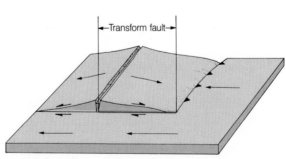

(B) Ridge-trench transform fault.

FIGURE 18.22

Transform faults can connect convergent and divergent plate boundaries in various combinations. Diagram **A** shows the common ridge–ridge transform fault. Note that relative motion occurs only along the boundary of the plates between the two segments of the ridge. Diagram **B** shows a ridge–trench transform fault, and diagram **C** shows a trench–trench transform fault. In all cases, the trend of a transform fault is parallel to the direction of relative motion between plates. This characteristic is helpful in determining the direction of plate motion.

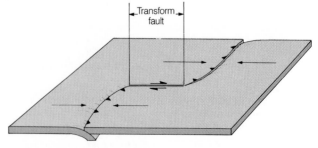

(C) Trench-trench transform fault.

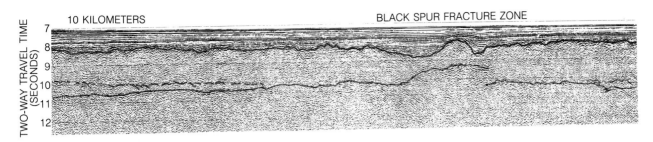

FIGURE 18.23
Seismic images of transform fault boundaries show the discontinuity along the fracture zone where the two plates slide past each other. The fracture zone may extend for thousands of kilometers. This profile shows the Black Spur fracture zone about 1000 km east of Florida.

have been offset, a transform fault connects the two divergent plate boundaries and creates a major topographic feature called a *fracture zone.* Fracture zones, however, are not what they might at first seem to be. The apparent offset of the oceanic ridge may suggest a simple strike-slip fault with displacements of thousands of kilometers, but one must keep in mind the relative motion of the plates at the spreading axis. Relative motion between the plates and seismic activity occur only in the area between the offset segments of the ridge (Figure 18.24). This zone is the only place where the fault forms a boundary between the plates. Beyond this zone, the plates on either side of the fracture are moving in the same direction and at the same rate and can be considered to be linked together. Note that the oceanic ridge is *not* being offset by motion along the transform fault. It was offset previously and may represent an old line of weakness in the rifted continental crust that preceded the development of oceanic crust. Moreover, volcanic activity usually is not associated with transform faults. As the plates slide past each other, their boundaries are fractured and broken. This fracturing produces parallel ridges and troughs along the fault zone.

Transform faults can also join ridges to trenches and trenches to trenches. In all cases, transform faults are parallel to the direction of relative plate motion, so that neither divergence nor convergence occurs along this type of boundary. The plates slide along the fracture system, and their movement produces only fracturing and seismic activity.

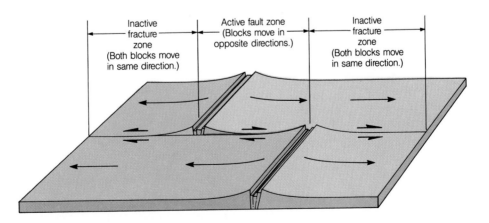

FIGURE 18.24
The relative movement of plates at a ridge–ridge transform fault changes along the trend of the fracture zone. The plates are moving away from the ridge, but the relative motion between the plates along the transform fault depends on the position of the spreading center. Along an active fault, between two segments of the ridge, plates on opposite sides of the fault move in opposite directions. Beyond the spreading centers, however, the plates move in the same direction on both sides of the fault, with no relative motion along the fault plane.

PLATE MOTION

The motion of a series of rigid plates on a sphere can be complex. Each plate moves as an independent unit, in different directions and at different velocities.

The geometry of a curved plate moving on a sphere was worked out 200 years ago by the Swiss mathematician Leonhard Euler (1707–1783) and now provides the basis for analyzing plate motion. The basic analysis of this type of motion is illustrated in Figure 18.25. In the figure, the motion of plate 1 with respect to plate 2 is a rotation around the axis A (called the *axis of plate rotation*), one pole of which is the point P (the *pole of rotation*). Note that the pole of the plate rotation is completely independent of Earth's spin axis and has no relation to the magnetic poles.

Several important facts about plate motion are immediately apparent from Figure 18.25. First, different parts of a plate move with different velocities. Maximum velocity occurs at the equator of rotation and minimum velocity at the poles of rotation. This may best be understood by considering a plate so large that it covers an entire hemisphere. All motion occurs around the axis of

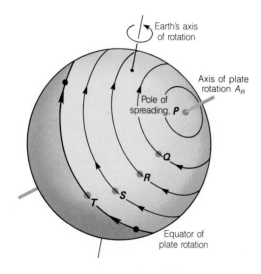

(A) Plate motion can be easily understood by considering a plate that covers an entire hemisphere. Each point on the plate would move along lines of latitude with respect to the pole of spreading (P).

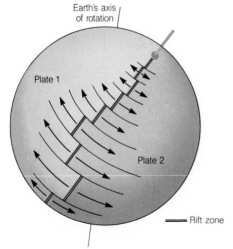

(B) The motion of plate 1 with respect to plate 2 also can be described as rotation around some imaginary axis extending through Earth. A segment of the oceanic ridge lies on a line of longitude that passes through the pole of that axis, and transform faults lie on lines of latitude with respect to that pole. The amount and rate of spreading are at a maximum at the equatorial line and are zero at the pole.

FIGURE 18.25
Plate motion on a sphere requires that the plates rotate around an axis of spreading, the pole of which is called a pole of spreading. Plates always move parallel to the fracture zones and along circles of latitude perpendicular to the spreading axis.

rotation. The pole of rotation has zero velocity because it is a fixed point around which the hemispheric shell moves. Points Q, R, and S have progressively higher velocities, with a maximum velocity at point T, which lies on the equator of rotation.

Note also that transform faults lie on lines of latitude relative to the pole of rotation. This condition holds for most transform faults in nature, as can be seen on a physiographic map of the Atlantic (see the inside covers of this book). We can thus use the orientation of transform faults to locate the pole of rotation for each plate.

Spreading ridges are linear and are usually perpendicular to plate motion. They are commonly oriented along lines of latitude relative to the plate's pole of rotation. It is important to understand that the poles of rotation do not necessarily lie on the plate in question (Figure 18.25).

What is unique about the motion of plates on a sphere?

The direction of movement of the major plates in relation to their neighbors can be determined in several ways. As we have seen, the trends of the oceanic ridge and the associated fracture zones are related to the location of the pole of rotation. Indications of movement are also drawn from seismic data (see page 469), from relative ages of different regions of the sea floor, and from the ages of chains of volcanic islands and seamounts (see page 497). From these data, geologists have determined the motion of the present tectonic plates. This motion can be summarized using Figure 18.14.

What geologic features indicate the direction of plate motion?

The Pacific plate is moving in a generally northwesterly direction, from the East Pacific rise toward the system of trenches in the western Pacific. It is bordered by several small plates along the subduction zone, so the relative motion at each of the trenches differs from the general trend. The American plates are moving westward from the mid-Atlantic ridge, converging with the Pacific, Cocos, and Nazca plates. The Australian plate is moving northward.

Africa and Antarctica, however, present a different situation. Both are nearly surrounded by ridges, and they have no associated subduction zones to accommodate the new lithosphere generated along the ridges. There is some northward movement of Africa toward the convergent boundary in the Mediterranean area, but this does not accommodate the east–west spreading from the Atlantic and Indian ridges. The African and Antarctic plates are apparently being enlarged as new lithosphere is generated at their margins. Without subduction zones, the ridges surrounding these plates must be moving outward in relation to Africa and Antarctica. The African and Antarctic plates illustrate a very important point: *Plate margins are not fixed but can move as much as the plates themselves can.* If two divergent plate margins are not separated by a subduction zone, new lithosphere is formed at each spreading axis, but none is destroyed between them. The plate between the ridges is continually enlarged, so the ridges themselves must move apart.

As shown in Figure 18.26, if a spreading axis moves toward a subduction zone with the migrating plate, both the ridge and the subduction zone are ultimately destroyed, and a single plate is formed where once there were three.

Subduction zones are also temporary features, for an active zone can be abandoned and a new one created in a different position (Figure 18.27). This process has apparently occurred several times in the western Pacific, where double trenches are common.

Another important change is in the lengths of plate margins. An oceanic ridge is essentially a fracture in the lithosphere. Besides growing larger with time, it can grow longer. A good example is the ridge in the Atlantic Ocean. It has grown and lengthened considerably since spreading began to separate South America from Africa.

From the foregoing discussion, it is apparent that plate margins are not permanent but can be displaced and can migrate to different positions. As a plate margin moves, the shape and configuration of the plate are changed, and in many cases, the plate itself is ultimately destroyed.

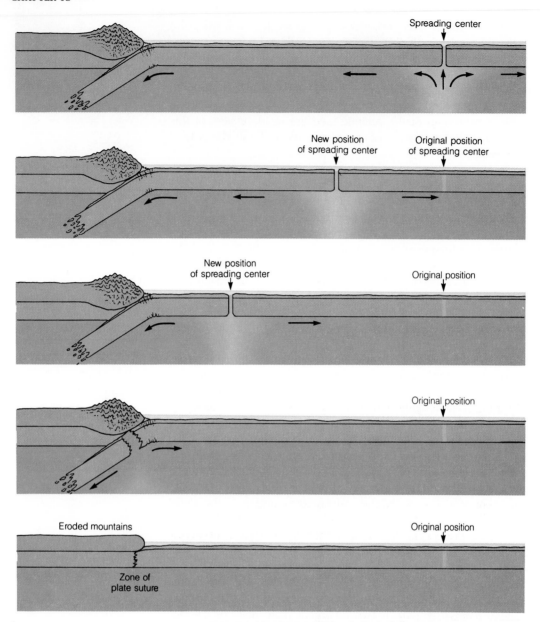

FIGURE 18.26
The migration of a spreading axis can result in the ultimate destruction of both the spreading
axis and a subduction zone. If a spreading axis descends into a subduction zone, it is consumed
with the descending plate. In the extreme case, the remaining plates are sutured together to form
a single plate where once there were three. In any case, the rate of subduction is greatly reduced.

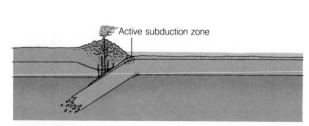

(A) Early stage.

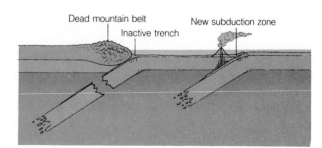

(B) Late stage.

FIGURE 18.27
A subduction zone can be abandoned and a new one created elsewhere as a result of
variations in stress in the lithosphere. This would leave an inactive trench, a dead mountain belt,
and an isolated fragment of the descending lithosphere.

RATES OF PLATE MOTION

Magnetic reversals on the ocean floor provide a timing mechanism to measure the relative velocity of plate motion. The results show that plates move at different rates ranging from 1.3 to 18.3 cm per year.

The velocity of movement of one plate relative to another can be measured by dating the magnetic reversals on the sea floor, using the time scale of magnetic stratigraphy (page 431). The timing system is the oscillation of Earth's magnetic field. As discussed on page 429, for at least 2 billion years Earth's magnetic field has switched back and forth between "normal" polarity (as exists today) and reverse polarity, in which the north and south poles are reversed. The switching back and forth between these two states has been at irregular intervals, some as short as 20 thousand years, others longer than 10 million years. The pattern of this "irregularly ticking clock" is known and calibrated by radiometric dates. Thus the pattern of magnetic reversals in the rocks on the sea floor can be used to establish magnetic time lines (isochrons), which can be used like tree rings in dating the age of the rocks on the sea floor.

The pattern of magnetic reversals has been worked out for most of the sea floor and shows an incredible record of plate motion (Figure 18.11). As seen in Figure 18.11, magnetic time lines are symmetric about the ridge that formed them. The distance from the ridge axis to any isochron indicates the amount of sea floor created during that interval of time. Thus, the wider the band, the faster the rate of plate movement.

It is apparent from Figure 18.28 that the plates are moving at significantly different rates. The Pacific, Nazca, Cocos, and Indian plates are moving at a higher rate than the North American, African, Eurasian, and Antarctic plates.

How fast is London moving away from New York?

How fast is New York moving away from Kansas City? How do we know?

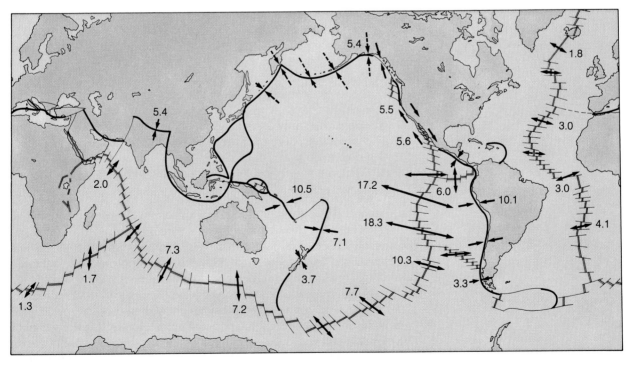

FIGURE 18.28
Relative velocities and directions of plate movement show how the major plates are currently interacting. The lengths of the arrows are proportional to the velocity of plate movement and the numbers represent velocity in centimeters per year.

You will note from the data shown in Figure 18.28 that the fastest-moving plates are those in which a large part of the margins is being subducted and that the slower-moving plates are those that lack subducting boundaries or that have large continental blocks imbedded in them. This has been interpreted by some geologists as evidence that the tectonic plates are part of the convection system of Earth and that plate motion is, to a large extent, a result of a cold, dense plate sinking into the mantle.

In addition to measuring plate velocities by magnetic reversals, current plate motion can be measured directly, using satellites and lasers. A narrow beam of light is emitted from an Earth-bound laser and bounced off an orbiting satellite whose position is known very accurately. The light is collected at the surface of Earth again, and the elapsed time is determined. This allows the location of the laser to be determined to within a centimeter. If the locations of several such stations on different plates are repetitively determined, the relative motion between plates can be accurately measured. The velocities measured in this fashion correlate well with those measured by geological means.

THE DRIVING MECHANISM

Forces that influence the motion of a plate include (1) slab-pull, (2) ridge-push, (3) basal drag, (4) friction along transform faults, and (5) friction between the converging slabs of lithosphere in a subduction zone.

In order to understand why the plates move, it is useful to examine the forces that act upon the plates. The most important of these are shown diagrammatically in Figure 18.29. Forces that influence motion of the plate include the following:

1. *Slab-pull,* a pull exerted on the plate as the slab descends into the asthenosphere in the subduction zone. The term *slab-pull,* however, does not mean that the lithosphere is under tension in the absolute sense, but rather that the slab sinks because it is more dense than the asthenosphere.
2. *Ridge-push,* the sum of the forces on or near the ridge; for example, the forces that are a result of the ridge being higher than the abyssal plains.
3. *Basal drag,* drag on the bottom of the plate caused by the movement of the lithosphere over the asthenosphere.
4. Friction along transform faults.
5. Friction between the converging slabs of lithosphere in a subduction zone.

Inasmuch as the plates have nearly constant velocities and are not accelerating, most researchers believe that the forces that drive the plates are approximately balanced by forces that resist their movement. Thus the driving forces provided by slab-pull and ridge-push are approximately balanced by basal drag, coupled with transform and collision resistance.

Although the basic description of plate tectonics was worked out in the 1960s, there is as yet no agreement about what moves the plates. This is because testing the actions of the possible driving forces is extremely difficult. Drive by slab-pull and ridge-push doubtless occurs, but the sense of the forces on the bases of the plates is more problematical. Undoubtedly, different forces act in different proportions on different plates. Many scientists conclude that the absolute velocity of a plate depends chiefly on the proportion of its margin

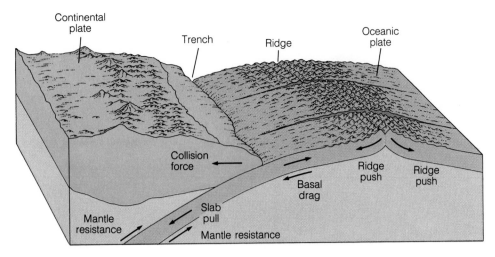

Continental plate — Trench — Ridge — Oceanic plate

Collision force — Ridge push — Ridge push

Basal drag

Mantle resistance — Slab pull — Mantle resistance

FIGURE 18.29
Forces active on the plates are indicated with arrows on the front of this block diagram. They include slab-pull, ridge-push, basal drag, and friction along transform faults and in the subduction zone.

that is subducting. For example, plates like the Pacific and Cocos plates, which have about 40% of their margins represented by subduction zones, have high plate velocities (greater than 5 cm/yr). Plates like the North American, which have smaller proportions of subducting margins, move more slowly (1 to 3 cm/yr). Hence, many scientists conclude that the major driving force is slab-pull. The correlation between the rate of plate motion and the proportion of subducting edge is so strong that other forces may be only minor. Ridge-push is believed to be a less important but still significant motivator of plate movement. The major retarding force must be resistance in the mantle to the motion of the subducting slab. The role of basal drag, either to drive the plates by asthenospheric convection or to retard their movement by viscous drag, is currently considered by most active investigators to be minor.

Presumably, the pattern of plate movement is determined largely by the fast plates, which are those with a large proportion of subducting edge. The remaining plates move around in the space between, under the influence of minor forces. This subject will occupy geologists for years to come.

What are ridge-push and slab-pull? Why are these forces important?

MANTLE CONVECTION

Convection in the mantle is believed to be the fundamental process responsible for plate motion, but there are two major schools of thought about the nature of the convection and where it occurs: (1) convection cells within the mantle carry the plates like a conveyor belt and (2) the plates themselves are an active part of the convecting process, not passive passengers.

One of the first models to explain the driving mechanism of plate tectonics suggested that *convection cells* within the mantle carried the plates and that the plates played little or no active part in the convection (Figure 18.30A). The rising limbs of the convecting cells in the mantle would therefore determine the positions of the oceanic ridges. The convecting mantle would cause the lithosphere to split, and the moving mantle would carry the lithosphere laterally toward the subduction zone. The descending cell would mark the location of the trench and would drag the lithosphere down into the mantle. Move-

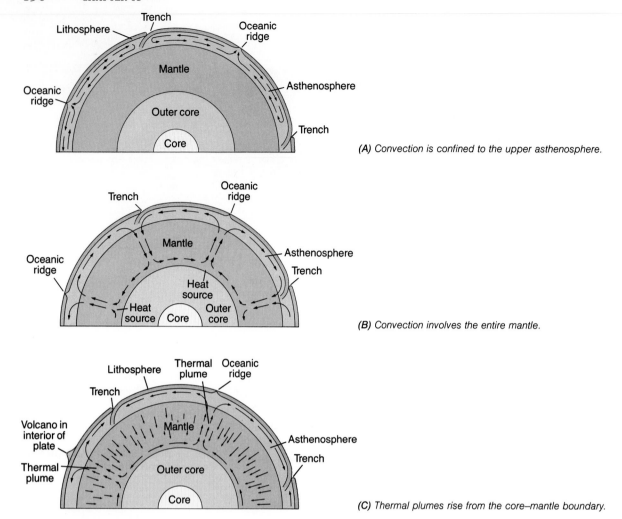

(A) Convection is confined to the upper asthenosphere.

(B) Convection involves the entire mantle.

(C) Thermal plumes rise from the core–mantle boundary.

FIGURE 18.30
Three suggested models of convection in the mantle show how flow in the asthenosphere
might move tectonic plates.

ments in the asthenosphere were thought to be strongly coupled to the lithosphere. In other words, convection cells in the mantle supposedly *cause*
ridges, trenches, and their effects. The distance between plate boundaries was
thought to be caused by the size of the convection cell.

Another model of convection theory considers the plates themselves to
be active participants in the convection process, not passive passengers on a
churning mantle. In this model the lithosphere is considered to be the cold
upper layer of the convection cell. Because of its greater density, the lithosphere tends to sink. Subduction occurs *not* because the plate is pulled down
by the descending mantle, but because the plate *is* the dense sinking limb of
the cell. If this is true, the plate may be driven by slab-pull. Mantle convection
may be driven by the plates and not vice versa. Thus the plate is removed by
forces largely independent of the mantle beneath.

Several examples demonstrate this process. If a large cauldron of molten
metal is allowed to cool, a skin of solid metal forms on the surface and, because
it is colder and more dense than the liquid, it founders and sinks into the
molten liquid, thereby stirring the melt. The same process has been observed
on a larger scale in lava lakes formed in the pit craters of Hawaiian volcanoes
(Figure 18.31). As molten lava cools, a solid crust forms over the lake, but the
crust splits into sheets and moves about. Because it is cooler and denser than
the underlying liquid, large slabs of the crust break up, split, and sink. Molten

FIGURE 18.31
Convection in a lava lake in Hawaii simulates convection and plate motion. As fresh, molten lava rises by convection, slightly older chilled lava (darker) is shoved aside to sink at some other (compressive) zone in the crater (out of view). Note transform fault near middle of the view and differential rate of spreading revealed by different widths of recently chilled lava on either side of spreading line. Front edge of view is about 50 m wide.

lava rises from below and creates a zone of new cooling crust. Many features of plate tectonics are exhibited in Figure 18.31. Spreading ridges, transform faults, and subduction zones are all observed. From this perspective it is more accurate to think of the plates and the mantle as forming a single, though complex, system, with each portion of the system affecting the other. In fact, however, convective flow in the mantle may have a radically different aspect than the motion of lithospheric plates seen at the surface of Earth.

The size and shape of the convection cells within the mantle are also a matter of considerable debate. The principal competing models are (1) layered mantle convection and (2) whole-mantle convection. There is evidence for and against each. The basic problem is that we do not understand enough about Earth's interior, and our current theories are based on what is happening at the surface and what geophysical and geochemical studies tell us about the deep interior.

In the first model, two separate convecting layers of the mantle are envisioned. The upper layer is confined largely to the asthenosphere and lithosphere (Figure 18.30A). Slabs of lithosphere are known to penetrate to depths of 700 km before becoming undetectable by their seismic activity, so a significant part of the upper mantle must be involved in convection. Below 700 km, the mantle is thought to convect independently of the upper mantle and probably at a very slow rate. The principal evidence for this model is the geochemical distinctiveness of midoceanic-ridge basalts, which come from the upper mantle. The lower mantle has not been processed by plate tectonics and

What are the three major models explaining the driving mechanism of plate tectonics? What do they all have in common?

is rarely sampled by magmas that reach the surface. There is thought to be little transfer of material across the boundary between the upper and lower mantle, but heat readily conducts across it.

The second model considers convection to involve the entire mantle. Heat for whole mantle convection is supplied from the outer core. The major difference in these models is the size of the convection cells (Figure 18.30B).

Another variety of mantle convection involves the rise of jetlike plumes of low-density material from the core–mantle boundary region. These plumes, or hotspots, are considered in the following section.

Until we are able to probe the interior of Earth more effectively and discover ways of mapping the fine details of its internal structures, we will have more questions than answers concerning convection systems in the mantle.

MANTLE PLUMES

Hotspots are believed to be surface manifestations of long, narrow columns of hot mantle material, called mantle plumes, which originate at depths greater than 700 km.

Although most tectonic activity is located along plate margins, a number of volcanic centers, such as Hawaii and Yellowstone National Park, exist far away from plate boundaries. These areas have been referred to as hotspots and are believed to be surface manifestations of mantle plumes—long, narrow columns of hot mantle material that originate at depths greater than 700 km. The locations of mantle plumes move very slowly with respect to one another and can be considered to be stationary and independent of plate motion above them. Thus, they supplement our knowledge of how plate tectonics works and they provide information about the deep mantle.

The idea of **hotspots** was established in 1963 from observations concerning the geology of the Hawaiian Islands. It was well known that the Hawaiian Islands grow progressively older in a line that begins at the southernmost island of Hawaii and runs northwestward.

The chain of active and recently extinct volcanoes is located atop a broad rise in the center of the Pacific plate. The island of Hawaii is an active shield volcano. Maui, an island about 80 km to the northwest, is barely extinct. Farther to the west, the chain of islands becomes progressively older (Figure 18.32). Other linear chains of volcanic islands and seamounts in the Pacific, Atlantic, and Indian oceans show similar trends, in which an active or young volcano is at one end of the chain and the series of volcanoes becomes progressively older toward the other end.

Why is there such an orderly and obviously meaningful pattern of volcanic activity in the interior of plates? We are not sure, but the simple idea of **mantle plumes** has generally become accepted as part of the plate tectonics theory.

What features are explained by mantle plumes that are not explained by moving plates?

Mantle plumes are believed to be isolated, long, slender columns of hot rock that originate deep inside Earth's mantle and rise slowly toward the surface, lifting the crust and forming volcanoes. They are thought to have diameters of 100 to 240 km and to rise at rates of perhaps 2 m per year. Some scientists think they originate at depths of at least 700 km, and perhaps at points as deep as the core–mantle boundary. Their position appears to be relatively stationary, so the lithospheric plates drift over them. The plumes are thus independent of the major tectonic elements of the crust, which are produced by plate movement. They rise up under continents and oceans alike, in the

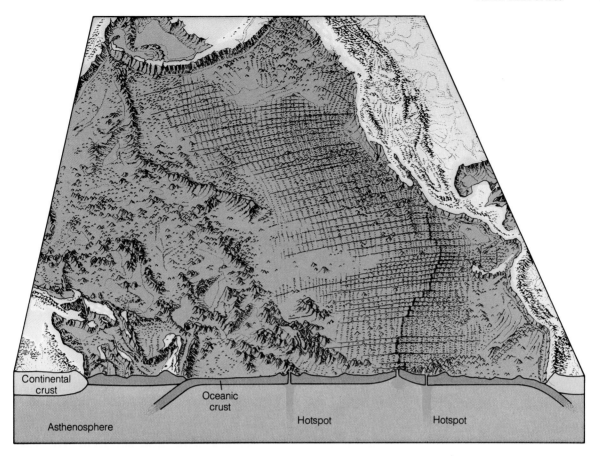

FIGURE 18.32
The motion of a plate over a fixed mantle plume produces a chain of volcanoes. The plumes originate deep in the mantle and the string of volcanoes produced reveals the path of the moving plates.

center of plates and along the oceanic ridges. The extra heat they bring to the lithosphere commonly produces domes up to 1000 km in diameter, with uplift ranging from 1 to 2 km.

It should be emphasized that this "plume model" is just that: a model. Plumes, of course, have not been observed directly, but the indirect evidence of mantle plumes is substantial. Geochemical studies show that the basalt erupted from hotspots is different from the basalt that forms from the upper mantle at spreading centers. This suggests that the lavas are derived from well below the asthenosphere and is an important argument in favor of a layered mantle. In addition, satellite measurements of Earth's gravitational field have shown that hotspots are located on sites of anomalously high gravity—areas of excess mass. The excess mass can be attributed to the bulges produced by the upwelling plumes. Recent advances in seismic studies of Earth's interior provide hope that in the near future we may be able to make a CAT-scan image of the plumes with sufficient resolution to determine their precise shapes and the locations of their roots (see Figure 19.17).

From evidence provided from studies of hotspots, it appears that the mantle plumes have a variety of shapes and sizes and that they originate at various depths. They may consist of hot mantle material rising as blobs rather than in a continuous streak. Melting of this material occurs as the pressure drops when the material rises to the surface. Moreover, mantle plumes appear to be temporary features that form and ultimately fade and die. Typical life spans are on the order of 100 million years.

Like the theory of plate tectonics itself, the idea of mantle plumes is a simple but powerful concept. It explains much of the geologic activity in the central parts of plates that never seemed to fit into any scheme of global tectonics. Further research will undoubtedly show that other geologic phenomena may be attributed to rising mantle plumes and their relation to moving plates. For example, as the plates move over a hotspot, so do plate boundaries. What happens when a plume is under a spreading axis? The answer may be found in Iceland and the Galápagos Islands, both of which are large buildups of basalt on the oceanic ridges. Can a row of hotspots initiate continental rifting? How do mantle plumes rise to the surface without being distorted by convection in the mantle? Are domal upwarps caused by plumes a primary cause for expansion and contraction of shallow seas over the continents? The idea of mantle plumes complements the theory of plate tectonics in that it helps to explain many features in the interiors of plates.

SUMMARY

1. The theory of drifting continents was proposed in the early 1900s. It was best developed by Alfred Wegener in his book *The Origin of Continents and Oceans.*

2. The theory was supported by various types of geologic evidence, including (a) matching geologic features on the margins of continents, (b) paleontologic findings, (c) Paleozoic glaciation, and (d) other paleoclimatic evidence. Without an understanding of the oceanic crust, however, a complete theory of Earth's dynamics could not be developed.

3. During the 1960s, new information about the topography and paleomagnetism of the ocean floor led to the development of the plate tectonics theory. Since then, the theory has been supported by a wide variety of geologic and geophysical data from both the continents and oceans.

4. Paleomagnetic studies show that the continents have changed their positions in relation to the magnetic poles and that each continent has followed a different path with respect to the poles throughout geologic time. These findings indicate that the continents have moved with respect to each other. Paleomagnetic reversals in the rocks of the sea floor occur in symmetrically matching sets on both sides of the oceanic ridge, a fact explained by sea-floor spreading.

5. The theory of plate tectonics explains Earth's crustal dynamics by the movement of rigid lithospheric plates. New lithosphere is created at oceanic ridges, where plates split and spread apart and basaltic magma wells up in the rift. Oceanic lithosphere is consumed where a plate descends into the mantle at a subduction zone.

6. The major structural features of Earth are formed along plate boundaries. At divergent plate boundaries, the lithosphere is under tension, and the major geologic processes are (a) rifting and block faulting, (b) the generation of basaltic magma and the formation of new oceanic lithosphere, and (c) rifting of continents. At convergent plate boundaries, the important geologic processes include (a) subduction, (b) generation of andesitic to silicic magma, and (c) metamorphism and mountain building. Transform fault boundaries are zones where plates slide passively past each other.

7. Each plate is part of a shell on a sphere, so different parts of the plate move at different velocities. The direction of plate motion can be determined from (a) trends of spreading centers and fracture zones, (b) seismic data, (c) ages of the sea floor on either side of the ridge, and (d) ages of island chains. Plates move from 1.3 to 18.3 cm per year.

8. A number of forces act upon the plates and influence their motion. The most important of these are (a) gravitational pulls in subduction zones, (b) gravitational sliding off elevated ridges, (c) frictional drag on the bottoms and sides of plates, and (d) collisional resistance forces between converging plates.

9. Convection of the mantle with the active participation of oceanic lithosphere is probably the fundamental driving mechanism responsible for plate motion, but details of the type of convection heat remain unknown.

10. Mantle plumes are isolated, long, slender columns of hot rock that originate deep inside Earth's mantle. Where they approach the surface, they create hotspots of high heat flow, volcanic activity, and broad crustal upwarps.

KEY TERMS

axis of plate rotation (p. 444)

basal drag (p. 448)

continental drift (p. 421)

convection cell (p. 449)

convergent plate boundary (p. 438)

divergent plate boundary (p. 437)

fracture zone (p. 443)

hotspot (p. 452)

magnetic anomaly (p. 429)

magnetic reversal (p. 429)

mantle plume (p. 452)

mid-Atlantic ridge (p. 431)

normal polarity (p. 428)

Pangaea (p. 421)

paleomagnetism (p. 427)

polarity epoch (p. 429)

polarity event (p. 429)

pole of rotation (p. 444)

reversed polarity (p. 429)

ridge-push (p. 448)

rift valley (p. 426)

sea-floor spreading (p. 426)

slab-pull (p. 448)

spreading axis (p. 437)

strike-slip fault (p. 442)

subduction (p. 438)

subduction zone (p. 438)

tectonic plate (p. 435)

transform fault (p. 442)

ADDITIONAL READINGS

Bird, J. M., ed. 1980. *Plate Tectonics*. Washington DC: American Geophysical Union.

Cloud, P. 1980. Beyond plate tectonics. *American Scientist* 68:381–87.

Condie, K. C. 1989. *Plate Tectonics and Crustal Evolution*, 3d ed. Elmsford, NY: Pergamon Press.

Cox, A., and R. B. Hart. 1986. *Plate Tectonics: How It Works*. Palo Alto, CA: Blackwell Scientific Pub.

Dewey, J. F. 1972. Plate tectonics. *Scientific American* 226(5):56–68.

Dewey, J. F., and J. M. Bird. 1970. Mountain belts and the new global tectonics. *Journal of Geophysical Research* 75(14):2625–2647.

Takeuchi, H., S. Uyeda, and H. Kanamouri. 1970. *Debate about the Earth*, rev. ed. San Francisco: Freeman, Cooper.

Van Andel, T. H. 1985. *New Views on an Old Planet*. New York: Cambridge University Press.

Weyman, D. 1981. *Tectonic Processes*. London: Geo. Allen and Unwin.

Wyllie, P. J. 1976. *The Way the Earth Works*. New York: Wiley.

York, D. 1975. *Planet Earth*. New York: McGraw-Hill.

REVIEW QUESTIONS

1. Briefly explain the theory of plate tectonics.
2. Describe the types of plate boundaries and give an example of each.
3. List the major processes that occur along each type of plate boundary.
4. Sketch a simple map of a part of the oceanic ridge and draw arrows to show the relative motion along ridge-to-ridge transform faults.
5. Describe the geometry of lithospheric plate motion over the plate.
6. How is the pattern of a series of transform faults governed by the pole of spreading?
7. Explain how plate margins, as well as the plate itself, can migrate.
8. Make a series of sketches showing how a subduction zone can be terminated.
9. Explain the origin of the following features in the context of plate tectonics: (a) Ural Mountains, (b) the Alps, (c) Iceland, (d) Hawaii, (e) San Andreas Fault, (f) Andes Mountains, and (g) volcanoes in Italy.
10. Draw a cross section showing a tectonic plate with divergent and convergent boundaries and label the major forces acting on the plate.
11. Explain the difference between (a) the convection model of plate motion, in which the mantle carries the plates, and (b) a model in which the plates themselves are part of the convecting process.
12. How fast are the plates moving? How do we determine rates of plate motion?
13. What are mantle plumes? Why do we believe they exist?

Earthquakes, perhaps more than any other phenomenon, demonstrate that Earth continues to be a dynamic planet, changing each day by internal tectonic forces. Most earthquakes occur along plate boundaries. As the plates move, these boundaries, spreading centers, subduction zones, and transform faults will be the sites of the most intense earthquake activity on Earth.

Earthquakes occur during sudden movements along faults. During long periods of slow deformation, elastic strain builds up between the rock bodies on opposite sides of a fault. Slip along the fault is prevented by friction until a threshold of strain is exceeded. Then the rocks snap past each other along the fault to release some of the stored energy.

The San Andreas Fault in California is a famous example. At the San Andreas fault zone the Pacific plate moves toward the northwest with respect to the North American plate, carrying with it the coastal area of California and Baja California. In this image of the San Francisco Bay area, it is easy to see why a major earthquake in the area is an event just waiting to happen. Several active faults can be traced across the area. The San Andreas Fault cuts across the peninsula and is marked by a linear lowland whose trace is emphasized by Crystal Springs Lake. The Hayward and Calaveras faults cut across the East Bay region.

Every year, more than 150,000 earthquakes are recorded by the worldwide network of seismic stations and are analyzed with the aid of computers at the earthquake data center in Boulder, Colorado. With this network, the exact location, depth, and magnitude of all detectable earthquakes are plotted on regional maps. Information about the direction of fault movement associated with the shock is also determined. As a result, we can literally monitor the details of present plate motion. But that is not all. Seismic waves also provide our most effective probe of Earth's interior, and they constitute the main method of collecting data upon which we base our present concepts of the internal structure of Earth.

19
Earth's Seismicity

MAJOR CONCEPTS

1. Seismic waves are vibrations in Earth caused by the rupture and sudden movement of rock.
2. Three types of seismic waves are produced by an earthquake shock: (a) P waves, (b) S waves, and (c) surface waves.
3. The primary effect of an earthquake is ground motion. Secondary effects include (a) landslides, (b) tsunami, and (c) regional or local uplift or subsidence.
4. Most earthquakes occur along plate boundaries. Divergent plate boundaries and transform fault boundaries produce shallow-focus earthquakes. Convergent plate boundaries produce a zone of shallow-focus, intermediate-focus, and deep-focus earthquakes.
5. Infrequent shallow-focus earthquakes occur in the interiors of the plates, away from the plate boundaries.
6. The velocities at which P waves and S waves travel through Earth indicate that Earth has a solid inner core, a liquid outer core, and a thick mantle with a soft asthenosphere and a rigid lithosphere.

CHARACTERISTICS OF EARTHQUAKES

Earthquakes are vibrations of Earth, caused by the rupture and sudden movement of rocks that have been strained beyond their elastic limits. Three types of seismic waves are generated by an earthquake shock: (1) primary waves, (2) secondary waves, and (3) surface waves.

Elastic-Rebound Theory

The origin of an *earthquake* can be illustrated by a simple experiment. Bend a stick until it snaps. Energy is stored in the elastic bending and is released if rupture occurs, causing the fractured ends to vibrate and send out sound waves. Detailed studies of active faults show that this model, known as the *elastic-rebound theory,* applies to all major earthquakes (Figure 19.1). Precision surveys across the San Andreas Fault in California show that railroads, fence lines, and streets are slowly deformed at first, as strain builds up, and are offset when movement occurs along the fault, releasing the elastic strain. The San Andreas Fault is the boundary between the Pacific and North American plates. Its movement is horizontal, with the Pacific plate moving toward the northwest. On a regional basis, the plates move quite steadily at a rate of roughly 5 cm per year. Along the fault, movement can be smooth or it can occur in a series of jerks, because sections of the fault can be "locked" together until enough strain accumulates to exceed the *elastic limit* of the rock and cause displacement. The elastic-rebound theory explains earthquakes as the result of either rupture and sudden movement along a fault or recurrent movement along existing fractures.

    The point within Earth where the initial slippage generates earthquake energy is called the *focus.* The point on Earth's surface directly above the focus is called the *epicenter* (Figure 19.2).

What causes earthquakes?

Types of Seismic Waves

Three major types of *seismic waves* are generated by an earthquake shock (Figure 19.3). Each type travels through Earth at a different speed, and each therefore arrives at a *seismograph* hundreds of kilometers away at a different

458

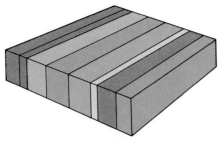

(A) Rocks remain undeformed in seismically inactive areas.

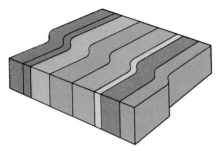

(B) Strain builds up in rocks in seismically active areas until the rocks rupture or move along preexisting fractures.

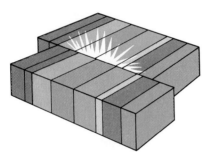

(C) Energy is released when the rocks rupture, and seismic waves move out from the point of rupture.

FIGURE 19.1
Earthquakes originate where rocks are strained beyond their elastic limits and rupture.

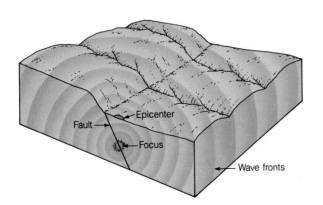

FIGURE 19.2
The relationship between an earthquake's focus, its epicenter, and seismic wave fronts is depicted in this diagram. The focus is the point of initial movement on the fault. Seismic waves radiate from the focus. The epicenter is the point on Earth's surface directly above the focus.

(A) Prior to seismic disturbance. A straight fence line provides a good reference marker for future movement.

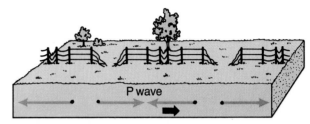

(B) Motion produced by a P wave. Particles are compressed and then are expanded in the line of wave progression. P waves can travel through any Earth material.

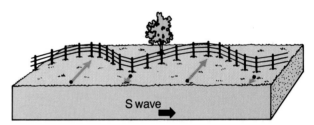

(C) Motion produced by an S wave. Particles move back and forth at right angles to the line of wave progression. S waves travel only through solids.

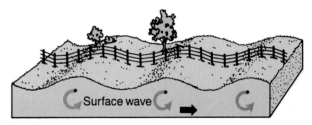

(D) Motion produced by a surface wave. Particles move in a circular path at the surface. The motion diminishes with depth, like that produced by surface waves in the ocean (see Figure 16.3).

FIGURE 19.3
Motion produced by the various types of seismic waves can be illustrated by the distortions they produce in a straight fence line.

time. The first waves to arrive are called *primary waves (P waves)*. These are a kind of *longitudinal wave,* identical in character to sound waves passing through a liquid or gas. The particles involved in these waves move forward and backward in the direction of wave travel, causing relatively small displacements. The next waves to arrive are called *secondary waves (S waves)*. In these, particles oscillate back and forth at right angles to the direction of wave travel. S waves cause strong movements to be recorded on a seismograph. The last waves to arrive are *surface waves,* which travel relatively slowly over Earth's surface. Particles involved in surface waves move in an orbit similar to that of particles in water waves.

Occurrence

The location of an earthquake focus is important in the study of plate tectonics because it indicates the depth at which rupture and movement occur. Although movement of material within Earth occurs throughout the mantle and core, earthquakes are concentrated in the upper 700 km.

Within the 700 km range, earthquakes can be grouped according to depth of focus. *Shallow-focus earthquakes* occur from the surface to a depth of 70 km. They occur in all seismic belts and produce the largest percentage of earthquakes. *Intermediate-focus earthquakes* occur between 70 and 300 km below the surface, and *deep-focus earthquakes* between 300 and 700 km. Both intermediate-focus and deep-focus earthquakes are limited in number and distribution. Generally, they are confined to convergent plate margins. The maximum energy released by an earthquake tends to become progressively smaller as the depth of focus increases. Also, seismic energy from a source deeper than 70 km is largely dissipated by the time it reaches the surface. Most large earthquakes therefore have a shallow focus, originating in the crust. The location of the focus of an earthquake is calculated from the time that elapses between the arrivals of the three major types of seismic waves.

The method of locating the epicenter of an earthquake is relatively simple and can be easily understood by reference to Figure 19.4. The P wave, traveling faster than the S wave, is the first to be recorded at the seismic station. The time interval between the arrival of the P wave and the arrival of the S wave is a function of the station's distance from the epicenter. By tabulating the travel times of P and S waves from earthquakes of known sources, seismologists have constructed time–distance graphs that can be used to determine the distance to the epicenter of a new quake. The seismic records indicate the distance, but not the direction, to the epicenter. Records from at least three stations are therefore necessary to determine the precise location of the epicenter.

Intensity

The *intensity,* or destructive power, of an earthquake is an evaluation of the severity of ground motion at a given location. It is measured in relation to the effects of the earthquake on human life. Generally, destruction is described in terms of the damage caused to buildings, dams, bridges, and other structures, as reported by witnesses. An intensity scale commonly used in the United States and a description of some of its criteria are presented in Table 19.1.

The intensity of an earthquake at a specific location depends on a number of factors. Foremost among these are (1) the total amount of energy released, (2) the distance from the epicenter, and (3) the type of rock and degree of consolidation. In general, wave amplitude and destruction are greater in soft, unconsolidated material than in dense, crystalline rock. The intensity is greatest close to the epicenter.

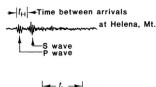
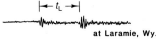
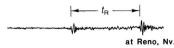

(A) *The greater the distance between the seismic event and a recording seismograph station, the more time it takes for the first wave to arrive. Also, the greater the distance, the longer the interval between the arrival of the P waves and the arrival of the S waves.*

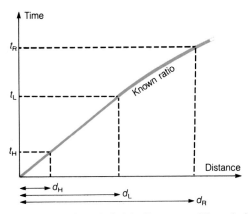

(B) *The time between the arrival of the P waves and the arrival of the S waves is correlated with the distance between the seismic event and the recording station. For example, time at Helena, t_H, yields distance from Helena, d_H.*

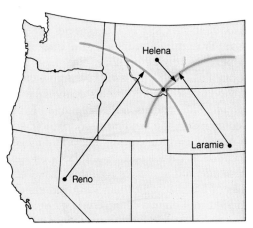

(C) *The direction of the event from any single station is not known, but by simply plotting the intersection of three arcs that have radii the respective distances from the three stations, a common point is found. That point lies at the epicenter of the seismic event.*

FIGURE 19.4
Locating the epicenter of an earthquake is accomplished by comparing the arrival times of P waves and S waves at three seismic stations.

TABLE 19.1
Scale of Earthquake Intensity

I	Not felt except by very few people under special conditions. Detected mostly by instruments.
II	Felt by a few people, especially those on upper floors of buildings. Suspended objects may swing.
III	Felt noticeably indoors. Standing automobiles may rock slightly.
IV	Felt by many people indoors, by a few outdoors. At night, some are awakened. Dishes, windows, and doors rattle.
V	Felt by nearly everyone. Many are awakened. Some dishes and windows are broken. Unstable objects are overturned.
VI	Felt by everyone. Many people become frightened and run outdoors. Some heavy furniture is moved. Some plaster falls.
VII	Most people are in alarm and run outside. Damage is negligible in buildings of good construction.
VIII	Damage is slight in specially designed structures, considerable in ordinary buildings, great in poorly built structures. Heavy furniture is overturned.
IX	Damage is considerable in specially designed structures. Buildings shift from their foundations and partly collapse. Underground pipes are broken.
X	Some well-built wooden structures are destroyed. Most masonry structures are destroyed. The ground is badly cracked. Considerable landslides occur on steep slopes.
XI	Few, if any, masonry structures remain standing. Rails are bent. Broad fissures appear in the ground.
XII	Virtually total destruction. Waves are seen on the ground surface. Objects are thrown in the air.

TABLE 19.2
Richter Scale of Earthquake Magnitude

Magnitude	Approximate Number per Year
1	700,000
2	300,000
3	300,000
4	50,000
5	6,000
6	800
7	120
8	20
>8	1 every few years

Magnitude

The ***magnitude*** of an earthquake is a measure of the amount of energy released. It is a much more precise measure than intensity. Earthquake magnitudes are based on direct measurements of the size (amplitude) of seismic waves, made with recording instruments, rather than on subjective observations of destruction. The total energy released by an earthquake can be calculated from the amplitude of the waves and the distance from the epicenter. Seismologists express magnitudes of earthquakes by using the Richter scale, which arbitrarily assigns 0 to the lower limits of detection. Each step on the scale represents an increase in amplitude by a factor of 10 (Table 19.2). The vibrations of an earthquake with a magnitude of 2 are therefore 10 times greater in amplitude than those of an earthquake with a magnitude of 1, and the vibrations of an earthquake with a magnitude of 8 are 1 million times greater in amplitude than those of an earthquake with a magnitude of 2.

The largest earthquake ever recorded had a magnitude of approximately 8.8 on the Richter scale. Significantly larger earthquakes are not likely to occur because rocks are not strong enough to accumulate more energy.

The primary effect of earthquakes is the violent ground motion accompanying movement along a fracture. This motion can shear and collapse buildings, dams, tunnels, and other rigid structures (Figure 19.5). Secondary effects include landslides, tsunami, and regional or local submergence of the land. The following are a few examples of well-documented earthquakes in historical times.

The earthquake that shook Peru on May 31, 1970, had a magnitude of 7.8. Claiming the lives of 50,000 people, it was the deadliest earthquake in Latin American history. Much of the damage to towns and villages was caused by the collapse of adobe buildings, which are easily destroyed by ground motion. Eighty percent of the adobe houses in an area of 65,000 km^2 were destroyed. Vibrations caused most of the destruction to buildings, but the massive landslides triggered by the quake in the Andes were a second major cause of fatalities. A huge debris flow buried 90% of the resort town of Yungay, with a population of 20,000, in a matter of a few minutes. Similar earthquake-triggered debris flows killed 41,000 in Ecuador and Peru in 1797 and, in 1939, killed 40,000 in Chile.

Another example of an earthquake at convergent plate margins is the 1960 earthquake that occurred in the Andes, in Chile, and caused extensive damage from ground motion, landslides, and flooding. It also produced a spectacular tsunami, which devastated seaports with a series of waves 7 m high. This tsunami crossed the Pacific at approximately 1000 km per hour and built up to 11 m high at Hilo, Hawaii. Twenty-two hours after the quake, it reached Japan, causing damage to property worth $70 million.

The earthquake that devastated southern Alaska late in the afternoon of March 27, 1964, was one of the largest tectonic events of modern times. Its magnitude was between 8.30 and 8.75, and its duration ranged from 3 to 4 minutes at the epicenter (by comparison, the San Francisco earthquake of 1906 lasted about 1 minute). The Alaskan earthquake was important to geologists because it had such marked effects at Earth's surface and because it was well documented. Despite its magnitude and severe effects, this quake caused far less property damage and loss of life than other national disasters (114 lives were lost and property worth $311 million was damaged) because, fortunately, much of the affected area was uninhabited. The crustal deformation associated with the Alaskan earthquake was the most extensive ever documented. The level of the land was changed in a zone 1000 km long and 500 km wide (an area of 500,000 km^2 was elevated or depressed). Submarine and terrestrial landslides triggered by the earthquake caused spectacular damage to communities, and the shaking spontaneously liquefied deltaic materials along the coast, causing slumping of the waterfronts of Valdez and Seward.

(A) Earthquake of June 16, 1964, Niigata, Japan. Tilted apartment houses produced by liquefaction of subsoil. Poor foundations were partly responsible for this type of damage. About one-third of the city subsided as much as 2 m as a result of sand compaction.

(B) Earthquake of October 17, 1989, Loma Prieta, California. Collapse of Nimitz Freeway, Oakland.

(C) Earthquake of September 19, 1985, Mexico City, Mexico. Collapse of 15-story reinforced concrete building.

(D) Earthquake of February 9, 1971, San Fernando, California. Lateral displacement of Southern Pacific railroad tracks near Los Angeles, California.

(E) Earthquake of March 27, 1964, Anchorage, Alaska. Subsidence at Government Hill School caused by ground deformation—slumping.

(F) Earthquake of December 7, 1988, Armenia, U.S.S.R. Collapse of a five-story concrete-frame apartment building in the town of Spitak.

FIGURE 19.5
The effects of historic earthquakes are dramatically displayed in the types of damage rendered to buildings and other structures.

One of the most famous earthquakes in history was the San Francisco earthquake of 1906. It lasted only a minute but had a magnitude of 8.3. The fire that followed caused most of the destruction (an estimated $400 million in damage and a reported loss of 700 lives). From a geological point of view, the earthquake was important because of the visible effects it produced along the San Andreas Fault. Horizontal displacement occurred over a distance of about 400 km and offset roads, fences, and buildings by as much as 7 m.

The Tangshan quake, which shook China on July 28, 1976, was probably the second most devastating earthquake in history. (The most destructive also occurred in China, in 1556.) The enormous shock registered 8.2 on the Richter scale. Although Chinese authorities have withheld many details about this disaster, scientists from Mexico and other countries were recently allowed to investigate the affected area. They say that Tangshan looked like Hiroshima after the explosion of the atomic bomb. The devastation of the city of 1 million was complete. The Tangshan quake killed 750,000 people.

In addition to the descriptions of awesome damage at Tangshan, the reports of the phenomenon itself have been interesting. Just before the earth began to shudder, residents were awakened by a brilliant incandescent light, which lit the early morning sky for hundreds of kilometers around. The glow reportedly was predominantly red and white. When the quake struck, many people were catapulted into the air, some as high as 2 m, by what was described as a violent, hammerlike, subterranean blow. Trees and crops were blown over as if by a colossal steamroller, and leaves were burned to a crisp. In the quake's aftermath, many bushes were scorched on one side only.

EARTHQUAKE PREDICTION

Effective earthquake prediction, which seemed so close at hand during the 1960s, is proving an elusive goal.

Why is earthquake prediction so difficult?

More than a million earthquakes occur around the world each year; 50 are large enough to cause property damage and loss of lives. Each earthquake disaster provides a vivid reminder of the threat earthquakes pose to the world's heavily populated areas, so considerable effort has been made to find methods of predicting earthquakes. Some people have tried to relate earthquakes to sunspots, tides, the alignment of planetary bodies, and other phenomena but have ignored many facts about Earth's seismicity. As a result, their predictions fail. Scientists have struggled with the problem, and during the 1970s significant strides were made toward measurements that might indicate an impending earthquake. Yet pinpoint earthquake prediction, which seemed so close at hand during the 1960s, is proving an elusive goal.

Chinese scientists claim to have been successful in predicting about 15 earthquakes in recent years. They rely heavily on the centuries-old idea that animals sense various underground changes prior to an earthquake and behave abnormally. Most of the bizarre behavior is simply increased restlessness. Cattle, sheep, and horses refuse to enter their corrals. Rats leave their hideouts and march fearlessly through houses. Shrimp crawl on dry land. Ants pick up their eggs and migrate en masse. Fish jump above the surface of the water, and rabbits hop aimlessly about.

Chinese scientists successfully predicted the Hai-cheng earthquake in February 1975 by means of unusual animal behavior. The most intriguingly bizarre behavior occurred in mid-December, when snakes came out of hibernation and froze to death on the icy ground, and groups of rats appeared and scurried about in the cold winter weather. These events were followed by a swarm of small earthquakes at the end of December. During January, Chinese scientists received thousands of reports of unusual animal behavior, especially in larger animals in the area that proved to be the quake epicenter.

One hypothesis offered to explain abnormal animal behavior before an earthquake is that certain animals are sensitive to small variations in Earth's magnetic field or to sounds produced by microfractures prior to the larger event. Animal sensors that detect light, sound, odor, touch, and temperature are well known, and they may have the ability to detect subtle changes in other physical phenomena.

Much of the work on earthquake prediction in the United States has been based on the dilatancy theory. Laboratory and field studies in recent years indicate that a rock subjected to stress swells just before it ruptures. This dilation is caused by the opening and extension of numerous tiny cracks, and it begins at levels of stress that are about half as great as those needed to break the rock. As rock dilates, changes occur in certain physical characteristics, such as electrical resistance, seismic wave velocities, and magnetic properties. Geologists therefore attempt to monitor uplift and tilting of the ground, electrical resistivity, the number of seismic events, and groundwater pressure.

For a prediction to benefit the populace, it must specify the time, the location, and the magnitude of the coming quake. Such accuracy is proving to be difficult to achieve. The problem is not so much that the dilatancy theory is wrong, but rather that it is inadequate. Different kinds of earthquakes apparently have different kinds of precursors. Instead of attempting to predict the time, place, and magnitude of an expected earthquake, geologists are now concentrating on the more modest goal of forecasting what areas of the world may be most susceptible to significant quakes. A major contribution to forecasting has been the compilation of a map showing the seismic potential of the world's major tectonic plate boundaries (Figure 19.6). This map essentially shows locations along the plate boundaries of the Pacific where major quakes

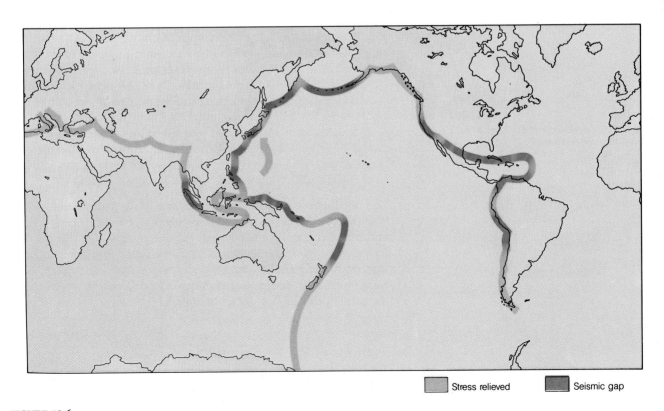

FIGURE 19.6
Seismic gaps are important in earthquake forecasting. Areas along plate margins that are not seismically active are believed to be building up stress and may be sites of significant seismic activity in the future. In the gray areas, earthquakes have relieved strain within the last 40 years, but in the red shaded areas, no large quakes have occurred and strain is still building up.

are most likely to occur in the near future. Along the plate margins are several gaps in seismic activity, where stress may be building up to a critical level. The most susceptible areas are those where major tremors have occurred in the past but have not occurred within the last 100 years. These include such heavily populated areas as southern California, central Japan, central Chile, Taiwan, and the west coast of Sumatra. These areas appear likely to experience a major earthquake (magnitude of 7.0 or greater) in the next few decades. One such seismic activity gap existed along the western coast of Mexico until a major quake struck the area on November 29, 1979.

If a reliable earthquake prediction system could give from 1 to 10 years' advance warning of a "killer" quake, what would be the appropriate social response? Would the usual flow of mortgage money be terminated? New earthquake insurance would certainly become unavailable. What about fire insurance, business expansion, unemployment, tax revenues, and demands on local government? The fear that false alarms would lead to adverse social response has led some to call for the withholding of predictions until prediction techniques are perfected to absolute certainty.

EARTHQUAKES AND PLATE TECTONICS

The distribution of earthquakes delineates plate boundaries. Shallow-focus earthquakes coincide with the crest of the oceanic ridge and with transform faults between ridge segments. Earthquakes at convergent plate margins occur in a zone inclined downward beneath the adjacent continent or island arc. Fault motion associated with earthquakes along plate margins shows the present direction of plate motion.

How is the distribution of earthquakes related to plate tectonics?

A worldwide network of 125 sensitive seismic stations was established in 1961 by the United States Coast and Geodetic Survey. Since then, the network has been expanded greatly, and the data received are processed by computers at the seismic data center in Boulder, Colorado. From this network, and from other seismic stations, seismologists have compiled an amazing amount of data concerning earthquakes and plate tectonics. Not only are the location and magnitude of thousands of earthquakes established and plotted on regional maps each year, but other information such as the direction of displacement on the earthquake faults is collected. The result is a new and important insight into the details of current plate motion. This is what has been found:

1. The distribution of earthquakes delineates plate boundaries.
2. Shallow-focus earthquakes coincide with the crest of the oceanic ridge and with transform faults between ridge segments.
3. Earthquakes at convergent plate margins occur in a zone inclined downward beneath the adjacent continent or island arc.
4. Fault motion associated with earthquakes along plate margins shows the present direction of plate motion.

Global Patterns of Earth's Seismicity

Tens of thousands of earthquakes have been recorded since establishment of the worldwide network of seismic observation stations. Their locations and depths are summarized in the seismicity map in Figure 19.7. From the standpoint of Earth's dynamics, this map is an extremely significant compilation because it shows where and how the crust of Earth is moving at the present time.

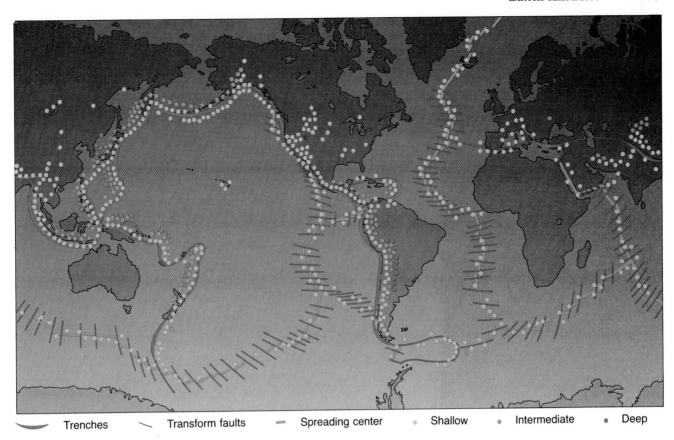

| Trenches | Transform faults | Spreading center | • Shallow | • Intermediate | • Deep |

FIGURE 19.7
Earth's seismicity is clearly related to plate margins. This map shows the locations of tens of thousands of earthquakes that occurred during a five-year period. Shallow-focus earthquakes occur at both divergent and convergent plate margins, whereas intermediate-focus and deep-focus earthquakes are restricted to the subduction zones of converging plates.

If a sufficient number of seismic stations record a quake, seismologists can determine the direction of movement along the fault when the earthquake occurred. This new insight into Earth's seismicity confirms, in a most remarkable way, the theory of plate tectonics because it shows the present-day movement of plates.

Seismicity at Divergent Plate Boundaries

The global patterns of Earth's seismicity show a narrow belt of shallow-focus earthquakes, coinciding almost precisely with the crest of the oceanic ridge and marking the boundaries between divergent plates. This zone is remarkably narrow in comparison to the zone of seismicity that follows the trends of young mountain belts and island arcs. The shallow earthquakes along divergent plate boundaries are less than 70 km deep and typically are small in magnitude. Although the zone looks like a nearly continuous line on regional maps, two types of seismic boundaries can be distinguished on the basis of fault motion. These are (1) spreading centers and (2) transform faults (Figure 19.8). Earthquakes associated with the crest of the oceanic ridge occur within or near the rift valley. They appear to be associated with normal faulting and intrusions of basaltic magmas. Locally, earthquakes occur in swarms. Detailed studies indicate that earthquakes associated with the ridge crest are produced by vertical faulting, a process that appears to be responsible for the ridge topography.

Shallow-focus earthquakes also follow the transform faults that connect offset segments of the ridge crest, but they generally are not associated with volcanic activity. Studies of fault motion indicate horizontal displacement in a

FIGURE 19.8
The distribution of earthquakes along divergent plate boundaries shows that seismicity along spreading centers results from normal faulting. Seismicity along transform faults results from strike-slip movement. No significant seismic activity occurs along the inactive fracture zone.

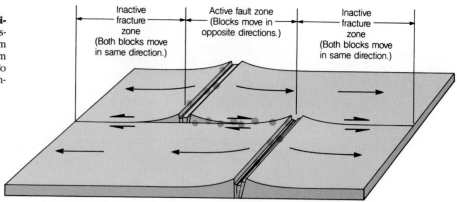

direction away from the ridge crest. Moreover, as is predicted by the plate tectonic theory, earthquakes are restricted to the active transform fault zone—the area between ridge axes—and do not occur in inactive fracture zones.

Seismicity at Convergent Plate Boundaries

The most widespread and intense earthquake activity on Earth occurs along subduction zones at convergent plate boundaries. This belt of seismic activity is immediately apparent from the world seismicity map (Figure 19.7), which shows a strong concentration of shallow, intermediate, and deep earthquakes coinciding with the subduction zones of the Pacific Ocean. The three-dimensional distribution of earthquakes in this belt defines a seismic zone that is inclined at moderate to steep angles from the trenches and extends down under the adjacent island arcs or continental borders. This distribution is well illustrated in the Tonga Trench in the South Pacific. The zone of seismicity there forms an inclined, nearly planar surface that plunges into the mantle to a depth of more than 600 km (see Figure 19.9).

Studies of fault motions from seismic waves generated in this zone indicate that the type of faulting varies with depth. Near the walls of the trench,

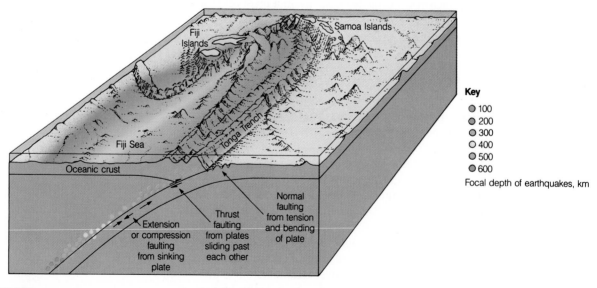

FIGURE 19.9
Earthquake foci in the Tonga region of the South Pacific occur in a zone inclined from the Tonga Trench toward the Fiji Islands. The top of the diagram shows the aerial distribution of foci, with focal depths represented by different-colored dots. The cross section on the front of the diagram shows how the seismic zone is inclined away from the trench. This seismic zone accurately marks the boundary of the descending plate in the subduction zone.

normal faulting is typical, resulting from tensional stresses generated by the initial bending of the plate. In the zone of the shallow earthquakes, thrust faulting dominates as the descending lithosphere slides beneath the upper plate. At intermediate depths, extension or compression can occur, depending on the specific characteristics of the subduction zone. Extension and normal faulting result when a descending slab, which is denser than the surrounding mantle, sinks under its own weight. Compression results if the mantle resists the downward motion of the descending plate. The zone of deep earthquakes shows compression within the descending slab of lithosphere, indicating that the mantle material at that depth resists the movement of the descending plate.

Intraplate Seismicity

Although most of the world's seismicity occurs along plate boundaries, the continental platforms also experience infrequent and scattered shallow-focus earthquakes. The zones of seismicity of East Africa and the western United States are most striking. They are probably associated with spreading centers, which can be projected into those regions. The minor shallow earthquakes in the eastern United States and Australia are more difficult to explain. Apparently, lateral motion of a plate across the asthenosphere involves slight vertical movement. Built-up stress can exceed the strength of the rocks within the lithospheric plate, causing infrequent faulting and seismicity. In contrast, the ocean floors beyond the spreading centers are seismically inactive, except for isolated earthquakes resulting from sudden rupture of the rock associated with stress produced by oceanic volcanoes.

Plate Motion as Determined from Seismicity

The data points in Figure 19.7 outline the seven major lithospheric plates. The present directions of plate movement can be deduced from the inclined zone of seismicity at deep-sea trenches and from the zone of shallow seismicity along the oceanic ridge. The Pacific plate, consisting almost entirely of oceanic crust, is moving northwestward, away from the spreading center along the eastern Pacific rise. The direction of movement along the convergent plate margins varies locally because the Pacific plate is bordered by several different plates (Cocos, Nazca, Antarctic, Australian, and so on). The width of the deep-focus earthquake zone indicates that the plate is descending into the mantle at an angle of roughly 45 degrees. The slabs of lithosphere extending to depths of 700 km at convergent boundaries indicate that the present plate motion has continued long enough for at least 1000 km of new crust to have been generated at the oceanic ridge.

The American plates move westward from the mid-Atlantic ridge, encountering the Pacific and adjacent plates along the trench on the West Coast.

The African plate is moving north, toward the convergent boundary in the Mediterranean region, driven by spreading from the ridge, which essentially surrounds it in the Atlantic and Indian oceans. There are no subduction zones between the spreading centers of the Atlantic and Indian oceans; therefore, the plate boundaries surrounding Africa are apparently moving in relation to the African plate and to each other. The same is probably true of the Antarctic plate, which is completely surrounded by the spreading center of the oceanic ridge.

The Himalayas and the Tibetan Plateau define a wide belt of shallow earthquakes. In this area, the convergence of plates has produced a collision of two continents. India moved in from the south and collided with Asia, which rode up and over the Indian plate to form a double thickness of continental rocks in the area. This convergence produced the wide zone of exceptionally high topography in the Himalayas and the Tibetan Plateau.

SEISMIC WAVES AS PROBES
OF EARTH'S INTERIOR

Seismic waves passing through Earth are refracted in ways that show distinct discontinuities within Earth's interior and provide the basis for the belief that Earth has (1) a solid inner core, (2) a liquid outer core, (3) a soft asthenosphere, and (4) a rigid lithosphere.

How can seismic waves "X-ray" the internal structure of Earth?

Speculations about the interior of Earth have stimulated the imagination of humans for centuries, but only after we learned how to use seismic waves to obtain an "X-ray" picture of Earth were we able to probe the deep interior of Earth and formulate models of its structure and composition. Seismic waves—both P waves and S waves—travel faster through rigid material than through soft or plastic material. The velocities of these waves traveling through a specific part of Earth thus give an indication of the type of rock there. Abrupt changes in seismic wave velocities indicate significant changes in Earth's interior.

Seismic waves are similar in many respects to light waves, and their paths are governed by laws similar to those of optics. Both seismic vibrations and light vibrations move in a straight line through a homogeneous body. If they encounter a boundary between different substances, however, they are reflected or refracted. Familiar examples are light waves reflected from a mirror or refracted (bent) as they pass from air to water.

If Earth were a homogeneous solid, seismic waves would travel through it at a constant speed. A *seismic ray* (a line perpendicular to the wave front) would then be a straight line like those shown in Figure 19.10. Early investigations, however, found that seismic waves arrive progressively sooner than was expected at stations progressively farther from the earthquake's source. The rays arriving at a distant station travel deeper through Earth than those reaching stations closer to the epicenter. Obviously, then, if the travel times of long-distance waves are progressively shortened as they go deeper into Earth, they must travel more rapidly at depth than they do near the surface. The significant conclusion drawn from these studies is that Earth is not a homogeneous, uniform mass but has physical properties that change with depth. As a result, seismic rays are believed to follow curved paths through Earth (Figure 19.11).

In 1906, scientists recognized that whenever an earthquake occurs, there is a large region on the opposite side of the planet where the seismic waves are not detectable. To better understand the nature and significance of this *shadow zone*, refer to Figure 19.12. For an earthquake at a particular spot (labeled "0°"), the shadow zone for P waves invariably exists between 103 and 143 degrees from the earthquake's focus. Evidently something deflects the waves from a linear path. The best explanation for this shadow zone is that Earth has a core through which P waves travel relatively slowly. Seismic rays traveling through the mantle follow a curved path from the earthquake's focus and emerge at the surface between 0 and 103 degrees from the focus (slightly more than a quarter of the distance around Earth). In Figure 19.12, ray 1 just misses the core and is received by a station located 103 degrees from the focus. Ray 2, however, being steeper than ray 1, encounters the core's boundary, where it is refracted. It travels through the core, is refracted again at the core's boundary, and is finally received at a station on the opposite side of Earth. Ray 3 is similarly refracted and emerges on the opposite side, 143 degrees from the focus. Other rays that are steeper than ray 1 are also refracted through the core and emerge between 143 and 180 degrees from the focus. Refraction at the boundary between the core and the mantle thus causes a P-wave shadow zone

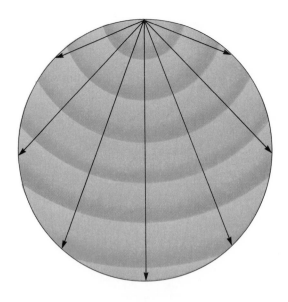

FIGURE 19.10
Seismic waves in a homogeneous planet would be neither re-flected nor refracted. Lines drawn perpendicular to the wave fronts (rays) would follow linear paths.

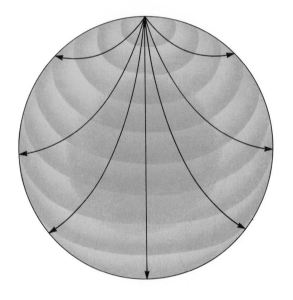

FIGURE 19.11
Seismic waves in a differentiated planet would pass through ma-terial that gradually increases in density with depth. As a result, wave velocities would increase steadily with depth, and rays would follow curved paths.

over part of Earth's surface, that is, a zone where no P-waves are received. The shadow zone extends as a band across the planet, as shown in Figure 19.13.

In 1914, Beno Gutenberg, a German seismologist, calculated that the depth to the surface of the core is 2900 km. Later analysis of more numerous and more reliable seismic data showed that Gutenberg's original estimate was remarkably accurate, with a probable error of less than two-thirds of 1%. More recent studies of the P-wave shadow zone show that some weak P waves of low amplitude are received in this zone. This suggests the presence of a solid inner core, which deflects the deep, penetrating P waves in the manner shown in Figure 19.13.

The surface of the core has an even more pronounced effect on S waves, but this effect cannot be explained by reflection or refraction. S waves simply do not pass through the core at all. They produce a huge shadow zone ex-tending almost halfway around Earth opposite the earthquake's focus (Figure 19.14). One difference between P waves and S waves is particularly significant. P waves pass through any substance—solid, liquid, or gas. S waves, however, are transmitted only through solids that have enough elastic strength to return to their former shapes after being distorted by the wave motion. They cannot be transmitted through a liquid. The fact that S waves will not travel through the core, therefore, is generally taken as evidence that the outer core is liquid.

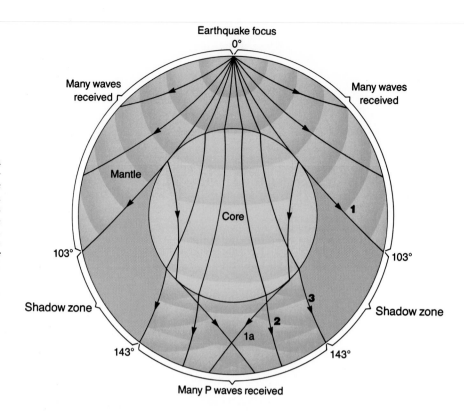

FIGURE 19.12
A P-wave shadow zone occurs in the area between 103 and 143 degrees from an earthquake's focus. The best way to explain the P-wave shadow zone is to postulate that Earth has a central core through which P waves travel relatively slowly. Ray 1 just misses the core and is received at a station located 103 degrees from the earthquake's focus. A steeper ray, such as ray 2, encounters the boundary of the core and is refracted. It travels through the core, is refracted again at the core's boundary, and is received at a station less than 180 degrees from the focus. Similarly, ray 3 is refracted and emerges at the surface 143 degrees from the focus. Other rays that are steeper than ray 1 are severely bent by the core, so that no P waves are directly received in the shadow zone. From shadow zones, seismologists calculate that the boundary of the core is 2900 km below the surface.

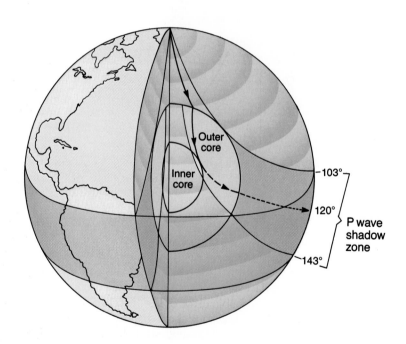

FIGURE 19.13
P waves are deflected by the inner core and are received in the shadow zone as weak, indirect signals. This deflection suggests that the inner core is solid.

Seismic Discontinuities

With the present worldwide network of recording stations, even minor variations in seismic wave velocities, called *seismic discontinuities,* can be determined with considerable accuracy. These data, summarized graphically in velocity-depth curves like the one in Figure 19.15, provide additional information about Earth's interior. The most striking variation occurs at the core's boundary, at a depth of 2900 km. There, S waves stop, and the velocity of P waves is drastically reduced.

Other discontinuities are apparent but are less striking. The first occurs between 5 and 70 km below the surface. This is called the *Mohorovičić dis-*

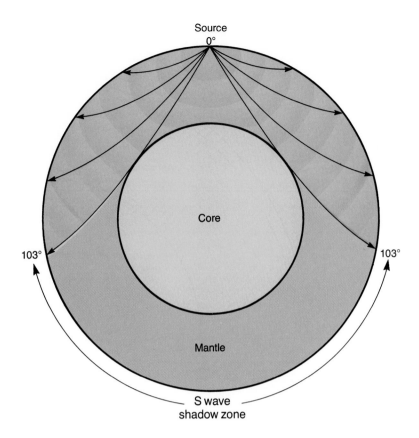

FIGURE 19.14
The shadow zone of S waves extends almost halfway around Earth opposite the earthquake's focus. This can be explained if the outer core of Earth is liquid. Because S waves cannot travel through liquid, they do not pass through the core.

continuity (or simply Moho), after Andrija Mohorovičić, the Yugoslavian seismologist who first recognized it. The discontinuity is considered to represent the base of the crust.

Perhaps the most significant discontinuity, however, is the low-velocity zone from 100 to 200 km below the surface (Figure 19.16). Gutenberg recognized this zone in the 1920s, but the discovery was viewed with skepticism by most other seismologists at that time. The normal trend is for seismic wave velocities to increase with depth. In this low-velocity zone, however, the trend is reversed, and seismic waves travel about 6% more slowly than they do in adjacent regions. More recent seismic data again confirmed Gutenberg's observations. Most seismologists are now convinced that this zone is more plastic than the areas above and below it and that it is capable of slow flow under long-term stress.

The generally accepted explanation for the low seismic wave velocities in this zone is that temperature and pressure cause part of the material, perhaps from 1% to 10%, to melt, so that a crystal–liquid mixture is produced. A small amount of liquid film around the mineral grains serves as a lubricant, increasing the plastic nature of the material. This low-velocity zone, as we will see in later chapters, plays a key role in the theories of motion of material at Earth's surface.

Other rapid changes in seismic wave velocities occur in the mantle at depths of about 350 km and 700 km. These are interpreted as the results of phase changes in the minerals, in which the atomic packing is rearranged in denser and more compact units.

Seismic Tomography

Over the past few years, scientists have utilized a new analytic technique called *seismic tomography,* which promises to greatly enhance our knowledge of Earth's internal structure, including the pattern of flow in the mantle.

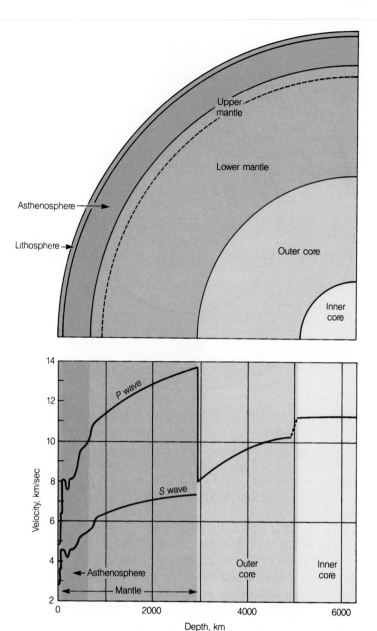

FIGURE 19.15
The internal structure of Earth is deduced from variations in the velocity of seismic waves at depth. The velocity of both P waves and S waves increases until they reach a depth of approximately 100 km. There the waves abruptly slow down, and their velocities continue to decrease until they have traveled to a depth of 200 km. This low-velocity layer is called the asthenosphere (see Figure 19.16). Below 200 km beneath the surface, the velocity of P waves and S waves increases until the waves reach a depth of about 2900 km, where both velocities change abruptly. S waves do not travel through the central part of Earth, and the velocity of the P waves decreases drastically. This variation is the most striking discontinuity and is considered to signify the boundary between the core and the mantle. Another discontinuity in P-wave velocity, at a depth of 5000 km, indicates the surface of the inner core.

Seismic tomography is like its medical analogue, the CAT scan (computer-assisted tomography; tomograph is based on a Greek word *tomos,* meaning section). In a CAT scan, X-rays that penetrate the body from all directions are used to construct an image of a slice (cross section) through the body. Bones, organs, and tumors are identified because they have different densities and absorb X-rays differently. With the aid of a computer, these images are stacked side by side to produce a three-dimensional view. In seismic tomography, seismic waves are used like X-rays. Seismologists analyze the velocity of hundreds of thousands of seismic waves as they pass through Earth in different directions. The results are images showing regions where the waves travel faster or slower than normal (Figure 19.17). Geophysicists know from laboratory studies and from observations near volcanoes that seismic waves are slowed down by unusually hot rock and are speeded up by cooler rock. The tomographs can thus be interpreted as temperature maps. The next step is to assume that hot parts of the mantle, being less dense than their surroundings, will rise, whereas the cool mantle rock will sink. Thus the tomograph can be used to outline the patterns of flow in the mantle.

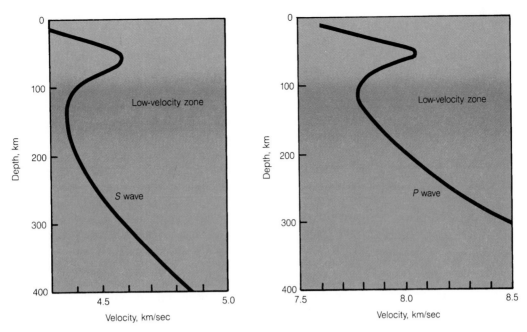

FIGURE 19.16
The rapid decrease in velocities of both S waves and P waves suggests a zone of low strength in the upper mantle, at depths between 100 and 200 km below the surface. This zone, the asthenosphere, is commonly referred to as the low-velocity layer of the upper mantle.

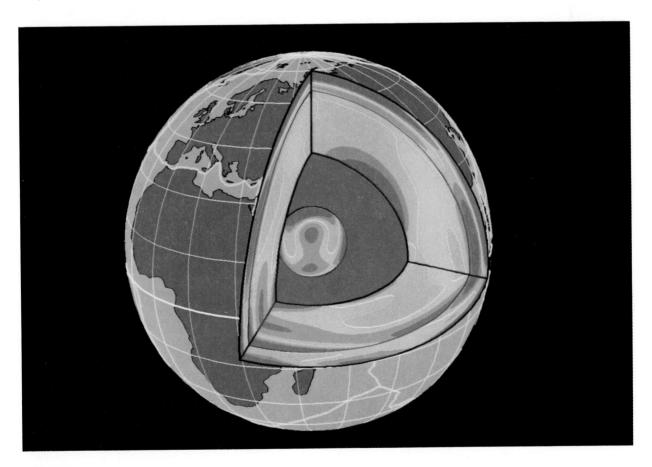

FIGURE 19.17
The internal structure of Earth as seen by seismic tomography. This illustration consists of three-dimensional plots of anomalies in seismic velocities. Seismic waves travel faster through cold, rigid rock and slower through warm, soft rock. The current view of Earth's interior is no longer a simple one of progressively hotter concentric layers of crust, mantle, and core. Instead, as this seismic tomograph shows, Earth's interior contains slowly swirling regions of hot and cooler rock. Below the thin lithosphere shades of orange delineate hot mantle rock. Blue indicates rigid cooler rock. The liquid outer core is homogeneous and has been omitted here to reveal the inner core of solid iron.

The three-dimensional view of the mantle obtained from seismic tomography confirms what one might expect from the plate tectonics theory. At a depth of 150 km, slow seismic zones occur under most of the volcanic regions, including the midocean ridges. In contrast, the shields of Canada, Brazil, Siberia, Africa, and Australia are all fast. At depths of 350 km, the midocean ridge system is no longer continuous but is broken up into isolated segments. At depths of 550 km, there is even less relation between mantle and surface features. This indicates that the midocean ridge system is not simply the surface expression of vertical upwelling currents. Instead, it must be fed by lateral transport of hot material from a few anomalies in the upper mantle.

Seismic tomography is also providing a more sophisticated view of the core. Early maps of the core–mantle boundary show that the surface of the core is not smooth, but is marked by broad swells and depressions with a difference in height of up to 20 km. The presence of any kind of topography on the core–mantle boundary bears on a number of fundamental geologic questions. A rough boundary would presumably disturb flow of the liquid iron in the outer core, much as a mountain influences the flow pattern of winds. The topography of the core also has important implications on the flow of energy in Earth. A rough core boundary would imply that a lot of energy escapes the core as heat, which might be the source of heat for convection in the mantle and the resulting tectonic plates.

Many other basic geologic questions must look to Earth's interior for answers. Projects planned for the future include increasing the number of seismic stations so that we might obtain sharper global images of the mantle and the hidden flow that shapes the surface of Earth.

SUMMARY

1. Earthquakes are vibrations in Earth caused by the sudden rupture of rocks that have been strained beyond their elastic limits.
2. Three types of seismic waves are generated by an earthquake shock: P waves, S waves, and surface waves.
3. The intensity of an earthquake is a measure of its destructive power. It depends on the total energy released, the distance from the quake's epicenter, and the nature of the rocks in the crust.
4. The magnitude of an earthquake is a measure of the total energy released.
5. The primary effect of earthquakes is ground motion. Secondary effects include (a) landslides, (b) tsunami, and (c) regional uplift and subsidence.
6. The establishment of a worldwide network of sensitive seismic stations has enabled seismologists to monitor the thousands of earthquakes that occur each year. This new insight into Earth's seismicity confirms, in a most remarkable way, the theory of plate tectonics because it shows the present-day movement of plates.
7. The distribution pattern of earthquakes dramatically outlines the plate margins.
8. Shallow earthquakes develop in a narrow zone along divergent plate margins, where the mantle rises and pulls the plate apart.
9. Where plates converge and one is thrust under the other, a zone of shallow, intermediate, and deep earthquakes is produced as the oceanic plate moves down into the mantle.
10. In the central parts of the plates there is little differential movement. Few earthquakes occur in these stable areas.
11. The velocities of P waves and S waves through Earth indicate that Earth has a solid inner core, a liquid outer core, a thick mantle, a soft asthenosphere, and a rigid lithosphere.

REVIEW QUESTIONS

1. Explain the elastic-rebound theory of the origin of earthquakes.
2. Describe the motion and velocity of the three major types of seismic waves.
3. Explain how the location of the epicenter of an earthquake is determined.
4. What secondary effects commonly accompany earthquakes?
5. Describe the global pattern of earthquakes.
6. How does the depth of earthquakes indicate (a) convergent plate margins and (b) divergent plate margins?
7. Draw a diagram showing the paths that would be followed by seismic rays through Earth if the core were only half the diameter shown in Figure 19.14.
8. What do seismic velocity-depth diagrams (see Figure 19.15) tell us about the internal structure of Earth?

ADDITIONAL READINGS

Asada, T., ed. 1982. *Earthquake Prediction Techniques*. Tokyo: University of Tokyo Press.

Bolt, B. A. 1978. *Earthquakes: A Primer*. San Francisco: Freeman.

Eckel, E. G. 1970. *The Alaskan Earthquake, March 27, 1964: Lessons and Conclusions*. U.S. Geological Survey Professional Paper 546. Reston, VA: U.S. Geological Survey.

Iacopi, R. 1964. *Earthquake Country*. Menlo Park, CA: Lane.

Nicholas, T. C. 1974. Global summary of human response to natural hazards: Earthquakes. In *Natural Hazards*, G. F. White, ed., pp. 274–84. New York: Oxford University Press.

Press, F. 1975. Earthquake prediction. *Scientific American* 232(5):14–23.

Wyllie, P. J. 1976. *The Way Earth Works*. New York: Wiley.

KEY TERMS

deep-focus earthquake (p. 460)

earthquake (p. 458)

earthquake intensity (p. 460)

earthquake magnitude (p. 462)

elastic limit (p. 458)

elastic-rebound theory (p. 458)

epicenter (p. 458)

focus (p. 458)

intermediate-focus earthquake (p. 460)

longitudinal wave (p. 460)

Moho (Mohorovičić discontinuity) (p. 472)

P wave (primary wave) (p. 460)

S wave (secondary wave) (p. 460)

seismic discontinuity (p. 472)

seismic ray (p. 470)

seismic tomography (p. 473)

seismic wave (p. 458)

seismograph (p. 458)

shadow zone (p. 470)

shallow-focus earthquake (p. 460)

surface wave (p. 460)

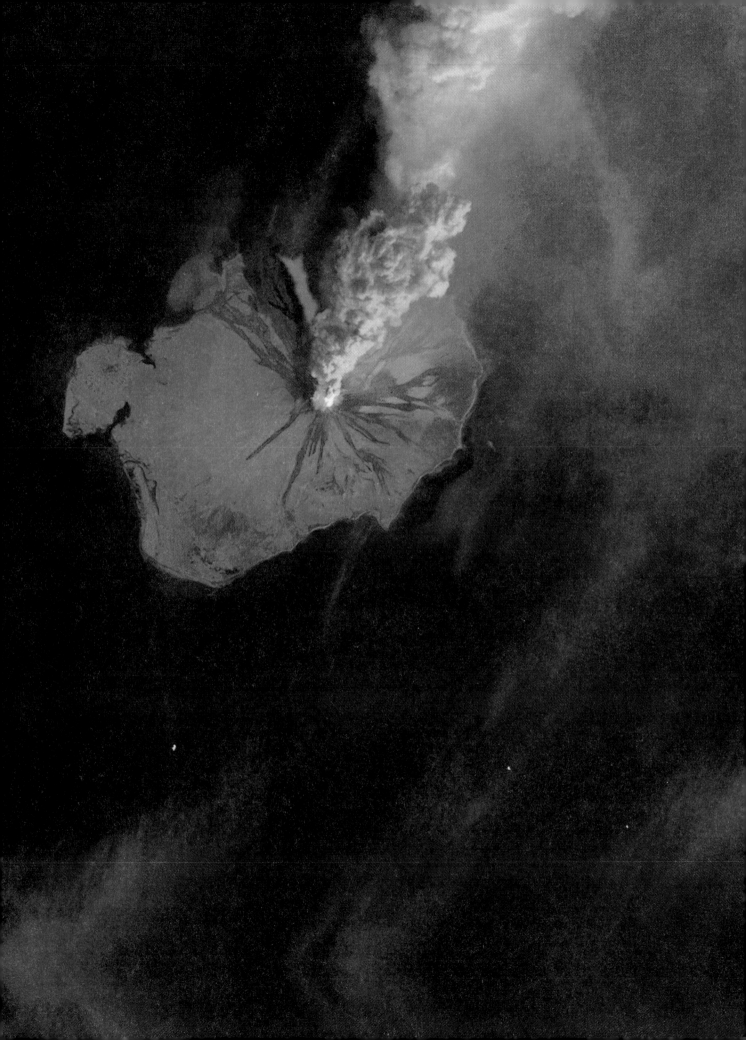

V olcanic eruptions are one of the most spectacular of all geologic phenomena, and for centuries they have caused dismay and terror to people who live nearby. Although they have always attracted attention because of their sometimes violent and catastrophic destruction, the significance of volcanic eruptions was only partly understood. It is now clear that volcanism is a key process in Earth's dynamics. It is not a rare or abnormal event. It has occurred on Earth throughout most of geologic history and undoubtedly will continue far into the future.

Augustine Island, a volcano in the Aleutian Island arc, erupted shortly after midnight on March 27, 1986. Airline pilots saw bright flashes of light, and a cloud of ash rose more than 11 km into the atmosphere. This image was recorded by Landsat 5 within hours after the initial eruption and was processed in false color to aid in scientific analysis. Red represents flowing hot ash, blue indicates snow, and gray-brown shows flows of cool ash. This type of explosive eruption represents a fundamental type of volcanism resulting from converging tectonic plates.

To appreciate the importance of volcanic activity in the dynamics of Earth, consider the volume of rock it produces. More than two-thirds of the face of Earth—the ocean floors—is underlain entirely by rocks derived from lava during the last 200 million years. Volcanic activity is also the most significant process in building island arcs. In addition, volcanism is important in many mountain chains, and vast floods of lava have constructed large lava fields along the margins of many continents.

Volcanic activity is significant for more than just the quantity of lava it produces. It is closely associated with the movements of tectonic plates, movements that also produce earthquakes, ocean basins, continents, and mountain belts. Moreover, volcanism provides a window into Earth's interior, giving us tangible evidence of processes operating far below the planet's surface.

20
Volcanism

MAJOR CONCEPTS

1. Most volcanism corresponds to active seismic zones and is clearly associated with plate boundaries.
2. The type of volcanic activity depends on the type of plate boundary.
3. Basaltic magma is generated at divergent plate boundaries by the partial melting of the asthenosphere. It is extruded mostly in quiet fissure eruptions.
4. Granitic magma is generated at convergent plate margins by the partial melting of the oceanic crust.
5. Intraplate volcanic activity probably represents local hotspots in the mantle.

GLOBAL PATTERNS OF VOLCANISM

Most volcanic activity coincides with the active seismic regions of the world and is clearly associated with plate boundaries. The type of volcanic activity depends on the type of plate boundary.

Volcanoes occur in many areas of the world, including the islands of the oceans and the young mountain ranges and plateaus of the continents. The distribution of volcanic activity, however, is not random. If the locations of active or recently active volcanoes* are plotted on a tectonic map (Figure 20.1), two important facts stand out:

1. Most volcanic activity coincides with the active seismic regions of the world and is clearly associated with plate boundaries.
2. The type of volcanic activity depends on the type of plate boundary.

As can be seen in Figure 20.1, a worldwide belt of volcanic activity occurs along divergent plate margins, but it is largely concealed beneath the ocean; there are probably many more volcanoes along the oceanic ridge than shown in Figure 20.1. Locally, enough lava is sometimes extruded along this zone to build a volcanic pile rising above sea level. Iceland, the best example, is built entirely of volcanic rock and continues to be volcanically active. Surtsey, a new island off Iceland's southern coast, was built by eruptions that started in 1963. A volcano erupted suddenly in 1973 on the tiny coastal island of Heimaey and buried much of Iceland's largest fishing port. More volcanism in this area can be expected in the future. South of Iceland, only a few submarine volcanoes rise above sea level, but submarine eruptions of gas and ash rising above the surface of the water have been seen by sailors in the area of the mid-Atlantic ridge. Other volcanoes located along divergent plate margins include those associated with the East African rift valleys and perhaps those along the margins of the Basin and Range province in the western United States.

The most notable belt of volcanic activity occurs along subduction zones and is intimately related to the intense seismicity that also occurs there. The most spectacular belt of volcanism practically surrounds the Pacific basin and has long been referred to as the "ring of fire." The volcanoes in this belt are unquestionably among Earth's great physical features. Their distribution is controlled by the subduction zones of the three major plates that make up the Pacific basin and the associated minor ones (such as the Caribbean and Philippine plates). Another similar belt of volcanic activity follows the convergent margin of the African plate. It extends through southern Europe to the Middle

What determines the location of active volcanoes on Earth?

---

* When geologists speak of recently active volcanoes, they are referring to those that have erupted during the last 10,000 years or so (which in a geologic time frame is recent).

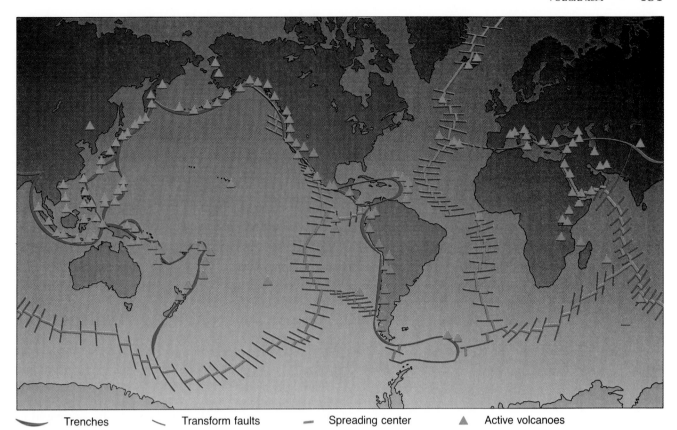

Trenches   Transform faults   — Spreading center   ▲ Active volcanoes

FIGURE 20.1
Active volcanoes are concentrated along plate boundaries. Basaltic volcanism and shallow intrusions occur all along the oceanic ridge, where plates are moving apart. Except for Iceland and a few small islands, however, the volcanic activity at spreading centers is concealed beneath the ocean. The most conspicuous volcanic activity occurs in the chains of andesitic volcanoes that form over subduction zones, at convergent plate margins. These include the ring of fire around the Pacific Ocean and the volcanoes of the Mediterranean and the Near East. Intraplate volcanism occurs mostly in the Pacific, where the movement of plates over hotspots in the mantle produces volcanic islands and seamounts.

East. Volcanic activity along the Australian plate boundary is concentrated in the island arcs of Indonesia.

Although most of the world's famous volcanoes are located along plate margins and correspond to the belt of intense seismic activity, some volcanoes occur in the middle of tectonic plates. Most of these are in the Pacific. The islands of Hawaii are the most notable example.

VOLCANISM AT DIVERGENT PLATE MARGINS

> *At divergent plate margins, basaltic magma is generated by the partial melting of the asthenosphere and is extruded largely as fissure eruptions.*

Until recently, little was known about the nature of volcanism along divergent plate margins because most of these areas are beneath the sea. Now, however, our understanding of this process has been broadened by data from three different lines of study:

1. Observations of the sea floor from submarine profiles, drilling, dredging, and submarine photography

2. Geologic studies in areas such as Iceland, where parts of the diverging plates are elevated above sea level
3. Studies of diverging plates along continental rift systems, where plateau basalts occur

These studies indicate that most volcanic activity along divergent plate margins takes the form of quiet eruptions of basaltic magma and that such eruptions have occurred there throughout all of geologic time. Magma extruded along divergent plate boundaries is believed to be generated by the partial melting of the mantle.

The Generation of Magma at Divergent Plate Margins

Why and how is magma generated along spreading centers rather than in some other place? The answer lies in the special characteristics of temperature and pressure in the asthenosphere and in their relationship to the melting of minerals in the mantle. The generation of magma along divergent plate margins is not the result of high temperature alone; it is also related to the effects of pressure and the temperature at which melting occurs.

As is shown in Figure 20.2, the balance of temperature, pressure, and composition in the asthenosphere (between 100 and 200 km below the surface) allows some melting to occur. In the overlying lithosphere, the temperature is too low for melting to occur. In the underlying mantle, the confining pressure is so great that the rocks are kept well under their melting points. The balance that permits some minerals to melt is reached in the asthenosphere.

The physical characteristics of the asthenosphere can be compared to those of slushy snow. It is a mixture of solid crystals and liquid, or melt. One factor particularly enhances the generation of magma along spreading centers. As the magma moves upward, the pressure is reduced, and the decrease in pressure lowers the temperature at which melting occurs. Magma is thus generated along spreading centers, in contrast to other zones, largely because of pressure. The diagram in Figure 20.3 illustrates the basic processes.

It should be reemphasized that each mineral in a rock has its own melting point. Magma produced by the partial melting of the asthenosphere does not have the composition of the original peridotite, which is a rock composed of the minerals olivine and pyroxene. Laboratory experiments on melting peridotite at high pressure indicate that if 10 to 30% of the rock melts, the resulting magma has a basaltic composition. Basaltic magma, because it is less dense than the solid mantle material, moves upward, into the fractures between the spreading plates, and is extruded on the ocean floor.

Why is volcanic activity at divergent plate margins different from that at convergent margins?

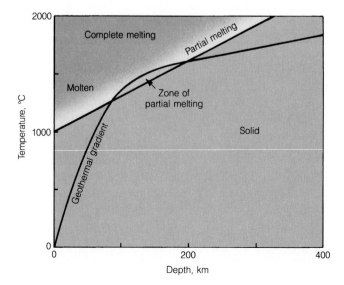

FIGURE 20.2
Temperature and pressure in the asthenosphere, at depths between 100 and 200 km, are in a critical balance at which melting occurs. From the surface to a depth of about 100 km (the lithosphere), the temperature is too low for melting. At depths between 100 and 200 km (the asthenosphere), the temperature curve passes into the zone in which melting occurs. Although the temperature continues to rise with increasing depth, pressures below the asthenosphere are too great for melting to occur.

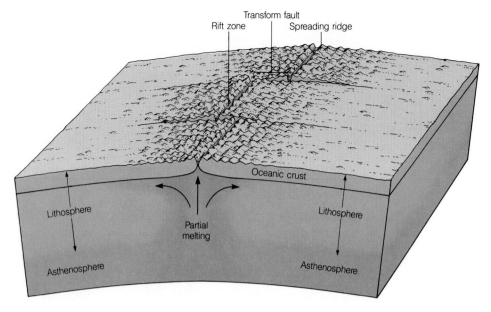

FIGURE 20.3
The origin of basaltic magma can be explained by the partial melting of mantle material that rises up at the spreading zone between two divergent plates. As mantle material moves upward along the spreading center, hot peridotite (the major rock in the mantle) partially melts because of the decrease in pressure. The resulting liquid forms a basaltic magma.

Direct Observations of the Sea Floor

In recent years, extensive studies of dredged samples and cores obtained from deep-sea drilling have improved our understanding of the processes operating along the oceanic ridge. Some of the most successful research came from a combined effort of French and American scientists, who studied and photographed parts of the mid-Atlantic ridge south of Iceland. This research program, known as Project FAMOUS (French-American Mid-Ocean Undersea Study), began in 1971. It culminated in 1974 with a series of 42 descents in deep-diving vessels capable of reaching the sea floor to collect samples and take photographs. Thousands of photographs show that the oceanic ridge, as expected, is composed of innumerable structures of *pillow lava* (Figure 20.4). Instead of forming a single unit, as basalt flows do on land, submarine flows characteristically form multitudes of bubblelike structures resembling a jumbled mass of pillows. Pillow basalt forms because underwater flows are chilled as soon as the liquid rock comes in contact with the water. This produces rounded, frozen skins on the lava while the interior remains liquid. A succession of swelling, pillow-shaped bodies forms at breaks in the crust of the parent flow. These bodies can become detached and come to rest on the sea floor while they are still hot and plastic. Accumulations of pillow basalt thus show unmistakable evidence of underwater formation.

Another highly significant discovery of Project FAMOUS was the presence of numerous open *fissures* in the crust along the oceanic ridge. More than 400 fractures, some as wide as 3 m (Figure 20.5) were observed in an area of 6 km². These are considered conclusive evidence that the oceanic crust is being pulled apart.

Geologic Studies of Iceland

Iceland is the best modern example of an area where the oceanic ridge rises above sea level. There geologists can examine in detail the surface expression of an active spreading center. The island is a plateau of basalt with a well-

FIGURE 20.4
Pillow basalt along the mid-Atlantic ridge was photographed at close range by scientists in the deep-diving submersible *Alvin*. Little or no sediment covers the basalt because this part of the sea floor is very young. The large elliptical structure is approximately 1 m long.

FIGURE 20.5
Open fissures along the mid-Atlantic ridge were photographed by scientists in the deep-diving submersible *Alvin*. Hundreds of fissures such as this were mapped. They clearly indicate that the rift zone in the oceanic ridge is pulling apart.

marked, troughlike rift extending through the center (see Figure 18.12). Although some cinder cones develop, the great floods of basalt have been extruded quietly along the fissures. The youngest rocks are located along the rift, with progressively older basalts occurring toward the east and west coasts. Of special importance is the presence of innumerable vertical basalt dikes, called ***sheet dikes.*** The aggregate thickness of these vertical dikes is about 400 km, which represents the total width of new crust created in Iceland during the last

65 million years. This information—together with data obtained from seismic studies, deep-sea drilling, and sea-floor observations made by Project FAMOUS—confirms that volcanism along the divergent plate margins occurs largely as fissure eruptions in which great floods of basaltic magma are extruded to form new crust.

Studies of Plateau Basalts

Where a spreading center passes beneath a continent, the continental crust is split, and great volumes of basalt commonly are extruded and spread over large areas near the rift system. These great floods of lava fill lowlands and depressions in the existing topography, and with subsequent crustal uplift, they erode into basalt plateaus (Figure 20.6). The deposits therefore are referred to as *flood basalts,* or *plateau basalts.* For example, in southern Brazil, more than 1,000,000 km^3 of basalt were extruded in a relatively short period of geologic time (10 million years). Similar floods have occurred in the Deccan Plateau of India, the Ethiopian Plateau of Africa, the Columbia Plateau of the western United States, and large areas of Siberia, Greenland, Antarctica, and northern Ireland. Much older flood basalts are found in northern Michigan and the Piedmont region of the eastern United States.

Plateau basalts are believed to represent the initial stage of continental rifting and provide direct evidence of the nature of volcanic activity along divergent plate margins. Figure 20.7 is a map of the southern continents prior to separation, which began in late Mesozoic time. The plateau basalts lie along the present-day continental margins, but they were originally extruded along rift systems that later developed into the oceanic ridge.

Among the best examples of plateau basalts are those of the Columbia Plateau, in eastern Washington and Oregon and western Idaho. These basalts lie along the northward extension of the eastern Pacific ridge, and they may well represent the initial stages of the breakup of western North America. The Columbia River Basalt covers an area of nearly 5,000,000 km^2 with a total thickness of between 1 and 2 km (Figure 20.6). This great accumulation of lava was not fed by central eruptions associated with a single volcano. Instead, the lava flowed to the surface through numerous fissures. Vast *dike swarms* now mark some of the fissures through which the volumes of lava were extruded.

FIGURE 20.6
Flood basalts of the Columbia Plateau were extruded during the last 30 million years. They provide important information about the style of volcanic activity along fissures and rift zones.

FIGURE 20.7
The major plateau basalts are located near the continental margins. They are believed to represent extrusions associated with the initial stage of continental rifting.

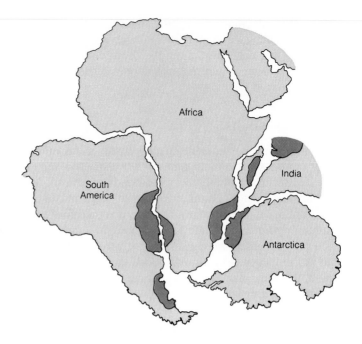

The lavas of the Columbia Plateau were extruded during the past 30 million years, with considerable activity during the last 1 million years. The most recent eruptions occurred on the Snake River plain in southern Idaho, where numerous flows have been extruded during the last few hundred years (Figure 20.8). There, the lava forms extensive flows with pahoehoe or aa surfaces. Locally, along the fracture system, spatter cones or piles of cinder accumulate instead of the pillow lava that forms beneath the ocean.

An important implication of these and other observations is that the style of volcanic activity along divergent plate margins is predominantly extrusion of basaltic lava along fissures. The fluid lava flows readily from cracks and fissures in the rift zone and tends to spread out laterally instead of building high volcanoes. Extrusions beneath the sea form pillow lava, and eruptions of basalt on land produce floods of aa or pahoehoe flows. Regardless of where the lava is extruded, fissure eruptions dominate volcanic processes along divergent plate margins.

The great volume of lava extruded along divergent plate margins is difficult to comprehend, because most of it is hidden beneath the ocean. The spreading centers, however, are the sites of the most extensive volcanism on Earth. To appreciate this fact, consider the amount of new oceanic crust created during the last 10 million years (Figure 20.9). Approximately 20 km^3 of basalt is extruded each year along the oceanic ridge.

FIGURE 20.8
Fissure eruptions in the Snake River plain, in Idaho, show the style of this type of volcanic activity. The basalt was extruded quietly along the zone of fissures and flowed rapidly across the surface as floods of lava.

FIGURE 20.9
New crust created from volcanic activity along divergent plate margins forms a broad swath along the crest and flanks of the oceanic ridge. This great volume of lava indicates that divergent plate margins are the sites of the most extensive volcanic activity on Earth.

VOLCANISM AT CONVERGENT PLATE MARGINS

At convergent plate margins, silicic magma is generated in the subduction zone by the partial melting of the oceanic crust. It is intruded in the upper plate as batholiths or is extruded as composite volcanoes in island arcs or mountain belts. Andesitic magma is viscous and erupts violently, commonly in the form of an ash flow.

Most volcanoes erupting above sea level are clearly associated with subduction zones at convergent plate margins (Figure 20.1). The geographic setting for volcanic activity along such zones depends on the type of plate interaction. Where two oceanic plates converge, an arcuate chain of islands forms on the edge of the overriding plate, parallel to the trench. Typical examples of this volcanic setting are Japan, the Aleutian Islands, and the Philippine Islands. Where a continent occurs on the active margin of the overriding plate, similar volcanic activity develops in the mountain belt. The Andes Mountains of South America are an example of this setting. In both cases, the close association of volcanism and the convergent plate margins is clear.

The type of volcanism along subduction zones, however, is quite different from the basaltic fissure eruptions that characterize spreading centers. The magma generated at a subduction zone is largely andesitic. It is somewhat richer in silica than basalt is, and thus it is more viscous. Entrapped gas cannot escape easily. This composition results in violent, explosive eruptions from central vents, which commonly produce ash flows, stratovolcanoes, and collapse calderas.

The Generation of Magma at Convergent Plate Margins

Figure 20.10 serves as a quick visual summary of the major factors involved in the generation of magma in a subduction zone. Here, magma originates by the partial melting of the basalts and sediments of the oceanic crust as the latter plunge diagonally into the hot asthenosphere. The key to understanding volcanic activity within this zone is the composition of the magma produced by the partial melting of the basaltic oceanic crust. The lithosphere, descending into a subduction zone, is subjected to progressively higher temperatures, and it begins to melt. The first material to melt is the layer of silica-rich sediment saturated with seawater. Then follow Na-plagioclase, amphibole, and, finally, pyroxene. The residue, containing minerals rich in magnesium and iron (olivine and some pyroxene), does not melt. It continues to sink and becomes assimilated in the mantle. The magma produced by the partial melting of the oceanic crust is thus enriched in silica, sodium, and potassium (compared with the basaltic magma generated from the partial melting of the upper mantle) and typically produces andesites and associated rocks. Because of its lower density, it migrates upward through the crust and can accumulate as granitic batholiths or be extruded as ash flows or andesitic lava.

What is the origin of the "ring of fire"?

As is shown in Figure 20.10, the depth of the descending plate increases away from the trench. This causes several systematic variations in volcanic activity across an *island arc.* (1) Volcanic activity typically begins abruptly along a line 200 or 300 km landward from the trench axis. This line is called the *volcanic front.* It occurs because 100 km is the critical depth at which partial melting produces enough magma to migrate to the surface. Between the trench and the island arc is a gap where no volcanic activity occurs. This area is called the *arc-trench gap.* The width of the gap is, of course, an expression of the angle of the descending plate. (2) The volume of magma that migrates to the surface and is erupted decreases rapidly behind the volcanic front,

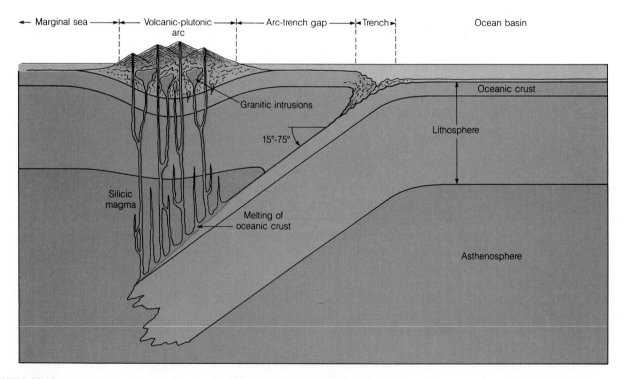

FIGURE 20.10
The generation of magma at convergent plate margins is primarily due to the slow heating of basalts and oceanic sediments as the descending plate slides down into the hot asthenosphere. At a critical depth, partial melting occurs in the descending slab of oceanic lithosphere. The melting generates andesitic magma, which rises buoyantly to form granitic batholiths or the volcanic rocks of an island arc.

possibly as a result of a depletion of water and minerals with low melting points. (3) The potassium content of the magma increases away from the trench, toward the back of the volcanic arc. This increase may be related to differences in the depth at which the magma originates.

Note that at both spreading centers and subduction zones the generation of magma by partial melting differentiates and segregates the materials of Earth. At spreading centers, low-density material, enriched in silica, aluminum, sodium, potassium, and calcium, separates from the iron- and magnesium-rich mantle and is concentrated in the oceanic crust. At subduction zones, magma is further enriched in silica by a second episode of partial melting and is concentrated in the island arcs or mountain belts of the continents. Because of its low density, this material cannot sink into the mantle but is concentrated in the continental crust. This *magmatic differentiation* is the principal method by which continents grow. We will say more about this process in Chapter 22.

There have been numerous historic volcanic eruptions, but most were not recorded, and many were not even observed. The record of the volcanic activity is preserved in the fresh, unweathered, uneroded cones and flows. A few eruptions, however, have had a major impact on human affairs and have therefore been described in minute detail. These accounts and studies of recent volcanic fields help us understand the nature of volcanic activity associated with converging plates.

Vesuvius, A.D. 79. An extraordinarily vivid and accurate eyewitness account of the eruption of Mount Vesuvius in A.D. 79 was recorded by Pliny the Younger, then a 17-year-old boy, who related details of how his famous uncle, Pliny the Elder, died in the destruction of Pompeii and Herculaneum.

Beginning in A.D. 63 and continuing for 16 years, earthquakes shook the west coast of Italy around what is now Naples. Then, on the morning of August 24, A.D. 79, Mount Vesuvius exploded with a devastating eruption of white-hot ash and gas. Within a few hours, it asphyxiated and buried the population of Pompeii. Many people, suffocated by sulfurous fumes from ash clouds, died in their homes or on the streets. The entire town and most of its 20,000 inhabitants were buried by ash and forgotten for more than a thousand years, until Pompeii was excavated in 1748 (Figure 20.11).

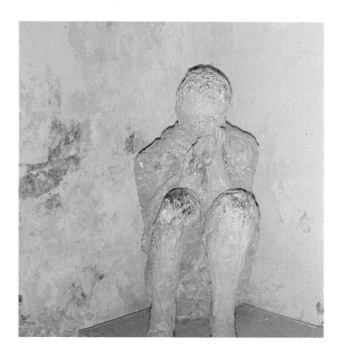

FIGURE 20.11
Pompeii, Italy, was completely covered with ash from the eruption of Vesuvius in A.D. 79. Excavations provide important insight into volcanic activity at convergent plate margins. Bodies of people asphyxiated by poisonous gas during the eruption were buried in ash. Eventually the bodies decomposed, leaving cavities in the ash. By filling these cavities with plaster, archeologists can make detailed casts.

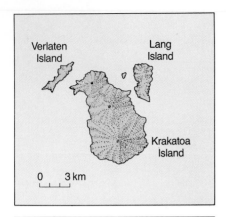

FIGURE 20.12
Maps of Krakatoa before *(top)* and after *(bottom)* the eruption of 1883 show the force of violent volcanic eruptions at convergent plate boundaries.

The ash fall that buried Pompeii is a first-class example of the type of violent eruption that is common in volcanoes along converging plates. The once-smooth and symmetrical cone of Mount Vesuvius was shattered by the explosion, which created a large caldera where a peak once existed.

Krakatoa, 1883. Krakatoa is a relatively small volcanic island west of Java, part of an island arc along the subduction zone associated with the Java Trench. After remaining dormant for two centuries, it began to erupt on May 20, 1883, and the eruption culminated in a series of four great explosions on August 26 and 27. One of them was heard in Australia, 4800 km away. The explosions are considered the greatest in recorded history. The whole northern part of the island, which stood about 600 m high, was blown off, forming a huge caldera 300 m below sea level (Figure 20.12). Tremendous quantities of ash were thrown high into the atmosphere. Some of the ash circled the globe for two years. Krakatoa was uninhabited, but more than 36,000 people were killed in Java and Sumatra by the huge tsunami produced by the explosion.

Mount Pelée, 1902. Ash flows are an important phenomenon associated with volcanism along converging plates. The great eruption of Mount Pelée, on the island of Martinique, in the West Indies, helped initiate an interest in and an understanding of this type of eruption. The eruption of Pelée was preceded by nearly a month of extrusions of steam and fine ash from the volcanic vent. Then, on May 8, 1902, a gigantic explosion blew ash and steam thousands of meters into the air. The denser hot ash moved as a body and swept down the slopes like an avalanche. In less than two minutes, the hot, incandescent ash flow moved 10 km from the side vent on Pelée and swept over the city of Saint Pierre. It annihilated the population of more than 30,000 people, except for one man, a prisoner, who was being held deep underground in the city jail. Every flammable object was instantly set aflame, and as the ash moved over the waterfront, all of the ships capsized.

The ash flow proper consisted of a mixture of hot glass shards, ash, pumice, and frothy volcanic glass, which flowed at the base of a billowing cloud of gas. The fundamental force that caused the ash to flow so rapidly was simply the pull of gravity. A mixture of hot gas and fragments of ash and lava is highly mobile and practically frictionless. Each particle is separated from its neighbors and from the surface over which it moves by a cushion of expanding gas.

Intermittent ash-flow eruptions continued on Mount Pelée for several months. By October, a bulbous dome of lava too thick to flow had formed in the crater. A spire of solidified lava was then slowly pushed up from a vent in the dome, like toothpaste from a tube. This Spire of Pelée was between 130 and 230 m in diameter, and it rose as much as 26 m per day, reaching a maximum height of 340 m above the crater floor. The spire repeatedly crumbled and grew again from the lava dome, which glowed red in the night.

The violent eruptions that characterize volcanoes at convergent plate margins result from the high silica content of the magma, which makes it thick and viscous. Dissolved gases cannot easily escape. As a result, tremendous pressure builds up in the magma, and when eruptions occur, they are highly explosive. The explosion produces huge quantities of ash, hot ash flows, and thick, viscous lava.

Mount Saint Helens, 1980. The best-documented example of the eruption of a composite volcano is the eruption of Mount Saint Helens on May 18, 1980, with an explosion estimated to have roughly 500 times the force of the atomic bomb that destroyed Hiroshima (Figure 20.13). Numerous indications of an impending eruption had drawn dozens of scientists to the site to monitor earthquakes, emissions of gas, and physical changes in the mountain. The event of May 18 is thus the best-documented volcanic eruption in history.

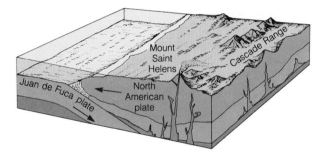

FIGURE 20.13
The eruption of Mount Saint Helens is well documented by aerial photographs and provides scientists with a rare opportunity to study details of volcanic eruptions. This photograph, taken about 3 hours after the initial blast on May 18, 1980, shows the ash cloud, which reached an elevation of about 18 km. Steam and ash were blown to the northeast across the continental United States.

Mount Saint Helens, in the state of Washington, is part of the Cascade Range, which extends about 1500 km in a north–south line from British Columbia to northern California and includes 15 major composite volcanoes (Figure 20.14). This is the North American part of the ring of fire, which encircles the Pacific Ocean. All volcanoes in the ring of fire have the same origin. They result from the subduction of oceanic plates in the vicinity of the deep-sea trenches that surround the Pacific. As the descending plate reaches regions of higher temperature and pressure, partial melting occurs, generating magma, which rises to the surface. Mount Saint Helens is the youngest volcano in the Cascade Range. Most of the existing mountain is only 2500 years old, but it overlies an older volcanic center with evidence of numerous previous eruptions. The history of eruptions extends back as far as 37,000 years.

Mount Saint Helens had been dormant for 123 years, but on March 20, 1980, it began to stir, with a series of small earthquakes. After a week of increasing local seismicity, it began to eject steam and ash. This was the first of

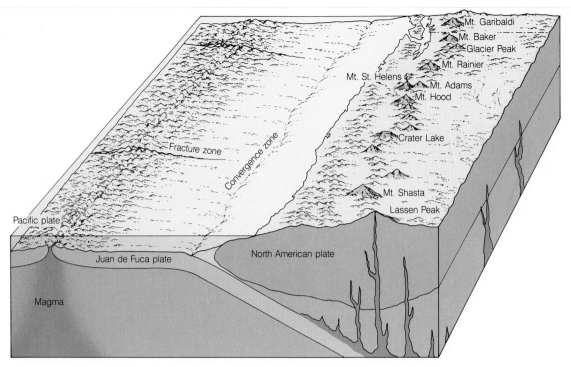

FIGURE 20.14
The Cascade Range contains 15 large composite volcanoes extending in a line from British Columbia to northern California. The volcanoes are formed by the subduction of the Gorda plate beneath the North American plate. Note the location of Mount Saint Helens.

a series of moderate eruptions that continued intermittently for the next six weeks. Within a few days after these first eruptions started, warnings were issued by the United States Geological Survey. During the weeks to come, the United States Forest Service and state officials closed all areas near the mountain, undoubtedly saving thousands of lives.

By the second week of activity, more than 30 geologists had gathered at the site to carry out a wide variety of studies. Much of their effort was directed toward monitoring the development of a large bulge on the north flank of the mountain. By the end of April, the bulge was 2 km long and 1 km wide and was expanding horizontally at a steady rate of 1.5 m per day. Clearly, the mountain was being inflated by magmatic intrusion. Sulfur dioxide (SO_2) was being released at a rate of 50 metric tons per day from March to May 18. Volcanoes that are actively erupting give off as much as 1000 metric tons of SO_2 per day, so an increase in the emission of SO_2 might suggest an increase in magmatic activity. Monitoring the seismicity, the bulge, and the emission of gases, geologists believed that they would detect some significant change to warn them of an impending large eruption, but no anomalous activity occurred. In fact, seismic activity decreased. Thirty-nine earthquakes were recorded on May 15 and only 18 on May 17.

On Sunday morning, May 18, the mountain was silent. Only minor plumes of steam rose from two vents. David Johnston, a 30-year-old geologist, was monitoring the emission of gases and making visual observations 8 km northwest of the volcano's crater at an observation post with a two-way radio. "Vancouver! Vancouver!" he cried. "This is it!" Moments later, Johnston vanished in the explosion of hot ash and gas as more than 4 km³ of material was thrown from the blast on the north side of the mountain.

The best way to understand the nature of the eruption is to study the sequence of diagrams in Figure 20.15. At 8:32 A.M., the mountain was shaken by an earthquake with a magnitude of approximately 5 on the Richter scale. The north slope began to undulate and then move downslope as a great debris

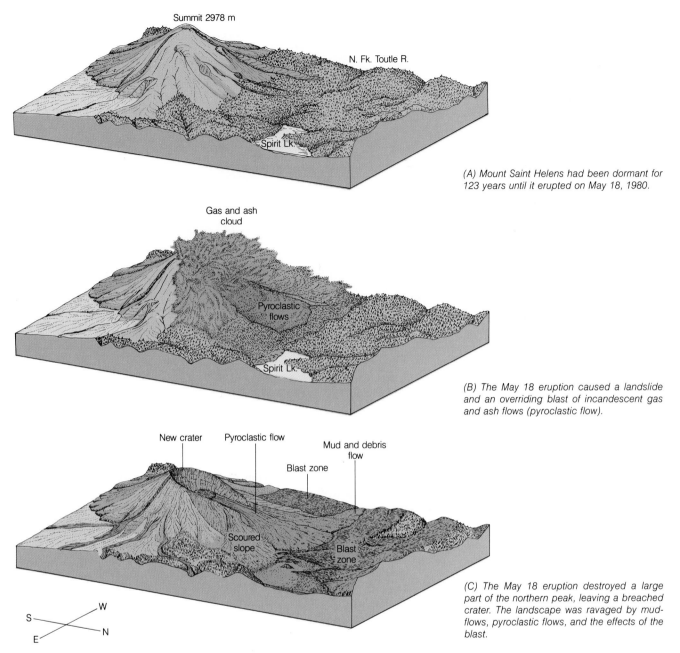

(A) Mount Saint Helens had been dormant for 123 years until it erupted on May 18, 1980.

(B) The May 18 eruption caused a landslide and an overriding blast of incandescent gas and ash flows (pyroclastic flow).

(C) The May 18 eruption destroyed a large part of the northern peak, leaving a breached crater. The landscape was ravaged by mud-flows, pyroclastic flows, and the effects of the blast.

FIGURE 20.15
The sequence of events in the eruption of Mount Saint Helens, as seen from the northeast, is depicted in this series of diagrams.

flow. As the north side of the volcano gave way, it uncapped the bottled-up gas and magma, and an eruption cloud blasted laterally, above, and over, the collapsing slope. This lateral blast of rock, ash, and gas caused most of the destruction and loss of life. The blast wave leveled the forest in an area 35 km wide and 23 km outward on the north flank of the mountain (see Figure 20.16). One man died in his truck still holding his camera. A young couple was buried in their tent by falling trees, their arms still around each other.

The eruption caused three separate but somewhat interrelated processes: (1) mudflows, (2) ash flows, and (3) ash falls. Mudflows originated largely from debris flows saturated with water on the upper slopes of the mountain. Most of the mudflows were hot. They moved rapidly downslope, and the most massive and devastating of them traveled many kilometers down the Toutle River. Their

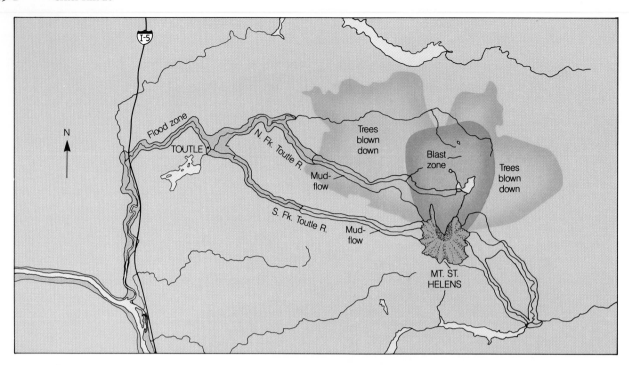

FIGURE 20.16
The effects of the eruption of Mount Saint Helens are depicted on this map, showing the blast area on the northern flank of the mountain, the area where trees were blown down, and the path of mudflows on the Toutle River.

rush downslope killed several people. At the height of the flow, the lower Toutle River swelled to 1.5 km wide and was heated to 90°C. Mudflows swept up 123 homes, as well as cars, logging trucks, and timber, and carried them downstream to batter and destroy bridges. So much sediment from the Toutle was carried into the Columbia River that the downstream depth of the Columbia was reduced from 12 m to 4 m within a day, and ships upstream were trapped.

An important part of the eruption was the extrusion of numerous ash flows. Traveling as fast as 130 km per hour, the incandescent ash and debris (at a temperature of about 500°C) extended north for a distance of 9 km. Some ash flows reached Spirit Lake, where, together with debris flows that were deposited earlier, they blocked the lake's outlet. Consequently, the level of water in the lake rose 60 m.

Immediately after the eruption, a vertical cloud of ash rose to heights of 18 km. The ash clouds then fanned out downwind (eastward), and ash began to settle like a soft, gray snow. Within two hours, it began to fall on Yakima, Washington, and by the next day the ash cloud was over Cheyenne, Wyoming, approximately 1400 km away. The cloud then moved in a broad arc across the United States and had completely circled Earth by June 5.

The particles of ash are composed of fine-grained crystals, fragments of volcanic glass, and rock material that once formed the summit dome. The ash is slightly acidic and is often coated with sulfur from the eruption. In just a few hours, Mount Saint Helens had thrown up almost as much ash as Vesuvius did in A.D. 79, when it buried Pompeii and Herculaneum.

On May 25, another eruption sent a cloud of steam and ash to elevations of 13 km. This event caused no structural changes in the mountain. Ash flows were barely large enough to travel to the base of the mountain, and mudflows were minor. Similar activity occurred again on June 12, with the extrusion of ash flows that moved downslope as far as Spirit Lake. This eruption was preceded by a significant increase in harmonic tremors (constant, rhythmic vibrations). The mountain erupted again on July 22 and on August 7, ejecting ash and steam to heights of approximately 13 km and extruding new ash flows over the northern flank. Both events were preceded by harmonic tremors and a noticeable decrease in the ratio of CO_2 and SO_2 emissions the day before the eruption.

It is clear from the geologic record that Mount Saint Helens has had a history of spasmodic explosive activity separated by intervals of dormancy lasting 200 or 300 years. The most recent explosions demonstrate that the show is far from over. Future eruptions will undoubtedly produce lava flows, domes, tephra, and ash flows—the typical *ejecta* of stratovolcanoes. Whatever Mount Saint Helens does next, geologists stand only to gain. By catching a volcanic eruption in the act, we are able to study many aspects of volcanism that are not preserved in the rocks from past eruptions. Geologists hope to learn how to assess potential hazards and even how to predict the time of major eruptions. A prime candidate as a predictive tool is the volcano's seismic activity: harmonic tremors have preceded several eruptions. Other clues may be revealed by studies of the emission of gases, such as SO_2, CO_2, and H_2 (hydrogen).

INTRAPLATE VOLCANIC ACTIVITY

Volcanic eruptions in the central parts of plates may be important as surface expressions of local thermal variations, or hotspots, in the mantle material. Most intraplate volcanic activity occurs on the floor of the South Pacific and has produced numerous submarine volcanoes and volcanic islands.

Volcanic eruptions in the central parts of plates, beyond the active margins, are trivial compared with those along spreading centers and subduction zones, but they may be important as surface expressions of local thermal variations, or hotspots, in the mantle material. Most intraplate volcanic activity occurs on the floor of the South Pacific, producing numerous submarine volcanoes and volcanic islands, both as isolated features and in linear chains.

Igneous activity within the continental platforms, in areas not associated with plate margins, is rare. Where it does occur, it commonly is limited to scattered extrusions and small dikes and sills. These are thought to be the result of mantle plumes (rising masses of hot mantle material, which may or may not be parts of a large convection cell).

Figure 20.17 is a map showing the distribution of major centers of intraplate volcanic activity in the Pacific. At first glance the distribution of intraplate volcanoes may appear to be random, but obvious linear trends, or chains, are soon apparent. Excellent examples include the Hawaiian–Emperor chain, the Tuamotu–Line chain, and the Austral–Marshall–Gilbert chain. The best data available suggest that volcanic chains are formed as a lithospheric plate moves over a *mantle plume,* or *hotspot* (Figure 20.18). Volcanism occurring over a hotspot produces a submarine volcano, which can grow into an island. If the hotspot's position in the mantle remains fixed for a long time, the moving lithosphere carries the volcano beyond the magma source. This volcano then becomes dormant, and a new one forms over the fixed hotspot. A continuation of this process would build one volcano after another, producing a linear chain of volcanoes parallel to the direction of plate motion.

Isolated volcanoes can result from small hotspots that do not endure long enough to produce a volcanic chain, or they can develop from minor pockets of magma carried with the moving asthenosphere.

If plate margins are the site of most geologic activity, why do some volcanoes occur in the central parts of plates?

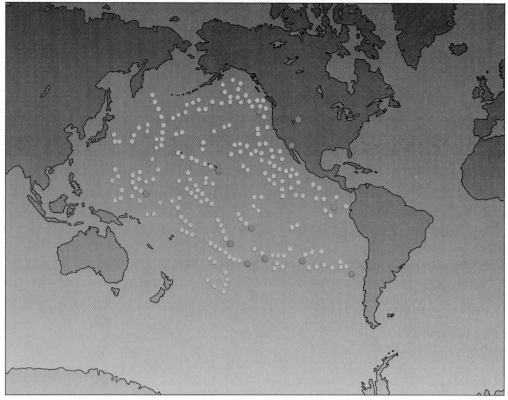

Volcanic island and seamounts   Hot spots

FIGURE 20.17
Intraplate volcanism occurs mostly in the Pacific plate, where hundreds of islands and sea-mounts have been formed by basaltic volcanism. These are believed to be produced by masses of hot mantle material, called mantle plumes, rising beneath the plate. Linear chains of volcanoes can be produced as a plate moves over a hotspot as indicated by red dots.

The type of volcanism in the interior of plates is similar, in most cases, to that along diverging plates. The products are largely basaltic lava extruded by quiet fissure eruptions. This basaltic lava is believed to be a derivative from the mantle, just like the lava that is found along the oceanic ridge.

Hawaii is one of the best known examples of volcanic activity over a hotspot (Figure 20.19). It is the active area of a series of volcanoes stretching across the Pacific sea floor from the Hawaiian islands to the Aleutian Trench. The amount of lava that has erupted from the hotspot is more than enough to cover the entire state of California with a layer a mile thick.

As can be seen in Figure 20.19, the island of Hawaii consists of five volcanoes, each built up by innumerable eruptions. Many of the more recent flows can be seen along the flanks of Mauna Loa extending as dark lines from the summit ridge toward the sea. The oval-shaped caldera was formed by collapse at Mauna Loa's summit.

To the right of Mauna Loa is Kilauea volcano where young lava flows erupted primarily from rift zones flow to the sea (Figure 20.20).

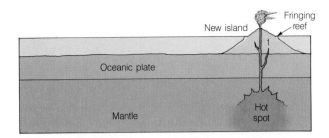

(A) A volcanic island or seamount is built up by extrusions from a fixed hotspot, or source of magma, in the mantle.

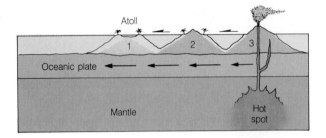

(B) As the plate moves, the volcano is carried away from the source of magma and becomes extinct. The surface of the island can then be eroded to sea level, and reefs can grow to form an atoll. A new island is then formed over the hotspot.

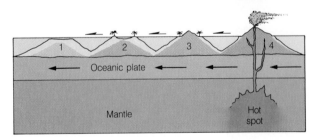

(C) Continued plate movement produces a chain of islands.

(D) The islands of the chain are progressively older away from the hotspot.

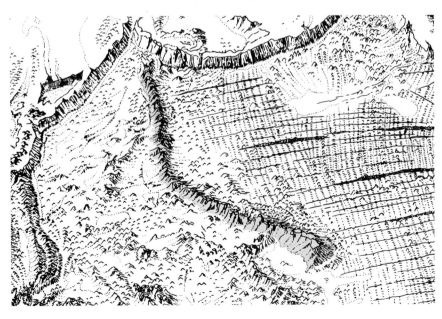

(E) An abrupt change in the direction of plate movement is indicated by a change in the direction of a chain of islands. The Emperor seamount chain began to form more than 40 million years ago, when the Pacific plate was moving northward. About 25 million years ago, the plate moved northwestward and started to form the Midway-Hawaiian chain.

FIGURE 20.18
The origin of linear chains of volcanic islands and seamounts is explained by the plate tectonics theory.

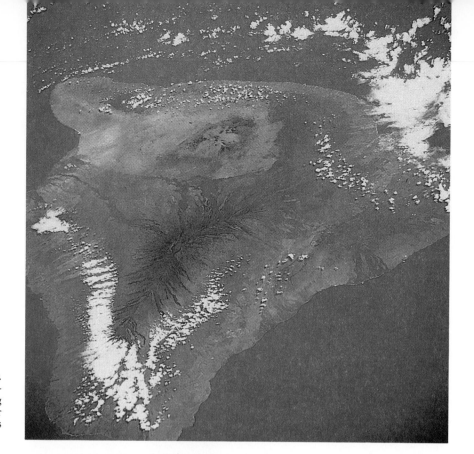

FIGURE 20.19
The island of Hawaii as seen from space. Mauna Loa is the large volcano in the foreground which has erupted many times during the last 150 years. The individual flows appear as thin dark lines extending from fissures along the southern flank.

FIGURE 20.20
Fissures along the southwest rift zone of Kilauea volcano.

SUMMARY

1. Most volcanic activity occurs in active seismic zones and is clearly associated with plate boundaries. The type of volcanism depends on the type of plate boundary.
2. At divergent plate margins, basaltic magma is generated by the partial melting of the asthenosphere and is extruded largely as fissure eruptions. Pillow lavas form on the sea floor, and plateau basalts are extruded where continents overlie rift zones.
3. At convergent plate margins, andesitic or silicic magma is generated in the subduction zone by the partial melting of the oceanic crust. It is intruded in the upper plate as batholiths or is extruded as composite volcanoes in island arcs or mountain belts. Andesitic magma is viscous and erupts violently, commonly in the form of an ash flow.
4. Minor intraplate volcanic activity probably indicates local hotspots in the mantle. The greatest intraplate volcanism occurs in the Pacific, where numerous shield volcanoes form islands and seamounts.

KEY TERMS

arc-trench gap (p. 488)

dike swarm (p. 485)

ejecta (p. 495)

fissure (p. 483)

flood basalt (p. 485)

hotspot (p. 495)

island arc (p. 488)

magmatic differentiation (p. 489)

mantle plume (p. 495)

pillow lava (p. 483)

plateau basalt (p. 485)

pumice (p. 490)

sheet dike (p. 484)

volcanic front (p. 488)

REVIEW QUESTIONS

1. Compare the locations of recently active volcanoes with the global patterns of seismicity. What appears to control the location of volcanic activity?
2. What is the ring of fire?
3. What volcanic features were observed by Project FAMOUS?
4. Describe the type of volcanic activity in Iceland. What is the significance of sheet dikes?
5. What does the study of plateau basalts tell us about volcanism along divergent plate margins?
6. Summarize the characteristics of volcanism along divergent plate margins.
7. Explain how magma is generated along divergent plate margins.
8. Describe the type of volcanic activity along convergent plate margins and cite several examples.
9. How does the composition of lava influence the style of volcanic eruption?
10. Why is there a gap between the trench and the volcanic front in a subduction zone?
11. What type of volcanic activity occurs within the central parts of tectonic plates, beyond the active margins?
12. Explain the origin of chains of volcanic islands and seamounts.
13. Explain why basaltic magma is not generally expected to be generated in the upper part of the continental crust.
14. Explain why there are numerous rhyolitic tuffs and ash-flow tuffs but few rhyolite lava flows.

ADDITIONAL READINGS

Bullard, F. 1984. *Volcanoes of the Earth,* 2nd rev. ed. Austin, TX: University of Texas Press.

Decker, R., and B. Decker. 1981. The Eruption of Mt. St. Helens. *Scientific American* 244(3):68–80.

Decker, R., and B. Decker. 1981. *Volcanoes.* San Francisco: Freeman.

Eiby, G. A. 1980. *Earthquakes.* New York: Van Nostrand Reinhold.

Francis, P. 1976. *Volcanoes.* New York: Penguin.

Green, J., and N. M. Short. 1971. *Volcanic Landforms and Surface Features: A Photographic Atlas and Glossary.* New York: Springer-Verlag.

Macdonald, G. A. 1983. *Volcanoes,* 2nd ed. Englewood Cliffs, NJ: Prentice-Hall.

Simkin, T., and R. S. Fiske. 1983. *Krakatoa, 1883. The Volcanic Eruption and Its Effects.* Washington, DC: Smithsonian Institution Press.

Tazieff, H. 1974. *The Making of the Earth. Volcanoes and Continental Drift.* New York: Saxon House.

21
Evolution of Oceans

The ocean floor is an alien landscape, nothing like that which we see on the continents. It was formed by volcanic activity and faulting, and much of its surface reflects these processes. In addition, sediment settling from the ocean blankets some areas to form large, flat, featureless surfaces. It is the ocean floor, not the continents, that is the most widespread surface of the planet.

Until recently we could not even dream of seeing and understanding the features on the ocean floor. They were almost as inaccessible as the stars. Now the surface of the ocean floor has been mapped from oceanographic vessels using special instruments that plot a profile of the sea floor, and most of our understanding of the ocean basin comes from these studies. The findings are as important as they are spectacular. A great "mountain range," called the oceanic ridge, extends as a continuous unit through all of the ocean basins, a distance of more than 64,000 km. The ridge is cut by long fracture zones, several thousand kilometers long. Flanking the oceanic ridge is the abyssal floor, above which rise numerous submarine volcanoes, called seamounts. The deepest parts of the ocean are the trenches, some of which descend to depths of 11,000 m. In addition, small ocean basins, commonly referred to as seas, have also been explored. These are mostly isolated from the major ocean basins.

This high-resolution image of the ocean floor shows the midoceanic ridge in the eastern Pacific Ocean south of Baja California. Note the linear trends of the topography parallel to the crest of the ridge and the volcanic cones rising as seamounts.

This new knowledge of the landforms on the ocean basin helped revolutionize geological thinking. For the first time, we have been able to consider the geology of the entire planet. No amount of research on the continents alone could have revealed what we now know from studies of the ocean basin and the continents together.

The Oceanic Ridge

The *oceanic ridge* is the most pronounced tectonic feature on Earth. If the ridge were not covered with water, it would be visible from as far away as the Moon. It is essentially a broad, fractured swell, generally more than 1500 km wide, with peaks rising as much as 3 km above the surrounding ocean floor. It covers nearly 23% of Earth's surface, almost as much as the surface of the continents. The remarkable characteristic of the ridge is that it extends as a continuous feature around the entire globe, like the seam of a baseball. It extends from the Arctic basin, down through the center of the Atlantic, into the Indian Ocean, and across the South Pacific, terminating in the Gulf of California, a total length of more than 64,000 km. Without question, it is the greatest "mountain" system on Earth. The "mountains" of the oceanic ridge, however, are nothing like the mountains of the continents, which were built largely of folded and metamorphosed sedimentary rocks. By contrast, the ridge is composed entirely of basalt and is not deformed by folding.

Many of the characteristics of the ridge are apparent in the seismic reflection profile reproduced in Figure 21.3. On a regional basis, the ridge is a broad segment of the ocean floor that is arched up and broken by numerous fault blocks, which form linear hills and valleys. The highest and most rugged topography is located along the axis, and a prominent *rift valley* marks the crest of the ridge throughout its length. As is shown in Figure 21.3, oceanic sediments are thickest down the flanks of the ridge but thin rapidly toward the crest.

Throughout most of its length, the oceanic ridge is cut by a series of transform faults (Figure 21.4). These are sites of continued seismic activity. Beyond an active transform fault, the fracture zone is expressed by an abrupt, steep cliff, which in places can be traced for several thousand kilometers.

Detailed studies of the axial spreading zone of the mid-Atlantic ridge were made in 1974, when scientists in deep-diving vessels sampled, observed, and photographed the ridge for the first time. Without doubt, this project made some of the most remarkable submarine discoveries of modern times. The photographs show extensive pillow basalts, so recent that little or no sediment covers the fine details of their surface textures (Figure 20.4). Numerous open fissures in the crust were also observed and mapped (Figure 20.5). In one small area of only 6 km², 400 open fissures were mapped, some of which are as wide as 3 m. These are considered conclusive evidence that the oceanic crust is being pulled apart. The eruption of lava from these fractures, which parallel the rift valley, would tend to create long, narrow ridges—a feature that does indeed characterize the morphology of the oceanic ridge.

What geologic features characterize the oceanic ridge?

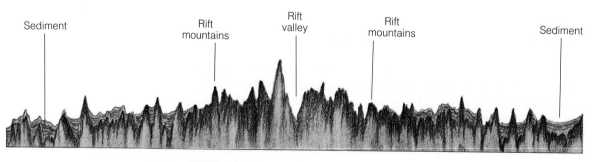

Sediment   Rift mountains   Rift valley   Rift mountains   Sediment

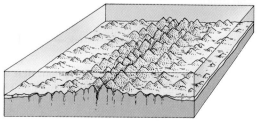

FIGURE 21.3
A seismic reflection profile across the mid-Atlantic ridge at 44 degrees north latitude shows that the crest of the ridge is marked by a deep rift valley, which can be traced along the entire length of the ridge. Sediment is thickest down the flanks of the ridge, but it thins rapidly near the crest. The idealized diagram of the ridge was based on a series of profiles.

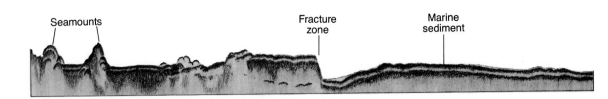

Seamounts   Fracture zone   Marine sediment

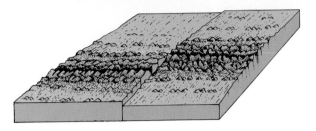

FIGURE 21.4
A seismic reflection profile across the Murray fracture zone, in the eastern Pacific Ocean, shows that the fracture is expressed by a pronounced vertical cliff, which separates areas of contrasting topography. On the left side of the fault, seamounts are abundant. To the right, the sea floor is relatively smooth and featureless. Note how this contrast in topography on the block diagram is produced by strike-slip faults. Note the seismic profile is parallel to the front of the idealized block diagram.

The general character of the oceanic ridge seems to be a function of the rate of plate separation. Where the rate of spreading is relatively low (less than 5 cm per year), the ridge is higher and more rugged and mountainous than it is where rates of spreading are more rapid. Moreover, rift valleys on slow-spreading ridges are prominent, whereas rift valleys in areas with high rates of spreading are more subdued.

The Abyssal Floor

Vast areas of the deep ocean consist of broad, relatively smooth surfaces known as the *abyssal floor.* This type of sea-floor topography was discovered in 1947 by oceanographic expeditions surveying the mid-Atlantic ridge. Subsequently, much of it has been mapped in detail with precision depth recorders, which are capable of measuring elevations on the ocean floor with relief as small as 2 m. The abyssal floor extends from the flanks of the oceanic ridge to the continental margins, generally lying at depths ranging from about 3 km to 5.5 km.

In most ocean basins, the abyssal floor can be subdivided into two sections: the abyssal hills and the abyssal plains. The *abyssal hills* are relatively small hills, rising from 75 to 900 m above the ocean floor (Figure 21.5). They are circular or elliptical and range from 1 to 8 km in width at the base. The hills are found along the flanks of the oceanic ridge and occur in profusion in parts of the ocean floor separated from land by trenches. In the Pacific, they cover between 80% and 85% of the ocean floor; thus, abyssal hills can be considered the most widespread landform on Earth.

The *abyssal plains* are exceptionally flat areas of the ocean floor where the abyssal hills are completely buried by sediment (Figure 21.5). Commonly, they are located near the margins of a continent, where sediment from a continental mass is transported by turbidity currents and spreads over the adjacent ocean floor.

The origin of the abyssal hills and abyssal plains can be traced to the oceanic ridge and the development of new crust at the spreading center. New lithosphere, created at the ridge crest, slowly recedes from the rift zone and is gradually modified in several important ways. The new crust cools and contracts, deepening the ocean basin as it moves away from the ridge. Fine-grained *pelagic sediments,* consisting of dust and the shells of marine organisms, slowly, but continually, settle over all of the sea floor. Linear hills, formed by volcanic activity, intrusions of magma, and block faulting, become the foundations of the abyssal hills. As plates move away from the spreading center, the superficial features of the landforms created at the ridge are gradually modified and concealed. Eventually, they may be completely buried in sediment, thus forming the flat abyssal plains. The outline of the buried rock surface can be traced in seismic reflection profiles such as that in Figure 21.5.

What controls the development of the abyssal floor?

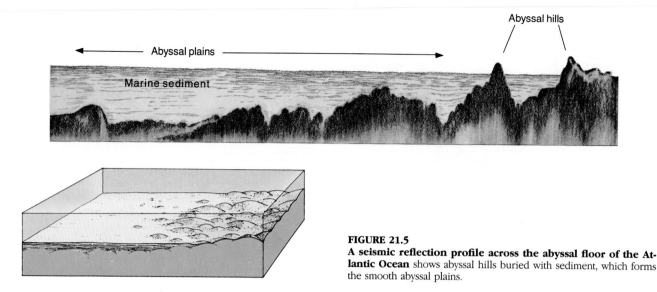

Abyssal hills

Abyssal plains

Marine sediment

FIGURE 21.5
A seismic reflection profile across the abyssal floor of the Atlantic Ocean shows abyssal hills buried with sediment, which forms the smooth abyssal plains.

The distribution of abyssal plains and abyssal hills substantiates this explanation of their origin. Abyssal plains occur only where the topography of the sea floor does not inhibit turbidity currents from spreading sediment from the continents over the sea floor. In the Atlantic Ocean, abyssal plains occur near the margins of the continents of North America, South America, Africa, and Europe. In the Pacific Ocean, by contrast, there are few abyssal plains because turbidity currents cannot flow past the deep trenches that lie along most of the continental margins. The inflowing sediment accumulates in the trenches, so most of the Pacific floor lacks abyssal plains and is covered instead with abyssal hills. The largest abyssal plains in the Pacific are found in the northeast, off the coast of Alaska and western Canada. This is the only significant segment of continental mass in the Pacific that is not bordered by deep trenches. In the North Pacific, between the Aleutian Trench and the continental shelf, the Bering abyssal plain covers most of the deep-sea floor north of the Aleutian Islands. This deep basin is underlain by an abnormally thick section of layered sediment, which is as deep as 2 km in places. As the physiographic map shows (see the inside covers of this book), the Aleutian ridge has cut off this corner of the Pacific basin, acting like a dam behind which sediments have rapidly accumulated. In the Indian Ocean, abyssal plains occur only along the margins of Africa and India. They are absent in the eastern part of the ocean, where the deep Java Trench, along the continental margins, acts as a sediment trap. Another major area of abyssal plains lies off the northern shore of Antarctica.

Large cone-shaped or fan-shaped deposits of sediment derived from the continents lie on the abyssal plains offshore from most of the world's great rivers. These are called *deep-sea fans* (such as the areas of the sea floor east and west of India). They resemble alluvial fans and deltas in that they are fan-shaped accumulations of sediment located at the mouths of rivers. They are different from the land deposits, however, because the sediment is transported and deposited primarily by turbidity currents. Most large fans are located at the base of the continental slope, with their apexes at the mouths of submarine canyons cut into the edge of the shelf. The main source of sediment for the fans is mud brought in by major rivers. Turbidity currents intermittently flush sediment through the canyons, building up depositional fans where the currents reach the lower gradient of the ocean floor.

Most, if not all, deep-sea fans are marked by one or more deep channels, which usually are the extensions of submarine canyons cut in the continental slope. As the slopes flatten, the channels develop natural levees, much like those formed by low-gradient streams on land.

The submarine fan of the Ganges River, in the northwestern Indian Ocean, is by far the largest deep-sea fan in the world. It is over 2800 km long and covers slightly more than 4,000,000 km². This accumulation represents about 70% of the debris derived from the erosion of the Himalayas. The remaining 30% is deposited on the floodplain and delta of the Ganges and by the Indus River, to the west.

Other large fans are the Indus fan, on the western side of the Indian peninsula; the Amazon and the Congo fans, in the South Atlantic; the Mississippi fan, in the Gulf of Mexico; and the Laurentian fan, in the North Atlantic. A number of smaller fans have been mapped off the Pacific coast of North America. There are no large fans in the Pacific because most major rivers drain into the Atlantic and Indian oceans and because deep-marine trenches trap most of the sediment that does flow into the Pacific.

Trenches

A subduction zone, where two plates converge and one slab of lithosphere plunges down into the mantle, is expressed topographically by a **trench**. We have seen in previous chapters that a subduction zone is characterized by intense volcanic activity and seismicity. It is also marked by a large **gravity anomaly** (abnormally high or low gravitational force within an area).

Trenches, some of which reach nearly 11,000 m below sea level, are the deepest parts of the ocean. As is shown in the seismic reflection profile in Figure 21.6, they typically are asymmetrical. A relatively steep slope lies on the landward side, along the continental landmass, and a more gentle slope lies on the side of the ocean basin. Individual trenches less than 100 km wide can form continuous features extending for several thousand kilometers across the deep-ocean floor.

What geologic features characterize the deep-sea trenches?

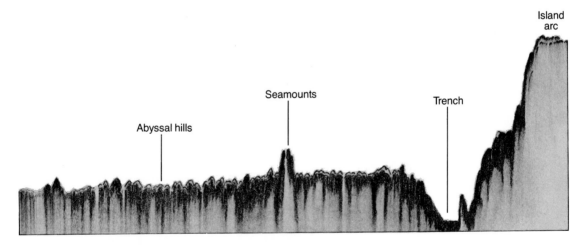

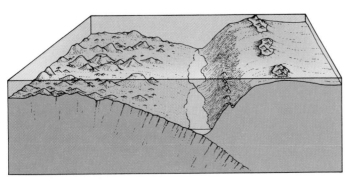

FIGURE 21.6
A seismic reflection profile across the central part of the Aleutian Trench shows the steep flank of the trench alongside the Aleutian island arc (right) and the gentle slope toward the ocean basin. The trench is the surface expression of a subducting plate.

The most striking examples occur in the western Pacific. A trench system there extends from the vicinity of New Zealand to Indonesia to Japan and then northeastward along the southern flank of the Aleutian Islands. Long trenches also occur along the western coast of Central America and South America, in the Indian Ocean west of Australia, in the Atlantic off the tip of South America, and in the Caribbean Sea.

Islands and Seamounts

Literally thousands of submarine volcanoes occur on the ocean floor, with the greatest concentration in the western Pacific. Some rise above sea level and form *islands,* but most remain submerged and are called *seamounts* (Figure 21.7). They often occur in groups or chains, with individual volcanoes being as much as 100 km in diameter and 1000 m high. A *guyot* is a special type of seamount with a flat, mesalike top surface, rather than a cone.

Chains of islands and seamounts are believed to develop as the sea floor moves over a hotspot in the upper mantle. The hotspot presumably forms from a huge column of upwelling lava known as a *mantle plume.* It lies in a fixed position under the lithosphere (see Figure 20.18).

What are the origin and significance of seamounts?

The origin of guyots, however, has been the subject of much debate. There is little doubt that they are ancient erosional platforms developed by stream and wave action on volcanic islands. Dredging samples show that some guyots are covered with volcanic debris that has been eroded and washed by wave action. Many are coated with coral, which can grow only in very shallow water. The big question raised by these findings is how the guyots became submerged from 1000 to 2000 m below sea level. Many atolls also are composed of coral reefs thousands of meters thick, all of which had to originate near sea level. These facts are accounted for by the theories of sea-floor spreading and plate tectonics. Initially, as a volcano forms near the crest of the oceanic ridge, it rises above sea level because that area of the ocean floor is high because it is warm and expanded by ascending convection currents. Wave action then erodes the top of the island to a flat surface, and coral reefs grow on the platform near sea level. As sea-floor spreading continues, the flat-topped island migrates off the swell and becomes submerged. If environmental con-

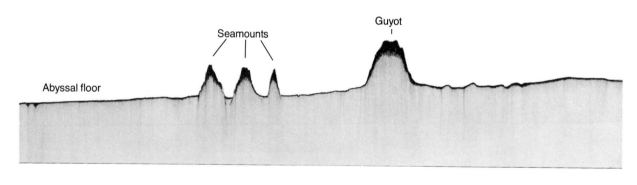

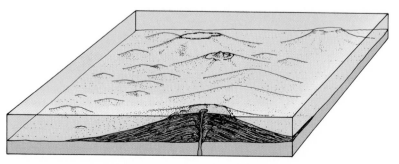

FIGURE 21.7
A seismic reflection profile across seamounts in the central Pacific Ocean shows the general configuration of typical seamounts rising above the ocean floor. Seamounts are submarine volcanoes, which usually occur in groups or chains. Some rise above sea level to form islands.

ditions are favorable, coral reefs will continue to grow upward as the seamount slowly moves off the flanks of the ridge and as the sea floor moves off the oceanic ridge into deeper water. Atolls such as Bikini Atoll can thus develop a reef rock over 1000 m thick as the island is slowly submerged.

Continental Margins

The *continental margins* are covered by the ocean but are not geologically part of the oceanic crust. They are composed of continental crust and sediment derived from erosion of the land.

This part of the sea floor can be divided into three major sections: the continental shelf, the continental slope, and the continental rise. The *continental shelf* is simply a submerged part of the shield or stable platform. The gently sloping shelf extends from the shoreline to the area where the continental margin begins its steep descent to the ocean floor. The shelf can be as much as 1500 km wide, and its depth ranges from 20 to 550 m at its outer edge. At present, the continental shelves comprise 18% of Earth's total continental area. At times in the geologic past, however, they were much larger because the oceans spread much farther over the continental platforms.

Characteristically, the continental shelf is smooth and flat, but its topography has been influenced greatly by changes in sea level. Large areas were once exposed as dry land and thus were subjected to subaerial processes, so the shelf topography can have features formed by both marine and nonmarine processes.

The *continental slope* descends from the outer edge of the continental shelf as a long, continuous slope to the deep-ocean basin (Figure 21.8). If the oceans were drained of water, the continental slope would appear as the most conspicuous boundary on Earth's surface. It would appear as a long, continuous slope rising from the abyssal floor to the high continental platform. It marks the edge of the continental granitic rock mass, the boundary between the continental crust and the oceanic crust. Continental slopes are found around the margins of every continent and around smaller pieces of continental crust such as Madagascar and New Zealand. Study the continental slopes shown on the physiographic map (see the inside covers of this book), especially those surrounding North America, South America, and Africa, and note that they form one of Earth's major topographic features. They are by far the longest and highest slopes on Earth. These long, straight margins are the topographic expression of the geologic difference between the continental crust and the oceanic crust, reflecting a fundamental difference in structure and rock type. Within this zone, from 20 to 40 km wide, the average relief above the sea floor is 4000 m. Along the marginal trenches, relief is as great as 10,000 m. In contrast to the shorelines of the continents, the edges of continental slopes are relatively straight over distances of thousands of kilometers.

The *continental rise* is the transition between the continent and the ocean basin. It has a gentle, inclined surface, rising from the abyssal plains to the continental slope. The continental rise is apparently formed by sediment deposited at the base of the continental slope.

Two types of continental margins can be recognized. Each is subjected to different stresses, and thus they develop different characteristics. The shelves on the *passive margin* of a continent are typically wide and are not subjected to strong compressive forces.

In contrast to a continent's passive margin, the *active margin* is subjected to much more stress because it impinges on another moving plate. Crustal deformation results. Where a continent converges with an oceanic plate, the oceanic plate moves down and under the lighter continental crust, and a trench develops along the continental margin.

What are the major geologic features that characterize the margins of continents?

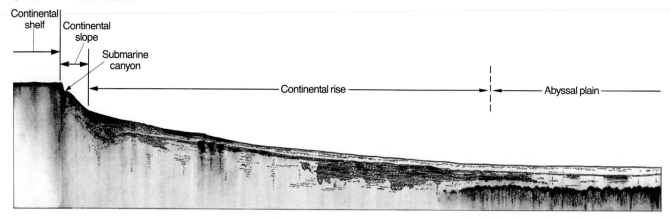

Continental shelf

Continental slope

Submarine canyon

Continental rise

Abyssal plain

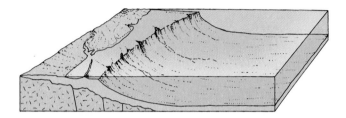

FIGURE 21.8
A seismic reflection profile across the western continental slope and continental rise of Africa shows the profile of several submarine canyons near the upper part of the slope and the thick accumulation of undeformed sediments on the continental margin. The continental slope merges into the adjacent abyssal plains, which cover the abyssal hills.

Submarine Canyons

Submarine canyons are common along the continental slope and have been studied for many years, long before the ocean floor was mapped. Some extend landward across the continental shelf for some distance. As is shown on the physiographic map, they typically cut through the edge of the continental shelf and terminate on the deep abyssal floor, some 5000 or 6000 m below sea level. The profile in Figure 21.8 crosses three canyons near the upper part of the continental slope. Submarine canyons have a V-shaped profile and a system of tributaries and thus closely resemble the great canyons cut by rivers on the continents (Figure 21.9). Many pioneer researchers therefore suggested that submarine canyons were also cut by rivers, but the problem of how this could happen remained unanswered for many years. Some canyons are 6000 m below sea level, and it was difficult to understand how sea level could change enough for rivers to cut to that depth. Some even suggested that the continents were uplifted thousands of meters, were dissected by streams to form canyons, and then subsided, so that the canyons became submerged. With our increased knowledge of the characteristics of the ocean floor and a better understanding of submarine processes, it now appears that submarine canyons are usually cut by turbidity currents flowing from the continental shelf to the abyssal floor. Turbidity currents can move at rates ranging up to 95 km per hour (Figure 6.11) and carry a large sediment load capable of vigorous erosion.

How do submarine canyons form?

Gravity Mapping of the Ocean Floor

A new map of the ocean floor has recently been created from satellite measurements and reveals many details either unknown or unconfirmed before (Figure 21.10). Using a pencil-thin beam of microwaves, a radar altimeter on board NASA's Seasat satellite measured the distance between the satellite and the ocean surface with an accuracy of about 10 cm. This survey confirms that the ocean surface is more than 60 m higher in some places than in others.

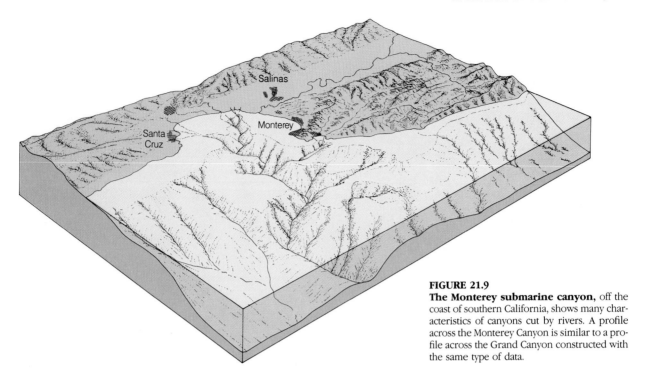

FIGURE 21.9
The Monterey submarine canyon, off the coast of southern California, shows many characteristics of canyons cut by rivers. A profile across the Monterey Canyon is similar to a profile across the Grand Canyon constructed with the same type of data.

The bulges and depressions occur because of gravitational differences that result from unequal distribution of mass in Earth's interior. Scientists also found much subtler variations in the ocean surface, and these corresponded with hills and valleys on the ocean floor. Where gravity is strong, such as around seamounts, water is attracted, creating a rise in sea level. This occurs because the added mass of rock in the seamount exerts a gravitational pull on the surrounding water. If, for example, the island of Hawaii were totally submerged, sea level would be about 30 m higher above the volcanic mountain than above the surrounding abyssal plains. In contrast, a trench has an absence of mass, so the ocean surface is depressed about 20 m above a deep trench. The ocean surface thus reflects the topography of the sea floor beneath it.

The map shown in Figure 21.10 was made by using a computer program to erase the regional bulges and depressions of the ocean surface caused by variations in Earth's internal mass. Only those variations resulting from the topography of the ocean floor remained. We are thus able for the first time to "see" a regional panorama of the ocean floor. We can recognize familiar features, originally discovered by sonar and plotted on physiographic maps, such as the mid-Atlantic ridge, the Mariana Trench, long fracture zones, and other landforms. The new map has also revealed details of unexplored areas and shows previously unknown seamounts, continuous submarine ridges, and areas where the oceanic crust has been buckled by compression. Most intriguing are several swells on the Pacific floor, which may provide direct evidence of convection under the tectonic plates.

In summary, the topography of the ocean floor has been mapped in detail that would have seemed impossible only a few years ago. Most of the landforms are relatively young and have been created either directly or indirectly by the tectonic system. The most spectacular feature is the oceanic ridge, which is a continuous feature throughout all the ocean basins. The ridge is cut by long fracture zones, some of which can be traced thousands of kilometers. Flanking the oceanic ridge is the abyssal floor, above which rise numerous submarine volcanoes called seamounts. The deepest parts of the oceans are the trenches, some of which descend to depths of 11,000 m.

How does the theory of plate tectonics explain the location and origin of the major landforms on the ocean floor?

Why is the topography of the ocean floor reflected in the elevation of sea level?

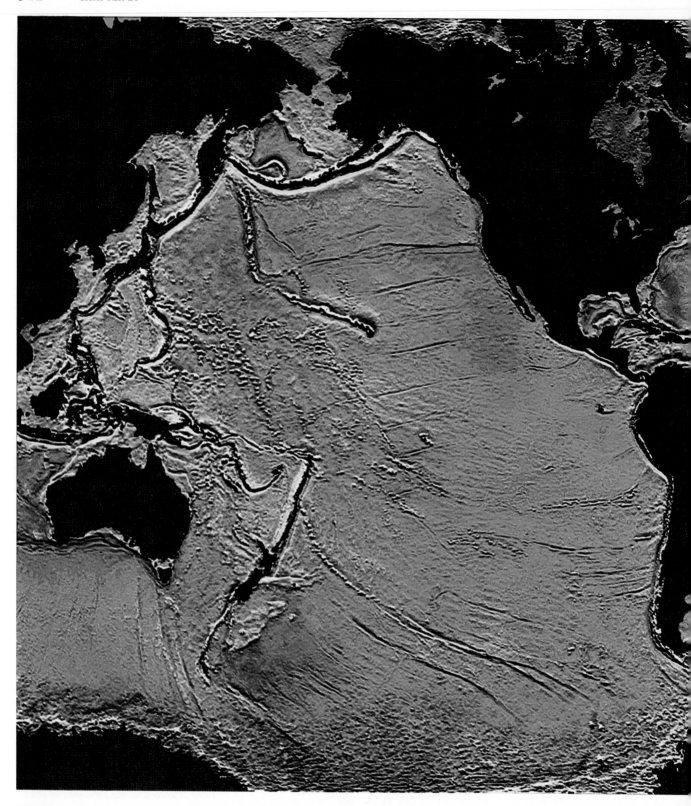

FIGURE 21.10
A map of the topography of the ocean floor was made from satellite measurements of the sea surface processed with sophisticated color computer graphics. This map not only confirms the data from sonar profiles but reveals many details that were previously either unknown or unsubstantiated. For example, new seamounts and fractures were discovered in the Indian Ocean that will help geologists better understand how Africa, India, and Australia have drifted northward from Antarctica.

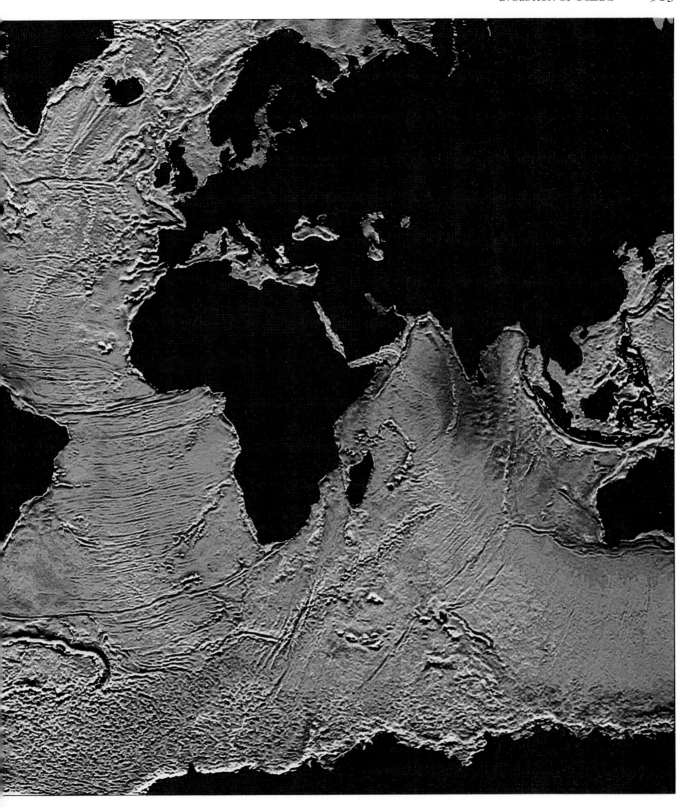

COMPOSITION AND STRUCTURE
OF THE OCEANIC CRUST

The oceanic crust is relatively thin (from 7 to 8 km) and is composed mostly of basalt. It consists of four major layers: (1) a surface layer of marine sediment, (2) an underlying layer of pillow basalt, (3) an older zone of sheeted dikes of basalt, and (4) a layer of gabbro. The oceanic crust is not deformed by tight folding or thrust faulting. It is geologically young; essentially all of it was formed during the last 200 million years.

With the growth of knowledge about the ocean basins came considerable efforts to determine the composition and structure of the oceanic crust. A variety of methods have been used, including seismic reflection, drilling, and studies of fragments of oceanic crust thrust up to the surface. Among the more significant facts we have learned about the oceanic crust are the following:

1. The oceanic crust is composed of four major layers: (a) a surface layer of marine sediment, (b) a layer of pillow basalt, (c) a zone of sheeted dikes, and (d) a layer of gabbro.
2. The oceanic crust and its topographic features are related in some way to igneous activity.
3. The rocks of the ocean basin have not been deformed by strong compression, so their simple structure contrasts markedly to the complex structure in the folded mountains and shields of the continents.
4. The rocks of the ocean basin are young in terms of geologic time. All appear to be less than 200 million years old, whereas the great bulk of continental rocks, the ancient rocks of the shields, are more than 700 million years old.

How does the oceanic crust differ from the continental crust?

The diagrams in Figure 21.11 illustrate the main elements of the composition and structure of the oceanic crust, as it is presently understood. At the top of the sequence (layer 1) is a relatively thin layer of sediment. This consists of calcareous and siliceous shells of microscopic marine organisms together with red clay, which is derived from the continents. Layer 2 consists of pillow basalts, which were originally fed by numerous dikes and intruded into vertical fractures. The pillow lava represents volcanic extrusions on the sea floor, and the dikes are feeder vents, which were produced as the tectonic plates split and moved apart. Layer 3 consists almost entirely of the dikes, so the individual basalt bodies are numerous vertical sheets. Layer 4 consists of rocks that are dominantly coarse-grained gabbro, which is believed to represent magma that was generated at a spreading center but cooled very slowly at some depth. Underlying the gabbro are peridotites, composed almost entirely of olivine and pyroxene. This material is considered part of the mantle. The boundary between the gabbro and the peridotite is the Moho.

Seismic reflection studies and drilling on the ocean floor provide important data concerning the upper layers of the oceanic crust. However, our most direct and detailed information is obtained from areas where large fragments of the oceanic crust have been incorporated in a folded mountain belt at a convergent plate margin and are available for direct observation. The most notable exposures are in Oman (Figure 21.12) Cyprus, Greece, New Guinea, Newfoundland, and California. They show the exact sequence illustrated in Figure 21.11.

A significant fact about the oceanic crust is that drilling samples brought up thus far are all much younger than the rocks that form the bulk of the continents. The oldest basalts retrieved from the ocean floor are only 50 million years old, and extensive drilling in recent years has produced no sediment older than 200 million years.

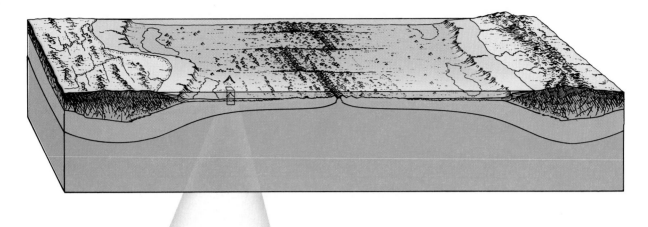

FIGURE 21.11
The structure of the oceanic crust consists of four distinct layers. Layer 1 is a relatively thin sequence of oceanic sediments, composed of the shells of microscopic marine organisms mixed with red clay. Layer 2 is pillow basalt fed by numerous dikes. Layer 3 is almost entirely composed of basalt dikes in vertical sheets. Layer 4 is gabbro, which is believed to represent magma that was generated at a spreading center and cooled slowly at depth.

1 Sediment
2 Pillow basalt
3 Basalt dikes
Crust
4 Gabbro
Moho
Peridotite
Upper mantle

THE OCEAN BASINS

The three major ocean basins—the Atlantic, Indian, and Pacific—are believed to have originated by sea-floor spreading. They differ from each other in size, shape, sea-floor topography, and age. The small ocean basins in the western Pacific, the Red Sea, the Caribbean, and the Mediterranean resulted from one of three processes: (1) the growth of island arcs, (2) initial rifting, or (3) the convergence of continental plates.

Although all of the ocean basins originated in the same manner (that is, by sea-floor spreading), they are quite different in size, shape, and topographic features. These differences are significant because they tell much about the ages and evolution of each individual ocean basin.

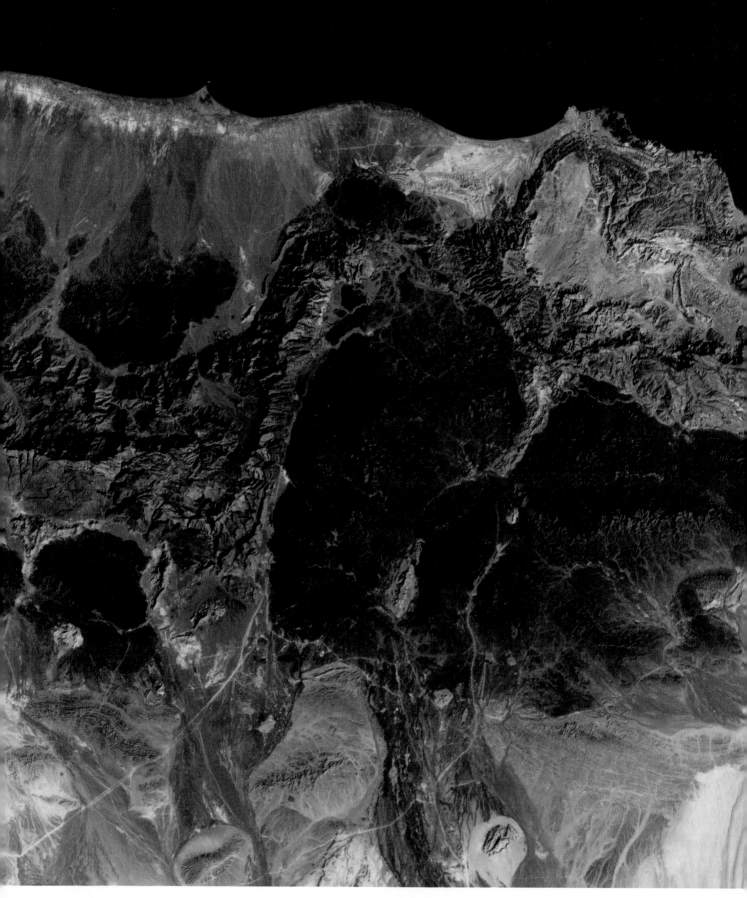

FIGURE 21.12
Oceanic crust in the Oman Mountains of northern Oman (southeast Arabian Peninsula) is shown in this satellite image as dark, massive rock with a highly dissected drainage pattern. It is a classic ophiolite sequence of oceanic crust anomalously superposed on top of continental crust. The lowermost units are peridotites, above which are massive gabbros, which in turn are overlain by sheeted dikes and pillow basalts. These oceanic rocks were formed at an oceanic ridge and were thrust over the continental crust by plate collision during Cretaceous time.

The Atlantic Ocean

The structure and regional topography of the Atlantic floor are basically simple, reflecting the initial opening of this young ocean basin. The Atlantic basin also shows remarkable symmetry in the distribution of the major features (see the physiographic map inside the covers of this book). The dominant feature is the mid-Atlantic ridge, which forms an S-shaped pattern down the center of the basin. It separates the ocean floor into two long, parallel sub-basins, trending north and south, which are characterized by abyssal plains. Abyssal hills occur alongside the ridge, and the plains occur along the margins of the continental platforms. The symmetry of the Atlantic basin extends to the continental margins: The outlines of Africa and Europe fit those of South America and North America.

The Arctic Ocean

Although the Arctic Ocean has been considered by some to be an extension of the Atlantic, recent exploration shows that it is unique in several respects: (1) It is nearly landlocked. If the basin beyond the continental slope is considered, there is only one inlet, the Lena Trough (between Spitsbergen and Greenland). (2) The continental shelf north of Siberia is the widest shelf in the world (over 1100 km across). Most of the shelf is deeper than 200 m, but very little exceeds 600 m in depth. (3) Two prominent submarine ridges trend roughly 140 degrees east and divide the deep Arctic Ocean into three major basins.

The Indian Ocean

The Indian Ocean is the smallest of the three great oceans. It connects with both the Atlantic and the Pacific through broad, open seas south of Africa and Australia. Like the Atlantic, its most conspicuous feature is the oceanic ridge, which continues from the Atlantic around southern Africa and splits near the center of the Indian Ocean to form a pattern similar to an inverted Y (see the physiographic map inside the covers). The northern segment of the ridge extends into the Gulf of Aden, apparently connecting with the African rift valleys and the Red Sea rift. The ridge thus divides the ocean basin into three major parts.

The topography of the Indian Ocean floor, unlike that of the other oceans, is dominated by scattered blocks and some remarkably linear plateaus called *microcontinents.* Most are oriented in a north–south direction. Prominent parallel fracture zones are numerous. The striking northward trend of the parallel fracture zones, together with the trend of the linear microcontinents, imparts a linear structural fabric to the Indian Ocean floor.

What is the difference in the size, shape, age, and history of the ocean basins?

The Pacific Ocean

The Pacific differs somewhat from the other oceans in that the oceanic ridge lies near its eastern margin. Its basin covers approximately half of the planet and is the largest region of oceanic crust. It is probably the oldest basin and lacks the symmetry of the Atlantic and Indian basins.

The oceanic ridge continues in a broad sweep from the Indian Ocean, trends eastward between New Zealand and Antarctica, and then turns northward along the American side of the ocean. The rifting associated with the ridge extends into the western United States at the head of the Gulf of California and is probably involved in the structure of the Basin and Range province in Utah and Nevada and the San Andreas Fault. It reappears off the coast of Oregon.

The floor of the western Pacific is studded with more seamounts, guyots, and atolls than all the other oceans combined. As is apparent from the physiographic map, many of the seamounts form linear chains that extend for considerable distances. The margins of the Pacific are also different from those of the other oceans. They generally are marked by a line of deep, arcuate trenches. In the eastern Pacific, the trenches lie along the margins of Central America and South America, paralleling the great mountain systems of the Andes and the Rockies. Local relief from the top of the Andes to the bottom of the trench is 14,500 m (nearly nine times the depth of the Grand Canyon). In the western Pacific, a nearly continuous line of trenches extends from the margins of the Gulf of Alaska, along the margins of the Aleutians, Japan, and the Philippines, and down to New Zealand.

Small Ocean Basins

A number of *small deep-ocean basins* are nearly isolated from the major oceanic plates. Nevertheless, they are considered a direct consequence of plate tectonic processes. Most of them are nearly landlocked and connect to the main ocean by narrow gaps (Figure 21.13). The Mediterranean Sea, the Red Sea, and the Sea of Japan are excellent examples. These small ocean basins originate in three ways: (1) by the growth of island arcs, (2) by the initial rifting of continental plates, and (3) by the convergence of continental plates and the destruction of larger ocean basins. The small ocean basins are temporary features. Like the major ocean basins, they grow but are ultimately destroyed.

The growth of island arcs created the small ocean basins of the Pacific. They formed along subduction zones that are not directly adjacent to a continent, so the associated volcanic arcs isolated segments of oceanic crust to form small independent basins. Several such basins extend from the Aleutian Islands to New Guinea. Most of them are shallower than the adjacent open ocean, partly because they trap the sediment washed from the continents. Sediment in the Bering basin even overflows through gaps in the Aleutian arc and spills into the Aleutian Trench.

The Red Sea and the Gulf of California are examples of another type of small ocean basin, which develops in the initial rift zone, where the continents split and begin to spread apart. Basalt from the upwelling mantle commonly fills the new opening during the early stages of rifting, and sediment eroded from the adjacent continents can help keep the basin filled to near sea level. If spreading is rapid, a long, narrow ocean basin develops. An interesting and potentially important feature of these new basins is the circulation of water through the still hot crust. Rising through the sediments on the ocean floor, it forms pools of hot brine with concentrations of rare metals.

As the sea floor continues to spread, the intervening sea develops the characteristics of a major ocean basin: an oceanic ridge, abyssal plains, and continental slopes. The Arctic Ocean is apparently in this transitional stage from sea to ocean.

The Mediterranean and Black seas, in contrast to those previously described, are basins formed by the closing of ocean basins as a result of converging continents, namely, Africa and Eurasia. These seas are remnants of the basin of the once vast Tethys Sea, which extended in an east–west direction and separated Pangaea into two large continental landmasses, *Laurasia* and *Gondwanaland,* 200 million years ago (Figure 21.14). As Africa and India moved northward, the Tethys Sea gradually closed, although it continued to link the Atlantic and Indian oceans for many millions of years. Compression due to northward movement of Africa and India produced the sinuous mountain belts of the Alps and the Himalayas, which are composed mostly of folded and uplifted deep-ocean sediments. If the present plate motion continues, the Mediterranean basin will eventually be completely closed, and Africa will be tucked under the Alps, much as India is thrust under the Himalayas.

Recent drilling in the Mediterranean basin indicates a fascinating history during the last 20 million years, the period when this sea became isolated from

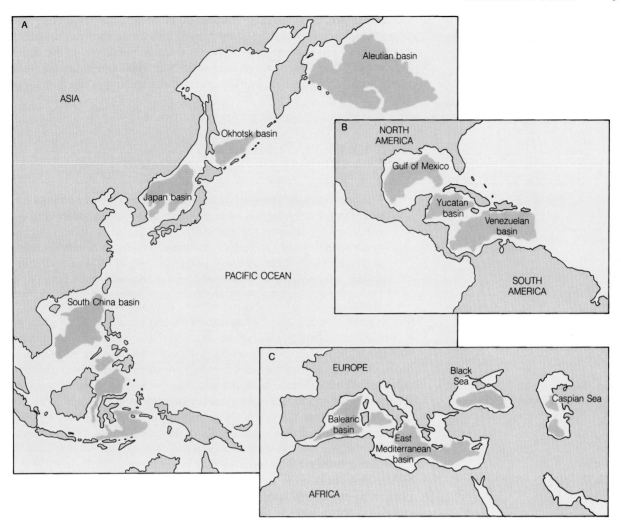

FIGURE 21.13
Small ocean basins originate in several ways. Those in the western Pacific and the western Atlantic developed when island arcs isolated part of the sea from the main ocean basins. The Mediterranean basin and the Black and Caspian seas represent remnants of the ancient Tethys Sea, which was closed by the convergence of India and Africa with Europe and Asia.

the major oceans. As is shown in Figure 21.13, the Mediterranean is almost landlocked, being connected with the Atlantic by a narrow gap through the Strait of Gibraltar. The deep basin is completely isolated, and this produces some very interesting hydrologic conditions. Evaporation removes more than 4000 km³ of water each year from the Mediterranean Sea, but less than 500 km³ are replaced by rain and surface runoff from Europe and Africa. Each year approximately 3500 km³ of water flow in through the Strait of Gibraltar to maintain the Mediterranean at sea level. If the strait were closed, the Mediterranean Sea would evaporate in about 1000 years.

Total evaporation of the Mediterranean actually happened between 5 million and 8 million years ago. Deep drilling by the *Glomar Challenger* has revealed that the Mediterranean basin was then a deep, desolate, dry ocean basin 3000 m below sea level. Core samples of windblown sand and salt deposits show that it was a stark desert region. Huge volcanoes (now islands) rose above the basin floor. The continental platforms of Africa and Europe surrounded the basin as huge plateaus. The area was undoubtedly devoid of life: temperatures must have reached 65°C.

The salt deposits and windblown sand that formed in the dry basin are now covered throughout by the deep-sea organic oozes, implying that the basin was flooded almost instantaneously about 5.5 million years ago. The inundation probably resulted from erosion of the barrier at the Strait of Gibraltar, so water from the Atlantic flowed into the Mediterranean basin over an enormous waterfall. Estimates based on fossils preserved in the sediments indicate that the flow through the strait was 1000 times greater than the present-day flow over Niagara Falls and that the basin was filled in about 100 years.

HISTORY OF PLATE MOVEMENT DURING THE LAST 200 MILLION YEARS

The considerable amount of data on plate motion enable us to reconstruct the position of continents and to trace plate movement with considerable certainty. They indicate that a large continental mass (Wegener's Pangaea) began to break up and drift apart about 200 million years ago. Dispersal and collision of the fragments have continued to the present time.

The tectonic system probably has operated during much of Earth's history, and it is believed to be responsible for the origin and evolution of continents as well as for the growth and destruction of ocean basins. Our understanding of the early history of plate movements comes mostly from evidence preserved in continental rocks. Ocean basins come and go because the ancient oceanic crust is consumed at subduction zones and replaced by newer oceanic crust created at spreading centers. Continents have drifted, with tectonic plates splitting and joining a number of times, but details of the patterns of ancient plate movements are scanty.

The considerable amount of data on plate motion during the last 200 million years enables us to reconstruct the position of continents and to trace plate movement with considerable certainty. The various geologic data discussed on pages 420–34 all suggest the same basic pattern of plate motion. They indicate that a large continental mass (Wegener's Pangaea) began to break up and drift apart about 200 million years ago. Dispersal and collision of the fragments have continued to the present time. The reconstruction of Pangaea in terms of absolute coordinates is now possible, and the directions and rates of plate movement have been determined.

The history of relative plate movement during the last 200 million years is shown in Figure 21.14. These maps, adapted from those prepared by C. Denham and C. Scotese, were constructed with cartographic precision using a computer to synthesize all available evidence, including the geologic fit of continents, paleomagnetic pole positions, and patterns of sea-floor spreading.

The initial event in the splitting of the continents was the extrusion of large volumes of basalt along the initial rift zone. Remnants of these basalts are found in the Triassic basins of the eastern United States and the plateau basalts of southwestern Africa, western India, and eastern Brazil. A northern rift split Pangaea along an east–west line slightly north of the equator and separated Laurasia from Gondwanaland, rotating Laurasia clockwise. A southern rift split South America and Africa away from the remaining Gondwanaland. Soon afterward, India was severed from Antarctica and moved rapidly northward. The plate containing Africa converged toward Eurasia, forming a subduction zone in the Tethyan Sea. By the end of the Cretaceous period, 65 million years ago after 135 million years of movement, the separation of South America from Africa was complete, and the South Atlantic Ocean had widened to at least 3000 km. All of the major continents were blocked out by this time, except for the connection between Greenland and Europe and between Australia and Antarctica. A new rift separated Madagascar from Africa, and India continued moving northward.

How have the oceans evolved? What major changes in the oceans and continents will likely occur in the next 10 million years?

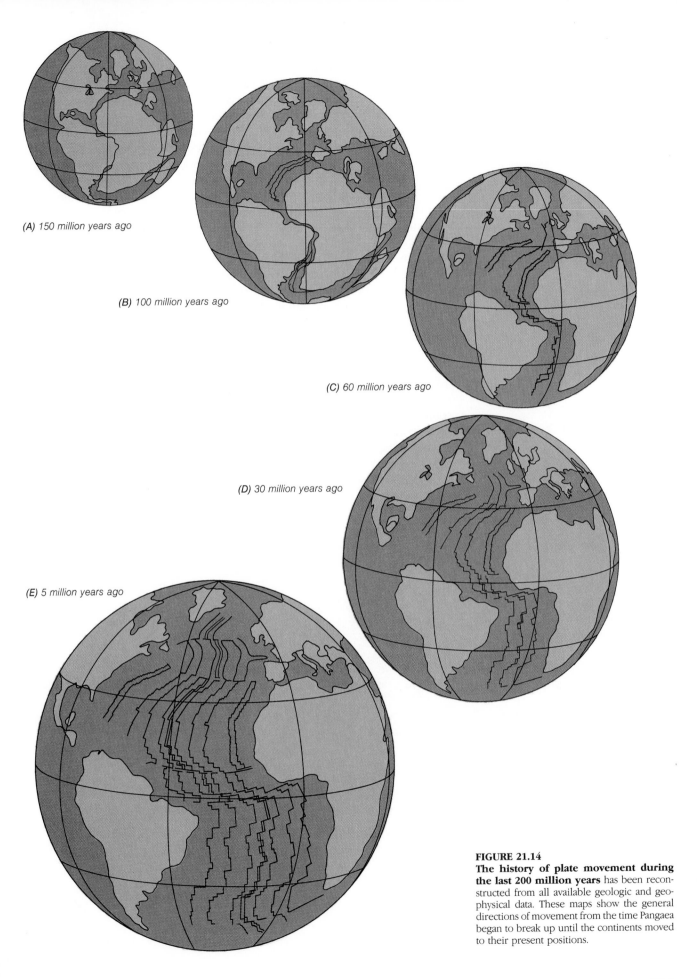

(A) 150 million years ago

(B) 100 million years ago

(C) 60 million years ago

(D) 30 million years ago

(E) 5 million years ago

FIGURE 21.14
The history of plate movement during the last 200 million years has been reconstructed from all available geologic and geophysical data. These maps show the general directions of movement from the time Pangaea began to break up until the continents moved to their present positions.

A north–south trench system must have existed in the Pacific (the South American–Central American trench) and consumed oceanic crust on the western edge of the rapidly westward-moving plates carrying North and South America. North America probably encountered this trench in late Jurassic time. The trench eventually was overridden by the continued westward drift of North America, causing the deformation of the Rocky Mountains. About the same time, the same trench was encountered by South America. This encounter developed the early Andean folded mountain belt.

Throughout the Cenozoic period (Figure 21.14), the mid-Atlantic ridge extended into the Arctic and finally detached Greenland from Europe. During that time, the two Americas were joined by the Isthmus of Panama, which was created by volcanism along the subduction zone. The Indian landmass completed its northward movement and collided with Asia, creating the Himalayas. Australia drifted northward from Antarctica.

Finally, a branch of the Indian rift system split Arabia away from Africa, creating the Gulf of Aden and the Red Sea. Then a spur of the rift meandered west and south to create the East African rift valleys. Less pronounced changes induced the partial closing of the Caribbean region and the continued widening of the South Atlantic.

Figure 21.15 shows the anticipated positions of plates 50 million years from now. Through the extrapolation of present directions and rates of plate movement, certain changes seem likely. The Atlantic and Indian oceans will continue to grow at the expense of the Pacific. Australia will move northward and encounter the Eurasian plate. The eastern part of Africa will split off along the rift valleys, but the northern part of the continent of Africa probably will move northward and close the Mediterranean Sea. New land will be created in the Caribbean by compressional uplift. Baja California and part of western California will be severed from North America and drift to the northwest.

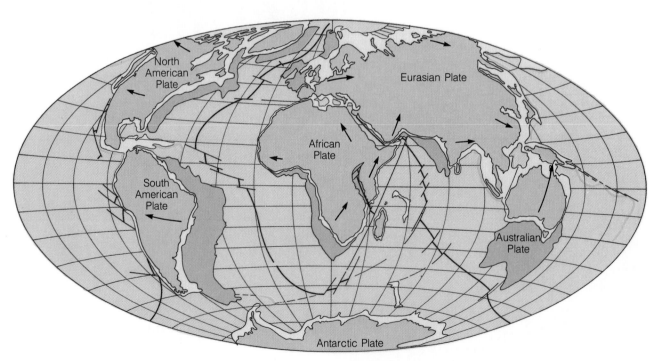

FIGURE 21.15
Anticipated plate movement during the next 50 million years.

SUMMARY

1. The oceanic ridge is a broad, fractured swell, extending as a continuous feature for more than 64,000 km. It has a narrow, axial rift valley and is cut by numerous transverse fractures.

2. Abyssal hills are formed at or near the ridge and move away from it with the drifting plates. As they move, they become covered with sediment. They can eventually be buried to form abyssal plains.

3. Trenches form at subduction zones where two plates converge and one is thrust down into the mantle. Typically, they are bordered by island arcs or young mountain ranges.

4. Volcanic islands, or seamounts, can originate near a spreading center and move with the spreading plates. They may also form over local hotspots in the mantle, producing linear chains of islands as the plates move over the hotspots.

5. The continental shelf and slope are not parts of the oceanic crust but are submerged parts of the continental platform. The continental slope is dissected by many deep submarine canyons. These are similar in size and depth to the great canyons cut on the continents by rivers.

6. The oceanic crust is relatively thin (from 7 to 8 km) and is mostly basalt. It is composed of four major layers: (1) a surface layer of marine sediment, (2) a layer of pillow basalt, (3) a zone of sheeted dikes of basalt, and (4) a basal layer of gabbro. The oceanic crust is *not* deformed by tight folding or thrust faulting. It is geologically young: Essentially all of it was formed during the last 200 million years.

7. The three major ocean basins—the Atlantic, Indian, and Pacific—are believed to have originated by sea-floor spreading. They differ from each other in size, shape, sea-floor topography, and age. The small ocean basins in the western Pacific, the Red Sea, the Caribbean, and the Mediterranean resulted from one of three processes: the growth of island arcs, initial rifting, or the convergence of continental plates.

8. We can outline in some detail the history of the ocean basins during the last 200 million years. The Atlantic, Indian, and Arctic oceans were formed by the rifting of a supercontinent (Pangaea). The major fragments of the supercontinent moved with the spreading plates, and as the ocean basins enlarged, the continents eventually arrived at their present positions.

KEY TERMS

abyssal floor (p. 505)

abyssal hill (p. 505)

abyssal plain (p. 505)

active plate margin (p. 509)

bathymetric chart (p. 503)

continental margin (p. 509)

continental rise (p. 509)

continental shelf (p. 509)

continental slope (p. 509)

deep-sea fan (p. 506)

Gondwanaland (p. 518)

gravity anomaly (p. 507)

guyot (p. 508)

island (p. 508)

Laurasia (p. 518)

mantle plume (p. 508)

microcontinent (p. 517)

oceanic ridge (p. 504)

passive plate margin (p. 509)

pelagic sediment (p. 505)

rift valley (p. 504)

seismic reflection profile (p. 503)

seamount (p. 508)

small deep-ocean basin (p. 518)

submarine canyon (p. 510)

trench (p. 507)

REVIEW QUESTIONS

1. Make a sketch map of the major types of landforms found on the ocean floor.
2. Study the seismic reflection profile in Figure 21.1 and label the following: (a) oceanic sediment, (b) bedrock, and (c) probable faults.
3. Explain the origin of the following: (a) the oceanic ridge, (b) fracture zone, (c) abyssal hills, (d) abyssal plains, (e) trenches, (f) seamounts, and (g) submarine canyons.
4. What is the major source of sediment that covers the abyssal plains?
5. What is the significance of the fact that the thickness of pelagic sediments increases with distance from the crest of the oceanic ridge?
6. Many submarine volcanoes (guyots) have flat tops. Explain their origin.
7. Compare and contrast the Atlantic Ocean and the Pacific Ocean basins with respect to shape, age, major structural features, and history.
8. Compare and contrast the Mediterranean Sea and the small ocean basins in the western Pacific with respect to such factors as structure, age, origin, and history.
9. Briefly outline the history of plate movement during the last 200 million years.

ADDITIONAL READINGS

Bishop, J. M. 1984. *Applied Oceanography.* New York: Wiley.

Gross, M. G. 1986. *Oceanography,* 3rd ed. Englewood Cliffs, NJ: Prentice-Hall.

Kennett, J. P. 1982. *Marine Geology.* Englewood Cliffs, NJ: Prentice-Hall.

Scientific American. 1983. The ocean. San Francisco: Freeman.

Scrutton, R. A., and M. Talwani, eds. 1982. *The Ocean Floor.* New York: Wiley International.

Stowe, K. 1983. *Ocean Science,* 2nd ed. New York: Wiley International.

Granite
Batholith

Metamorphic rocks

Granite

Granite
Batholith

Metamorphic rocks

Granite

Although the theory of plate tectonics was developed largely from new studies of the sea floor, a record of plate motion during most of geologic time is preserved only in continental rocks. The oceanic crust is a temporary feature, continually created along the oceanic ridge and then destroyed at subduction zones. Plate movement can be studied and measured in the relatively young oceanic rocks, but essentially all of the oceanic crust is less than 200 million years old. By contrast, the rocks of the continents are as old as 3.8 billion years, and a record of plate movement during the early history of Earth is preserved only in the ancient metamorphic and igneous rocks of the continental shields.

On the continental shields, geologists find evidence of a long and complex history of Earth's dynamics. They also find evidence of how the continents originated and evolved. One area that has been especially interesting is the shield of western Australia. This remarkable Landsat image shows large elliptical masses of granite (light yellowish tones) intruded into the older greenish metamorphic rocks. These intrusions and their country rocks stand out as perhaps the best exposed examples of ancient granitic and metamorphic terrain anywhere on Earth. The dark green metamorphic rocks were originally basalt, andesite, and rhyolite, with some interbedded sedimentary rocks. These intrusions occurred about 3 billion years ago and represent an early style of tectonism, which created the nuclei of modern continents. No fragments of original crust have ever been found. Why did continents evolve on our planet? Did Earth always have plate tectonics? Are continents still evolving?

In this chapter, we will consider these questions and how deformation, metamorphism, igneous activity, erosion, sedimentation, and isostatic adjustment all play a role in continental evolution.

22
Evolution of Continents

MAJOR CONCEPTS

1. The continents are made up of three basic structural components: (a) shields, (b) stable platforms, and (c) young, folded mountain belts.
2. Orogenesis (mountain building) occurs at convergent plate margins.
3. The major features of orogenesis are (a) intensive crustal deformation, (b) metamorphism, and (c) igneous activity.
4. The characteristics of a mountain belt and the sequence of events that produced it depend on the rock types that are involved and the types of interactions at convergent plate boundaries.
5. Three major convergent plate interactions are recognized: (a) convergence of two oceanic plates, (b) convergence of a continental and an oceanic plate, and (c) convergence of two continental plates.
6. Continents grow by accretion as new crustal material forms in an orogenic belt.

THE CONTINENTAL CRUST

The continents are made up of three basic structural components: (a) shields, (b) stable platforms, and (c) young, folded mountain belts.

The continental crust has been studied for more than 150 years, but only recently have we been able to synthesize the vast amount of complex data concerning continental geology and develop a reasonably clear concept of what continents really are. The more important geologic facts about continents can be summarized as follows:

1. Continents are composed of huge slabs of "granitic" rock. This does not mean that all of the rocks of the continents are igneous, however. Much of the continental crust is metamorphic rock of roughly the same composition as granite. Seismic data suggest that there is a gradation from the silicic composition observed near the surface of a continent to a more basaltic composition near its base.
2. Continents range in thickness from 30 to 50 km. The thickest portions are beneath folded mountain belts.
3. Continents cover roughly one-third of the planet. The volume of continental crust appears to have grown throughout most of geologic time.

What makes the continental crust unique?

4. The continents contain the oldest rocks on Earth, which range in age to 3.8 billion years. By contrast, all of the oceanic crust is less than 200 million years old.
5. Planetary studies indicate that Earth is the only terrestrial planet of the solar system with continental crust.
6. Continental crust results from *planetary differentiation,* in which the materials in a planetary body are separated according to density into a layered body with a core, a mantle, and a crust.
7. Although each continent is unique, they all have three basic components:
   a. A large area of exposed basement complex known as the shield.
   b. Broad, flat stable platforms, where the igneous and metamorphic rocks of the basement complex are covered with a veneer of sedimentary rocks.
   c. Young, folded mountain belts along the continental margins.
8. Geologic differences among continents are mostly in the size, shape, and proportions of the three basic components.

Shields

The continental *shields* are the key to modern theories of the origin and evolution of continents. Take a moment to study Figures 22.1 and p. 132.

The most striking characteristic seen in these photographs is the vast expanse of the low, relatively flat surface of the shield. Throughout an area of thousands of square kilometers, this surface lies within a few hundred meters of sea level. The only features that stand out in relief are resistant rock formations that rise a few tens of meters above the surrounding, less-resistant rocks. On a regional basis, shields are flat and almost featureless.

A second fundamental characteristic of shields is their structure and composition. Shields are composed of a highly deformed sequence of metamorphic rocks and granitic intrusions, known as a **basement complex.** Their structural complexities are shown by patterns of erosion, the alignment of lakes, and differences in the tones of photographs. Faults and joints are common and are expressed at the surface by linear depressions, some of which can be traced for hundreds of kilometers. Most of the rocks are Precambrian, and most were formed under high temperature and highly compressive stresses several kilometers below the surface. In Figure 22.1, metamorphic rocks appear in tones of dark gray. Granitic intrusions, which appear in lighter tones, have a more massive texture. Figure 22.1 shows evidence that the shields have

FIGURE 22.1
Complex metamorphic rocks (dark tones) are intruded by granitic rocks (light tones) to form the complex structural characteristics of a shield. These rocks were formed in the roots of an ancient folded mountain belt during Precambrian time, approximately 1.8 billion years ago. On a regional basis, the shield is a broad surface of low relief, eroded close to sea level.

been intensely deformed, so the deeper sedimentary and volcanic rocks were converted to complex metamorphic rocks. They were also intruded by silicic magmas. Subsequently, erosion removed the upper cover of the sedimentary and metamorphic terrain, exposing what we now see at the surface.

From these facts our present understanding of shields can be summarized graphically in Figure 22.2. The basic structure of complex igneous and metamorphic rocks, eroded to a flat surface near sea level, forms the nucleus of every continent.

The belts of metamorphic rocks, together with igneous intrusions, indicate that the shields are composed of a series of zones that were once highly mobile and tectonically active. These facts have long been known. For many years, geologists have considered the shields to consist of a series of ancient mountain belts that have been eroded down to their roots and have since remained stable.

Why don't Mars, Mercury, and the Moon have continents?

The Canadian Shield of North America is relatively well known. The map in Figure 22.3 summarizes many of its structural details derived from data compiled from years of fieldwork, geophysical studies, and radiometric dating. In a general way, North America is quite symmetrical. The oldest rocks, recording the first clearly recognizable geologic events, are in the Superior, Wyoming, and Slave provinces. They consist of volcanic flows and volcanic-derived sediments that are similar to those forming today in the island arcs of the Pacific. These sediments were metamorphosed and engulfed by granitic intrusions. The granite-forming event, as determined by radiometric dating, occurred between 2.5 and 3.4 billion years ago. Earth has been estimated to be a little more than 4.5 billion years old, so this part of the crust apparently formed from 1.1 to 2.0 billion years after Earth's origin. Generally speaking, the

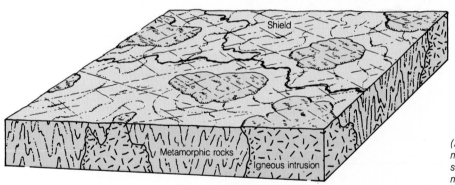

(A) The shield consists of a complex of metamorphic rocks and a variety of igneous intrusions. The upper surface is flat and is commonly eroded down to near sea level.

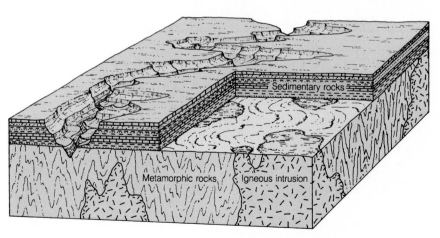

(B) Throughout most of the interior of the United States, the basement complex is covered with a veneer of horizontal sedimentary rock. In some places, such as the Grand Canyon, erosion has cut through the sedimentary veneer to expose the basement complex below.

FIGURE 22.2
The general characteristics of the continental crust are depicted in these block diagrams.

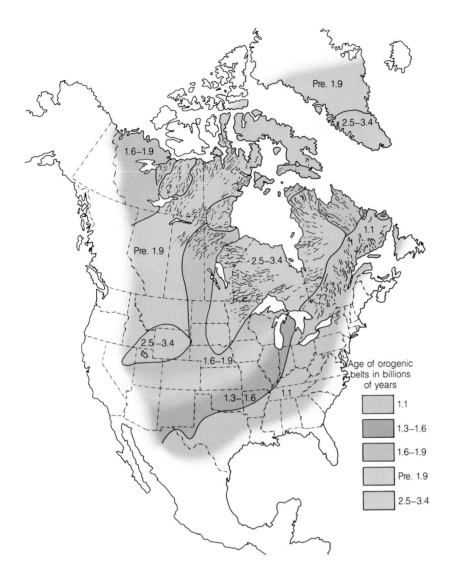

FIGURE 22.3
Major structural trends and radiometric dates of the basement rocks in North America show several geologic provinces, each representing a mountain-building event. The numbers refer to the ages, in billions of years, of the major granitic intrusions, and the lines represent the trends of the folds and the structural trends in the metamorphic rocks. The shield apparently grew by accretion as new mountain belts formed along its margins.

composition of this ancient continental crust (prior to granitic intrusion) was closer to basalt than to granite. Granitic intrusions now constitute three-fourths of the area. The metamorphosed sediments and volcanic rocks display structural trends in a northward direction, as is shown in Figure 22.3.

This region of the crust is significant because it is thought to represent the first stable and resistant granitic crust in North America. Since its formation, the upper surface has been eroded down almost to sea level and has persisted as a stable unit. It has sometimes been partly submerged beneath a shallow sea and has sometimes rested slightly above sea level, but it was never again the site of mountain-building processes. This block of crust was probably once 20% to 60% larger than it is now because the trends of the metamorphic structures are terminated abruptly by younger provinces. We can assume that the original continental mass was split and fragmented by plate movement, with each segment subsequently acting as a center for future continental growth.

Surrounding the Superior province to the south, west, and north is a vast area of gneiss and granite from 1.9 to 1.6 billion years old. The rocks in this younger province are also metamorphosed, but the original sediment was quite different from the sediment that formed the rocks of the Superior province. These younger rocks contain less lava, and they show a distinct increase in quartz-rich sandstone and limestone, which originally formed on a continental shelf. In addition, the structural trends are oriented in a different direction.

Why do we believe that the shields consist of a series of belts formed in the roots of ancient mountain systems?

To the southeast, the rocks are younger still, and the associated granitic intrusions are approximately 1.1 billion years old. The rocks are mostly quartzites, marbles, and schists derived from well-sorted sandstone, shale, and limestone. They also include volcanic rocks that are richer in silica than the older volcanics of the Superior province.

The concentric pattern of the provinces in North America is considered strong evidence that the continent grew by the accretion of material around its margins during a series of orogenic events. Each province represents a mountain-building event during which sediments were deformed into a mountain range by converging plates. Subsequently, erosion removed the upper part of the deformed belt, exposing the deeper, highly deformed metamorphic rocks and the associated granitic intrusions.

Stable Platforms

Large areas of the basement are covered with a series of horizontal sedimentary rocks, as is shown in Figure 22.2. The *stable platforms* form much of the broad, flat lowlands of the world and are known locally as plains, steppes, and low plateaus (Figure 22.4). The relationship between the sedimentary rocks of the stable platform and the underlying igneous and metamorphic complex (the basement complex) is known from thousands of wells that penetrate the sedimentary cover and from seismic studies, which reveal the rock structure beneath the surface. Although locally the rocks appear almost perfectly horizontal, on a regional basis they are warped into broad, shallow domes and basins.

The flat-lying sedimentary rocks that cover parts of the basement are predominantly sandstone, shale, and limestone, which were deposited in ancient shallow seas. These flat-lying marine sediments, preserved on all continents, show that large areas of the *cratons* have periodically been flooded by the sea and have then reemerged as dry land. At present, more than 11% of the continental crust (namely, the continental shelf) is covered with water. At various periods in the past, however, shallow seas spread over a much greater percentage of the land surface.

There are several reasons for the periodic flooding of large areas of the shields during various periods of geologic time. The shields are broad, flat areas eroded down to within a few tens of meters of sea level. Uplifts or downwarps, which occur as the tectonic plates move across the globe, thus cause expansion or contraction of shallow seas over the continents. For example, rapid convection in the mantle would arch up the oceanic ridge and displace water from the ocean basins to the continents. If the convection rate decreased, the oceanic ridge would subside and the seas would recede. As a tectonic plate moves with the convecting asthenosphere, it can rise and fall slightly and thus cause expansion and contraction of shallow seas over the cratons. In addition, hotspots beneath a continental mass could cause it to arch up, which would produce a general withdrawal of the sea.

Mountain Belts

The great, linear *folded mountain belts,* which typify the continental margins, are one of Earth's most distinctive tectonic features. The rocks within these relatively narrow zones are compressed and folded and in many places are broken by thrust faults. As a result, the crust is shortened by as much as 30% (Figure 22.5).

The young, active mountain belts of today occur along convergent plate margins, coinciding with zones of intense seismicity and andesitic volcanism (Figure 22.6). Two major mountain belts are still in the process of growing. They are (1) the Cordilleran belt, which includes the Rockies and the Andes, and (2) the Himalaya–Alpine belt, which extends across Asia and western Eu-

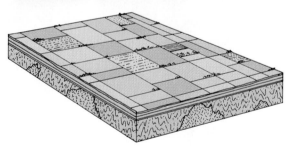

FIGURE 22.4
Stable platforms are areas where the basement complex is covered with a veneer of sedimentary rocks. They constitute much of the world's broad, flat lowlands, known locally as plains, steppes, and low plateaus.

rope and into North Africa. Older mountain ranges can still be expressed by significant topographic relief, although deformation ceased long ago. Examples include the Appalachian Mountains of the eastern United States, the Great Dividing Range of eastern Australia, and the Ural Mountains of Russia.

The location of young mountains in long, narrow belts along continental margins is significant because it implies that mountains do not result from a uniform worldwide force evenly distributed over Earth. They must be the result of forces concentrated along the margins of continents.

Another important aspect of their location is that many of the older mountain belts extend to the ocean and abruptly terminate at the continental margin. The northern Appalachians of the eastern United States, the Atlas Mountains of Africa, and the mountains of Great Britain are excellent examples. The abrupt termination of the folded structures suggests that the mountain systems were once much more continuous and have been separated by continental rifting.

What is distinctive about mountain belts?

FIGURE 22.5
The structure of a folded mountain belt in western Pakistan is clearly visible in this satellite photograph. The ridges are plunging anticlines, many of which are in the initial stage of erosion. Some are partly breached, and exposures of older rocks can be seen in their cores. Young mountain belts such as this occur along continental margins as a result of the motion of converging plates.

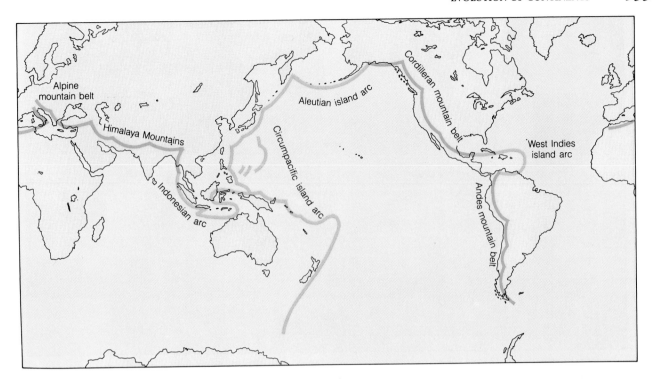

FIGURE 22.6
Active mountain belts of Earth occur along convergent plate margins and coincide with zones of intense seismicity and andesitic volcanism. The structural trends of the mountains extend into the oceans as island arcs.

Shields, stable platforms, and young, folded mountain belts are the three basic structural components of all continents. The major geologic differences among continents are mostly in the size, shape, and proportions of these three components. Let us briefly consider the continents individually and note their similarities and differences. Figure 22.7 shows the distribution of shields, platforms, and young, folded mountain belts on each continent and provides important background for subsequent discussions of the origin and evolution of continents.

North America.   North America has a large shield, most of which is in Canada. It extends from the Arctic islands southward to the Great Lakes area and westward to the plains of western Canada. Most of the Canadian Shield is less than 300 m above sea level. The stable platform extends through the central United States and western Canada and is covered with sedimentary rocks, slightly warped into broad domes and basins. The Appalachians are an old mountain belt, a large part of which was covered with Cretaceous and Tertiary sediments. The Rocky Mountains, a part of the Cordilleran folded mountain belt, were deformed during Cretaceous time. Parts of this belt are still active today.

South America.   South America consists of a broad shield in Brazil and Venezuela and stable platforms in the Amazon basin and along the eastern flanks of the Andes Mountains. The Andes Mountains are part of the Cordilleran folded mountain belt, which extends from Alaska to the southern tip of South America. The continent has no mountain belts on the eastern margin.

Australia.   Australia is remarkable in many ways. Most of the continent is extremely flat and low, barely above sea level. Shields are exposed in the north, west, and south; and in the central part of the continent, a veneer of sedimentary rocks covers the older, complex igneous and metamorphic terrain. An old mountain range occurs on the eastern margins, and a younger, tectonically active mountain belt occurs in New Guinea along the converging plate margin. Australia has been isolated from the rest of the world since the breakup of Pangaea and has evolved a unique community of plants and animals

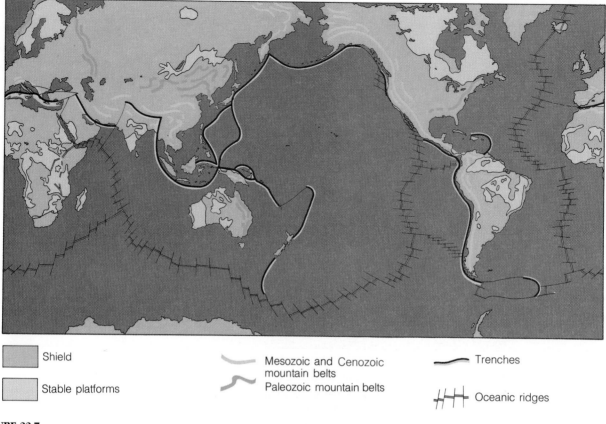

FIGURE 22.7
Major tectonic features of Earth provide a record of global dynamics during the last 3.8 billion years. The shields formed more than 600 million years ago and were eroded down to sea level. The stable platforms provide a history of sedimentation in the shallow seas that periodically covered the cratons during the past 600 million years. Paleozoic mountain belts represent ancient convergent plate margins, which existed more than 200 million years ago. Deep-sea trenches, young folded mountain belts, and the oceanic ridge outline the boundaries of the plates that are active at the present time.

found nowhere else in the world. It is now situated between 12 and 16 degrees south latitude, so that much of the continent is a great desert. At present, it is moving slowly to the north.

Africa. Africa consists of several extensive shields, with sedimentary rocks of the stable platform preserved in a series of isolated elliptical basins. The only folded mountain belts occur as narrow bands along the southern and northern margins of the continent. In terms of plate tectonics, Africa is distinctive because it is nearly surrounded by spreading centers. The African plate is thus subject to compression on all margins except the one to the north. This may help explain why the continent is abnormally high. A large rift system extends from the Red Sea and the Gulf of Aden southward almost 4000 km. This is the longest rift zone on any continent. Africa sits astride the equator and spans several climatic zones—the tropics in the central belt, flanked by the low-latitude Sahara and Kalahari deserts.

Asia. Asia has a much larger area of folded mountains—the Himalayas—which constitute a wide east–west range extending from Europe to southeast Asia. Most of the Asian basement complex is covered with sedimentary rocks, which form the vast, low stable platforms of central Asia. Igneous and metamorphic rocks of the shield are exposed in parts of Siberia. The Ural Mountains mark the margin between the continents of Asia and Europe, and the Himalaya Mountains form the margin between the continent of Asia and the subcontinent of India.

Europe. The European shield is exposed in the Baltic area, but most of the central and southern parts of the continent are a large stable platform. In central Europe, the sedimentary rocks have been warped into broad domes and basins. The eroded edges of these structures, such as the Paris basin, can be seen on a regional physiographic map. The Alps are folded mountains along the southern margin of the continent.

India. India is a small block of continental crust that moved rapidly northward at the time of the breakup of Pangaea and collided with Asia to form the great Himalaya Mountains. The Himalayas are high because part of the Indian continent is thrust under Asia. Most of the Indian continent is a shield with only limited areas of stable platform. The west-central part of India is distinctive in that it is covered with horizontal basaltic flows and sedimentary rocks that are related to those in South America, both being formed before the continents drifted apart.

This brief overview of the continents serves to emphasize an important fact. From the point of view of geologic structure and mode of origin, all continents are similar. They are all composed of shields, platforms, and folded mountain belts. They differ mostly in the size, shape, and proportions of these structural features. The style of landforms that develop on each of these major structural components can be quite distinctive because of the influence climate has on the development of landforms. The Canadian shield is similar to the shields in Australia and Brazil, yet because of climate, the landforms of each have unique characteristics. The Canadian shield has been glaciated and its river system obliterated; there are many surface features formed by glacial ice. Australia is a desert. Sand dunes cover vast areas. The shields are modified by desert landforms. In Brazil, tropical weathering is extreme and has developed a soil up to 700 feet thick, which covers and obscures the structure of the underlying bedrock.

All three shields are fundamentally the same, but climate has exerted a major control on development of the landscape. And so it is with the stable platform and mountain belts. They are quite similar in basic geologic structure but distinctive in details of their landscapes.

OROGENESIS (MOUNTAIN BUILDING)

Mountain belts are formed at convergent plate margins by compressive forces that result from plate collision. They therefore provide a record of ancient plate motion and a history of Earth's dynamics. The factors that appear to be most important in the mountain-building process are (1) rock sequences, (2) structural deformation, (3) metamorphism, (4) igneous activity, and (5) erosion and isostatic adjustment of the crust.

The highly deformed rocks in mountain belts have attracted the attention of geologists for over 150 years, but the question of why Earth has mountains has not been satisfactorily answered until recently. The question becomes more important with the recognition that folded mountain belts have formed on Venus but not on other planets that have been explored in the solar system.

We now know that mountain belts are formed at convergent plate margins by compressive forces that result from plate collision. Mountains, therefore, provide a record of ancient plate motion and a history of Earth's dynamics.

The process of mountain building is called *orogenesis*. The factors that appear to be most important in this process are the following:

1. Rock sequences
2. Structural deformation
3. Metamorphism
4. Igneous activity
5. Erosion and isostatic adjustment of the crust, which occur after orogenesis and continue until the mountain belt is eroded to near sea level

Rock Sequences and Orogenesis

Over 100 years ago, James Hall (1811–1898), a noted American geologist, recognized that the total thickness of strata in the Appalachian Mountains is approximately 8 or 10 times the thickness of sedimentary rocks of equivalent age in the stable platform. He also noted that, in both areas, most of the rocks throughout the entire sequence had been deposited in shallow water. This is indicated by shallow-marine fossils, mud cracks, and interbedded layers of coal. From these observations, Hall concluded that the present mountain range was once a region of the crust that gradually subsided much more than the rest of the continent. The gradual subsidence permitted a great thickness of shallow-marine sediments to accumulate at rates that were roughly equivalent to the rates of subsidence. The elongate subsiding trough is called a *geosyncline.* After receiving a critical thickness of sediments (approximately 14 km), the geosyncline was compressed into a folded mountain range, uplifted, and eroded.

Why are there specific types of rock sequences in different mountain belts?

The geosynclinal theory of mountain building was further developed by James Dana (1813–1895), another American geologist, who also worked in the Appalachians. He proposed that mountain building involves a three-phase cycle consisting of (1) geosynclinal sedimentation and contemporaneous subsidence, (2) compression and deformation, and (3) uplift and erosion.

Studies of mountain belts in different parts of the world indicated that the Hall–Dana geosynclinal theory was too simple. Instead of a single belt, parallel belts of different types of geosynclinal sedimentation were recognized. One belt, called a *miogeosyncline,* is adjacent to the stable platform and is underlain by continental crust. It consists of clean, well-sorted, shallow-water sandstone, limestone, and shale, with no volcanic rocks. The other belt, called a *eugeosyncline,* consists of sediments deposited in deep-marine environments and commonly includes volcanic rocks. Modern theories of geosynclinal sedimentation and mountain building use new information about sedimentation along continental margins and in the adjacent deep-marine basins and relate the formation of geosynclines to the plate tectonics theory.

The theory of plate tectonics proposes that orogenesis occurs along convergent plate margins. Unlike the geosynclinal theory, it does not require specific rock types and events to occur in the same order in all *orogenic belts.* Indeed, a variety of rock sequences is deposited along continental margins and on the ocean floor. When they are deformed at convergent plate margins, different sequences produce different styles of mountain belts. Modern oceanographic studies have determined that the classical miogeosynclinal sequence of sediments accumulates along the continental margins and that the eugeosynclinal suites of rocks form in the deep water beyond. Probably the best known example is the present Atlantic continental margin of the eastern United States (Figure 22.8). The sequence of thick sediments has been called a *geocline* by some geologists because it is not a two-sided trough but is open toward the ocean.

Two distinctly different rock sequences are recognized. In *miogeoclines,* which lie in shallow water along the continental margins, the sequence consists of clean, well-sorted sandstone, shale, and limestone derived from erosion of the continents. In *eugeoclines,* in the deep water off the continental mar-

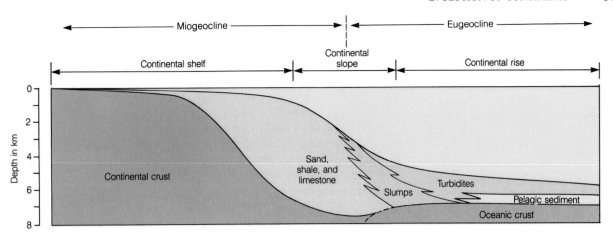

FIGURE 22.8
Two types of rock sequences accumulate along continental margins. Clean, well-sorted sandstone, shale, and limestone, deposited in shallow water, accumulate on the continental shelf. Poorly sorted, dirty sandstone and shale are deposited by turbidity currents in the deep water beyond the continental margins.

gins, the sequence consists of poorly sorted sandstone and shale deposited by turbidity currents, submarine slump blocks, and rock debris from submarine landslides. As is shown in Figure 22.8, the turbidites of the eugeoclinal sequence grade seaward into deep-marine organic oozes.

Shallow-water miogeoclinal sediments also can form behind an island arc, and deep-marine eugeoclinal assemblages can form in trenches or on the seaward side of arcs.

A third rock assemblage, which is common in some mountain belts, is called an *ophiolite* sequence. It consists of peridotite, gabbro, pillow basalt, and the deep-marine sediment that forms on the oceanic crust. These rocks are believed to have been scraped off a subducting plate and plastered against the upper plate as a chaotic mélange, with complex folding and thrust faulting.

One feature of geoclinal sequences that always has been difficult to explain is the great thickness (as much as 15 km) of shallow-marine sediments. For shallow-water sediments to accumulate to such depths, the rate at which the crust subsides must be essentially the same as the rate at which the sediment accumulates. What causes this gradual subsidence? The question has not been completely answered, but this subsidence may be related to vertical movements of the crust following extension and rifting of a continent. As the continent begins to split apart and a new ocean basin begins to form, the continental crust is uparched, extended, and split. The edges of the rift zone, which eventually become new continental margins and the sites of geoclinal sedimentation, are soon leveled by erosion. As the continent moves off from the uparched spreading center, it begins to subside, and sediment derived from erosion of the continent accumulates along the margins. Gradual subsidence of the continental margins occurs for two reasons: (1) the continent cools and moves off from the hot rising mantle that underlies the ridge, and (2) the weight of the deposited sediment causes the crust to be depressed.

In summary, the sedimentary rocks commonly involved in orogenesis are (1) shallow-marine miogeoclinal sediments, (2) deep-marine eugeoclinal sediments, and (3) ophiolite sequences of the oceanic crust. To study the history of mountain building and the plate movements that produce it, we must go beyond the direct results of orogenesis, such as deformation, metamorphism, and igneous activity. Geologists must also study the details of sedimentary rock assemblages that were deposited before plate collision. These assemblages provide insight into the nature of plate interactions.

Structural Deformation

Is there an organized structure in mountains?

The single most distinctive feature of a mountain belt is the structural deformation of the rocks. The nature of this deformation was discussed in Chapter 8 (refer again to Figure 8.1). Deformed structures result from compression, and the scale of deformation ranges from wrinkled grains or fossils in the rock to folds tens of kilometers wide.

From detailed field mapping, the structure of an entire mountain range can be determined and illustrated by geologic maps and cross sections. Figure 22.9 includes three well-known examples.

In the Canadian Rockies, a major type of deformation is thrust faulting, in which large slices of rock have been thrust over others in a belt 60 km wide (Figure 22.9A). The orientation of the faults and the direction of displaced beds indicate that the rocks were thrust from the margin of the continent toward the interior.

A cross section of the Appalachian Mountains shows a different style and magnitude of deformation (Figure 22.9B). The major structural feature is a series of tight folds. Deformation is most intense near the continental margins, dying out toward the continental interior.

The structure of the Alps is even more complicated (Figure 22.9C). Great overturned folds called **nappes** (French, "tablecloths") show enormous amounts of crustal shortening. The rocks are so intensely deformed that spherical pebbles have been stretched into rods as much as 30 times longer than the

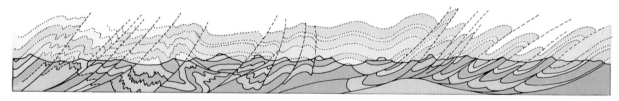

(A) The Canadian Rockies contain both folds and thrust faults.

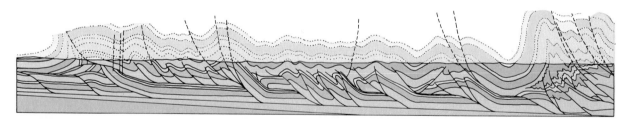

(B) The Appalachian Mountains consist of tight folds and thrust faults that have been eroded down to within 1000 to 3000 m of sea level. Resistant sandstones form the mountain ridges.

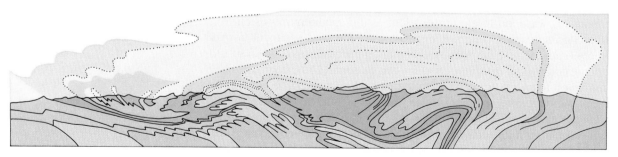

(C) The Alps are complex folds, many of which are overturned.

FIGURE 22.9
The structure of folded mountain belts reflects intense compression at convergent plate boundaries, but each range can have its own structural style.

original diameters of the pebbles. Most of these structures can be explained only in terms of compressive forces.

Note that, in each mountain belt, the internal structures result from strong horizontal compressive forces. Similar deformation exists in older rocks in places where erosion has reduced the topographic relief.

Metamorphism

In the deeper parts of the orogenic belt, intensive plastic deformation and recrystallization at elevated temperature and pressure metamorphose the original sedimentary and volcanic rocks into schists and gneisses. The horizontal stress generated by converging plates causes the recrystallization of many minerals and develops foliation that is perpendicular to the direction of stress. Slaty cleavage, schistosity, and gneissic layering in the deeper parts of a mountain range are thus characteristically vertical or dip at a high angle. Abnormally high heat in local areas within a mountain belt can produce a system of metamorphic zones.

In the deeper parts of a mountain belt, metamorphism can become intense enough to produce granitic *migmatite*. Migmatite is a complex mixture of thin layers of granitic material between sheets of schist or gneiss. It develops largely from the partial melting of preexisting rocks. Apparently, the magma generated by the compression and heat does not migrate far and is mixed with the unmelted material. In these zones, high temperatures and pressures soften the entire rock body, which behaves like a highly viscous liquid if it is subjected to stress. Metamorphic rocks in deeper parts of orogenic belts consequently exhibit complex flow structures (see Figure 7.4).

Igneous Activity

Igneous activity associated with mountain building is part of the fundamental process of differentiation by which Earth's materials are separated and concentrated into layers according to density. In particular, this igneous activity is responsible for *magmatic differentiation.*

The concentration of Earth's lighter material in the continental crust occurs in two steps. The first phase begins at a spreading center where partial melting of peridotite in the upper mantle generates a basaltic magma, which rises to form oceanic crust. Basalt is richer than peridotite in the lighter elements, particularly silicon and oxygen. In the second phase, partial melting of the oceanic crust forms a silica-rich magma, which is then emplaced in the mountain belt as granitic intrusions and andesitic volcanic products. This process further separates the lighter elements, especially silicon and oxygen, and concentrates them in the continental crust. The granitic continental crust is less dense than the mantle and the oceanic crust, and its buoyancy prevents it from being consumed at subduction zones. Once it is formed, the continental crust remains on the outer surface of Earth.

The generation of silica-rich magma at a subduction zone is shown in Figure 22.10. The magma conceivably can result from partial melting in three different regions of the subduction zone: (1) in the subducting oceanic crust, (2) in the overlying mantle, as hot fluids percolate upward, and (3) near the base of the continental crust, as upward-migrating magma raises the temperature. Water in the pore spaces of the rock and in chemical combination in many minerals of the oceanic crust plays an important role in the type of igneous activity at convergent plate margins. As the crust is subducted and heated, this water is driven out. It percolates upward and enhances melting at higher levels in the mantle.

Why are metamorphic processes associated with mountain building?

Why do andesitic and silicic magmas form in mountain belts?

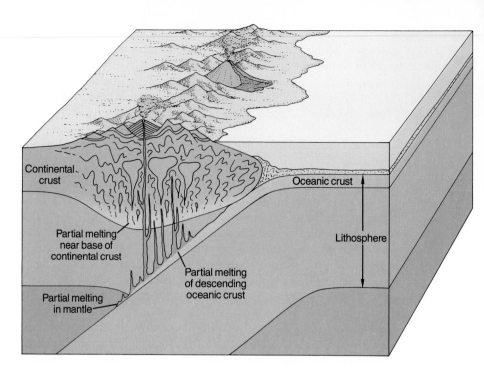

FIGURE 22.10
The generation of silica-rich magmas can occur by partial melting in three different regions of a subduction zone. The process can involve (1) partial melting of the descending oceanic crust, (2) partial melting of the upper mantle, and (3) partial melting of the lower part of the continental crust.

Continental crust

Oceanic crust

Partial melting near base of continental crust

Lithosphere

Partial melting of descending oceanic crust

Partial melting in mantle

The Evolution of a Mountain Belt

What is the origin of mountains?

The general concept of how a mountain belt evolves into a new region of the continental basement complex is shown in Figure 22.11. Folding and thrusting occur at relatively shallow depths, metamorphism occurs deeper, and partial melting occurs at still greater depths. Silicic magma initially forms deep within the crust, at points where the rock begins to melt. The magma is then injected into the foliation of the adjacent metamorphic rock, so that a migmatite is formed. Much of the magma migrates upward because it is less dense than the solid rock. As it rises, it forms teardrop-shaped bodies, which collect into larger and larger masses. The boundaries of the body of magma are generally parallel to the broad zones of foliated metamorphic rock. The rising silicic magma cuts across the upper folded strata, which are not metamorphosed but are only deformed by folding and faulting. The magma can cool within a few kilometers of the surface, forming a batholith, or it can be extruded as volcanic material.

Erosion and Isostatic Adjustment

The history of a mountain belt does not end with deformation, metamorphism, and igneous activity. After the orogenic activity is terminated (presumably because of shifts of the convection cells in the mantle), deformation ceases, but erosion and *isostasy* combine to modify the orogenic belt. As is illustrated in Figure 22.11, the entire crust is deformed by the orogeny, so that a mountain root, composed of the most intensely deformed rocks, extends down into the mantle. The high mountain range and its deep roots are in isostatic balance. As erosion wears away the high mountains, this balance is destroyed. Isostasy then causes the roots of the mountain belt to rise in a broad upwarp to reestablish the balance. In the early stages of erosion, the removal of 500 m of rock is accompanied by an isostatic uplift of approximately 400 m, with a net lowering of the surface of only 100 m. The rate of net lowering is believed to decrease as the mountain belt is eroded down and approaches a new state of isostatic equilibrium. Uplift continues as long as erosion removes material from the mountain range. Eventually, however, a balance is reached, when the mountainous topography is eroded to near sea level and the deep mountain root is isostatically adjusted.

Why is simple isostatic adjustment important in the evolution of a mountain belt?

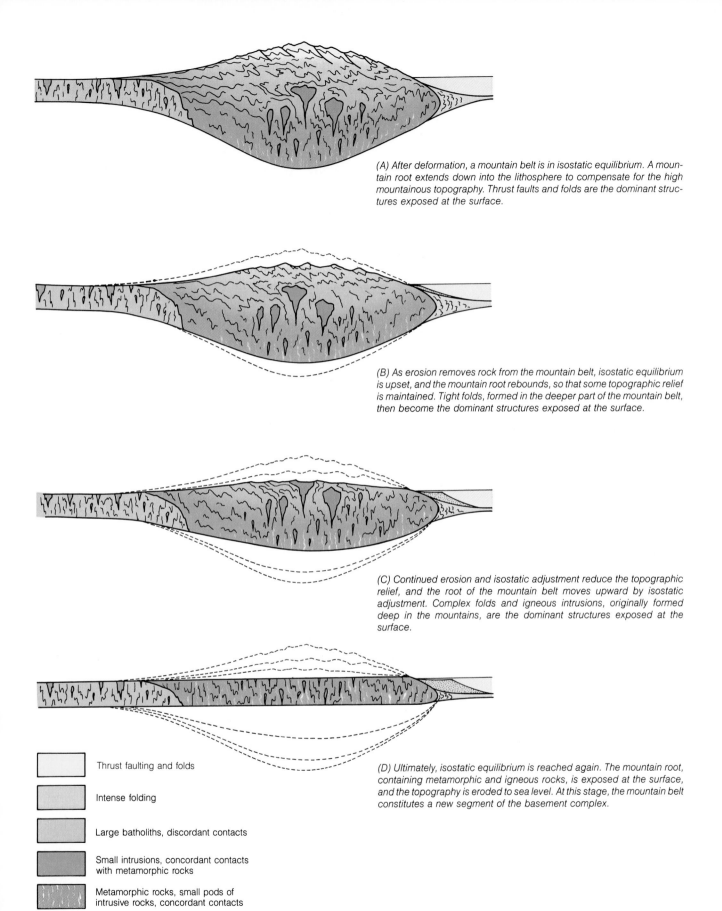

(A) After deformation, a mountain belt is in isostatic equilibrium. A mountain root extends down into the lithosphere to compensate for the high mountainous topography. Thrust faults and folds are the dominant structures exposed at the surface.

(B) As erosion removes rock from the mountain belt, isostatic equilibrium is upset, and the mountain root rebounds, so that some topographic relief is maintained. Tight folds, formed in the deeper part of the mountain belt, then become the dominant structures exposed at the surface.

(C) Continued erosion and isostatic adjustment reduce the topographic relief, and the root of the mountain belt moves upward by isostatic adjustment. Complex folds and igneous intrusions, originally formed deep in the mountains, are the dominant structures exposed at the surface.

(D) Ultimately, isostatic equilibrium is reached again. The mountain root, containing metamorphic and igneous rocks, is exposed at the surface, and the topography is eroded to sea level. At this stage, the mountain belt constitutes a new segment of the basement complex.

Thrust faulting and folds

Intense folding

Large batholiths, discordant contacts

Small intrusions, concordant contacts with metamorphic rocks

Metamorphic rocks, small pods of intrusive rocks, concordant contacts

FIGURE 22.11
Erosion and isostatic adjustment combine to transform a newly deformed mountain belt into a segment of the basement complex of the continent.

541

It is important to note that the rocks exposed at the surface when isostatic balance is established are the metamorphic and igneous rocks formed at great depths in the orogenic belt. These rocks are tectonically stable and become part of the basement complex.

TYPES OF OROGENIC ACTIVITY

The characteristics of a mountain belt and the sequence of events that produced it depend largely on the type of crust carried by converging plates and the type of sediments involved in the deformation. Three distinctive types of convergence are recognized: (1) the convergence of two oceanic plates, (2) the convergence of a continental plate and an oceanic plate, and (3) the convergence of two continental plates.

Why are there three different types of orogenesis?

In the preceding section, we saw that mountain building is the result of plate convergence and that it involves intensive deformation, metamorphism, and igneous activity. The characteristics of a mountain belt and the sequence of events in development can vary. They depend on the types of interactions at convergent plate margins and the types of rock sequences that are involved in the deformation. There are three fundamentally different types of convergence:

1. Convergence of two oceanic plates
2. Convergence of continental and oceanic plates
3. Convergence of two continental plates

Convergence of Two Oceanic Plates

What typical sequence of rock would be produced in ocean-to-ocean orogenesis?

The main types of orogenic activity involving the convergence of two oceanic plates are shown in Figure 22.12. The major result is an arc of volcanic islands above the subduction zone. In the first stages of convergence, the process can be relatively simple and restricted to volcanic activity on the overriding plate. It does not involve widespread metamorphism or granitic intrusion. The Tonga Islands are an example of a simple island arc (volcanic arc) in the present oceans (Figure 22.12A).

In more complex island arcs (such as Japan), crustal deformation, metamorphism, and igneous activity combine to produce distinctive rock associations and deformational patterns. Their major features, illustrated in Figure 22.12B, are the following:

1. Rock sequences consist of the oceanic sediments and pillow basalts of the oceanic crust, in addition to sediment derived from the erosion of the volcanic arc.
2. Partial melting of rocks in or above the subduction zone produces a silica-rich magma, which rises to form granitic intrusions and volcanic products in the island arc.
3. A zone of high-pressure, low-temperature metamorphism develops at the outer margins of the overriding plate.
4. An inner zone of higher-temperature metamorphism develops, with associated granitic intrusions.
5. Crustal deformation of the volcanic arc results from the converging plates and the intrusion of silicic magma.

The zone of high-pressure, low-temperature metamorphism that develops on the outer margin of the overriding plate is easily understood in light of its dynamic setting. High pressure is produced because this zone is the focal point of the plates' convergence. Low temperature results because the cold

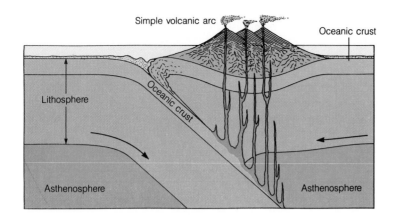

(A) Simple ocean-to-ocean orogenesis is restricted largely to volcanic activity and does not involve widespread metamorphism or granitic intrusions. An example is the Tongan island arc in the South Pacific.

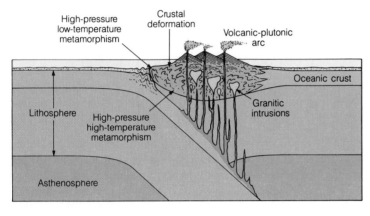

(B) Complex ocean-to-ocean orogenesis involves crustal deformation, metamorphism, and granitic intrusions, as well as volcanic activity, producing a complex island arc. An example is the Japanese arc.

FIGURE 22.12
Ocean-to-ocean orogenesis

oceanic plate descends at rates between 5 and 15 cm per year. These rates exceed the rate of heat flow from the mantle through the descending lithosphere. The rocks that are metamorphosed consist of sediment derived from erosion of the adjacent volcanic arc, oceanic sediments, and pillow basalts of the oceanic crust. This complex mixture is squeezed downward along the top of the descending plate at temperatures that generally are lower than 300 °C. The mixture of pillow basalts, oceanic sediments (ophiolite complex), and erosional debris from the island arc is metamorphosed into a distinctive, fine-grained schistose rock characterized by high-pressure, low-temperature mineral assemblages. The rock typically develops a bluish color because of the presence of the mineral glaucophane and is thus referred to as a ***blueschist***. There are no contemporaneous igneous intrusions in this zone because magma is generated only where the descending plate reaches a greater depth.

Blueschists are found in Japan, California, and New Zealand and have been recognized in the Alps, but they represent only a small fraction of the metamorphic rocks exposed on the continents. Their small proportion can be explained by the fact that these high-pressure, low-temperature rocks lose their distinctive character if they are subsequently heated above 300 °C. Such heating often occurs due to granitic intrusions that develop in continental orogenic belts.

In deeper parts of the subduction zone, magma generated by partial melting of the descending oceanic plate rises to form large granitic batholiths. These intrusions make up the deeper parts of the island arc. They also produce a surrounding zone of high temperature, which develops a very different assemblage of metamorphic minerals. At shallow depths, high-pressure, low-temperature metamorphism occurs. At deeper levels, high-pressure, high-temperature metamorphism (high-grade metamorphism) occurs. The original rocks are the volcanic flows and the sediments that formed within the arc system proper. The resulting metamorphic rocks are quite distinct from blueschist, which is formed near the trench.

As a result of the collision of converging plates, the older rocks in the volcanic arc are compressed, folded, and broken by thrust faults. In addition, deformation results from the emplacement of granitic batholiths. The fold axes and thrust faults trend parallel to the long axis of the arc and the linear belts of granitic batholiths.

Orogenesis caused by the convergence of two oceanic plates is distinctive in (1) rock sequence (andesitic and volcanic sediments), (2) metamorphism (blueschist), and (3) zonation of ophiolite and volcanic rock types.

Orogenesis in island arcs is significant because the deformed andesitic volcanic rocks (and sediments derived from them) are differentiated from oceanic basalt. They represent the initial stage in the development of new continental crust in areas where only oceanic crust previously existed. This process provides important insight into the origin and evolution of continents.

Convergence of Continental and Oceanic Plates

What sequence of rocks develops in continent-to-ocean orogenesis?

The style of mountain building produced by the collision of a continental plate and an oceanic plate resembles that produced by the convergence of two oceanic plates, but it is distinctive in two main respects.

1. Thick sequences of geoclinal sediments, derived from the continent, commonly occur along the continental margins and are quite different from the rock sequences deposited around a volcanic arc. The convergence of the plates produces a mountain belt characterized by folded sedimentary rocks with roots of granitic intrusions and metamorphosed sedimentary rocks.
2. Silica-rich magma generated by the partial melting of the descending oceanic crust is placed upon or within the continental crust, rather than upon the ocean floor. Consequently, a high topography is produced, which is subject to rapid erosion. Volcanic material is soon stripped off the mountain range, and granitic batholiths commonly are ultimately exposed.

A schematic cross section showing the major features produced by the convergence of continental and oceanic plates is shown in Figure 22.13. Before the continental crust becomes involved in the subduction zone, a considerable thickness of sandstone, shale, and limestone can accumulate in the miogeocline along the continental margin. During this same period, deep-marine sediments accumulate in the deep-ocean basin. When the plates converge, the buoyant granitic mass of the continental crust always overrides the adjacent oceanic plate. As the continental plate approaches the subduction zone, some of the deep-marine sediments on the oceanic plate can be crumpled and deformed. Slabs of oceanic crust shear off and are incorporated in the chaotic mass. This material, like that developed by the convergence of two oceanic plates, is subjected to high-pressure, low-temperature metamorphism.

The thick sequence of geoclinal sediments along the continental margin is then compressed and deformed. Thrust faults occur in the shallow zone of the mountain belt, where brittle rocks fail by rupture. At intermediate depths, plastic flow develops tight folds and nappes. Intense metamorphism occurs within the deeper zones, where temperature and pressure are relatively high. The partial melting of the descending lithosphere generates silica-rich magma, which rises to form granitic intrusions in the deformed sediments in the upper plate or is extruded as volcanic material at the surface.

With continued compression, deformation of the geoclinal sediments and the resulting crustal thickening progress landward, and the edge of the deformed continental margin rises above sea level. Erosion of the mountain

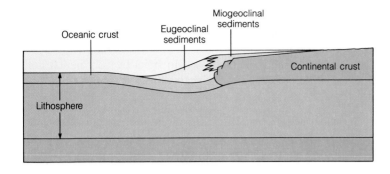

Oceanic crust   Eugeoclinal   Miogeoclinal
sediments   sediments

Continental crust

Lithosphere

(A) Rock sequences before orogenesis

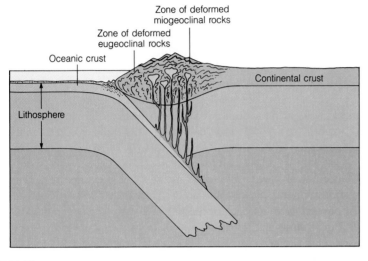

Zone of deformed
miogeoclinal rocks

Zone of deformed
eugeoclinal rocks

Oceanic crust

Continental crust

Lithosphere

(B) Orogenesis

FIGURE 22.13
Continent-to-ocean orogenesis involves the deformation of the thick sequence of sediment that accumulates along continental margins. This material is deformed into a folded mountain belt with miogeoclinal sediments (clean, well-sorted, shallow-marine sand, shale, and limestone) located near the continent interior and the eugeoclinal deep-marine turbidites located near the ocean margin. Granitic batholiths and metamorphosed sediments occur in the deeper zones of the orogenic belt. High topography is produced, and andesitic volcanism is common.

belt begins and continues contemporaneously with subsequent deformation. Sediments eroded from the growing mountain range can be shed in both directions, and thick sequences of coarse conglomerates and sandstone can be deposited as alluvial fans or deltas in front of the rising mountain belt.

As the orogenic belt evolves, resistance builds up. Convergence may stop, or the subduction zone may move seaward to form an ocean-to-ocean orogenic belt. After compression terminates, erosion continues to wear down the mountain range. Isostatic rebound occurs until, ultimately, the mountain roots are exposed at the surface, and the topography is eroded to near sea level.

There are many excellent examples of orogenic belts formed by the convergence of continental and oceanic plates. The Rocky Mountains of western North America were deformed during late Mesozoic and early Tertiary time. The Andes Mountains of South America have been deformed from the Tertiary to the present. The Appalachian Mountains of the eastern United States were deformed during the late Paleozoic. All represent continent-to-ocean convergence, although the structure and topography of each is unique in some respects.

Convergence of Two Continental Plates

In the tectonic system, virtually all oceanic crust is destined to descend into the mantle by the subduction processes at convergent plate margins. As continents, carried by the plates, move toward a subduction zone, the ocean basin is continually reduced in size. It is eliminated altogether if two continents collide.

The convergence of two predominantly continental plates generates an orogenic belt with several characteristics that are quite different from those of the other types of orogeny.

Although the generation of mountain belts by continental collision is exceedingly complex in detail, the major events are similar to those outlined in Figure 22.14.

1. Geoclinal sedimentation occurs along the margin of each continent (Figure 22.14A).
2. Before the actual continental collision, the wedge of sediments along the margin of the continent above the subduction zone is deformed (Figure 22.14B). The oceanic lithosphere is consumed at the subduction zone, and the ocean basin decreases in size.
3. As the continents approach collision, segments of the remaining oceanic crust are deformed by overthrusting and eventually are squeezed between the converging plates.
4. As the continental crust moves into the subduction zone, its buoyancy prevents it from descending into the mantle more than perhaps 40 km below its normal depth. It can be thrust under the overriding plate, however, so that a double layer of low-density continental crust is produced. This layer rises buoyantly to create a wide belt of deformed rock with an adjacent high plateau (Figure 22.14C).
5. Alternatively, the continental masses can become welded together, and fragments of ophiolite assemblage (oceanic crust) can be caught between them and squeezed upward.
6. The oceanic slab of lithosphere, descending into the mantle, ultimately becomes detached and sinks independently. When the slab has been consumed, the volcanic activity and earthquakes it generated cease.
7. Eventually convergence stops as resisting forces build up, and the mountain belt is eroded and adjusts isostatically.
8. The welding together of two continents produces a single large continental mass with an internal mountain range (Figure 22.14D).

The Himalaya Mountains are an example of orogenesis due to continental collision. They were formed during the last 100 million years as India moved northward and destroyed the oceanic lithosphere that formerly separated it from Asia. As two continents collided, India was thrust under the Asian plate, and the Himalaya Mountains and the extensive highlands of the Tibetan Plateau were formed. Earthquakes are frequent in the region, but they are shallow and occur in a broad, diffuse zone because there is no descending oceanic plate.

The Alps and the Ural Mountains are other examples. The European Alps result from the convergence of the African and Eurasian plates. In many ways, Africa, moving northward against Eurasia, is like India, but it has not evolved to the point at which the oceanic crust (the Mediterranean Sea) is completely consumed. The Urals were formed much earlier, during late Paleozoic time, when the Siberian continental mass collided with Europe. They are not tectonically active, but erosion and isostatic adjustments maintain a mountainous topography.

In summary, mountain building is a fundamental process in the differentiation of Earth. Orogenesis occurs along convergent plate boundaries. Processes of compressional deformation, metamorphism, and igneous activity are involved in mountain building. The style of the orogenesis and the specific events can vary significantly, depending on the interactions that occur along the convergent plate boundaries. The three fundamental types of convergence are ocean-to-ocean, ocean-to-continent, and continent-to-continent collisions. Each produces its own distinctive style of mountain building.

What sequence of rocks would be found in a mountain belt formed by the collision of two continents?

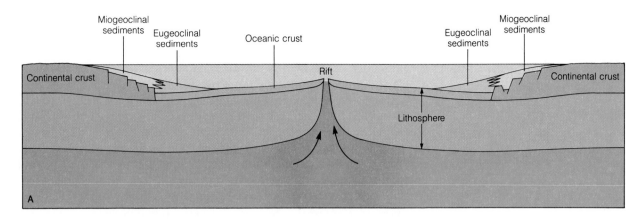

Miogeoclinal sediments Eugeoclinal sediments Oceanic crust Eugeoclinal sediments Miogeoclinal sediments

Continental crust Rift Continental crust

Lithosphere

A

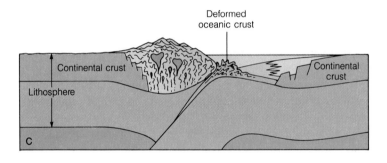

Deformed geoclinal sediments Oceanic crust Eugeoclinal sediments Miogeoclinal sediments

Continental crust Lithosphere Continental crust

B

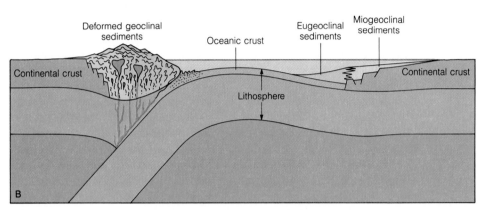

Deformed oceanic crust

Continental crust Continental crust

Lithosphere

C

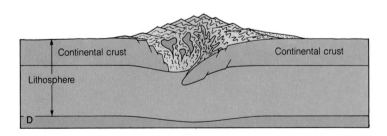

Continental crust Continental crust

Lithosphere

D

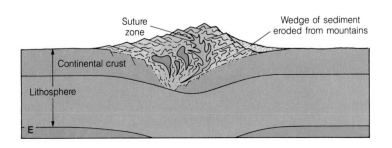

Suture zone Wedge of sediment eroded from mountains

Continental crust

Lithosphere

E

FIGURE 22.14
Continent-to-continent orogenesis involves the deformation of oceanic and geoclinal sediments and commonly produces a complex, high mountain range. As two continents converge, the oceanic crust is caught between them and deformed. A double layer of continental crust can be produced, resulting in abnormally high topography. The continents are welded together. The descending oceanic plate becomes detached in the subduction zone and sinks independently. When this slab is consumed, volcanic activity and deep earthquakes end.

547

Accreted Terranes

Recent studies in western North America reveal a number of segments or blocks of diverse origins within the Rocky Mountain orogenic belt that appear to be juxtaposed in an unorganized manner. The blocks or terranes (the term terrane is used with reference to a region or group of rocks that has a common age, structure, stratigraphy, and origin) are variable in size, and their rocks, fossils, histories, and magnetic properties contrast sharply with those of adjacent provinces. Each terrane is unique and completely unrelated to the others, except that they are now together in one place. These exotic segments of the orogenic belt are referred to as accreted terranes. Fossils indicate that the rocks in a series of *accreted terranes* were formed at different times and in different environments, and paleomagnetic data show that the various terranes originated at different latitudes.

The accreted terranes are believed to be small crustal fragments, island arcs, or seamounts that have been transported thousands of kilometers by the moving oceanic plate and added to a large continental mass at the subduction zone. The present oceans contain many such features that move with the plates; they will resist subduction and will eventually be sutured onto a continent. These features stand out on a physiographic map of the oceans (see the inside covers). For example, in the Indian Ocean, Madagascar, the Seychelles Island platform, and Ninety East Ridge are small fragments of continental crust (microcontinents) that will eventually become accreted terrane. New Zealand, the Philippines, Taiwan, and Japan, plus the numerous chains of seamounts, are similar examples in the Pacific. Others are being produced today as Baja California is split away from Mexico and southern California is moving northward along the San Andreas Fault.

Accreted terranes were first recognized in the orogenic belt of western North America (Figure 22.15). In this region numerous independent segments of the crust, each with its own internal structure, rock types, and fossils, contrast sharply with adjacent provinces. One accreted terrane contains remnants of ancient seamounts; another, segments of limestone platforms like the Bahamas; still others are pieces of old volcanic arcs. Some are crustal fragments of metamorphosed basement complexes. Based on paleomagnetic properties and fossils, it is known that many of these crust segments originated in what is now the South Pacific and have traveled thousands of kilometers to eventually become part of the Rocky Mountains. The Appalachian Mountains, from Newfoundland to Alabama, contain slices of ancient Europe, Africa, and oceanic islands. Accreted terranes have also been found in Southeast Asia, China, and Siberia, but the original locations of these fragments have not yet been determined. The present distribution of oceanic plateaus (Figure 22.16) gives some idea of accreted terranes that will likely be welded onto continental blocks in the future.

ORIGIN AND EVOLUTION OF CONTINENTS

Continental crust began to form at an early stage in Earth's history and has grown by accretion as mountain belts evolved into segments of the shields. Major pulses of continental growth appear to have occurred in three phases, roughly 1, 2 and 3 billion years ago.

The precise way in which continents first began to evolve from the primitive crust is uncertain, but it undoubtedly resulted from strong convective currents in the mantle, which produced cycles of melting and cooling of crust, subsidence, and renewed melting. Weathering, erosion, and sedimentation helped

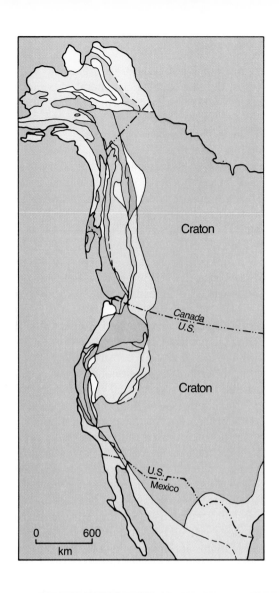

FIGURE 22.15
Accreted terranes in western North America are composed of pre-Mesozoic rocks. Prior to reaching their present positions these rocks were microcontinents or oceanic islands thousands of kilometers away in the South Pacific. The basement complex of North America is shown in tan and the accreted terranes in bright colors. Green represents mostly Paleozoic microcontinent rocks, blue represents rock sequences formed in island arcs, red represents oceanic crust. Each terrane has distinct stratigraphy, fossil assemblages, and/or volcanic rock types. Apparently all originated in widely separated areas and were jammed against the North American continent at different times.

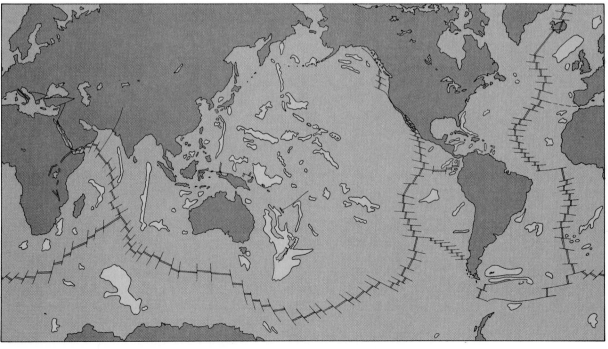

FIGURE 22.16
Oceanic plateaus are concentrated in the western Pacific and Indian oceans. They are composed of fragments of continental crust, chains of seamounts, and segments of island arcs.

concentrate quartz, feldspars, and other minerals of the "granitic" crust. By these processes, continental crust developed a special chemical composition (high in quartz and feldspars), entirely different from that of the "primitive" crust and the oceanic crust.

Major pulses of continental growth appear to have occurred in three phases, roughly 1, 2, and 3 billion years ago.

In the beginning, the surface of Earth was much like that of the Moon, Mercury, and other primitive planetary bodies. Its entire surface was pock-marked with impact craters. The temperature at the surface may have exceeded the boiling point of water. There were no continents or ocean basins, no land or sea, and no life. Some 4 billion years later, at the end of the Precambrian, the planet Earth looked much like it does today (see Figure 1.2). Blue seas filled with diverse plants and complex animals covered much of the planet. There were continents, ocean basins, and mountain chains. The land was still devoid of life, but river systems carved a landscape that has proven to be unique in the solar system.

No one knows precisely how the landscape changed so drastically, but we can gain some insight by studying other planetary bodies (see Chapter 24). We can also use principles of physics and chemistry to construct theoretical models that point the way to a better understanding. Continents evolve because of planetary differentiation, and some form of continental crust appears to have formed early in Earth's history. The oldest rocks in Earth's crust, collected from the shields of Greenland, South Africa, and Australia, are continental rocks roughly 3.8 billion years old; they formed during the rapid decline of impact cratering that followed the accretion of Earth. They are not part of the primitive crust, but possess an advanced continental composition. Because many are metamorphosed sedimentary rocks, they show evidence of active weathering and erosion and deposition in water. They are thus a record of some type of ocean and atmosphere. This fact illustrates that the oceans had already condensed by this time from the gases released by volcanoes. Moreover, these sediments demonstrate that some high-standing continental platforms evolved very early in Earth's history. The mineralogy and chemistry of some of these Archean metamorphic rocks show that they recrystallized at about 550 to 800 °C at 5 to 8 kilobars. These great pressures show that these rocks were at depths of up to 27 km and were probably underlain by an additional 35 km of crust. It thus appears that some type of continental crust 50 to 70 km thick existed on Earth by 3.8 billion years ago. These continental blocks were probably small, the configuration of the ocean basins complex, and convection in the mantle more vigorous.

No matter how one looks at the origin and evolution of our planet, Earth has been cooling since it first became differentiated. It was thus hotter during Archean time than it is now (Figure 22.17). Higher temperatures in the early stages of Earth's history resulted from the following:

1. Kinetic energy released by the impact of meteorites and asteroids
2. Release of gravitational energy during the separation of the core
3. Radioactive decay of isotopes

The greater amount of heat within Earth would undoubtedly affect the style and rate of convection in the mantle and the style of Earth's tectonic system. In all probability, the hot mantle was less viscous, flowed more rapidly, and convected more vigorously than it does today. In the primitive Earth there were probably numerous hotspots rooted in the mantle and distributed more randomly than the linear pattern of oceanic ridges and rift systems we see today. Because the production and cooling of basaltic oceanic crust is the principal means by which heat is lost from today's Earth, the primitive, hot Earth may have been overwhelmed by volcanism. It is possible that eruptions occurred almost entirely beneath a global ocean condensed from the vast clouds of steam released during accretion and, later, volcanism.

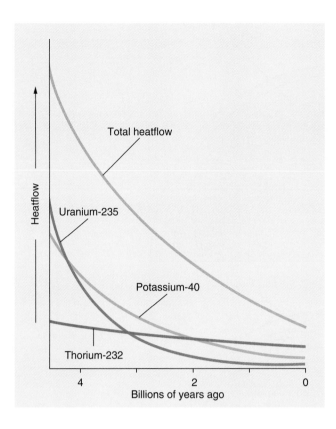

FIGURE 22.17
The amount of heat generated by the decay of radioactive isotopes was much larger in the early stages of Earth's history than now. Many of the short-lived isotopes have completely vanished. Long-lived isotopes have passed through several half-lives so that in many cases less than 1/64th the original amount remains. With a decrease in the amount of radioactive elements, there is a decrease in the amount of heat produced.

It is therefore likely that the first continental crust did not form from "classical" plate tectonics but developed from huge volcanic islands built above the rising and descending convection currents, such as is illustrated in Figure 22.18. Partial melting above a rising convection current in the mantle may have formed magma of basaltic composition, which accumulated upon the floor of the ancient ocean and eventually rose above sea level. These plates were probably short-lived, and most of them were recycled back into the mantle.

In the zone where two convecting currents met and moved downward or where loading occurred, this primitive crust was probably too hot and too light to be subducted and was instead compressed and became thickened. It is likely that the base of the downward bulge partially melted, producing a silicic magma that, because of its low density, was injected into the rocks above. No new material from the mantle needed to be added, so that differentiation by partial melting of the primitive crust proceeded toward silicic (continental) compositions. Part of the lower basaltic crust may have been swept back into the mantle. These embryonic continents consisted of metamorphosed basaltic lavas (greenstones) and are marked by large granitic intrusions. They are preserved in the oldest shields as deformed granite–greenstone belts. In this manner, primitive crust became differentiated into more silicic continental crust and was ultimately stopped from possible recycling back into the mantle. In addition, weathering and erosion produced wedges of sediment on the submerged flanks of these continental nuclei.

This hypothetical model explains some of the relationships of early Precambrian rocks of Australia and South Africa (page 534), but the precise way in which continents originated is still hotly debated.

It seems likely that, as soon as the small continents with a significant proportion of low-density silicic rocks were generated, the oceanic lithosphere would be subducted beneath them and the system would evolve toward modern plate tectonics (Figure 22.19), with long, continuous zones of subduction and the eruption of andesites found in the Proterozoic and younger rocks. It

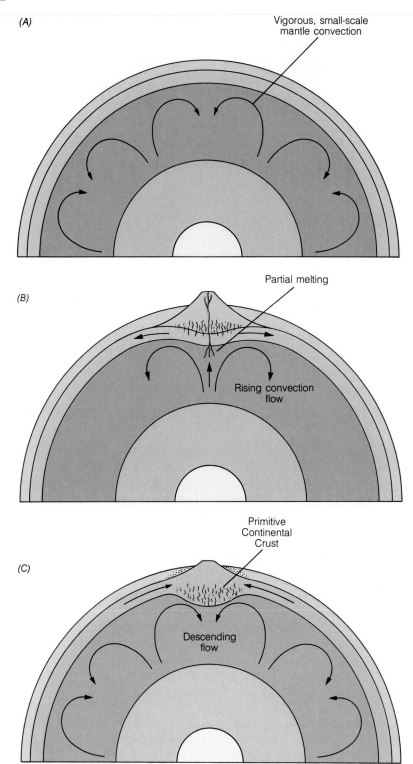

FIGURE 22.18
The origin of the first continents is a matter of conjecture, but we can speculate that embryonic continents were formed by vigorous small-scale mantle convection (*diagram A*). Above a rising current (*diagram B*) partial melting of the mantle would produce volcanic material that was relatively high in silica and consequently would be less dense than the mantle. It would therefore resist assimilation back into the mantle. Above descending currents (*diagram C*) the crust would be compressed, thickened, and metamorphosed, thus changing toward a true continental crust. Small landmasses emerging above the sea would be weathered and eroded. This would produce sediment that would be deposited adjacent to the island and that would be high in silica and other light minerals. In this way small islands of metamorphosed andesitic volcanics, and sediment derived from them, would form small islands of embryonic continental crust.

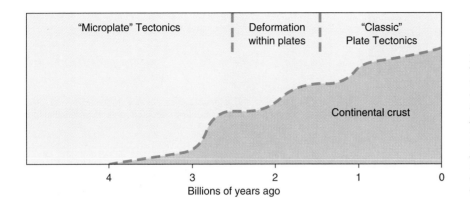

FIGURE 22.19
The increase in the volume of continental crust during Precambrian time probably occurred during three major growth phases as shown in this diagram. Isotopic data and geochemical consideration suggest that the major periods of growth were between 3 and 2.5, just after 2, and about 1 billion years ago. As heat within Earth diminished and continents grew larger, the style of tectonics also changed from "microplate" tectonics to the classic style we see today.

appears that this critical threshold was crossed sometime in the Proterozoic era, more than 1 billion years ago.

The oceans play a vital role in the origin of the continents and the eventual development of plate tectonics. Water lowers the melting temperature of lithospheric rocks compared to dry rock. In addition, wet partial melting is critical for the development of silicic magma compositions from rocks richer in iron and magnesium. Once subduction of wet oceanic crust occurred, partial melting in the subduction zone produced magmas that were richer in silica and that resulted in andesitic volcanoes or formed granitic intrusions. These low-density rocks further stabilized the continental nuclei. Sediment derived from erosion of the land further separated the low-density material and concentrated it in sedimentary rocks, which are less dense than the basaltic oceanic crust. Their buoyancy prevented any significant consumption of continental materials in subduction zones.

The newly added volcanic material and sediment were ultimately deformed into orogenic belts by subsequent collision with other plates. Metamorphism occurred in the mountain roots, and granitic intrusions resulted from partial melting in the lower crust and along subduction zones (Figure 22.20A). Eventually, erosion and isostatic adjustment produced new segments of the stable shields, in which igneous and metamorphic rocks formed deep in the mountain roots were exposed at the surface (Figure 22.20B). Sediment derived from the erosion of these orogenic belts was deposited along the margins of the small continents. Repetition of this process deformed the geoclinal sediments along the continental margins into new orogenic belts. New igneous material was added to the continents from the subduction zone (Figure 22.20C). Continents continued to grow by accretion and reworking of material along their margins during each subsequent orogenic event (Figure 22.20D).

Thus, during the period of plate tectonics, the major processes of continental evolution have been (1) igneous activity and extraction of material from the mantle, with the emplacement of new material in an orogenic belt, (2) mountain building along convergent plate margins, involving deformation and metamorphism of geoclinal sediments, and (3) erosion and isostatic adjustment of the mountain belt to produce a new section of a shield, which is eroded to a flat surface near sea level. Broad upwarps or slight vertical movements of the craton or ocean, caused by changes in spreading rate, permit the ocean to expand and contract over the continent, thus depositing a thin but vastly important sediment cover, forming the stable platform.

At any stage, rifts can form within stable continental blocks as new spreading centers develop, presumably as a result of shifting patterns of plate movements and attendant convection in the mantle. The continent then splits, and its fragments drift apart with the spreading lithospheric plates (Figure 22.21). Each fragment may than act as a separate center for future continental growth.

What processes are involved in the evolution of continents?

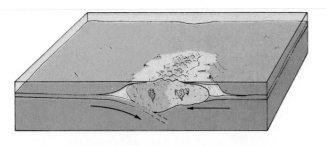

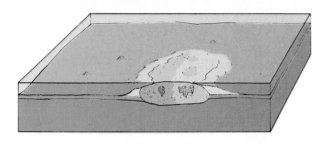

(A) The sediment and andesitic volcanics are deformed by compression at converging plates. Mountain belts and metamorphic rocks form an embryonic continent.

(B) The erosion of mountains concentrates the light minerals (quartz, clay, and calcite) as geoclinal sediments along the margins of the small continent.

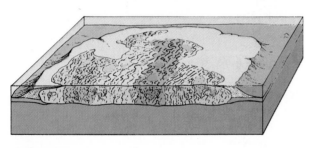

(C) The geoclinal and orogenic cycle is repeated. New material is added to the continental mass in the form of granitic batholiths and andesitic flows.

(D) The continent continues to grow by accretion.

FIGURE 22.20
Continents evolve by accretion of low-density rock material during a series of orogenic events. They begin as island arcs and grow by the addition of new material, which is metamorphosed into a segment of continental crust with each orogenic event.

We need to emphasize one additional point. Radiometric dates of Precambrian rocks suggest that continental growth occurred in several major episodes separated by longer periods of quiescence and stability. Approximately 3.5 billion years ago there was much less continental crust than at present; 90 to 95% of the crust has been formed since that time. Most continental crust was formed during Precambrian time during three major periods of growth (Figure 22.19). The first major period occurred between 2.8 and 2.5 billion years ago. At the end of this time, about 75% of the present volume of continental crust existed. Subsequent phases of growth, from 1.9 to 1.7 billion years ago and from 1.2 to 0.9 billion years ago, added most of the rest. Phanerozoic additions at plate margins produced only a small proportion of continental crust.

In the next chapter we will consider the environmental effects of processes operating in and on Earth and the production of natural resources by these processes.

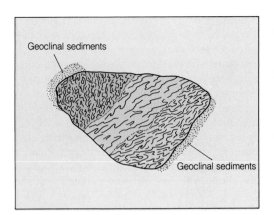

(A) The original continental mass is composed of orogenic belts welded together into a stable craton.

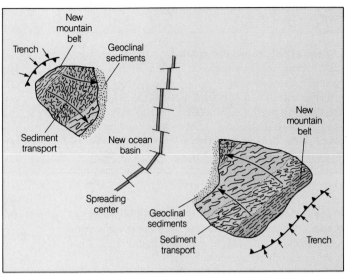

(B) A spreading center splits the continent, and the fragments drift apart. A subduction zone ultimately develops along the leading edge of the plate, and each fragment can grow as an independent continental mass.

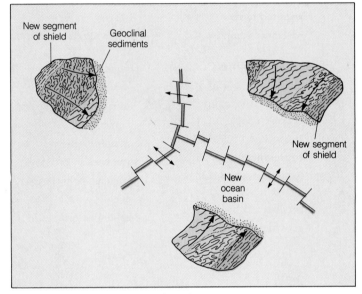

(C) Further fragmentation can occur with each fragment acting as a separate continental nucleus.

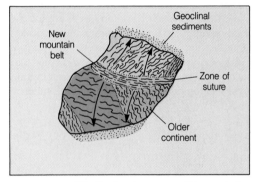

(D) Continents can collide and become welded together.

FIGURE 22.21
Fragmentation of continents can occur at any time during the evolution of a continental mass.
Each fragment then acts as a nucleus for further continental accretion.

SUMMARY

1. Geologic mapping shows that continents consist of three major components: (1) shields, (2) stable platforms, and (3) young mountain belts.

2. The characteristics of a mountain belt and the sequence of events that produced it depend largely on the type of crust carried by converging plates and the type of sediments involved in the deformation. Three distinctive types of convergence are recognized: (1) convergence of two oceanic plates, (2) convergence of a continental plate and an oceanic plate, and (3) convergence of two continental plates.

3. Convergence of two oceanic plates produces island arcs with andesitic volcanism, metamorphism of ophiolite rock sequences, and emplacement of granitic batholiths in the deeper parts of the orogenic belt.

4. Convergence of a continental plate and an oceanic plate involves deformation of a thick geoclinal sequence of sediment, andesitic volcanism, and emplacement of granitic batholiths in the deformed mountain belt.

5. Convergence of two continental plates produces a continental suture or a double layer of continental crust with high, mountainous topography and limited igneous activity.

6. After orogenesis, erosion and associated isostatic adjustment reduce the mountainous topography to a flat surface near sea level. These processes expose the metamorphic rock of the deep mountain roots as a new segment of the shield.

7. In orogenic processes, silica-rich, low-density material is concentrated on the outer part of the crust through the partial melting of the oceanic plate in the subduction zone.

8. Continents thus grow by accretion, as new silica-rich material is added to an orogenic belt through igneous activity at convergent plate margins. This low-density crust resists any further subduction and remains on the outer lithosphere. Subsequent rifting can split and separate continental blocks, but each block then acts as a separate center for future continental growth.

9. Continental growth is simply part of the process of planetary differentiation, in which lighter materials are separated and concentrated in the outer layers of Earth. The process of differentiation has produced the ocean basins, the continental platforms, and the surface fluids (air and water). Differentiation thus is directly or indirectly responsible for the origin of all surface features and the internal structure of the planet.

KEY TERMS

accreted terrane (p. 548)

basement complex (p. 527)

blueschist (p. 543)

continental accretion (p. 553)

craton (p. 530)

eugeocline (p. 536)

eugeosyncline (p. 536)

folded mountain belt (p. 530)

geocline (p. 536)

geosyncline (p. 536)

isostasy (p. 540)

magmatic differentiation (p. 539)

migmatite (p. 539)

miogeocline (p. 536)

miogeosyncline (p. 536)

nappe (p. 538)

ophiolite (p. 537)

orogenesis (p. 535)

orogenic belt (p. 536)

shield (p. 527)

stable platform (p. 530)

REVIEW QUESTIONS

1. What are the composition and structure of the continental crust?
2. Describe the Canadian shield. How does it differ from the Ethiopian shield (of Africa) or the Australian shield?
3. How does a stable platform differ from a shield?
4. How are young mountain belts related to island arcs?
5. Study the tectonic map of the world (Figure 22.7) and label (a) shields, (b) stable platforms, (c) Mesozoic and Cenozoic mountain belts, (d) oceanic ridges, and (e) trenches.
6. What is orogenesis? What factors are most important in this process?
7. Describe and diagram the two types of rock sequences that accumulate along continental margins.
8. Describe the styles of structural deformation that occur during orogenesis.
9. Explain the processes of metamorphism and igneous activity that accompany orogenesis.
10. What role does isostatic adjustment play in the evolution of a mountain belt and the development of a new segment of the shield?
11. Sketch a series of diagrams to show the type of orogenesis that involves the convergence of two oceanic plates.
12. Compare and contrast orogenesis involving two oceanic plates and orogenesis involving a continental and an oceanic plate.
13. How does the rock sequence in mountains formed by continent–island arc convergence differ from the sequence in continent–continent convergence?
14. Describe orogenesis involving two continental plates. How can this process be recognized in the rock record?
15. Outline the steps in the evolution of a continent.

ADDITONAL READINGS

Condie, K. C. 1989. *Plate Tectonics and Crustal Evolution,* 3d ed. Elmsford, NY: Pergamon Press.

Dietz, R. 1972. Geosynclines, mountains, and continent building. *Scientific American* 226(3): 30–38.

Engel, A. E. J., and C. G. Engel. 1964. Continental accretion and the evolution of North America. In *Advancing Frontiers in Geology and Geophysics: A Volume in Honour of M. S. Krishman,* eds. A. P. Subramanian and S. Balakrishna, pp. 17–37e. Hyderabad: Indian Geophysical Union. Reprinted in *Adventures in Earth History,* ed. P. E. Cloud. 1970. pp. 293–312. San Francisco: Freeman.

Hallam, A. 1973. *A Revolution in Earth Sciences.* Oxford: Clarendon Press.

McElhinny, M. W., and D. A. Valencio. 1981. *Paleoreconstruction of the Continents.* Geodynamic Series, Vol. 2. Boulder, Colorado: Geological Society of America.

Motz, L. 1979. *The Rediscovery of Earth.* New York: Van Nostrand Reinhold.

Sullivan, W. 1974. *Continents in Motion: The New Earth Debate.* New York: McGraw-Hill.

Van Andel, T. H. 1985. *New Views on an Old Planet.* New York: Cambridge Univ. Press.

Wyllie, P. J. 1976. *The Way Earth Works.* New York: Wiley.

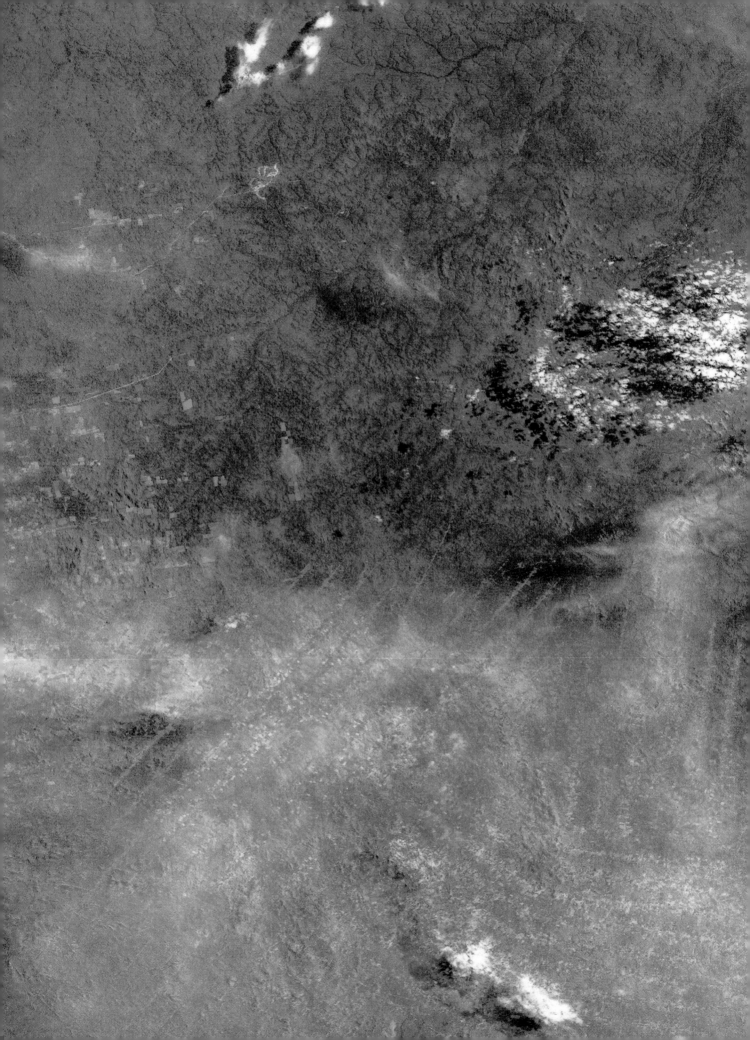

23
Environment

In the preceding chapters, we have studied Earth as a dynamic system that produces a constantly changing environment. Some rapid changes, such as earthquakes, volcanoes, and floods, may have catastrophic effects upon society. Most changes, however, are so slow by human standards that they are barely noticeable. Perhaps there is comfort in realizing that most geologic processes are of little concern to us as individuals. Still, history records the rise and fall of communities, cultures, and civilizations in which environment and resources played a major role. An example is the decline of Mesopotamia 1500 years ago. As the people began to irrigate their arid lands, salt became concentrated in the soil and drastically reduced agricultural production. This led to the decline of their civilization. Today many third-world countries have their own energy crisis: a crisis of firewood. Overpopulation and the resulting massive consumption of forests for fuel are major factors in the famine in Ethiopia. Deforestation destroys the root system, which holds the soil together; erosion accelerates, and within a decade or two the topsoil is gone and agriculture rapidly declines.

What is the fate of our civilization? Will the Mississippi River ultimately shift its course on its deltaic plain and abandon New Orleans? Will the major reservoirs behind our marvelous dams soon become full of mud? Most of our resources are finite and nonrenewable. During the next 10 years, we will use more oil, gas, iron, and other mineral resources than were consumed throughout all of history.

The Landsat image on the facing page is a scene from Rondônia, a state in western Brazil. In 1973 the rainforest was nearly pristine. By 1987 farm plots (shown in light blue) gashed much of the tropical forest. This abrupt change in the landscape resulted from government encouragement of thousands of poor Brazilians to settle the interior. In the 1980s a fifth of the rainforest was stripped away, but the tropical soil was unproductive, so most farmers were forced to leave after a few seasons. When we alter our environment, we upset the balance of the natural systems that were slowly established over thousands of years. Adjustments to these changes we have imposed cannot always be anticipated.

1. Our environment consists of natural systems that have operated in a delicate balance for a long period of time.
2. We can manipulate many natural systems, such as rivers and shorelines.
3. Natural systems adjust to artificial changes in ways that cannot always be anticipated.
4. Waste disposal always involves natural systems. Waste products cannot be "thrown away." They remain in some form in Earth's system.
5. Mineral resources are concentrated by geologic processes operating in the hydrologic and tectonic systems. These resources are finite and nonrenewable.
6. There are limits to growth.

ENVIRONMENTAL GEOLOGY

Our environment consists of natural systems that have operated in a delicate balance for a long period of time. Although we can manipulate many natural systems, there are commonly many unforeseen consequences. Natural systems adjust to artificial changes in ways that cannot always be anticipated.

We are a part of nature, one of the many species of animals adapted to the present natural environment. We are so well adapted that we dominate all other species, and our numbers are so great that we have become an effective agent of physical, chemical, and biological change. At present, we are capable of significantly modifying the natural systems in which we live. Many of our modifications, unfortunately, conflict with the normal evolution of Earth's environment. For example, since the development of an agrarian culture, we have drastically changed large areas of Earth's surface from natural wilderness to controlled, cultivated land. This has greatly strained the food supply of many animal populations, which previously had established a degree of balance. The advancements of science and technology have telescoped thousands, if not millions, of years of normal evolutionary changes into days. Many species simply cannot adapt over such a short period of time.

We have also (1) diverted and manipulated drainage systems, (2) altered the quality and flow of water, both on the surface and in the subsurface, (3) changed the atmosphere, and (4) modified coastal and oceanic waters. The changes we institute, almost without exception, are intended to produce improvements and advantages for society. Frequently, the results are quite the opposite. At best, some are detrimental on a short-term basis. Others are catastrophic and irreversible in the long run. We are thus forced to adapt to a rapidly changing environment, one that we have in part created. Certainly some changes are necessary, and at times we must tolerate undesirable side effects (anticipated or unanticipated) as the "necessary price" of change.

Clearly, a decision to alter our environment should be made carefully. First, we should have a sufficient geological and biological understanding of how natural systems operate to gauge how they will be affected by artificial modifications. If we decide to dam a river, we should understand the side effects of altering part of a natural river system. We should ask how the damming will affect erosion and sedimentation downstream. How will the groundwater system respond? How will lakes be altered? How will the beaches along shores far removed from the dam be affected? How will marine and terrestrial life be changed? If we develop a great volume of waste, we should know how it will be assimilated into Earth's systems. How will it affect the quality of surface and ground water, the atmosphere, and the oceans? If we build cities, we should fully understand how this development will affect the terrain, the atmosphere, and the hydrosphere. In too many cases, when we spend time, energy, and money to correct or alter a natural condition to suit our needs, we upset a natural balance. We then have to spend more money and energy to reestablish the balance.

The human race exists by geologic consent. The basis of our entire environment is the geologic system. It is therefore incumbent on us to understand the system and to live within it.

Modification of Natural Systems

Modification of River Systems. Water is perhaps our most important natural resource. As the population has grown, we have increasingly modified river systems to better use this finite resource. Most of our attempts to modify river systems have involved the construction of dams and canals, but urbanization also affects river systems. The important point to be considered in all of these manipulations is that a river is a system that has approached some degree of equilibrium among a large number of variables over a long period of time. If this equilibrium is upset, a number of rapid adjustments occur. Recall the example of the Aswan High Dam in Egypt (see pages 243–44) and how the Nile adjusted to it. Whenever a dam is built, several major geologic adjustments occur:

1. Sediment is trapped in the reservoir, and the clear water discharged downstream is capable of accelerated erosion.
2. Deltas are deprived of their source of sediment, and coastal erosion results.
3. Groundwater systems are modified by artificial lakes created behind the dam.

A more universal problem is the subtle modification of a river system by urbanization (see Figure 12.11). Where cities are constructed, the surface run-off is modified by streets, sidewalks, parking lots, and roofs of buildings, which make large areas of the surface impermeable. This change produces two major effects. First, the volume of surface runoff is greatly increased and is channeled into gutters and storm drains. As a result, flooding increases in intensity and frequency. Second, groundwater systems are altered by reduced infiltration. The water table is lowered; the aquifers are not replenished.

Modification of Slope Systems. Slopes are also dynamic systems that have some degree of balance (see Figure 11.13). Construction on hills and slopes modifies the system, causing an increase in the magnitude and frequency of mass movement. A landslide in northern Italy that resulted from slope modification associated with the Vaiont Reservoir was cited earlier (page 217). The problem exists wherever uncontrolled hillside construction occurs, especially where the slope is underlain by unconsolidated rock material. Deforestation is also a problem in slope erosion (see Figure 11.2) and in the modification of the atmosphere (Figure 23.1).

Modification of Groundwater Systems. The use of groundwater resources is constantly increasing. As we drill wells and pump groundwater, we create a new and unnatural groundwater discharge. Subsidence is one of the more important results. Mexico City is an excellent example: Many of its buildings have subsided more than 3 m (see Figure 14.26). The encroachment of saltwater into the lens-shaped body of fresh water on an island or a peninsula also commonly results from excessive pumping (see Figure 14.23).

Alteration of the water table can result from the modification of surface infiltration. An example is the construction of a canal system in southern

How has human activity modified the major geologic systems? What effect has this had on the natural systems?

FIGURE 23.1
Deforestation in Brazil is apparent in the clearing and development patterns of this space photograph.

Florida, which has caused a variety of problems in the Everglades (see Figure 14.24). In contrast, the irrigation of arid lands causes the water table to rise. This condition can produce unstable slopes and accelerated mass movement, such as occurred in the Pasco basin in Washington (see page 322).

Modification of Shorelines. Perhaps even more dramatic are the results of modifying shorelines. In Santa Barbara, California, the process of longshore drift was manipulated by constructing a breakwater to form a harbor (see Figure 16.8). If longshore drift is disrupted, the supply of sand is eliminated or reduced, and beach erosion results. Oil spills are also becoming a major problem in modifying shoreline ecology (Figure 23.2).

These and other examples illustrate a great central theme: Geologic systems are complex, dynamic systems in which balance has been established among a number of variables. Manipulating these systems destroys the balance, often resulting in many serious side effects—some detrimental over the short term, others catastrophic and irreversible in the long run.

FIGURE 23.2
Environmental impact of an oil spill in Alaska.

Geologic Problems with Waste Disposal

Our industrialized society produces an ever-increasing variety and quantity of toxic waste. Traditionally, people have used fresh water to remove and dilute solid and liquid wastes and have used the atmosphere to dilute the gaseous waste products of combustion. Until recently, however, people generally have been unaware that a local natural system of waste disposal can become saturated, so that an unhealthy environment is created. We cannot take our waste "away," as has been suggested by some politicians. The waste products are in Earth's natural systems and will remain in them. The problem is made particularly acute by such business practices as planned obsolescence, the use of throwaway containers, and the hard sell of new models of old products. In addition, high labor costs often make it uneconomical to repair, reclaim, and recycle used items, so the volume of waste grows unnecessarily, at a staggering rate. The replacement of products, of course, greatly reduces natural resources. Unfortunately, waste is not just a by-product. Eventually, it is a product itself (Figure 23.3).

Waste disposal has many geologic ramifications. If waste is buried, the quality of groundwater is threatened. If it is dumped into streams and rivers, it accumulates on beaches and in estuaries, altering the environment of the oceans. Previous methods of elimination have not been "waste disposal"; they have been "waste dispersal." Any significant solution to the problem of elimination must consider what kinds of waste disposal and dispersal a given geologic environment can accommodate without critical alterations in geologic and biological conditions.

Solid Wastes. *Solid wastes* are disposed of in many ways, including landfill, incineration, composting, open dumping, animal feeding, fertilizing, and disposal in oceans. The geologic consequences include changes in the surface of the land where the waste is deposited and changes in the environment (rivers, lakes, oceans, and groundwater) where the mass of waste is concentrated. The major problems with solid waste disposal involve the hydrologic characteristics of the site. These include the porosity and permeability

FIGURE 23.3
Solid waste produced by metropolitan regions commonly increases the volume of sediment eroded by river systems. This great volume of material impacts on most natural systems operating at or near the surface.

FIGURE 23.4
Air pollution has reached a critical level in many metropolitan and industrial regions of the world.

of the rock in which the fill is located and whether or not the waste deposit intersects the water table. The altered topography associated with dumps and landfills is also critical because it can change the drainage and groundwater conditions. Perhaps the most critical contamination problem is created as water passes through a landfill, dissolves organic and inorganic compounds, and incorporates them into the groundwater reservoirs.

Liquid Wastes.   Traditionally, *liquid wastes* have been discharged into surface drainage systems and diluted. They accumulate ultimately in lakes and oceans, where they are stored. As the volume of liquid waste increases, the capacity of the natural water system to dilute it is overwhelmed, and the drainage system becomes a system of moving waste.

One very subtle type of liquid pollutant is the hot water created by cooling systems in power plants and factories. Although the water itself is not contaminated, the temperature alone is enough to alter the biological conditions in the streams and lakes into which it flows. Such pollution is called thermal pollution.

Gaseous Wastes (Air Pollution).   The population explosion, with the consequent industrial expansion, has produced a variety of *gaseous wastes* and pollutants in the form of minute liquid and solid particles that are suspended in the atmosphere. In the past, pollutants were expelled into the air with the reasonable assurance that normal atmospheric processes would disperse and dilute them to a harmless, unnoticeable level. In many heavily industrialized areas, however, the atmosphere's capacity for absorption and dispersal has been exceeded, and the composition of the air has been radically altered (Figure 23.4). The problem is so severe in some areas that rain is made more acid than normal by pollutants, particularly oxides of sulfur and nitrogen, and is called acid rain. If the troposphere (the lower part of the atmosphere, which is involved in most human activities) extended indefinitely into space, air pollution would not pose a problem. The troposphere, however, extends only to an altitude of 10 or 15 km, and few pollutants move out of it into the overlying stratosphere for any great length of time. A steadily increasingly volume of pollutants is thus concentrated mostly in the lower part of the troposphere.

A dramatic example of air pollution is the oilfield fires in Kuwait that resulted from the 1991 Persian Gulf War. The Landsat images in Figure 23.5 graphically illustrate the sequence of events. On January 5, before fighting began, the wells were intact. On February 15, nearly a month into the air war and a week before the ground war, a number of wells were afire. By March 3, at the conclusion of the ground campaign, more than 600 wells had been ignited, sending clouds of thick, black smoke into the atmosphere. A ground

(A) January 6, 1991, Kuwait before fighting began

(B) February 15, 1991, Kuwait one month after fighting began

(C) March 3, 1991, Kuwait at the conclusion of the ground campaign

FIGURE 23.5
Landsat images of oil well fires in Kuwait show the magnitude of this man-made environmental catastrophe.

FIGURE 23.6
A ground view of the Kuwait oil well fires.

view of this catastrophe is shown in Figure 23.6. The fires were expected to last for at least another year; even the most optimistic said that this could prove to be the worst man-made atmospheric pollution event in history. The fires reportedly consumed several million barrels of oil per day and spewed 500,000 tons of particulates into the air each week. In the spring of 1991, air-sampling instruments at the Mauna Loa observatory in Hawaii recorded numerous "spikes" of soot, five times higher than normal.

The regional and worldwide effects of the Kuwaiti oil fires are not yet clear. Preliminary computer modeling predicted that acid rain could affect areas as far as 2000 km from Kuwait. By mid-1991, unprecedented acid rain in southern Russia was announced by Soviet scientists. In addition, smoke and darkened snow in Pakistan and northern India were visible in satellite images. Because soot-covered snow absorbs more heat, rapid melting could result, causing floods or damage to crops. Some scientists even suggested a possible connection between the smoke and the unusual intensity of the typhoon that struck Bangladesh on May 1, 1991, killing more than 100,000 people.

Radioactive Wastes. All industries face waste-disposal problems, but none are greater than those of the nuclear energy industry. The generation of nuclear energy creates numerous radioactive isotopes—some with short half-lives, others with very long ones. Nuclear waste is extremely hazardous itself, but another nuclear waste product is a large amount of heat. Any disposal system must therefore be capable of removing the waste while completely isolating it from the biological environment. In addition, containment must be maintained for exceptionally long periods. Compared to the waste produced by many other industries, the volume of *radioactive waste* is not large, but the hazards and the heat that are generated are considerable.

One of the more promising methods of radioactive waste disposal involves storage in thick salt formations. Salt deposits are desirable because they are essentially impermeable and are isolated from circulating groundwater. In addition, salt yields to plastic flow, so it is unlikely to fracture and make contact with leaching solutions over extended periods of time. Salt also has a high thermal conductivity and thus can absorb heat from the waste, and it has approximately the same shielding properties as concrete. In theory, radioactive wastes would be solidified and sealed in containers from 15 to 60 cm in diameter and as much as 3 m in length. The containers would then be shipped to salt mines in the stable interior of the continent, where seismic activity is minimal. There, they would be placed in holes drilled in a salt formation deep in a mine. When filled with waste, the hole would be packed with crushed salt and closed.

Mining Wastes. The waste products from mining operations include (1) tailings and dumps, (2) altered terrain (due to open-pit mining and *strip mining*), (3) changes in the composition of the surface, and (4) solid, liquid, and gaseous wastes produced by refining.

In the United States, approximately 3 billion metric tons of rock are mined each year. About 85% comes from open-pit and strip mines, which require the removal of an additional 6 billion metric tons of rock as overburden. Such surface mining operations have affected about 12,000 km^2 of land in the United States. The principal geologic problem arises from the alteration of the terrain by the creation of open pits and artificial mounds and hills of tailings.

An additional problem arises if mine tailings enter the drainage system. They can choke a stream channel, increasing the flood hazards. Alteration of a stream system also can be produced from placer mining, in which the movement of large quantities of sediment upsets the balance of the stream.

RESOURCES

Mineral resources have been concentrated by a variety of geologic processes related to the plate tectonics and hydrologic systems. Most mineral resources are therefore finite and nonrenewable.

The important minerals on which modern civilization depends constitute an infinitesimally small part of Earth's crust, whereas the rock-forming minerals (such as feldspar, quartz, calcite, and clay) are abundant and widely distributed. Copper, tin, gold, and other metallic minerals occur in quantities measured in parts per million (and, in most cases, a very few parts per million). The important question, then, is how are these very small quantities of important minerals concentrated into deposits large enough to be used? The basic answer is simple: They are deposited by the various geologic processes operating in Earth's system.

It may surprise you to learn that essentially every geologic process—including igneous activity, metamorphism, sedimentation, weathering, and deformation of the crust—plays a part in the genesis of some valuable mineral deposits. The occurrence or absence of most mineral deposits, therefore, is controlled by the specific geologic conditions of a region.

It is critical to understand that the processes that form these minerals operate so slowly (by human standards) that the rates of replenishment are infinitesimally small in comparison to rates of consumption. We must clearly understand that mineral deposits are finite and therefore are exhaustible and nonrenewable. If the approximate extent of a deposit and its rate of consumption are known, we can predict how long it will last. Our resources are like a checking account that will never receive another deposit. The faster we withdraw or the larger the check we write, the sooner the account will be depleted. Moreover, today relatively few areas of potential mineral deposits are still

How are natural resources formed by geologic processes?

unexplored. Most of the continents have been mapped and studied extensively, so the inventory of natural resources is nearly complete. We basically know the extent of our mineral resources and the rates of consumption. Projecting how long they will last is not difficult.

Most natural resources are finite and nonrenewable. To best appreciate this fact, we will consider some of the principles that govern the concentration of rare minerals in ore deposits and the origin of some nonmetallic resources. These principles are complex and diversified, but the origin of most mineral deposits is somehow related to the various tectonic and hydrologic processes. We can therefore recognize major groups of mineral deposits formed by (1) igneous processes, (2) metamorphic processes, (3) sedimentary processes, and (4) weathering processes.

Igneous Processes

Many metallic ores are concentrated by magmatic processes, much as silicate minerals are. That is, concentration results from differences in crystal structure, order of crystallization, density, and other factors. Magmatic intrusions are themselves concentrations, on a regional scale, of silicate minerals and some metals. They characteristically are generated at plate boundaries. On a local scale, metallic minerals are concentrated in specific areas of intrusion in a variety of ways. One process is direct *magmatic segregation,* in which heavy mineral grains that crystallize early sink down through the fluid magma and accumulate in layers near the base of the igneous body. Deposits of chromite, nickel, and magnetite are good examples of this type of concentration (Figure 23.7). Late crystallization is another process by which rare minerals are concentrated. Many elements that occur in amounts of only a few parts per million in the original magma do not fit readily into the crystalline structure of the silicate rock-forming minerals. These are concentrated in the residual liquid, and thus, as a magma cools, rare elements, such as gold, silver, copper, lead, and zinc, become concentrated in the last remaining fluid. These late-stage, metal-rich solutions can then be squeezed into the fractures of the surrounding rock. As they cool, the rare elements are precipitated as veins of metallic minerals, which differ little from ordinary dikes and sills except in their chemistry and mineralogy. The residual hot solutions of the magma contain much water, so the deposits are known as *hydrothermal deposits* (Figure 23.7).

Hydrothermal solutions are not necessarily injected into the surrounding rock to form veins of mineral deposits. They can also permeate the early-formed crystals in the granitic rock and disseminate metallic minerals throughout the body. This process produces large deposits of low-grade ore because practically the entire igneous body is mineralized. If the intrusion is exposed at the surface, these low-grade deposits can be mined profitably by special equipment and open-pit mining techniques. An important example of this type of concentration is the disseminated copper in porphyritic intrusions. These deposits, called *porphyry coppers,* currently account for more than 50% of the world's copper production.

FIGURE 23.7
The origin of metallic mineral deposits is commonly associated with igneous intrusions. The concentration of minerals by magmatic segregation occurs as heavy, early-formed crystals sink through the fluid magma and accumulate in layers near the base of the magma chamber. The chromite deposits of South Africa are an example. The concentration of minerals in hydrothermal vein deposits occurs during the late stage of crystallization. Many rare minerals, such as gold, silver, copper, lead, and zinc, do not fit into the crystalline structure of silicate minerals, so they are concentrated in residual fluids of the magma. This material can then be squeezed into fractures in the surrounding rock, and the rare metallic elements are precipitated as vein deposits. The concentration of ore deposits by contact metamorphism occurs as fluids from the cooling magma replace parts of the surrounding rock. Iron deposits commonly are formed by this process.

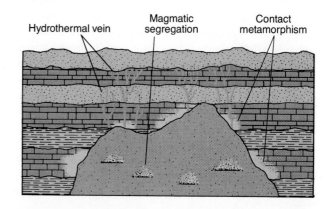

Porphyry copper and many other metallic minerals are associated with granitic intrusions, which in turn are associated with convergent plate boundaries. The metals are believed to derive from partial melting of the oceanic plate in the subduction zone, and their final concentration and emplacement result from the differentiation of magma as it cools. Mineral deposits, therefore, are products of the tectonic system. The plate tectonics theory promises to be an important key to a better understanding of the genesis of minerals and to future exploration for minerals.

Metamorphic Processes

Ore deposits are also formed by *contact metamorphism,* or metamorphism along the contact between an igneous intrusion and the surrounding rocks. In this process, variations in heat and pressure and the presence of chemically active fluids in the cooling magma alter the adjacent rock by adding or replacing certain components. Limestone that surrounds granitic intrusions is particularly susceptible to alteration and replacement by chemicals in the hot, acid magma solutions. For example, large volumes of calcium can be replaced by iron to form a valuable ore deposit (Figure 23.7). Regional metamorphism changes the texture and mineralogy of rocks and in the process forms deposits of nonmetallic minerals, such as asbestos, talc, and graphite.

Sedimentary Processes

A significant result of the erosion, transportation, and deposition of sediments is the segregation and concentration of material according to size and density. As was emphasized in Chapter 6, soluble minerals are transported in solution, silt and clay-size particles are transported in suspension, and sand and gravel are moved mostly as bed load by strong currents. As a result, sand and gravel are concentrated in river bars and beaches. These deposits, both modern and ancient, are a valuable resource of the construction industry. In the United States alone, over $1 billion worth of sand and gravel is mined each year, making this operation the largest mineral industry in the country not associated with fuel production.

Sedimentary processes also concentrate other valuable materials, such as gold, diamonds, and tin oxide. Originally formed in veins, volcanic pipes, and intrusions, these minerals are eroded and transported by streams. Because they are heavy, they are deposited and concentrated where current action is weak, such as on the insides of meander bends or on protected beaches and bars (Figure 23.8). Such layers and lenses of valuable minerals are known as *placer* deposits, and large operations mine them along both modern and ancient rivers and beaches.

Sedimentary processes concentrated the great bulk of iron ore mined today. Banded iron formations of Precambrian age are especially significant because of their abundance. These deposits consist of alternating layers of iron oxide and chert formed during a unique period of Earth's history, from 1.8 billion to 2.2 billion years ago, when oxidation conditions on Earth were much different and iron was removed from solution in seawater.

Another way in which sedimentary processes concentrate valuable minerals is the evaporation of saline waters in large lakes and restricted embayments of the ocean. As evaporation proceeds, dissolved minerals are concentrated and eventually precipitated. These *evaporite* deposits include minerals of important commercial value, such as potassium, sodium, and magnesium salts; gypsum; sulfates; borate; and nitrates.

Weathering Processes

The simple process of weathering also concentrates minerals. It removes soluble material from rocks and leaves the insoluble material as a residue. Weath-

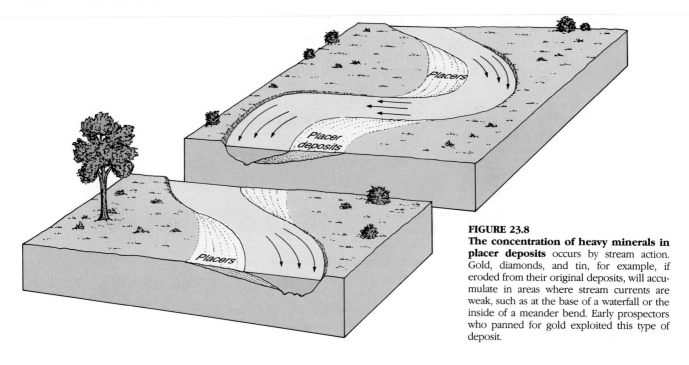

FIGURE 23.8
The concentration of heavy minerals in placer deposits occurs by stream action. Gold, diamonds, and tin, for example, if eroded from their original deposits, will accumulate in areas where stream currents are weak, such as at the base of a waterfall or the inside of a meander bend. Early prospectors who panned for gold exploited this type of deposit.

ering, therefore, can enrich ore deposits that were originally formed by other processes (Figure 23.9), or it can concentrate in the regolith material a mineral that was originally dispersed throughout a rock body. For example, extensive weathering of granite in tropical and semitropical zones commonly concentrates relatively insoluble metallic oxides in the thick regolith as it removes the more soluble material. Deposits of aluminum, nickel, iron, and cobalt are formed in this way.

The Nonrenewability of Mineral Deposits

The few examples cited here show that the world's valuable mineral deposits were formed in a systematic way by the major geologic processes during various periods in the geologic past. Their formation required long intervals of time and occurred under specific geologic conditions. Some deposits were formed in such restricted geologic settings that they approach uniqueness. For example, 40% of the world's reserves of molybdenum is in one igneous intrusion in Colorado, 77% of the tungsten reserves are in China, more than 50% of the tin reserves are in Southeast Asia, and 75% of the chromium reserves are in South Africa. If resources are depleted, we cannot "just go out and find some more." There simply are no more. Fortunately, most metals, unlike fossil fuels, can be recycled. We obviously must emphasize recycling to conserve metals.

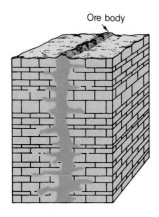

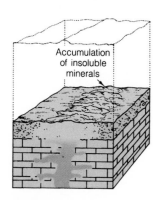

FIGURE 23.9
The concentration of ore deposits by weathering processes occurs as soluble rock, such as limestone, is removed in solution, leaving insoluble minerals concentrated as a residue.

ENERGY

> *Renewable energy resources include solar energy, hydropower, tidal energy, and geothermal energy. Together they cannot be expected to fulfill our energy needs. Nuclear energy may become a major energy source in the future.*

The technological progress and standard of living of modern society are intimately related to energy consumption. Until recently, energy resources and the capacity for growth seemed unlimited. Now, for the first time, people all over the world are experiencing a serious energy shortage, one that is sure to become more acute in the future. Understanding the sources of energy and how they can be used most effectively are among the more pressing problems of the twentieth century.

If we consider Earth's entire dynamic system, we see that our sources of energy are found in both renewable and nonrenewable forms. A renewable energy source is either one that is available in unlimited amounts, for all practical purposes, or one that will not be appreciably diminished in the foreseeable future. Solar energy, tidal energy, and geothermal energy are the most important examples. In contrast, nonrenewable energy sources, like mineral resources, are finite and exhaustible. They cannot be replaced once they are consumed. Coal and petroleum, the fossil fuels on which modern culture relies so much, are nonrenewable. These energy sources have been concentrated and preserved by geologic processes that operated over the vast periods of geologic time. Although the same processes may function today, they operate too slowly to replenish our supplies of these fuels.

An important aspect of today's energy picture is that over 90% of the energy we use is produced from nonrenewable fossil fuels. The exponential growth in consumption of the world's fossil fuels has brought on the present energy crisis. Few technical analysts, whether they be economists, geologists, or engineers, doubt that the problem exists. They see the trends and events that indicate, in the relatively near future, difficulties ranging from an awkward situation to disaster and economic peril. By contrast, few of the general public—if we are to accept the results of public opinion polls—believe that a problem exists. They see only gas pumps and energy bills, OPEC, and oil glut, not the statistics of the analysts.

The energy crisis, however, was accurately predicted. In 1956, M. King Hubbert, an eminent research geologist for Shell Oil Company, analyzed the reserves, production, and rate of consumption of petroleum. From these data, he predicted a continuous decline in production beginning in 1970 (Figure 23.10). Look at these curves carefully and note that they show a peak in production in about 1975. If our actual reserves exceeded this estimate by a third, the whopping increase would postpone the "day of reckoning" by only five years. Hubbert's projection caused some alarm in the petroleum industry more than 30 years ago, but it was mostly ignored by the government. It is estimated that petroleum production will decline to near exhaustion by the year 2070. Coal is expected to replace petroleum as the main hydrocarbon resource, with

Why is there an energy crisis?

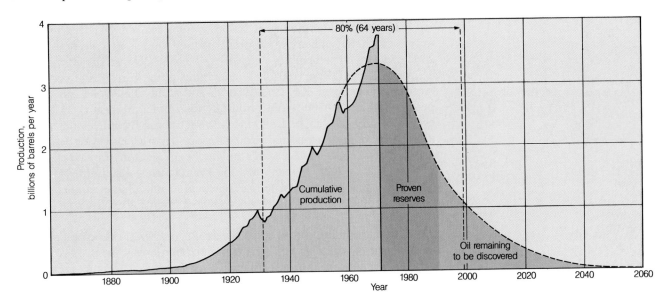

FIGURE 23.10
Projected rates of petroleum consumption were calculated in 1956 by M. King Hubbert. His analysis of reserves, production, and consumption rates predicted a decrease in production beginning in 1970, with supplies declining to near exhaustion by the year 2070. The major oil crisis that occurred in 1974 certainly substantiates his remarkable projections.

peak production by the year 2100 (plus or minus 100 years). It is quite clear, therefore, that in the very near future our reliance on fossil fuels must be greatly curtailed. Fossil fuels will have to be replaced by other energy sources.

Renewable Energy Sources

Solar Energy. Solar radiation is the most important *renewable energy source,* or sustained-yield source. It has the added benefits of being clean, constant, and reliable. The major problem, of course, is that solar energy is distributed over a broad area, so that in order to be used, it must first be concentrated in a small control center, where it can be converted to electricity and distributed. Panels for collecting solar radiation have been mounted on buildings to provide heating or cooling, but collecting systems for large-scale use of solar energy, although technically feasible, are not at present economical. For example, to satisfy the present need for electrical energy in the United States, a collecting system covering 25,000 km^2 (one-tenth the size of the state of Nevada) would be required. Large-scale use of solar energy, therefore, is a long way in the future, although local use in individual homes and buildings would help alleviate the need for other forms of energy.

Hydropower. Hydropower is another sustained source of energy. It has been developed in the United States to approximately 25% of its maximum capacity. With full development by the year 2000, hydropower would still provide only 10% or 15% of the energy needed in the United States. A problem with hydropower is that, when a river system is modified by a dam, unforeseen side effects can occur. Moreover, the reservoir behind the dam is a temporary feature, destined to become filled with sediment. The useful life expectancy for most large reservoirs is only 100 to 200 years, so this source of energy is, in reality, limited.

Tidal Energy. Another sustained energy source is the ocean tides. Tidal power can be harnessed by a dam built across the mouth of a bay. At the narrowed entrance of the bay, the rise and fall of the tides produce a strong tidal current flow, which can be used to turn turbines to generate electricity. Even with maximum development, however, tidal power could supply only 1% of the energy needed in the United States.

Geothermal Energy. Earth has its own internal source of heat, which is expressed on the surface by hot springs, geysers, and active volcanoes (Figure 23.11). In general, temperature increases systematically with depth, at a rate of approximately 3 °C per 100 m. Temperatures at the base of the continental crust can range from 200 °C to 1000 °C, and at the center of Earth they are perhaps as high as 4500 °C. Unfortunately, most of Earth's heat is far too deep ever to be artificially tapped, and the heat we can reach by drilling is typically too diffuse to be of economic value. Like ore deposits, however, geothermal energy can be concentrated locally and has been used for years in Iceland and in areas of Italy, New Zealand, and the United States. Geothermal energy is concentrated along active plate margins or in mantle plumes, and concentrations are most likely to occur in regions where magma is relatively close to the surface.

Estimates indicate that, at its maximum worldwide development, geothermal energy would yield 10 times as much power as it did in 1969, but this would still amount to only a small fraction of the world's total energy requirements. Locally, however, geothermal energy can be a significant source.

Fossil Fuels

Coal, petroleum, and natural gas commonly are referred to as *fossil fuels* because they contain solar energy preserved from past geologic ages. The idea

FIGURE 23.11
Geothermal energy in New Zealand

that we currently use energy released by the sun more than 200 million years ago may seem remarkable, but the basic process of storing solar energy is quite simple. Heat from the sun is converted by biological processes into combustible, carbon-rich substances (plant and animal tissues). These are subsequently buried by sediment and preserved (Figure 23.12).

Coal.   Extensive coal deposits originate from plant material that flourished in ancient swamps typically found in low-lying coastal plains. A modern example of such an area is the present Great Dismal Swamp, along the coast of Virginia and North Carolina. In this area, the lush growth of vegetation has produced a layer of *peat* more than 2 m thick, covering an area of over 5000 km². In such an environment, the layer of peat is covered with sand and mud from the adjacent lagoon and beach as sea level slowly rises (Figure 23.13). Under pressure from the overlying sediment, water and organic gases (volatiles) are squeezed out, and the percentage of carbon increases. By this process, peat is compressed and is eventually transformed into coal.

If sea level rises and falls repeatedly, a series of coal beds can develop, interbedded with beach sand and nearshore mud. The essential requirements for the development of coal deposits, therefore, are a luxuriant growth of vegetation and relatively rapid burial by sediment to prevent decomposition. Coal deposits are thus restricted to the latter part of the geologic record, when plant life became plentiful. The most important coal-forming periods in Earth's history were the Pennsylvanian and Permian periods. Great swamps and forests then covered large parts of most of the continents. (The Western industrial nations, such as Great Britain, Germany, and the United States, developed with energy from these coals.) Other important periods of coal formation were the Cretaceous and Tertiary.

With the completion of at least a reconnaissance geologic mapping of most of the continents, all of the major coalfields are believed to have been discovered, and a reasonably accurate inventory of the world's coal reserves can therefore be made. Most of the reserves are found in the United States and Russia, so the developing nations are not likely to grow by vast amounts of coal.

Coal is important because there are reasonably large reserves of it. With the depletion of oil and gas, we are already witnessing an important shift to reliance on coal. The real pinch on oil and gas is now upon us and will put tremendous pressure on the coal industry, which undoubtedly will bring serious environmental problems.

How are coal, petroleum, and natural gas formed? Why are they called fossil fuels?

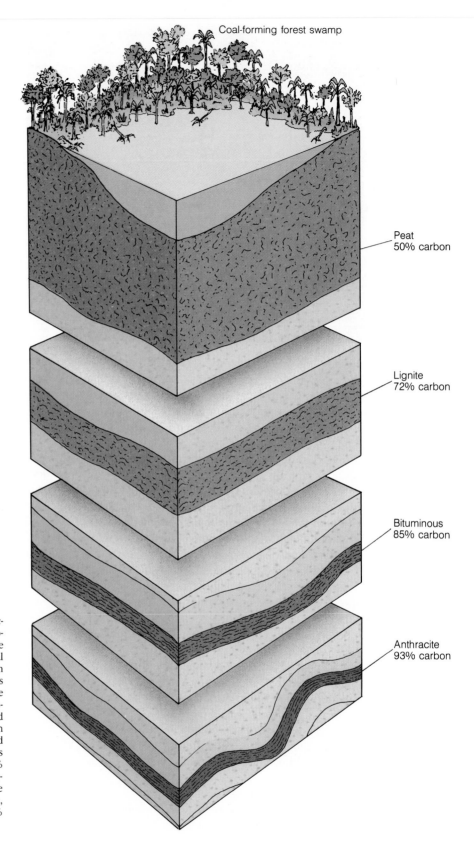

Coal-forming forest swamp

Peat
50% carbon

Lignite
72% carbon

Bituminous
85% carbon

Anthracite
93% carbon

FIGURE 23.12
The origin of coal involves burial, compaction, and induration of plant material. The process begins in extensive swamps such as the Great Dismal Swamp of Virginia. Plant material produced in the swamp decomposes to form peat (about 50% carbon). Subsidence causes the peat to be buried with sediment, and the resulting increase in temperature and pressure compacts the peat, expelling water and gases and thus forming lignite and brown coals (about 72% carbon). With continued subsidence and deeper burial, the lignite is compressed into bituminous coal (about 85% carbon). Further compression (commonly induced by tectonism) drives out most of the remaining hydrogen, nitrogen, and oxygen, producing anthracite coal, which is about 93% carbon.

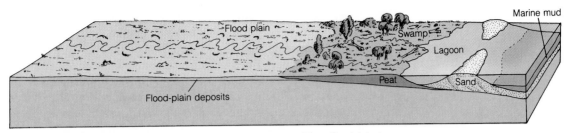

(A) The sequence of sedimentary environments along a coast grades seaward from floodplain, to swamp and lagoon, to barrier bar, to offshore mud.

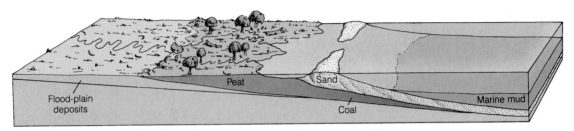

(B) As the sea expands inland, each sedimentary environment shifts landward. Swamp vegetation is deposited over the adjacent floodplain, sand of the barrier island is deposited over the previous swamp (peat) muck, and marine mud is deposited over the sand. The heat and pressure of the overlying sediment change the peat to coal.

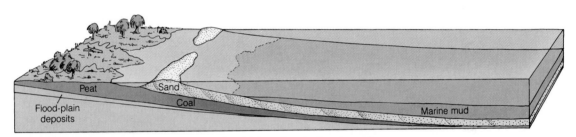

(C) Continued expansion of the sea superposes coal over floodplain sediments, beach sand over coal, and mud over beach sand.

(D) As the sea recedes, the sequence is reversed, that is, sand of the barrier island is deposited over offshore mud, coal is deposited over the sand, and floodplain sediments are deposited over coal. By expansion and contraction of the sea, layers (lenses) of sediments and coal are thus deposited in an orderly sequence.

FIGURE 23.13
Coal deposits are commonly formed by the expansion and contraction of a shoreline and involve the associated movement of a swamp and barrier bar.

Petroleum and Natural Gas. Petroleum and natural gas are hydrocarbons (molecules composed only of hydrogen and carbon in various combinations). In contrast to coal, the hydrocarbons forming oil and gas deposits originate largely from microscopic organisms that once lived in the oceans or in large lakes. The remains of these organisms accumulated with mud on the sea floor. Because of their rapid burial, they escaped complete decomposition.

Deposits of oil and gas form if four basic conditions are met.

1. The source beds must have sufficient organic material in the fine-grained sediments.
2. The beds must be buried deep enough (usually at least 500 m) for heat and pressure to compress rock and cause the chemical transformations that break down organic debris into hydrocarbons.
3. Once formed, the hydrocarbons must migrate upward from the source beds into more porous and permeable rock (usually sandstone or porous limestone or dolomite) called reservoir beds.
4. As the oil and gas migrate through the reservoir beds, a trap, or a barrier, must cause the deposits to accumulate.

If the reservoir beds provide an unobstructed path to the surface, the oil and gas seep out at the surface and are lost. This is one of the reasons why most oil and gas deposits are found in relatively young rocks. In older rocks there has been more time for erosion and Earth movements to provide a means for the oil and gas to escape. Barriers, or traps, in the general path of upward migration can result from a variety of geologic conditions, such as those shown in Figure 23.14. Exploration for oil and gas, therefore, is based on finding sequences of sedimentary rocks that provide good source and reservoir beds and then locating an effective trap.

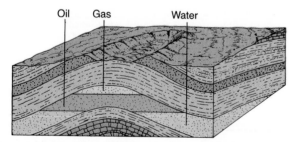

(A) Anticline. Oil, being lighter than water, migrates up the dip of permeable beds and can be trapped beneath a relatively impermeable shale bed in the crest of an anticline

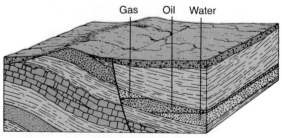

(B) Fault trap. Impermeable beds can be displaced against a permeable stratum and then trap the oil as it migrates updip.

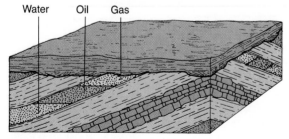

(C) Unconformity. An impermeable layer can cap inclined strata to form a seal that traps the upward-migrating oil.

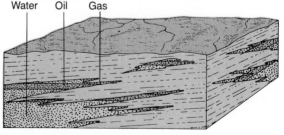

(D) Impermeable barrier. Shale surrounding a sandstone lens can form and prevent the oil from escaping.

FIGURE 23.14
The accumulation of oil and gas requires (1) a reservoir rock (a permeable formation, such as a porous sandstone) into which the petroleum can migrate and (2) a barrier, an impermeable cap rock, to trap the fluids. Some of the geologic structures that trap oil and gas are shown here.

In some instances, hydrocarbons can remain as solids in the shale in which the organic debris originally accumulated. These deposits are known as *oil shales.* They are reservoirs of oil that may become important in the future. The problem with oil shale is that it must be mined and heated to extract the oil. This process requires considerable energy and is not economically feasible at present.

Oil and gas are convenient forms of energy because they are easy to handle and transport. Unfortunately, at the present rate of consumption, the known reserves will be rapidly depleted. If the current trends continue, we will soon be forced to begin large-scale gasification and liquefaction of coal and oil shale deposits and to rely more on nuclear and solar energy. Clearly, we can expect to pay a great deal more for petroleum in the future and should use alternate sources of energy whenever possible.

Nuclear Energy

The ever-increasing demands for energy and the decreasing supply of fossil fuels naturally put the spotlight on nuclear power as the answer to our energy requirements. The technology of nuclear energy production is well developed. Modern society, however, has hesitated to move toward large-scale production of nuclear energy because of the possibility of serious long-range environmental problems. Radiation hazards, problems of waste disposal, and thermal pollution of fresh and marine waters, as well as potential terrorist activities, are among the greatest concerns.

The key element in the development of nuclear energy is uranium. The average uranium content in the rocks of Earth's crust is only 2 parts per million. It is concentrated in deposits by late magmatic segregation, and it occurs in veins associated with intrusive igneous rocks. As the veins oxidize, the water-soluble uranium is leached out and transported by surface water and groundwater. It can later be deposited in permeable sedimentary rocks as it is absorbed by clay minerals and organic matter. The rich uranium deposits in the Colorado Plateau are concentrated in ancient stream channels, especially where fossil wood and bones are found. Important uranium deposits occur in Canada, the United States, South Africa, and Russia. If the hazards and environmental problems can be solved, nuclear energy may indeed become a more important source of energy.

LIMITS TO GROWTH AND CONSUMPTION

Rapid population growth and the associated industrial expansion cause consumption of natural resources to increase at an exponential rate. We are finding that there are limits to growth. These limits probably will be reached through the depletion of natural resources.

The consumption of natural resources is proceeding at a phenomenal rate. A moment's reflection shows that the rapid population and industrial growth that have prevailed during the last few hundred years are not normal. Indeed, this period is one of the most abnormal phases of human history. Current rates of growth and the associated consumption of natural resources present one of our most serious problems. The problem is basically one of changing from a period of growth to a period of nongrowth. This change will require a fundamental revision of current popular economic and social thinking, which is based on the assumption that growth must be permanent and the even more basic assumption that growth must occur for society to prosper.

Throughout this book the recurring central theme has been that Earth is a dynamic planetary system. Mineral and energy resources are systematically formed by specific geologic processes operating in the system. Resources do not occur haphazardly, nor are they distributed evenly throughout the continents. Some continents and nations have natural resources and some do not. Iceland and Hawaii, for example, being built exclusively of basaltic lava, cannot be considered good potential sources of petroleum and natural gas, no matter how much those areas might be explored. In addition, mineral deposits are nonrenewable resources. Once a deposit is mined out and used, it is gone forever.

Modern industrial civilization has developed principally during the last century. It differs from all previous civilizations in the amounts of energy and resources it uses and in its rate of growth. It is important, therefore, to consider how resources are used and the rates at which they are being depleted.

Perhaps the most critical point to make is that the rate of resource consumption is not static. It increases exponentially. The exponential rate results both from population growth and from the growth of the average annual consumption per person, which increases yearly. In other words, growth is exponential because of an increasing population and a rising standard of living.

The limits of growth in the world system probably will not be imposed by pollution. The limits will be set by the depletion of natural resources. The projected interaction of some of the major variables as the world system grows to its ultimate limits is shown in Figure 23.15. Food, industrial output, and population will continue to grow exponentially until the rapid depletion of resources forces a sharp decline in industrial growth.

In our finite Earth's system, unlimited exponential growth is impossible. In fact, the transition to a stable or declining phase has already begun. This does not pose insurmountable technological, biological, or social problems. It does require some fundamental adjustments in our present growth culture. If we can achieve appropriate cultural adjustments, the steady-state system could foster a social environment that would be conducive to the flowering of one of humanity's greatest intellectual advances. The alternative could be catastrophic.

Can a finite Earth sustain unlimited growth? How much growth is enough?

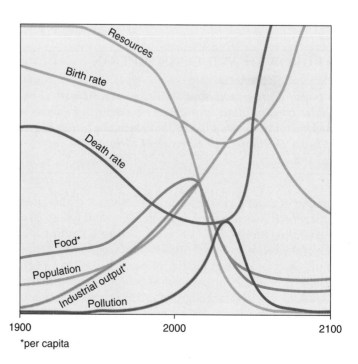

FIGURE 23.15
A computer model of resource consumption and its influence on other variables assumes no major changes in the physical, economic, and social relationships that historically have governed the development of the world system. All variables plotted here follow historical values from 1900 to 1970. Food, industrial output, and population grow exponentially until the rapidly diminishing resource base forces a slowdown in industrial growth. Because of natural delays in the system, both population and pollution continue to increase for some time after the peak of industrialization. Population growth is finally halted by a rise in the death rate due to decreased food and medical supplies.

1900   2000   2100

*per capita

SUMMARY

1. Earth's natural systems are intricately related, and they maintain a delicate balance with one another. We are capable of modifying and manipulating most natural systems, including rivers, slopes, groundwater, and shorelines. Such systems often adjust to artificial modifications with many unanticipated side effects.

2. Waste disposal is a major environmental problem. Matter cannot be created or destroyed. It can only change from one state to another or from one chemical combination to another. Problems with waste disposal involve solid, liquid, gaseous, and radioactive wastes.

3. Mineral resources have been concentrated by a variety of geologic processes. These processes are related to the plate tectonics and hydrologic systems. Most resources are therefore finite and nonrenewable.

4. Energy resources are classified as renewable and nonrenewable forms. Renewable energy sources include (1) solar energy, (2) hydropower, (3) tidal energy, and (4) geothermal energy. None of these can be expected to supply a large percentage of the energy that will be required worldwide in the near future.

5. Nonrenewable energy sources include fossil fuels and nuclear energy. Most industry is based on fossil fuels, particularly oil and gas. Fossil fuels are essentially solar energy from past geologic ages stored in the form of organic matter.

6. Nuclear energy could become a major energy source in the future. There are some serious environmental concerns, however, such as radiation hazards, waste disposal, and misuse by irresponsible persons or groups.

7. Rapid population growth and the associated industrial expansion cause consumption of natural resources to increase at an exponential rate. We are finding that there are limits to growth. These limits probably will be reached through the depletion of natural resources.

KEY TERMS

contact metamorphism (p. 569)

evaporite (p. 569)

fossil fuel (p. 572)

gaseous waste (p. 564)

hydrothemal deposit (p. 568)

hydrothermal solution (p. 568)

liquid waste (p. 564)

magmatic segregation (p. 568)

mining waste (p. 567)

oil shale (p. 577)

peat (p. 573)

placer (p. 569)

porphyry copper (p. 568)

radioactive waste (p. 566)

renewable energy source (p. 572)

solid waste (p. 563)

strip mining (p. 566)

REVIEW QUESTIONS

1. List some of the problems that result from artificial modifications of river systems and cite several examples.
2. What are some of the major problems associated with artificial modifications of the groundwater system?
3. What natural geologic systems are likely to become contaminated by waste disposal? Consider all types of waste.
4. How does mining waste affect the environment?
5. List the ways in which mineral resources are concentrated. Can they be considered renewable?
6. Explain how magmatic segregation concentrates ores such as chromium.
7. How are minerals concentrated by streams?
8. Explain how minerals are concentrated by weathering processes.
9. What energy sources are renewable?
10. What are fossil fuels?
11. Explain how coal originates.
12. Describe various kinds of petroleum traps. What kinds of strata serve as (a) source beds and (b) reservoir beds?
13. Why is oil rarely found in the axis of a syncline?
14. What are the major problems with nuclear energy?
15. Discuss the factors that limit the growth of population and industrialization.

ADDITIONAL READINGS

Canadian Department of Environment. 1984. *The Acid Rain Story.*

Crandall, R. W. 1983. *Controlling Industrial Pollution.* Washington, DC: The Brookings Institution.

Edelen, G. W., Jr. 1981. Hazards from floods. In *Facing Geological Hydrologic Hazards, Earth-Science Considerations.* W. W. Hays, ed. U.S. Geological Survey Professional Paper 1240–B:39–52.

Elliot, J. 1980. Lessons from Love Canal. *Journal of the American Medical Association* 240:2033–34, 2040.

Fleming, R. W., and F. A. Taylor. 1980. Estimating the cost of landslide damage in the United States. U.S. Geological Survey Circular 832.

Griggs, G. B., and J. A. Gilchrist. 1983. *Geologic Hazards, Resources, and Environmental Planning.* 2nd ed. Belmont, CA: Wadsworth.

Hanks, T. C. 1985. The national earthquake hazards reduction program: Scientific status. U.S. Geological Survey Bulletin 1659.

Hubbert, M. K. 1962. *Energy Resources: A Report to the Committee on Natural Resources.* Washington, DC: National Academy of Sciences-National Research Council.

Hubbert, M. K. 1971. The energy resources of the earth. *Scientific American* 224(3):60–84. Reprint no. 663. San Francisco: Freeman.

Hunt, C. B. 1983. How safe are nuclear waste sites? *Geotimes* 28(7):21–22.

International Institute for Applied Systems Analysis. 1981. *Energy in a Finite World.* Cambridge, MA: Ballinger.

Pye, U. I., and R. Patrick. 1983. Ground water contamination in the United States. *Science* 221:713–718.

U.S. Environmental Protection Agency. 1980. *Acid Rain.*

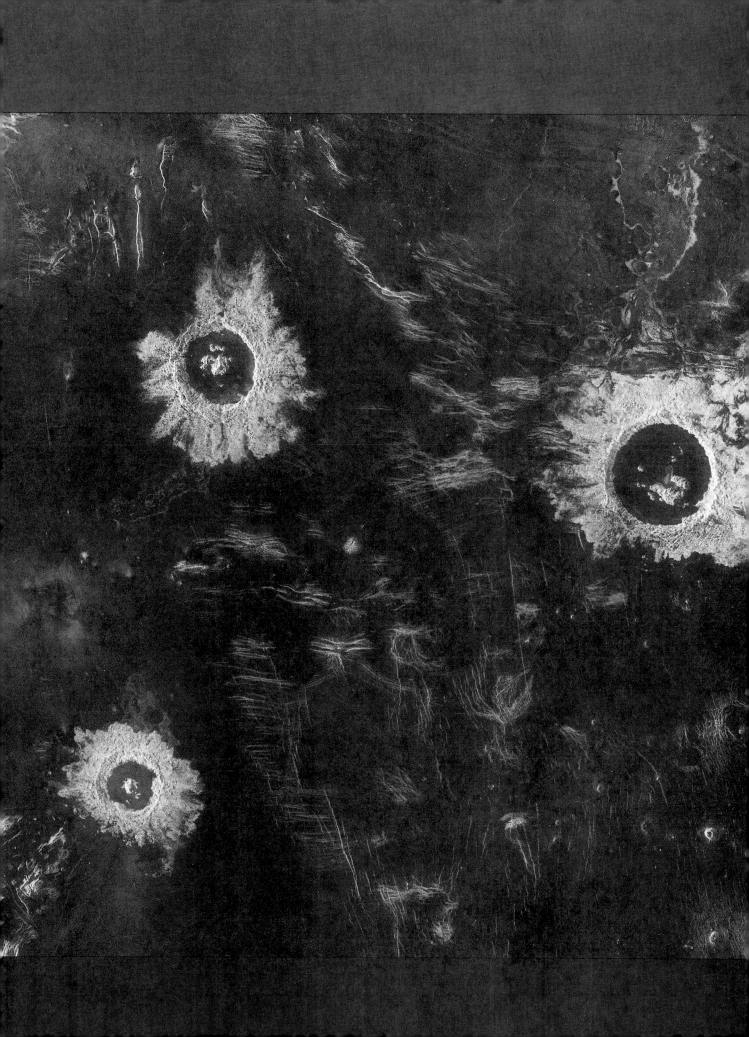

The geologic exploration of the solar system is one of the most exciting adventures ever experienced by geologists, because for the first time, whole new worlds can be compared geologically with Earth. The study of planetary geology, however, does much more than merely satisfy scientific curiosity. By comparing in detail the geologic nature and evolution of different planets, we can better recognize those principles and processes that are fundamental to the geology of Earth and those that are of secondary importance.

A sample of what we have discovered is shown in this radar image of Venus. Three large, bright impact craters occur on the smooth lava plains that are cut by a gridlike system of fractures. Several shield volcanoes with summit craters can be seen near the lower right corner. Venus has many features similar to those on Earth but has a tectonic style of its own. Why?

We also wonder why the Moon has no atmosphere, no folded mountain belts, and no recent volcanism. Why does Mars have such large shield volcanoes and an immense canyon? How did stream channels develop, and how did catastrophic flooding occur on Mars? Why did the Moon, Mercury, Venus, Mars, and some asteroids all have major thermal events involving outpourings of basalt early in their histories? Why are each of the moons of Jupiter so different, and what are the origin and history of the small, icy satellites of Saturn? What do the moons of Uranus and Neptune tell us about volcanism and tectonics? To gain an insight into these and other questions about the geology and evolution of the planets, we will venture into a brief study of planetology.

24
Other
Planets

MAJOR CONCEPTS

1. Impact cratering was the dominant geologic process in the early history of all planetary bodies in the solar system.
2. Earth, the Moon, Mercury, Venus, and Mars form a family of related planets, known as the terrestrial planets, which probably experienced similar sequences of events in their early histories.
3. Both the Moon and Mercury are primitive bodies, and their surfaces have not been modified by hydrologic and tectonic systems.
4. Mars had a more eventful geologic history involving crustal uplift, volcanism, stream erosion, and eolian activity.
5. The surface of Venus is dominated by volcanism and such tectonic features as faults and folded mountain belts. The crust of Venus does not appear to be broken into tectonic plates, however, and tectonic motion does not appear to have played a dominant, ongoing role in altering the surface.
6. Cratering on the icy moons of Jupiter, Saturn, Uranus, and Neptune suggests that the period of intense bombardment affected the entire solar system more than 4 billion years ago.
7. Most of the icy moons of Uranus and Neptune show evidence of geologic activities such as volcanic extrusions of slushy ice and rifting. Miranda, a tiny moon of Uranus, has a unique terrain that possibly represents fragmentation and then reassembling of the fragments.
8. The planets formed in a thermal gradient around the sun. The inner planets are thus rich in silicates and iron, which are stable at high temperature, and the outer planetary bodies have large amounts of ice, which is stable at low temperature.
9. The geologic evolution of a planet depends on its source of heat energy, its size, and its composition.

THE TERRESTRIAL PLANETS

The terrestrial planets are composed of rocky material that condensed near the sun. The Moon and Mercury were too small to generate enough heat to sustain a tectonic system and ceased to be active after their first major thermal event. Mars, being larger, developed a more prolonged period of tectonism. Venus, nearly as large as Earth, has continent-like land masses, volcanic activity, and folded mountain belts.

The Moon

With the development of the space program the Moon has become one of the best understood planetary bodies in the solar system. As curious as it may seem, less than two decades after putting astronauts on the Moon, we probably understand the earliest history of the Moon better than that of Earth, for Earth lacks a rock record of the first 800 million years of its history. One of the most important results of the exploration of the Moon is the discovery that cratering from the impact of meteorites and comets is a fundamental and universal process in planetary development. The Moon is pockmarked with billions of **craters,** which range in size from microscopic pits on the surface of rock specimens to huge, circular **basins** hundreds of kilometers in diameter. Craters are also abundant on Mercury and Mars and to a lesser extent on Venus (in spite of its dense atmosphere), the asteroids, and many of the satellites of the outer planets. Indeed, impact cratering was undoubtedly the dominant geologic process on the surface of Earth during the early stages of its evolution.

Probably Earth once looked much like the Moon does today. How are impact craters formed? What is their geologic significance?

Impact Craters. Impact craters are the dominant landform on most planetary bodies of the solar system, so some understanding of the impact process is essential before one can begin to understand the origin and history of these planetary bodies. Impact processes are nearly instantaneous, but their formation can be studied in the laboratory with high-speed motion pictures and can be simulated by detonation of strong explosives. Conceptually, the process is relatively simple, as is illustrated in Figure 24.1. As a *meteorite*

What is formed when a meteorite impacts on a planetary surface?

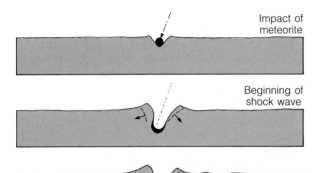

(A) Impact of meteorite causes the rock to be instantly fractured, fused, and partly metamorphosed.

(B) Shock wave is propagated downward and outward from point of impact.

(C) Shock wave expands.

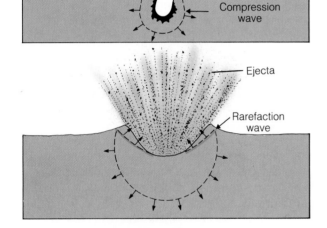

(D) Shock wave is reflected back toward the surface. Crater begins to form and material is fragmented similar to that produced by an explosion.

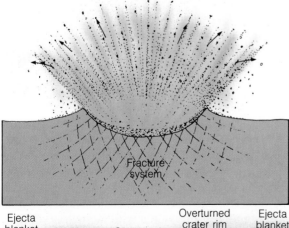

(E) Fragmented material is cycled upward and outward. Solid bedrock is fractured and forced upward to form crater rim.

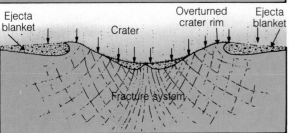

FIGURE 24.1
Hypothetical stages in the formation of a meteorite impact crater are depicted in this series of diagrams. The kinetic energy of the meteorite is almost instantly transferred to the ground as a shock wave, which moves out, compressing the rock. At the point of impact, the rock is intensely fractured, fused, and partly vaporized by shock metamorphism. The shock wave is reflected back as a rarefaction wave, which throws out large amounts of fragmental debris, and the solid bedrock is forced upward to form the crater rim. A large amount of fragmental material falls back into the crater.

strikes the surface, its kinetic energy is almost instantaneously transferred to the ground as a shock wave, which moves downward and outward from the point of impact. This initial compression wave is followed by a rarefaction wave, which rebounds in the opposite direction and causes material to be ejected from the surface and thrown out along ballistic trajectories. This fragmented material accumulates around the crater, forming an *ejecta blanket* and a system of splashlike rays. Such a crater is referred to as a **rayed crater.** A central peak on the crater floor can result from the rebound, and rocks in the crater rim can be overturned. Many large craters (more than 300 km in diameter) contain a series of concentric ridges and depressions and hence are called **multiring basins.** Meteorite impacts, like other rock-forming processes that operate on the surface of a planetary body, create new landforms (craters) and new rock bodies (ejecta blankets) so that a record of the event is preserved.

How is an impact crater modified with time?

After a crater is formed, it is subject to certain types of modification. The steep crater walls rapidly slump and move downslope by mass movement. This process results in partial filling of the depression and the formation of concentric terraces (slump blocks) inside the crater rim. The rays gradually become obliterated by subsequent bombardment. Isostatic adjustment can arch up the crater floor to compensate for the removal of material in the formation of the crater. Further modification results from three main processes: (1) subsequent meteorite impact can partly destroy or obliterate the crater and ejecta blanket, (2) the crater can be covered with ejecta from the formation of younger craters, and (3) the crater and ejecta can be buried by lava flows.

The Surface of the Moon. Utilizing thousands of satellite photographs of the Moon, geologists have been able to map it with greater accuracy than we could map Earth only a few decades ago. We have analyzed over 380 kg of lunar rock samples, determined their ages, and received data from a variety of geophysical instruments. As it turns out, the Moon is truly a whole new world.

Study the surface of the Moon (Figure 24.2) and you will see two contrasting types of landforms, which reflect two major periods in its history. The bright, densely cratered highland resulted from the intense bombardment of meteorites, most of which impacted more than 4 billion years ago. The dark, smooth areas, called **maria,** mostly occupy low regions, such as the circular interiors of impact basins. We know from rock samples brought back from the *Apollo* missions that the maria resulted from great floods of basaltic lava, which filled many large craters and spread out over the surrounding area. Most of this volcanic activity therefore occurred after the formation of the densely cratered terrain. Radiometric dates on samples brought back from the Moon indicate that most of the lavas are over 3 billion years old.

Tectonic Activity. Almost no global tectonic activity has occurred on the Moon during the last 3 billion years, and very little has occurred during its entire history. We know this because the thousands of craters that cover the lunar surface provide an excellent reference system for even the most subtle structural deformation. Like lines on graph paper, the circular craters would show the effects of deformation of the crust by compression, tension, or shear stresses and would record, by changes in their shapes, even the slightest disturbances. Because the network of craters is essentially undeformed, the lunar crust appears to have been fixed throughout time. We find no evidence of intense folding or thrust faulting and no indication of major rifts. The main lunar features that can be attributed to structural deformation are narrow grabens, formed by minor extension, and wrinkle ridges, formed by minor compression. Nor has the lunar surface been modified by wind, water, or glaciers. Without an atmosphere, it has no hydrologic system, and its surface is strikingly different from that of Earth.

FIGURE 24.2

The surface of the Moon shows two contrasting types of landforms: densely cratered highlands, called terrae, and dark, smooth areas of lava plain, called maria. We know from rock samples brought back by the *Apollo* missions that the maria resulted from great floods of lava, which filled many large craters and spread out over the surrounding area. The volcanic activity thus occurred after the formation of the densely cratered terrain. These relationships between surface features imply that the Moon's history involved three major events: (1) a period of intense bombardment by meteorites, (2) a period of volcanic activity, and (3) a subsequent period of relatively light meteorite bombardment (resulting in young, bright-rayed craters). The lunar surface has a very low level of erosion and has not been modified by wind, water, or glaciers.

The Lunar Time Scale. One of the most significant results of geologic exploration of the Moon has been the construction of a lunar geologic time scale. This was done by using the same principles of superposition and cross-cutting relations that were devised in the early nineteenth century by geologists studying Earth. Look carefully at Figure 24.2 and you can read the record of lunar history yourself. A period of intense bombardment is recorded in the densely cratered terrain. This was followed, or accompanied, by the impact of large asteroid-sized bodies that produced the huge multiring basins such as Imbrium basin. A subsequent thermal event is evident in the floods of lava that fill the lowlands and spread over parts of the densely cratered terrain. A period of light bombardment by meteorites followed, which formed craters upon the lava flows of the maria. Radiometric dates of lunar rock samples provide benchmarks of absolute time, and the major events in lunar history have been outlined as shown in Figure 24.3.

How did scientists develop a geologic time scale for the history of the Moon?

Perhaps the most important aspect of the Moon's geologic evolution is that the most dynamic events occurred during the early history of the solar system, before the oldest rock on Earth was formed. The Moon thus provides important insights into planetary evolution that are unobtainable from studies of Earth.

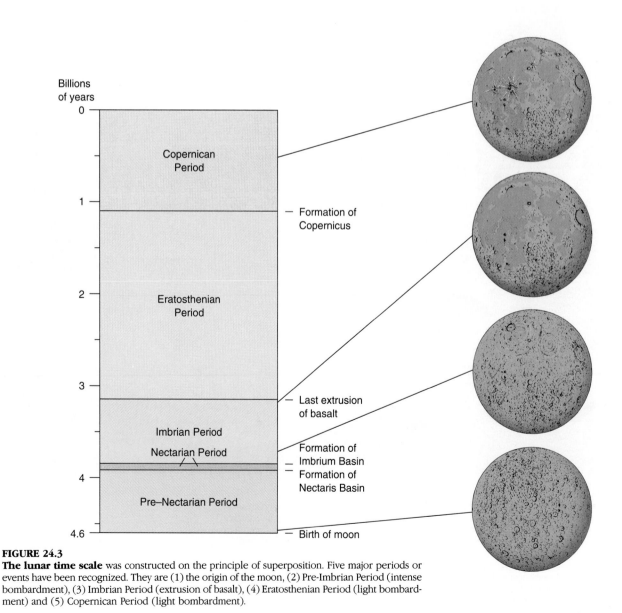

FIGURE 24.3
The lunar time scale was constructed on the principle of superposition. Five major periods or events have been recognized. They are (1) the origin of the moon, (2) Pre-Imbrian Period (intense bombardment), (3) Imbrian Period (extrusion of basalt), (4) Eratosthenian Period (light bombardment) and (5) Copernican Period (light bombardment).

Mercury

The surface features of Mercury are strikingly similar to those of the Moon, as is evident from the mosaics shown in Figure 24.4. Most of Mercury's surface is nearly saturated with craters that have a range in size similar to that of the craters on the Moon. Some are obviously old because they are battered by the impact of other meteorites; others are relatively young and have bright rays extending out from the point of impact.

The largest impact structure seen on Mercury is the Caloris basin (Figure 24.4), a multiring basin similar in size and form to the Imbrium basin on the Moon. Caloris's ejecta radiate from the main ring, forming elongate hills and valleys that extend outward over a distance roughly equal to the diameter of the basin. Other multiring basins on Mercury are smaller.

Smooth plains material covers the floor of the Caloris basin, as well as much of the lowlands beyond. The basin's similarity to the lunar maria suggests that it also was formed by the extrusion of fluid basaltic lava.

Employing the same methods and techniques used to study the Moon, geologists have been able to establish a preliminary geologic time scale for Mercury and have developed a working hypothesis for its geologic evolution. Mercury's surface features indicate a sequence of four major events, broadly similar to the sequence of events recorded on the Moon: (1) accretion, planetary differentiation, and intense meteorite bombardment, (2) formation of multiring basins, (3) flooding of basins, probably by extrusion of basaltic lava, and (4) light meteorite bombardment. Because there is no atmosphere or water on Mercury, its surface has not been modified by a hydrologic system.

What were the major events in the history of Mercury?

FIGURE 24.4
Photomosaic images of Mercury, made from photographs taken at a distance of 234,000 km, show that Mercury and the Moon are strikingly similar. Each has a densely cratered terrain, multiring basins, a younger sea of dark plains (maria), and young rayed craters.

Mars

Mars is a planet of fascinating geologic phenomena. Almost every geologic feature is gigantic—immense volcanoes, enormous canyons, and giant landslides. There is evidence not only of stream action but of catastrophic flooding in which sheets of water possibly 300 meters deep surged over the landscape, eroding deep channels. Wind action is also an important process on Mars; it has modified most of the planet's surface features to some extent. In addition, the polar regions are covered with alternating layers of ice and windblown sediment, and details of the advancing and retreating polar caps have been recorded by a type of time-lapse photography. It is obvious from these observations that the geologic agents operating on Mars have varied both from place to place and from time to time and that Mars has a fascinating record of geologic events quite different from those of the Moon, Mercury, and Earth. Mars has intrigued science-fiction writers for decades; now that it has been explored, it is much more fascinating than anyone could ever imagine.

FIGURE 24.5
The planet Mars, as photographed from the *Viking* spacecraft, shows a surface somewhat similar to that of the Moon, yet significant differences exist. Unlike the Moon, Mars has many features indicating that its surface has been modified by atmospheric processes, recent volcanic activity, and crustal deformation.

Impact Craters. Although a variety of landforms are evident on the surface of Mars, cratering still appears to have been a dominant process, especially during the planet's early history (Figure 24.5). The craters on Mars record impacts from a population of meteorites similar to those that bombarded the surfaces of the Moon and Mercury. The Martian craters, however, have many distinctive features, which reflect the planet's particular gravitational attraction, surface materials, and erosional history. Most of the craters are shallow, flat-floored depressions. They show evidence of much more erosion and modification by sedimentation than do craters on the Moon and Mercury. Rayed craters are rare because the Martian winds can easily erode the loose ray material.

But most of the young craters on Mars are unique in a very surprising way. The ejecta blankets from the craters appear to have flowed over the Martian surface like avalanches and mudflows instead of accumulating as free-falling particles. A spectacular example is the crater Yuty, shown in Figure 24.6. Tongues of ejecta debris, like huge splashes of mud, have flowed outward from the crater rim. Near the right side of Figure 24.6, the ejecta flow can be seen overlapping the older eroded debris from the large adjacent crater, and it actually flows up and over the cliff formed on the older, eroded ejecta blanket.

The flow of ejecta material is believed to result from the presence of ground ice in the soil and rock of Mars, a geologic condition believed to be similar to that at the polar regions of Earth. Upon impact, the ground ice instantly melts, so that material ejected from craters moves like great globs of mud.

An important finding is that the cratering history of Mars parallels that of the Moon and Mercury. Each planet experienced an early period of intense bombardment and a subsequent period of impact by asteroid-size bodies, which produced large multiring basins. This was followed by extrusions of lava that formed broad, smooth plains. Since then, the rate of impact on all of these planets has rapidly decreased.

How does Mars differ from the Moon and Mercury?

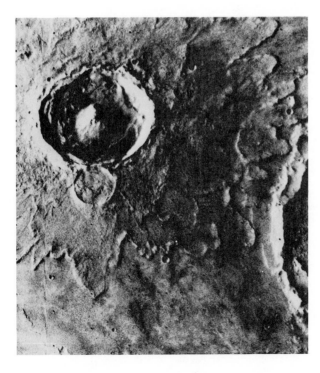

FIGURE 24.6
Ejecta debris from many craters on Mars has flowed like a mudflow. This suggests that the ejecta were fluidized at the time of impact, probably by the melting of near-surface ground ice.

Volcanic Features.   Some of the most spectacular features observed on the surface of Mars are the giant volcanoes in its northern hemisphere. They are much larger than any seen on Earth, and their freshness suggests that Mars has just been "turned on" volcanically. Studies of the available photographs also reveal a number of older volcanoes and volcanic plains that are greatly eroded. It thus appears that Mars has had a long and interesting volcanic history, much more complex than that of Mercury or the Moon. The interior of Mars must have remained hot much longer than the interiors of smaller planetary bodies such as the Moon and Mercury.

The enormous volcanoes on Mars are at least twice the size of the largest volcano on the Earth. Olympus Mons, the largest, is 700 km in diameter and rises 23 km above the surrounding plains (Figure 24.7). This is nearly half again the distance from the deepest spot on Earth—the Mariana Trench—to the top of Mount Everest. At the crest of Olympus Mons is a huge, complex, collapsed caldera, similar to those on the volcanoes of Hawaii. Other Martian volcanoes resemble Olympus Mons but differ in size and detail.

Vast areas of Mars are also covered by volcanic plains, similar to the plains of the lunar maria. Some lava plains formed during the intense bombardment of Mars more than 4 billion years ago. Eruptions have continued until at least fairly recent times, however, as seen by the vast aprons of lava that surround Olympus Mons and nearby volcanoes.

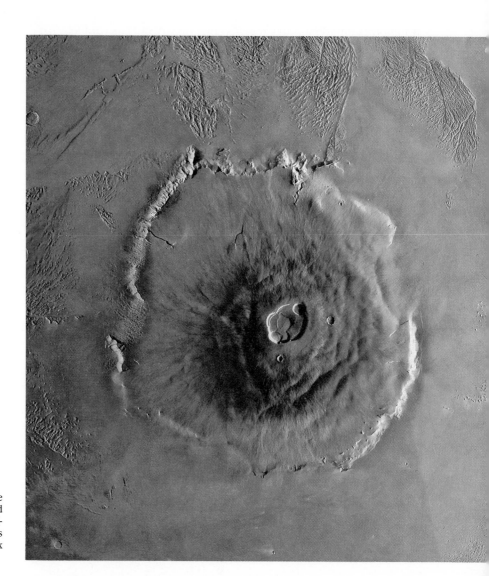

FIGURE 24.7
Volcanoes in the Tharsis region include huge structures, much larger than any found on the Earth. Olympus Mons, the largest volcano on Mars, is shown here. It is 700 km across at the base and 23 km high. The complex caldera at the summit is 65 km in diameter.

The Martian volcanoes are important because they indicate that the planet has been thermally active during much of its early history and possibly is still active today. Although volcanic activity on the Moon and Mercury terminated after the period of extensive extrusion of basalt, it has continued on the larger planets, Mars and Earth.

Tectonic Features. The presence of undeformed craters across the surface of Mars clearly indicates that its crust is not broken into a series of tectonic plates that move about like those on Earth, nor has it been subjected to extensive compression. There are no continents, ocean basins, or folded mountain belts. Yet, there is clear evidence of tensional stresses that have deformed the Martian crust. These, together with evidence of prolonged volcanic activity, clearly indicate that Mars has a tectonic style of its own. Large domes form bulges in the crust roughly 5000 km in diameter and 7 km high. In addition, systems of faults displace the crust to form long narrow grabens that have influenced the development of Valles Marineris, the "Grand Canyon" of Mars (Figure 24.8). This huge canyon is about 5000 km long and, in places, 6 km deep. The walls are eroded, with deep side canyons with branching tributaries

What is the origin of the "Grand Canyon" on Mars?

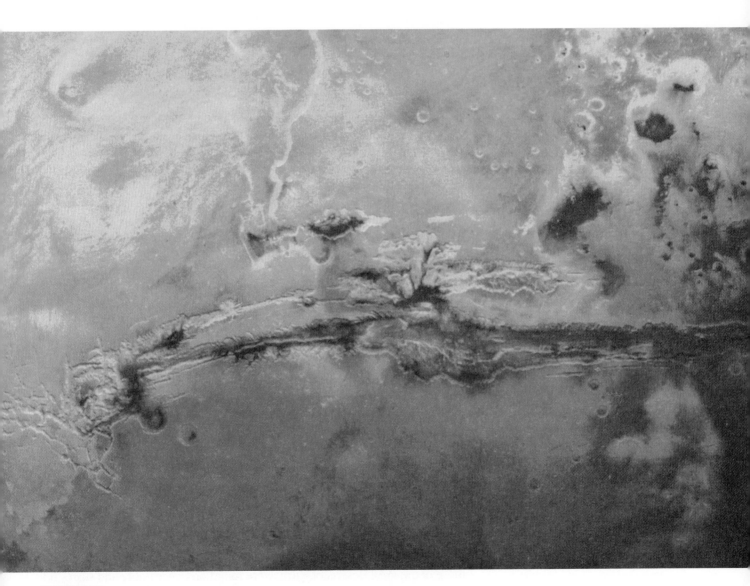

FIGURE 24.8
Valles Marineris, the Grand Canyon of Mars, extends across an area longer than the United States. This huge chasm is about 5000 km long and at least 6 km deep. The Grand Canyon of the Colorado River is about the size of a minor tributary of the canyon on Mars.

FIGURE 24.9
Stream channels on Mars are similar in many respects to dry river beds in arid regions of Earth. Some have typical braided patterns, others meander. This photomosaic taken in 1976 by a *Viking* orbiter shows a drainage system more than 700 km long. The system of branching tributaries is short and stubby, suggesting they developed primarily from seepage rather than from runoff of precipitation.

similar to the features of canyons cut by running water on Earth. The configuration of the canyon walls indicates that erosion has considerably widened and modified the original rift valleys. The dominant erosional processes were probably landslides, debris flows, wind, and possibly running water. The features on Earth that are most comparable in size to Valles Marineris are the Red Sea and the rift valleys of east Africa. Like the rift system on Earth, Valles Marineris started where the crust was pulled apart and the interior block subsided. Subsequent erosion sculptured the rim of the rift valley.

Fluvial Features. Perhaps the most startling result of the *Mariner 9* project (1971) was the discovery on Mars of numerous channels that closely resemble dry river beds on Earth (Figure 24.9). Many channels originate in the southern highlands, near the erosional escarpment, and empty into the low plains to the north. If they were located on Earth, no one would hesitate to call them dry river beds. Their presence on Mars, however, raises some of the most intriguing and perplexing questions about the planet. The existence of water in the liquid state depends on the temperature and pressure regime at the planet's surface. The present Martian atmosphere is characterized by low pressures and low temperatures. As a result, water in the liquid state cannot exist in large amounts at the surface. It will either freeze or evaporate. At present, most of the water on Mars is locked up either as ice in the polar regions or as ground ice beneath the surface. The widespread river channels on Mars, however, imply that not only was running water once present but that, in places, large-scale flooding occurred. When did large volumes of water flow on Mars? Did life evolve on the Red Planet? Why did Mars enter its present period of deep freeze?

We cannot answer these questions at the present time, but we are quite certain that the dry river channels indicate that the surface of Mars was eroded by running water at some earlier stage in its history. The change from earlier fluvial periods to the present dry and cold period may be similar to the change from glacial to interglacial periods on Earth.

Mass Movement. High-resolution photography from the *Viking* orbiters provides dramatic evidence that mass movement is an important process in the evolution of the Martian landscape, especially in the enlargement of canyons. The importance of mass movement can be seen in the oblique view in Figure 24.10. The canyon shown in this photograph is approximately 5 km deep. Multiple massive landslides and debris flows are visible on both walls.

Eolian Features. Wind is at present the dominant geologic agent operating on the surface of Mars, but here again its scale of activity is enormous when compared to wind activity on Earth.

When *Mariner 9* entered orbit around Mars in 1971, a raging dust storm completely obscured the entire planet. The storm, which had begun two months before the encounter and continued for several months after the spacecraft was in orbit, could be seen through telescopes from Earth. The magnitude of a global storm on Mars is hard to imagine. Winds comparable to a strong hurricane on Earth raged continually for several months, with gusts of up to 500 km per hour. The storm blew dust high into the atmosphere, so high that only the tops of the highest volcanoes could be seen as dark spots above the global cloud of dust. When the storm cleared, a variety of features resulting from the wind activity were photographed. Behind the wind shadows of craters, dark streaks covered broad areas. Large dune fields, remarkably similar to those of the great deserts of Earth, were found on crater floors and surrounding areas. In addition, surface photographs taken by the *Viking* landers show many wind-generated features, such as small dunes, ventifacts, lag deposits, and sand streaks behind boulders (Figure 24.11).

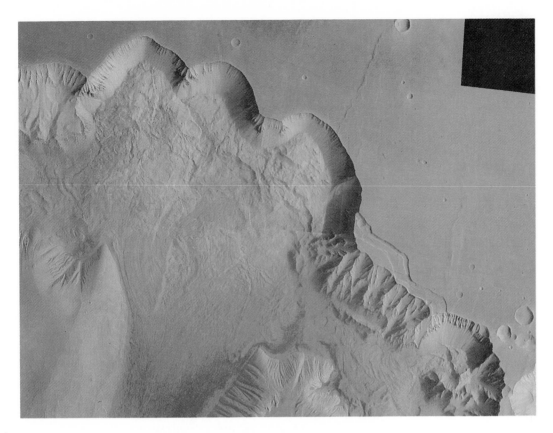

FIGURE 24.10
Landslides in Valles Marineris are shown in this photo. The landslide on the far wall has two components—an upper blocky portion, which is probably disrupted caprock, and a finely striated lobate extension, which is probably debris derived from the old cratered terrain exposed in the lower canyon walls. Similar lineations are found on terrestrial landslides and show the direction of movement. This part of Valles Marineris is about 5 km deep.

FIGURE 24.11
View of the Martian surface photographed by a *Viking* lander. The large boulder in the left foreground is approximately 2 m across. Many of the features shown here indicate the importance of wind activity in forming details of the Martian landscape. The gravel surface was formed as lag deposits resulting from deflation.

593

The Geologic History of Mars. The relative ages of the major rock bodies and terrain types on Mars can be determined by using the basic principles of superposition. A geologic time scale for Mars has been developed that can be tentatively correlated with the major periods in the time scales of other planetary bodies in the solar system. The major events in the history of Mars are outlined as follows:

1. Accretion, planetary differentiation, an intense meteorite bombardment
2. Crustal upwarping and faulting; development of volcanoes
3. Continued extrusion of lava to form volcanic plains
4. Continued volcanism and eolian activity

Venus

The surface of Venus is totally obscured by a thick atmosphere composed mostly of carbon dioxide with clouds of sulfuric acid. Nonetheless, Soviet and American spacecraft using radar instruments have revealed an outline of the planet's most important surface features. Volcanic plains, mountain belts, volcanoes, and two high "continents" that rise several kilometers above vast rolling lowlands show that Venus has a surface similar in some ways to the surface of Earth with its continents and ocean basins. However, Venus has almost no water, and temperatures at its surface (almost 750 K) are higher than on Mercury. Now, with the U.S. *Magellan* mission (1991), an orbiting satellite is able to map the surface of Venus in detail, showing features as small as 120 m in diameter. With this global high-resolution imagery, scientists are able to compare the geology of Venus with that of other planetary bodies. This is what has been found.

What are the major products of volcanic activity on Venus?

Volcanic Features. Images of the first 20% to 30% of the surface of Venus made by *Magellan* show that volcanic features dominate its landscape. Volcanic plains with a wide variety of surface features make up over 80% of the planet observed so far. The plains are built by sequences of thin fluid lava flows, the youngest of which can be seen emerging from several sources along fracture systems and then converging and ponding in the low parts of the plains (Figure 24.12A). In addition, thousands of small shield volcanoes generally 2 to 8 km in diameter with summit craters are scattered across the plains and are concentrated into clusters locally. These features are similar to seamounts on Earth. Commonly, the locations of small shields are controlled by fracture systems on the plains. Some lava flows occur in extremely narrow, sinuous channels, 0.5 to 1.5 km wide, that extend for hundreds of kilometers and appear to empty into low-lying areas of the volcanic plains. In some areas eruptions have built up larger volcanoes, as much as 225 km in diameter, that have radial flow units. These structures may indicate the presence of broad mantle upwelling or hotspots like those on Earth.

Most of the volcanism on Venus appears to be the result of fluid basaltic (?) eruptions, although some steep-sided volcanic domes suggest the presence of lavas with higher viscosities and possibly more-silicic compositions (Figure 24.12B). The general morphology of these domes is similar to that of the silicic lava domes on Earth, but they are somewhat larger. The relatively large diameters may be due to the high surface temperature that would retard the cooling of magma and enhance its flow.

Tectonic Features. The surface of Venus has a wide variety of tectonic features that are commonly displayed in a spectacular fashion because they are unmodified by weathering and erosion. Some areas are completely laced with polygonal patterns of faults and fractures that are extremely regular in pattern and orientation (Figure 24.12C, D). Some form a grid pattern with two sets of fractures that have different characteristics (Figure 24.12E). In other areas deformation of the plains occurs in linear belts of narrow ridges similar to the wrinkle ridges on the Moon but wider and longer (Figure 24.12F). The most spectacular tectonic features on Venus are the orogenic belts of western Ishtar Terra, one of the continent-like highlands. The belts are over 1000 km long, several hundred km wide, and several km high. The high-resolution *Magellan* data reveal a complex structure within the mountain system (Figure 24.12F). Folds and thrust sheets form parallel ridges and troughs that clearly represent compressional deformation. The presence of these folded mountain belts clearly implies significant crustal shortening—a phenomenon that on Earth resulted from plate tectonics. They form the steepest slopes yet observed on Venus and suggest that the forces forming them have been active in recent times and are perhaps still active today.

How do the mountain belts on Venus compare with those on Earth?

Impact Features.   Venus does not have a heavily cratered terrain dominated by ancient impact structures as do the Moon, Mercury, and Mars, but it does have a number of relatively young craters. Those larger than 10 km in diameter show familiar features such as central peaks, terraced rims, and radial ejecta. Yet the impact structures also show some remarkably unique features. Many of the craters are asymmetrical, suggesting that the projectile hit the surface at an oblique angle (Figure 24.12G). Many smaller craters have complex interiors that appear to represent impact of several projectiles simultaneously. These clusters are interpreted to represent a family of smaller projectiles derived from a larger meteorite that broke up as it entered the very dense atmosphere of Venus. One of the most distinctive characteristics of the ejecta of craters on Venus is the striking contrast in roughness compared to the smooth volcanic plains upon which they lie. Rough surfaces appear white on radar images; smooth surfaces appear dark. Thus, there is a clear demarcation between the margins of the ejecta blanket and the smooth plains. Note also that the margins of the ejecta blankets are commonly lobate, suggesting that the ejecta flowed like an avalanche or mud flow. This may be due to the extreme heat on the surface of Venus. The heat would tend to partially melt the ejecta, causing the material to move as a flow rather than fall as discrete particles. These lobes are similar, at least in general appearance, to those around the impact craters on Mars.

How do the impact features on Venus differ from those on Mercury, the Moon, and Mars?

Many large craters show evidence of distinctive bright flow features extending tens to hundreds of kilometers beyond the edges of the ejecta blankets. These features may be due to flows of impact melt or to ashflow-like material composed of fine-grained ejecta heated by the impact of meteorites and the high surface temperature of Venus.

Erosional Features.   Although erosional and depositional features dominate the surface of Earth and are common on Mars, they appear to be relatively insignificant on Venus. The lack of sediment supply (no soil-forming process or fluvial erosion exists) limits the effectiveness of sedimentary processes on the surface of Venus to eolian and mass movement processes. In a few localized areas, however, distinct wind streaks and dunes have been observed near isolated supplies of sediment, such as fault scarps and ejecta, but these are limited. On Earth's Moon, soil, or regolith, is produced by micrometeorite bombardment. On Earth, the atmosphere screens out micrometeorites, and aqueous physical and chemical processes dominate. On Venus, weathering processes are different still. The dense atmosphere screens out micrometeorites, and because of the constant high temperatures and lack of water on the surface, rocks are not subject to freeze-thaw cycles and aqueous physical and chemical erosion processes as they are on Earth. Tectonic and volcanic structures and sequences of events are thus remarkably well preserved on Venus, allowing study of important processes and relations that are commonly eroded and obscured on Earth.

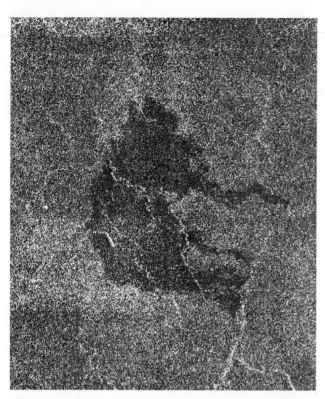

(A) Lava flows (dark) emerging from several sources along a fracture (bright discontinuous line) merge to form a lava lake in the low part of the plains.

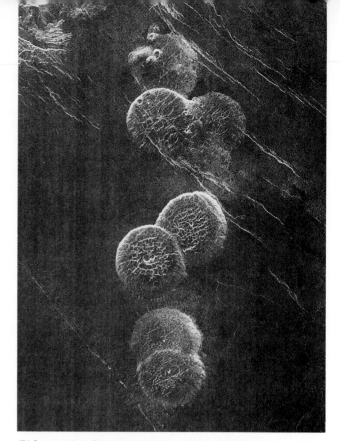

(B) Steep-sided, flat-topped domes similar to silicic domes on Earth. Their structure and morphology suggest they were formed by viscous magma.

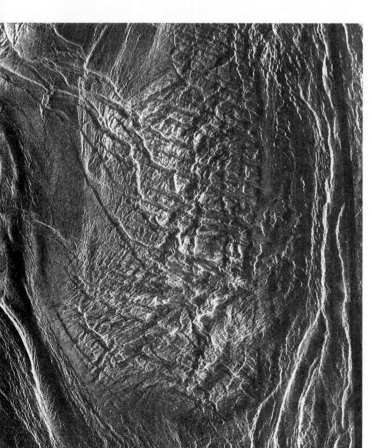

(C) A large structural dome cut by a complex fault system and flanked by a series of folds. The width of the image is about 125 km.

FIGURE 24.12
The surface of Venus as revealed by *Magellan* radar images.

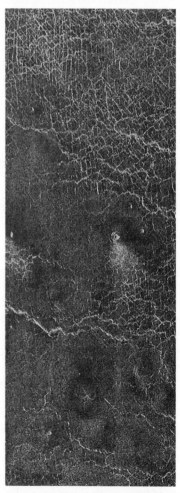

(D) Intensely fractured plains produce a complex polygonal pattern. The small bright spots are believed to be volcanic vents.

596

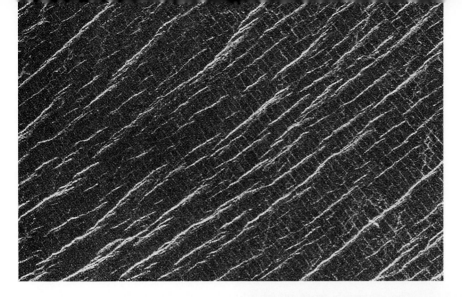

(E) Volcanic plains cut by intersecting fracture systems form a remarkable grid pattern. The fractures in each set have different characteristics of size and shape and represent different stages of deformation.

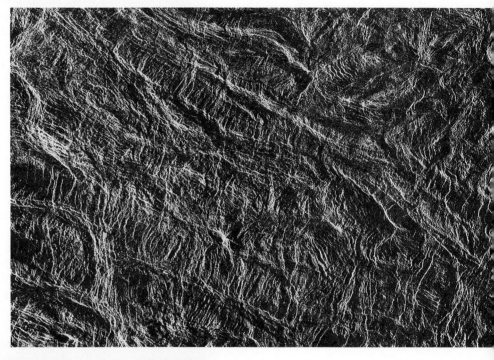

(F) Intense folding and faulting show that the crust of Venus has been deformed by tectonic processes a number of times. Without stream erosion the structures have a distinctive surface expression.

(G) Impact craters on Venus are unique because of the high surface temperature and dense atmospheric pressure. This crater is about 31 km in diameter. The asymmetrical, radial-lobate ejecta pattern suggests the ejected material was fluid like a mudflow. Note the bright flows extending to the upper right as part of the continuous ejecta deposit.

It appears that Venus has had a long and especially eventful geologic history. Venus, like its sister planet Earth, does lack a heavily cratered terrain that must have formed early in the history of the solar system. Billions of years of volcanism and tectonism have erased any vestige of its battered crust. Nonetheless, it does not appear that Venus has developed a system of plate tectonics to recycle its lithosphere and rid its interior of heat. Instead, Venus, like Mars, seems to be losing heat by hotspot development. The studies presently under way may tell us if Venus has remained geologically active until this day.

PLANETARY BODIES IN THE OUTER SOLAR SYSTEM

The materials from which planetary bodies in the outer solar system were formed consist mostly of the lighter elements: hydrogen, helium, and oxygen. Hence, the satellites of the giant planets are composed mostly of ice. They are relatively small and did not generate enough internal heat to sustain geologic activity much beyond the period of intense bombardment. Io, Europa, Ganymede, Enceladus, Miranda, and Triton are notable exceptions in that each has distinctive tectonic styles resulting from unique energy systems.

Jupiter and Its Satellites

Jupiter and its moons constitute a planetary system of incredible beauty and intrigue. The giant planet has a volume 1300 times greater than Earth's, but it is composed mostly of gas and liquid swirling in complex patterns. Jupiter has no solid surface, hence no record of a geologic history. But its moons are solid planetary bodies that contain geologic wonders far beyond those anticipated only twenty years ago. The four large moons of Jupiter (Io, Europa, Ganymede, and Callisto), called the Galilean satellites because they were discovered by Galileo in 1610, have compositions dramatically different from those of the rocky inner planets. Three of these moons have surfaces composed mostly of water ice. Each of Jupiter's moons shows a diverse landscape resulting from impact, volcanism, and surface fracturing (Figure 24.13).

Io. Io is the innermost major satellite of Jupiter. It is only slightly larger and denser than Earth's Moon, but the exploration of Io has proved to be one of the most significant results of the *Voyager* missions, which explored the outer solar system in the last decade. Like the inner planets, Io must be composed mostly of silicate minerals. However, unlike the Moon Io is volcanically active, and much of its topography is influenced in some way by the molten sulfur spewed from huge volcanoes scattered across its surface (Figure 24.14A). The volcanic eruptions on Io are like huge fountains; they spray ash much as a sprinkler system waters a lawn. The form of the spray depends on the size and shape of the vent. Some are in the form of huge umbrellas; others are jetlike streaks. Io appears to be the most volcanically active body in the solar system and probably has been throughout much of geologic time. The question is why? The Moon and Mercury, which are about the same size as Io, cooled quickly because of their small sizes and have been volcanically dead for almost 3 billion years. The reason for Io's volcanism may be found in the internal friction generated in the satellite as it circles Jupiter. Europa and Ganymede, as well as Jupiter, exert a gravitational pull on Io, forcing it into an eccentric orbit. This eccentricity causes variations in the amplitude of the tidal force due to Jupiter's gravitational field. Io is therefore constantly wrenched

FIGURE 24.13
Jupiter and its major moons were photographed early in March 1979 by *Voyager 1* and assembled into this composite picture. The satellites are not to scale but are in their proper relative positions. Io, orange in color (upper left) is nearest Jupiter; next are Europa (center), Ganymede, and Callisto (lower right). Not visible are Jupiter's faint rings, seen for the first time by *Voyager 1*.

The Galilean satellites are roughly the size of Earth's Moon and show remarkable differences in surface features and composition. Io, the innermost of the satellites, is rocky and has active volcanoes. Its surface is young, and all ancient impact craters are covered and obliterated. Sulfur volcanoes and lava flows dominate its surface. Europa has a thick outer shell of fractured ice. Ganymede has large expanses of ancient cratered terrain cut by younger grooved terrain formed by tectonic activity and volcanic extrusions. Callisto is heavily cratered and has a large multiring basin.

about by variable gravitational forces. These conditions cause the interior of the satellite to yield somewhat, producing friction and heat. Indeed, Io may be molten at a depth of only 20 km.

Europa.   Europa has a density of 3.03 g per cm^3 and must therefore be composed mostly of rock, but spectroscopic measurements indicate that Europa is surrounded by a frozen ocean of ice. Its surface is distinctive in that it is essentially free from craters and is almost perfectly smooth, suggesting that Europa is geologically active. The major surface features on Europa are sets of tan streaks or bands that appear to be similar to fractured ice packs in the polar regions of Earth (Figure 24.14B). Most appear to be shallow valleys up to 200 km wide and up to 3000 km long. Internal heat apparently formed an exotic form of lava, a watery slush that erupted through cracks and fissures in the crust and coated the surface with fresh new ice. The near-absence of impact craters on Europa suggests that the surface is very young, formed after the early periods of heavy meteorite bombardment that scarred the ancient surfaces of Mercury and the Moon. Resurfacing by eruption of watery lavas must have continued until relatively recent times.

Ganymede.   Jupiter's largest satellite is Ganymede, with a diameter of approximately 1.5 times that of Earth's Moon. It is even larger than the planet Mercury. It has a bulk density of only 1.93 g per cm^3 and consists of large amounts of water in the form of ice, a conclusion supported by the planet's high reflectivity. This evidence suggests that Ganymede is composed of approximately 50% water. A preliminary model of Ganymede's internal structure suggests a silicate rock core surrounded by a mantle of liquid water or slushy ice and a rigid ice crust.

The view shown in Figure 24.14C reveals that Ganymede has a cratered surface with two contrasting terrain types. The older terrain is nearly saturated with craters. It is dark and is believed to be composed of "dirty" ice, containing fragments of dust and particles from outer space. This older crust appears to have been fractured and split apart, and many of the fragments have shifted about. The younger terrain is brighter and consists of a series of grooves and complex strips, or bands, which are closely spaced and roughly parallel. Crater density on the grooved terrain is much less than that on the darker surface, so the grooved terrain is believed to be much younger. Its features probably result from tectonic processes acting on an ice crust. Apparently the old, cratered terrain was fractured and split apart at some time late in the period of intense bombardment, and cleaner ice from below was extruded into the fractures to form the grooved terrain. This suggests the breaking and movement of crustal fragments, a type of plate tectonics on a frozen world with a lithosphere of ice.

Callisto.   Callisto is the outermost Galilean satellite; like Ganymede, it is believed to consist of a rocky core surrounded by a thick mantle of ice. The images sent back from *Voyager* spacecraft show that Callisto, in contrast to the other Galilean satellites, is saturated with craters, somewhat like the highlands of the Moon (Figure 24.14D). The general surface of Callisto is dark, dirty ice, similar to the older terrain of Ganymede, but many craters have bright rays and ejecta blankets. The bright material is probably clean melted ice from below the dirty crust, ejected onto the surface from impact. Thus, the surface of Callisto is believed to be very old, recording events during the early history of the solar system. Aside from the densely cratered terrain, the most striking feature on Callisto is a large multiring basin. This feature is reminiscent of the multiring basins on the planets of the inner solar system. Important differences in numbers of rings, their spacing, and their elevation exist, however, probably because the response of an ice crust to the shock of impact is different from that of a rocky crust.

What geologic features are unique on each of the Galilean satellites of Jupiter?

(A) Io

(B) Europa

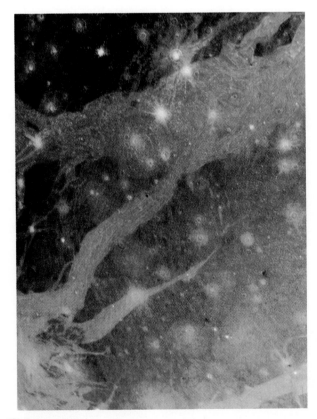

(C) Ganymede

FIGURE 24.14
The Galilean moons of Jupiter

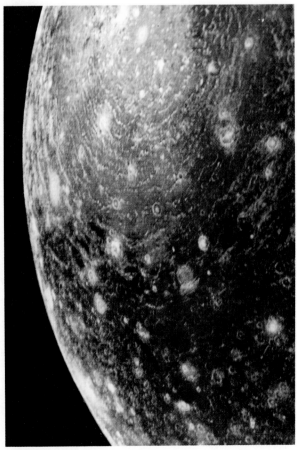

(D) Callisto

Saturn and its Satellites

Saturn is similar to Jupiter in many ways. It is a gigantic ball of gas, mostly hydrogen and helium, and is the center of a miniature planetary system with an elaborate family of satellites. Its atmosphere is not as colorful as Jupiter's but is marked by dark bands alternating with lighter zones. The rings of Saturn, of course, have long been considered its most dramatic feature; they have intrigued astronomers for over 300 years (Figure 24.15). Now that we have seen them close up they are even more astonishing. They extend over a distance of 40,000 km and yet are only a few kilometers thick. The rings are probably made up of billions of particles of ice and ice-covered rock ranging from a few micrometers to a meter or more in diameter. Rarely are particles in the rings more than a kilometer in diameter. Each particle moves in its independent orbit around Saturn, producing an extraordinarily complex ring structure.

With the exception of the satellite Titan, the seven largest moons of Saturn are small, icy bodies that appear through a telescope only as points of light. They range from 390 to 1530 km in diameter, 1/2 to 1/10 the diameter of the Moon.

What is the composition of the satellites of Saturn?

One might think that such small icy bodies would have little interest to the geologist because of their cold origins, primitive compositional character, and apparent lack of an internal source of heat for any geologic activity. As it turns out, however, the moons of Saturn experienced many events, now preserved in the morphology of their surface features, that help us better understand the geology of the entire solar system.

What geologic activity occurred on Enceladus?

The surfaces of most of the icy satellites of Saturn are saturated with impact craters; some of these are gigantic compared to the size of the planetary body (Figure 24.15). Others have large fracture systems and strange surface markings, probably resulting from an exotic type of volcanism formed on these small bodies by the melting of their icy interiors. One satellite, Enceladus, has a grooved terrain with fractures and fissures and smooth plains that may have been produced by "lavas" of slushy water that erupted from the fissures. The heating of Enceladus is probably related to the same type of tidal heating that warms Io. Titan, Saturn's largest moon, has a thick atmosphere composed mainly of nitrogen, a characteristic it shares with Earth.

FIGURE 24.15
Saturn and its major moons assembled as a composite picture from *Voyager* photographs. This artist's view shows Dione in the forefront, Saturn rising behind, Tethys and Mimas fading to the distance to the right, Enceladus and Rhea off Saturn's rings to the left, and Titan in its distant orbit at the top. The icy moons of Saturn are a geologically diverse group of mostly small satellites. They show an amazing variety of young and old surfaces, impact craters, evidence of icy volcanism, and global fracture systems.

Enceladus is one of the most interesting and is shown close-up on the facing page. Although it is a tiny satellite, it shows a remarkable record of geologic activity. The smooth uncratered surface is geologically young, which indicates that Enceladus has experienced a relatively recent thermal event and an exotic form of volcanic activity in which floods of water and icy slush were extruded.

Uranus and Its Satellites

Uranus is a unique world. Like Jupiter and Saturn, it has no solid surface but is enveloped by a thick atmosphere of hydrogen and helium. In contrast to all other planets in the solar system, it is tipped and spins on its side; that is, its axis of rotation lies nearly in the plane of its orbit. Thus, it rolls like a ball as it moves on its orbital path around the sun, whereas other planets spin like tops. Uranus has five major moons that can be seen from telescopic observations on Earth. Each moon occupies a nearly circular orbit lying in the plane of Uranus's equator. Their orbits share the unusual axial inclination of the planet itself. Oberon, Titania, Ariel, and Umbriel are quite similar in size (1100 to 1600 km in diameter) and are approximately the size of the intermediate moons of Saturn (Tethys, Dione, and Rhea). Their surfaces are nearly saturated with craters (Figure 24.16).

How is Miranda different from all other planetary bodies in the solar system?

Miranda is a different story! *Voyager* flew within 29,000 km of the tiny moon and photographed its surface features in remarkable detail. Most scientists expected Miranda to be a small sphere of ice, like Saturn's tiny moon Mimas, little changed since its birth more than 4.6 billion years ago. What the pictures of Miranda revealed was a strange planetary body with a surface no one had ever imagined. Miranda is a small, icy body with barely enough gravitational strength to pull itself into a sphere, yet its surface consists of a complex and alien terrain. The surface of Miranda consists of two strikingly different terrain types (Figure 24.16). One terrain is an old, densely cratered surface similar to that on other densely cratered planetary bodies and is therefore probably more than 4 billion years old. The other is completely different, consisting of three roughly circular or oval areas of complex structures, characterized by parallel sets of alternating light and dark bands and by cliffs and ridges. But this is not all. Miranda also has a series of enormous faults that can be traced across the globe. Some are older than the complex terrain; others are younger. One fault forms a huge cliff 20 km high, which is more than twice as deep as the Grand Canyon on Earth.

Why does Miranda have such a strange and unusual surface? Some scientists believe that shortly after it was formed Miranda collided with a large comet or meteorite that shattered it into several large fragments that remained in the same general orbit as the original satellite. Gravitational attraction caused the fragments to reassemble, and some of the large fragments that had been near the core were now exposed at the surface.

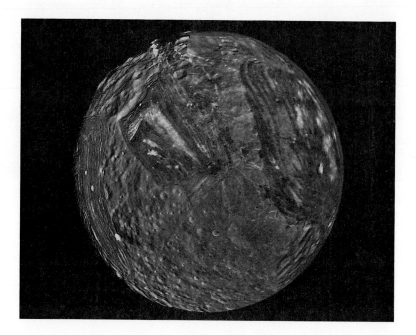

FIGURE 24.16
Uranus and its five major satellites assembled in a composite picture from *Voyager* photographs. Note that the axes of rotation of Uranus and its moons are nearly perpendicular to that of the rest of the planets in the solar system. A close-up view of Miranda is shown on the facing page.

Miranda is about the size of Saturn's tiny moon Mimas (470 km in diameter). The view on the facing page shows evidence of significant crustal mobility and resurfacing after the period of heavy bombardment, plus a huge fault scarp (at the top of the image). Miranda may have been fragmented by collision with a large comet and then reassembled.

Neptune and Its Satellites

Neptune is only slightly smaller than Uranus and is similar to its neighbor in its composition. Both planets, called the twins of the outer solar system, are thought to have large cores of water ice and rock surrounded by thick atmospheres of hydrogen, helium, and minor methane. Only two moons (Figure 24.17) were known to orbit Neptune before the *Voyager* spacecraft passed it in August of 1989. At that time six additional tiny moons were discovered. Triton, the largest moon, is only slightly smaller than Earth's Moon. It has an extremely tenuous atmosphere of nitrogen and methane, and its exotic landscape is formed from ices of those gases. Triton has a surprisingly large variety of geologic features, including ice caps, fractured terrain, "lava" lakes, and volcanic or geyser eruptions (Figure 24.17). Its constantly changing ice caps, together with ices of nitrogen and methane found on its surface, mark it as one of the most distinctive planetary bodies in the solar system.

What geologic activity was observed on Triton?

FIGURE 24.17
Neptune and its largest satellite, Triton, as photographed by *Voyager 2.* Neptune is a giant gaseous planet with a banded blue atmosphere decorated with brilliant white clouds of methane ice. Because Triton is so far from the sun, its surface temperature is so low (37 K) that nitrogen is frozen solid to form a large ice cap. A close-up view of Triton is shown on the facing page.

Triton has complex surface features that show incredible geologic processes. A fractured terrain with many crisscrossed linear features appears to be the result of rifting. Floods of lava (probably mixtures of water, nitrogen, and methane) formed smooth plains and lava lakes. Dark streaks are formed from geyserlike volcanic eruptions.

THE ORIGIN OF THE SOLAR SYSTEM

The solar system probably formed by the gravitational collapse of a huge cloud of gas and dust. The reason for differences in planetary composition appears to be related to distance from the sun. Bodies close to the sun (the terrestrial planets) were formed from rocks and metals that crystallized at high temperatures, whereas the outer planets were formed from elements that form solids (condense) at low temperatures.

Most scientists believe that the universe began about 15 billion years ago in what has become known as the Big Bang. This gigantic explosion caused matter to expand to form the billions of swirling galaxies and, in time, the stars and their planets. It is generally believed that our solar system was spawned in a cold diffuse cloud of gas and dust, deep within the spiral arm of the Milky Way galaxy. The huge cloud was made up largely of the two lightest elements, hydrogen and helium. The cloud not only rotated slowly about a central concentration of mass, but it contained a system of complicated eddies. Under the force of gravity, the giant cloud began to collapse and assume the shape of a rotating disk, with an increasingly hot and dense mass at the center (Figure 24.18).

During the collapse, much of the cloud's matter swirled toward the dense central core to form the sun. The outer part of the cloud was naturally the coldest, so that substances there, such as water, ammonia, and methane, solidified. Nearer the sun, those materials remained as vapor, but only silicon, iron, aluminum, and similar materials could crystallize at high temperature into solids and formed rocky material. Thus, early in the history of the solar system, there was a separation and differentiation of material. Denser particles were concentrated in the central region whereas lighter icy materials were concentrated near the fringes of the cloud. Over a relatively short period of time (possibly as short as 100,000 years) the small particles in the embryonic solar system accreted into larger and larger particles until asteroid-sized bodies of rock and ice called planetesimals were formed. As the planetesimals traveled in orbit around the infant sun, the larger bodies grew by gravitational attraction because they were able to sweep up much of the smaller material near their orbital paths. These planetesimals became the principal planets. The size and composition of a planet were therefore determined to a considerable degree by its distance from the sun. In the high-temperature regions near the sun, only materials such as metals and silicates could crystallize into solids and accrete to form planets. Proceeding outward toward cooler and cooler temperatures, materials with lower melting points such as water and methane could also become solid (ices). In the coldest outer regions, virtually all materials except gaseous hydrogen, helium, and neon were solid and formed the planets and their satellites. Gases were gravitationally anchored on the huge icy cores of the giant planets. The general variation in planetary composition follows this sequence with regard to distance from the sun: small planets of rock and metal near the sun, giant planets of gas and ice far from the sun in the outer solar system.

With time, the gravitational field of the largest body (the ancestral sun) imposed more structure and order to the system. Most of the material of the nebula swirled inward where it compacted into a solar core. The intense pressure raised the core temperature to a point where it ignited, becoming a vast nuclear furnace—a new star, the sun. The principal planets and their satellites orbited the sun, sweeping up most of the remaining debris in their orbital paths. This final stage of planetary formation is clearly recorded on the surfaces of the Moon, Mercury, Mars, and most other planetary bodies as

(A) A slowly rotating portion of a large nebula becomes a distinct globule as a mostly gaseous cloud collapses by gravitational attraction.

(B) Rotation of the cloud prevents collapse of the equatorial disk while a dense central mass forms.

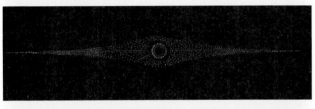

(C) A protostar "ignites" and warms the inner part of the nebula, possibly vaporizing preexisting dust. As the nebula cools, condensation produces solid grains that settle to the central plane of the nebula.

(D) The dusty nebula clears either by dust aggregation into larger particles (planets or planetesimals) or by ejection during a T-Tauri stage of the star's evolution. A star energized by fusion and a system of cold bodies remains. Gravitational accretion of these small bodies eventually leads to the development of a small number of major planets.

FIGURE 24.18
The evolution of a dusty nebula with a surrounding system of orbiting planets is shown in this schematic diagram.

densely cratered terrain. The subsequent history of a planet would depend upon its own internal thermal energy and upon the thermal energy reaching the planet's surface from the sun.

All planetary bodies were heated to some degree as a result of the impact of the numerous planetesimals that formed them. If heated sufficiently, much of the planet melted and the constituent materials became differentiated; that is, denser materials were separated and concentrated in the core and lighter materials were concentrated near the surface. This process is called *planetary differentiation.*

The subsequent thermal history and resulting geologic activity of a planet are largely dependent on the planet's size. Small bodies tend to cool down rapidly because they have large surface areas compared to their masses. Larger bodies tend to retain their heat longer, and as a result, have prolonged periods of internal geologic activity such as volcanism and crustal deformation. An additional source of heat is the radioactive decay of elements (long-lived isotopes of uranium, thorium, and potassium continue to heat Earth today, but other short-lived isotopes could have been important in the past). But other factors are also important, such as tidal forces, the presence or absence of an atmosphere and water, the amount of solar radiation, and the influence of a biosphere.

Concluding Statement

This has been an extraordinary period of geologic exploration. The *Voyager* missions, as well as others, have explored all of the planets except Pluto. The space program has permitted us to see whole new worlds and, for the first time, to compare them with our own. All of the planetary bodies in the solar system are fascinating but none has more intrigue than the planet Earth. It is unique in our solar system. In exploring the worlds in our solar neighborhood, we are beginning to understand why Earth is so different: why we have an atmosphere, climates, continents, and ocean basins; why we alone have life.

The evidence we have accumulated suggests that all nine planets and their many moons were born at the same time from the vast amount of matter moving about the sun. The terrestrial planets, Mercury, Venus, Earth, and Mars, are composed of rocky material that formed at the high temperature that prevailed near the sun. They all probably started out pretty much alike, but being different in size and in their distance from the sun, they evolved along separate paths. Mercury and the Moon were too small to generate enough internal heat to sustain long-lived tectonic systems, and after their first major thermal events, they ceased to be geologically active. They thus remain "fossils" of the early stages in planetary development and provide the first chapters of the history of our corner of space. We are now reading stories once thought to be lost forever because Earth's "genesis" rocks, formed during this infancy period, were destroyed by subsequent cycles of weathering, erosion, and tectonism. Now, on the surfaces of the Moon and Mercury, we can read a record of the events that occurred soon after Earth was born—a chapter of our beginnings that we knew nothing about only a few decades ago.

Mars, being larger and cooling more slowly than the Moon and Mercury, developed a more prolonged period of tectonism. Its surface is marked by huge volcanic features, major rifting, and associated landforms modified by running water. It is far enough away from the sun to have incorporated significant amounts of water. Moreover, it is large enough to hold an atmosphere and, as a result, has had a type of hydrologic system.

Venus is more like Earth, yet it is different. We don't know all of the reasons why. It has continent-like land masses, volcanic activity, and folded mountains. Its atmosphere is dense and quite unlike that of Earth. Apparently the high surface temperature of Venus never dipped low enough for liquid water to form. As a result, all of the CO_2 in the atmosphere remained gaseous, being unable to be incorporated into sedimentary rocks. This sent the planet into a hellish greenhouse condition.

The materials from which the planetary bodies in the cold outer solar system were formed consisted mostly of the lighter elements: hydrogen, helium, and oxygen. Hence, the solid planetary bodies in this part of the solar system are composed mostly of ice. They are all relatively small and did not generate enough internal heat to sustain geologic activity much beyond the period of intense bombardment. Io, Europa, Ganymede, Enceladus, and Miranda are notable exceptions in that each has evolved distinctive surface features resulting from a unique energy system.

And so, by studying other planetary bodies, we gain a greater understanding of our own Earth. Its size and composition are just right for the development of a tectonic system that recycles the lithosphere, creates continents and ocean basins, and concentrates ores and minerals. Earth's gravitational field is strong enough to hold an atmosphere. We are just the right distance from the sun so that water can exist as solid, liquid, and vapor and can move in a hydrologic cycle. If we were a little closer to the sun, our oceans would evaporate; if we were farther from the sun, they would freeze solid.

The study of other planets has taught us that Earth is a small place, a world that is an oasis in space, a home we are still learning to understand.

SUMMARY

1. A vast amount of new knowledge has been obtained about the planets of the solar system as a result of the space program. This new knowledge permits geologists, for the first time, to study details of other planets and compare the planets with Earth. Some of the more important discoveries and some hypotheses that explain them are summarized here.

   a. **Craters.** Cratering has been the dominant geologic process in the solar system, and it may have been the dominant process on Earth during the first half billion years of its history.

   b. **Volcanism.** Most of the planetary bodies in the solar system experienced a thermal event early in their history, after the period of intense bombardment. Mars and Venus have experienced more-recent volcanic activity, as is indicated by their great shield volcanoes. The volcanic styles of Earth and of Io reflect their distinctive tectonic systems.

   c. **Tectonic features.** The circular shape of craters on the planetary bodies of the solar system indicates that the crusts of many of these planets have not been deformed by strong compressive forces. Mercury has a global pattern of thrust faults. Mars and Venus have upwarps and graben systems. None of the planets, however, has developed an active tectonic system like Earth's. Ganymede, Europa, and Enceladus present some distinctive tectonic styles involving ice tectonics. Major grabens also occur on some of the moons of Saturn and Uranus, possibly the result of freezing water ice and expansion on a planetary scale.

   d. **Atmospheres.** The surfaces of the Moon and Mercury have not been changed by atmospheric processes. Mars, in contrast, has a thin atmosphere. Its surface has been modified by stream erosion and deposition, mass movement, wind, ice caps, and ground ice. Venus has a dense atmosphere, which influences impact processes, weathering, erosion, and volcanic extrusions in surprising ways.

   e. **Internal structure.** To various degrees, all of the large planetary bodies appear to have experienced differentiation as a result of accretionary heating.

   f. **Planetary composition.** The composition (silicate vs. ice and gas) and size (small vs. large) of the planets were largely controlled by their distance from the sun. The planets formed 4.5 billion years ago when a cloud of gas and dust condensed to form solids around a hot star. High-temperature silicates and metals condensed near the sun and eventually formed the terrestrial planets. Ices of water, ammonia, and methane also condensed at great distances from the sun and formed the large outer planets. Thick cloaks of hydrogen and helium were then attached to the icy cores.

   g. **Geologic history.** A geologic time scale for events on most of the planetary bodies has been established by the principles of superposition and crosscutting relations. These provide a framework of relative time in which the major events of their geologic history can be arranged chronologically.

2. The early histories of all planetary bodies in the solar system are similar in that they experienced intense bombardment, which produced densely cratered terrain, and, somewhat later, they experienced the impact of asteroid-sized bodies, which formed multiring basins. A subsequent thermal event is also evident in most planetary bodies—an event that formed smooth volcanic plains (probably basalt on the inner planets and water ice on the outer satellites).

3. The Moon and Mercury and many satellites of the outer planets remain primitive bodies in that their surfaces have not been changed by atmospheric erosion or by continuing convection in the planet's interiors.

4. Mars has had a major eventful geologic history involving (1) the formation of densely cratered terrain, (2) uplift of segments of the crust and radial fracturing, (3) widespread extrusion of lava, which formed parts of the northern plains, (4) renewed uplift in the Tharsis region and the formation of Valles Marineris, and (5) volcanism and eolian activity.

5. In planetary development, Venus appears to be more like Mars than any other planetary body, and the history of Venus is probably similar to that of Mars. These two planets appear to have reached an intermediate stage of planetary development between that of a primitive, impact-dominated body (such as the Moon and Mercury) and a tectonically active, water-dominated planet (such as Earth).

6. Io seems to be the most volcanically active body in the solar system. Its surface is continually being renewed by volcanism.

7. The small, icy satellites of Saturn probably experienced intense meteorite bombardment during an early period of their history, as did the other planetary bodies of the solar system.

8. Earth, unlike the other planets, has had an active tectonic system throughout its history. Its unique atmosphere and abundant supply of water have also played an important role in the development of the planet's surface features. Earth's dynamic systems—its tectonic system and its hydrologic system, both acting to modify its surface—are responsible for the features that distinguish Earth from other planetary bodies.

KEY TERMS

basin (p. 582)

crater (p. 582)

ejecta blanket (p. 584)

maria (p. 584)

meteorite (p. 583)

multiring basin (p. 584)

planetary differentiation (p. 609)

rayed crater (p. 584)

ADDITIONAL READINGS

Betty, J. K., B. O'Leary, and A. Chaikin. 1990. *The New Solar System.* 3rd ed. New York: Cambridge University Press.

Greeley, R. 1985. *Planetary Landscapes.* London: Allen and Unwin.

Hamblin, W. K., and E. H. Christiansen. 1990. *Exploring the Planets.* New York: Macmillan Publishing Co.

Morrison, D., and T. Owen. 1987. *The Planetary System.* Reading MA: Addison-Wesley.

REVIEW QUESTIONS

1. What geologic process has been most significant in modifying the surfaces of the Moon, Mercury, and Mars?
2. Outline the stages in the production of a crater by the impact of a meteorite. What geologic features are produced by impact?
3. How are craters modified with time?
4. Explain how a geologic time scale was developed for events in the Moon's history.
5. Outline the major events in lunar history.
6. Compare and contrast the geology of Mercury with that of the Moon.
7. Describe the craters on Mars. How do they differ from those formed on the Moon?
8. Describe the volcanoes on Mars.
9. What tectonic features are found on Mars?
10. Describe the fluvial features on the surface of Mars. How do they compare with fluvial features on Earth?
11. Describe Valles Marineris. How did it originate?
12. What evidence indicates "catastrophic" flooding on Mars?
13. Describe the surface features generated by wind on Mars.
14. Outline the geologic history of Mars. How is it similar to that of the Moon?
15. Compare and contrast the surface features of Venus with those on Earth and Mars.
16. Compare and contrast the surface features of the moons of Jupiter.
17. Explain the origin of volcanism on Io.
18. What is the significance of the major surface features on the Saturnian moon Enceladus?
19. How is Earth geologically unique among the planetary bodies of the solar system?
20. What is unique about Miranda?

Glossary

Aa flow A lava flow with a surface typified by angular, jagged blocks (see diagram). Contrast with *pahoehoe flow*.

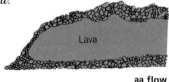

aa flow

ablation Reduction of a glacier by melting, evaporation, iceberg calfing, or deflation.

abrasion The mechanical wearing away of a rock by friction, rubbing, scraping, or grinding.

absolute time Geologic time measured in a specific duration of years (in contrast to relative time, which involves only the chronologic order of events).

abyssal Pertaining to the great depths of the oceans, generally 1000 fathoms (2000 m) or more below sea level.

abyssal floor The deep, relatively flat surface of the ocean floor located on both sides of the oceanic ridge. It includes the abyssal plains and the abyssal hills.

abyssal hills The part of the ocean floor consisting of hills rising as much as 1000 m above the surrounding floor. They are found seaward of most abyssal plains and occur in profusion in basins isolated from continents by trenches, ridges, or rises (see diagram).

abyssal plains Flat areas of the ocean floor, having a slope of less than 1:1000. Most abyssal plains lie at the base of a continental rise and are simply areas where abyssal hills are completely covered with sediment (see diagram).

active margin (plate tectonics) The leading edge of a lithospheric plate bordered by a trench.

aftershock An earthquake that follows a larger earthquake. Generally, many aftershocks occur over a period of days or even months after a major earthquake.

agate A variety of cryptocrystalline quartz in which colors occur in bands. It is commonly deposited in cavities in rocks.

aggradation The process of building up a surface by deposition of sediment.

A horizon The topsoil layer in a soil profile.

alcove A large niche or recession formed in a steep cliff.

alluvial fan A fan-shaped deposit of sediment built by a stream where it emerges from an upland or a mountain range into a broad valley or plain (see diagram). Alluvial fans are common in arid and semiarid climates but are not restricted to them.

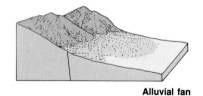

Alluvial fan

alluvium A general term for any sedimentary accumulations deposited by comparatively recent action of rivers. It thus includes sediment laid down in river beds, flood plains, and alluvial fans.

alpine glacier A glacier occupying a valley. Synonymous with *mountain glacier, valley glacier.*

amorphous solid A solid in which atoms or ions are not arranged in a definite crystal structure. Examples: glass, amber, obsidian.

amphibole An important rock-forming mineral group of ferromagnesian silicates. Amphibole crystals are constructed from double chains of silicon-oxygen tetrahedra. Example: hornblende.

amphibolite A metamorphic rock consisting mostly of amphibole and plagioclase feldspar.

andesite A fine-grained igneous rock composed mostly of plagioclase feldspar and from 25 to 40% amphibole and biotite, but no quartz or K-feldspar. It is abundant in mountains bordering the Pacific Ocean, such as the Andes Mountains of South America, from which the name was derived. Andesitic magma is believed to originate from fractionation of partially melted basalt.

andesite line The boundary in the Pacific Ocean separating volcanoes of the inner Pacific basin, which discharge only basalt, from those near the continental margins, which discharge both andesite and basalt.

angular unconformity An unconformity in which the older strata dip at a different angle (generally steeper) than the younger strata (see diagram).

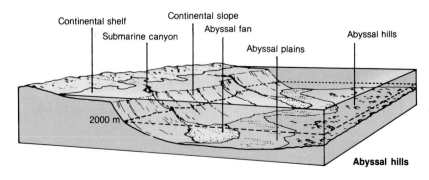

Continental shelf
Submarine canyon
Continental slope
Abyssal fan
Abyssal plains
Abyssal hills
2000 m
Abyssal hills

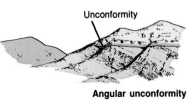

Unconformity
Angular unconformity

anomaly A deviation from the norm or average.

anorthosite A coarse-grained intrusive igneous rock composed primarily of calcium-rich plagioclase feldspar.

anticline A fold in which the limbs dip away from the axis. After erosion, the oldest rocks are exposed in the central core of the fold (see diagram).

Anticline

aphanitic texture A rock texture in which individual crystals are too small to be identified without the aid of a microscope. In hand specimens, aphanitic rocks appear to be dense and structureless.

aquifer A permeable stratum or zone below the Earth's surface through which groundwater moves (see diagram).

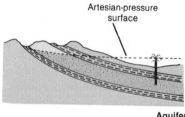

Aquifer

arc-trench gap The geographic area in an island arc deep-sea trench system that separates the arc of volcanoes from the trench. In most cases, the gap is about 100 km wide.

arête A narrow, sharp ridge separating two adjacent glacial valleys.

arkose A sandstone containing at least 25% feldspar.

artesian basin A geologic structural feature in which groundwater is confined and is under artesian pressure.

artesian-pressure surface The level to which water in an artesian system would rise in a pipe high enough to stop the flow.

artesian water Groundwater confined in an aquifer and under pressure great enough to cause the water to rise above the top of the aquifer when it is tapped by a well.

ash Volcanic fragments the size of dust particles.

ash flow A turbulent blend of unsorted pyroclastic material (mostly fine-grained) mixed with high-temperature gases ejected explosively from a fissure or crater.

ash-flow tuff A rock composed of volcanic ash and dust, formed by deposition and consolidation of ash flows.

asteroid A small, rocky planetary body orbiting the sun. Asteroids are numbered in the tens of thousands. Most are located between the orbit of Mars and the orbit of Jupiter. Their diameters range downward from 770 km.

asthenosphere The zone in the Earth directly below the lithosphere, from 70 to 200 km below the surface. Seismic velocities are distinctly lower in the asthenosphere than in adjacent parts of the Earth's interior. The material in the asthenosphere is therefore believed to be soft and yielding to plastic flow.

astrogeology The study of extraterrestrial bodies by the application of geologic methods and knowledge.

asymmetric fold A fold (anticline or syncline) in which one limb dips more steeply than the other (see diagram).

Asymmetric fold

atmosphere The mixture of gases surrounding a planet. The Earth's atmosphere consists chiefly of oxygen and nitrogen, with minor amounts of other gases. Synonymous with *air*.

atoll A ring of low coral islands surrounding a lagoon (see diagram).

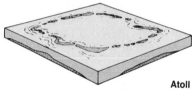

Atoll

atom The smallest unit of an element. Atoms are composed of protons, neutrons, and electrons.

attitude The three-dimensional orientation of a bed, fault, dike, or other geologic structure. It is determined by the combined measurements of the dip and the strike of a structure.

axial plane With reference to folds, an imaginary plane that intersects the crest or trough of a fold so as to divide the fold as symmetrically as possible (see diagram).

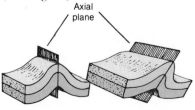

Axial plane

axis 1 (crystallography) An imaginary line passing through a crystal around which the parts of the crystal are symmetrically arranged. 2 (fold) The line where folded beds show maximum curvature. The line formed by the intersection of the axial plane with the bedding surface (see diagram).

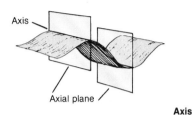

Axis

Backswamp The marshy area of a floodplain at some distance beyond and lower than the natural levees that confine the river.

backwash The return sheet flow down a beach after a wave is spent.

badlands An area nearly devoid of vegetation and dissected by stream erosion into an intricate system of closely spaced, narrow ravines.

bajada The surface of a system of coalesced alluvial fans.

bar An offshore, submerged, elongate ridge of sand or gravel built on the sea floor by waves and currents.

barchan dune A crescent-shaped dune, the tips or horns of which point downwind. Barchan dunes form in desert areas where sand is scarce.

barrier island An elongate island of sand or gravel formed parallel to a coast (see diagram).

Barrier island

barrier reef An elongate coral reef that trends parallel to the shore of an island or a continent, separated from it by a lagoon (see diagram).

Barrier reef

basalt A dark-colored, aphanitic (fine-grained) igneous rock composed of plagioclase (over 50%) and pyroxene. Olivine may or may not be present. Basalt and andesite represent 98% of all volcanic rocks.

base level The level below which a stream cannot effectively erode. Sea level is the ultimate base level, but lakes form temporary base levels for inland drainage systems.

basement complex A series of igneous and metamorphic rocks lying beneath the oldest stratified rocks of a region (see diagram). In shields, the basement complex is exposed over large areas.

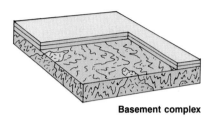

Basement complex

basin 1 (structural geology) A circular or elliptical downwarp. After erosion, the youngest beds are exposed in the central part of the structure. 2 (topography) A depression into which the surrounding area drains.

batholith A large body of intrusive igneous rock exposed over an area of at least 100 km^2 (see diagram).

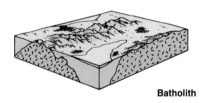

Batholith

bathymetric chart A topographic map of the Earth's surface underlying a body of water (such as the ocean floor).

bathymetry The measurement of ocean depths and the charting (mapping) of the topography of the ocean floor.

bauxite A mixture of various amorphous or crystalline hydrous aluminum oxides and aluminum hydroxides, commonly found as a residual clay deposit in tropical and subtropical regions. Bauxite is the principal commercial source of aluminum.

bay (coast) A wide, curving recess or inlet between two capes or headlands.

baymouth bar A narrow, usually submerged ridge of sand or gravel deposited across the mouth of a bay by longshore drift. Baymouth bars commonly are formed by extension of spits along embayed coasts.

beach A deposit of wave-washed sediment along a coast between the landward limit of wave action and the outermost breakers.

beach drift The migration of sediment along a beach caused by the impact of waves striking the shore at an oblique angle.

bed A layer of sediment 1 cm or more in thickness.

bedding plane A surface separating layers of sedimentary rock.

bed load Material transported by currents along the bottom of a stream or river by rolling or sliding, in contrast to material carried in suspension or in solution.

bedrock The continuous solid rock that underlies the regolith everywhere and is exposed locally at the surface. An exposure of bedrock is called an outcrop.

Benioff zone A zone of deep-focus earthquakes that dips away from a deep-sea trench and slopes beneath the adjacent continent or island arc.

berm A nearly horizontal portion of a beach or backshore formed by storm waves. Some beaches have no berms; others have several.

B horizon The solid zone of accumulation underlying the *A* horizon of a soil profile. Some of the material dissolved by leaching in the *A* horizon is deposited in the *B* horizon.

biologic material A general term for material originating from organisms. Examples: fossils (shells, bones, leaves), peat, coal.

biosphere The totality of life on or near the Earth's surface.

biotite "Black mica." An important rock-forming ferromagnesian silicate with silicon-oxygen tetrahedra arranged in sheets.

bird-foot delta A delta with distributaries extending seaward and in map view resembling the claws of a bird. Example: the Mississippi Delta.

block faulting A type of normal faulting in which segments of the crust are broken and displaced to different elevations and orientations.

blowout A basin excavated by wind erosion.

blueschist A fine-grained schistose rock characterized by high-pressure, low-temperature mineral assemblages, and typically blue in color.

boulder A rock fragment with a diameter of more than 256 mm (about the size of a volleyball). A boulder is one size larger than a cobble.

bracketed intrusion An intrusive rock that was once exposed at the surface by erosion and was subsequently covered by younger sediment. The relative age of the intrusion thus falls between, or is bracketed by, the ages

of the younger and older sedimentary deposits.

braided stream A stream with a complex of converging and diverging channels separated by bars or islands. Braided streams form where more sediment is available than can be removed by the discharge of the stream.

breaker A collapsing water wave.

breccia A general term for sediment consisting of angular fragments in a matrix of finer particles. Examples: sedimentary breccias, volcanic breccias, fault breccias, impact breccias.

butte A somewhat isolated hill, usually capped with a resistant layer of rock and bordered by talus (see diagram). A butte is an erosional remnant of a formerly more extensive slope.

Butte

Calcite A mineral composed of calcium carbonate (CaCO$_3$).

caldera A large, more or less circular depression or basin associated with a volcanic vent. Its diameter is many times greater than that of the included vents. Calderas are believed to result from subsidence or collapse and may or may not be related to explosive eruptions.

calving The breaking off of large blocks of ice from a glacier that terminates in a body of water.

capacity The maximum quantity of sediment a given stream, glacier, or wind can carry under a given set of conditions.

capillary A small, tubular opening with a diameter about that of a human hair.

capillary action The action by which a fluid (such as water) is drawn up into small openings (such as pore spaces in rocks) due to surface tension.

capillary fringe A zone above the water table in which water is lifted by surface tension into openings of capillary size.

carbonaceous Containing carbon.

carbonate mineral A mineral formed by the bonding of carbonate ions (CO$_3^{2-}$) with positive ions. Examples: calcite (CaCO$_3$), dolomite [CaMg(CO$_3$)$_2$].

carbonate rock A rock composed mostly of carbonate minerals. Examples: limestone, dolomite.

carbon 14 A radioisotope of carbon. Its half-life is 5730 years.

catastrophism The belief that geologic history consists of major catastrophic events involving processes that were far more intense than any we observe now. Contrast with *uniformitarianism.*

cave A naturally formed subterranean open area, chamber, or series of chambers, commonly produced in limestone by solution activity.

cement Minerals precipitated from groundwater in the pore spaces of a sedimentary rock and binding the rock's particles together.

Cenozonic The era of geologic time from the end of the Mesozoic era (65 million years ago) to the present.

chalcedony A general term for fibrous cryptocrystalline quartz.

chalk A variety of limestone composed of shells of microscopic oceanic organisms.

chemical decomposition Synonymous with chemical weathering.

chemical weathering Chemical reactions that act on rocks exposed to water and the atmosphere so as to change their unstable mineral components to more stable forms. Oxidation, hydrolysis, carbonation, and direct solution are the most common reactions. Synonymous with *decomposition.*

chert A sedimentary rock composed of granular cryptocrystalline silica.

C horizon The zone of soil consisting of partly decomposed bedrock underlying the *B* horizon. It grades downward into fresh, unweathered bedrock.

cinder A fragment of volcanic ejecta from 0.5 to 2.5 cm in diameter.

cinder cone A cone-shaped hill composed of loose volcanic fragments.

cirque An amphitheater-shaped depression at the head of a glacial valley, excavated mainly by ice plucking and frost wedging (see diagram).

clastic 1 Pertaining to fragments (such as mud, sand, and gravel) produced by the mechanical breakdown of rocks. 2 A sedimentary rock composed chiefly of consolidated clastic material.

clastic texture The texture of sedimentary rocks consisting of fragmentary particles of minerals, rocks, and organic skeletal remains (see diagram).

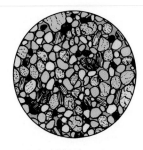

clay Sedimentary material composed of fragments with a diameter of less than 1/256 mm. Clay particles are smaller than silt particles.

clay minerals A group of fine-grained crystalline hydrous silicates formed by weathering of minerals such as feldspar, pyroxene, or amphibole.

cleavage The tendency of a mineral to break in a preferred direction along smooth planes.

cobble A rock fragment with a diameter between 64 mm (about the size of a tennis ball) and 2567 mm (about the size of a volleyball). Cobbles are larger than pebbles but smaller than boulders.

columnar jointing A system of fractures that splits a rock body into long prisms, or columns (see diagram). It is characteristic of lava flows and shallow intrusive igneous flows.

competence The maximum size of particles that a given stream, glacier, or wind can move at a given velocity.

composite volcano A large volcanic cone built by extrusion of alternating layers of ash and lava (see diagram). Synonymous with *stratovolcano.*

compression A system of stresses that tends to reduce the volume of or shorten a substance.

conchoidal fracture A type of fracture that produces a smooth, curved surface. It is characteristic of quartz and obsidian.

concretion A spherical or ellipsoidal nodule formed by accumulation of mineral matter after deposition of sediment.

cone of depression A conical depression of the water table surrounding a well after heavy pumping (see diagram).

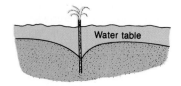

conglomerate A coarse-grained sedimentary rock composed of rounded fragments of pebbles, cobbles, or boulders.

consequent stream A stream that has a course determined by, or directly resulting from, the original slope on which it developed.

contact The surface separating two different rock bodies.

contact metamorphism Metamorphism of a rock near its contact with a magma.

continent A large landmass, from 20 to 60 km thick, composed mostly of granitic rock. Continents rise abruptly above the deep-ocean floor and include the marginal areas submerged beneath sea level. Examples: the African continent, the South American continent.

continental accretion The theory that the continents have grown by incorporation of deformed sediments along their margins.

continental crust The type of crust underlying the continents, including the continental shelves. The continental crust is commonly about 35 km thick. Its maximum thickness is 60 km, beneath mountain ranges. Its density is 2.7 g/cm^3, and the velocities of primary seismic waves traveling through the crust are less than 6.2 km/sec. Synonymous with *sial.* Contrast with *oceanic crust.*

continental drift The theory that the continents have moved in relation to one another.

continental glacier A thick ice sheet covering large parts of a continent. Present-day examples are found in Greenland and Antarctica.

continental margin The zone of transition from a continental mass to the adjacent ocean basin. It generally includes a continental shelf, continental slope, and continental rise.

continental rise The gently sloping surface located at the base of a continental slope (see diagram for *abyssal plains*).

continental shelf The submerged margin of a continental mass extending from the shore to the first prominent break in slope, which usually occurs at a depth of about 120 m.

continental slope The slope that extends from a continental shelf down to the ocean deep. In some areas, such as off eastern North America, the continental slope grades into the more gently sloping continental rise.

convection Movement of portions of a fluid as a result of density differences produced by heating (see diagram).

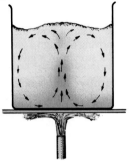

Convection

convection cell The space occupied by a single convection current.

convection current A closed system in which material is transported as a result of thermal convection. Convection currents are characteristic of the atmosphere and of bodies of water. They are believed also to be generated in the interior of the Earth. In the plate tectonic theory, convection within the mantle is thought to be responsible for the movement of tectonic plates.

convergent plate boundary The zone where the leading edges of converging plates meet. Convergent plate boundaries are sites of considerable geologic activity and are characterized by volcanism, earthquakes, and crustal deformation. See also *subduction zone.*

Copernican period The period of lunar history during which rayed craters, such as Copernicus, and their associated rim deposits were formed (from 2 billion years ago to the present).

Copernican system The youngest system of rocks on the Moon, formed during the Copernican period.

coquina A limestone composed of an aggregate of shells and shell fragments (see diagram).

Coquina

coral A bottom-dwelling marine invertebrate organism of the class Anthozoa.

core The central part of the Earth below a depth of 2900 km.

Coriolis effect The effect produced by a Coriolis force, namely, the tendency of all particles of matter in motion on the Earth's surface to be deflected to the right in the Northern Hemisphere and to the left in the Southern Hemisphere.

country rock A general term for rock surrounding an igneous intrusion.

covalent bond A chemical bond in which electrons are shared between different atoms so that none of the atoms has a net charge.

crater An abrupt circular depression formed by extrusion of volcanic material, by collapse, or by the impact of a meteorite (see diagram).

Crater

craton The stable continental crust, including the shield and stable platform areas, most of which have not been affected by significant tectonic activity since the close of the Precambrian era.

creep The imperceptibly slow downslope movement of material (see diagram).

Creep

crevasse **1** (glacial geology) A deep crack in the upper surface of a glacier. **2** (natural levee) A break in a natural levee.

cross-bedding Stratification inclined to the original horizontal surface upon which the sediment accumulated. It is produced by deposition on the slope of a dune or sand wave (see diagram).

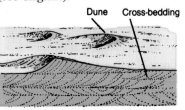

Dune   Cross-bedding

Cross-bedding

crosscutting relations, principle of The principle that a rock is younger than any rock across which it cuts.

crust (planetary structure) The outermost layer, or shell, of the Earth (or any other differentiated planet). The Earth's crust is generally defined as the part of the Earth above the Mohorovičić discontinuity. It represents less than 1% of the Earth's total volume (see diagram). See also *continental crust, oceanic crust.*

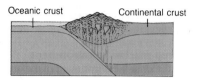

Oceanic crust   Continental crust

Crust

crustal warping Gentle bending (upwarping or downwarping) of sedimentary strata.

cryptocrystalline texture The texture of rocks composed of crystals too small to be identified with an ordinary microscope.

crystal A solid, polyhedral form bounded by naturally formed plane surfaces resulting from growth of a crystal lattice.

crystal face A naturally formed smooth plane surface of a crystal.

crystal form The geometric shape of a crystal (see diagram).

Crystal form

crystal lattice A systematic, symmetrical network of atoms within a crystal.

crystalline texture The rock texture resulting from simultaneous growth of crystals.

crystallization The process of crystal growth. It occurs as a result of condensation from a gaseous state, precipitation from a solution, or cooling of a melt.

crystal structure The orderly arrangement of atoms in a crystal.

cuesta An elongate ridge formed on the tilted and eroded edges of gently dipping strata.

Daughter isotope An isotope produced by radioactive decay of its parent isotope. The quantity of a daughter isotope continually increases with time.

debris flow The rapid downslope movement of debris (rock, soil, and mud).

fan A fan-shaped deposit of sediment. See also *alluvial fan, deep-sea fan.*

fault A surface along which a rock body has broken and been displaced.

fault block A rock mass bounded by faults on at least two sides (see diagram).

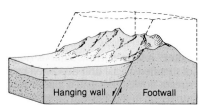

Hanging wall / Footwall

Fault block

fault scarp A cliff produced by faulting.

faunal succession, principle of The principle that fossils in a stratigraphic sequence succeed one another in a definite, recognizable order.

feldspar A mineral group consisting of silicates of aluminum and one or more of the metals potassium, sodium, or calcium. Examples: K-feldspar, Ca-plagioclase, Na-plagioclase.

felsite A general term for light-colored aphanitic (fine-grained) igneous rocks. Example: rhyolite.

ferromagnesian minerals A variety of silicate minerals containing abundant iron and magnesium. Examples: olivine, pyroxene, amphibole.

fiord A glaciated valley flooded by the sea to form a long, narrow, steep-walled inlet.

firn Granular ice formed by recrystallization of snow. It is intermediate between snow and glacial ice. Sometimes referred to as *névé.*

fissure An open fracture in a rock.

fissure eruption Extrusion of lava along a fissure (see diagram).

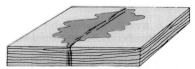

Fissure eruption

flint A popular name for dark-colored chert (cryptocrystalline quartz).

flood basalt An extensive flow of basalt erupted chiefly along fissures. Synonymous with *plateau basalt.*

floodplain The flat, occasionally flooded area bordering a stream.

fluvial Pertaining to a river or rivers.

fluvial environment The sedimentary environment of river systems.

focus The area within the Earth where an earthquake originates.

fold A bend, or flexure, in a rock (see diagram).

Fold

folded mountain belt A long, linear zone of the Earth's crust where rocks have been intensely deformed by horizontal stresses and generally intruded by igneous rocks. The great folded mountains of the world (such as the Appalachians, the Himalayas, the Rockies, and the Alps) are believed to have been formed at convergent plate margins.

foliation A planar feature in metamorphic rocks, produced by the secondary growth of minerals. Three major types are recognized: slaty cleavage, schistosity, and gneissic layering.

footwall The block beneath a dipping fault surface.

foreshore The seaward part of the shore or beach lying between high tide and low tide.

formation A distinctive body of rock that serves as a convenient unit for study and mapping.

fossil Naturally preserved remains or evidence of past life, such as bones, shells, casts, impressions, and trails.

fossil fuel A fuel containing solar energy that was absorbed by plants and animals in the geologic past and thus is preserved in organic compounds in their remains. Fossil fuels include petroleum, natural gas, and coal.

fracture zone 1 (field geology) A zone where the bedrock is cracked and fractured. 2 (oceanography) A zone of long, linear fractures on the ocean floor, expressed topographically by ridges and troughs. Fracture zones are the topographic expression of transform faults.

fringing reef A reef that lies alongside the shore of a landmass (see diagram).

Fringing reef

frost heaving The lifting of unconsolidated material by the freezing of subsurface water.

frost wedging The forcing apart of rocks by the expansion of water as it freezes in fractures and pore spaces.

Gabbro A dark-colored, coarse-grained rock composed of Ca-plagioclase, pyroxene, and possibly olivine, but no quartz.

gas The state of matter in which a substance has neither independent shape nor independent volume. Gases can readily be compressed and tend to expand indefinitely.

geocline An elongate prism of sedimentary rock deposited in a subsided part of the continental margins and adjacent oceanic crust. A modern example is the continental margin of the eastern United States. See also *eugeocline, miogeocline.*

geode A hollow nodule of rock lined with crystals; when separated from the rock body by weathering, it appears as a hollow, rounded shell partly filled with crystals (see diagram).

Geode

geologic column A diagram representing divisions of geologic time and the rock units formed during each major period.

geologic cross section A diagram showing the structure and arrangement of rocks as they would appear in a vertical plane below the Earth's surface.

geologic map A map showing the distribution of rocks at the Earth's surface.

geologic time scale The time scale determined by the geologic column and by radiometric dating of rocks.

geosyncline A subsiding part of the lithosphere in which thousands of meters of sediment accumulate. See also *eugeosyncline, miogeosyncline.*

geothermal Pertaining to the heat of the interior of the Earth.

geothermal energy Energy useful to human beings that can be extracted from steam and hot water found within the Earth's crust.

geothermal gradient The rate at which temperature increases with depth.

geyser A thermal spring that intermittently erupts steam and boiling water.

glacial environment The sedimentary environment of glaciers and their meltwaters.

glacier A mass of ice formed from compacted, recrystallized snow that is thick enough to flow plastically.

glass 1 A state of matter in which a substance displays many properties of a solid but lacks crystal structure. 2 An amorphous igneous rock formed from a rapidly cooling magma.

glassy texture The texture of igneous rocks in which the material is in the form of natural glass rather than crystal.

global tectonics The study of the characteristics and origin of structural features of the Earth that have regional or global significance.

Glossopteris flora An assemblage of late Paleozoic fossil plants named for the seed fern *Glossopteris,* one of the plants in the assemblage. These flora are widespread in South America, Africa, Australia, India, and Antarctica, and provide important evidence for the theory of continental drift.

gneiss A coarse-grained metamorphic rock with a characteristic type of foliation (gneissic layering), resulting from alternating layers of light-colored and dark-colored minerals. Its composition is generally similar to that of granite.

gneissic layering The type of foliation characterizing gneiss, resulting from alternating layers of the constituent silicic and mafic minerals.

Gondwanaland The ancient continental landmass that is thought to have split apart during Mesozoic time to form the present-day continents of South America, Africa, India, Australia, and Antarctica (see diagram).

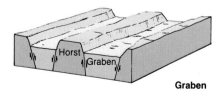

Gondwanaland

graben An elongate fault block that has been lowered in relation to the blocks on either side (see diagram).

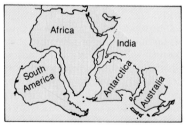

Graben

gradation Leveling of the land due to erosion by such agents as river systems, groundwater, glaciers, wind, and waves.

graded bedding A type of bedding in which each layer is characterized by a progressive decrease in grain size from the bottom of the bed to the top (see diagram).

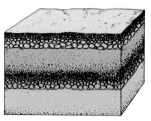

Graded bedding

graded stream A stream that has attained a state of equilibrium, or balance, between erosion and deposition, so that the velocity of the water is just great enough to transport the sediment load supplied from the drainage basin, and neither erosion nor deposition occurs.

gradient (stream) The slope of a stream channel measured along the course of the stream.

grain A particle of a mineral or rock, generally lacking well-developed crystal faces.

granite A coarse-grained igneous rock composed of K-feldspar, plagioclase, and quartz, with small amounts of ferromagnesian minerals.

granitization Formation of granitic rock by metamorphism without complete melting.

gravity anomaly An area where gravitational attraction is greater or less than its normal value.

graywacke An impure sandstone consisting of rock fragments and grains of quartz and feldspar in a matrix of clay-size particles.

groundmass The matrix of relatively fine-grained material between the phenocrysts in a porphyritic rock.

ground moraine Glacial deposits that cover an area formerly occupied by a glacier; they typically produce a landscape of low, gently rolling hills.

groundwater Water below the Earth's surface. It generally occurs in pore spaces of rocks and soil.

guyot A seamount with a flat top (see diagram).

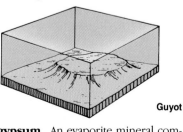

Guyot

gypsum An evaporite mineral composed of calcium sulfate with water ($CaSO_4 \cdot 2H_2O$).

Half-Life The time required for half of a given sample of a radioactive

isotope to decay to its daughter isotope.

halite An evaporite mineral composed of sodium chloride (NaCl).

hanging valley A tributary valley with the floor lying ("hanging") above the valley floor of the main stream or shore to which it flows (see diagram). Hanging valleys commonly are created by deepening of the main valley by glaciation, but they can also be produced by faulting or rapid retreat of a sea cliff.

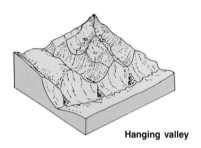

Hanging valley

hanging wall The surface or block of rock that lies above an inclined fault plane (see diagram).

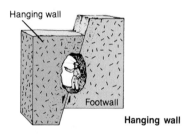

Hanging wall

hardness 1 (mineralogy) The measure of the resistance of a mineral to scratching or abrasion. 2 (water) A property of water resulting from the presence of calcium carbonate and magnesium carbonate in solution.

headland An extension of land seaward from the general trend of the coast; a promontory, cape, or peninsula.

headwater erosion Extension of a stream headward, up the regional slope of erosion.

heat flow The flow of heat from the interior of the Earth.

high-grade metamorphism Metamorphism that occurs under high temperature and high pressure.

hogback A narrow, sharp ridge formed on steeply inclined, resistant rock.

horizon 1 (geologic) A plane of stratification assumed to have been originally horizontal. 2 (soil) A layer of soil distinguished by characteristic physical properties. Soil horizons generally are designated by letters (for example, *A* horizon, *B* horizon, *C* horizon).

horn A sharp peak formed at the intersection of the headwalls of three or more cirques (see diagram).

Horn

hornblende A variety of the amphibole mineral group.

hornfels A nonfoliated metamorphic rock of uniform grain size, formed by high-temperature metamorphism. Hornfelses typically are formed by contact metamorphism around igneous intrusions.

horst An elongate fault block that has been uplifted in relation to the adjacent rocks (see diagram).

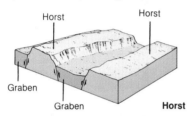

Horst

hot spot The expression at the Earth's surface of a mantle plume, or column of hot, buoyant rock rising in the mantle beneath a lithospheric plate.

hummock A small, rounded or cone-shaped, low hill or a surface of other small, irregular shapes. A surface that is not equidimensional or ridgelike.

hydraulic Pertaining to a fluid in motion.

hydraulic head The pressure exerted by a fluid at a given depth beneath its surface. It is proportional to the height of the fluid's surface above the area where the pressure is measured.

hydrologic system The system of moving water at the Earth's surface (see diagram).

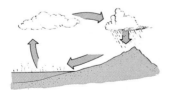

Hydrologic system

hydrolysis Chemical combination of water with other substances.

hydrosphere The waters of the Earth, as distinguished from the rocks (lithosphere), the air (atmosphere), and living things (biosphere).

hydrostatic pressure The pressure within a fluid (such as water) at rest, exerted on a given point within the body of the fluid.

hydrothermal deposit A mineral deposit formed by high-temperature groundwater. The high temperature commonly is associated with emplacement of a magma.

Ice sheet A thick, extensive body of glacial ice that is not confined to valleys. Localized ice sheets are sometimes called *ice caps.*

ice wedging A type of mechanical weathering in which rocks are broken by the expansion of water as it freezes in joints, pores, or bedding planes. Synonymous with *frost wedging.*

igneous rock Rock formed by cooling and solidification of molten silicate minerals (magma). Igneous rocks include volcanic and plutonic rocks.

Imbrian period The period of lunar history during which the large multiringed basins, such as Mare Imbrium, were formed and the mare basalts extruded (from 3.9 billion to 3.1 billion years ago).

Imbrian system The system of rocks formed on the Moon during the Imbrian period.

inclination, magnetic The angle between the horizontal plane and a magnetic line of force.

inclusion A rock fragment incorporated into a younger igneous rock.

intermediate-focus earthquake An earthquake with a focus located at a depth between 70 and 300 km.

intermittent stream A stream through which water flows only part of the time.

internal drainage A drainage system that does not extend to the ocean.

interstitial Pertaining to material in the pore spaces of a rock. Petroleum and groundwater are interstitial fluids. Minerals deposited by groundwater in a sandstone are interstitial minerals.

intrusion 1 Injection of a magma into a preexisting rock. 2 A body of rock resulting from the process of intrusion.

intrusive rock Igneous rock that, while it was fluid, penetrated into or between other rocks and solidifed. It can later be exposed at the Earth's surface after erosion of the overlying rock.

inverted valley A valley that has been filled with lava or other resistant material and has subsequently been eroded into an elongate ridge.

ion An atom or combination of atoms that has gained or lost one or more electrons and thus has a net electrical charge.

ionic bond A chemical bond formed by electrostatic attraction between oppositely charged ions.

ionic substitution The replacement of one kind of ion in a crystalline lattice by another kind that is of similar size and electrical charge.

island A landform smaller than a continent and completely surrounded by water.

island arc A chain of volcanic islands. Island arcs are generally convex toward the open ocean. Example: the Aleutian Islands.

isostasy A state of equilibrium, resembling flotation, in which segments of the Earth's crust stand at levels determined by their thickness and density (see diagram). Isostatic equilibrium is attained by flow of material in the mantle.

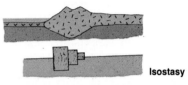

Isostasy

isotope One of the several forms of a chemical element that have the same number of protons in the nucleus but differ in the number of neutrons and thus differ in atomic weight.

Joint A fracture in a rock along which no appreciable displacement has occurred.

Kame A body of stratified glacial sediment. A mound or an irregular ridge deposited by a subglacial stream as an alluvial fan or a delta.

karst topography A landscape characterized by sinks, solution valleys, and other features produced by groundwater activity (see diagram).

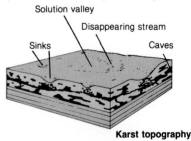

Solution valley

Disappearing stream

Sinks

Caves

Karst topography

kettle A closed depression in a deposit of glacial drift formed where a block of ice was buried or partly buried and then melted.

Laccolith A concordant igneous intrusion that has arched up the strata into which it was injected, so that it forms a pod-shaped or lens-shaped body with a generally horizontal floor (see diagram).

Laccolith

lag deposit A residual accumulation of coarse fragments that remains on the surface after finer material has been removed by wind.

lagoon A shallow body of seawater separated from the open ocean by a barrier island or reef.

lamina (*pl. laminae*) A layer of sediment less than 1 cm thick.

laminae Plural of *lamina*.

laminar flow A type of flow in which the fluid moves in parallel lines. Contrast with *turbulent flow*.

landform Any feature of the Earth's surface having a distinct shape and origin. Landforms include major features (such as continents, ocean basins, plains, plateaus, and mountain ranges) and minor features (such as hills, valleys, slopes, drumlins, and dunes). Collectively, the landforms of the Earth constitute the entire surface configuration of the planet.

landslide A general term for relatively rapid types of mass movement, such as debris flows, debris slides, rockslides, and slumps.

lateral moraine An accumulation of till deposited along the side margins of a valley glacier. It accumulates as a result of mass movement of debris on the sides of the glacier.

laterite A soil that is rich in oxides of iron and aluminum formed by deep weathering in tropical and subtropical areas.

Laurasia The ancient continental landmass that is thought to have split apart to form Europe, Asia, North America, and Greenland.

lava Magma that reaches the Earth's surface.

leach To dissolve and remove the soluble constituents of a rock or soil.

leachate A solution produced by leaching. Example: water that has seeped through a waste disposal site and thus contains in solution various substances derived from the waste material

leaching The process by which ground water dissolves and transports soluble components of a rock or soil.

lee slope The part of a hill, dune, or rock that is sheltered or turned away from the wind. Synonymous with *slip face*.

levee, natural A broad, low embankment built up along the banks of a river channel during floods.

limb The flank, or side, of a fold.

limestone A sedimentary rock composed mostly of calcium carbonate ($CaCO_3$).

lineament A topographic feature or group of features having a linear configuration. Lineaments commonly are expressed as ridges or depressions or as an alignment of features such as stream beds, volcanoes, or vegetation.

liquid The state of matter in which a substance flows freely and lacks crystal structure. Unlike a gas, a liquid retains the same volume independent of the shape of its container.

lithification The processes by which sediment is converted into sedimentary rock. These processes include cementation and compaction.

lithosphere The relatively rigid outer zone of the Earth, which includes the continental crust, the oceanic crust, and the part of the mantle lying above the softer asthenosphere.

load The total amount of sediment carried at a given time by a stream, glacier, or wind.

loess Unconsolidated, wind-deposited silt and dust.

longitudinal dune An elongate sand dune oriented in the direction of the prevailing wind.

longitudinal profile The profile of a stream or valley drawn along its length, from source to mouth.

longitudinal wave A seismic body wave in which particles oscillate along lines in the direction in which the wave travels. Synonymous with *P wave*.

longshore current A current in the surf zone moving parallel to the shore. Longshore currents occur where waves strike the shore at an angle. The waves push water and sediment obliquely up the beach, and the backwash returns straight down the beach face, so the water and sediment follow a zigzag pattern, with net movement parallel to the shore.

longshore drift The process in which sediment is moved in a zigzag pattern along a beach by the swash and backwash of waves that approach the shore obliquely.

low-grade (low-rank) metamorphism Metamorphism that is accomplished under low or moderate temperature and low or moderate pressure.

Mafic rock An igneous rock containing more than 50% ferromagnesian minerals.

magma A mobile silicate melt, which can contain suspended crystals and dissolved gases as well as liquid.

magmatic differentiation A general term for the various processes by which early-formed crystals or early-formed liquids are separated and removed from a magma to produce a rock with composition different from that of the original magma. Early-crystallized ferromagnesian minerals commonly are separated by gravitational settling, so that the parent magma is left enriched in silica, sodium, and potassium.

magmatic segregation Separation of crystals of certain minerals from a magma as it cools. For example, some minerals (including certain valuable metals) crystallize while other components of the magma are still liquid. These early-formed crystals can settle to the bottom of a magma chamber and thus become concentrated there, forming an ore deposit.

magnetic anomaly A deviation of observed magnetic inclination or intensity (as measured by a magnetometer) from a constant normal value.

magnetic reversal A complete 180-degree reversal of the polarity of the Earth's magnetic field.

mantle The zone of the Earth's interior between the base of the crust (the Moho discontinuity) and the core.

mantle plume A buoyant mass of hot mantle material that rises to the base of the lithosphere. Mantle plumes commonly produce volcanic activity and structural deformation in the central part of lithospheric plates.

marble A metamorphic rock consisting mostly of metamorphosed limestone or dolomite.

mare (pl. *maria*) Any of the relatively smooth, low, dark areas of the Moon. The lunar maria were formed by extrusion of lava.

maria Plural of *mare*.

mascon A concentration of mass in a local area beneath a lunar mare. The term is derived from *mass concentration*.

mass movement The transfer of rock and soil downslope by direct action of gravity without a flowing medium (such as a river or glacial ice). Synonymous with *mass wasting*.

matrix The relatively fine-grained rock material occupying the space between larger particles in a rock. See also *groundmass*.

meander A broad, looping bend in a river (see diagram).

Meander

mechanical weathering The breakdown of rock into smaller fragments by physical processes such as frost wedging. Synonymous with *disintegration*.

medial moraine A ridge of till formed in the middle of a valley glacier by the junction of two lateral moraines where two valley glaciers converge (see diagram).

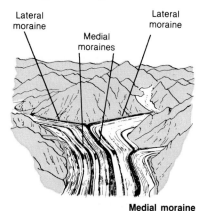

Medial moraine

melt A substance altered from the solid state to the liquid state.

mesa A flat-topped, steep-sided highland capped with a resistant rock formation. A mesa is smaller than a plateau but larger than a butte.

Mesozoic The era of geologic time from the end of the Paleozoic era (225 million years ago) to the beginning of the Cenozoic era (65 million years ago).

metaconglomerate A metamorphosed conglomerate.

metallic bond A chemical bond in which shared electrons move freely among the atoms.

metamorphic Pertaining to the processes or products of metamorphism.

metamorphic rock Any rock formed from preexisting rocks within the Earth's crust by changes in temperature and pressure and by chemical action of fluids.

metamorphism Alteration of the minerals and textures of a rock by changes in temperature and pressure and by a gain or loss of chemical components.

meteorite Any particle of solid matter that has fallen to the Earth, the Moon, or another planet from space.

mica A group of silicate minerals exhibiting perfect cleavage in one direction.

microcontinent A relatively small, isolated fragment of continental crust. Example: Madagascar.

mid-Atlantic ridge The mountain range extending from north to south down the central part of the Atlantic Ocean floor.

migmatite A mixture of igneous and metamorphic rocks in which thin dikes and stringers of granitic material interfinger with metamorphic rocks.

mineral A naturally occurring inorganic solid having a definite internal structure and a definite chemical composition that varies only within strict limits. Chemical composition and internal structure determine its physical properties, including the tendency to assume a particular geometric form (crystal form).

miogeocline (plate tectonics) A geocline situated near a continental margin containing a thick sequence of well-sorted clastic and chemical sediments derived from the continent.

miogeosyncline A geosyncline situated near a continental margin and receiving well-sorted clastic and chemical sediments from the continent, not associated with volcanism.

mobile belts Long, narrow belts in the continents that have been subjected to mountain-building processes.

Mohorovičić discontinuity The first global seismic discontinuity below the surface of the Earth. It lies at a depth varying from about 5 to 10 km beneath the ocean floor to about 35 km beneath the continents. Commonly referred to as the Moho.

monadnock An erosion remnant rising above the peneplain (see diagram).

Monadnock

monocline A bend or fold in gently dipping horizontal strata (see diagram).

Monocline

moraine A general term for a landform composed of till.

mountain A general term for any landmass that stands above its surroundings. In the stricter geological sense, a mountain belt is a highly deformed part of the Earth's crust that has been injected with igneous intrusions and the deeper parts of which have been metamorphosed. The topography of young mountains is high, but erosion can reduce old mountains to flat lowlands.

mud crack A crack in a deposit of mud or silt resulting from the contraction that accompanies drying.

mudflow A flowing mixture of mud and water (see diagram).

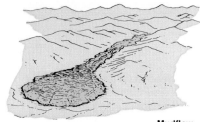

Mudflow

multiringed basin A large crater (more than 300 km in diameter) containing a series of concentric ridges and depressions. Example: the Orientale basin on the Moon.

Nappe Faulted and overturned folds.

natural arch An arch-shaped landform produced by weathering and differential erosion.

névé Granular ice formed by recrystallization of snow. Synonymous with *firn*.

nodule A small, irregular, knobby, or rounded rock that is generally harder than the surrounding rock.

nonconformity An unconformity in which stratified rocks rest on eroded granitic or metamorphic rocks (see diagram).

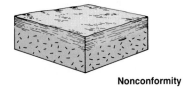

Nonconformity

normal fault A steeply inclined fault in which the hanging wall has moved downward in relation to the footwall (see diagram). Synonymous with *gravity fault*.

Footwall

Hanging wall

Normal fault

nuée ardente A hot cloud of volcanic fragments and superheated gases that flows as a mass, because it is denser than air (see diagram). Upon cooling, it forms a type of rock called tuff, including ash-flow tuff and welded tuff.

Nuée ardente

Obsidian A glassy igneous rock with a composition equivalent to that of granite.

ocean basin A low part of the lithosphere lying between continental masses. The rocks of an ocean basin are mostly basalt with a veneer of oceanic sediment.

oceanic crust The type of crust that underlies the ocean basins. It is about 5 km thick, composed predominantly of basalt. Its density is 3.0 g/cm³. The velocities of compressional seismic waves traveling through it exceed 6.2 km/sec. Compare with *continental crust.*

oceanic ridge The continuous ridge, or broad, fractured topographic swell, that extends through the central part of the Arctic, Atlantic, Indian, and South Pacific oceans. It is several hundred kilometers wide, and its elevation above the ocean floor is 600 m or more. It thus constitutes a major structural and topographic feature of the Earth.

offshore The area from low tide seaward.

oil shale Shale that is rich in hydrocarbon derivatives. In the United States, the chief oil shale is the Green River Formation, in the Rocky Mountain region.

olivine A silicate mineral with magnesium and iron but no aluminum [$(Mg,Fe)_2SiO_4$].

oolite A limestone consisting largely of spherical grains of calcium carbonate in concentric spherical layers (see diagram).

Oolite

ooze (marine geology) Marine sediment consisting of more than 30% shell fragments of microscopic organisms.

ophiolite A sequence of rocks characterized by ultramafic rocks at the base and (in ascending order) gabbro, sheeted dikes, pillow lavas, and deep-sea sediments. The typical sequence of rocks constituting the oceanic crust.

orogenesis The processes of mountain building.

orogenic Pertaining to deformation of a continental margin to the extent that a mountain range is formed.

orogenic belt A mountain belt.

orogeny A major episode of mountain building.

outcrop An exposure of bedrock.

outlet glacier A tonguelike stream of ice, resembling a valley glacier, that forms where a continental glacier encounters a mountain system and is forced to move through a mountain pass in large streams.

outwash Stratified sediment washed out from a glacier by meltwater streams and deposited in front of the end moraine.

outwash plain The area beyond the margins of a glacier where meltwater deposits sand, gravel, and mud washed out from the glacier (see diagram).

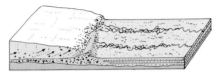

Outwash plain

overturned fold A fold in which at least one limb has been rotated through an angle greater than 90 degrees (see diagram).

Overturned fold

oxbow lake A lake formed in the channel of an abandoned meander.

oxidation Chemical combination of oxygen with another substance.

Pahoehoe flow A lava flow with a billowy or ropy surface. Contrast with *aa flow.*

paleocurrent An ancient current, which existed in the geologic past, with a direction of flow that can be inferred from cross-bedding, ripple marks, and other sedimentary structures.

paleogeography The study of geography in the geologic past, including the patterns of the Earth's surface, the distribution of land and ocean, and ancient mountains and other landforms.

paleomagnetism The study of ancient magnetic fields, as preserved in the magnetic properties of rocks. It includes studies of changes in the position of the magnetic poles and reversals of the magnetic poles in the geologic past.

paleontology The study of ancient life.

paleowind An ancient wind, existing in the geologic past, the direction of which can be inferred from patterns of ancient ash falls, orientation of cross-bedding, and growth rates of colonial corals.

Paleozoic The era of geologic time from the end of the Precambrian (600 million years ago) to the beginning of the Mesozoic era (225 million years ago).

Pangaea A hypothetical continent from which the present continents originated by plate movement from the Mesozoic era to the present.

parabolic dune A dune shaped like a parabola with the concave side toward the wind.

partial melting The process by which minerals with low melting points liquefy within a rock body as a result of an increase in temperature or a decrease in pressure (or both) while other minerals in the rock are still solid. If the liquid (magma) is removed before other components of the parent rock have melted, the composition of the magma can be quite different from that of the parent rock. Partial melting is believed to be important in the generation of basaltic magma from peridotite at spreading centers and in the generation of granitic magma from basaltic crust at subduction zones.

passive margin (plate tectonics) A lithospheric plate margin at which crust is neither created nor destroyed. Passive plate margins generally are marked by transform faults. Contrast with *active margin.*

peat An accumulation of partly carbonized plant material containing approximately 60% carbon and 30% oxygen. It is considered an early stage, or rank, in the development of coal.

pebble A rock fragment with a diameter between 2 mm (about the size of a matchhead) and 64 mm (about the size of a tennis ball).

pediment A gently sloping erosion surface formed at the base of a receding mountain front or cliff. It cuts across bedrock and can be covered with a veneer of sediment (see diagram). Pediments characteristically form in arid and semiarid climates.

Pediment

pelagic sediment Deep-sea sediment composed of fine-grained detritus that slowly settles from surface waters. Common constituents are clay, radiolarian ooze, and foraminiferal ooze.

peneplain An extensive erosion surface worn down almost to sea level (see diagram). Subsequent tectonic activity can lift a peneplain to higher elevations.

Peneplain

peninsula An elongate body of land extending into a body of water.

perched water table The upper surface of a local zone of saturation that lies above the regional water table (see diagram).

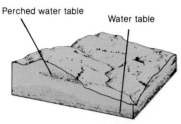

Perched water table

peridotite A dark-colored igneous rock of coarse-grained texture, composed of olivine, pyroxene, and some other ferromagnesian minerals, but with essentially no feldspar and no quartz.

permafrost Permanently frozen ground.

permanent stream A stream or reach of a stream that flows continuously throughout the year. Synonymous with *perennial stream.*

permeability The ability of a material to transmit fluids.

phaneritic texture The texture of igneous rocks in which the interlocking crystals are large enough to be seen without magnification.

phenocryst A crystal that is significantly larger than the crystals surrounding it (see diagram). Phenocrysts form during an early phase in the cooling of a magma when the magma cools relatively slowly.

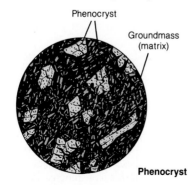

Phenocryst

Groundmass (matrix)

Phenocryst

physiographic map A map showing surface features of the Earth.

physiography The study of the surface features and landforms of the Earth.

pillar A landform shaped like a pillar.

pillow lava An ellipsoidal mass of igneous rock formed by extrusion of lava underwater.

pinnacle A tall, tower-shaped or spire-shaped pillar of rock.

placer A mineral deposit formed by the sorting or washing action of water. Placers are usually deposits of heavy minerals, such as gold.

plagioclase A group of feldspar minerals with a composition range from $NaAlSi_3O_8$ to $CaAl_2Si_2O_8$.

planetary differentiation The processes by which the materials in a planetary body are separated according to density, so that the originally homogeneous body is converted into a zoned or layered (shelled) body with a dense core, a mantle, and a crust.

plastic deformation A permanent change in a substance's shape or volume that does not involve failure by rupture.

plate (tectonics) A broad segment of the lithosphere (including the rigid upper mantle, plus oceanic and continental crust) that floats on the underlying asthenosphere and moves independently of other plates.

plateau An extensive upland region.

plateau basalt Basalt extruded in extensive, nearly horizontal layers, which, after uplift, tend to erode into great plateaus (see diagram). Synonymous with *flood basalt.*

Plateau basalt

plate tectonics The theory of global dynamics in which the lithosphere is believed to be broken into individual plates that move in response to convection in the upper mantle. The margins of the plates are sites of considerable geologic activity.

platform reef An organic reef with a flat upper surface developed on submerged segments of a continental platform.

playa A depression in the center of a desert basin, the site of occasional temporary lakes (see diagram).

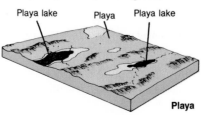

Playa lake   Playa   Playa lake

Playa

playa lake A shallow temporary lake formed in a desert basin after rain (see diagram).

Pleistocene The epoch of geologic time from the end of the Pliocene epoch of the Tertiary period (about 2 million years ago) to the beginning of the Holocene epoch of the Quaternary period (about 10,000 years ago). The major event during the Pleistocene was the expansion of continental glaciers in the Northern Hemisphere. Synonymous with *glacial epoch, ice age.*

plucking (glacial geology) The process of glacial erosion by which large rock fragments are loosened by ice wedging, become frozen to the bottom surface of the glacier, and are torn out of the bedrock and transported by the glacier as it moves. The process involves the freezing of subglacial meltwater that seeps into fractures and bedding planes in the rock.

plunge The inclination, with respect to the horizontal plane, of any linear structural element of a rock. The plunge of a fold is the inclination of the axis of the fold.

plunging fold A fold with its axis inclined from the horizontal.

plutonic rock Igneous rock formed deep beneath the Earth's surface.

pluvial lake A lake that was created under former climatic conditions, at a time when rainfall in the region was more abundant than it is now. Pluvial lakes were common in arid regions during the Pleistocene.

point bar A crescent-shaped accumulation of sand and gravel deposited on the inside of a meander bend (see diagram).

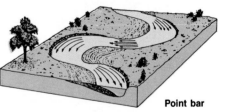

Point bar

polarity epoch A relatively long period of time during which the Earth's magnetic field is oriented in either the normal direction or the reverse direction.

polarity event A relatively brief interval of time within a polarity epoch; during a polarity event, the polarity of the Earth's magnetic field is reversed with respect to the prevailing polarity of the epoch.

polar wandering The apparent movement of the magnetic poles with respect to the continents.

pore fluid A fluid, such as groundwater or liquid rock material resulting from partial melting, that occupies pore spaces of a rock.

pore space The spaces within a rock body that are unoccupied by solid material. Pore spaces include spaces between grains, fractures, vesicles, and voids formed by dissolution.

porosity The percentage of the total volume of a rock or sediment that consists of pore space.

porphyritic texture The texture of igneous rocks in which some crystals are distinctly larger than others.

porphyry copper Deposits of copper disseminated throughout a porphyritic granitic rock.

pothole A hole formed in a stream bed by sand and gravel swirled around in one spot by eddies (see diagram).

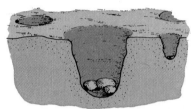

Pothole

Precambrian The division of geologic time from the formation of the Earth (about 4.5 billion years ago) to the beginning of the Cambrian period of the Paleozoic era (about 600 million years ago). Also, the rocks formed during that time. Precambrian time constitutes about 90% of the Earth's history.

Pre-Imbrian period The earliest period of lunar history, extending from the formation of the planet (about 4.5 billion years ago) to the formation of the multiringed basins (about 3.9 billion years ago).

Pre-Imbrian system The system of rocks formed on the Moon during the pre-Imbrian period. It includes most of the material on the lunar highlands.

pressure ridge An elongate uplift of the congealing crust of a lava flow, resulting from the pressure of underlying and still fluid lava.

primary coast A coast shaped by subaerial erosion, deposition, volcanism, or tectonic activity.

primary sedimentary structure A structure of sedimentary rocks (such as cross-bedding, ripple marks, or mud cracks) that originates contemporaneously with the deposition of the sediment (in contrast to a secondary structure, such as a joint or fault, which originates after the rock has been formed).

primary wave See *P wave.*

pumice A rock consisting of frothy natural glass.

P wave (primary seismic wave) A type of seismic wave, propagated like a sound wave, in which the material involved in the wave motion is alternately compressed and expanded.

pyroclastic Pertaining to fragmental rock material formed by volcanic explosions.

pyroclastic texture The rock texture of igneous rocks consisting of fragments of ash, rock, and glass produced by volcanic explosions.

pyroxene A group of rock-forming silicate minerals composed of single chains of silicon-oxygen tetrahedra. Compare with *amphibole*, which is composed of double chains.

Quartz An important rock-forming silicate mineral composed of silicon-oxygen tetrahedra joined in a three-dimensional network. It is distinguished by its hardness, glassy luster, and conchoidal fracture.

quartzite A sandstone recrystallized by metamorphism.

Radioactivity The spontaneous disintegration of an atomic nucleus with the emission of energy.

radiocarbon A radioactive isotope of carbon, ^{14}C, which is formed in the atmosphere and is absorbed by living organisms.

radiogenic heat Heat generated by radioactivity.

radiometric dating Determination of the age in years of a rock or mineral by measuring the proportions of an original radioactive material and its decay product. Synonymous with *radioactive dating.*

ray crater A meteorite crater that has a system of rays extending like splash marks from the crater rim (see diagram).

Ray crater

recessional moraine A ridge of till deposited at the margin of a glacier during a period of temporary stability in its general recession.

recharge Replenishment of the groundwater reservoir by the addition of water.

recrystallization Reorganization of elements of the original minerals in a rock resulting from changes in temperature and pressure and from the activity of pore fluids.

reef A solid structure built of shells and other secretions of marine organisms, particularly coral.

regolith The blanket of soil and loose rock fragments overlying the bedrock.

rejuvenated stream A stream that has had its erosive power renewed by uplift or lowering of the base level or by climatic changes.

relative age The age of a rock or an event as compared with some other rock or event.

relative dating Determination of the chronologic order of a sequence of events in relation to one another without reference to their ages measured in years. Relative geologic dating is based primarily on superposition, faunal succession, and cross-cutting relations.

relative time Geologic time as determined by relative dating, that is, by placing events in chronologic order without reference to their ages measured in years.

relief The difference in altitude between the high and the low parts of an area.

reverse fault A fault in which the hanging wall has moved upward in relation to the footwall; a high-angle thrust fault (see diagram).

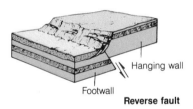

Hanging wall
Footwall
Reverse fault

rhyolite A fine-grained volcanic rock composed of quartz, K-feldspar, and plagioclase. It is the extrusive equivalent of a granite.

rift system A system of faults resulting from extension.

rift valley 1 A valley of regional extent formed by block faulting in which tensional stresses tend to pull the crust apart. Synonymous with *graben.* 2 The downdropped block along divergent plate margins.

rill A very small stream.

rille An elongate trench of cracklike valleys on the Moon's surface. Rilles can be sinuous and meandering or relatively linear structural depressions.

rip current A current formed on the surface of a body of water by the convergence of currents flowing in opposite directions. Rip currents are common along coasts where longshore currents move in opposite directions.

ripple marks Small waves produced on a surface of sand or mud by the drag of wind or water moving over it.

river system A river with all of its tributaries.

roche moutonnée An abraded knob of bedrock formed by an over-riding glacier. It typically is striated and has a gentle slope facing the up-stream direction of ice movement (see diagram).

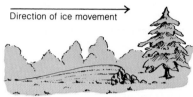

Direction of ice movement

Roche moutonnée

rock An aggregate of minerals that forms an appreciable part of the lithosphere.

rockfall The most rapid type of mass movement, in which rocks ranging from large masses to small fragments are loosened from the face of a cliff.

rock flour Fine-grained rock particles pulverized by glacial erosion.

rock glacier A mass of poorly sorted, angular boulders cemented with interstitial ice. It moves slowly by the action of gravity.

rockslide A landslide in which a newly detached segment of bedrock suddenly slides over an inclined surface of weakness (such as a joint or bedding plane).

runoff Water that flows over the land surface.

Sag pond A small lake that forms in a depression, or sag, where active or recent movement along a fault has impounded a stream.

saltation The transportation of particles in a current of wind or water by a series of bouncing movements (see diagram).

Saltation

salt dome A dome produced in sedimentary rock by the upward movement of a body of salt (see diagram).

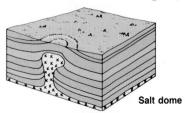

Salt dome

saltwater encroachment Displacement of fresh groundwater by

salt water in coastal areas, due to the greater density of salt water.

sand Sedimentary material composed of fragments ranging in diameter from 0.0625 to 2 mm. Sand particles are larger than silt particles but smaller than pebbles. Much sand is composed of quartz grains, because quartz is abundant and resists chemical and mechanical disintegration, but other materials, such as shell fragments and rock fragments, can also form sand.

sandstone A sedimentary rock composed mostly of sand-size particles, usually cemented by calcite, silica, or iron oxide.

sand wave A wave produced on a surface of sand by the drag of air or water moving over it. Sand waves include dunes and ripple marks.

scarp A cliff produced by faulting or erosion.

schist A medium-grained or coarse-grained metamorphic rock with strong foliation (schistosity) resulting from parallel orientation of platy minerals, such as mica, chlorite, and talc.

schistosity The type of foliation that characterizes schist, resulting from the parallel arrangement of coarse-grained platy minerals, such as mica, chlorite, and talc.

scoria An igneous rock containing abundant vesicles.

sea arch An arch cut by wave erosion through a headland (see diagram).

Sea arch

sea cave A cave formed by wave erosion.

sea cliff A cliff produced by wave erosion.

sea-floor spreading The theory that the sea floor spreads laterally away from the oceanic ridge as new lithosphere is created along the crest of the ridge by igneous activity.

seamount An isolated, conical mound rising more than 1000 m above the ocean floor (see diagram). Seamounts are probably submerged shield volcanoes.

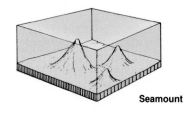

Seamount

sea stack A small, pillar-shaped, rocky island formed by wave erosion through a headland near a sea cliff (see diagram).

Sea stack

secondary coast A coast formed by marine processes or the growth of marine organisms.

secondary wave See *S wave.*

sediment Material (such as gravel, sand, mud, and lime) that is transported and deposited by wind, water, ice, or gravity; material that is precipitated from solution; deposits of organic origin (such as coal and coral reefs).

sedimentary differentiation The process in which distinctive sedimentary products (such as sand, shale, and lime) are generated and progressively separated from a rock mass by means of weathering, erosion, transportation, and deposition.

sedimentary environment A place where sediment is deposited and the physical, chemical, and biological conditions that exist there. Examples: rivers, deltas, lakes, shallow-marine shelves.

sedimentary rock Rock formed by the accumulation and consolidation of sediment.

seep A spot where groundwater or other fluids (such as oil) is discharged at the Earth's surface.

seif dune A longitudinal dune of great height and length.

seismic Pertaining to earthquakes or to waves produced by natural or artificial earthquakes.

seismic discontinuity A surface within the Earth at which seismic wave velocities abruptly change.

seismic ray The path along which a seismic wave travels. Seismic rays are perpendicular to the wave crest.

seismic reflection profile A profile of the configuration of the ocean floor and shallow sediments on the floor obtained by reflection of artificially produced seismic waves.

seismic wave A wave or vibration produced within the Earth by an earthquake or artificial explosion.

seismograph An instrument that records seismic waves.

settling velocity The rate at which suspended solid material subsides and is deposited.

shadow zone (seismology) An area where there is very little or no direct reception of seismic waves from a given earthquake because of refraction of the waves in the Earth's core. The shadow zone for *P waves* is between about 103 and 143 degrees from the epicenter.

shale A fine-grained clastic sedimentary rock formed by consolidation of clay and mud.

shallow-focus earthquake An earthquake with a focus less than 70 km below the Earth's surface.

shallow-marine environment The sedimentary environment of the continental shelves, where the water is usually less than 200 m deep.

sheeting A set of joints formed essentially parallel to the surface. It allows layers of rock to spall off as the weight of overlying rock is removed by erosion. It is especially well developed in granitic rock.

shield An extensive area of a continent where igneous and metamorphic rocks are exposed and have approached equilibrium with respect to erosion and isostasy. Rocks of the shield are usually very old (that is, more than 600 million years old).

shield volcano A large volcano shaped like a flattened dome and built up almost entirely of numerous flows of fluid basaltic lava. The slopes of shield volcanoes seldom exceed 10 degrees, so that in profile they resemble a shield or broad dome.

shore The zone between the waterline at high tide and the waterline at low tide. A narrow strip of land immediately bordering a body of water, especially a lake or an ocean.

sial A general term for the silica-rich rocks that form the continental masses.

silicate A mineral containing silicon-oxygen tetrahedra, in which four oxygen atoms surround each silicon atom.

silicon-oxygen tetrahedron The structure of the ion SiO_4^{-2}, in which four oxygen atoms surround a silicon atom to form a four-sided pyramid, or tetrahedron.

sill A tabular body of intrusive rock injected between layers of the enclosing rock.

silt Sedimentary material composed of fragments ranging in diameter from 1/265 to 1/16 mm. Silt particles are larger than clay particles but smaller than sand particles.

siltstone A fine-grained clastic sedimentary rock composed mostly of silt-size particles.

sima A general term for the magnesium-rich igneous rocks (basalt, gabbro, and peridotite) of the ocean basins.

sinkhole A depression formed by the collapse of a cavern roof (see diagram).

Sinkhole

slate A fine-grained metamorphic rock with a characteristic type of foliation (slaty cleavage), resulting from the parallel arrangement of microscopic platy minerals, such as mica and chlorite.

slaty cleavage The type of foliation that characterizes slate, resulting from the parallel arrangement of microscopic platy minerals, such as mica and chlorite. Slaty cleavage forms distinct zones of weakness within a rock, along which it splits into slabs.

slip face See *lee slope.*

slope retreat Progressive recession of a scarp or the side of a hill or mountain by mass movement and stream erosion.

slump A type of mass movement in which material moves along a curved surface of rupture.

snowline The line on a glacier separating the area where snow remains from year to year from the area where snow from the previous season melts.

soil The surface material of the continents, produced by disintegration of rock. Regolith that has undergone chemical weathering in place.

soil profile A vertical section of soil showing the soil horizons and parent material.

solid The state of matter in which a substance has a definite shape and volume and some fundamental strength.

solifluction A type of mass movement in which material moves slowly downslope in areas where the soil is saturated with water. It commonly occurs in permafrost areas.

solution valley A valley produced by solution activity, either by dissolution of surface materials or by removal of subsurface materials such as limestone, gypsum, or salt.

sorting The separation of particles according to size, shape, or weight. It occurs during transportation by running water or wind.

spatter cone A low-steep-sided volcanic cone built by accumulation of splashes and spatters of lava (usually basaltic) around a fissure or vent.

specific gravity The ratio of the weight of a substance to the weight of an equal volume of water.

spheroidal weathering The process by which corners and edges of a rock body become rounded as a re-

Spheroidal weathering

sult of exposure to weathering on all sides, so that the rock acquires a spheroidal or ellipsoidal shape (see diagram).

spit A sandy bar projecting from the mainland into open water. Spits are formed by deposition of sediment moved by longshore drift (see diagram).

Spit

splay A small deltaic deposit formed on a floodplain where water and sediment are diverted from the main stream through a crevasse in a levee.

spreading axis The imaginary axis through the Earth about which a set of tectonic plates moves. The motion of a diverging plate can be described as rotation around a spreading axis.

spreading center A plate boundary formed by tensional stress along the oceanic ridge. Synonymous with *divergent plate boundary, spreading edge.*

spreading pole A pole of the imaginary axis about which a set of tectonic plates moves. The spreading poles are the two points at which a spreading axis intersects the Earth's surface.

spring A place where groundwater flows or seeps naturally to the surface.

stable platform The part of a continent that is covered with flatlying or gently tilted sedimentary strata and underlain by a basement complex of igneous and metamorphic rocks. The stable platform has not been extensively affected by crustal deformation.

stack See *sea stack.*

stalactite An icicle-shaped deposit of dripstone hanging from the roof of a cave (see diagram on next page).

stalagmite A conical deposit of dripstone built up from a cave floor (see diagram on next page).

star dune A mound of sand with a high central point and arms radiating in various directions.

stock A small, roughly circular intrusive body, usually less than 100 km² in surface exposure.

strata Plural of *stratum.*

stratification The layered structure of sedimentary rock.

stratovolcano A volcano built up of alternating layers of ash and lava flows. Synonymous with *composite volcano.*

stratum (pl. *strata*) A layer of sedimentary rock.

streak The color of a powdered mineral.

stream load The total amount of sediment carried by a stream at a given time.

stream order The hierarchical number of a stream segment. The smallest tributary has the order number of 1, and successively larger tributaries have progressively higher numbers.

stream piracy Diversion of the headwaters of one stream into another stream. The process occurs by headward erosion of a stream having greater erosive power than the stream it captures (see diagram).

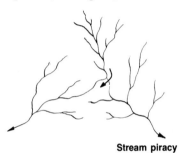

Stream piracy

stream terrace One of a series of level surfaces in a stream valley representing the dissected remnants of an abandoned floodplain, stream bed, or valley floor produced in a previous stage of erosion or deposition.

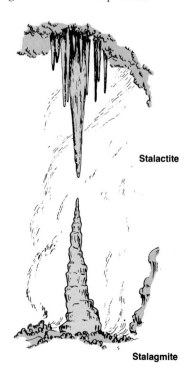

Stalactite

Stalagmite

stress Force applied to a material that tends to change its dimensions or volume; force per unit area.

striation A scratch or groove produced on the surface of a rock by a geologic agent, such as a glacier or stream.

strike The bearing (compass direction) of a horizontal line on a bedding plane, a fault plane, or some other planar structural feature (see diagram).

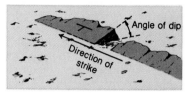

Strike

strike-slip fault A fault in which movement has occurred parallel to the strike of the fault (see diagram).

Strike-slip fault

strike valley A valley that is eroded parallel to the strike of the underlying nonresistant strata.

strip mining A method of mining in which soil and rock cover are removed to obtain the sought-after material.

subaerial Occurring beneath the atmosphere or in the open air, with reference to conditions or processes (such as erosion) that occur on the land. Contrast with *submarine* and *subterranean.*

subaqueous sand flow A type of mass movement in which saturated sand or silt flows beneath the surface of a lake or an ocean.

subduction Subsidence of the leading edge of a lithospheric plate into the mantle.

subduction zone An elongate zone in which one lithospheric plate descends beneath another. A subduction zone is typically marked by an oceanic trench, lines of volcanoes, and crustal deformation associated with mountain building. See also *convergent plate boundary.*

submarine canyon A V-shaped trench or valley with steep sides cut into a continental shelf or continental slope.

subsequent stream A tributary stream that is eroded along an underlying belt of nonresistant rock after the main drainage pattern has been established (see diagram).

Subsequent stream

subsidence A sinking or settling of a part of the Earth's crust with respect to the surrounding parts.

superposed stream A stream with a course originally established on a cover of rock now removed by erosion, so that the stream or drainage system is independent of the newly exposed rocks and structures. The stream pattern is thus superposed on, or placed upon, ridges or other structural features that were previously buried (see diagram).

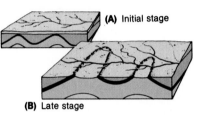

(A) Initial stage

(B) Late stage

Superposed stream

superposition, principle of The principle that, in a series of sedimentary strata that has not been overturned, the oldest rocks are at the base and the youngest are at the top.

surface creep Slow downwind movement of large sand grains by rolling or sliding along the surface due to the impact of smaller, saltating grains.

surface wave (seismology) A seismic wave that travels along the Earth's surface. Contrast with *P waves* and *S waves,* which travel through the Earth.

suspended load The part of a stream's load that is carried in suspension for a considerable period of time without contact with the stream bed. It consists mainly of mud, silt, and sand. Contrast with *bed load* and *dissolved load.*

swash The rush of water up onto a beach after a wave breaks.

S wave (secondary seismic wave) A seismic wave in which particles vibrate at right angles to the direction in which the wave travels. Contrast with *P wave.*

symmetrical fold A fold in which the two limbs are essentially mirror images of each other.

syncline A fold in which the limbs dip toward the axis. After erosion, the youngest beds are exposed in the central core of the fold.

Talus Rock fragments that accumulate in a pile at the base of a ridge or cliff (see diagram).

Talus

tectonic creep Slow, apparently continuous movement along a fault (as opposed to the sudden rupture that occurs during an earthquake).

tectonics The branch of geology that deals with regional or global structures and deformational features of the Earth.

tension Stress that tends to pull materials apart.

tephra A general term for pyroclastic material ejected from a volcano. It includes ash, dust, bombs, and other types of fragments.

terminal moraine A ridge of material deposited by a glacier at the line of maximum advance of the glacier (see diagram).

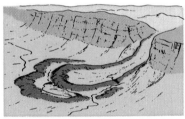

Terminal moraine

terra (pl. *terrae*) A densely cratered highland on the Moon.

terrace A nearly level surface bordering a steeper slope, such as a stream terrace or wave-cut terrace (see diagram).

Terrace

terrae Plural of *terra.*

texture The size, shape, and arrangement of the particles that make up a rock.

thin section A slice of rock mounted on a glass slide and ground to a thickness of about 0.03 mm.

thrust fault A low-angle fault (45 degrees or less) in which the hanging wall has moved upward in relation to the footwall (see diagram). Thrust faults are characterized by horizontal compression rather than by vertical displacement.

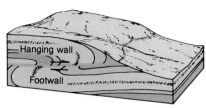

Thrust fault

tidal bore A violent rush of tidal water.

tidal flat A large, nearly horizontal area of land covered with water at high tide and exposed to the air at low tide. Tidal flats consist of fine-grained sediment (mostly mud, silt, and sand).

till Unsorted and unstratified glacial deposit.

tillite A rock formed by lithification of glacial till (unsorted, unstratified glacial sediment).

tombolo A beach or bar connecting an island to the mainland (see diagram).

Tombolo

topography The shape and form of the Earth's surface.

transform fault A special type of strike-slip fault forming the boundary between two moving lithospheric plates, usually along an offset segment of the oceanic ridge (see diagram). See also *passive plate margin.*

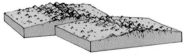

Transform fault

transpiration The process by which water vapor is released into the atmosphere by plants.

transverse dune An asymmetrical dune ridge that forms at right angles to the direction of prevailing winds.

travertine terrace A terrace formed from calcium carbonate deposited by water on a cave floor.

trellis drainage pattern A drainage pattern in which tributaries are arranged in a pattern similar to that of a garden trellis (see diagram).

Trellis drainage pattern

trench (marine geology) A narrow, elongate depression of the deep-ocean floor oriented parallel to the trend of a continent or an island arc.

tributary A stream flowing into or joining a larger stream.

tsunami A seismic sea wave; a long, low wave in the ocean caused by an earthquake, faulting, or a landslide on the sea floor. Its velocity can reach 800 km per hour. Tsunamis are commonly and incorrectly called tidal waves.

tuff A fine-grained rock composed of volcanic ash.

turbidity current A current in air, water, or any other fluid caused by differences in the amount of suspended matter (such as mud, silt, or volcanic dust). Marine turbidity currents, laden with suspended sediment, move rapidly down continental slopes and spread out over the abyssal floor (see diagram).

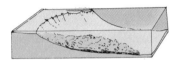

Turbidity current

turbulent flow A type of flow in which the path of motion is very irregular, with eddies and swirls.

Ultimate base level The lowest possible level to which a stream can erode the Earth's surface; sea level.

ultramafic rock An igneous rock composed entirely of ferromagnesian minerals.

unconformity A discontinuity in the succession of rocks, containing a gap in the geologic record. A buried erosion surface. See also *angular unconformity, nonconformity.*

uniformitarianism The theory that geologic events are caused by natural processes, many of which are operating at the present time.

upwarp An arched or uplifted segment of the crust.

Valley glacier A glacier that is confined to a stream valley. Synonymous with *alpine glacier, mountain glacier.*

varve A pair of thin sedimentary layers, one relatively coarse-grained and light-colored, and the other relatively fine-grained and dark-colored, formed by deposition on a lake bottom during a period of one year (see diagram). The coarse-grained layer is formed during spring runoff, and the fine-grained layer is formed during the winter when the surface of the lake is frozen.

Varve

ventifact A pebble or cobble shaped and polished by wind abrasion (see diagram).

Ventifact

vesicle A small hole formed in a volcanic rock by a gas bubble that became trapped as the lava solidified (see diagram).

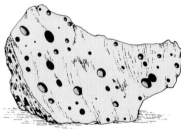

Vesicle

viscosity The tendency within a body to resist flow. An increase in viscosity implies a decrease in fluidity, or ability to flow.

volatile 1 Capable of being readily vaporized. 2 A substance that can readily be vaporized, such as water or carbon dioxide.

volcanic ash Dust-size particles ejected from a volcano.

volcanic bomb A hard fragment of lava that was liquid or plastic at the time of ejection and acquired its form and surface markings during flight through the air (see diagram). Volcanic bombs range from a few millimeters to more than a meter in diameter.

Volcanic bomb

volcanic front The line in a volcanic arc system (parallel to a trench) along which volcanism abruptly begins.

volcanic neck The solidified magma that originally filled the vent or neck of an ancient volcano and has subsequently been exposed by erosion (see diagram).

Volcanic neck

volcanism The processes by which magma and gases are transferred from the Earth's interior to the surface.

Wash A dry stream bed (see diagram).

Wash

water gap A pass in a ridge through which a stream flows (see diagram).

Water gap

water table The upper surface of the zone of saturation (see diagram).

Water table

wave base The lower limit of wave transportation and erosion, equal to half the wavelength.

wave-built terrace A terrace built up from wave-washed sediments. Wave-built terraces usually lie seaward of a wave-cut terrace.

wave crest The highest part of a wave.

wave-cut cliff A cliff formed along a coast by the undercutting action of waves and currents.

wave-cut platform A terrace cut across bedrock by wave erosion. Synonymous with *wave-cut terrace*.

wave-cut terrace See *wave-cut platform*.

wave height The vertical distance between a wave crest and the preceding trough.

wavelength The horizontal distance between similar points on two successive waves, measured perpendicular to the crest.

wave period The interval of time required for a wave crest to travel a distance equal to one wavelength; the interval of time required for two successive wave crests to pass a fixed point.

wave refraction The process by which a wave is bent or turned from its original direction. In sea waves, as a wave approaches a shore obliquely, part of it reaches the shallow water near the shore while the rest is still advancing in deeper water; the part of the wave in the shallower water moves more slowly than the part in the deeper water. In seismic waves, refraction results from the wave encountering material with a different density or composition.

wave trough The lowest part of a wave, between successive crests.

weathering The processes by which rocks are chemically altered or physically broken into fragments as a result of exposure to atmospheric agents and the pressures and temperatures at or near the Earth's surface, with little or no transportation of the loosened or altered materials.

welded tuff A rock formed from particles of volcanic ash that were hot enough to become fused together.

wind gap A gap in a ridge through which a stream, now abandoned as a result of stream piracy, once flowed (see diagram).

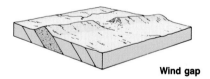

Wind gap

wind shadow The area behind an obstacle where air movement is not capable of moving material.

wrinkle ridge A sinuous, irregular segmented ridge on the surface of a lunar mare, believed to be a result of deformation of the lava.

X-ray diffraction In mineralogy, the process of identifying mineral structures by exposing crystals to a beam of X-rays and studying the resulting diffraction patterns.

Yardang An elongate ridge carved by wind erosion.

yazoo stream A tributary stream that flows parallel to the main stream for a considerable distance before joining it. Such a tributary is forced to flow along the base of a natural levee formed by the main stream.

Zone of aeration The zone below the Earth's surface and above the water table, in which pore spaces are usually filled with air (see diagram).

zone of saturation The zone in the subsurface in which all pore spaces are filled with water (see diagram).

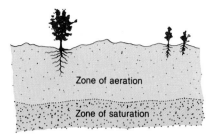

Zone of aeration

Illustration Credits

Chapter 13, page 270 Image copyright by Earth Satellite Corp., Chevy Chase, Maryland, under GEOPIC trademark.

Figures 13.2 and 13.9 Modified from A. N. Strahler, 1981, *Physical geology* (New York: Harper & Row), pp. 437, 449.

Figure 13.11B, C After A. N. Strahler, 1951, *Physical geography* (New York: John Wiley and Sons).

Chapter 14, page 294 Image copyright by Earth Satellite Corp., Chevy Chase, Maryland, under GEOPIC trademark.

Figure 14.11 Modified from Time-Life *Planet Earth* series, 1982, *Underground worlds* (Alexandria, Va.: Time-Life Books), p. 69.

Figure 14.14 Courtesy of John Shelton.

Figure 14.17B Courtesy of David Herron.

Figure 14.19 U. S. Department of the Interior, Washington, D. C.

Figure 14.22 Modified from W. J. Schneider, 1970, *Hydrologic implications of solid waste disposal,* U. S. Geological Survey Circular 601-F.

Figure 14.24 Modified from F. Ward, 1972, The imperiled Everglades, *National Geographic* 141(1):1.

Chapter 15, page 326 Courtesy of U. S. Geological Survey.

Figure 15.2 Courtesy of Petroconsultants, S. A., Geneva, Switzerland.

Figure 15.5 Modified from Lake Gillian and Conn Lake topographic maps, Northwest Territories, Geological Survey of Canada.

Figure 15.6 Courtesy of J. D. Ives.

Figure 15.7 Modified from R. F. Flint, 1957, *Glacial and Pleistocene geology* (New York: John Wiley and Sons).

Figure 15.8 After W. W. Atwood, 1940, *Physiographic provinces of North America* (Boston: Ginn and Company).

Figure 15.10 Courtesy of JPL, Pasadena, California.

Figure 15.15 Courtesy of Petroconsultants, S. A., Geneva, Switzerland.

Figure 15.16 After A. N. Strahler, 1951, *Physical geography* (New York: John Wiley and Sons).

Figure 15.17 National Air Photo Library, Department of Energy, Mines, and Resources, Canada.

Figure 15.21 Compiled from glacial map of the U. S. east of the Rocky Mountains, G. S. A.

Figure 15.22 After R. F. Flint, 1971, *Glacial and Quaternary geology* (New York: John Wiley and Sons).

Figure 15.23 Compiled from R. J. W. Douglas, 1970, *Geology and economic minerals of Canada* (Ottawa: Department of Energy, Mines, and Resources, Geological Survey of Canada), p. 691, and R. F. Flint, 1971, *Glacial and Quaternary geology* (New York: John Wiley and Sons), p. 234.

Figure 15.24 National Air Photo Library, Department of Energy, Mines, and Resources, Canada.

Figure 15.25 After J. L. Hough, 1958, *Geology of the Great Lakes* (Urbana: University of Illinois Press), and R. J. W. Douglas, 1970, *Geology and economic minerals of Canada* (Ottawa: Department of Energy, Mines, and Resources, Geological Survey of Canada), pp. 714–25.

Figure 15.26 After Teller, Lake Agassiz, Canadian Geological Survey.

Figure 15.27 After R. F. Flint, 1971, *Glacial and Quaternary geology* (New York: John Wiley and Sons), p. 447.

Figure 15.30 Courtesy of John Shelton.

Chapter 16, page 366 Image copyright by Earth Satellite Corp., Chevy Chase, Maryland, under GEOPIC trademark.

Figure 16.6 U. S. Geological Survey.

Figures 16.7B and 16.8D Courtesy of National Atmospheric and Oceanic Administration.

Figure 16.14 Courtesy of Alan L. Mayo, Geophoto Publishing Co. Photo by Steven R. Lower.

Figures 16.16 and 16.18 U. S. Department of Agriculture, ASCS Western Aerial Photo Lab., Salt Lake City, Utah.

Figure 16.23 Courtesy of John Shelton.

Figure 16.24 Courtesy of Bruce Coleman Inc.

Chapter 17, page 398 Image copyright by Earth Satellite Corp., Chevy Chase, Maryland, under GEOPIC trademark.

Figure 17.6 U. S. Geological Survey.

Figure 17.7 EROS Data Center, Sioux Falls, South Dakota.

Figure 17.9 Courtesy of Bruce Coleman.

Figure 17.14 Courtesy of Earth Satellite Corp., Washington, D. C.

Chapter 18, page 418 Image copyright by Earth Satellite Corp., Chevy Chase, Maryland, under GEOPIC trademark.

Figure 18.1 After A. Wegener, 1915, *The origin of continents and oceans,* English translation of the 4th edition, 1929 (New York: Dover), Figure 1.

Figure 18.2 Modified from L. Motz, 1979, *The rediscovery of the earth* (New York: Van Nostrand, Reinhold), p. 255, Figure 17.2.

Figure 18.3 After P. M. Hurley, 1969, The confirmation of continental drift, *Scientific American* 218(4):52–64.

Figure 18.6 After American Association of Petroleum Geologists, 1928, *Theory of continental drift: A symposium* (Tulsa, Okla.: American Association of Petroleum Geologists), Figure 2.

Figure 18.9 After A. Cox, G. B. Dalrymple, and R. R. Doell, 1967, Reversals of the earth's magnetic field, *Scientific American* 216(2):44–54.

Figure 18.13 Modified from P. J. Wyllie, 1976, *The way the earth works* (New York: John Wiley and Sons).

Figure 18.15 After B. Isacks, J. Oliver, and L. R. Sykes, 1968, Seismology and the new global tectonics, *Journal of Geophysical Research* 73(18):5855–99.

Chapter 19, page 456 Image copyright by Earth Satellite Corp., Chevy Chase, Maryland, under GEOPIC trademark.

Figure 19.3 Modified from J. H. Zumberge and C. A. Nelson, 1972, *Elements of geology,* 3rd ed. (New York: John Wiley and Sons).

Figure 19.5 U. S. Geological Survey.

Figure 19.6 After B. Isacks, J. Oliver, and L. R. Sykes, 1968, Seismology and the new global tectonics, *Journal of Geophysical Research* 73(18):5855–99.

Figure 19.9 Modified from L. R. Sykes, 1966, The seismicity and deep structure of island arcs, *Journal of Geophysical Research* 71(12):2981–3006.

Figure 19.17 Courtesy of Adam M. Dzieworski, Harvard University.

Chapter 20, page 478 Courtesy of NASA.

Figures 20.4 and 20.5 Courtesy of Woods Hole Oceanographic Institute.

Figure 20.7 After I. G. Gass, P. J. Smith, and R. C. L. Wilson, eds., 1971, *Understanding the earth* (Cambridge, Mass.: MIT Press).

Figure 20.13 U. S. Geological Survey.

Figure 20.15 Modified from Geo-Graphics, Portland, Oregon.

Figures 20.19 and 20.20 Courtesy of J. D. Griggs, U. S. Geological Survey.

Chapter 21, page 500 Courtesy of Ken C. Macdonald, University of California, Santa Barbara.

Figure 21.1 Courtesy of K. C. Macdonald.

Figure 21.9 Compiled from data in F. P. Shepard, 1963, *Marine geology* (New York: Harper & Row), pp. 321–22.

Figure 21.10 Courtesy of Woods Hole Oceanographic Institute.

Figure 21.12 Image copyright by Earth Satellite Corp., Chevy Chase, Maryland, under GEOPIC trademark.

Figure 21.14 Courtesy of C. R. Scotese.

Figure 21.15 After R. S. Dietz and J. C. Holden, 1970, *Scientific American* 223(4):30–41.

Chapter 22, page 524 Courtesy of NASA.

Index

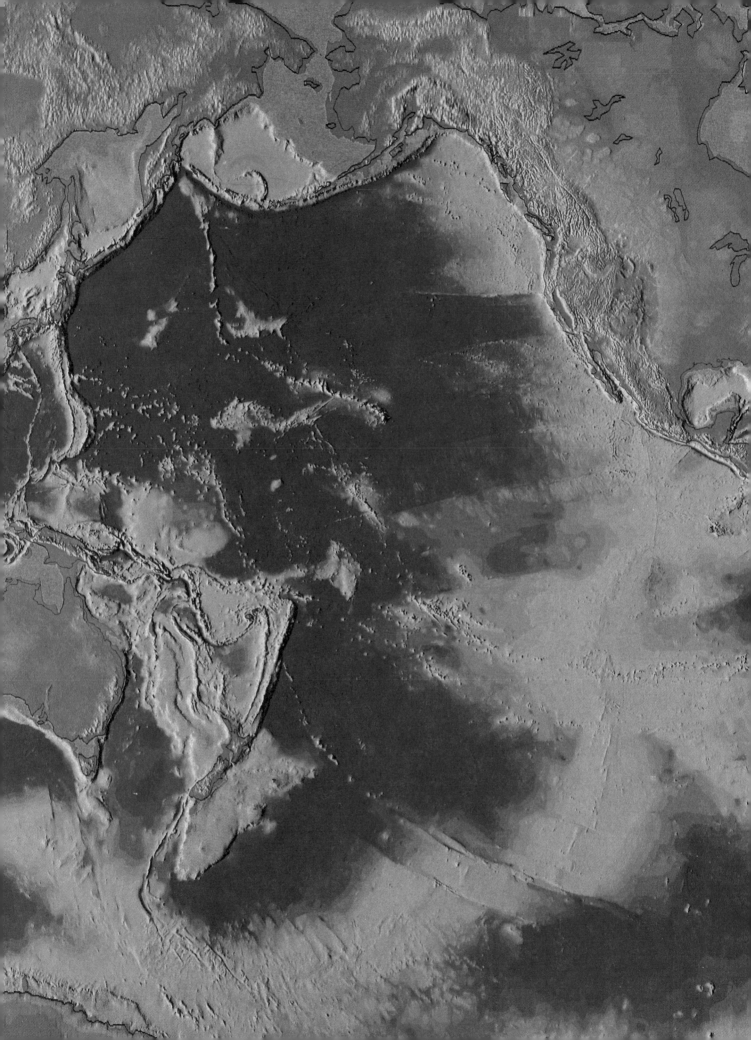